Petri Nets for Systems Engineering

Springer
Berlin
Heidelberg
New York
Hong Kong
London
Milan
Paris
Tokyo

Claude Girault
Rüdiger Valk

Petri Nets for Systems Engineering

A Guide to Modeling, Verification, and Applications

With 190 Figures and 9 Tables

Claude Girault
University of Paris VI
Lab. LIP6 (CNRS-UMR 7606)
4 Place Jussieu
75252 Paris Cedex 05
France
Claude.Girault@lip6.fr

Rüdiger Valk
University of Hamburg
Department of Computer Science
Vogt-Kölln-Str. 30
22527 Hamburg
Germany
valk@informatik.uni-hamburg.de

Library of Congress Cataloging-in-Publication Data applied for

Die Deutsche Bibliothek – CIP-Einheitsaufnahme
Girault, Claude; Valk, Rüdiger: Petri Nets for Systems Engineering. A Guide to Modeling, Verification, and Applications – Berlin; Heidelberg; New York; Hong Kong; London; Milan; Paris; Tokyo: Springer, 2003

ACM Computing Classification (1998): D.1.5, D.2.2–6, F.1.2, F.3.1–2, F.4.1, D.3.1, D.4, B.1.2, B.6, C.1, C.2.4, I.2.11, J.1, J.4

ISBN 978-3-642-07447-9

Springer-Verlag Berlin Heidelberg New York,
a member of BertelsmannSpringer Science+Business Media GmbH

Printed in Germany

Cover Design: KünkelLopka, Heidelberg

Printed on acid-free paper

Preface

Nowadays, computer-based systems are indispensable for almost all areas of modern life. As has been frequently stated, they suffer, however, from being insufficiently correct or reliable. Software development projects fail their completion deadlines and financial frames. Though computer systems surpass in size all traditional complex systems ever produced, the discrepancy in quality standards seems to be unbridgeable. It has been argued frequently that traditional engineering methods and standards should be adapted to software development needs and so the field of software engineering was created. Though some progress is observable in this area, modelling, analysis, and implementation techniques lack a powerful modelling method connecting all these areas. Moreover, though graphic-based modelling techniques are of increasing interest, there are very few that are founded on formal methods.

This book intends to bridge the gap between Petri nets, which fulfil many desirable requirements, and the systems modelling and implementation process. Petri nets are introduced from its basics and their use for modelling and verification of systems is discussed. Several application domains are selected to illustrate the method.

The book has been written in the context of the MATCH project. We wish to thank all participants for their contributions to the concept and design of the book. It took several meetings and numerous sessions to decide on its structure and contents. In particular, we gratefully acknowledge the contribution of portions of text, mutual proof reading, the supplying of useful comments and suggestions, and – last but not least – the perseverance in maintaining the complex process of communication which led to this result. To the European Community we are deeply indebted for providing the financial base to organize all these meetings.

Rainer Mackenthun put a lot of effort into organizing a common bibliography. The completion of the book would have been unthinkable without the hard and careful work of Berndt Farwer who solved the problem of compiling the chapters and sections and removed many inconsistencies. It is a pleasure for us to acknowledge these efforts.

We are grateful for the numerous suggestions and the encouragement which came from the participants of the summer school in Jaca, Spain, and students in our universities, who worked with early versions of the text. Our thanks are also due to Hans Wössner of Springer-Verlag for his unique editorial help and suggestions.

Paris and Hamburg, September 2002 Claude Girault and Rüdiger Valk

Contents

Part II. Modelling

Part III. Verification

List of Authors and Affiliations

The authors of the present volume are listed here in alphabetical order together with their current affiliation and e-mail address. An updated list can be found at http://www.informatik.uni-hamburg.de/TGI/service/MATCHbook/frame.html.

Name	e-mail
Wil van der Aalst[b]	w.m.p.v.d.aalst@tm.tue.nl
Patrick Barril†[e]	(see footnote on page 3)
Twan Basten[c]	a.a.basten@tue.nl
José-Manuel Colom[f]	jm@posta.unizar.es
Alioune Diagne[e]	Alioune.Diagne@lip6.fr
Claude Dutheillet[e]	Claude.Dutheillet@lip6.fr
William El Kaim[h]	william.elkaim@thalesgroup.com
Joaquin Ezpeleta[f]	ezpeleta@posta.unizar.es
Berndt Farwer[d]	farwer@informatik.uni-hamburg.de
Marie Pierre Gervais[e]	mpg@src.lip6.fr
Claude Girault[e]	Claude.Girault@lip6.fr
Mike van de Graaf[i]	mvdgraaf@bakkenist.nl
Serge Haddad[g]	Serge.Haddad@lamsade.dauphine.fr
Jean Michel Ilié[e]	Jean-Michel.Ilie@lip6.fr
Fabrice Kordon[e]	Fabrice.Kordon@lip6.fr
Rainer Mackenthun[d]	Rainer.Mackenthun@isst.fhg.de
Daniel Moldt[d]	moldt@informatik.uni-hamburg.de
Denis Poitrenaud[e]	Denis.Poitrenaud@lip6.fr
Manuel Silva[f]	silva@posta.unizar.es
Mark-Oliver Stehr[d]	stehr@informatik.uni-hamburg.de
Enrique Teruel[f]	eteruel@posta.unizar.es
Rüdiger Valk[d]	valk@informatik.uni-hamburg.de
Isabelle Vernier-Mounier[e]	Isabelle.Vernier-Mounier@lip6.fr
Marc Voorhoeve[a]	wsinmarc@win.tue.nl

a Eindhoven University of Technology
Department of Mathematics and Computing Science
P.O. Box 513, 5600 MB Eindhoven, The Netherlands

b Eindhoven University of Technology
Department of Information and Technology
P.O. Box 513, 5600 MB Eindhoven, The Netherlands

c Eindhoven University of Technology
Department of Electrical Engineering
P.O. Box 513, 5600 MB Eindhoven, The Netherlands

d University of Hamburg
Department of Computer Science
Vogt-Kölln-Str. 30, 22527 Hamburg, Germany

e Université Paris VI
Lab. LIP6
4, Place Jussieu, 75252 Paris Cedex 05, France

f University of Zaragoza
Departamento de Informatica e Ingeniería de Systemas
María de Luna 3, 50015 Zaragoza, Spain

g Université Paris-Dauphine
Centre des Ressources Informatiques en Commun
Place du Maréchal de Lattre de Tassigny, 75775 Paris Cedex 16, France

h THALES Research and Technology
Software Architecture Group
Domaine de Corbeville, 91404 Orsay Cedex, France

i Bakkenist Management Consultants
P.O. Box 23103, Wisselwerking 46, 1100 XP Amsterdam Zuidoost/Diemen, The Netherlands

Introduction: Purpose of the Book

This book intends to show how Petri nets fill many of the needs of systems modelling, their verification and implementation, as mentioned in the preface. It first introduces Petri nets in such a way that only those features necessary for system engineers are presented and then introduces important fields such as modelling concepts and verification techniques.

The advantages of Petri nets for the modelling of systems are well-known:

- They provide a graphically and mathematically founded modelling formalism. This is in contrast to many similar techniques, where only one of these properties is well developed and the other is added in a less systematic way. These two sides of the coin are of high importance as the system development process needs graphical as well as algorithmic tools.
- To date there exists a huge variety of algorithms for the design and analysis of Petri nets and powerful computer tools have been developed to aid this process. To give just one example, we would like to mention reachability analysis as a subfield of model checking.
- Abstraction and hierarchical design is crucial for the effective design of large scale and complex systems. Petri nets provide mechanisms for abstraction and refinement that are well integrated into the basic model.
- There is a huge number of commercial or university tools for the design, simulation, and analysis of Petri-net-based systems. Many of them achieve industrial standards.
- Petri nets have been used in many different application areas. As a result there is a high degree of expertise in the modelling field.
- Different variants of Petri net models have been developed that are all related by the basic net formalism which they build upon. This allows them to meet the needs in different application domains on the one hand, but on the other hand gives facilities for communication and the transfer of methods and tools from one field to another. Currently, besides the basic model, there are extensions such as timed, stochastic, high-level, and object-oriented Petri nets, meeting the specific needs for (almost) every applications area that comes to mind.

After a general introduction the contents of the book are oriented towards the software and hardware development process. The modelling, validation, and execution phases in the software life cycle for which there are very few scientific results to be found in the literature are covered in some detail here. On the other hand in fields like verification a considerable depth of research has been reached. This is reflected in the book by some in-depth studies of these issues. Despite the fact that the scientific maturity of the covered fields varies, a holistic approach has been chosen. As a result some parts present genuine research results while others are restricted to an overview of the field, mostly referring to the literature.

Following the introduction is Part I, *Petri Nets – Basic Concepts.* It introduces essential features such as locality versus concurrency, graphical versus algebraic representation and, refinement versus composition. Then in an intuitive manner, arc-constant, place/transition, and coloured nets are introduced using a running example. Chapters 4 and 5 give the essentials of formal definitions, in particular the incidence matrices and some basic properties such as reachability graph, linear invariants, liveness, and reversibility. The last chapter of Part I presents an outline of the more advanced topics covered in Parts II to V.

Part II, *Modelling*, gives an introduction to the construction methods of systems using Petri nets. This part starts with a chapter giving some introductory and more complex examples. Here the reader can obtain some deeper knowledge about the specific potential of Petri nets. The examples cover elementary nets, place/transition nets, and coloured nets. Subsequently, design methodologies are presented in a more systematic way. The bottom-up method starts with building simple nets, that are combined into more and more complex nets until the desired model is obtained. This approach is contrasted to the top-down method of decomposing nets into smaller parts. In practical work both methods have to be used in a mixed form. Different communication mechanisms between parts are studied as well as interconnecting techniques of different parts. A state-oriented style is contrasted to event-oriented modelling. The systematic approaches are illustrated by three case studies.

Part III, *Verification*, consists of an overview of the main approaches to verification of Petri net models. There are chapters covering the exploration of the state space and model checking, structural methods – such as invariants, linear algebraic techniques and reductions – and some advanced methods using deduction and process algebra. This part is meant to reflect the current state of the art in verification which, of course, cannot be exhaustive in a book like this. The reader will nevertheless acquire a thorough knowledge of many verification issues and will be guided to the wealth of further literature.

A part of the book that will be especially interesting for system engineers with a more practical background is Part IV, *Validation and Execution.* This part is concerned not only with the software life cycle but gives detailed analysis of the practical use of Petri net models for the development of large-scale systems. This is manifested in the possibilities of execution of and code generation from such abstract models. This part also includes an overview of tools available to support the process of software and system development.

The final part of the book (Part V) is dedicated to three in-depth studies of different *Application Domains.* The coverage takes account of all phases of the respective development phases for such diverse domains as flexible manufacturing systems, workflow management systems, and telecommunications systems.

The book has been written in the scope of the project MATCH (Modelling and Analysis of Time Constrained and Hierarchical Systems) which was a Human Capital and Mobility initiative, sponsored by the European Union. Among its objectives there was the organisation of two complementary advanced summer schools and two books, one focusing on performance modelling and evaluation, and the present one on modelling and verification.

This book presents the results from the cooperation of four Petri net research groups from the universities of Eindhoven, Hamburg, Paris VI, and Zaragoza. The contributing authors are (in alphabetical order, see also p. XV):

W. van der Aalst (Chapter 25),
P. Barril[†] (Chapter 20),
T. Basten (Chapter 16),
J.-M. Colom (Chapters 5 & 15),
A. Diagne (Chapters 10 & 11),
C. Dutheillet (Chapter 14),
W. El Kaim (Chapter 21),
J. Ezpeleta (Chapter 24),
B. Farwer (Chapter 16),
M.-P. Gervais (Chapter 26),
C. Girault (editorial parts),
M. van de Graaf (Chapter 25),
S. Haddad (Chapters 13 & 15),
J.-M. Ilié (Chapter 14),
F. Kordon (Chapters 19 & 21),
R. Mackenthun (Chapters 10 & 11),
D. Moldt (Chapter 19),
D. Pointrenaud (Chapter 14),
M. Silva (Chapter 5),
M.-O. Stehr (Chapter 16),
E. Teruel (Chapters 5 & 15),
R. Valk (editorial parts, Chapters 1–4 & 8),
I. Vernier-Mounier (Chapter 14),
M. Voorhoeve (Chapters 9, 10 & 11).

[†] We very much regret that our honoured colleague Patrick Barril died during the period of writing this book.

A comprehensive index and bibliographical information can be found at the end of the book. References for further reading given in the preceding chapters may be found here. The index gives quick access to the main definitions and keywords.

Part I

Petri Nets – Basic Concepts

1. Introduction

Due to their numerous features and various applications there are many ways to introduce Petri nets. In this part, we first focus on the modelling of actions. In general, actions depend on a limited set of conditions, restrictions, etc. which could be called the local environment. Petri nets model actions by the change of their local environment. This *principle of locality* is the basis of the superiority of Petri nets in modelling concurrency. It is, however, a widespread misunderstanding that Petri nets should not be used when the application systems do not exhibit any concurrent behaviour. There are other features, such as graphical and textual representation of Petri nets, refinement, and abstraction, that can contribute to a well-structured and reliable system construction. Many of these features as well as methods and tools for system analysis will be presented later on in this book.

In Chapter 2, we restrict our attention to a basic set of such principles, namely *locality and concurrency*, *graphical and algebraic/textual representation*, *conflict and confusion*, *refinement and composition*. Refinement will not only be used for structuring application systems, but also to transform Petri nets between different degrees of abstraction. The notion of refinement is strongly related to the concept of *net morphisms* which are instances of morphisms in general algebraic structures, where *quotient* is the mathematical term for abstraction. In a first reading, Section 2.5 *Net Morphisms* may be skipped since all definitions with respect to refinement and abstraction are already dealt with in the preceding Section 2.4 *Refinement and Composition*.

While the notion of a Petri net is introduced in Chapter 2, its behaviour is explained in a very elementary way. In Chapter 3, the basic models of *place/transition nets* and *coloured nets* are introduced in a more specific but still intuitive style. As a starting point, the model of an *arc-constant net* is introduced, since its properties nicely connect the former models.

Chapter 4 provides formal definitions of the three previously studied models, however, in a different order: first place/transition nets, then arc-constant nets and finally coloured nets as this reflects their relation by increasing complexity. The method of introducing these three models is similar: after the definition of the net model, the incidence matrices are explained and the corresponding transition occurrence rules are given. Due to its complexity more emphasis is given to the subject of the incidence matrix in the case of coloured Petri nets. Also different ways of representing their incidence matrices are compared.

Chapters 2 to 4 are connected by a common example of a simple starting procedure for a car race. This example illustrates communicating sequential processes in general.

Some important properties of Petri nets are introduced in Chapter 5 such as *boundedness*, *liveness*, and *reversibility*. The motivation is that these are also important properties of real systems. Hence, the question of validating such properties in large-scale systems arises. The reachability graph for place/transition nets is introduced which in principle allows for the checking of such properties. Due to the high complexity of the reachability graph method in general, more structural methods are desirable. Chapter 5 also gives some first hints on how linear algebra methods can be used for verification.

An example of such a verification technique is given for a Petri net modelling a production cell. Almost all methods of this chapter will be treated in further detail in subsequent chapters, particularly in Part III.

Chapter 6 gives a detailed overview of the remainder of the book. This chapter is intended to give the more experienced reader the opportunity to decide on a further reading strategy, while the novice is briefly introduced to the aspects and problems that are treated in the following parts of the book.

Petri nets were introduced by Carl Adam Petri in his Ph.D. thesis in 1962 [Pet62]. Today, there are thousands of papers and monographs on the topic. Many of them are referenced in this book and most of them can be found in the *Petri Net Bibliography* maintained at `http://www.informatik.uni-hamburg.de/TGI/pnbib/index.html`. The notion of place/transition net was introduced in 1980 (see [JV80]) to distinguish this model from nets without annotation (in the sense of Definition 2.2.1) and other net models. High-level Petri nets were introduced in 1981 in the form of predicate/transition nets ([GL81]) and coloured nets ([Jen81]). A three-volume monograph on coloured nets was published in the 1990s: [Jen92b], [Jen94], and [Jen97].

2. Essential Features of Petri Nets*

In this chapter a basic model is introduced by extracting information from an intuitive example, given in this section. By doing this some essential features of Petri nets will appear, namely the principles of *locality*, *concurrency*, *graphical* and *algebraic representation*. For illustration we choose an example where several objects are subject to a coordination procedure. Similar examples can be found in various other fields, e.g. from computer integrated manufacturing to office automation. In our example, a race among a number of cars is started. When the starter receives ready signs from all cars, he gives the starting signal and the cars begin the race. For simplicity, we restrict the example to one starter and two racing cars (Figure 2.1).

Fig. 2.1. Starting two racing cars

Suppose that for a computer application (e.g. simulation, race-control) the following essential conditions and actions have been identified:

* Author: R. Valk

a) List of conditions:

p_1: car a; preparing for start
p_2: car a; waiting for start
p_3: car a; running
p_4: ready sign of car a
p_5: start sign for car a
p_6: starter; waiting for ready signs
p_7: starter; start sign given
p_8: ready sign of car b
p_9: start sign for car b
p_{10}: car b; preparing for start
p_{11}: car b; waiting for start
p_{12}: car b; running

b) List of actions:

t_1: car a; send ready sign
t_2: car a; start race
t_3: starter; give start sign
t_4: car b; send ready sign
t_5: car b; start race

2.1 Locality and Concurrency

Identifying and, in particular, *separating* passive elements (such as conditions) from active elements (such as actions) is a very important step in the design of systems. This duality is strongly supported by Petri nets. We formulate the first principle: *the principle of duality.* Whether an object is seen as active or passive may depend on the context or the point of view of the system. For instance, a statement of a programming language can be modelled as an active element if its execution is modelled. Alternatively, it may be seen as passive if it is subject to an operation in a compiler.

I	The Principle of Duality for Petri Nets

There are two disjoint sets of elements: *P-elements* (state elements, places) and *T-elements* (transition elements, transitions).
Entities of the real world, interpreted as passive elements, are represented by P-elements (conditions, places, resources, waiting pools, channels etc.).
Entities of the real world, interpreted as active elements, are represented by T-elements (events, transitions, actions, executions of statements, transmission of messages etc.).

To build an operational model for our example, we select truth values TRUE and FALSE to hold at the beginning. The initial state m_1 is characterised by the conditions that car a and car b are preparing for start (i.e.

$p_1 = p_{10} = T$ (TRUE)) and that the starter is waiting for ready signs (i.e. $p_6 = T$). Hence, we obtain the following global state vector m_1:

$$m_1 = [p_1 = T, p_2 = F, p_3 = F, p_4 = F, p_5 = F, p_6 = T,$$
$$p_7 = F, p_8 = F, p_9 = F, p_{10} = T, p_{11} = F, p_{12} = F]$$

Two of the actions, namely t_1 and t_4, may occur in this initial state. By action t_1 car a gives the ready sign and therefore stops preparing for start ($p_1 = F$). Then it is waiting for start ($p_2 = T$) having given the ready sign ($p_4 = T$). The resulting state is denoted by m_2 and is given by:

$$m_2 = [p_1 = F, p_2 = T, p_3 = F, p_4 = T, p_5 = F, p_6 = T,$$
$$p_7 = F, p_8 = F, p_9 = F, p_{10} = T, p_{11} = F, p_{12} = F]$$

The first observation we make is that most conditions are untouched by any of these actions; hence, only a few conditions are relevant. This property is called *locality of action.* In Figure 2.2, for action t_1 the affected conditions are marked by surrounding circles and are connected by arcs with the action, which is represented by a rectangle. Restricting actions to those parts of the global state vector which are in some causal dependency leads to a considerable simplification, i.e. for the complexity of description as well as for new conceptional approaches, such as the notion of concurrency. Thus, we formulate a principle of locality.

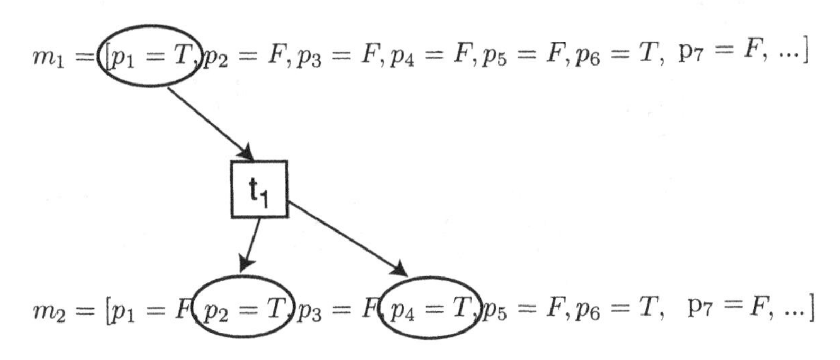

Fig. 2.2. Locality of action t_1

Action t_1 may occur if p_1 holds TRUE and p_2 and p_4 hold FALSE. Adding to the set t_1 its set of conditions $\{p_1, p_2, p_4\}$, where p_1 is called a *pre-condition* and p_2, p_4 are *post-conditions* of t_1, we obtain the locality of t_1. The notion of locality will be used in Section 2.5 of this chapter to characterise net morphisms.

II	The Principle of Locality for Petri Nets
The behaviour of a transition exclusively depends on its *locality*, which is defined as the totality of its input and output objects (pre- and post-conditions, input and output places, ...) together with the element itself.	

The second action that may occur in m_2 is t_4 transforming m_2 into m_3 with:

$$m_3 = [p_1 = F, p_2 = T, p_3 = F, p_4 = T, p_5 = F, p_6 = T,$$
$$p_7 = F, p_8 = T, p_9 = F, p_{10} = F, p_{11} = T, p_{12} = F].$$

The locality of t_4 is $\{t_4, p_{10}, p_8, p_{11}\}$. Therefore, t_1 and t_4 share no conditions in the marking m_1 and may occur completely independently. This is the principle of *concurrency*: actions with disjoint localities may occur independently. In Figure 2.3, the occurrences of t_1 and t_4 are represented as a common step leading from marking m_1 to marking m_3. Note that the notion of concurrency is different from parallelism. Parallel action may be synchronised by a central clock, whereas concurrent events are not connected by any causality.

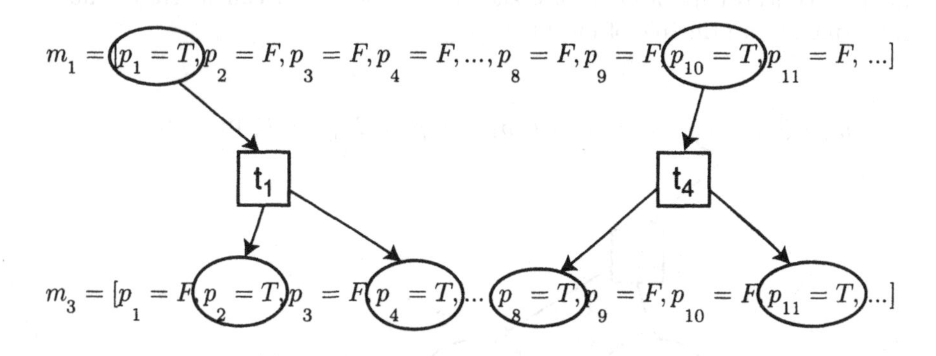

Fig. 2.3. Concurrent actions t_1 and t_4

III	The Principle of Concurrency for Petri Nets
Transitions having disjoint locality occur independently (concurrently).	

2.2 Graphical and Algebraic Representation

In Figure 2.4, actions t_1 and t_4 are drawn with their pre- and post-conditions. In this formal form they are called *transitions*. Conditions are represented by circles which are called *places*. In addition, Figure 2.4 shows *all* transitions of the example. Places, transitions, and arcs together form a *net*. Fusing places bearing identical names, we obtain the net of Figure 2.5. Some places contain *tokens* that mark the initial conditions. They will be explained later on.

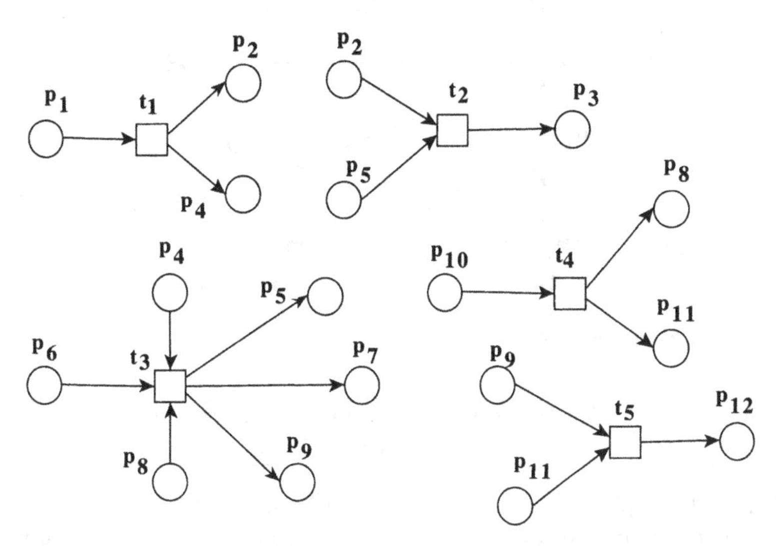

Fig. 2.4. Actions represented by transitions

This leads to a principle of graphical representation.

IV	The Principle of Graphical Representation for Petri Nets
P-elements are represented by rounded graphical symbols (circles, ellipses,...) (round like the top of the letter P). T-elements are represented by edged graphical symbols (rectangles, bars,...) (edged like the top of the letter T). Arcs connect each T-element with its locality, which is a set of P-elements. Additionally, there may be inscriptions such as names, tokens, expressions, guards.	

For many purposes – for instance, listing, analysis, and mathematical description – an algebraic description of Petri nets is useful. In most cases it is equivalent to the graphical representation.

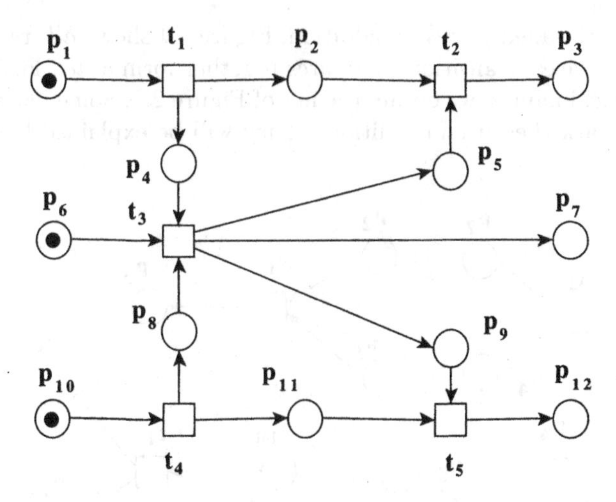

Fig. 2.5. Net $\mathcal{N}$ of the example

V	**The Principle of Algebraic Representation for Petri Nets**
For each graphical representation there is an algebraic representation containing equivalent information. It contains the set of places, transitions, and arcs, and additional information such as inscriptions.	

The base of all Petri net models is the definition of a net ([Pet96]).

Definition 2.2.1. *A* net *is a triple* $\mathcal{N} = (P, T, F)$ *where*

- P *is a set of* places,
- T *is a set of* transitions, *disjoint from* P, *and*
- F *is a* flow relation $F \subseteq (P \times T) \cup (T \times P)$ *for the set of arcs.*

If P *and* T *are finite, the net* $\mathcal{N}$ *is said to be finite.*

Sometimes, instead of P the letter S is used, coming from the notion of state element (S-element). For the example net we obtain $P = \{p_1, \ldots, p_{12}\}$, $T = \{t_1, \ldots, t_5\}$, $F = \{(p_1, t_1), (t_1, p_2), (t_1, p_4), \ldots\}$.

The holding of a condition is represented by a token in the corresponding place. In the net $\mathcal{N}$ of Figure 2.5 such tokens show the initial state m_1. The occurrence rule for transitions is illustrated in Figure 2.6 using the example of transition t_3 from Figure 2.5. Transition t_3 "may occur" or "is activated" if all pre-conditions hold (are marked by a token) and no post-condition holds. With the occurrence of t_3 all tokens are removed from the pre-conditions (input places) and are added to the post-conditions (output places).

In Figure 2.7, all possible occurrences of transitions are shown. Observe in particular the concurrent occurrence of t_1, t_4 and t_2, t_5.

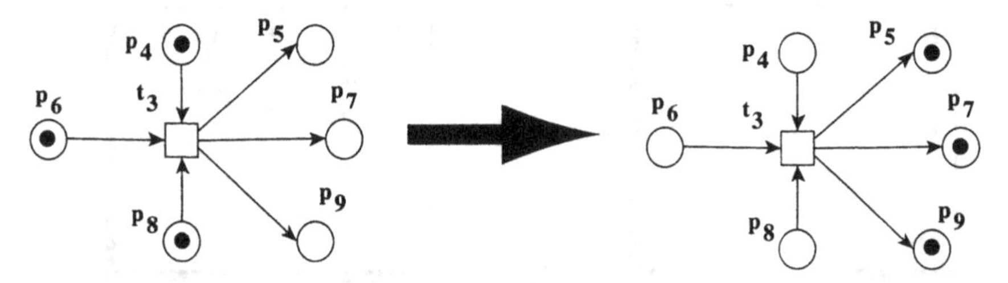

Fig. 2.6. Transition occurrence rule

To denote the places connected to a transition (and vice versa), the following standard notation is used. Given an element $x \in P \cup T$, then $^\bullet x := \{y \in P \cup T \mid (y, x) \in F\}$ denotes the set of all *input elements* of x, and $x^\bullet := \{y \in P \cup T \mid (x, y) \in F\}$ denotes the set of all *output elements* of x. If x is a place, then $^\bullet x$ and $x^\bullet$ denote the set of *input* and *output transitions* respectively. The corresponding notion holds for transitions. It is convenient to extend this definition to hold for a set $A \subseteq P \cup T$ by $^\bullet A := \{y \mid \exists x \in A . (y, x) \in F\}$ and $A^\bullet := \{y \mid \exists x \in A . (x, y) \in F\}$.

To give an example, for $A = \{t_1, p_5, p_{11}\}$ in the net of Figure 2.5 we obtain $^\bullet A = \{p_1, t_3, t_4\}$ and $A^\bullet = \{p_2, p_4, t_2, t_5\}$. The notion of *locality* of a transition was used in Section 2.1 to introduce concurrency of two transitions. Now, it can formally be defined as follows: $loc(t) := \{t\} \cup {^\bullet t} \cup t^\bullet$. Hence, t_1 and t_2 are concurrent if $loc(t_1) \cap loc(t_2) = \emptyset$. In a similar way we also define the locality of a place $p \in P$ by $loc(p) := \{p\} \cup {^\bullet p} \cup p^\bullet$.

2.3 Concurrency, Conflict, and Confusion

Contrary, in some sense, to concurrency is the notion of conflict. To illustrate this, we extend our initial example in such a way that after the start phase the cars begin to proceed independently (Figure 2.8).

Now consider a marking such as $\{p_3, p_7, p_{12}\}$, where the starting phase is over for both cars. Then transitions t_6, t_7 and t_8, t_9 may occur independently. To be more precise, the pairs of transitions (t_6, t_8), (t_7, t_8), (t_6, t_9), and (t_7, t_9) may occur concurrently. In a similar way, conditions such as p_{13} and p_{15} may hold independently. To introduce the notion of conflict we now assume that the race road becomes so narrow at some point that only one car can pass at a time. Assume t_6 and t_7 to be the entry and exit events of this narrow section for car a, and t_8 and t_9 the same for car b. Then neither (t_6, t_8), (t_7, t_8), (t_6, t_9) nor (t_7, t_9) can occur concurrently any longer, and moreover p_{13} and p_{15} cannot hold simultaneously. The effect of this modification could be implemented using the extension of the net in Figure 2.9. Here, a new place *conf* with one token imposes the condition that after the occurrence of one of the transitions t_6 or t_8 the other one is not enabled until

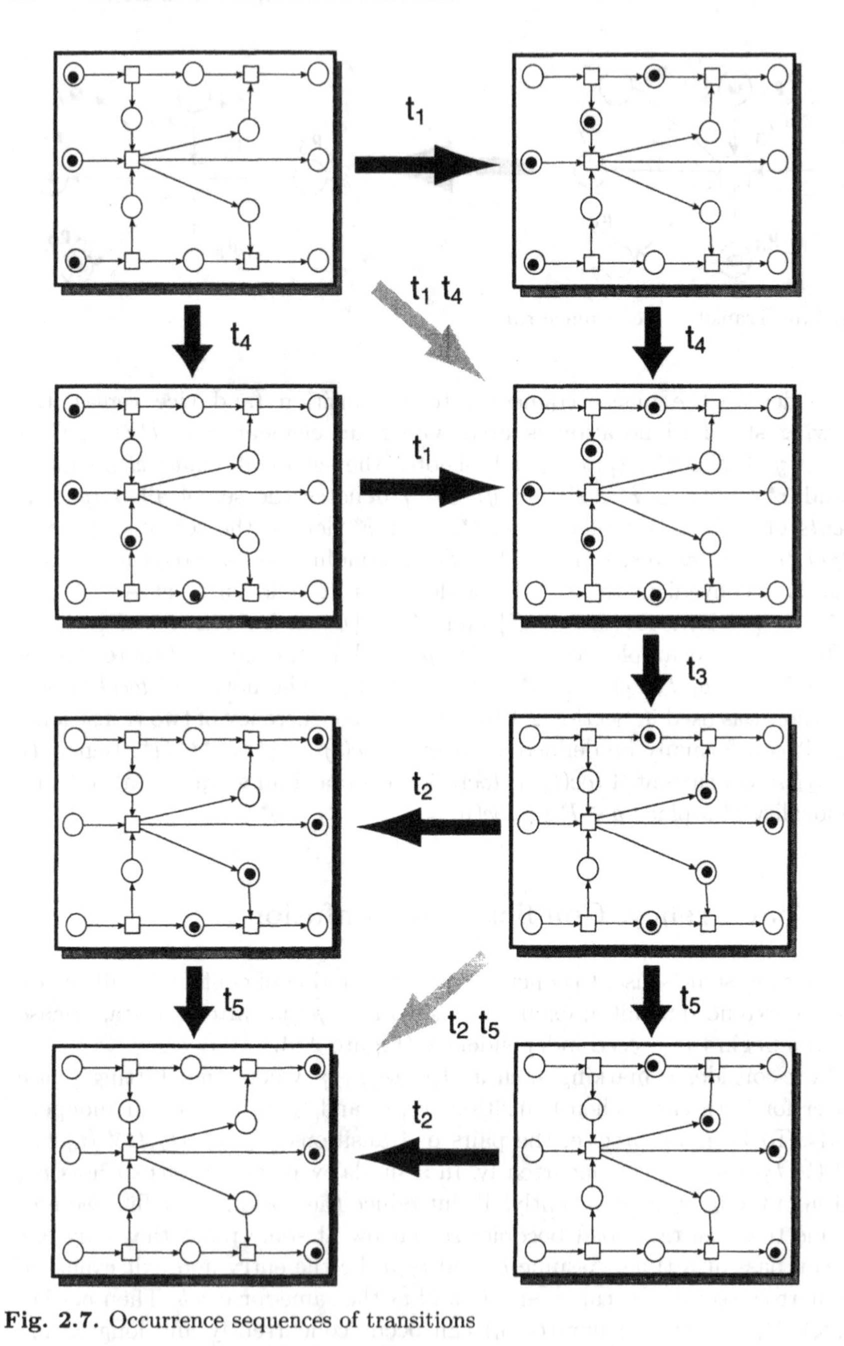

Fig. 2.7. Occurrence sequences of transitions

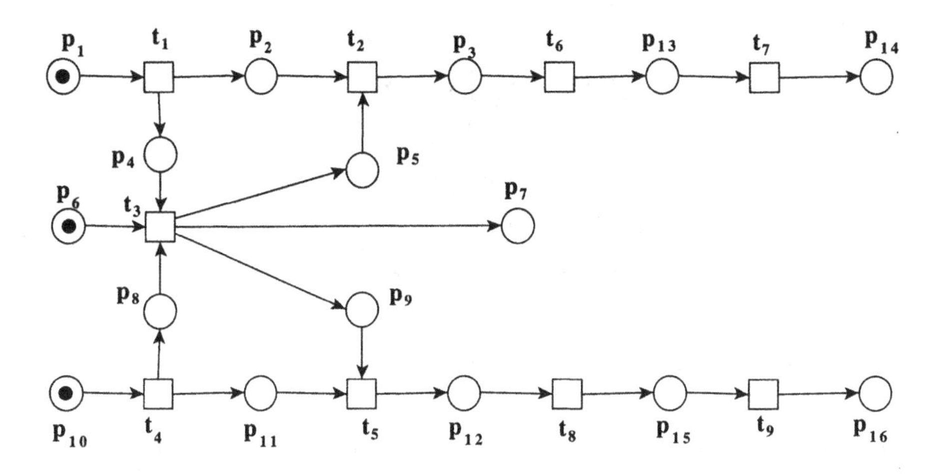

Fig. 2.8. Extension of the net from Figure 2.5

t_7 or t_9 respectively have occurred. Hence, a marking containing p_{13} and p_{15} simultaneously is impossible. This is known as *mutual exclusion*. The place *conf* contains two output transitions and is therefore called a *structural conflict* place. In the marking $\{p_3, p_7, p_{12}, conf\}$ also a *behavioural conflict* occurs since both transitions t_6 and t_8 are enabled, but only one of them can occur.

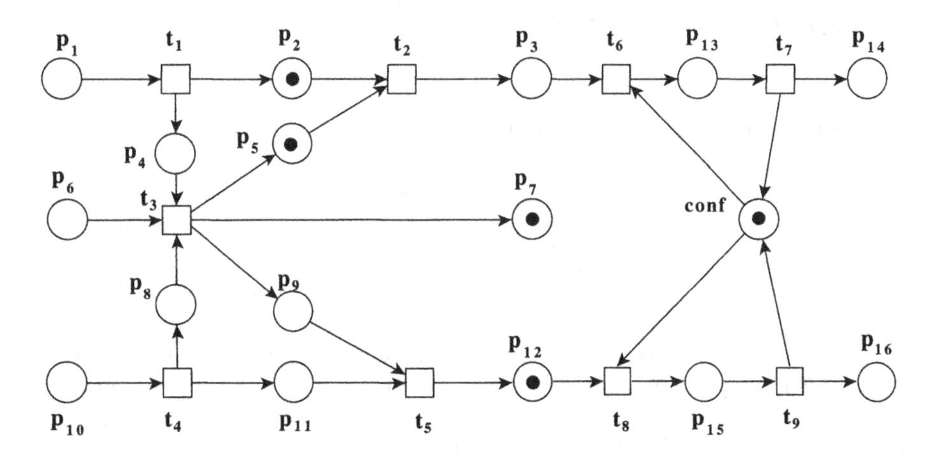

Fig. 2.9. Car race with conflict and confusion

A similar situation is shown in Figure 2.10. The conflict place *conf* can be seen as a resource shared by the actions t_6 and t_8, such that these transitions

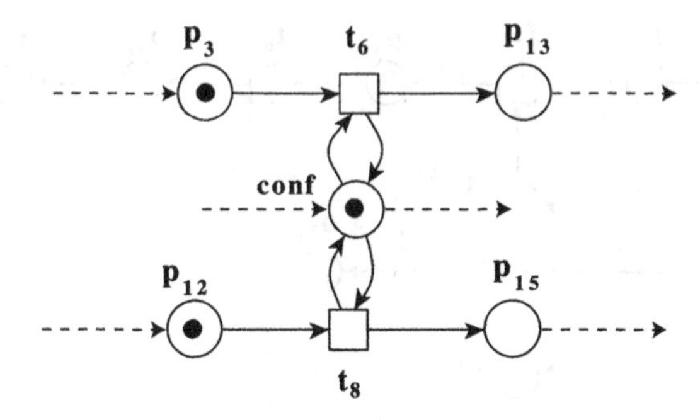

Fig. 2.10. Common resource

cannot occur simultaneously. They have a non-disjoint locality and are not concurrent.

Concurrent transitions behave independently and should not have any impact on each other. However, as observed by C. A. Petri, this is not always true. Transitions t_2 and t_8 in the net of Figure 2.9 have disjoint localities (see Section 2.1) and may occur concurrently in the given marking. In this marking there is no (behavioural) conflict for t_8. After the occurrence of t_2, however, there is a transition which conflicts with t_8, namely t_6. Hence, the occurrence of a concurrent transition can change the situation of t_8 with respect to behavioural conflicts. Such situations showing the sophisticated interaction of concurrency and conflicts are called *confusions*. For a formal definition of confusion see [Thi87].

2.4 Refinement and Composition

Building hierarchies by abstraction or refinement is an important technique in system design. Petri nets support such approaches by abstraction techniques that are inherently compatible with the structure of the model. We start by defining the border of a set, which will be the interface of a part to be considered.

Let $\mathcal{N} = (P, T, F)$ be a net, $X := P \cup T$, and $Y \subseteq X$ a set of elements. Then $\partial(Y) := \{y \in Y \mid \exists x \notin Y . x \in loc(y)\}$ is the *border* of the set Y. Y is called *place-bordered* or *open*[1] if $\partial(Y) \subseteq P$, and *transition-bordered* or *closed*[1] if $\partial(Y) \subseteq T$. In order to define a well-structured abstraction, place- and transition-bordered sets may be replaced by a single element. Note that a set Y can be open *and* closed at the same time, e.g. $Y := P \cup T$. In such a case the context of the application determines whether Y is to be replaced by a place or a transition.

[1] Open and closed sets define a topology for a net, which formalises the notion of vicinity of elements with respect to the graphical structure.

The set $Y = \{p_3, p_4, t_2, t_3, t_4\}$ of the net in Figure 2.11 is transition-bordered and can be abstracted to a transition t_Y such that the result is again a net, as shown in Figure 2.12. This operation will now be formalised.

Let $\mathcal{N} = (P, T, F)$ be a net and Y a non-empty transition-bordered set of elements. Then $\mathcal{N}[Y] = (P[Y], T[Y], F[Y])$ is said to be a *simple abstraction* of $\mathcal{N}$ with respect to Y if: $P[Y] = P \backslash Y$, $T[Y] = (T \backslash Y) \cup \{t_Y\}$ where t_Y is a new element, $F[Y] = \{(x, y) \mid x \notin Y \wedge y \notin Y \wedge (x, y) \in F\} \cup \{(x, t_Y) \mid x \notin Y \wedge \exists y \in Y \,.\, (x, y) \in F\} \cup \{(t_Y, x) \mid x \notin Y \wedge \exists y \in Y \,.\, (y, x) \in F\}$. $P[Y]$ contains all places with the exceptions of those in Y. $T[Y]$ contains all transitions with the exceptions of those in Y and a new element t_Y. $F[Y]$ is the union of three sets of arcs, namely (1) those having no end point in Y, (2) those leading from outside of Y to t_Y and (3) those leading from t_Y to the outside of Y.

Analogously, if Y is a non-empty place-bordered set, then $\mathcal{N}[Y] = (P[Y], T[Y], F[Y])$ is obtained by defining $P[Y] = (P \backslash Y) \cup \{p_Y\}$ where p_Y is a new element, $T[Y] = T \backslash Y$, $F[Y] = \{(x, y) \mid x \notin Y \wedge y \notin Y \wedge (x, y) \in F\} \cup \{(x, p_Y) \mid x \notin Y \wedge \exists y \in Y \,.\, (x, y) \in F\} \cup \{(p_Y, x) \mid x \notin Y \wedge \exists y \in Y \,.\, (y, x) \in F\}$.

The definition of $\mathcal{N}[Y]$ is ambiguous if Y is place- and transition-bordered at the same time. Then we write $\mathcal{N}[Y^{(p)}]$ if Y is used as a place-bordered set and $\mathcal{N}[Y^{(t)}]$ if it is used as a transition-bordered set.

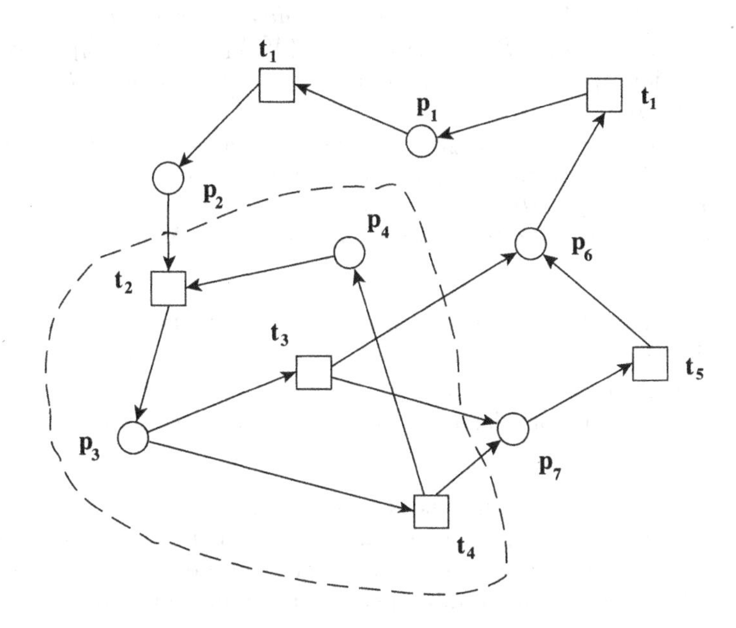

Fig. 2.11. A transition-bordered set

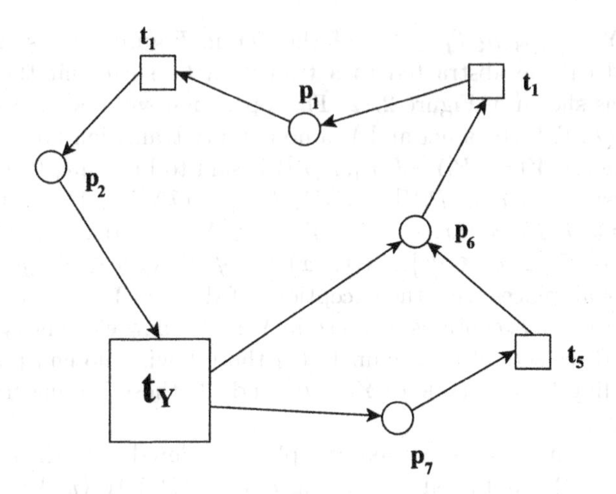

Fig. 2.12. Abstraction from the net of Figure 2.11

Definition 2.4.1.

a) *If* $\mathcal{N}_2 = \mathcal{N}_1[Y]$ *is a simple abstraction of* $\mathcal{N}_1$ *for some place- or transition-bordered set* Y*, then* $\mathcal{N}_1$ *is said to be a* simple refinement *of* $\mathcal{N}_2$*. If there is a set* $\{Y_1, Y_2, \dots, Y_n\}$ *of pairwise disjoint place- or transition-bordered subsets of* $P_1 \cup T_1$ *then* $\mathcal{N}_2 = (\dots((\mathcal{N}_1[Y_1])[Y_2])\dots[Y_n])$ *is called an* abstraction *of* $\mathcal{N}_1$*, and* $\mathcal{N}_1$ *is a* refinement *of* $\mathcal{N}_2$ *and is denoted by* $\mathcal{N}_2 = \mathcal{N}_1[Y_1, Y_2, \dots, Y_n]$.

b) *An abstraction* $\mathcal{N}_2 = \mathcal{N}_1[Y_1, Y_2, \dots, Y_n]$ *of* $\mathcal{N}_1$ *is called* strict *if every* Y_i *is either a set of places, i.e.* $Y_i \subseteq P_1$*, or a set of transitions, i.e.* $Y_i \subseteq T_1$*. In the definition of a strict abstraction* Y_i *will be replaced by a place* p_{Y_i} *in the first case and by a transition* t_{Y_i} *in the second.* $\mathcal{N}_1$ *is called a* strict refinement *of* $\mathcal{N}_2$.

Again, by the following convention ambiguity of the notation can be removed: if in a) or b) a set Y_i $(1 \le i \le n)$ is both place- and transition-bordered, the abstraction is denoted by $\mathcal{N}_2 = \mathcal{N}_1[Y_1, \dots, Y_i^{(d)}, \dots, Y_n]$ where $d = p$ or $d = t$ if Y_i is considered a place- or transition-bordered set respectively.

Remark 2.4.2. In the literature and sometimes also in this book strict abstractions are called *foldings*. The notion of folding is related to the definition of *net morphisms*. In Section 2.5, the equivalence of strict abstractions and (epi-)foldings is formally established by a theorem. This is to justify the interchangeable use of these terms.

It is easy to verify that the abstraction of a net is again a net. Not every abstraction, however, has a meaningful interpretation. To give an example, consider the fragment of a net in Figure 2.13a. Intuitively the abstraction in

Figure 2.13d has a corresponding behaviour ("a token is passed through"). Figure 2.13d is also an abstraction of Figure 2.13b, but now the behaviour is different. In Figure 2.13d a token can pass, but not so in Figure 2.13b. Note that the set Y to be abstracted is not necessarily connected as a subgraph. However, Figure 2.13d can be interpreted as a merge of two places. This operation is called a *fusion of places* or place fusion. It will be used in Part II of this book to create larger nets from smaller ones. This kind of abstraction will be represented graphically as shown in Figure 2.13c. The dual situation for a transition-bordered set is shown in Figures 2.13e–h. Thus Figures 2.13f, g and h describe a *fusion of transitions* or transition fusion.

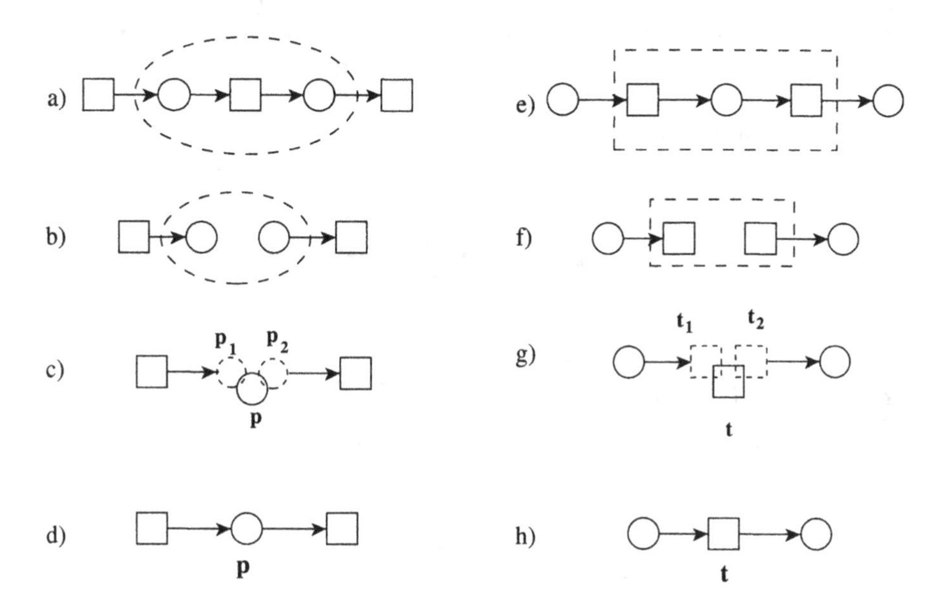

Fig. 2.13. Abstraction and fusion

In Figure 2.14 two nets are given that can be seen as a decomposition of the extended race-example net of Figure 2.9. By fusion of the places p'_3, p''_3 and p'_{12}, p''_{12}, the original net from Figure 2.9 is obtained.

To present a larger and more meaningful example of a simple abstraction, we consider the net $\mathcal{N}$ from Figure 2.9 and the transition-bordered set $Y = \{t_1, t_2, t_3, t_4, t_5, p_2, p_4, p_5, p_8, p_9, p_{11}\}$. The abstraction $\mathcal{N}[Y]$ describes a system where the starting phase of the race is modelled by a single atomic action t_Y, as depicted in Figure 2.15.

If Y_1 and Y_2 are disjoint sets then the iterated abstractions $(\mathcal{N}[Y_1])[Y_2]$ and $(\mathcal{N}[Y_2])[Y_1]$ are of interest. The resulting nets are isomorphic and denoted by $\mathcal{N}[Y_1, Y_2]$. With Y as in the preceding example and $Y_1 = \{t_6, p_{13}, t_7\}$ and $Y_2 = \{t_8, p_{15}, t_9\}$, from the net in Figure 2.9 we obtain the abstraction

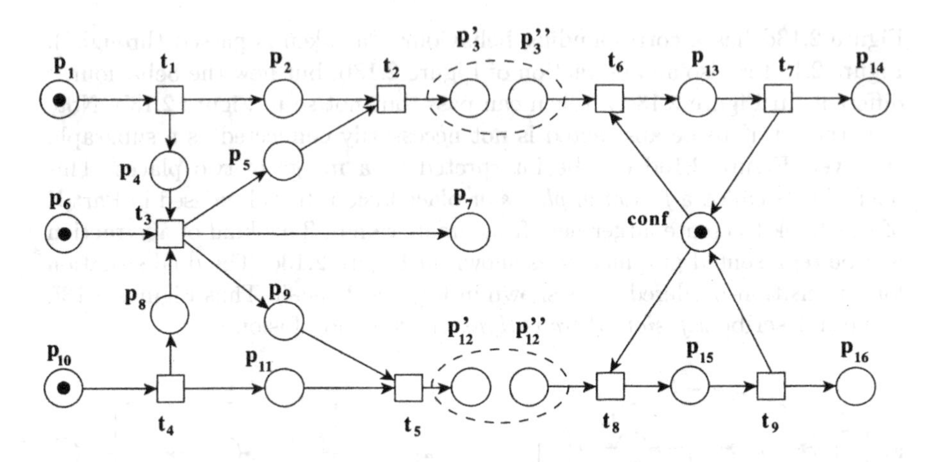

Fig. 2.14. Composition by fusion

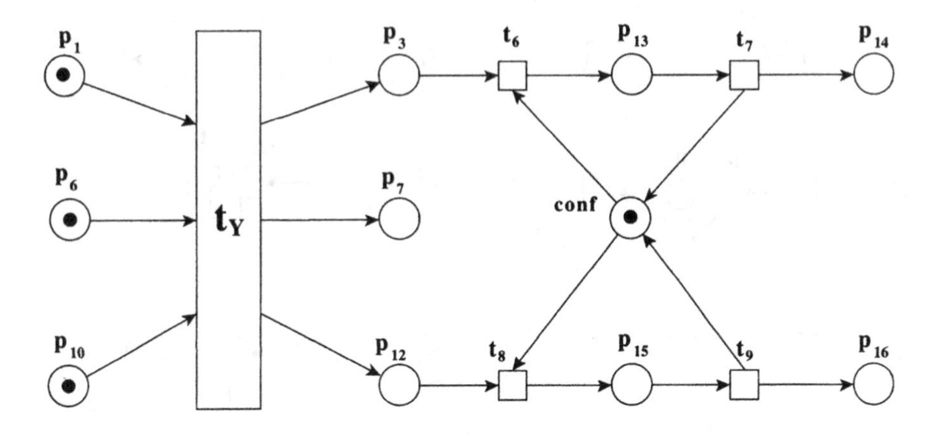

Fig. 2.15. Abstraction of the starting phase (Figure 2.9)

$\mathcal{N}[Y, Y_1, Y_2]$ in Figure 2.16b. In Figure 2.16a the sets to be abstracted are represented by dashed rectangular lines. This abstraction has a behavioural interpretation. After the starting phase t_Y of the race, considered as an atomic event, the next actions of the two cars are given, where they pass the critical section in mutual exclusion. These subsequent actions are also represented as indivisible actions.

An example of a strict abstraction is given in Figure 2.19. In fact, all sets to be abstracted contain either places or transitions. A semantical interpretation will be given in Chapter 3.

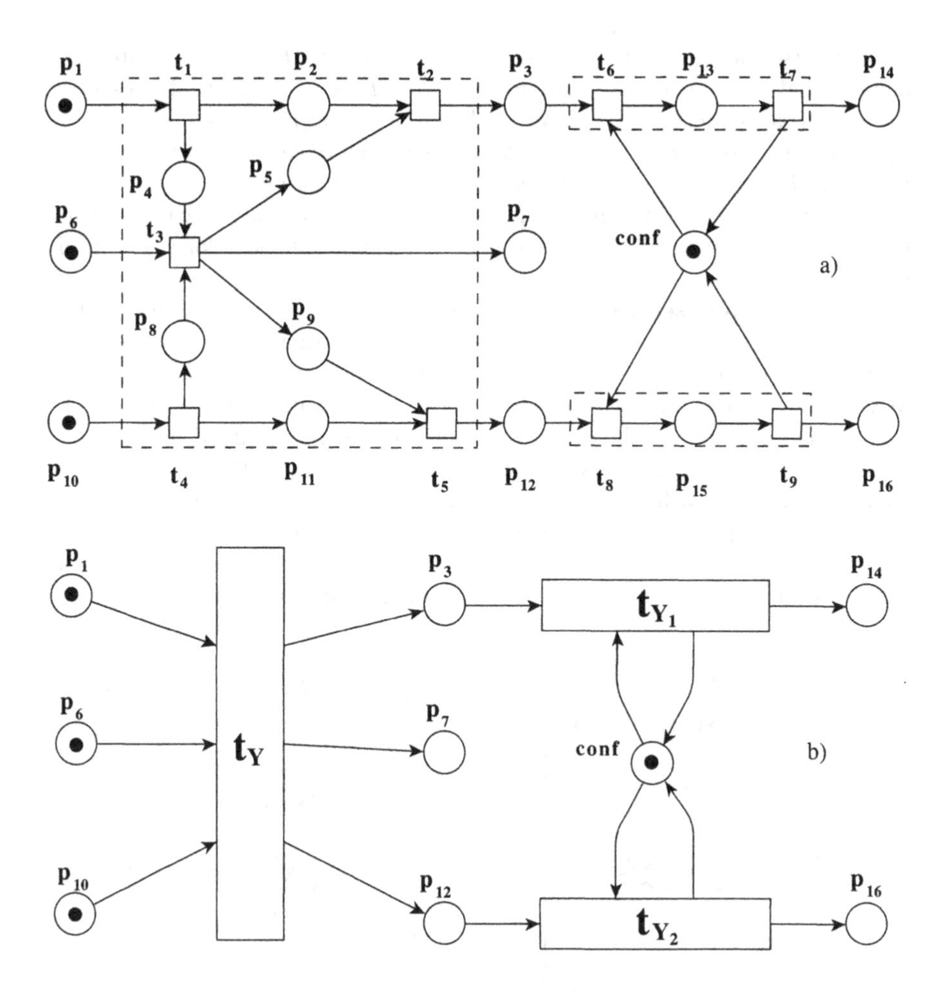

Fig. 2.16. Iterated abstraction

2.5 Net Morphisms

As for algebraic representations in general, structure-preserving mappings known as *morphisms* play an important role. They are the basis of many operations on nets, such as abstractions, composition, hierarchy, and foldings. This concept is also fundamental for the creation of tools. For these reasons we introduce in this section some basic notions of net morphisms. From an application point of view however, the notion of abstraction, as given in Section 2.4, is sufficient. Therefore, at the end of this section a theorem will formally state the equivalence of these two concepts. The mathematically disinclined reader may skip this section on a first reading.

Given two nets $\mathcal{N}_1 = (P_1, T_1, F_1)$ and $\mathcal{N}_2 = (P_2, T_2, F_2)$, in a first step we define a *net morphism* as a mapping $\varphi : P_1 \cup T_1 \to P_2 \cup T_2$ that preserves the graphical structure given by the F-relation: $\forall x, y \in P_1 \cup T_1 . (x, y) \in F_1 \Rightarrow (\varphi(x), \varphi(y)) \in F_2 \vee \varphi(x) = \varphi(y)$ and is therefore called *F-preserving.* In Figure 2.17, two F-preserving maps from a net $\mathcal{N}_1$ to a net $\mathcal{N}_2$ are shown by dashed arrows. However, whereas φ describes an abstraction in case b), this is not true in case a). Therefore this property is not sufficient for a notion of net morphism.

This problem arises because some place-bordered sets are the image of a set not having that property (e.g. the set $\{p'\}$). The same holds for transition-bordered sets (e.g. $\{t'\}$). In this section place- and transition-bordered sets will be denoted open and closed respectively (see the definition in Section 2.4). Using this terminology, the case just described can be excluded if open and closed sets are required to have respectively open and closed inverse images.[2] By this property it will be impossible for a net arrow $(p, t) \in F$ to be mapped to an arrow $(f(p), f(t)) \in F'$ where $f(p)$ is not a place or $f(t)$ is not a transition. In the full definition of a net morphism, the *at-relation* A has been introduced to express this as a condition called *A-preservation* ([Pet96]).

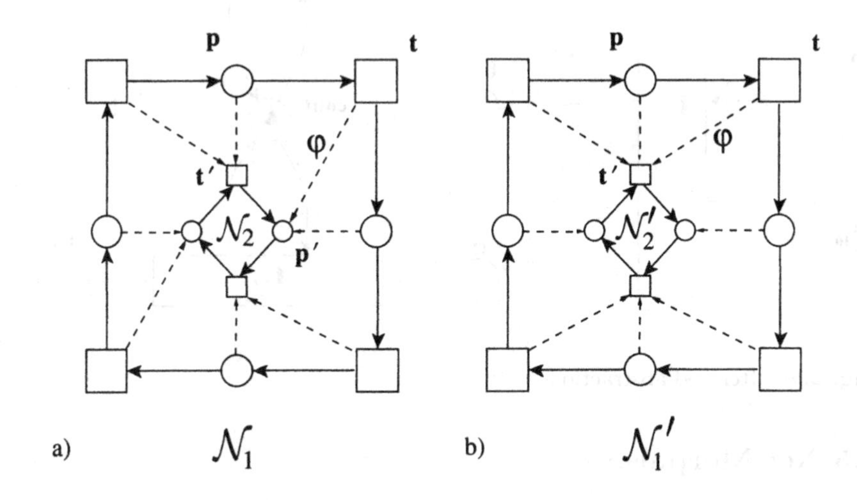

Fig. 2.17. Two F-preserving mappings

Definition 2.5.1. *Let $\mathcal{N}_1 = (P_1, T_1, F_1)$ and $\mathcal{N}_2 = (P_2, T_2, F_2)$ be two nets and $\varphi : X_1 \to X_2$ a mapping where $X_i = P_i \cup T_i$. Furthermore, define the* at-relation $A_i := (F_i \cup F_i^{-1}) \cap (P_i \times T_i)$, *where $i \in \{1, 2\}$ in both cases.*

[2] The reader knowing the mathematical definition of a topology will recognise the property of a continuous mapping.

a) φ *is said to be a* net morphism *(or a* net mapping*) if:*
1. $\forall x, y \in P_1 \cup T_1 \,.\, (x,y) \in F_1 \Rightarrow (\varphi(x), \varphi(y)) \in F_2 \vee \varphi(x) = \varphi(y)$ *("F-preservation"),*
2. $\forall x, y \in P_1 \cup T_1 \,.\, (x,y) \in A_1 \Rightarrow (\varphi(x), \varphi(y)) \in A_2 \vee \varphi(x) = \varphi(y)$ *("A-preservation", "continuity").*

b) A net morphism φ *is a* folding *if* $\varphi(P_1) \subseteq P_2$ *and* $\varphi(T_1) \subseteq T_2$*.*

c) A net morphism φ *is an* epimorphism *if* φ *is surjective and for every* $(x_2, y_2) \in F_2$ *there is an arc* $(x_1, y_1) \in F_1$ *such that* $(x_2, y_2) = (\varphi(x_1), \varphi(y_1))$*. A folding which is an epimorphism is called an* epifolding*.*

d) A net morphism φ *is a* net isomorphism *if* φ *is a bijection and* φ^{-1} *is also a net morphism.*

Recall that the at-relation describes the "proximity" of elements in a net, and A-preservation is equivalent to the property of continuity with respect to a topology on the net given by open (place-bordered) sets. The mapping φ of Figure 2.17a) is F-preserving but not A-preserving, since $(p,t) \in A$, but $(\varphi(p), \varphi(t)) = (t', p') \notin A'$ and $\varphi(p) \neq \varphi(t)$. We now give a different but equivalent definition of morphism that avoids the introduction of the A-relation ([DM96]):

A mapping $\varphi : X_1 \rightarrow X_2$ is called a *net morphism* if the following holds:

1. $(x,y) \in F_1 \cap (P_1 \times T_1) \Rightarrow (\varphi(x), \varphi(y)) \in F_2 \cap (P_2 \times T_2) \vee \varphi(x) = \varphi(y)$ and
2. $(x,y) \in F_1 \cap (T_1 \times P_1) \Rightarrow (\varphi(x), \varphi(y)) \in F_2 \cap (T_2 \times P_2) \vee \varphi(x) = \varphi(y)$.

The following lemma gives a characterisation of continuous mappings by locally-defined properties that will first be defined. They have been studied in [DM96] under the name *vicinity-respecting properties.*

Definition 2.5.2. *Let* φ *be a mapping as in Definition 2.5.1. Then* φ *is said to be*

a) locally closed *iff* $p \in P_1$ *and* $\varphi(p) = t' \in T_2$ *implies* $\varphi(loc(p)) = \{t'\}$ *and*
b) locally open *iff* $t \in T_1$ *and* $\varphi(t) = p' \in P_2$ *implies* $\varphi(loc(t)) = \{p'\}$*.*

The following lemma will be used in the proof of the main theorem of this section.

Lemma 2.5.3. *If* φ *is A-preserving then* φ *is both locally closed and locally open. The converse implication holds if, in addition,* φ *is assumed to be F-preserving.*

Proof. To prove that φ is locally closed, assume $p \in P_1$ and $\varphi(p) = t' \in T_2$. If $t \in loc(p)$, we have to show that $\varphi(t) = t'$. In fact, if $t \in {}^\bullet p$ and $(t,p) \in F$, then $(p,t) \in F^{-1}$, and $(p,t) \in A_1$. If $t \in p^\bullet$ and $(p,t) \in F$, then also $(p,t) \in A_1$. Hence, by the property of A-preservation, $(\varphi(p), \varphi(t)) \in A_2$ or $\varphi(p) = \varphi(t)$. But $(\varphi(p), \varphi(t)) \in A_2$ cannot hold since $\varphi(p) \in T_2$, and so $\varphi(p) = \varphi(t)$ is proved. In a similar way, φ is proved to be locally open.

To prove the second part of the statement, assume that $(x, y) \in A_1$ and $(\varphi(x), \varphi(y)) \notin A_2$. We have to show that $\varphi(x) = \varphi(y)$.

There are three cases: a) $\varphi(x) \notin P_2$, b) $\varphi(y) \notin T_2$, and c) $(\varphi(x), \varphi(y)) \notin F_2 \cup F_2^{-1}$. By $(x, y) \in A_1$ we have $x \in P_1$, $y \in T_1$, and $y \in loc(x)$. Since φ is locally closed, case a) implies $\varphi(x) = \varphi(y)$. Since φ is locally open, case b) also implies $\varphi(x) = \varphi(y)$. Thus case c) remains. By the assumption that φ is F-preserving, $y \in loc(x)$ implies $(\varphi(x), \varphi(y)) \in F_2 \cup F_2^{-1}$ and so case c) leads to a contradiction.

To illustrate the lemma, consider again the example from Figure 2.17a. In this case φ is neither locally closed nor locally open, and is not A-preserving, whereas in part b) of this figure all properties of the lemma hold. Figure 2.18 gives an example where φ is locally open and closed, but neither F-preserving nor A-preserving.

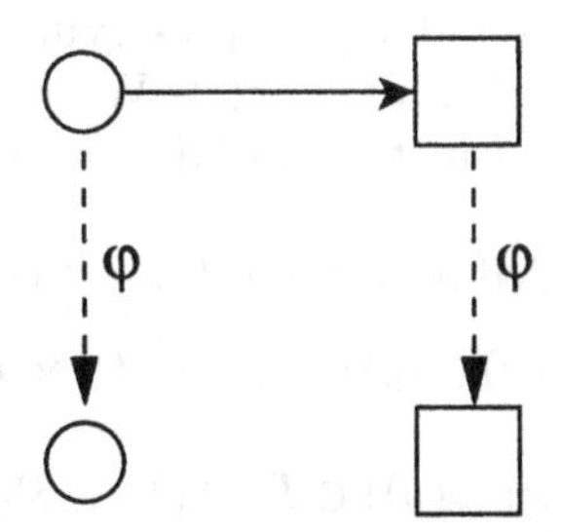

Fig. 2.18. A mapping that is not F-preserving

We are now ready to formulate the main theorem of this section, namely the characterisation of refinements by morphisms. Such a morphism has to be surjective in a strong sense, i.e. all elements $x \in X_2$ *and* all arcs $(x, y) \in F_2$ in $\mathcal{N}_2$ have to be images under φ. Such a morphism has been introduced as an epimorphism in Definition 2.5.1.

Theorem 2.5.4. *Let $\mathcal{N}_1 = (P_1, T_1, F_1)$ and $\mathcal{N}_2 = (P_2, T_2, F_2)$ be finite nets.*

a) $\mathcal{N}_2$ is an abstraction of $\mathcal{N}_1$ if and only if there is an epimorphism from $\mathcal{N}_1$ to $\mathcal{N}_2$.

b) $\mathcal{N}_2$ is a strict abstraction of $\mathcal{N}_1$ if and only if there is an epifolding from $\mathcal{N}_1$ to $\mathcal{N}_2$.

Proof. To simplify the proof of the theorem, we also consider the trivial case of an abstraction $\mathcal{N}_1[Y]$ where Y is a singleton set, i.e. a set $Y = \{x\}, x \in P_1 \cup T_1$. Here, in the construction of an abstraction, $x \in Y$ is simply replaced by x_Y. This allows us to consider any abstraction as $\mathcal{N}_1[Y_1, \ldots, Y_k]$ where $Y = \{Y_1, \ldots, Y_k\}$ is a finite partition of $P_1 \cup T_1$ (i.e. a finite covering by disjoint sets).

To prove a), let φ be an epimorphism from $\mathcal{N}_1$ to $\mathcal{N}_2$. Then we define $Y = \{Y_1, \dots, Y_k\} := \{\varphi^{-1}(x) \mid x \in P_2 \cup T_2\}$.

Since φ is a mapping and $P_1 \cup T_1$ is finite, Y is a finite partition. By the assumption that φ is F- and A-preserving it follows from Lemma 2.5.3 that $Y_i = \varphi^{-1}(x)$ is open or closed. It remains to be shown that $\mathcal{N}_1[Y_1, \dots, Y_k]$ is "isomorphic" to $\mathcal{N}_2$, i.e. there is a net isomorphism ψ from $\mathcal{N}_1[Y_1, \dots, Y_k]$ to $\mathcal{N}_2$. The mapping defined by $\psi(x_{Y_i}) := \varphi(y)$ for some $y \in Y_i$ is well-defined and bijective. Furthermore it is a morphism. In fact, if (x_{Y_i}, x_{Y_j}) is an F-edge in $\mathcal{N}_1[Y_1, \dots, Y_k]$, then by construction of $\mathcal{N}_1[Y_1, \dots, Y_k]$ there are elements $x' \in Y_i$ and $y' \in Y_j$ such that $(x', y') \in F_1$; hence, since $Y_i \neq Y_j$ we also have that $\varphi(x') \neq \varphi(y')$ and $(\varphi(x'), \varphi(y')) = (\psi(x_{Y_i}), \psi(x_{Y_j})) \in F_2$. By this ψ is F-preserving. Furthermore, $\{x_{Y_i}\}$ is open $\Leftrightarrow$ Y_i is open $\Leftrightarrow$ $\psi(x_{Y_i})$ is open and $\{x_{Y_i}\}$ is closed $\Leftrightarrow$ Y_i is closed $\Leftrightarrow$ $\psi(x_{Y_i})$ is closed. By this, ψ is also A-preserving.

Conversely, if $(x_2, y_2) \in F_2$ then, since φ is an epimorphism, there are elements $(x', y') \in F_1$ such that $(\varphi(x'), \varphi(y')) = (x_2, y_2)$. If $Y_i = \varphi^{-1}(x_2)$ and $Y_j = \varphi^{-1}(y_2)$ then $x' \in Y_i$, $y' \in Y_j$, $x_{Y_i} = \psi^{-1}(x_2)$ and $x_{Y_j} = \psi^{-1}(y_2)$ and there is an F-edge (x_{Y_i}, x_{Y_j}) in $\mathcal{N}_1[Y_1, \dots, Y_k]$. Thus, we also have that ψ^{-1} is a morphism and ψ an isomorphism.

To finish the proof of a), assume that $\mathcal{N}_2$ is an abstraction $\mathcal{N}_1[Y_1, \dots, Y_k]$ of $\mathcal{N}_1$ (where $\{Y_1, \dots, Y_k\}$ is a partition of $P_1 \cup T_1$). Then it is easy to verify that the map φ_1, defined by $\varphi_1(x_1) = x_Y :\Leftrightarrow x_1 \in Y$, is an epimorphism. Recall, that by the definition of an abstraction each of the sets Y_i is non-empty.

To prove part b), it is sufficient to observe that every open or closed set Y_i is a subset of places or a subset of transitions, and that the same holds for each $\varphi^{-1}(x)$ with $x \in P_2 \cup T_2$.

The condition that the nets are finite can be dropped if abstractions are defined with respect to arbitrary partitions. To illustrate the theorem, consider the abstraction in Figure 2.16b of the net from Figure 2.16a. Then the epimorphism φ is given by $p_1 \mapsto p_1$, $p_2 \mapsto t_Y$, $p_3 \mapsto p_3$, $\dots$, $p_{12} \mapsto p_{12}$, $p_{13} \mapsto t_{Y1}$, $\dots$, $p_{15} \mapsto t_{Y2}, \dots$, $t_1 \mapsto t_Y$, $t_2 \mapsto t_Y, \dots$, $t_6 \mapsto t_{Y1}$, $t_7 \mapsto t_{Y1}$, $t_8 \mapsto t_{Y2}$, $t_9 \mapsto t_{Y2}$. The mapping φ of Figure 2.17b is another such example.

Net morphisms are of particular interest when different net classes are considered. In the next chapter, coloured nets will be shown to be foldings of place/transition nets. Figure 2.19 shows a strict abstraction, which can be interpreted by Theorem 2.5.4b as an epifolding. A semantical interpretation will be given in Section 3.2.

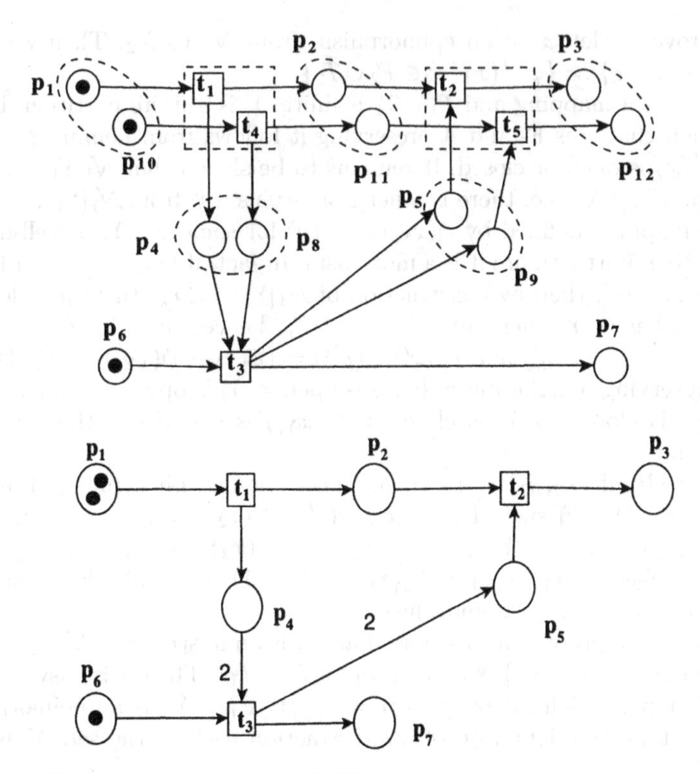

Fig. 2.19. Strict abstraction and epifolding

3. Intuitive Models*

This chapter introduces the most frequently used models of Petri nets, namely place/transition nets and coloured Petri nets, in an intuitive manner. While place/transition nets are a slight generalisation of the nets introduced in Chapter 2, coloured Petri nets allow much more compact models of systems. For bibliographic references see Chapter 1.

3.1 Arc-Constant Nets

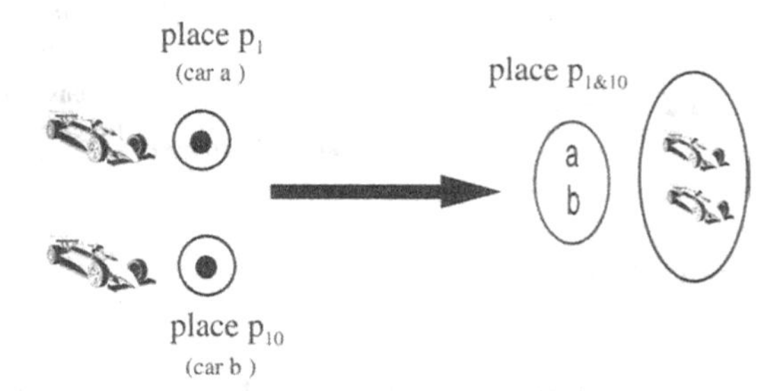

Fig. 3.1. From tokens distinguished by places to individual tokens

In the nets of Chapter 2 a token on a place indicated that the condition associated with that place was satisfied. Such tokens can also be seen as objects in a pool or resources in a storage facility. Two tokens, however, are not distinguishable from each other. The tokens representing *car a* and *car b* in the net of Figure 2.5 are distinguished by their places p_1 and p_{10}. A different but more compact and natural way is to represent them in one place, say $p_{1\&10}$, but by individual tokens a and b (see Figure 3.1).

Distinguishable tokens are said to be *coloured*. Usually coloured tokens are divided into different types which are called *colours*. Hence, colours can be thought of as data types. Since each place may contain objects of a specific

* Author: R. Valk

colour, for each place p a *colour domain* or *colour set* $cd(p)$ is defined. In our case $cd(p_{1\&10}) = \{a, b\}$. Other examples of colour domains are the set of integers or boolean values.

For a transition t with respect to an input place it must be indicated which of the individual tokens should be removed. In the net of Figure 3.2 this is done by inscriptions on the corresponding arc. For instance, transition t_1 can occur if there is an object a in the place $p_{1\&10}$. When this is the case, on the occurrence of t_1 the token a is removed and added to the place $p_{2\&11}$ and an (indistinguishable) token is added to p_4. We suppose that the places $p_{1\&10}$, $p_{2\&11}$, and $p_{3\&12}$ have the colour domain $cars = \{a, b\}$ denoting *car a* and *car b*. The starter is modelled by the token s, hence places p_6 and p_7 have the colour domain $starter = \{s\}$. As shown in Figure 3.2 colour domains are represented by lower case italics near the place symbols.

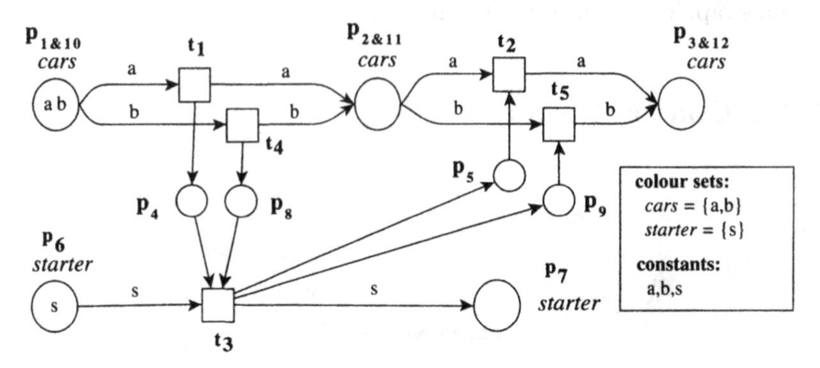

Fig. 3.2. Arc-constant CPN $\mathcal{N}_1$

The places p_4, p_8, p_5, and p_9 are supposed to hold an indistinguishable token and therefore have the colour domain $token = \{\bullet\}$, which is assumed to hold by default. Also, by default, arcs without any inscription (such as (t_1, p_4)) are assumed to bear the constant "$\bullet$". This example of a coloured net can be understood as a *folding* of the net $\mathcal{N}$ in Figure 2.5. To underline this aspect, names such as $p_{1\&10}$ are chosen since this place plays the role of the former places p_1 and p_{10}. The reader should note that the (ordinary) net $\mathcal{N}$ of Figure 2.5 and the (coloured) net $\mathcal{N}_1$ contain the same information and have a very similar behaviour. Since $\mathcal{N}_1$ contains objects from a colour domain, it is called a *coloured Petri net (CPN)*. To distinguish this model from general CPNs, we call it an *arc-constant coloured Petri net* (ac-CPN), since the inscriptions on the arcs are constants (and not variables).

For the next step in our introduction, we assume that the modeller of the example system also wants to represent the messages *"ready sign of car a"* (rsa), *"ready sign of car b"* (rsb), *"start sign for car a"* (ssa), and *"start sign for car b"* (ssb) explicitly by tokens with identifiers as given. The correspond-

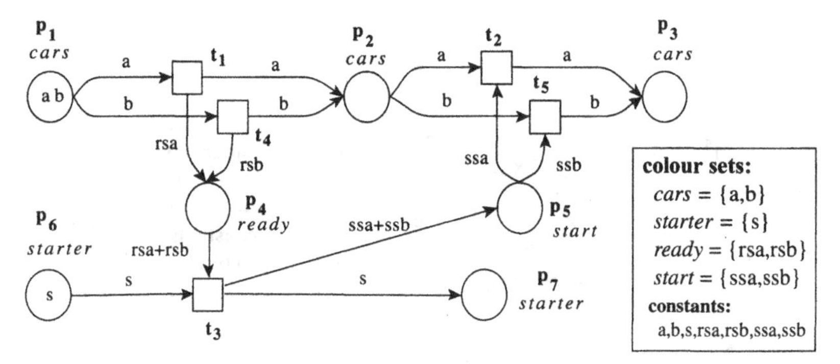

Fig. 3.3. Arc-constant CPN $\mathcal{N}_2$

ing net $\mathcal{N}_2$ is shown in Figure 3.3. As a new feature of this net, transition t_3 has to remove both signals rsa and rsb from place p_4. Thus the new expression $rsa + rsb$ denotes the set $\{rsa, rsb\}$. By such an inscription, t_3 is enabled if both rsa and rsb are in p_4 and on the occurrence of this transition both tokens are removed. In the general case, *bags* (multisets) will be used instead of sets where multiple copies of elements are allowed. A formal definition of bags will be given in Section 4.2. In the present section an intuitive understanding is sufficient.

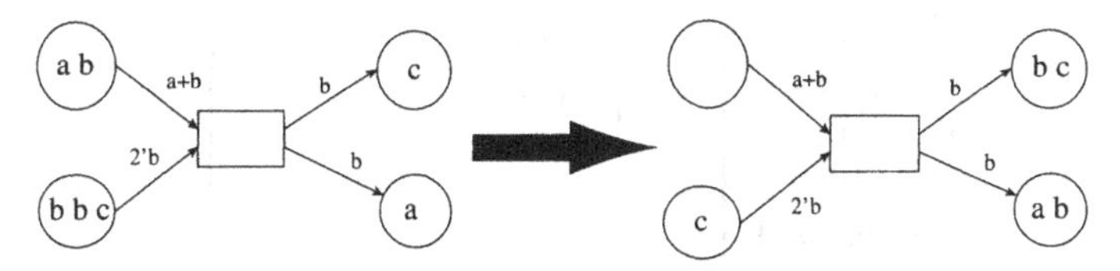

Fig. 3.4. Occurrence rule for ac-CPN

The transition-occurrence rule for arc-constant coloured nets is illustrated in Figure 3.4: all input places contain at least as many individual tokens as specified by the corresponding arcs. Thus the transition is enabled. In the successor marking they are removed and tokens are added to the output places, as indicated by the arc inscriptions.

The initial marking of $\mathcal{N}_2$ (Figure 3.3) can be formalised as a vector $(\{a, b\}, \emptyset, \emptyset, \emptyset, \emptyset, \{s\}, \emptyset)$ where entry i gives the bag of coloured tokens in place p_i. The entire occurrence sequence, i.e. the initial marking and all successor markings with respect to the transition sequence t_4, t_1, t_3, t_2, t_5, is given in the following vectors.

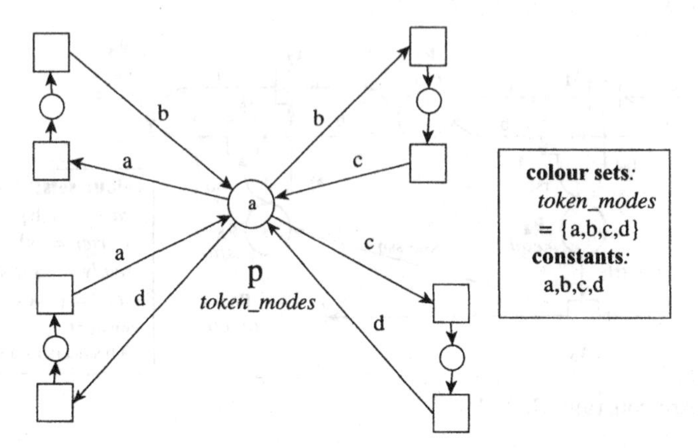

Fig. 3.5. Ring bus as ac-CPN

$$\begin{pmatrix} \{a,b\} \\ \emptyset \\ \emptyset \\ \emptyset \\ \emptyset \\ \{s\} \\ \emptyset \end{pmatrix} \xrightarrow{t_4} \begin{pmatrix} \{a\} \\ \{b\} \\ \emptyset \\ \{rsb\} \\ \emptyset \\ \{s\} \\ \emptyset \end{pmatrix} \xrightarrow{t_1} \begin{pmatrix} \emptyset \\ \{a,b\} \\ \emptyset \\ \{rsa, rsb\} \\ \emptyset \\ \{s\} \\ \emptyset \end{pmatrix} \xrightarrow{t_3}$$

$$\begin{pmatrix} \emptyset \\ \{a,b\} \\ \emptyset \\ \emptyset \\ \{ssa, ssb\} \\ \emptyset \\ \{s\} \end{pmatrix} \xrightarrow{t_2} \begin{pmatrix} \emptyset \\ \{b\} \\ \{a\} \\ \emptyset \\ \{ssb\} \\ \emptyset \\ \{s\} \end{pmatrix} \xrightarrow{t_5} \begin{pmatrix} \emptyset \\ \emptyset \\ \{a,b\} \\ \emptyset \\ \emptyset \\ \emptyset \\ \{s\} \end{pmatrix}$$

Figure 3.5 contains a different example of an arc-constant net. It shows a net where a token circulates through four functional units. This is however not organised by an elementary cycle, but by a bus structure where the central place represents the bus.

3.2 Place/Transition Nets

Now before introducing general CPNs we give an example in Figure 3.6 of a so-called place/transition net (P/T net) $\mathcal{N}_3$. It can be seen as an abstraction obtained from $\mathcal{N}_2$ by removing the individuality of tokens. In fact, the individual tokens a and b are replaced by two anonymous tokens "•".

As a consequence there is no need to distinguish transitions t_1 and t_4. Instead of the expression *rsa+rsb* for the set $\{rsa, rsb\}$ we obtain the number *2* indicating that two tokens have to be removed. In general, arc inscriptions in P/T nets are natural numbers denoting the number of tokens to be moved, as shown in Figure 3.7.

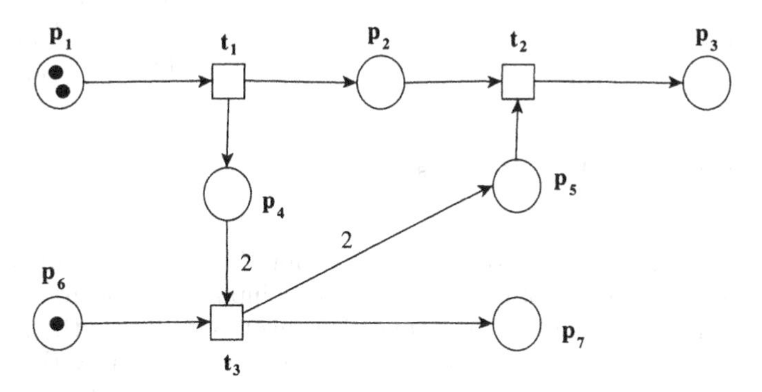

Fig. 3.6. Place/transition net $\mathcal{N}_3$

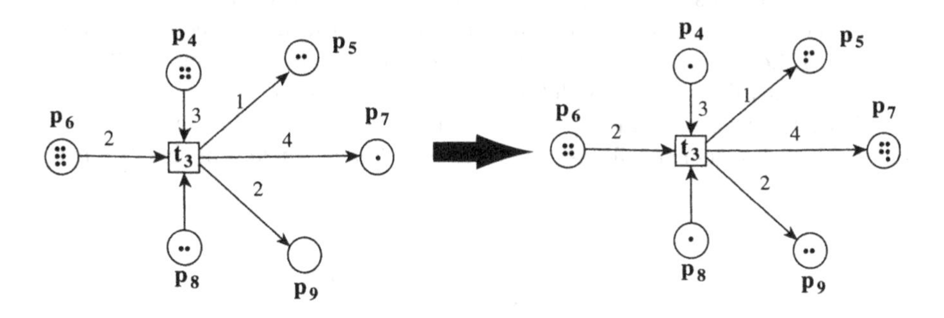

Fig. 3.7. Occurrence rule for P/T nets I

It is convenient to define *markings* as vectors of integers, assuming a total ordering of the places. Thus the initial marking of $\mathcal{N}_3$ in Figure 3.6 is the vector $\mathbf{m}_0 = (2, 0, 0, 0, 0, 1, 0)$. Alternatively, a marking may be given as a mapping $\mathbf{m}_0 : P \to \mathbb{N}$, i.e. in our example $\mathbf{m}_0(p_1) = 2$, $\mathbf{m}_0(p_6) = 1$ and $\mathbf{m}_0(p_i) = 0$ in all other cases. Sometimes it is also convenient to write a marking as a sequence of place names having exponents that give the number of tokens in that place. The place is omitted if this number is zero, i.e. $\mathbf{m}_0 = p_1^2 p_6$ or $\mathbf{m}_0 =< p_1^2 p_6 >$ (further variants of this notation are $\mathbf{m}_0 = p_1^2 + p_6$ and $\mathbf{m}_0 = < p_1^2 + p_6 >$).

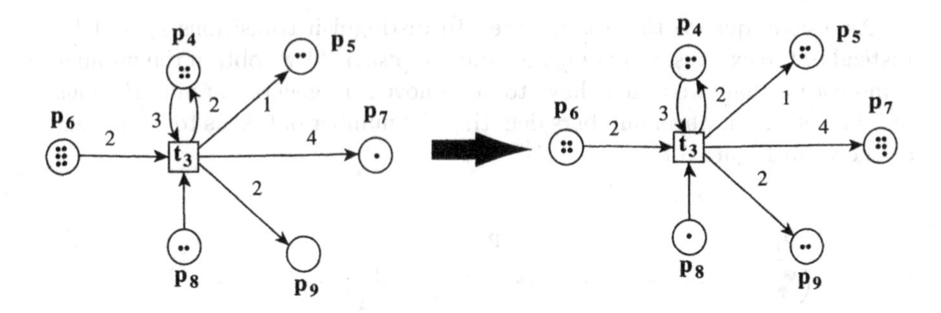

Fig. 3.8. Occurrence rule for P/T nets II

Figure 3.7 illustrates the occurrence rule for P/T nets. Transition t_3 is enabled since the number of tokens in any of its input places is greater than or equal to the number on the arc connecting this input place with t_3. When t_3 occurs, the number of tokens in the input place is reduced by this value. For instance, $\mathbf{m}(p_6) = 6$ becomes $\mathbf{m}_1(p_6) = 6 - 2 = 4$ in the successor marking $\mathbf{m}_1$. The number of tokens in the output places is increased by the corresponding value (e.g. $\mathbf{m}_1(p_7) = \mathbf{m}(p_7) + 4 = 1 + 4 = 5$). The occurrence rule extends easily to cases where places are both input *and* output places of a transition, as shown for p_4 in Figure 3.8. In this case the token number in the successor marking is computed by both operations of decreasing *and* increasing, i.e. $\mathbf{m}_1(p_4) = \mathbf{m}(p_4) - 3 + 2 = 4 - 3 + 2 = 3$. The occurrence rule for P/T nets will be formally defined in the next chapter.

Using the vector notation, a transition sequence with markings is given below. Such a sequence will be called the *occurrence sequence* of the P/T net $\mathcal{N}_3$.

$$\begin{pmatrix}2\\0\\0\\0\\0\\1\\0\end{pmatrix} \xrightarrow{t_1} \begin{pmatrix}1\\1\\0\\1\\0\\1\\0\end{pmatrix} \xrightarrow{t_1} \begin{pmatrix}0\\2\\0\\2\\0\\1\\0\end{pmatrix} \xrightarrow{t_3} \begin{pmatrix}0\\2\\0\\0\\2\\0\\1\end{pmatrix} \xrightarrow{t_2} \begin{pmatrix}0\\1\\1\\0\\1\\0\\1\end{pmatrix} \xrightarrow{t_2} \begin{pmatrix}0\\0\\2\\0\\0\\0\\1\end{pmatrix}$$

3.3 Coloured Nets

Let us return to coloured Petri nets. Now consider net $\mathcal{N}_4$ in Figure 3.9 compared with net $\mathcal{N}_2$ in Figure 3.3. The most significant difference of these nets is that transitions t_1, t_4 and t_2, t_5 of $\mathcal{N}_2$ are folded together and renamed in $\mathcal{N}_4$ as t_1 and t_2 respectively. To preserve the original behaviour, transitions t_1 and t_2 are considered to occur in two "occurrence modes" *mode1* and *mode2*. Depending on the mode, different elements in the arc inscriptions have to be selected. To this end arc expressions are replaced by *arc vectors* of such expressions, as shown in Figure 3.9. Each entry of an arc vector uniquely corresponds to an occurrence mode of the transition.

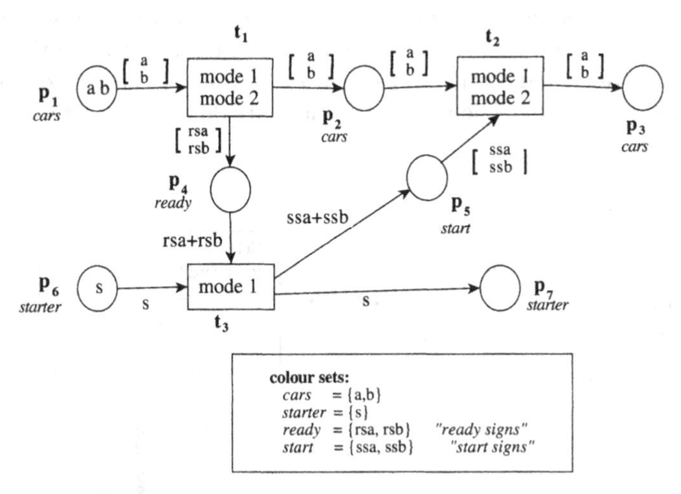

Fig. 3.9. CPN $\mathcal{N}_4$ with transition modes

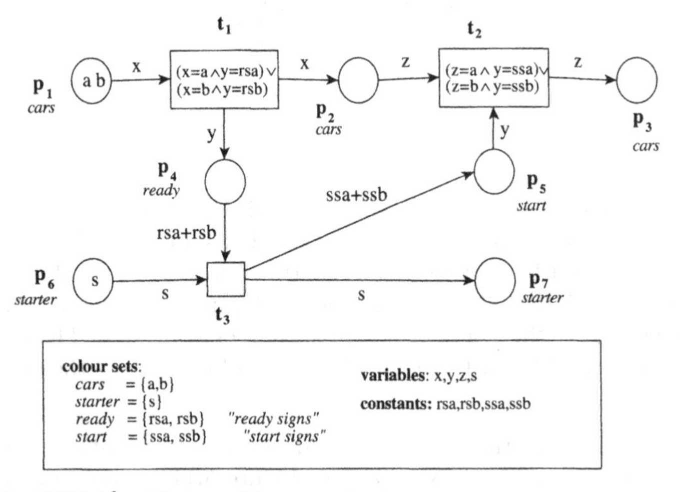

Fig. 3.10. CPN $\mathcal{N}_5$ with transition guards

The occurrence rule is changed as follows:

1. Select an occurrence mode of the transition.
2. Temporarily replace arc vectors by the component corresponding to this mode.
3. Apply to this transition the occurrence rule of arc-constant CPN.

Usually in the literature the association of values with different occurrence modes is represented by expressions containing variables. For each transition a finite set of variables is defined which is strictly local to this transition. These variables have types or colour domains which usually are the colours of the places connected to the transition. In Figure 3.10 the set of variables of transition t_1 is $\{x, y\}$. The types of x and y are dom(x) = *cars* and dom(y) = *ready*, respectively.

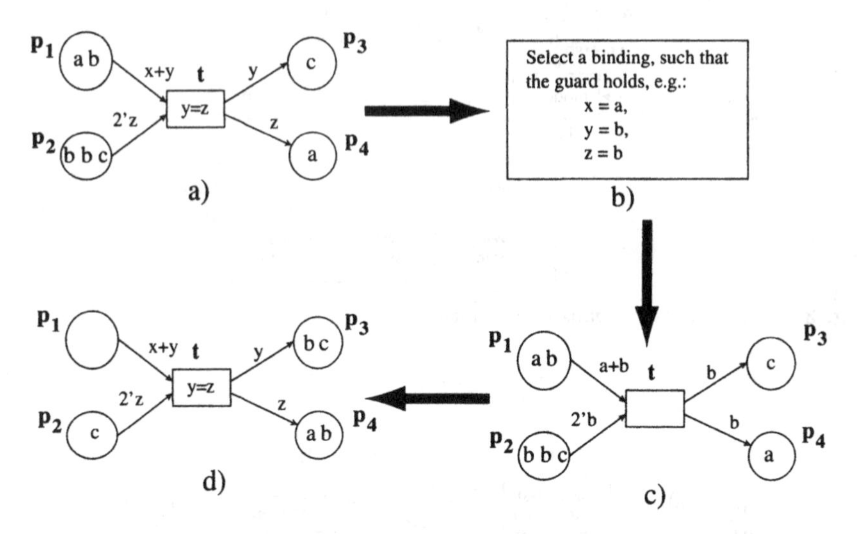

Fig. 3.11. Occurrence rule for CPN

An assignment of values to variables is called a *binding*. Clearly, not all possible bindings can be allowed for a correctly behaving net. The appropriate restriction is defined by a predicate at the transition which is called a *guard*. The occurrence rule for CPNs is explained in Figure 3.11, where all places p are assumed to have the colour set $cd(p) = objects = \{a, b, c\}$. The colour (or domain) of all variables in this example is also *objects*. Now the occurrence rule is:

1. Select a binding such that the guard holds, i.e. associate with each variable a value of its colour, as shown in b) of Figure 3.11 for the binding $\beta_1 = [x = a, y = b, z = b]$.
2. Temporarily replace variables by the associated constants, as shown in c).

3. Apply to this transition the ac-CPN occurrence rule from Figure 3.4, as shown in d).

The occurrence of a transition should be understood as a single step from a) to d). If in b) a binding such as $\beta_2 = [x = a, y = b, z = c]$ is selected, then the transition is *not* activated in this binding (or mode) since the guard is not satisfied. For a binding such as $\beta_3 = [x = b, y = b, z = b]$ the guard holds, but there are insufficient tokens in the input places, namely there are not two copies of b in p_1. The selection of a binding is local to a transition. Hence, the variable z in $\mathcal{N}_5$ can be replaced by x without changing the behaviour of the net.

Applying the occurrence rule to the CPN $\mathcal{N}_5$, we obtain the same behaviour as for $\mathcal{N}_4$. In particular, the two possible bindings of t_1 correspond to the two occurrence modes *mode1* and *mode2* of t_1 in $\mathcal{N}_4$. The representation by variables becomes essential if there is a large number of occurrence modes. To give an example, the places and the transition of the net in Figure 3.12 have the colour set *integer*. Hence, there is an infinity of occurrence modes for the transition. (The expression *x+1* in this net denotes an arithmetic operation.) The successor marking has the integers 4 and 3 in p_1 and p_2 respectively.

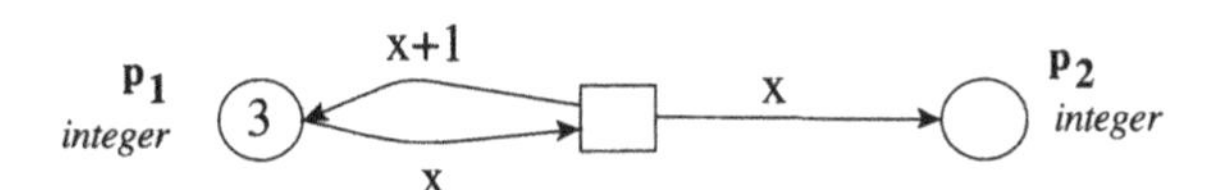

Fig. 3.12. CPN with colour domain integer

An *occurrence sequence* for the coloured net of Figure 3.10 can be represented as follows. The entries of the vectors are sets of colour elements instead of integers. The transitions are given together with the binding used, in particular: $\beta_a = [x = a, y = rsa]$, $\beta_b = [x = b, y = rsb]$ and $\hat{\beta}_a = [x = a, y = ssa]$, $\hat{\beta}_b = [x = b, y = ssb]$. For transition t_3 no binding is necessary since all arcs have constants as inscriptions. β is the "empty binding".

$$\begin{pmatrix} \{a,b\} \\ \emptyset \\ \emptyset \\ \emptyset \\ \emptyset \\ \{s\} \\ \emptyset \end{pmatrix} \xrightarrow{(t_1,\beta_b)} \begin{pmatrix} \{a\} \\ \{b\} \\ \emptyset \\ \{rsb\} \\ \emptyset \\ \{s\} \\ \emptyset \end{pmatrix} \xrightarrow{(t_1,\beta_a)} \begin{pmatrix} \emptyset \\ \{a,b\} \\ \emptyset \\ \{rsa, rsb\} \\ \emptyset \\ \{s\} \\ \emptyset \end{pmatrix} \xrightarrow{(t_3,\beta)}$$

$$\begin{pmatrix} \emptyset \\ \{a,b\} \\ \emptyset \\ \emptyset \\ \{ssa, ssb\} \\ \emptyset \\ \{s\} \end{pmatrix} \xrightarrow{(t_2,\hat{\beta}_a)} \begin{pmatrix} \emptyset \\ \{b\} \\ \{a\} \\ \emptyset \\ \{ssb\} \\ \emptyset \\ \{s\} \end{pmatrix} \xrightarrow{(t_2,\hat{\beta}_b)} \begin{pmatrix} \emptyset \\ \emptyset \\ \{a,b\} \\ \emptyset \\ \emptyset \\ \emptyset \\ \{s\} \end{pmatrix}$$

Note that the introduced semantics of coloured nets preserve all principles from Chapter 2:

I) The principle of duality for Petri nets: Places are clearly distinguished from transitions.
II) The principle of locality for Petri nets: The behaviour of a transition depends only on its locality (which should be understood here as all places, arcs, and guards connected with this transition).
III) The principle of concurrency for Petri nets: Transitions with disjoint localities behave concurrently.
IV) The principle of graphical representation for Petri nets: obvious.
V) The principle of algebraic representation for Petri nets: see the next chapter.

3.4 Foldings

The techniques of abstraction, refinement, and composition apply to coloured nets as well. The example of Figure 2.15 could be modified for coloured nets in the style of $\mathcal{N}_5$ in Figure 3.10.

As shown by the intuitive introduction of different models, particular net morphisms, namely foldings, are useful to relate higher-level nets to lower-level nets. In this section the informal notion will be related to the formal introduction of a folding in Definition 2.5.1. In this definition a folding was defined as a net morphism φ that maps places to places and transitions to transitions, i.e. $\varphi(P_1) \subseteq P_2$ and $\varphi(T_1) \subseteq T_2$. Furthermore, the F- and the A-relations are preserved: $(x,y) \in F_1 \Rightarrow (\varphi(x),\varphi(y)) \in F_2$ and $(x,y) \in A_1 \Rightarrow (\varphi(x),\varphi(y)) \in A_2$. In contrast with to general net morphisms, arcs are preserved and not hidden by abstraction. By part b) of Theorem 2.5.4, (epi-)foldings are represented by strict abstractions.

To give an example, the arc-constant net $\mathcal{N}_2$ of Figure 3.3 can be seen as a folding of its first version, the net $\mathcal{N}$ in Figure 2.5. By Theorem 2.5.4 foldings can be represented as strict abstractions as shown in Figure 3.13. In part a) of this figure the net from Figure 2.5 is given in a different layout and with some open sets marked by dashed lines. Since there are no F-arcs remaining within these sets, the abstraction is strict and the corresponding

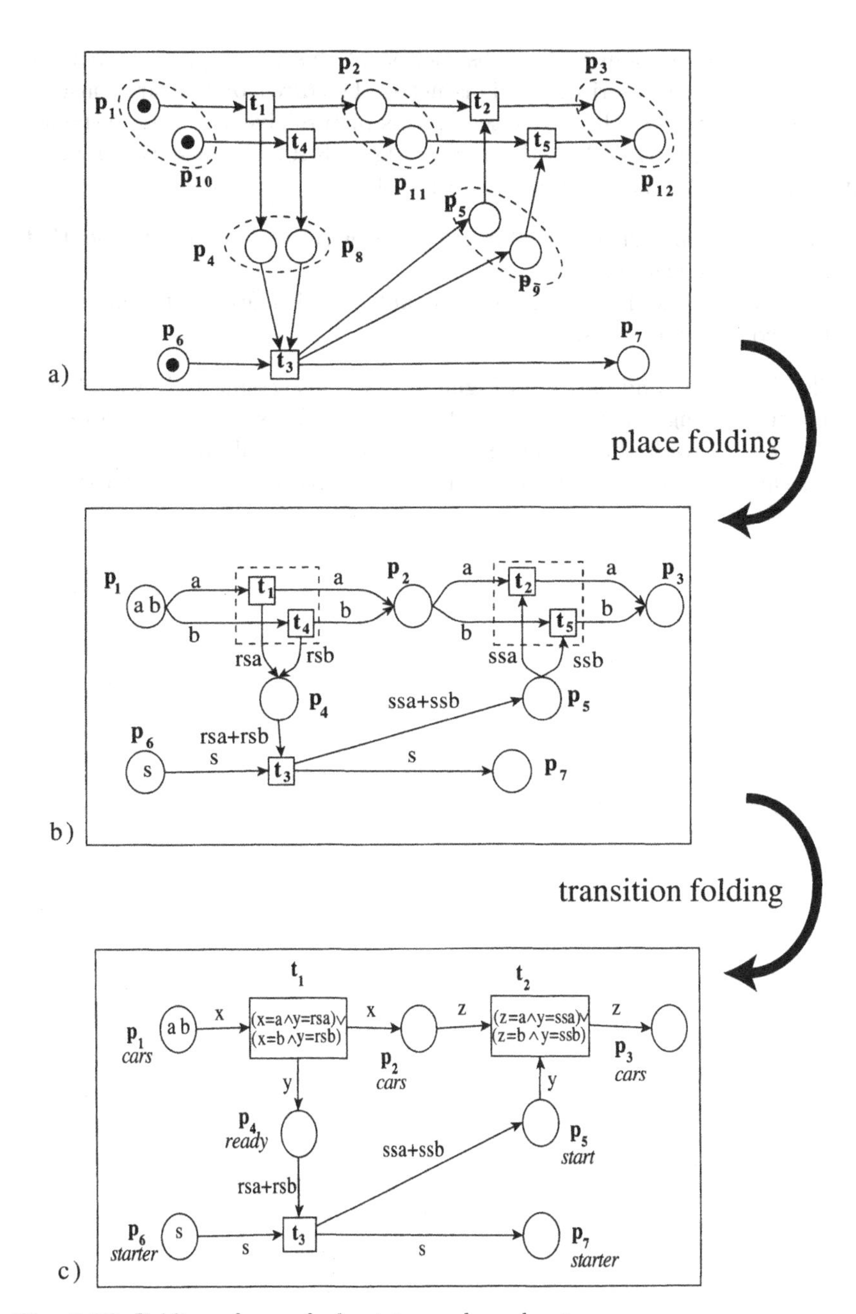

Fig. 3.13. Folding of a marked net to a coloured net

morphism is an epifolding. The abstraction of this net is given in part b) together with some inscriptions. This net is the corresponding arc-constant CPN from Figure 3.3. The closed sets in this picture give a folding to the net in part c), which is the coloured net from Figure 3.10. The observation made with this example leads to a general principle:

I) An arc-constant CPN is a folding of a net with respect to open sets that contain no transitions.
II) A CPN is the folding of an arc-constant net with respect to closed sets that contain no places.

Clearly, part a) and part c) of Figure 3.13 are also connected by a folding, namely the folding shown in Figure 2.19. The reader should note, however, that the notions of folding and strict abstraction refer only to the net underlying a particular model and not to the inscriptions related to the particular net models.

4. Basic Definitions*

In this chapter a formal definition of place/transition nets and coloured Petri nets is given. Although the former are special cases of the latter, for didactic reasons we start with place/transition nets.

4.1 Formal Definition of Place/Transition Nets

Place/transition nets are nets in the sense of Definition 2.2.1 together with a definition of arc weights.

Definition 4.1.1. *A* place/transition net *(P/T net) is defined by a tuple* $\mathcal{N} = \langle P, T, \mathbf{Pre}, \mathbf{Post} \rangle$, *where*

- P *is a finite set (the* set of places *of* $\mathcal{N}$*),*
- T *is a finite set (the* set of transitions *of* $\mathcal{N}$*), disjoint from* P*, and*
- $\mathbf{Pre}, \mathbf{Post} \in \mathbb{N}^{|P|\times|T|}$ *are matrices (the* backward *and* forward incidence matrices *of* $\mathcal{N}$*).* $\mathbf{C} = \mathbf{Post} - \mathbf{Pre}$ *is called the* incidence matrix *of* $\mathcal{N}$*.*

There is an arc with weight $n > 0$ from a place $p \in P$ to some transition $t \in T$ iff $\mathbf{Pre}[p, t] = n$ with $n > 0$ and there is an arc with weight $n > 0$ from a transition $t \in T$ to some place $p \in P$ iff $\mathbf{Post}[p, t] = n > 0$. Hence, $F := \{(p, t) \in P \times T \mid \mathbf{Pre}[p, t] > 0\} \cup \{(t, p) \in T \times P \mid \mathbf{Post}[p, t] > 0\}$ is the set of arcs of $\mathcal{N}$.

This leads to the following alternative but equivalent definition of place/transition nets, which is closer to the graphical representation.

Definition 4.1.2. *A* place/transition net *(P/T net) is defined by a tuple* $\mathcal{N} = \langle P, T, F, W \rangle$, *where*

- $\langle P, T, F \rangle$ *is a net (Definition 2.2.1) with finite sets* P *and* T*, and*
- $W : F \to \mathbb{N}\backslash\{0\}$ *is a function (*weight function*).*

Example 4.1.3. For the P/T net $\mathcal{N}_3$ of Figure 3.6 we have:
$P = \{p_1, \ldots, p_7\}$, $T = \{t_1 \ldots, t_3\}$, $F = \{(p_1, t_1), (t_1, p_2), \ldots, (p_4, t_3), \ldots\}$, $W(p_1, t_1) = 1, W(t_1, p_2) = 1, \ldots, W(p_4, t_3) = 2, \ldots,$

---

* Author: R. Valk

$\mathbf{Pre}[p_1, t_1] = 1, \mathbf{Post}[p_2, t_1] = 1, \ldots, \mathbf{Pre}[p_4, t_3] = 2, \ldots$
The complete matrices **Pre**, **Post**, and **C** are given in Table 4.1. Empty entries denote the value zero.

Table 4.1. The incidence matrices of the P/T net $\mathcal{N}_3$

Pre	t_1	t_2	t_3	**Post**	t_1	t_2	t_3	**C**	t_1	t_2	t_3
p_1	1			p_1				p_1	−1		
p_2		1		p_2	1			p_2	1	−1	
p_3				p_3		1		p_3		1	
p_4			2	p_4	1			p_4	1		−2
p_5		1		p_5			2	p_5		−1	2
p_6			1	p_6				p_6			−1
p_7				p_7			1	p_7			1

Definition 4.1.4. *A* marking *of a P/T net* $\mathcal{N} = \langle P, T, \mathbf{Pre}, \mathbf{Post} \rangle$ *is a vector* $\mathbf{m} \in \mathbb{N}^{|P|}$. $\mathcal{N}$ *together with a marking* $\mathbf{m_0}$ *(initial marking) is called a* P/T net system $\mathcal{S} = \langle \mathcal{N}, \mathbf{m_0} \rangle$ *or* $\mathcal{S} = \langle P, T, \mathbf{Pre}, \mathbf{Post}, \mathbf{m_0} \rangle$. *A transition* $t \in T$ *is* enabled *in a marking* $\mathbf{m}$ *if* $\mathbf{m} \geq \mathbf{Pre}[\bullet, t]$. *In this case the* successor-marking relation *is defined by* $\mathbf{m} \xrightarrow{t} \mathbf{m}' \Leftrightarrow \mathbf{m} \geq Pre[\bullet, t] \wedge \mathbf{m}' = \mathbf{m} + \mathbf{Post}[\bullet, t] - \mathbf{Pre}[\bullet, t] = \mathbf{m} + C[\bullet, t]$. $\mathbf{Pre}[\bullet, t]$ *denotes the t-column vector* $\mathbf{Pre}[\bullet, t] = (\mathbf{Pre}[p_1, t], \ldots, \mathbf{Pre}[p_{|P|}, t])$ *of the* $|P| \times |T|$ *matrix* **Pre**. *The same holds for* $\mathbf{Post}[\bullet, t]$ *with respect to* **Post**.[1]

For the net $\mathcal{N}_3$ from Figure 3.6 and Example 4.1.3 the initial marking is $\mathbf{m_0} = (2, 0, 0, 0, 0, 1, 0)$. Since $Pre[\bullet, t_1] = (1, 0, 0, 0, 0, 0, 0)$, transition t_1 is enabled in $\mathbf{m_0}$ and the successor marking is $\mathbf{m}' = \mathbf{m_0} + C[\bullet, t_1] = (2, 0, 0, 0, 0, 1, 0) + (-1, 1, 0, 1, 0, 0, 0) = (1, 1, 0, 1, 0, 1, 0)$.

Definition 4.1.5. *The successor-marking relation of Definition 4.1.4 is extended to hold for sequences of transitions (i.e. elements from the set* T^* *of word over the alphabet* T*) by*

- $\mathbf{m} \xrightarrow{w} \mathbf{m}'$ *if* w *is the empty word* λ *and* $\mathbf{m} = \mathbf{m}'$ *and*
- $\mathbf{m} \xrightarrow{wt} \mathbf{m}'$ *if* $\exists \mathbf{m}'' . \mathbf{m} \xrightarrow{w} \mathbf{m}'' \wedge \mathbf{m}'' \xrightarrow{t}$ *for* $w \in T^*$ *and* $t \in T$.

For a net system $\mathcal{S} = \langle \mathcal{N}, \mathbf{m_0} \rangle$ *the set* $\mathrm{RS}(\mathcal{S}) = \mathrm{RS}(\mathcal{N}, \mathbf{m_0}) := \{\mathbf{m} \mid \exists w \in T^* . \mathbf{m_0} \xrightarrow{w} \mathbf{m}\}$ *is the* reachability set. *It can be denoted by* $\mathrm{RS}(\mathbf{m_0})$ *if* $\mathcal{N}$ *is obvious from the context.* $FS(\mathcal{S}) := \{w \in T^* \mid \exists \mathbf{m} . \mathbf{m_0} \xrightarrow{w} \mathbf{m}\}$ *is the* set of occurrence-transition sequences *(or* firing-sequence set*) of* $\mathcal{S}$.

As mentioned in Chapter 3, it is sometimes convenient to define the set $Occ(\mathcal{S})$ of *occurrence sequences* to be the set of all sequences of the form

[1] Note that sometimes the notations $\mathbf{Pre}[P, t]$ and $\mathbf{Post}[P, t]$ are used instead of $\mathbf{Pre}[\bullet, t]$ and $\mathbf{Post}[\bullet, t]$ respectively. The intended meaning is the same in both cases.

$$\mathbf{m_0}, t_0, \mathbf{m_1}, t_1, \mathbf{m_2}, t_2, \ldots, t_{n-1}, \mathbf{m_n} \qquad (n \geq 1)$$

such that $\mathbf{m_i} \xrightarrow{t_i} \mathbf{m_{i+1}}$ for $i \in \{0, \ldots, n-1\}$. An example can be found in Section 3.2.

4.2 Formal Definition of Arc-Constant Nets

For a coloured net we have to specify the colours and, for all places and transitions, particular colour sets (colour domains). Since arc inscriptions may contain different elements or, as in the case of P/T nets, multiple copies of an element, bags (multisets) are used.

A *bag* (or *multiset*) 'bg', over a non-empty set A, is a function $bg : A \to \mathbb{N}$, sometimes denoted as a formal sum $\sum_{a \in A} bg(a)'a$. Bag(A) denotes the set of all bags over A. Extending set operations to Bag(A) we define the *sum* (+) and *difference* (−) as follows:

If bg, bg_1 and bg_2 are bags over A, then:

- $bg_1 + bg_2 := \sum_{a \in A}(bg_1(a) + bg_2(a))'a$
- $bg_1 \leq bg_2 :\Leftrightarrow \forall a \in A \,.\, bg_1(a) \leq bg_2(a)$
- $bg_1 - bg_2 := \sum_{a \in A}(bg_1(a) - bg_2(a))'a$ if $bg_2 \leq bg_1$ and
- $|bg| := \sum_{a \in A} bg(a)$ is the size of bg and $\emptyset$ denotes the *empty bag* (with $|bg| = 0$).

An example of a bag over $A = \{a, b, c, d\}$ is $bg_1 = \{a, a, b, b, b, d\}_b$ (where the index distinguishes set brackets from bag brackets) or, equivalently, $bg_1 = 2'a + 3'b + d$. With $bg_2 = a + 2'b$ we obtain $bg_1 + bg_2 = 3'a + 5'b + d$ and $bg_1 - bg_2 = 1'a + 1'b + 1'd = a + b + d$.

Definition 4.2.1. *An* arc-constant coloured Petri net *(ac-CPN) is defined by a tuple* $\mathcal{N} = \langle P, T, \mathbf{Pre}, \mathbf{Post}, \mathcal{C}, cd \rangle$, *where*

- P *is a finite set (the* set of places *of* $\mathcal{N}$*),*
- T *is a finite set (the* set of transitions *of* $\mathcal{N}$*), disjoint from* P*,*
- $\mathcal{C}$ *is the set of* colour classes,
- cd*:* $P \to \mathcal{C}$ *is the* colour domain mapping, *and*
- $\mathbf{Pre}, \mathbf{Post} \in \mathcal{B}^{|P| \times |T|}$ *are matrices (the* backward *and* forward incidence matrices *of* $\mathcal{N}$*) such that* $\mathbf{Pre}[p,t] \in \mathrm{Bag}(cd(p))$ *and* $\mathbf{Post}[p,t] \in \mathrm{Bag}(cd(p))$ *for each* $(p,t) \in P \times T$*.* $\mathbf{C} = \mathbf{Post} - \mathbf{Pre}$ *is called the* incidence matrix.

Note that in this definition $\mathcal{B}$ can be taken as the set Bag(A), where A is the union of all colour sets from $\mathcal{C}$. The difference operator in $\mathbf{C} = \mathbf{Post} - \mathbf{Pre}$ is a formal one here, i.e. the difference is not computed as a value.

Example 4.2.2. The matrices **Pre** and **Post** of the arc-constant CPN $\mathcal{N}_2$ of Figure 3.3 are given in Table 4.2. To improve clarity, the colour domains of the places are also contained in these matrices. Bags are represented by formal expressions, e.g. $2'a + b$ denotes the bag containing a twice and b once.

Table 4.2. The incidence matrices of the ac-CPN $\mathcal{N}_2$ from Figure 3.3

Pre	t_1	t_2	t_3	t_4	t_5
$p_1 : cars$	a			b	
$p_2 : cars$		a			b
$p_3 : cars$					
$p_4 : ready$			$rsa+rsb$		
$p_5 : start$		ssa			ssb
$p_6 : starter$			s		
$p_7 : starter$					

Post	t_1	t_2	t_3	t_4	t_5
$p_1 : cars$					
$p_2 : cars$	a			b	
$p_3 : cars$		a			b
$p_4 : ready$	rsa			rsb	
$p_5 : start$			$ssa+ssb$		
$p_6 : starter$					
$p_7 : starter$			s		

colour sets:
$cars = \{a, b\}, starter = \{s\}$
$ready = \{rsa, rsb\}$ *"ready signs"*
$start = \{ssa, ssb\}$ *"start signs"*
constants:
$a, b, s, rsa, rsb, ssa, ssb$

Definition 4.2.3. *A marking of an ac-CPN* $\mathcal{N} = \langle P, T, \mathbf{Pre}, \mathbf{Post}, \mathcal{C}, cd\rangle$ *is a vector* $\mathbf{m}$ *such that* $\mathbf{m}[p] \in \mathrm{Bag}(cd(p))$ *for each* $p \in P$. $\mathcal{N}$ *together with a marking* $\mathbf{m}_0$ *(initial marking) is called an* ac-CPN system *and is denoted by* $\mathcal{S} = \langle \mathcal{N}, \mathbf{m}_0\rangle$ *or* $\mathcal{S} = \langle P, T, \mathbf{Pre}, \mathbf{Post}, \mathcal{C}, cd, \mathbf{m}_0\rangle$. *A transition* $t \in T$ *is* enabled *in a marking* $\mathbf{m}$ *if* $\mathbf{m} \geq \mathbf{Pre}[\bullet, t]$ *(denoted by* $\mathbf{m}\xrightarrow{t}$*). In this case the* successor-marking relation *is defined by* $\mathbf{m}\xrightarrow{t}\mathbf{m}' \Leftrightarrow \mathbf{m} \geq \mathbf{Pre}[\bullet, t] \wedge \mathbf{m}' = \mathbf{m} + \mathbf{Post}[\bullet, t] - \mathbf{Pre}[\bullet, t] = \mathbf{m} + \mathbf{C}[\bullet, t]$. *(The bag operations* $+$, $-$ *and the bag relation* $\leq$ *are extended to vectors.)*

Definition 4.2.4. *The successor-marking relation of Definition 4.2.3 is extended to hold for sequences of transitions by*

- $\mathbf{m}\xrightarrow{w}\mathbf{m}'$ *if* w *is the empty word* λ *and* $\mathbf{m} = \mathbf{m}'$ *and*
- $\mathbf{m}\xrightarrow{wt}\mathbf{m}'$ *if* $\exists \mathbf{m}'' . \mathbf{m}\xrightarrow{w}\mathbf{m}'' \wedge \mathbf{m}''\xrightarrow{t}\mathbf{m}'$ *for* $w \in T^*$ *and* $t \in T$.

For a net system $\mathcal{S} = \langle \mathcal{N}, \mathbf{m_0}\rangle$ *the set* $\mathrm{RS}(\mathcal{S}) = \mathrm{RS}(\mathcal{N}, \mathbf{m_0}) := \{\mathbf{m} \mid \exists w \in T^* . \mathbf{m_0}\xrightarrow{w}\mathbf{m}\}$ *is the* reachability set. *It can be denoted by* $\mathrm{RS}(\mathbf{m_0})$, *if* $\mathcal{N}$ *is obvious from the context.* $FS(\mathcal{S}) := \{w \in T^* \mid \exists \mathbf{m} . \mathbf{m_0}\xrightarrow{w}\mathbf{m}\}$ *is the* set of occurrence-transition sequences *(or* firing sequence set*) of* $\mathcal{S}$.

For the net $\mathcal{N}_2$ from Figure 3.3 and Example 4.2.2 the initial marking is $\mathbf{m}_0 = (a + b, \emptyset, \emptyset, \emptyset, \emptyset, s, \emptyset)$. Since $Pre[\bullet, t_1] = (a, \emptyset, \emptyset, \emptyset, \emptyset, \emptyset, \emptyset)$, transition t_1 is enabled in $\mathbf{m}_0$ and the successor marking is

$$\begin{aligned}
\mathbf{m}' &= \mathbf{m}_0 + C[\bullet, t_1] \\
&= \mathbf{m}_0 + \mathbf{Post}[\bullet, t_1] - \mathbf{Pre}[\bullet, t_1] \\
&= (a + b, \emptyset, \emptyset, \emptyset, \emptyset, s, \emptyset) + (\emptyset, a, \emptyset, rsa, \emptyset, \emptyset, \emptyset) - (a, \emptyset, \emptyset, \emptyset, \emptyset, \emptyset, \emptyset) \\
&= (b, a, \emptyset, rsa, \emptyset, s, \emptyset).
\end{aligned}$$

4.3 Formal Definition of Coloured Nets

In a coloured Petri net the incidence matrices cannot be defined over $\mathcal{B} = \text{Bag}(A)$ as for arc-constant CPNs. For a transition the different modes or bindings of a transition t have to be represented. These are called *colours*, and are denoted by $cd(t)$. Therefore the colour domain mapping cd is extended from P to $P \cup T$. In the entries of the incidence matrices for each transition colour, a multiset has to be specified. This is formalised by a mapping from $cd(t)$ into the bags of colour sets over $cd(p)$ for each $(p,t) \in P \times T$.

Definition 4.3.1. *A* coloured Petri net *(CPN) is defined by a tuple* $\mathcal{N} = \langle P, T, \mathbf{Pre}, \mathbf{Post}, \mathcal{C}, cd\rangle$, *where*

- P *is a finite set (the* set of places *of* $\mathcal{N}$*),*
- T *is a finite set (the* set of transitions *of* $\mathcal{N}$*), disjoint from* P*,*
- $\mathcal{C}$ *is the set of* colour classes*,*
- $cd\colon P \cup T \to \mathcal{C}$ *is the* colour domain mapping*, and*
- $\mathbf{Pre}, \mathbf{Post} \in \mathcal{B}^{|P| \times |T|}$ *are matrices (the* backward *and* forward incidence matrices *of* $\mathcal{N}$*) such that* $\mathbf{Pre}[p,t] : cd(t) \to \text{Bag}(cd(p))$ *and* $\mathbf{Post}[p,t] : cd(t) \to \text{Bag}(cd(p))$ *are mappings for each pair* $(p,t) \in P \times T$.

$\mathcal{B}$ *can be taken as the set of mappings of the form* $f : cd(t) \to \text{Bag}(cd(p))$. *Again,* $\mathbf{C} = \mathbf{Post} - \mathbf{Pre}$ *is called the* incidence matrix.

As introduced in Chapter 3 and specified in Definition 4.3.1, the mapping

$$\mathbf{Pre}[p,t] : cd(t) \to \text{Bag}(cd(p))$$

Table 4.3. The incidence matrices of $\mathcal{N}_4$ and $\mathcal{N}_5$ in vector form

Pre	t_1	t_2	t_3
	$mode_1$ $mode_2$	$mode_1$ $mode_2$	$mode_1$
p_1 : *cars*	a b		
p_2 : *cars*		a b	
p_3 : *cars*			
p_4 : *ready*			$rsa+rsb$
p_5 : *start*		ssa ssb	
p_6 : *starter*			s
p_7 : *starter*			

Post	t_1	t_2	t_3
	$mode_1$ $mode_2$	$mode_1$ $mode_2$	$mode_1$
p_1 : *cars*			
p_2 : *cars*	a b		
p_3 : *cars*		a b	
p_4 : *ready*	rsa rsb		
p_5 : *start*			$ssa+ssb$
p_6 : *starter*			
p_7 : *starter*			s

colour sets:
$cars = \{a, b\}$, $starter = \{s\}$
$ready = \{rsa, rsb\}$ "ready signs"
$start = \{ssa, ssb\}$ "start signs"

defines for each transition colour (or occurrence mode) $\beta \in cd(t)$ of t a bag $\mathbf{Pre}[p,t](\beta) \in \text{Bag}(cd(p))$ denoting the token bag to be removed from p when t occurs in colour β. In a similar way, $\mathbf{Post}[p,t](\beta)$ specifies the bag to be added to p when t occurs in colour β. Hence, the overall effect of the action performed on the occurrence of a transition t is given by a tuple corresponding to the arcs connected with t. To be more precise, let t be a transition having $q = |{}^\bullet t|$ input places. Then for a transition colour $\beta \in cd(t)$, the effect of an occurrence of t with respect to these input places ${}^\bullet t = \{p_1, \ldots, p_q\}$ is given by the q-tuple of bags:

$$(bg_1, .., bg_q) = (\mathbf{Pre}[p_1,t], \ldots, \mathbf{Pre}[p_q,t]) \in \text{Bag}(cd(p_1)) \times \ldots \times \text{Bag}(cd(p_q)).$$

The same holds with respect to $t^\bullet$ and **Post**.

The colours (i.e. modes) of a transition t can be seen as particular subsets of tuples $cd(t) \subseteq \text{Bag}(cd(p_1)) \times \ldots \times \text{Bag}(cd(p_{|P|}))$, i.e. vectors having an entry for each place. However, as discussed before this can be an arbitrary set as well. In applications and for analytical tools effective representations of this set are necessary. In the following section we will discuss four classical ways to denote the mappings $\mathbf{Pre}[p,t]$ and $\mathbf{Post}[p,t]$, namely:

a) by vectors,
b) by projections,
c) by functions, and
d) by terms with functions and variables.

a) Representation of the incidence matrix entries by vectors: Assuming a predefined ordering $\beta_1 < \beta_2 < \beta_3 < \ldots$ of $cd(t) = \{\beta_1, \beta_2, \beta_3 \ldots\}$, the mapping $\mathbf{Pre}[p,t]$ is represented by the vector:

$$(\mathbf{Pre}[p,t](\beta_1), \mathbf{Pre}[p,t](\beta_2), \mathbf{Pre}[p,t](\beta_3) \ldots)$$

If the set $cd(t)$ is finite, a finite vector is obtained. To give an example, Table 4.3 shows such a representation for the CPN $\mathcal{N}_4$ of Figure 3.9. The transition colours β_1 and β_2 are denoted by $mode_1$ and $mode_2$ respectively, near the transition names.

b) Representation of the incidence matrix entries by projections: This representation is directly based on the transition colour set

$$cd(t) \subseteq \text{Bag}(cd(p_1)) \times \ldots \times \text{Bag}(cd(p_q))$$

assuming an ordering on a subset of places ($q \leq |P|$). Then the appropriate bag for each arc with respect to a transition colour is obtained by a projection function $pr_i : (bg_1, \ldots, bg_q) \mapsto bg_i$, returning the i-th component of the tuple. An example is given in Figure 4.1. Here the inscriptions on the arcs of the CPN $\mathcal{N}_6$, as well as the entries of the incidence matrices, are projections on the transition colour sets, which are given in the lower part of the figure. To give some examples, with $\beta_1 = (a, rsa)$ for the arcs (p_1, t_1) and (t_1, p_4) we obtain $\mathbf{Pre}[p_1, t_1](\beta_1) = pr_1(\beta_1) = a$ and $\mathbf{Post}[t_1, p_4](\beta_1) = pr_2(\beta_1) = rsa$

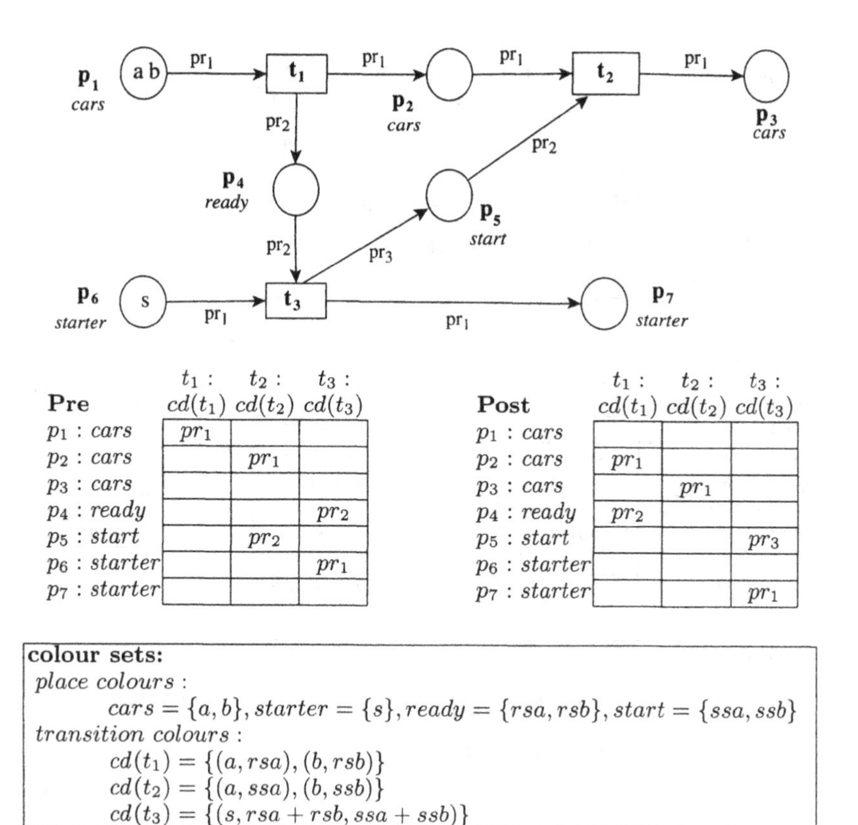

Pre	$t_1 : cd(t_1)$	$t_2 : cd(t_2)$	$t_3 : cd(t_3)$
$p_1 : cars$	pr_1		
$p_2 : cars$		pr_1	
$p_3 : cars$			
$p_4 : ready$			pr_2
$p_5 : start$		pr_2	
$p_6 : starter$			pr_1
$p_7 : starter$			

Post	$t_1 : cd(t_1)$	$t_2 : cd(t_2)$	$t_3 : cd(t_3)$
$p_1 : cars$			
$p_2 : cars$	pr_1		
$p_3 : cars$		pr_1	
$p_4 : ready$	pr_2		
$p_5 : start$			pr_3
$p_6 : starter$			
$p_7 : starter$			pr_1

colour sets:
place colours :

$$cars = \{a, b\}, starter = \{s\}, ready = \{rsa, rsb\}, start = \{ssa, ssb\}$$

transition colours :

$$cd(t_1) = \{(a, rsa), (b, rsb)\}$$
$$cd(t_2) = \{(a, ssa), (b, ssb)\}$$
$$cd(t_3) = \{(s, rsa + rsb, ssa + ssb)\}$$

Fig. 4.1. CPN $\mathcal{N}_6$ with incidence matrix, both in projection form

respectively. Note that some tuples are excluded by the definition of the subset $cd(t) \subseteq \text{Bag}(cd(p_1)) \times \ldots \times \text{Bag}(cd(p_q))$. In the limit this subset can contain a single element as for $cd(t_3)$ in the present example.

c) Representation of the incidence matrix entries by functions: In the place of projections other kinds of functions with range $\text{Bag}(cd(p_i))$ can be used. This is of particular importance when there is a "most important" token class among the objects to be moved by an occurring transition and the other tokens functionally depend on it. The colour set *car* in our example is an example of such a "main token class" with respect to transitions t_1 and t_2. By a function *sign* with $sign(a) = rsa$ and $sign(b) = rsb$, the corresponding start sign is associated with the token. The same is done by the function *signal* with $signal(a) = ssa$ and $signal(b) = ssb$, which specifies the start sign[2] to be received by the corresponding car. As shown in Figure 4.2, for instance, transition t_1 now has *cars* as its transition colour set (instead of

[2] The function is named *signal* in order to have unique names.

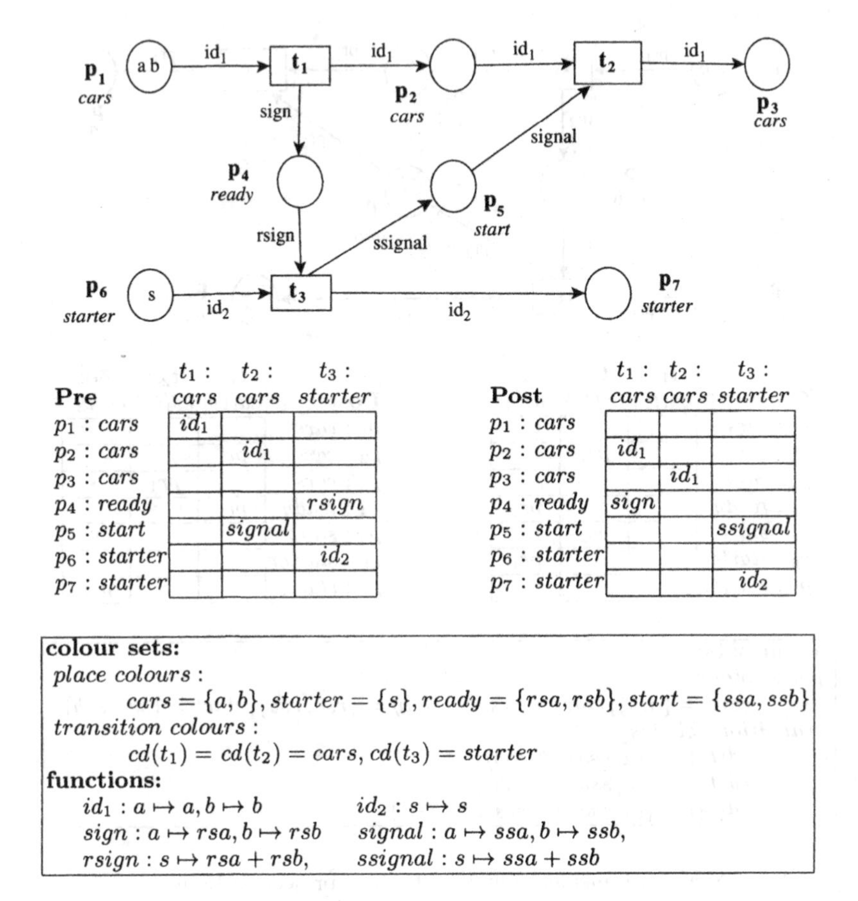

Pre	t_1 : *cars*	t_2 : *cars*	t_3 : *starter*
p_1 : *cars*	id_1		
p_2 : *cars*		id_1	
p_3 : *cars*			
p_4 : *ready*			*rsign*
p_5 : *start*		*signal*	
p_6 : *starter*			id_2
p_7 : *starter*			

Post	t_1 : *cars*	t_2 : *cars*	t_3 : *starter*
p_1 : *cars*			
p_2 : *cars*	id_1		
p_3 : *cars*		id_1	
p_4 : *ready*	*sign*		
p_5 : *start*			*ssignal*
p_6 : *starter*			
p_7 : *starter*			id_2

colour sets:
place colours :
$cars = \{a, b\}, starter = \{s\}, ready = \{rsa, rsb\}, start = \{ssa, ssb\}$
transition colours :
$cd(t_1) = cd(t_2) = cars, cd(t_3) = starter$
functions:
$id_1 : a \mapsto a, b \mapsto b$ $\quad id_2 : s \mapsto s$
$sign : a \mapsto rsa, b \mapsto rsb$ $\quad signal : a \mapsto ssa, b \mapsto ssb,$
$rsign : s \mapsto rsa + rsb,$ $\quad ssignal : s \mapsto ssa + ssb$

Fig. 4.2. CPN $\mathcal{N}_7$ with incidence matrix, both in function form

a subset of $cars \times ready$ as in Figure 4.1) and the functions pr_1 and pr_2 of Figure 4.1 are replaced by the functions *id* (for the identity mapping) and *sign* respectively.

d) Representation of the incidence matrix entries by terms:
Probably the most general form to represent the incidence matrix entries is by bags of terms from a Σ-algebra including function symbols and variables. This is used in "algebraic Petri nets" to define generic inscriptions. Here we consider interpreted terms, i.e. expressions over variables and interpreted functions. In the example given in Figure 4.3, however, the terms consist of single variables such as x or y. A different example would be $sign(x)$ on the arc (t_1, p_4), where the function *sign* is defined as in Figure 4.2. A more general example of a term bag would be $3'sign(x) + 2'signal(y)$ which is, however, not included in the example. See Figure 8.12 for a term containing a case-operation. The approach is attractive since, in contrast to the function

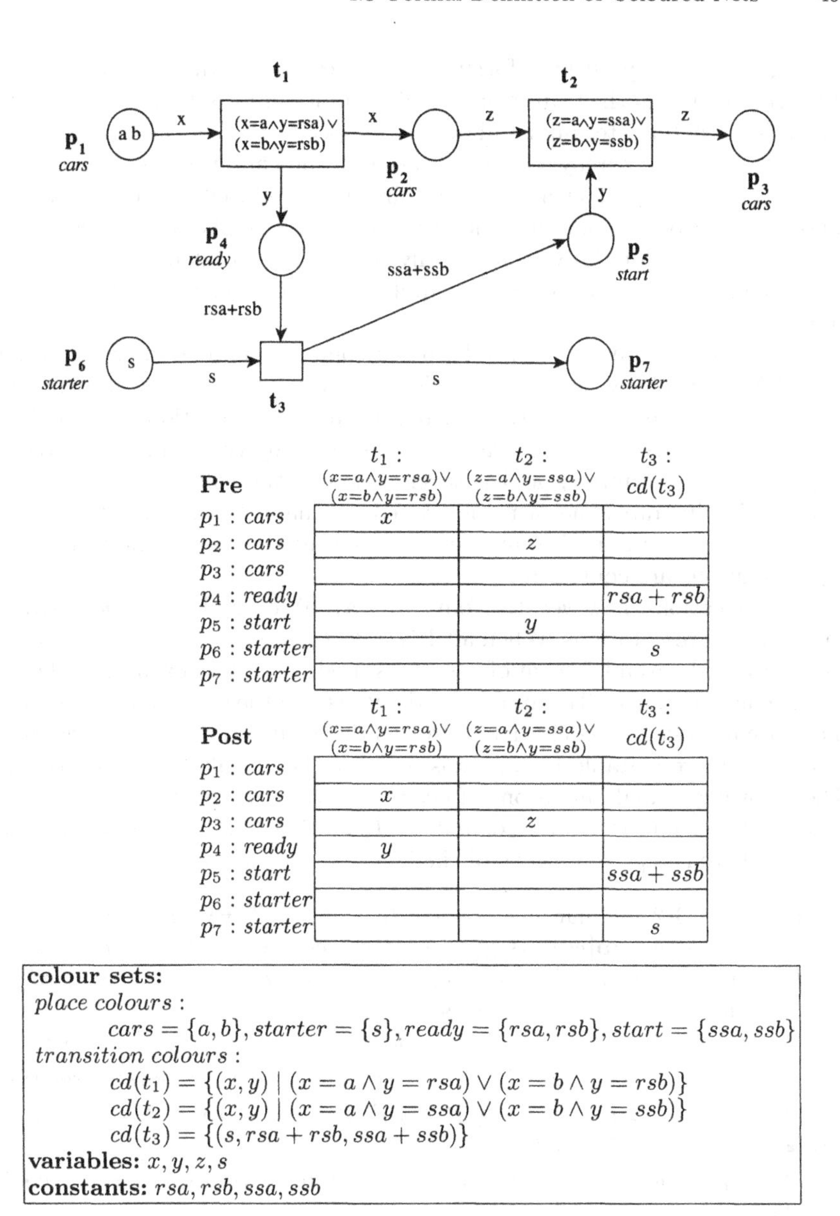

Pre

	t_1 : $(x{=}a{\wedge}y{=}rsa){\vee}$ $(x{=}b{\wedge}y{=}rsb)$	t_2 : $(z{=}a{\wedge}y{=}ssa){\vee}$ $(z{=}b{\wedge}y{=}ssb)$	t_3 : $cd(t_3)$
p_1 : *cars*	x		
p_2 : *cars*		z	
p_3 : *cars*			
p_4 : *ready*			$rsa + rsb$
p_5 : *start*		y	
p_6 : *starter*			s
p_7 : *starter*			

Post

	t_1 : $(x{=}a{\wedge}y{=}rsa){\vee}$ $(x{=}b{\wedge}y{=}rsb)$	t_2 : $(z{=}a{\wedge}y{=}ssa){\vee}$ $(z{=}b{\wedge}y{=}ssb)$	t_3 : $cd(t_3)$
p_1 : *cars*			
p_2 : *cars*	x		
p_3 : *cars*		z	
p_4 : *ready*	y		
p_5 : *start*			$ssa + ssb$
p_6 : *starter*			
p_7 : *starter*			s

colour sets:

place colours :

$cars = \{a, b\}, starter = \{s\}, ready = \{rsa, rsb\}, start = \{ssa, ssb\}$

transition colours :

$cd(t_1) = \{(x, y) \mid (x = a \wedge y = rsa) \vee (x = b \wedge y = rsb)\}$

$cd(t_2) = \{(x, y) \mid (x = a \wedge y = ssa) \vee (x = b \wedge y = ssb)\}$

$cd(t_3) = \{(s, rsa + rsb, ssa + ssb)\}$

variables: x, y, z, s

constants: rsa, rsb, ssa, ssb

Fig. 4.3. CPN $\mathcal{N}_5$ with incidence matrix, both in term form

form, guards are expressed as formulas over these terms. In most cases, as for the formulas within the transition box of $\mathcal{N}_5$ in Figure 4.3, these formulas are quantifier-free. This technique allows us to specify the transition colour set within the net and thereby reduces the need to study additional specifications. The occurrence rule of such nets is, as already introduced by Figure 3.11, as follows: (1) Choose a binding β for the variables. (2) Check that the guard evaluates to true for β. (3) Temporarily replace the variables by the chosen values and evaluate the term. (4) Apply the transition occurrence rule of ac-CPNs.

Note that the four examples discussed contain complete information for the CPN in graphical as well as in algebraic form. By "graphical form" we mean the Petri net graph with inscriptions and a specification of the colour sets and functions. The "algebraic form" is given by the incidence matrices together with inscriptions and a specification of the colour sets and functions. This illustrates the principle stated in Chapter 2, namely that Petri nets always have a graphical and an algebraic representation and that these representations are equivalent.

In applications, however, the four forms a), b), c), and d) are not used in their pure representations. When a CPN is designed for human readers, often the "term and variable" form of case d) is used, since this form is similar to concurrent programs with variables and guards. On the other hand, when the incidence matrices are constructed, frequently the function-form is preferred, since it is more suitable for analysis, e.g. for the calculation of invariants. The following formal definitions of occurrence rule, reachability set, etc. are uniquely based on the abstract definition (Definition 4.3.1) and do not depend on the different representations of the incidence matrix entries.

Definition 4.3.2. *A* marking *of a CPN* $\mathcal{N} = \langle P, T, \mathbf{Pre}, \mathbf{Post}, \mathcal{C}, cd \rangle$ *is a vector* $\mathbf{m}$ *such that* $\mathbf{m}[p] \in \mathrm{Bag}(cd(p))$ *for each* $p \in P$. $\mathcal{N}$ *together with a marking* $\mathbf{m}_0$ *(initial marking) is called a CPN* system *and is denoted by* $\mathcal{S} = \langle \mathcal{N}, \mathbf{m}_0 \rangle$ *or* $\mathcal{S} = \langle P, T, \mathbf{Pre}, \mathbf{Post}, \mathcal{C}, cd, \mathbf{m}_0 \rangle$.

A transition $t \in T$ *is* enabled *for binding* β *in a marking* $\mathbf{m}$ *iff* $\mathbf{m} \geq \mathbf{Pre}[\bullet, t](\beta)$ *(denoted by:* $\mathbf{m} \xrightarrow{t,\beta}$*). In this case the* successor-marking relation *is defined by* $\mathbf{m} \xrightarrow{t,\beta} \mathbf{m}' :\Leftrightarrow \mathbf{m} \geq \mathbf{Pre}[\bullet, t](\beta) \wedge \mathbf{m}' = \mathbf{m} + \mathbf{Post}[\bullet, t](\beta) - \mathbf{Pre}[\bullet, t](\beta) = \mathbf{m} + \mathbf{C}[\bullet, t](\beta)$.

The bag operations $+$, $-$ *and the bag relation* $\leq$ *are extended to vectors as before.*

If the particular binding is not important, the notation $\mathbf{m} \xrightarrow{t} \mathbf{m}' :\Leftrightarrow \exists \beta \,.\, \mathbf{m} \xrightarrow{t,\beta} \mathbf{m}'$ *is used.*

Definition 4.3.3. *The successor-marking relation of Definition 4.3.2 is extended to hold for sequences of transitions by*

- $\mathbf{m} \xrightarrow{w} \mathbf{m}'$ *if* w *is the empty word* λ *and* $\mathbf{m} = \mathbf{m}'$ *and*
- $\mathbf{m} \xrightarrow{wt} \mathbf{m}'$ *if* $\exists \mathbf{m}'' \,.\, \mathbf{m} \xrightarrow{w} \mathbf{m}'' \wedge \mathbf{m}'' \xrightarrow{t} \mathbf{m}'$ *for* $w \in T^*$ *and* $t \in T$.

For a net system $\mathcal{S} = \langle \mathcal{N}, \mathbf{m_0} \rangle$ *the set* $\mathrm{RS}(\mathcal{S}) = \mathrm{RS}(\mathcal{N}, \mathbf{m_0}) := \{\mathbf{m} \mid \exists w \in T^* . \mathbf{m_0} \xrightarrow{w} \mathbf{m}\}$ *is the* reachability set. *It can be denoted by* $\mathrm{RS}(\mathbf{m_0})$ *if* $\mathcal{N}$ *is obvious from the context.* $FS(\mathcal{S}) := \{w \in T^* \mid \exists \mathbf{m} . \mathbf{m_0} \xrightarrow{w} \mathbf{m}\}$ *is the* set of occurrence-transition sequences *(or* firing-sequence set*) of* $\mathcal{S}$. *As in the case of P/T nets it is sometimes convenient to define the set* $Occ(\mathcal{S})$ *of* occurrence sequences *to be the set of all sequences of the form*

$$\mathbf{m_0}, t_0, \mathbf{m_1}, t_1, \mathbf{m_2}, t_2, \ldots, t_{n-1}, \mathbf{m_n} \qquad (n \geq 1)$$

such that $\mathbf{m_i} \xrightarrow{t_i} \mathbf{m_{i+1}}$ *for* $i \in \{0, \ldots, n-1\}$. *If in an occurrence sequence the individual bindings are of importance, the notion is extended to*

$$\mathbf{m_0}, (t_0, \beta_0), \mathbf{m_1}, (t_1, \beta_1), \mathbf{m_2}, (t_2, \beta_2), \ldots, (t_{n-1}, \beta_{n-1}), \mathbf{m_n} \qquad (n \geq 1)$$

such that $\mathbf{m_i} \xrightarrow{t_i, \beta_i} \mathbf{m_{i+1}}$ *for* $i \in \{0, \ldots, n-1\}$.

Table 4.4. The incidence matrices of the *CPN* from Figure 3.11 in term form

Pre	$t : \{(x,y,z) \in \{a,b,c\}^3 \mid y=z\}$
$p_1 : \{a,b,c\}$	$x+y$
$p_2 : \{a,b,c\}$	$2'z$
$p_3 : \{a,b,c\}$	
$p_4 : \{a,b,c\}$	

Post	$t : \{(x,y,z) \in \{a,b,c\}^3 \mid y=z\}$
$p_1 : \{a,b,c\}$	
$p_2 : \{a,b,c\}$	
$p_3 : \{a,b,c\}$	y
$p_4 : \{a,b,c\}$	z

An example of an occurrence sequence for the current CPN has been given in Section 3.3. The definition of the successor marking is illustrated by the example from Figure 3.11. To begin with, recall that for this example the colour domains of all places are assumed to be $cd(p) = \{a, b, c\}$ and the colour domain of the transition t is assumed to be $cd(t) = \{(x, y, z) \in \{a, b, c\}^3 \mid y = z\}$. The corresponding incidence matrices are given in Table 4.4 (in term form).

For the binding $\beta = [x = a, y = b, z = b] \in cd(t)$ and the marking $\mathbf{m} = (a + b, 2'b + c, c, a)$ from Figure 3.11 we calculate the successor marking $\mathbf{m}'$ as follows. For a term τ and a binding β we denote by $\tau[\beta]$ the value returned after the evaluation of τ with β.

$$\mathbf{m}' = \mathbf{m} - \begin{pmatrix} \mathbf{Pre}[p_1, t](\beta) \\ \mathbf{Pre}[p_2, t](\beta) \\ \mathbf{Pre}[p_3, t](\beta) \\ \mathbf{Pre}[p_4, t](\beta) \end{pmatrix} + \begin{pmatrix} \mathbf{Post}[p_1, t](\beta) \\ \mathbf{Post}[p_2, t](\beta) \\ \mathbf{Post}[p_3, t](\beta) \\ \mathbf{Post}[p_4, t](\beta) \end{pmatrix}$$

$$= \begin{pmatrix} a+b \\ 2'b+c \\ c \\ a \end{pmatrix} - \begin{pmatrix} (x+y)[x=a, y=b, z=b] \\ (2'z)[x=a, y=b, z=b] \\ \emptyset \\ \emptyset \end{pmatrix} + \begin{pmatrix} \emptyset \\ \emptyset \\ (y)[x=a, y=b, z=b] \\ (z)[x=a, y=b, z=b] \end{pmatrix}$$

$$= \begin{pmatrix} a+b \\ 2'b+c \\ c \\ a \end{pmatrix} - \begin{pmatrix} a+b \\ 2'b \\ \emptyset \\ \emptyset \end{pmatrix} + \begin{pmatrix} \emptyset \\ \emptyset \\ b \\ b \end{pmatrix} = \begin{pmatrix} \emptyset \\ c \\ b+c \\ a+b \end{pmatrix}$$

5. Properties*

The construction of Petri net models from informal requirement specifications is not a trivial task, and requires a great deal of modelling experience, as well as knowledge of the techniques aiding in model construction. As a result, a Petri net model may differ considerably from its original specification. This is especially true when large Petri net models of complex systems are involved.

Therefore, a critical issue in the use of formal methods (in our case Petri nets) for problem solving is the construction of a *good* model, in particular a model that is *correct* with respect to some kind of logical specification. The existence of a correspondence between an original specification and its Petri net representation provides feedback to the designers who can, in many instances, clarify their perception of the system.

Different concepts of correctness exist. Basically, a system is said to be correct when two models, namely the *specification* and the *implementation*, are *equivalent* ([BGV91, PRS92]), or when the system exhibits a set of *desirable properties*, either expressed as formulas of a *logic* ([CES86, MP89]), or selected from a given *kit* (e.g. boundedness, liveness); see the following textbooks or surveys on basic Petri net theory and applications: [Pet81, Bra83, Sil85, Rei85a, ABC+95, Mur89]. These *desirable properties*, when interpreted in the context of the modelled system, allow the system designer to identify the presence or absence of the application-domain-specific functional properties of the system under design.

Broadly speaking, analysis methods for Petri net models can be classified as *enumeration-* or *net-driven*. The first step in enumeration techniques is the computation of the *reachability graph* (totally or partially). If the system is bounded, this can be used as the computational model for a proof system, which verifies formulas from a given (temporal) logic, or for decision procedures and tools for automatic verification ([CES86, MP89]).

It is also possible to deal with unbounded systems by using the *coverability graph* ([Fin93]), but this particular construction does not allow the deciding of important properties such as liveness and reachability ([Pet81]).

The basic idea in net-driven approaches is to obtain useful information about the behaviour, reasoning from the structure of the net and the initial marking. Crucial advantages of this approach are the deep understanding

* Authors: J.M. Colom, M. Silva, and E. Teruel

of the system behaviour that is gained, and the efficiency of the resulting algorithms. Two different approaches are ([Mur89, DHP+93]):

1) Net *transformations* (typically reductions), preserving the properties to be analysed ([Ber86]). It is expected that the transformed system is easier to analyse.
2) *Structure theory* results, based on graph theory and/or linear algebra. Typically, results for general Petri net systems are obtained, some of which become increasingly powerful when restricted to particular subclasses (see [TS94] for a recent survey).

In this chapter we introduce a basic kit of properties of Petri nets. These can be considered to be good properties which a Petri net modelling a system should satisfy. We then illustrate the analysis method based on the reachability graph and the structural analysis method based on the linear invariants. This kind of structural analysis is based on linear algebra.

5.1 Basic Properties

Only a few qualitative properties will be considered in this chapter. They are general in the sense that they are meaningful for any concurrent system, not only for those modelled with Petri nets. Nevertheless their statements, using Petri net concepts and objects, make them especially easy to understand in many cases. The properties to be considered are:

1) *Boundedness*, characterising finiteness of the state space.
2) *Liveness*, related to potential fireability in all reachable markings. *Deadlock-freeness* is a weaker condition in which global infinite activity (i.e. fireability) of the net system model is guaranteed, but some parts of it may not work at all.
3) *Reversibility*, characterising recoverability of the initial marking from any reachable marking.
4) *Mutual exclusion*, dealing with the impossibility of simultaneous *submarkings* (p-mutex) or *firing concurrency* (t-mutex).

Consider the net in Figure 5.1.a. Firing t_2 leads to $\mathbf{m} = p_3 + p_4$. Subsequently firing t_4, $\mathbf{m}_1 = p_1 + p_3$ is reached. Repeating k times the sequence $t_2 t_4$ the marking $\mathbf{m}_k = p_1 + k\, p_3$ is reached. Since the marking of p_3 can be arbitrarily large, place p_3 is said to be *unbounded*. In practice, the capacity of the physical element represented by p_3 will be finite, so an *overflow* can appear, which is a pathological situation.

The maximum number of tokens a place may contain is its (marking) *bound*. A place is bounded if its bound is finite. A net system is bounded if each place is bounded. System boundedness (i.e. all places bounded) is a generally required behavioural property.

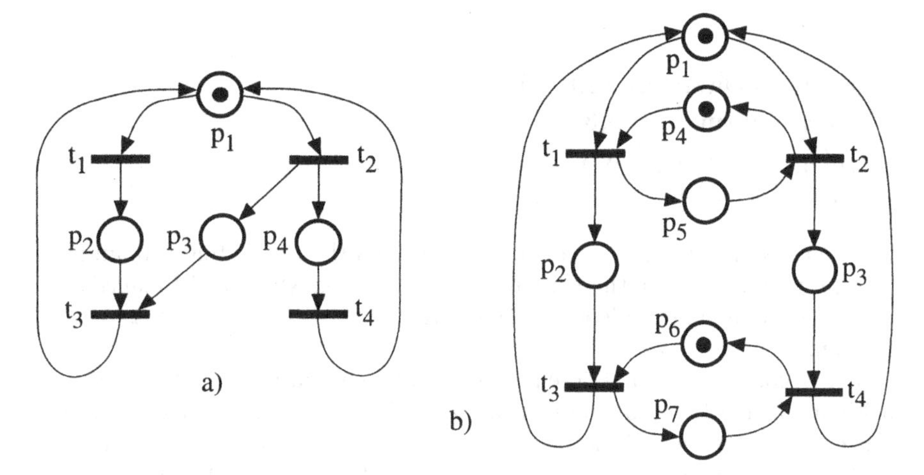

Fig. 5.1. On qualitative pathological behaviours: a) an unbounded, deadlockable (non-live), non-reversible net system; b) a live net system that by increasing the initial marking (e.g. $\mathbf{m_0}[p_5] = 1$) can reach a deadlock state!

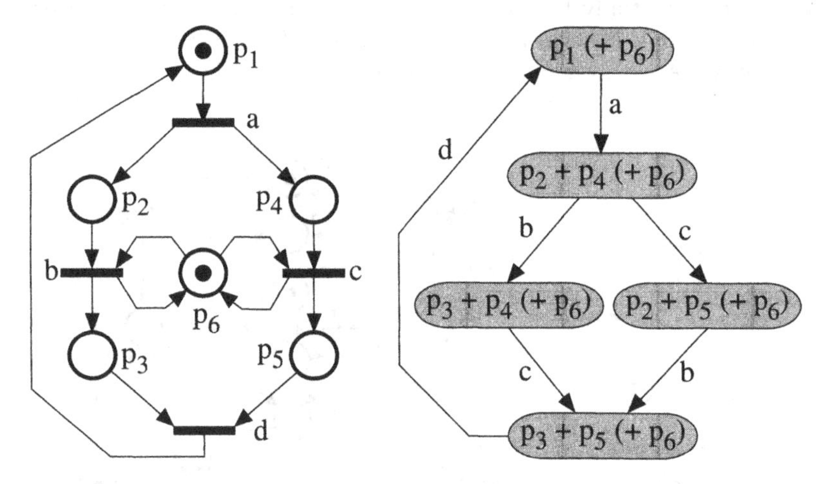

Fig. 5.2. Bounded, live, and reversible system and its reachability graph

For any initial marking we can define on the net structure of Figure 5.2a the following token conservation laws:

$$\begin{aligned}
\mathbf{m}[p_1] + \mathbf{m}[p_2] + \mathbf{m}[p_3] &= \mathbf{m_0}[p_1] + \mathbf{m_0}[p_2] + \mathbf{m_0}[p_3] = k_1(\mathbf{m_0}) \\
\mathbf{m}[p_1] + \mathbf{m}[p_4] + \mathbf{m}[p_5] &= \mathbf{m_0}[p_1] + \mathbf{m_0}[p_4] + \mathbf{m_0}[p_5] = k_2(\mathbf{m_0}) \\
\mathbf{m}[p_6] &= \mathbf{m_0}[p_6] = k_3(\mathbf{m_0})
\end{aligned}$$

where $\mathbf{m_0}$ is the initial marking and $\mathbf{m}$ any reachable marking. Therefore:

$$\mathbf{m}[p_1] \le \min(k_1(\mathbf{m_0}), k_2(\mathbf{m_0}))$$
$$\mathbf{m}[p_i] \le k_1(\mathbf{m_0}); i = 2, 3$$
$$\mathbf{m}[p_j] \le k_2(\mathbf{m_0}); j = 4, 5$$
$$\mathbf{m}[p_6] = k_3(\mathbf{m_0})$$

The above inequalities mean that *for any* $\mathbf{m_0}$ the net system is bounded. This property, stronger than boundedness, is called *structural boundedness* because it holds independently of the initial marking (only finiteness of $\mathbf{m_0}$ is assumed).

Let us consider now a different scenario where we fire t_1 from the marking in Figure 5.1a. After that, no transition can be fired: a *total deadlock* situation has been reached. A net system is said to be *deadlock-free* if from any reachable marking at least one transition can always occur. A stronger condition than deadlock-freeness is *liveness*. A transition t is potentially fireable at a given marking $\mathbf{m}$ if there exists a transition firing sequence σ leading to a marking $\mathbf{m}'$ in which t is enabled (i.e. $\mathbf{m} \xrightarrow{\sigma} \mathbf{m}' \ge \mathbf{Pre}[P, t]$). A transition is *live* if it is potentially fireable in all reachable markings. In other words, a transition is live if it never loses the possibility of firing (i.e. of performing some activity). A net system is live if all transitions are live.

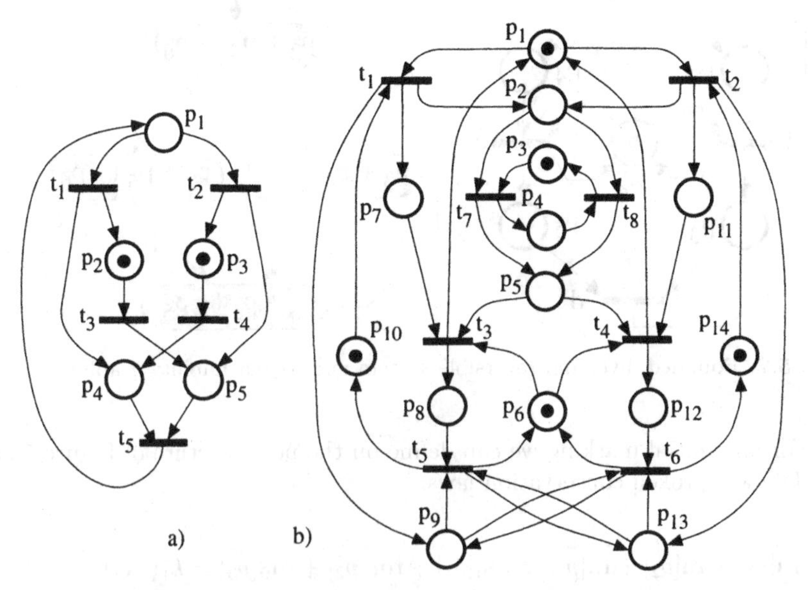

Fig. 5.3. On home states: a) The initial marking is not a home state, but all successor markings are home states. b) Net system that presents two livelocks, so there are no home states.

For any initial marking we can define on the net structure in Figure 5.1a, non-liveness holds (in fact, a total deadlock can always be reached). Non-liveness for arbitrary initial markings reflects a pathology of the net structure: *structural non-liveness.* A net is structurally live if there exists at least one live initial marking.

A paradoxical behaviour of concurrent systems is the following: At first glance it may be accepted as intuitive that increasing the initial marking (e.g. increasing the number of resources) of a net system helps to make it live. The net system in Figure 5.1b shows that increasing the number of resources can lead to deadlock situations: Adding a token to p_5, t_2 can be fired and a deadlock is reached! In other words, in general, liveness is not *monotonic* with respect to the initial marking. Note however that liveness can be marking-monotonic on certain net subclasses.

A marking is a *home state* if it is reachable from any other reachable marking. The initial marking of the net system in Figure 5.3a is not a home state: after the firing of transition t_3 or t_4 it is not possible to reach this initial marking. Nevertheless, each of the markings reachable from the initial one is a home state. For some subclasses of net systems the existence of home states is guaranteed ([Vog89, TS96]), but in general the existence of home states does not hold. The net system in Figure 5.3.b ([BV84]) is live and bounded but there are no home states. In fact, in the reachability graph there exist two different terminal strongly connected components, each one containing all transitions (thus live). Therefore, the markings of one of these components are unreachable from the markings belonging to the other component. Each one of these terminal strongly connected components is called a *livelock.* The set of home states of a net system is called the *home space.* The existence of a home space for a net system is a desirable property because it is strongly related to properties such as ergodicity, of crucial importance in the context of performance evaluation or system simulation.

In the particular case that the initial marking is a home state, the net system is reversible, so it is always possible to return to the initial marking.

Liveness, boundedness, and reversibility are just three different "good" (often required) behavioural properties that may be interesting to study in a net system. Figure 5.4 shows that they are independent of each other, giving examples of the eight cases we may have.

The last basic property we introduce in this section is *mutual exclusion.* This property captures constraints such as the impossibility of a simultaneous access by two robots to a single store. Two places (transitions) are in mutual exclusion if they can never be simultaneously marked (fired). For instance, in the net system in Figure 5.2 we can write: $\mathbf{m}[p_1] + \mathbf{m}[p_2] + \mathbf{m}[p_3] = 1$, so p_1, p_2, and p_3 are in mutual exclusion.

Table 5.1 summarises the definitions of the different properties we have introduced in this section.

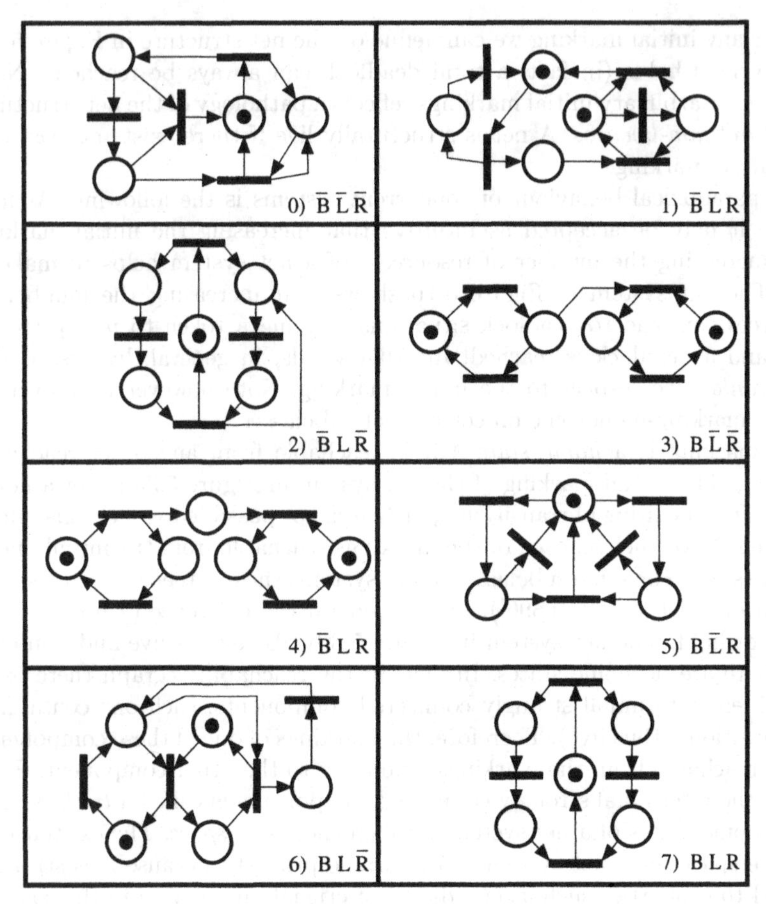

Fig. 5.4. Boundedness (B), liveness (L), and reversibility (R) are independent properties.

5.2 An Introduction to the Analysis

Conventionally, analysis techniques for Petri nets are classified as: (1) Enumeration, (2) Transformation, and (3) Structural analysis. Simulation methods have also been applied to study systems modelled with P/T nets. They proceed by playing the token game (firing enabled transitions) on the net system model under certain strategies. In general, simulation methods do not allow us to prove properties, but they may be of great help for understanding the modelled system or for fixing the problems manifested during the simulation. Simulation methods are extremely useful when time is associated with the net evolution (timed systems), or when we wish to know the response of a system described with a net in an environment which is also defined by simulation (see Part IV). In this section we do not consider simulation

Table 5.1. Summarising some basic logical properties

(1)	Bound of place p in $\langle \mathcal{N}, \mathbf{m_0} \rangle$ $\mathbf{b}(p) = \sup\{\mathbf{m}[p] \mid \mathbf{m} \in \mathrm{RS}(\mathcal{N}, \mathbf{m_0})\}$
(2)	p is bounded in $\langle \mathcal{N}, \mathbf{m_0} \rangle$ iff $\mathbf{b}(p) < \infty$
(3)	$\langle \mathcal{N}, \mathbf{m_0} \rangle$ is bounded if all places are bounded
(4)	$\langle \mathcal{N}, \mathbf{m_0} \rangle$ is deadlock-free iff $\forall \mathbf{m} \in \mathrm{RS}(\mathcal{N}, \mathbf{m_0})$. $\exists t \in T$ such that t is fireable at $\mathbf{m}$
(5)	t is live in $\langle \mathcal{N}, \mathbf{m_0} \rangle$ iff $\forall \mathbf{m} \in \mathrm{RS}(\mathcal{N}, \mathbf{m_0})$. $\exists \sigma$ such that $\mathbf{m} \xrightarrow{\sigma t} \mathbf{m}'$
(6)	$\langle \mathcal{N}, \mathbf{m_0} \rangle$ is live if all transitions are live
(7)	$\mathbf{m} \in \mathrm{RS}(\mathcal{N}, \mathbf{m_0})$ is a home state iff $\forall \mathbf{m}' \in \mathrm{RS}(\mathcal{N}, \mathbf{m_0})$. $\exists \sigma$ such that $\mathbf{m}' \xrightarrow{\sigma} \mathbf{m}$
(8)	$\langle \mathcal{N}, \mathbf{m_0} \rangle$ is reversible iff $\forall \mathbf{m} \in \mathrm{RS}(\mathcal{N}, \mathbf{m_0})$. $\exists \sigma$ such that $\mathbf{m} \xrightarrow{\sigma} \mathbf{m_0}$
(9)	Mutual exclusion in $\langle \mathcal{N}, \mathbf{m_0} \rangle$: p_i and p_j are in marking mutual exclusion iff $\nexists\ \mathbf{m} \in \mathrm{RS}(\mathcal{N}, \mathbf{m_0})$ such that $(\mathbf{m}[p_i] > 0)$ and $(\mathbf{m}[p_j] > 0)$ t_i and t_j are in firing mutual exclusion iff $\nexists\ \mathbf{m} \in \mathrm{RS}(\mathcal{N}, \mathbf{m_0})$ such that $\mathbf{m} \geq \mathbf{Pre}[P, t_i] + \mathbf{Pre}[P, t_j]$
(10)	Structural properties (abstractions of behavioural properties): $\mathcal{N}$ is structurally bounded iff $\forall \mathbf{m_0}$ (finite) $\langle \mathcal{N}, \mathbf{m_0} \rangle$ is bounded $\mathcal{N}$ is structurally live iff $\exists \mathbf{m_0}$ (finite) making $\langle \mathcal{N}, \mathbf{m_0} \rangle$ a live system

methods and we will only overview the three previously mentioned analysis techniques on P/T nets without interpretation.

Enumeration methods are based on the construction of a *reachability graph* (RG) which represents each net marking and the single transition firings between them. If the net system is bounded, the reachability graph is finite and the different qualitative properties can be verified easily. If the net system is unbounded, the RG is infinite and its construction is not possible. In this case, finite graphs known as *coverability graphs* can be constructed (see for example [Pet81, Rei85a, Fin93]). In spite of its power, enumeration is often difficult to apply, even in small nets, due to its computational complexity (it is strongly combinatorial).

Analysis by transformation proceeds by transforming a net system $\mathcal{S} = \langle \mathcal{N}, \mathbf{m_0} \rangle$ into a net system $\mathcal{S}' = \langle \mathcal{N}', \mathbf{m_0}' \rangle$ preserving the set of properties Π to be verified (i.e. $\mathcal{S}'$ satisfies the properties Π iff $\mathcal{S}$ satisfies them). The goal is to verify the properties Π in $\mathcal{S}'$ more easily than in $\mathcal{S}$. The state space of $\mathcal{S}'$ may be bigger than that of $\mathcal{S}$, but $\mathcal{S}'$ may belong to a subclass for which state enumeration can be avoided.

Reduction methods are a special class of transformation methods in which a sequence of net systems preserving the properties to be studied is constructed. The construction is done in such a way that the net system $\langle \mathcal{N}_{i+1}, \mathbf{m_0}_{i+1} \rangle$ is "smaller" (i.e. markings that have fewer tokens) than the previous in the sequence, $\langle \mathcal{N}_i, \mathbf{m_0}_i \rangle$.

The applicability of reduction methods is limited by the existence of irreducible net systems. Practically speaking, the reductions obtained are normally considerable, and can allow the desired properties to be verified directly. Because of the existence of irreducible systems, this method must be complemented by others.

Finally, *structural analysis techniques* investigate the relationships between the behaviour of a net system and its structure (hence their name), while the initial marking acts, basically, as a parameter. In this last class of analysis techniques, we can distinguish two subgroups:

1) *Linear algebra/Linear programming based techniques*, which are based on the net state equation. In certain analyses they permit a fast diagnosis without the necessity of enumeration.
2) *Graph based techniques*, in which the net is seen as a bipartite graph and some *ad hoc* reasonings are applied. These methods are especially effective in analysing restricted subclasses of ordinary nets.

The three groups of analysis techniques outlined above are by no means exclusive, but complementary, e.g. the conclusions obtained from a structural analysis of a given net model may simplify or accelerate a further enumeration analysis of this net model; or the application of reduction methods to a given net model, preserving the properties to be verified, may be the only way to obtain a manageable reachability graph to verify the properties. Normally the designer can use them according to the needs of the ongoing analysis process. Obviously, although we have distinguished between reduction and structural analysis methods, it must be pointed out that most popular reduction techniques act basically on the net-structure level and thus can also be considered as structural techniques.

5.2.1 Verification Based on the Reachability Graph

Given a net system $\mathcal{S} = \langle \mathcal{N}, \mathbf{m_0} \rangle$, its reachability graph is a directed graph $\mathrm{RG}(\mathcal{S}) = (V, E)$, where $V = \mathrm{RS}(\mathcal{S})$ and $E = \{\langle \mathbf{m}, t, \mathbf{m}' \rangle | \mathbf{m}, \mathbf{m}' \in \mathrm{RS}(\mathcal{S})$ and $\mathbf{m} \xrightarrow{t} \mathbf{m}'\}$ are the sets of nodes and edges, respectively.

If the net system $\mathcal{S} = \langle \mathcal{N}, \mathbf{m_0} \rangle$ is bounded, the $\mathrm{RG}(\mathcal{S})$ is finite and it can be constructed, for example, by Algorithm 5.1. It finishes when all the possible firings from the reachable markings have been explored. The tagging scheme in step 2.1 ensures that no marking is visited more than once. Each marking visited is tagged (step 2.1), and step 2.2.3 ensures that the only markings added to V are those that have not previously been added. When a marking is visited, only those edges representing the firing of an enabled transition are added to the set E in step 2.2.4.

The construction of the reachability graph is a very hard problem from a computational point of view. This is because the size of the state space may grow more than exponentially with respect to the size of the Petri net model (measured, for example, by the number of places). In [Val92] the reader can find a discussion of the size of the reachability graph obtained from a Petri net, the rôle of concurrency in the state space explosion problem, and some methods to obtain reduced representations of the state space.

Let us consider, for example, the net system in Figure 5.2 without the place p_6, and its reachability graph obtained by applying the Algorithm 5.1. The net system has five markings, thus it is bounded. It is also easy to conclude that all the places are 1-bounded. A closer look allows us to state that p_1, p_2, and p_3 (p_1, p_4, and p_5) are in mutual exclusion. Moreover, considering RS and the net structure (the pre-function), firing concurrency between transitions b and c can be decided. Observe at this point that introducing p_6 into our net system does not change the graph structure of the reachability graph, but transitions b and c are now in firing mutual exclusion. This example shows that the RG obtained is a *sequentialised observation* of the net system behaviour, and therefore it is not possible to use it to determine if two transitions can be fired concurrently from the same marking or if they are in a conflict. To avoid this problem, other reachability graphs capturing the true concurrency can be constructed. The basic idea is to increase the number of edges of the conventional RG to represent the concurrent firing of transitions from each marking.

For unbounded net systems $\mathcal{S}$, RS($\mathcal{S}$) is not a finite set and therefore the construction of RG($\mathcal{S}$) never ends. Karp and Miller [KM69] showed how to detect unboundedness of a net system by means of the following condition (incorporated in step 2.2.2 of algorithm 5.1 as a break condition): the system $\mathcal{S} = \langle \mathcal{N}, \mathbf{m_0} \rangle$ is unbounded iff there exists $\mathbf{m}'$ reachable from $\mathbf{m} \in \text{RS}(\mathcal{S})$, $\mathbf{m} \xrightarrow{\sigma} \mathbf{m}'$, such that $\mathbf{m} \lneq \mathbf{m}'$ (the repetition of σ allows a conclusion of unboundedness because the occurrence of σ strictly increases the content of tokens of the starting marking $\mathbf{m}$).

Algorithm 5.1 (Computation of the Reachability Graph)

Input - The net system $\mathcal{S} = \langle \mathcal{N}, \mathbf{m_0} \rangle$
Output - The directed graph $\text{RG}(\mathcal{S}) = (V, E)$ for bounded net systems

1. Initialise $\text{RG}(\mathcal{S}) = (\{\mathbf{m_0}\}, \emptyset)$; $\mathbf{m_0}$ is untagged;
2. **while** there are untagged nodes in V **do**
   2.1 Select an untagged node $\mathbf{m} \in V$ and tag it
   2.2 **for** each enabled transition, t, at $\mathbf{m}$ **do**
      2.2.1 Compute $\mathbf{m}'$ such that $\mathbf{m} \xrightarrow{t} \mathbf{m}'$;
      2.2.2 **if** there exists $\mathbf{m}'' \in V$ such that $\mathbf{m}'' \xrightarrow{\sigma} \mathbf{m}'$ and $\mathbf{m}'' \lneq \mathbf{m}'$
         then the algorithm fails and exits;
            (the unboundedness condition of $\mathcal{S}$ has been detected)
      2.2.3 **if** there is no $\mathbf{m}'' \in V$ such that $\mathbf{m}'' = \mathbf{m}'$
         then $V := V \cup \{\mathbf{m}'\}$; ($\mathbf{m}'$ is an untagged node)
      2.2.4 $E := E \cup \{\langle \mathbf{m}, t, \mathbf{m}' \rangle\}$
3. The algorithm succeeds and RG($\mathcal{S}$) is the reachability graph

---

Coverability graphs allow us to obtain finite representations of the RG of unbounded net systems ([KM69, Pet81, Rei85a, Fin93]). Roughly speaking, in a coverability graph the set of nodes is a finite set of marking vectors (called the *coverability set*) that covers all the markings of the reachability set. There is an edge, representing the firing of a transition t, between two nodes $\mathbf{m}$ and $\mathbf{m}'$ if and only if t is fireable from $\mathbf{m}$ and a marking covered by $\mathbf{m}'$ is reached. The loss of information in the computation of a coverability graph means that many important properties (e.g. marking reachability or deadlock-freeness) cannot be decided using it.

In order to analyse a given property in a bounded net system, the reachability graph is used as the basis for the corresponding *decision procedure*. It allows us to decide whether the net system satisfies a given property. All procedures are, in general, of exponential complexity in the size of the net (measured, for example, by the number of places) but they are of polynomial complexity in the size of the reachability graph (measured, for example, by the number of nodes and edges). The focus of the remainder of this section is on two general decision procedures.

In what follows we will define a *marking predicate* to be a propositional formula whose atoms are inequalities of the form: $\sum_{p \in A} k_p \mathbf{m}[p] \leq k$, where k_p and k are rational constants and A is a subset of places. Let us consider a net system $\mathcal{S} = \langle \mathcal{N}, \mathbf{m_0} \rangle$.

The first group of properties are the so-called *marking invariance properties*. A given marking predicate Π must be satisfied for all reachable markings (hence the name marking invariance property): $\forall \mathbf{m} \in \mathrm{RS}(\langle \mathcal{N}, \mathbf{m_0} \rangle)$, $\mathbf{m}$ satisfies Π. Examples of this are:

1) *k-boundedness of place p*: $\forall \mathbf{m} \in \mathrm{RS}(\mathcal{S}) \,.\, \mathbf{m}[p] \leq k$.
2) *Marking mutual exclusion between p and p'*: $\forall \mathbf{m} \in \mathrm{RS}(\mathcal{S}) \,.\, ((\mathbf{m}[p] = 0) \vee (\mathbf{m}[p'] = 0))$.
3) *Deadlock-freeness*: $\forall \mathbf{m} \in \mathrm{RS}(\mathcal{S}) \,.\, \exists t \in T \,.\, \mathbf{m} \geq \mathbf{Pre}[P, t]$.

Algorithm 5.2 (Decision procedure for marking invariances)

Input - The reachability graph $\mathrm{RG}(\mathcal{N}, \mathbf{m_0})$. The property Π.
Output - TRUE if the property is verified.

1. Initialise all elements of $\mathrm{RS}(\mathcal{S})$ as untagged.
2. **while** there is an untagged node $\mathbf{m} \in \mathrm{RS}(\mathcal{S})$ **do**
   2.1 Select an untagged node $\mathbf{m} \in \mathrm{RS}(\mathcal{S})$ and tag it
   2.2 **if** $\mathbf{m}$ does not satisfy Π
   then return FALSE (the property is not verified).
3. Return TRUE

Marking invariance properties can be decided through Algorithm 5.2, which is linear in the size of $\mathrm{RG}(\mathcal{S})$: each node is visited no more than once. If the algorithm succeeds, then all reachable markings from $\mathbf{m_0}$ satisfy Π.

If the algorithm fails at step 2.2, there is a path in the RG($\mathcal{S}$) from $\mathbf{m_0}$, containing at least one marking that does not satisfy Π.

Example 5.2.1 (Analysis of marking invariance properties). Let us consider the net system in Figure 5.2 for which RS($\mathcal{S}$) $= \{p_1 + p_6,\ p_2 + p_4 + p_6,\ p_3 + p_4 + p_6,\ p_2 + p_5 + p_6,\ p_3 + p_5 + p_6\}$. The execution of Algorithm 5.2 to verify the mutual exclusion property between places p_5 and p_6 ($\forall \mathbf{m} \in \mathrm{RS}(\mathcal{S}) \,.\, (\mathbf{m}[p_5] = 0) \vee (\mathbf{m}[p_6] = 0)$) starts by initialising all elements of RS($\mathcal{S}$) as untagged (step 1). Then the markings are visited one by one (e.g. in the previous order) until $p_2 + p_5 + p_6$ is visited, where the predicate Π is false; hence the algorithm stops and returns FALSE.

The second group of properties are the so-called *liveness invariance properties.* For each reachable marking $\mathbf{m}$ of a net system there exists at least one reachable marking from it satisfying the property Π: $\forall \mathbf{m} \in \mathrm{RS}(\mathcal{S}) \,.\, \exists \mathbf{m}' \in \mathrm{RS}(\mathcal{N}, \mathbf{m}) \,.\, (\mathbf{m}' \text{ satisfies } \Pi)$. Examples of this are:

1) *Liveness of t*: $\forall \mathbf{m} \in \mathrm{RS}(\mathcal{S}) \,.\, \exists \mathbf{m}' \in \mathrm{RS}(\mathcal{N}, \mathbf{m}) \,.\, \mathbf{m}' \geq \mathbf{Pre}[P, t]$.
2) $\mathbf{m}_H$ *is home state*: $\forall \mathbf{m} \in \mathrm{RS}(\mathcal{S}) \,.\, \exists \mathbf{m}' \in \mathrm{RS}(\mathcal{N}, \mathbf{m}) \,.\, \mathbf{m}' = \mathbf{m}_H$.
3) *Reversibility*: $\forall \mathbf{m} \in \mathrm{RS}(\mathcal{S}) \,.\, \exists \mathbf{m}' \in \mathrm{RS}(\mathcal{N}, \mathbf{m}) \,.\, \mathbf{m}' = \mathbf{m_0}$.

Algorithm 5.3 (Decision procedure for liveness invariances)

Input - The reachability graph RG($\mathcal{N}, \mathbf{m_0}$). The property Π.
Output - TRUE if the property is verified.

1. Decompose RG($\mathcal{N}, \mathbf{m_0}$) into its strongly connected components $C_1, \ldots, C_r$
2. Obtain the graph $\mathrm{RG}^c(\mathcal{S}) = (V_c, E_c)$ by shrinking $C_1, \ldots, C_r$ to a single node, i.e. $V_c = \{C_1, \ldots, C_r\}$. $\langle C_i, t, C_j \rangle \in E_c$ iff there exists $\langle \mathbf{m}, t, \mathbf{m}' \rangle \in E$, such that $\mathbf{m}$ is in the SCC C_i, $\mathbf{m}'$ is in the SCC C_j, and $i \neq j$.
3. Compute the set F of terminal strongly connected components from $\mathrm{RG}^c(\mathcal{S})$
4. **while** there is a $C_i \in F$ **do**
   4.1 **if** C_i does not contain a $\mathbf{m}'$ satisfying Π
   then return FALSE
   4.2 Remove C_i from F
5. Return TRUE

These properties cannot be verified by an exclusive linear inspection of the reachability set (as in Algorithm 5.2). The verification requires finding a reachable marking satisfying Π from each of the markings in RS($\mathcal{S}$). In order to verify the property we will classify the markings of RS($\mathcal{S}$) into subsets of mutually reachable markings through the concept of strongly connected components of a directed graph. Therefore, the property will be easily verified by checking that each terminal strongly connected component contains at least one marking satisfying Π. We now recall some basic concepts.

A *path* in a reachability graph $\mathrm{RG}(\mathcal{S})$ is any sequence of nodes $\mathbf{m}_1 \ldots \mathbf{m}_i \mathbf{m}_{i+1} \ldots \mathbf{m}_k$ of $\mathrm{RG}(\mathcal{S}) = (V, E)$ where all successive nodes $\mathbf{m}_i$ and $\mathbf{m}_{i+1}$ in the path satisfy $\langle \mathbf{m}_i, t, \mathbf{m}_{i+1} \rangle \in E$ for some t. The reachability graph $\mathrm{RG}(\mathcal{S})$ is *strongly connected* (SC) iff there is a path from each node in V to any other node in V. A *strongly connected component* (SCC) of a reachability graph is a maximal strongly connected subgraph. A strongly connected component of a graph will be called *terminal* if no node in the component has an edge leaving the component. The strongly connected components of a digraph (V, E) can be found in order $(|V| + |E|)$ steps (e.g. [Meh84]).

When computing the SCCs $C_1, \ldots, C_r$ of a reachability graph $\mathrm{RG}(\mathcal{S}) = (V, E)$, a new graph $\mathrm{RG}^c(\mathcal{S}) = (V_c, E_c)$ is induced by shrinking the strongly connected components to a single node, i.e. $V_c = \{C_1, \ldots, C_r\}$. For each edge $\langle \mathbf{m}, t, \mathbf{m}' \rangle \in E$ such that $\mathbf{m}$ is in a SCC C_i, and $\mathbf{m}'$ is in a different SCC C_j, there is an induced edge $\langle C_i, t, C_j \rangle \in E_c$. The graph $\mathrm{RG}^c(\mathcal{S})$ is an acyclic digraph. Therefore the terminal SCCs of $\mathrm{RG}(\mathcal{S})$ can be computed with linear complexity in the size of $\mathrm{RG}^c(\mathcal{S})$. This fact is exploited in Algorithm 5.3 for liveness invariance checking.

Algorithm 5.3 allows us to decide liveness invariance properties. The algorithm is of linear complexity in the size of $\mathrm{RG}(\mathcal{S})$. If the algorithm succeeds, all terminal SCCs contain at least one marking satisfying the property Π, and therefore for all reachable markings there exists at least one successor marking satisfying the property Π. If the algorithm fails, there exists at least one terminal SCC that does not contain markings satisfying the property Π, and therefore the reachable markings belonging to this SCC (at least) do not satisfy the liveness invariance property.

Remark 5.2.2. It is possible to design more specific (efficient) decision procedures for the analysis of a property if we know, a priori, some characteristics of the property to be verified or we know some other properties of the net system to be analysed. In the first case we can consider as an example the reversibility property. It is easy to see that if a net system is reversible then all terminal SCCs must contain the initial marking, i.e. the reachability graph must be strongly connected. In the second case we may know, for example, that the net system is reversible; then liveness of a transition t can be decided by checking the existence of an edge in the reachability graph labelled t (since the reachability graph is SC and therefore it is always possible to reach the marking from which t can be fired).

Example 5.2.3 (Analysis of liveness invariance properties). Let us consider the net system in Figure 5.3b for which the reachability graph is depicted in Figure 5.5. The execution of Algorithm 5.3 to verify the liveness property of this net system $(\forall \mathbf{m} \in \mathrm{RS}(\mathcal{S}) \,.\, \forall t \in T \,.\, [\exists \mathbf{m}^t \in \mathrm{RS}(\langle \mathcal{N}, \mathbf{m} \rangle) \,.\, \mathbf{m}^t \geq \mathbf{Pre}[P, t]])$ requires the computation of the strongly connected components of the $\mathrm{RG}(\mathcal{S})$ (step 1). In this case, there are three SCCs depicted in Figure 5.5 and named C_1, C_2, and C_3. The SCCs C_2 and C_3 are the terminal ones. Step 4 of the

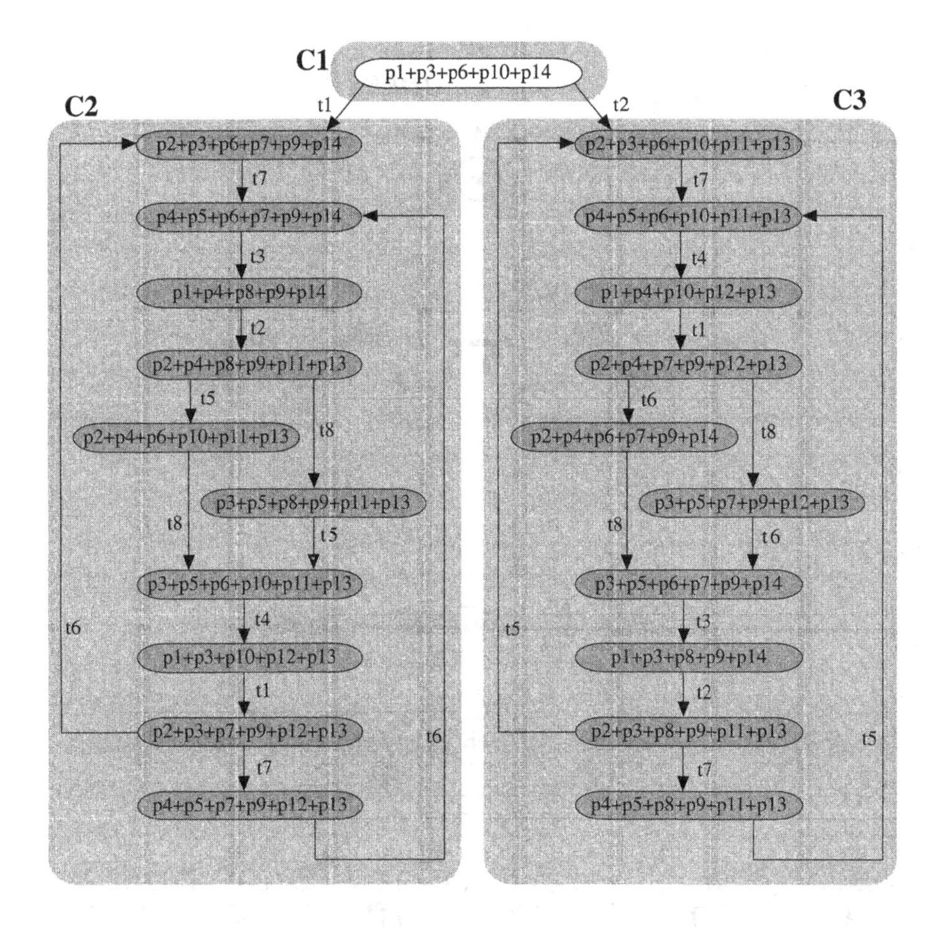

Fig. 5.5. Reachability graph of the net system in Figure 5.3b

algorithm will verify that each of these two SCCs contains for each transition t a marking $\mathbf{m}^t$ satisfying $\mathbf{m}^t \geq \mathbf{Pre}[P, t]$ (equivalently, contains edges labelled with all the transitions of the net). The reader can observe by inspection of the figure that all the transitions appear in some edge of C_2 and C_3, therefore the result of the algorithm will be TRUE.

The execution of Algorithm 5.3 to verify that the marking $\mathbf{m}_H = p_2 + p_3 + p_6 + p_7 + p_9 + p_{14}$ is a home state $(\forall \mathbf{m} \in \text{RS}(\mathcal{S}) \,.\, \exists \mathbf{m}' \in \text{RS}(\mathcal{N}, \mathbf{m})\,.$ such that $\mathbf{m}' = \mathbf{m}_H)$ gives the result FALSE, because the terminal SCC C_2 contains the marking $\mathbf{m}_H$, but the terminal SCC C_3 does not. Therefore, step 3.1 returns FALSE.

From a practical point of view, it is commonly accepted today that systems are too complex to be verified by hand. As a result, analysis is increas-

a)

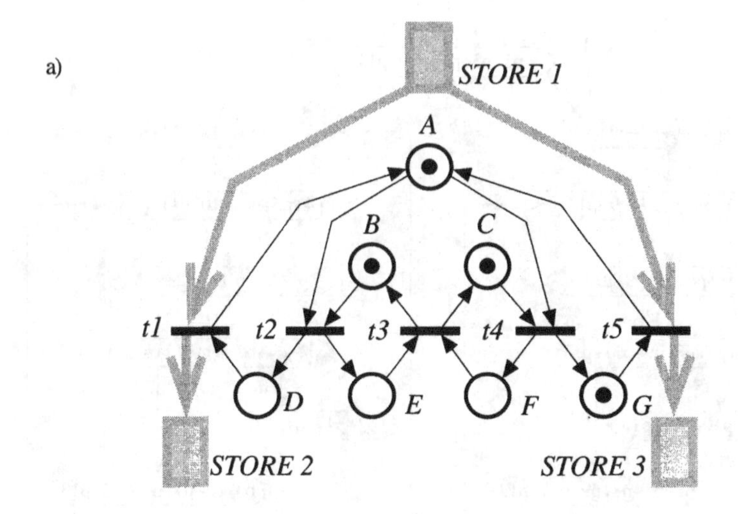

b)

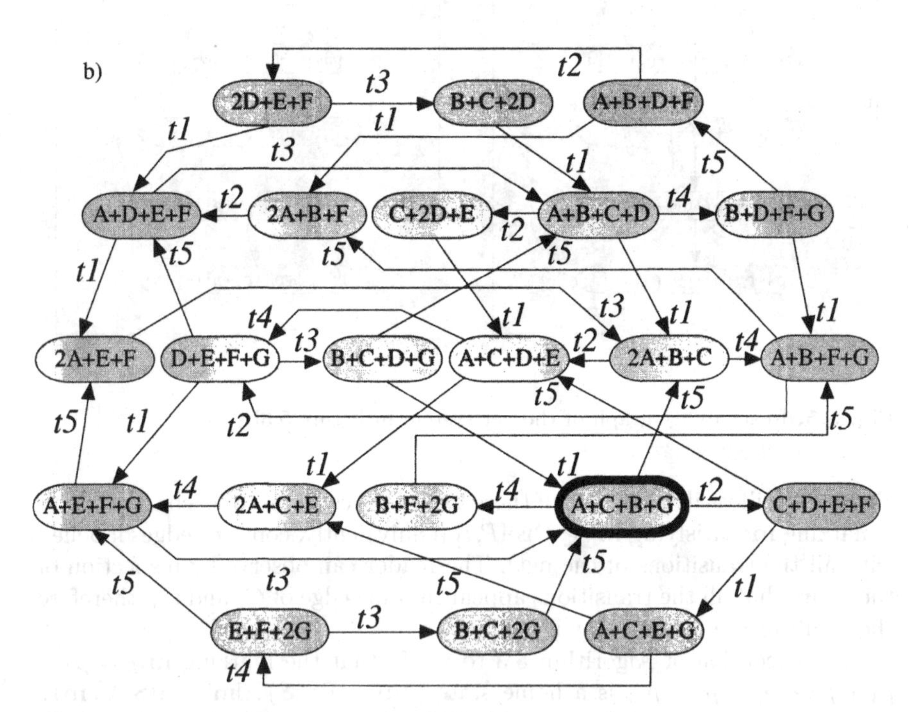

Fig. 5.6. Parts of *STORE 1* are sent to *STORE 2* and *STORE 3* according to the strategy defined by the subnet generated by $\{B, C, E, F\}$: (a) the net system; (b) the reachability graph.

ingly becoming synonymous with *computer-aided verification.*[1] Computer-aided verification means using a computer, for increased speed and reliability, to perform the analysis steps. For instance, the following example considers the analysis of a property in the group of so-called *synchronic properties* ([SC88]), demonstrating that an analysis by hand can be very difficult.

Example 5.2.4. Figure 5.6 shows a very simple net system: Parts are sent from *STORE 1* to *STORE 2* and *STORE 3.* The subnet generated by places $\{B, C, E, F\}$ imposes some restrictions on the way parts are distributed to the destination stores (i.e. the distribution is partially scheduled). The reachability graph is, even though it has been structured to clarify the presentation, difficult to understand and manage. The reader can try to check via the reachability graph (!) that the imposed distribution strategy is: parts are sent in a 1:1 relation to the destination stores, but sometimes allowing up to four consecutive deliveries to a given store (i.e. locally adjusting the possible demand, but maintaining a fair overall distribution).

Summarising, analysis techniques based on the reachability graph are only theoretically possible for bounded systems. They are very simple from a conceptual point of view. The problem that makes this approach not practical in many cases is its computational complexity: *the state-space explosion problem.*

On the other hand, it must be pointed out that the reachability/coverability graphs are computed for a given initial marking. This means that a parametric analysis of a net system (needed in earlier phases of the system design) where the initial marking of some places (e.g. representing the number of resources in the system) is a parameter, is not possible since for each value of the parameter a (possibly completely different) new reachability graph must be computed. Moreover, the reachability graph presents some difficulties in order for the analysis of properties where the distinction between conflict and concurrency plays a fundamental role (recall the net in Figure 5.2, in which the reachability graph is the same with place p_6 and without it). This is because the reachability graph gives a sequentialised view of the behaviour of the net system.

Although these analysis techniques have the drawbacks mentioned above, for bounded net systems they are the more general techniques and, in some cases, provide the only way to verify a given property.

[1] The International Conference on Computer-Aided Verification is the main forum for new results in this area. The proceedings to date have been published as *Lecture Notes in Computer Science (LNCS)* **407** (1989), **531** (1990), **575** (1991), **663** (1992), **697** (1993), **818** (1994), **939** (1995), **1102** (1996), **1254** (1997), **1427** (1998), **1633** (1999), **1855** (2000), **2102** (2001), **2404** (2002).

5.2.2 Verification Based on Linear Invariants

A *p-flow* (*t-flow*) is a vector $\mathbf{y} : P \to \mathbb{Q}$ such that $\mathbf{y} \cdot \mathbf{C} = 0$ ($\mathbf{x} : T \to \mathbb{Q}$ such that $\mathbf{C} \cdot \mathbf{x} = 0$), where $\mathbf{C}$ is the incidence matrix of the net. The set of p-flows (t-flows) is a vector space, orthogonal to the space of the rows (columns) of $\mathbf{C}$. Therefore, the flows can be generated from a *basis* of the space. Natural (i.e. non-negative integer) p-flows (t-flows) are called *p-semiflows* (*t-semiflows*): vectors $\mathbf{y} : P \to \mathbb{N}$ such that $\mathbf{y} \cdot \mathbf{C} = 0$ ($\mathbf{x} : T \to \mathbb{N}$ such that $\mathbf{C} \cdot \mathbf{x} = 0$). The following terminology is used with semiflows ([MR80]): The *support of a p-semiflow (t-semiflow)*, $\mathbf{y}$ is ($\mathbf{x}$): $||\mathbf{y}|| = \{p \in P | \mathbf{y}[p] > 0\}$ ($||\mathbf{x}|| = \{t \in T | \mathbf{x}[t] > 0\}$). A p-(t-)semiflow is *canonical* iff the g.c.d. of its non-null elements is equal to one. A net is *conservative* (*consistent*) iff there exists a p-semiflow (t-semiflow) such that $||\mathbf{y}|| = P$ ($||\mathbf{x}|| = T$).

The set of canonical p-semiflows (t-semiflows) of a given net can be infinite, since the sum of any two p-semiflows (t-semiflows) is also a p-semiflow (t-semiflow). Consider now the case of p-semiflows. A *generator set* of p-semiflows, $\Psi = \{\mathbf{y}_1, \mathbf{y}_2, \ldots, \mathbf{y}_q\}$, is made up of the least number of them which will generate any p-semiflow, $\mathbf{y} : P \to \mathbb{N}$, as follows: $\mathbf{y} = \sum_{\mathbf{y}_j \in \Psi} k_j \cdot \mathbf{y}_j$, $k_j \in \mathbb{Q}$ and $\mathbf{y}_j \in \Psi$. The p-semiflows of Ψ are said to be *minimal*. The following result characterises the generator set of each of the sets of semiflows (p-semiflows and t-semiflows) of a net.

Proposition 5.2.5. *A p-(t-)semiflow is minimal iff it is canonical and its support does not strictly contain the support of any other p-(t-)semiflow. Moreover, the set of minimal p-(t-)semiflows of a net is finite and unique.*

From the above result, the number of minimal p-(t-)semiflows is less than or equal to the number of incomparable vectors of dimension k ($k = |P|$ or $k = |T|$): Number of minimal p-(t-)semiflows $\leq \binom{k}{\lceil k/2 \rceil}$, where $\binom{\star}{\star}$ denotes binomial coefficient and $\lceil \star \rceil$ denotes rounding up to an integer. In practice the number of minimal p-(t-)semiflows is much smaller than the previously stated upper bound.

Algorithm 5.4 presents a simple way to compute the set of minimal p-semiflows Ψ from the incidence matrix of the net. Each row of matrix Ψ memorises the coefficients of the positive linear combination of rows of matrix $\mathbf{C}$ which generated the row of $\mathbf{A}$ with the same index. In step 3 of the algorithm, the rows of $\mathbf{A}$ are null and therefore each row $\Psi[i]$ is a p-semiflow: $\Psi[i] \cdot \mathbf{C} = 0$. The same algorithm can be used to compute the set of minimal t-semiflows if the input of the algorithm is the transpose of the incidence matrix.

The computation of minimal p-semiflows ($\mathbf{y}$) and minimal t-semiflows ($\mathbf{x}$) has been extensively studied. An *exponential number* of such minimal p- and t-semiflows with respect to the number of places (transitions) may appear, and therefore the time complexity of this computation cannot be polynomial.

Algorithm 5.4 (Computation of the minimal p-semiflows)

Input - The incidence matrix $\mathbf{C}$. A fixed but arbitrary order in P is supposed.
Output - The p-semiflow matrix Ψ, where each row is a minimal p-semiflow.

1. $\mathbf{A} = \mathbf{C}$; $\Psi = \mathbf{I_n}$; { $\mathbf{I_n}$ is an identity matrix of dimension n }
2. **for** $i = 1$ **to** m **do** { $m = |T|$ }
   2.1 Add to the matrix $[\Psi|\mathbf{A}]$ all rows which are natural linear combinations of pairs of rows of $[\Psi|\mathbf{A}]$ and which annul the i-th column of $\mathbf{A}$
   2.2 Eliminate from $[\Psi|\mathbf{A}]$ the rows in which the i-th column of $\mathbf{A}$ is non-null.
3. Transform the rows of Ψ into canonical p-semiflows and remove all non-minimal p-semiflows from Ψ using the characterisation of Proposition 5.2.5.

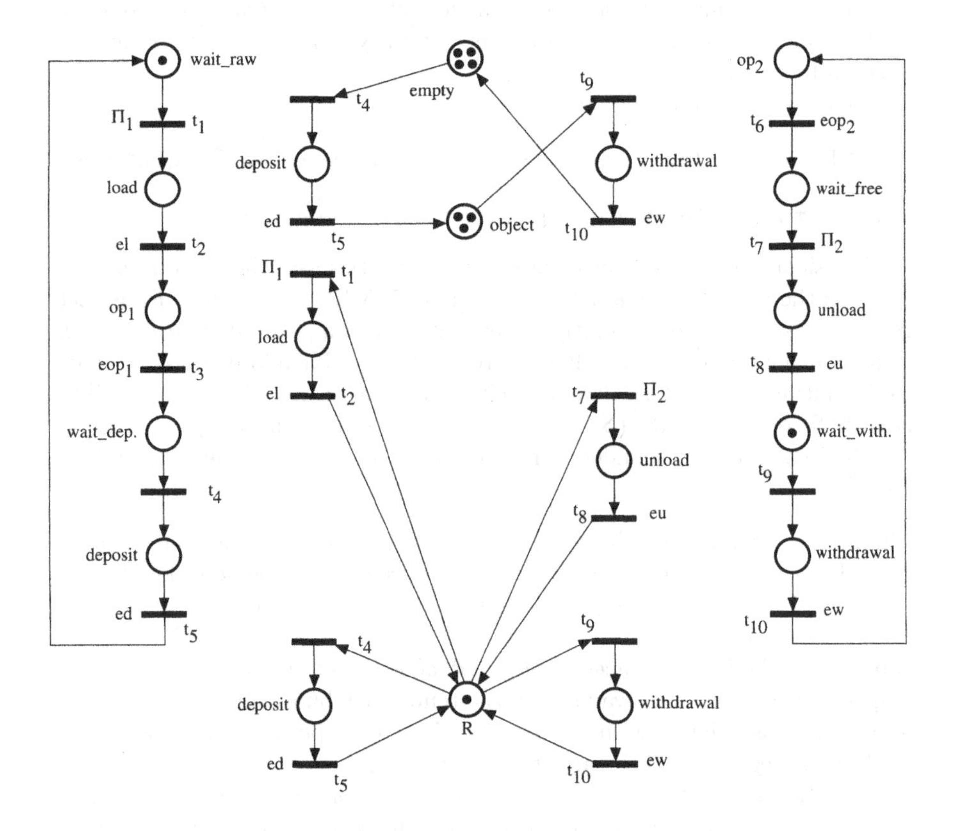

Fig. 5.7. A decomposed view of the net system in Figure 5.8

In [CS90a] a study is done merging traditional techniques in convex geometry with those developed within Petri nets.

p- and t-semiflows are dual structural objects (left or right annullers of the incidence matrix of the net) from which it is possible to obtain linear invariant laws on the reachable states or cyclic occurrence sequences respectively. These invariant laws arise from the structure of the net, and the initial marking plays the role of a parameter specifying a particular behaviour for the net. The following two classes of linear invariants can be obtained:

1) From p-semiflows: (token conservation law)

$$\mathbf{y} \in \mathbb{N}^n \wedge \mathbf{y} \cdot \mathbf{C} = 0 \Longrightarrow \forall \mathbf{m_0} \,.\, \forall \mathbf{m} \in \mathrm{RS}(\mathcal{N}, \mathbf{m_0}) \,.\, [\mathbf{y} \cdot \mathbf{m} = \mathbf{y} \cdot \mathbf{m_0}].$$

This marking invariant specifies that for all markings reachable from the initial one, the weighted sum of tokens at $\mathbf{m}$, $\mathbf{y} \cdot \mathbf{m}$, remains constant and equal to $\mathbf{y} \cdot \mathbf{m_0}$.

2) From t-semiflows: (cyclic behaviour law)

$$\mathbf{x} \in \mathbb{N}^m \wedge \mathbf{C} \cdot \mathbf{x} = 0 \Longrightarrow \exists \mathbf{m_0} \,.\, \exists \text{occurrence sequence } \sigma \,.\, [\mathbf{m_0} \xrightarrow{\sigma} \mathbf{m_0} \wedge \boldsymbol{\sigma} = \mathbf{x}],$$

where $\boldsymbol{\sigma}$ is the Parikh mapping of the occurrence sequence σ.

Classical reasoning for proving logical properties uses these *linear invariants* on the behaviour of a net system ([Lau87, MR80]). The key idea is: Let $\mathcal{S}$ be a net system and Ψ a matrix where each row is a p-semiflow: $\Psi[i] \cdot \mathbf{C} = 0$. If $\mathbf{m}$ is reachable from $\mathbf{m_0}$, then $\Psi \cdot \mathbf{m} = \Psi \cdot \mathbf{m_0}$. Therefore the set of natural solutions $\mathbf{m}$ of this equation defines a linearisation of the reachability set $\mathrm{RS}(\mathcal{S})$ denoted $\mathrm{LRS}^{\Psi}(\mathcal{S})$. This set can be used to analyse properties. It usually leads only to semidecision algorithms because, in general, $\mathrm{RS}(\mathcal{S}) \subset \mathrm{LRS}^{\Psi}(\mathcal{S})$.

Example 5.2.6 (Analysis based on linear invariants). The local controller attached to the production cell depicted in Figure 5.8a can be described by the given Petri net model. The places *wait_raw*, *load*, op_1, *wait_dep.*, and *deposit* represent the possible states of *MACH 1*; the place R is marked when the robot is available; the places *empty* and *object* contain as many tokens as empty slots or parts are available in the temporary buffer, etc. In this model actions are associated with places, e.g. *MACH 2* performs its operations while place op_2 is marked, and transitions represent atomic instantaneous changes of state. External inputs (from plant sensors) condition these possible changes of state, e.g. a load operation is initiated (transition t_1 is fired) when *MACH 1* is waiting for a raw part (*wait_raw* marked), the robot is available (R marked), and a raw part is detected by the sensor Π_1 (Π_1 is true). The model reflects the synchronisation constraints imposed by the use of the temporary buffer: a deposit operation cannot be initiated unless an empty slot is available (represented by an arc from the *empty* place to the t_4 transition); and a withdrawal operation cannot be initiated unless a part

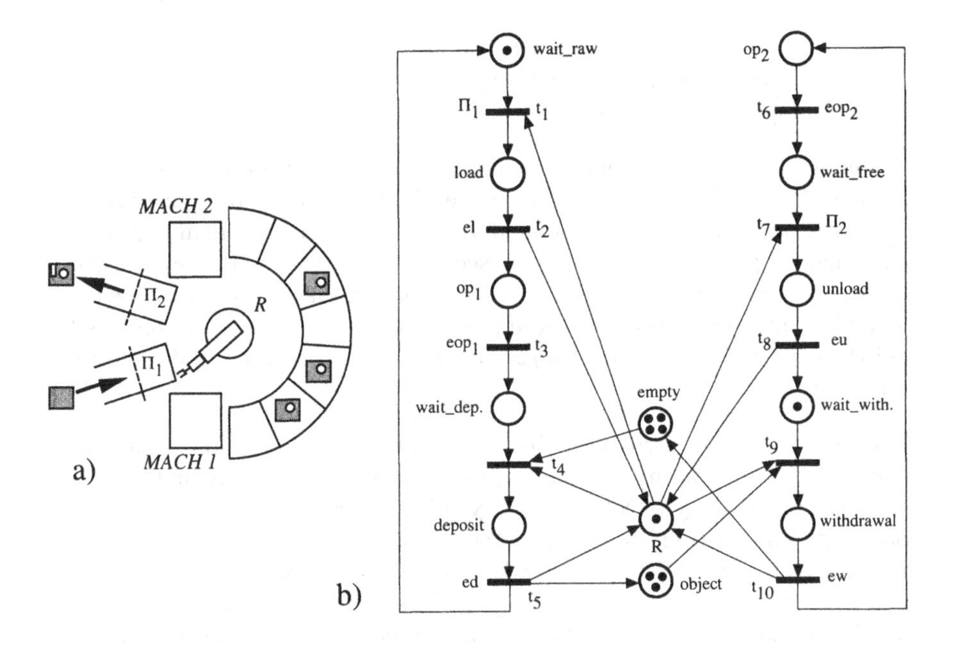

Fig. 5.8. a) A production cell with two machines, one robot, and a store b) Net system specifying its behaviour

is available in the buffer(represented by an arc from the *object* place to the t_9 transition). The arcs from the t_5 transition to the *object* place and from the t_{10} transition to the *empty* place represent the updating of the number of parts in the buffer after a deposit operation, and the number of empty slots after a withdrawing operation respectively. If the synchronisation constraints previously described were deleted from the model, the physical system could reach a deadlock situation, e.g. *MACH 2* withdraws a part when there are none available but *MACH 1* cannot deposit any because the robot is busy.

The marking linear invariants induced by the minimal p-semiflows of the net system in Figure 5.8 are the following:

$$\mathbf{m}[\text{wait\_raw}]+\mathbf{m}[\text{load}]+\mathbf{m}[\text{op}_1]+\mathbf{m}[\text{wait\_dep.}]+\mathbf{m}[\text{deposit}]=1 \quad (5.1)$$

$$\mathbf{m}[\text{op}_2]+\mathbf{m}[\text{wait\_free}]+\mathbf{m}[\text{unload}]+\mathbf{m}[\text{wait\_with.}]+\mathbf{m}[\text{withdrawal}]=1 \quad (5.2)$$

$$\mathbf{m}[\text{empty}]+\mathbf{m}[\text{deposit}]+\mathbf{m}[\text{object}]+\mathbf{m}[\text{withdrawal}]=7 \quad (5.3)$$

$$\mathbf{m}[\text{R}]+\mathbf{m}[\text{load}]+\mathbf{m}[\text{unload}]+\mathbf{m}[\text{deposit}]+\mathbf{m}[\text{withdrawal}]=1 \quad (5.4)$$

Because markings are non-negative integers (i.e. $\forall p \in P \,.\, \mathbf{m}[p] \geq 0$), the following can be easily deduced from the previous equalities:

1. Bounds:
   $\forall p_i \in P \setminus \{\text{empty}, \text{object}\} \,.\, (\mathbf{m}[p_i] \leq 1 \wedge \mathbf{m}[\text{empty}] \leq 7 \wedge \mathbf{m}[\text{object}] \leq 7)$.
2. The places in each of the following sets are in marking mutual exclusion:

a) {wait_raw, load, op_1, wait_dep., deposit}
b) {op_2, wait_free, unload, wait_with., withdrawal}
c) {R, load, unload, deposit, withdrawal}

Finally, from a conceptual point of view, the consideration of p-semiflows provides *decomposed views* of the structure of the net model. In Figure 5.7 the decomposition induced by the minimal p-semiflows of the system in Figure 5.8 is graphically presented. The decomposed view of a net system is even useful to derive an *implementation.* For example, the net system in Figure 5.8 can be implemented using two sequential processes (for *Machine1* and *Machine2*) and three semaphores (*object*, *empty* and *R*), where *R* is a mutual-exclusion semaphore.

Remark 5.2.7. Other structural objects similar to p- or t-semiflows have been defined ([MR80]) leading to other types of linear invariants. A first type to consider are vectors $\mathbf{y} \in \mathbb{N}^n$ such that $\mathbf{y} \cdot \mathbf{C} \lneq 0$. Such a vector $\mathbf{y}$ leads to the following marking law: $\forall \mathbf{m_0} \,.\, \forall \mathbf{m} \in \mathrm{RS}(\mathcal{N}, \mathbf{m_0}) \,.\, (\mathbf{y} \cdot \mathbf{m} \leq \mathbf{y} \cdot \mathbf{m_0})$. A second type are vectors $\mathbf{x} \in \mathbb{N}^m$ such that $\mathbf{C} \cdot \mathbf{x} \gneq 0$. In this case, a vector $\mathbf{x}$ of this kind leads to: $\exists \mathbf{m_0} \,.\, \exists \sigma \in \mathrm{L}(\mathcal{N}, \mathbf{m_0}) \,.\, \mathbf{m_0} \xrightarrow{\sigma} \mathbf{m} \geq \mathbf{m_0} \wedge \boldsymbol{\sigma} = \mathbf{x}$. These linear invariants (expressed as inequalities) can be used for analysis purposes in the same way as that presented previously for linear invariants obtained from semiflows.

6. Overview of the Book

This chapter is intended to give a more detailed overview of the book than the introduction in Chapter 1. Having read Chapters 1 to 5 the reader should have acquired some intuitive understanding of Petri nets as well as some familiarity with basic formal definitions and properties. At this point the reader should be in a position to understand most of the presentations in the remainder of the book. Although, for the novice it may be beneficial to read the book from beginning to end, more experienced readers should have no problem skipping chapters that are not of foremost importance for them.

In the preceding chapters Petri nets have been introduced using the example of modelling the starting phase of a car race together with some further nets for illustrating basic properties. A production cell with two machines was modelled towards the end of the preceding chapter. Although illustrative for introductory purposes, these examples teach the reader few principles for modelling real cases. Part II of the book is thus devoted to a thorough treatment of skills and methods for modelling systems by Petri nets from more or less formal specifications. To begin with, some more complex examples are given in Chapter 8. They range from simple P/T nets (task execution) to resource management by P/T nets (the banker's problem) to a coloured net model (alternating bit protocol).

Today, the modelling of information systems can be viewed as an art. It very much depends on the personal skills and styles of the people involved in building the models. This situation is however not satisfactory from a systematic or commercial point of view. Modelling of real-world cases to a large extent involves abstraction from those features classified as less important, in a consistent way. Here a formal model such as Petri nets can help a great deal: when thinking in terms of places and transitions the modeller is smoothly lead to create an operational model which is formal and intuitive at the same time. Such a phase is followed by the addition of details and technicalities.

An important contribution of the Petri net modelling technique is the clear concept of refinement and abstraction. It was introduced in Chapter 2 from a formal and conceptual point of view. Chapter 9 discusses its practical use in modelling, while Chapter 10 presents three more methodical approaches for the modelling of systems. The first is state-oriented, starting from specifications of properties of reachable states. The second is event-oriented, combining modules of a system in a bottom-up approach by synchronising events. The third approach introduces object-oriented methods for Petri nets, e.g. by considering nets as token objects or by introducing the class concept to coloured Petri nets. These methods are illustrated in Chapter 11 by applying them to a common case study on mutual exclusion.

Modern computer-based systems still do not achieve the quality standards that are necessary to meet requirements with respect to security or reliability. Deficiencies of software applications can even decrease commercial profitability to the point of total failures of projects. A system cannot be verified if the desired properties are unknown or ambiguous. Furthermore, a formal specification of the system itself is desirable.

Since the modelling of systems using Petri nets supports both aspects there have been vast research activities in the development of methods for the verification of systems. This is reflected by the extensive treatment of this topic in Part III. Chapter 13 gives a thorough introduction to the aspects dealt with in this part. It discusses the nature of verifying properties, the classes of nets that are considered (restrictions, extensions, abbreviations, and parametrisations). Furthermore the verification process itself and the methods are classified (graph theory, linear algebra, state-based, binary decision diagrams, on-the-fly verification, partial-order methods).

The analysis of the state space, also known as model checking, is perhaps the most frequently used approach in industrial verification. This is because of its conceptual simplicity. It can be applied to all kinds of formal modelling techniques bearing a notion of global state. The method is limited, however by its huge complexity in general. The so-called state-space explosion problem applies, in particular, to systems with significant inherent concurrency. There are many approaches to overcome or reduce these limitations. The Petri net model allows some special techniques since the structural information given by a net can be exploited. State-space analysis means the verification of some formal properties by inspection of the state space, i.e. the reachability graph of the Petri net. For this reason, Chapter 14 begins with the introduction of a formal specification language based on temporal logic. As illustrated by a mutual exclusion example, it allows the specification of so-called safety properties as well as liveness properties in a very compact way. Some general methods for the verification of temporal formulae are discussed as well as more specific topics, such as fairness assumptions, on-the-fly methods, and partial-order approaches. The latter are efficient techniques to reduce the time and space complexity of state-based methods via the concepts of so-called stubborn and sleep sets. Finally, symbolic and parametrised approaches are presented that reduce the size of the reachability graph representation by grouping states into classes. They apply in particular to so-called well-formed nets, a net class that introduces a structure which allows more efficient analysis techniques. For instance, a symbolic construction of the reachability graph permits the exploitation of the intrinsic symmetries of the model. The theory is discussed in detail and implementation issues are presented at the end of this chapter.

The state-space explosion problem is also tackled in Chapter 15, but here the state space is neither constructed nor inspected. All information on the net behaviour is deduced from its structure, e.g. from the structure of the graphical or algebraic representation. The techniques are in part derived from deep results from graph theory, linear algebra, and convex geometry. The results are quite impressive since the exponential complexity of state-space analysis is reduced to a complexity which is polynomial or less in many cases. The price to be paid for this improvement is that the Petri net has to satisfy some structural properties which are not granted for all applications. Typical restrictions aim at limiting the interplay between synchronisations and conflict. Hence the designer must find a compromise between the modelling power and the availability of powerful analysis tools.

A set of reduction rules is given that allows the elimination of places and transitions while important properties such as boundedness and liveness are preserved. With these reductions the state space is also reduced even before it is constructed. For these rules the notion of implicit places is introduced, which is also important in a different context (see the first example of Chapter 8). Linear algebra is used to deduce conditions for structural properties. These properties hold independently of the chosen initial marking. The notions of siphons and traps lead to structurally defined subclasses such as equal conflict and free-choice Petri nets, which are easier to analyse. At the end of Chapter 15 extensions of the definitions of invariants and reductions are discussed for the case of coloured nets.

Chapter 16 presents connections to important fields in the formal methods area: logic, algebraic specification, assertional reasoning, and process algebra. These connections are important as Petri net users cannot stay isolated from these areas where continual research is done and important developments take place. Algebraic specifications allow for a domain-independent definition of systems, which is not only important for improving correctness and reliability but is the basis for very high-level systems development. Algebraic Petri nets provide algebraically specified systems with a semantics of change and all the advantages of ordinary Petri nets. They can be treated as a subcase of the broader theory of rewrite specifications. Furthermore, this theory is the basis of a continually growing number of important tools that will most likely gain importance for systems designers in the near future. The chapter gives an introduction into the field, connecting it with Petri net terminology. Also some non-trivial distributed algorithms are verified within this framework.

Section 2 of Chapter 16 relates Petri nets to a standard technique in verification, namely verification by assertions and temporal logic. Invariants and leads-to-operators are used in a UNITY-like logical-rule system to verify properties of concurrent systems. Unlike many other contributions these notions are used for the verification of high-level nets in the form of coloured Petri nets. Proof rules of safety and liveness assertions are given, and illustrated using a simple coloured net. Also the important topic of compositionality is addressed, i.e. the problem of composing proofs while composing net components.

To argue about computations, the standard temporal logic is extended to a so-called logic of enablement in the next section. Reductions of complex nets to much simpler test nets are studied to apply preservation and reflection results. This result is a logical counterpart to the syntactical net reductions in Chapter 15. Linear logic is a quite surprising application of logic to systems design. Deductions in linear logic are very similar (in some cases isomorphic) to the occurrence sequences of Petri nets and therefore allow for a translation of results between the two fields. Linear logic is formulated for coloured nets for the first time and even for object nets. Through these results the power and the flexibility of the approach is demonstrated. In the last section of Chapter 16 an example is given of the combined application of two formal methods, namely Petri nets and process algebra for design and verification purposes. The key aspect is that system design is integrated with system verification. The integrated method is illustrated by the development of a simple production unit.

The aspects of validation and execution are treated in Part IV. By validation we mean the matching of the modelled system with the expectations of the user, client, or customer. Obviously, the frontiers to verification are not sharp. While the former deals with precisely stated properties, the latter has its focus on less formal steps of executing, simulating, animating, inspecting, testing, debugging, observing, or checking the system. This list could easily be extended and it is impossible to give a complete presentation of the topic in this framework. Here we concentrate on the major areas of Petri net validation, namely prototyping, net execution, and code generation. Because of their well-defined semantics, expressiveness, and graphical representation, Petri-net-based models are suitable for supporting different steps in the software life-cycle process as discussed in software engineering. Some of these aspects are presented, including prototyping and animation. The success of the Petri net modelling technique depends to a large extent on the availability of suitable tools. In this respect some important progress can be observed, and to date there is a large number of products that meet strict industrial criteria. They incorporate many of the aspects discussed in this part. However, as the field is changing very fast, concrete tools are not discussed here in detail.

Observation plays an important part in the phases of requirement analysis and design. A chapter on net execution is concerned with these aspects. The generation of code can also derive benefits from a Petri net model of a system. The model can provide important information on implementation with respect to specific computer architectures, such as centralised, distributed, or hybrid settings. Petri net partitioning algorithms rely on structural properties such as place invariants and place refinements. The results are illustrated by an implemented high-level formalism for code generation: H-COSTAM.

There are few modelling concepts that possess such a multitude of extensively elaborated application domains as do Petri nets. Therefore it is essential to include at least some of them in this book. Three such domains have been selected for a detailed presentation in Part V, namely flexible manufacturing systems, workflow management systems, and telecommunications systems. (For a collection of other application domains see Section 23.2.4.) From a conceptual view the first two of these domains seem to be very similar. In both cases sequences of task executions are modelled under the restriction of limited resources. However, on a closer look, they turn out to be rather different. Of these two, the study of flexible manufacturing systems has a longer tradition and has been developed within the discipline of mechanical engineering. Flexible manufacturing systems (Chapter 24) are characterised by flexible, concurrently operating, and mainly automated elements, such as a production controller, a machine, an automated guided vehicle, and a conveyor. Petri nets allow for the modelling of resource sharing, conflicts, mutual exclusion, concurrency, and non-determinism, which characterise critical elements of such manufacturing processes.

The second domain, workflow management (Chapter 25), is much younger. It arose from the study of information systems and is usually constituted as a generic software tool which allows for the definition, execution, registration, and control of commercial and administrative processes. Petri nets are a good candidate to become the foundation of a unified, vendor-independent workflow theory providing algorithms for verification and analysis. They can also be used by users without any engineering or programming background.

Telecommunications systems on the other hand have become a dominant factor of modern societies. They consist of a transport and a processing subsystem. The complexity of these systems easily exceeds all kinds of artificial and planned systems. The processing system includes software for the control and management of the transport network and the communication software (protocols). Protocol engineering has emerged as a specialisation of software engineering inheriting the problems of the more general domain, but having specific requirements in particular with respect to correctness and performance. Many analysis techniques for Petri nets have been successfully applied here and specific new algorithms have been developed. An introduction to the implications of building telecommunications systems with Petri nets is given in Chapter 26.

Part II

Modelling

7. Introduction

The Art of Modelling

The systems engineer often uses models to investigate properties of his system. In the context of this book, the models will be Petri nets, and the chapters to come are filled with various interesting properties to investigate (analyse, verify, validate) in this context. However, the models have to be built first, which is the topic of this part. In fact, the construction of formal models (such as Petri nets) is valuable in itself, as it enforces a full understanding of the aspects treated.

There exist two schools in Petri net modelling. The direct approach (cf. [Jen92b, Jen94]) views Petri nets as a user-friendly graphical modelling technique. It advocates models that are easy to grasp and close to the problem domain. The indirect approach (cf. [BG96]) views Petri nets as a powerful but low-level formalism. Models are constructed in another formalism and translated to Petri nets for analysis.

We think that both approaches are valid. The chapters in this section can be studied as a guideline for directly creating Petri net models or as a guideline for translating models in other formalisms into Petri nets.

Modelling is an art; there is no standard recipe for it. Different expert modellers will build different models for the same problem. Each model will have its merits; it is hard to choose the "best" one. On the other hand some models will be rejected by all experts. A mistake commonly made is making the model too detailed. Much effort is expended in doing so, and the model turns out to be too complex to analyse.

The example in Figure 7.1 illustrates this principle. The figure shows a part of a coloured net with a place of numeric colour. Two transitions have this place as input: one with a condition $x < 10$, and the other with $x > 5$. The analysis tools have to be quite powerful to determine that the conditions allow consumption of every possible token. This problem becomes even more apparent for more complex colour sets. The modeller should seriously consider whether the colour extension is needed or a place-transition net model would not suffice.

Modelling is thus, first and foremost, choosing what aspects to model and what to leave out. A novice has the tendency, understandably, to avoid such choices. As seen above, this can be a serious mistake. One must learn and

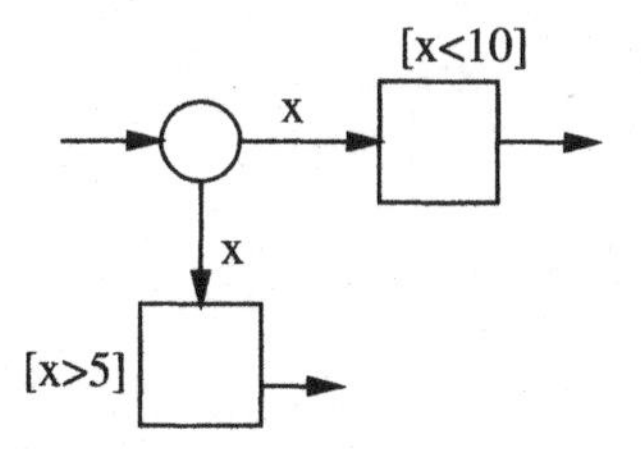

Fig. 7.1. Coloured net model

dare to choose and making a wrong choice is far less costly than not making any choice at all.

The aspects that benefit the most from Petri net models are related to the decomposition of a system into largely independent subsystems. These subsystems may differ, e.g. a computer system consisting of a CPU and input and output devices, or they may be the same, such as the nodes in a data communication network. They should cooperate, exchanging information from time to time. If such aspects are of little importance to a project, it is questionable whether Petri nets should be used at all.

Modelling Approaches

In the chapters to come, expert Petri net model builders show the tricks of their trade. In the earlier chapters, models are given without explaining how they were constructed. The reader can appreciate the end product and develop a sense of quality. The models are kept simple to aid the reader's understanding.

In the later chapters, methods for constructing models are given. There exist several substantially different methods, which may nevertheless lead to quite similar end results. All methods have a way of stating properties of the models to be constructed and verifying whether the constructed models satisfy them. The properties involved may deal with either *states* or *events*.

A typical state-oriented statement is: "There are ten cars in this parking lot." This fact is *static* and lasts for a certain period of time. Events, e.g. the arrival or departure of cars, may alter the truth of this statement. An event-oriented statement is: "I parked my car in the wrong place this morning." Parking one's car is an action, with a short duration. There may be a state witness of the event, i.e. my car with a parking ticket.

One may construct models by looking at the state and analysing how it changes. Another approach is looking at the events: how, when, and in what order they occur. Finally, one can look at both at the same time. All three approaches are represented in the following chapters.

Overview of Part II

In Chapter 8, we introduce a few example models. In Chapter 9, the basic ideas underlying modular modelling strategies are developed. The idea behind these strategies is to model a few aspects at the time and combine these smaller models into a large one covering all aspects. In Chapter 10, three such modelling strategies are unfolded, one that is initially based on *states*, one that is based on *events*, and an object-oriented approach that combines both aspects. Finally, in Chapter 11 the three strategies are applied to a common case study. It has to be noted that, although the approaches differ, the final results are very similar.

8. Modelling and Analysis Techniques by Example*

In the literature, and even more in unpublished sources such as lecture notes, there is a treasury of fine examples of using Petri nets for system design. Many such examples are given in this book to illustrate particular definitions, results, or methods. However, some very typical and interesting examples are not included, sometimes due to their size. This chapter is meant to partially bridge the gap and to give a deeper insight into the modelling potential of Petri nets. The examples are not chosen randomly but rather to cover different areas of applications using different Petri net model classes. In particular, nets, refinement, and abstraction of nets will be used for an example on task execution in a system of functional units. A well-known resource allocation problem will be modelled by place/transition nets, and state-space representation and place-invariants will be used to illustrate the problem. The alternating-bit protocol will be modelled by a coloured net. Different layers will again be connected by the concept of net refinement.

8.1 Nets, Refinement, and Abstraction

Example 1: Task Execution

This example models a task execution sequence by a number of machines: an object in a production line has to be processed, first by some machine M_1 and then by machines M_2 or M_3. The process is then repeated, which is very natural in the context of manufacturing systems. Besides the machines, operators for the machines are a second type of limited resource: operator O_1 can be operating M_1 or M_2, but not both at the same time. The same holds for O_2 with respect to M_1 and M_3.

In a first step, a designer can conceive the design of the system ignoring the existence of resources and can concentrate on the operation sequences of the parts to be processed. From the perspective of manufacturing systems, this means that the designer concentrates on the feasibility of the operation sequences according to the routing possibilities that the layout of the plant

* Author: R. Valk

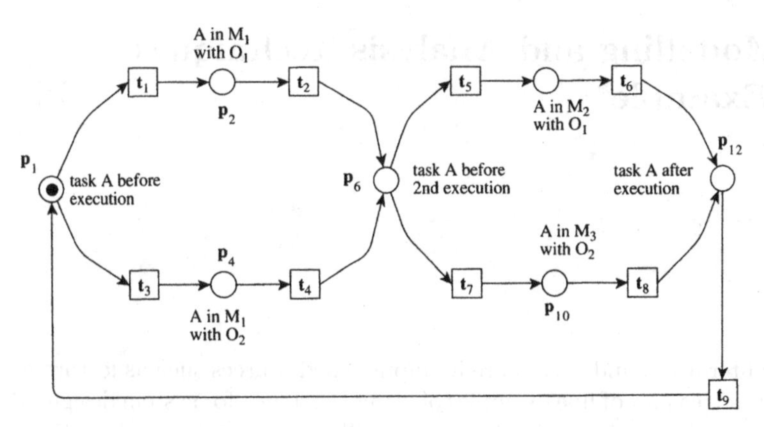

Fig. 8.1. Machine example without constraints

imposes. The resulting net is given in Figure 8.1 with p_1 initially the only place containing a token. This net is a monomarked state machine, where each circuit corresponds to a t-invariant. *T-invariants* describe sets of transitions which reproduce the initial marking when occurring in a suitable order. Therefore the operation sequences are correct. There are four elementary t-invariants (also known as t-semiflows, cf. page 68):

a) $t_1 t_2 t_5 t_6 t_9$
(process working only with operator 1)

b) $t_1 t_2 t_7 t_8 t_9$
(process with operator 1 working first, and operator 2 afterwards)

c) $t_3 t_4 t_7 t_8 t_9$
(process working only with operator 2)

d) $t_3 t_4 t_5 t_6 t_9$
(process with operator 2 working first, and operator 1 afterwards).

This reflects the operation sequences allowed by the specification. Next, the designer can incorporate the constraints imposed by the resources which are to be understood in a broad sense. This is to verify the feasibility of the operation sequences with a predefined number of resources (the availability of the machines M1, M2, M3 and the operators O1, O2 in this case). The resulting net is presented in Figure 8.2.

Before the execution all machines and operators are idle. Therefore the initial state corresponds to the given initial marking $\mathbf{m_0}$ with $\mathbf{m_0}[p_1] = \mathbf{m_0}[p_3] = \mathbf{m_0}[p_5] = \mathbf{m_0}[p_7] = \mathbf{m_0}[p_8] = \mathbf{m_0}[p_{11}] = 1$ corresponding to the local conditions:

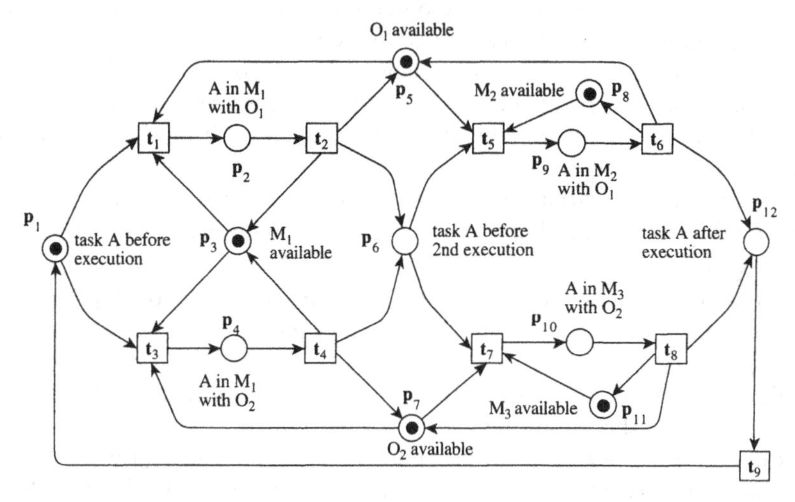

Fig. 8.2. Machine example with constraints

p_1:	task A before a new execution,
p_3:	M_1 available,
p_5:	O_1 available,
p_7:	O_2 available,
p_8:	M_2 available, and
p_{11}:	M_3 available.

On the occurrence of transition t_1, the beginning of a task is represented by machine M_1 with O_1 operating it. If O_2 is chosen to work instead, transition t_3 has to occur first. The processing of the task on M_1 by O_1 or O_2 ends with the occurrence of transitions t_2 and t_4 respectively. The execution of tasks on M_2 and M_3 can be described in a similar way. The example is taken from [Pet81], [Rei83] and [JV87].

The following effects can be observed:

a) The occurrence of transitions (actions) may be in conflict (e.g. t_5 and t_7). The procedure of a decision for resolving the conflict may not be visible since the model is omitting from this detail.
b) The occurrence of transitions may be causally dependent upon other transitions, e.g. t_2 not before t_1, t_5 not before t_3 (in the same cycle of operation). Note that t_2 cannot occur without a preceding occurrence of t_1, whereas t_5 can occur without a preceding occurrence of t_3.
c) Limited resources are modelled in a fairly direct way (e.g. O_1 cannot operate M_1 and M_2 at the same time) or in a less direct way (e.g. O_1 and O_2 cannot operate M_1 at the same time).
d) There are resources that do not restrict the behaviour of the net (e.g. p_8: M_2 available). This may be an indication that the net is an extract from

a more complex net where the resource may be critical (e.g. an extension also modelling the possibility that M_2 may not be available).

e) Transitions (actions) may occur independently ("concurrently"), e.g. with 2 tokens in p_1 (interpreted as two tasks to be processed), after some time one of them may reach p_6 while the other one is still in p_1. Then t_1 and t_5 may occur concurrently.

f) There are linear invariant equations[1] (place-invariants, p-semiflows) holding in all reachable markings $\mathbf{m}$. Consider the following examples:
   1. $\mathbf{m}[p_2] + \mathbf{m}[p_5] + \mathbf{m}[p_9] = 1$
      Either O_1 is idle or operating one of the machines M_1 or M_2, but only one of these cases can occur at the same time.
   2. $\mathbf{m}[p_2] + \mathbf{m}[p_3] + \mathbf{m}[p_4] = 1$
      Either M_1 is idle or operated by O_1 or by O_2, but only one of these cases can occur at the same time.
   3. $\mathbf{m}[p_1] + \mathbf{m}[p_2] + \mathbf{m}[p_4] + \mathbf{m}[p_6] + \mathbf{m}[p_9] + \mathbf{m}[p_{10}] + \mathbf{m}[p_{12}] = c$
      Exactly c tasks are in the production line, where $c = 1$ in Figure 8.2 and $c = 2$ in the modification of case e).

The t-invariants of this net system remain the same as in the first design. An interesting analysis with respect to the resources is that all added places representing these resources are implicit (redundant) in the net system of Figure 8.2. The linear invariant equations that define the implicitness property[2] of these places are:

g)
   1. $\mathbf{m}[p_3] = \mathbf{m}[p_1] + \mathbf{m}[p_6] + \mathbf{m}[p_9] + \mathbf{m}[p_{10}] + \mathbf{m}[p_{12}]$
   2. $\mathbf{m}[p_5] = \mathbf{m}[p_1] + \mathbf{m}[p_6] + \mathbf{m}[p_{12}] + \mathbf{m}[p_4] + \mathbf{m}[p_{10}]$
   3. $\mathbf{m}[p_7] = \mathbf{m}[p_1] + \mathbf{m}[p_6] + \mathbf{m}[p_{12}] + \mathbf{m}[p_2] + \mathbf{m}[p_9]$
   4. $\mathbf{m}[p_8] = \mathbf{m}[p_1] + \mathbf{m}[p_2] + \mathbf{m}[p_4] + \mathbf{m}[p_6] + \mathbf{m}[p_{10}] + \mathbf{m}[p_{12}]$
   5. $\mathbf{m}[p_{11}] = \mathbf{m}[p_1] + \mathbf{m}[p_2] + \mathbf{m}[p_4] + \mathbf{m}[p_6] + \mathbf{m}[p_9] + \mathbf{m}[p_{12}]$

This redundancy of the resource places indicates that there are sufficient resources in the sense that they do not constrain the operation sequences of the preliminary design of Figure 8.1. To give an example for this statement, consider the initial marking $\mathbf{m}_0$ of the net system of Figure 8.2, where $\mathbf{m}_0[p_1] = 1$. Then by the first equation of g), $\mathbf{m}_0[p_3] = 1$ and both transitions t_1 and t_3 are activated.

When introducing more than one part into the system, i.e. $\mathbf{m}_0[p_1] > 1$, the previous implicitness property is no longer true. In fact, the different parts now compete in order to obtain the resources of the system. Nevertheless, we can remove these competition relations (all of them or only some of them) from the net system if we increase the number of resources, i.e. if we increase the number of tokens at the initial marking in the places representing the

[1] See also page 70.

[2] See also Section 15.1.2 for a discussion of implicit places.

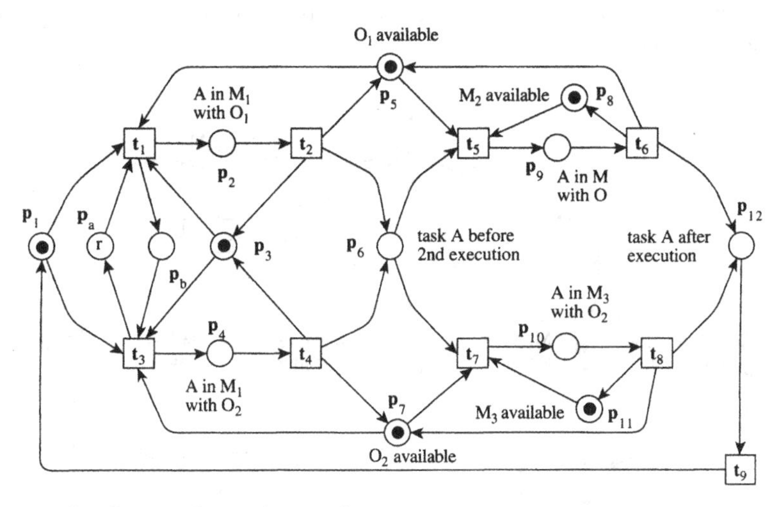

Fig. 8.3. Conflict resolution by regulation circuit

resources. The way in which we must increase the number of resources is governed by the above equations defining the implicitness property of the resource places (e.g. $\mathbf{m}_0[p_3] = k$ in the case considered above).

From a modelling point of view, the net system presented in Figure 8.2 is a good abstraction of the physical system considered. Nevertheless, for a designer of manufacturing systems, an important refinement introduces fairness constraints in order to reduce the indeterminism appearing at the conflicts. Schedulers should be introduced to impose a deterministic ratio between the occurrences of the transitions in the conflict.

For example, by the introduction of two places p_a and p_b a *regulation circuit* can be achieved as shown in the net system of Figure 8.3. A number of $r > 0$ tokens in p_a introduces a so-called *finite synchronisation distance* of size r. This means that one of the transitions t_1 and t_3 cannot occur more than r times without an interleaving occurrence of the other. In case of $r = 1$ strict alternation of the occurrences is obtained.

This example allows the application of the concept of abstraction for Petri nets in a meaningful way. If the action of a machine is considered as an indivisible step, then the two different transitions for its start and termination are combined, including the place connecting them. Figure 8.4 shows the result of such an abstraction of the net from Figure 8.2: the closed (cf. Section 2.4) sets $\{t_1, p_2, t_2\}, \{t_3, p_4, t_4\}, \{t_5, p_9, t_6\}$, and $\{t_7, p_{10}, t_8\}$ are replaced by transitions t_1, t_3, t_5 and t_7 respectively. The behaviour of this abstraction corresponds to the behaviour of the refined version, e.g. the occurrence sequence $t_1 t_2 t_5 name t_6$ of the net from Figure 8.2 corresponds to $t_1 t_5$ in the abstraction of Figure 8.4. Here the use of resources M_i and O_j is an indivisible action and the corresponding places are "side conditions" that never become unmarked.

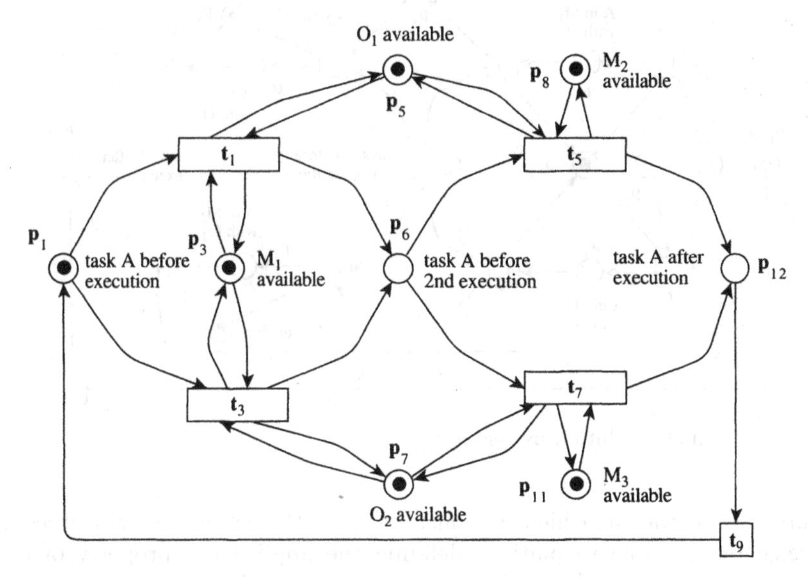

Fig. 8.4. First abstraction of the machine example

In the definition of steps, "simultaneous" concurrent transitions connected by such a common side condition are excluded. Therefore, even when there are two tokens (tasks) in the place p_1, transitions t_1 and t_3 cannot occur in the same step because of the common side condition p_3.

Further abstractions of this net are shown in Figure 8.5. The net in Figure 8.5a abstracts from the two modes of operation of machine M_1. Now the transitions are named M_i, since their occurrence represents an entire action of machine M_i on a task. The abstraction includes places p_3, p_8, and p_{11} (M_i available), so these places are no longer visible. In Figure 8.5b the places p_5 (O_1 available) and p_7 (O_2 available) are in addition merged by abstraction to a new place p_a.

The sub-net obtained by omitting this place restricts the view of the system to the machines without representing the operators (Figure 8.5c). All the nets in a, b, or c of this figure can be abstracted to the net in d, where the set of actions of all machines is modelled by a single transition t.

In these examples, the abstractions have a meaningful dynamic behaviour that is related to the behaviour of the original refined net. (At least a convincing interpretation for such a behaviour can be given.) However, the reader should be warned that this cannot be guaranteed in all cases. If for instance, as shown in Figure 8.5e, the three places p_5 (O_1 available), p_7 (O_2 available) and p_6 (from Figure 8.5a) are abstracted to one place p_x, then the initial

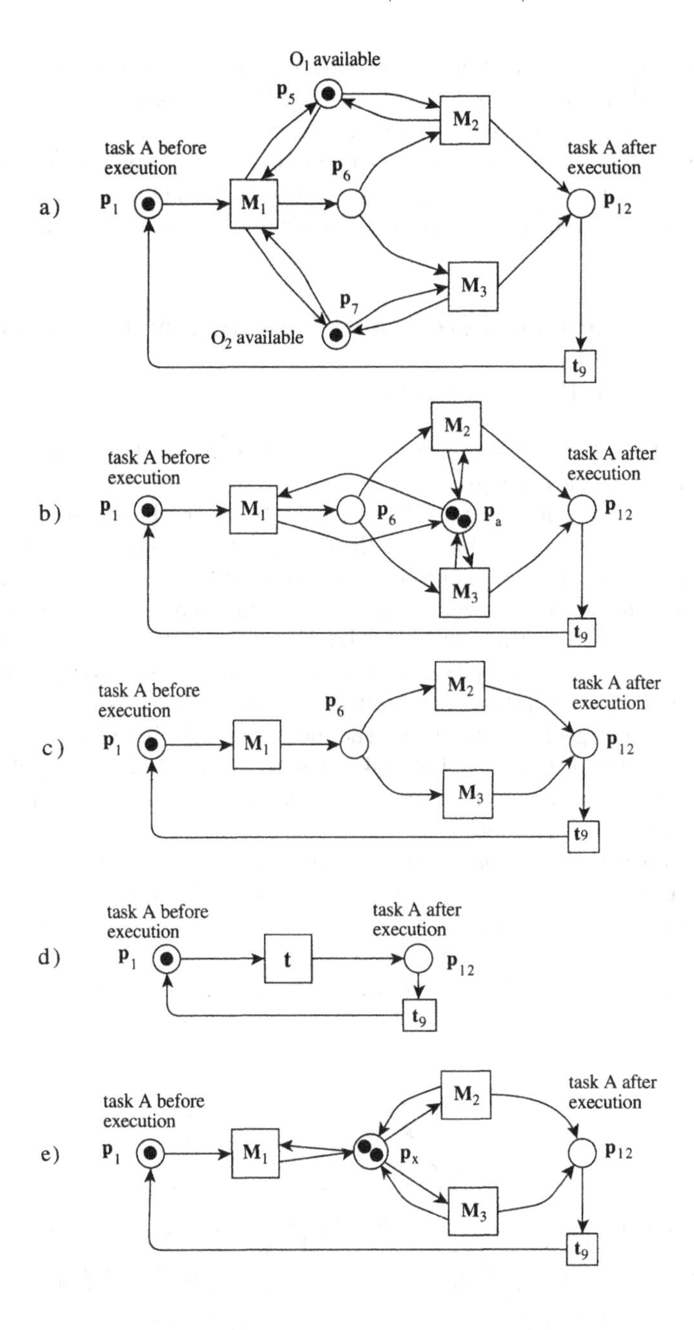

Fig. 8.5. Further abstractions of the machine example

marking is not predefined in a unique way. All conceivable solutions (0, 1 or 2 tokens) result in strange behaviour. In fact, the first case with no token in p_x does not work at all. If there are one or two tokens, M_2 may occur before M_1 which should not be allowed. Therefore, abstraction and refinement should be understood as well-defined operations on the net structure, whereas behaviour is preserved by this operation consistently only in particular cases.

8.2 Place/Transition Nets and Resource Management

Example 2: The Banker's Problem

The "Banker's Problem" was given by E.W. Dijkstra [Dij68] as an example of a resource sharing problem:

A banker has n clients, and a fixed capital g. Each client requires a predetermined amount, say f_i for the ith client, for his project. He does not need all the money at the beginning, but periodically he requests a unit of capital from the bank until his requirement is fulfilled. Some time later he returns his full loan to the bank. The banker may satisfy a given request if he has the money available, but he may choose not to do so. In that case the client has to wait until his request is satisfied. The banker's problem is to develop a strategy for distributing the money which will eventually satisfy all the clients' requirements. The banker has to avoid situations in which he has insufficient money but there are clients' requests still outstanding. These situations are called deadlocks.

An instance $\iota = (n, f, g)$ of the problem is characterised by a positive integer n, an n-tuple $f = (f_1, \ldots, f_n)$ and a number g. All amounts are positive integers. Given a particular problem instance, a state is an n-tuple $r = (r_1, \ldots, r_n)$ representing the amount required but not yet received by each client. Initially, $r = f$. A state is safe if it does not necessarily lead to a deadlock.

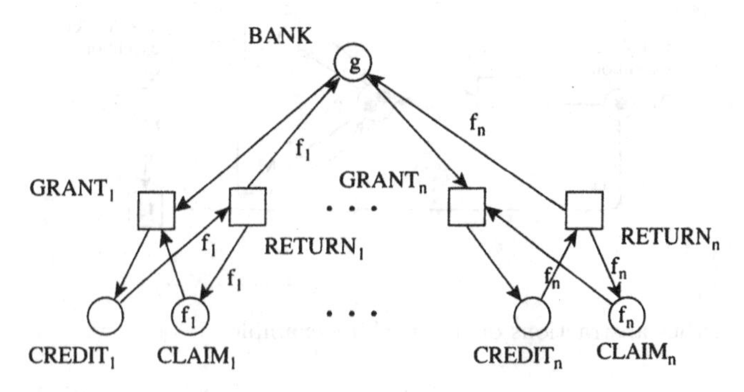

Fig. 8.6. The Banker's Problem with n clients

The place/transition net in Figure 8.6 represents the Banker's Problem as described above. The place *BANK*, holding the banker's cash, initially contains g units (tokens) of money. $CREDIT_i$ and $CLAIM_i$ stand for the loan and the remaining claim of client i respectively. Through the transition $GRANT_i$ this client obtains one unit (token) of money as often as $GRANT_i$ fires. $RETURN_i$ returns all the money back to the banker. $RETURN_i$ cannot fire unless the banker has fulfilled the maximal claim f_i of the client. By the same transition this claim is restored in $CLAIM_i$.

In this example we will study two instances of the problem, namely $\iota_1 = (2,(8,6),10)$, as given in Figure 8.7, and $\iota_2 = (3,(8,3,9),10)$, as given in Figure 8.9. These instances show how Petri net representations of a well-known problem can give a good intuitive understanding and allow for the application of formal techniques such as linear invariants and reachability graph analysis.

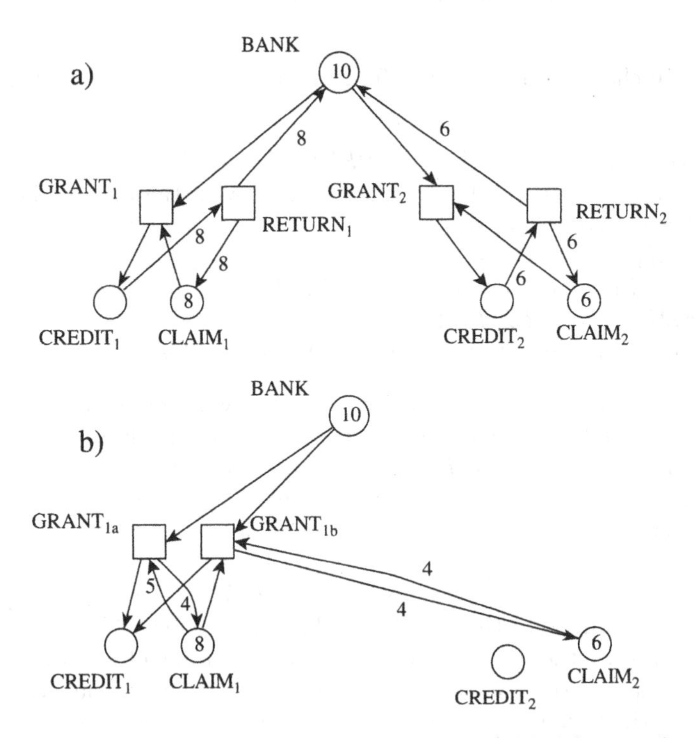

Fig. 8.7. a) The Banker's Problem with 2 clients. b) Modification for deadlock avoidance

Let us start with the instance $\iota_1 = (2,(8,6),10)$ from Figure 8.7a. Each state of the problem is representable by a marking, which is a vector of dimension 5 (since there are 5 places). By the following three linear invariant equations (one for the bank and two for the two clients) only two compo-

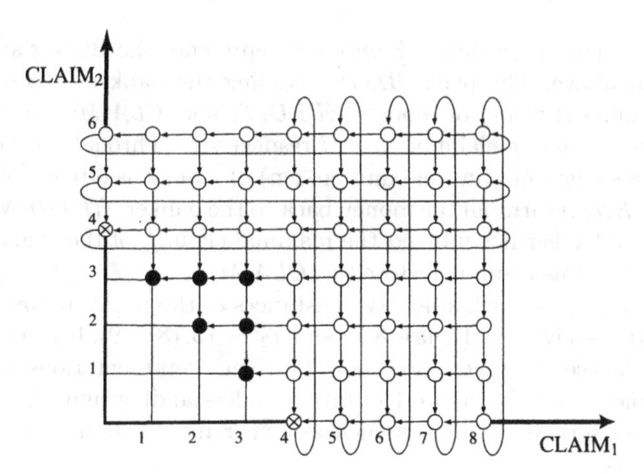

Fig. 8.8. State space of the net from Figure 8.7

nents can be chosen independently from the five components of a reachable marking:

- $\mathbf{m}[BANK] + \mathbf{m}[CREDIT_1] + \mathbf{m}[CREDIT_2] = 10$
  (The capital is with the bank or as a credit with the clients. The total amount is always 10 units.)
- $\mathbf{m}[CLAIM_1] + \mathbf{m}[CREDIT_1] = 8$
  (The sum of the current claim and credit of client 1 is always 8 units.)
- $\mathbf{m}[CLAIM_2] + \mathbf{m}[CREDIT_2] = 6$
  (The sum of the current claim and credit of client 2 is always 6 units.)

Hence, any reachable marking is completely specified by two components, say $(CLAIM_1, CLAIM_2)$. Then the other three components $BANK, CREDIT_1$, and $CREDIT_2$ can be computed from the three place-invariants. This allows us to represent the reachability set of the place/transition net from Figure 8.7 in the two-dimensional plane, as shown in Figure 8.8. In this figure the initial marking $\mathbf{m}_0 = (10, 0, 8, 0, 6)$ (assuming the following ordering of the places: ($BANK$, $CREDIT_1$, $CLAIM_1$, $CREDIT_2$, $CLAIM_2$)) is reduced to the pair $(\mathbf{m}_0(CLAIM_1), \mathbf{m}_0(CLAIM_2)) = (8, 6)$. It corresponds to the upper right node in the graph of Figure 8.8. All paths starting at this node correspond to occurrence sequences. Arcs to the left, right, down, and up represent the occurrence of transitions $GRANT_1, RETURN_1, GRANT_2$, and $RETURN_2$ respectively. It is clear that these occurrence sequences cause no problems when one client is completely served before the second one. On the other hand, interleaved serving can lead to one of the three deadlocks (1,3), (2,2) and (3,1).

It was shown by Dijkstra that there are further critical states to be avoided, namely those markings that inevitably lead to a deadlock, such as $(3, 3)$. He called these *unsafe states*. Markings representing such unsafe states are represented by black nodes. As shown in [HV87] and [VJ85], the *safe states* (represented by white nodes) can be characterised by their *minimal elements*: $(0,4)$ and $(4,0)$, marked by a cross in the node.

How can the net be modified in such a way that deadlocks are avoided? As can be seen from Figure 8.8 this would be the case if transition $GRANT_1$ were activated only in markings greater than or equal to $(4,0)$ or $(0,4)$. we can thus replace transition $GRANT_1$ by two modified copies $GRANT_{1a}$ and $GRANT_{1b}$, as shown in Figure 8.7b. These two transitions have the same effect as the original one, but possess a higher "activation level", according to the additional condition to be satisfied. An analogous construction should be applied to transition $GRANT_2$. The general procedure is given in [VJ85].

The second instance $\iota_2 = (3,(8,3,9),10)$ is an example from [BH73]. Its net representation is shown in Figure 8.9 and has similar properties to that of the previous example. There are now 7 places and 4 linear invariant equations. Hence, the reachable markings can be represented within 3 dimensions (Figure 8.11). One interesting property is the size of the reachability set, compared with the first instance. There are now 197 reachable markings. The subset of 137 safe states (white nodes) is generated by 10 minimal elements (white nodes with cross). A general method for computing these is presented in [HV87]. All the 60 black nodes represent unsafe states.

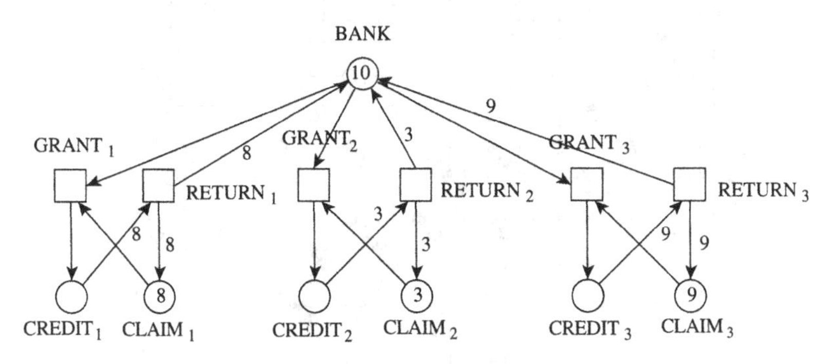

Fig. 8.9. Instance of the Banker's Problem with 3 clients

This second instance of the banker's problem also shows a feature not present in the first one: there are markings that do not necessarily lead to a deadlock, but are not safe in the sense that some transactions cannot be terminated. From the marking represented by the point $(4,3,6)$, for instance, the second client can continue arbitrarily many transactions, while clients 1 and 3 cannot even finish their current transactions. Therefore the definition of safe markings has to be extended. Following Dijkstra we might call a marking safe if the initial marking is reachable. Alternatively, a marking could be defined to be safe if an infinite occurrence sequence, in which all transitions

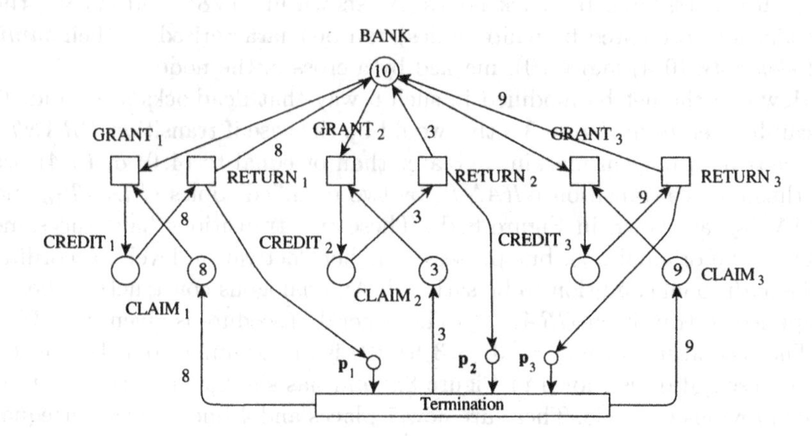

Fig. 8.10. Banker's Problem with 3 clients and termination

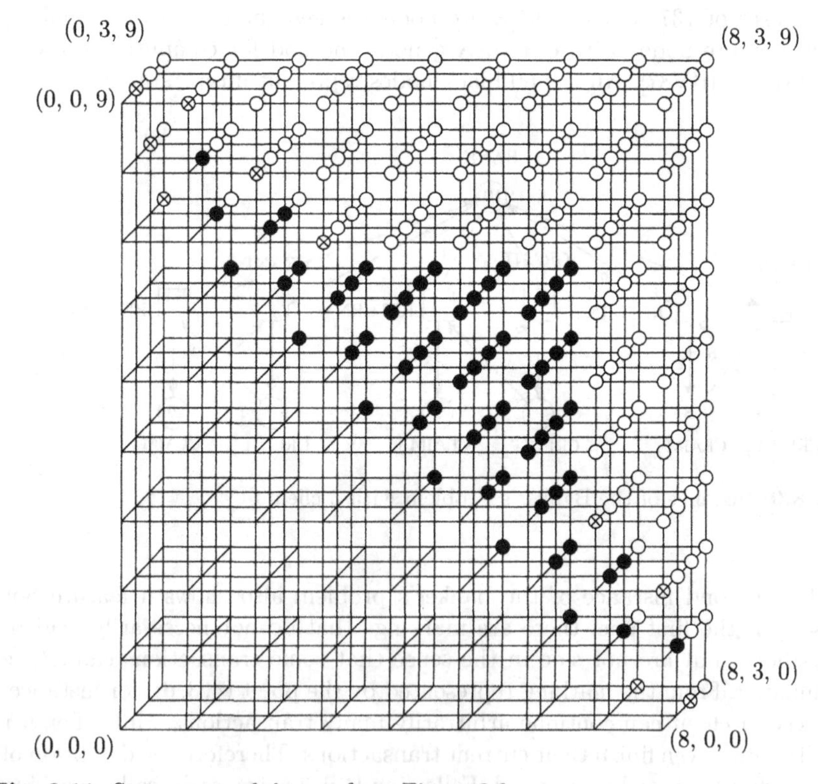

Fig. 8.11. State space of the net from Figure 8.9

occur an infinite number of times, can start in this marking. In the example the two definitions are equivalent.

If in each round each client has to finish exactly one transaction, then a transition called *TERMINATION* could be introduced, as shown in Figure 8.10. After having completed his own transaction client i marks the place p_i as a flag. When p_1, p_2, and p_3 are marked, transition *TERMINATION* can occur and reproduce the initial marking. By the place-invariants it can easily be proved that indeed the place *BANK* contains 10 tokens after the occurrence of *TERMINATION*.

With respect to the net model presented here, Dijkstra's notion of a safe state can be formalised in (at least) three different ways:

- A marking **m** is safe if the initial marking is reachable.
- A marking **m** is safe if there is an infinite occurrence sequence from **m**, containing all transitions an infinite number of times.
- A marking **m** is safe if transition *TERMINATION* can be fired from **m** within a finite number of steps (for the net of Figure 8.10 only).

We close the discussion of this example by a folding to a coloured net. The folding of the place/transition net from Figure 8.9 is given in Figure 8.12. For instance, the remaining total claim is a bag in the place $CLAIM$, giving the individual claims of each client a, b, and c by the number of individual tokens a, b and c respectively. The arc inscription on $(CREDIT, RETURN)$ is in the form of a case-statement, returning for a given binding such as $\beta = [y = c]$ the appropriate bag $W(CREDIT, RETURN)(\beta) = 9'c = \{c, c, c, c, c, c, c, c, c\}_b$. (The value of x is irrelevant for this transition and therefore omitted in β.) The colour set $cd(RETURN)$ of the transition $RETURN$ is defined by the colour set $clients = \{a, b, c\}$ which is motivated by the case. However, a different choice such as $cd(RETURN) = \{1, 2, 3\}$ or $cd(RETURN) = \{CLIENT_1, CLIENT_2, CLIENT_3\}$ would satisfy the formal definition as well. In the latter case the arc inscription on $(CREDIT, RETURN)$ should be modified to the statement
case y **of** $[CLIENT_1 \rightarrow 8'a \mid CLIENT_2 \rightarrow 3'b \mid CLIENT_3 \rightarrow 9'c]$

8.3 Coloured Nets, Abstraction, and Unfolding

Example 3: Alternating-Bit Protocol

In this example the well-known alternating-bit protocol will be modelled by a coloured net. The upper part of Figure 8.13 describes the transmission environment corresponding to the data link layer of the ISO/OSI reference model: a host X produces sequences of data units d (*data frames*). Each data unit is delivered to the entry of the protocol (transition A). After data transmission through the network, the data item is passed to the receiving host Y by transition B. All the places have a capacity of one data item.

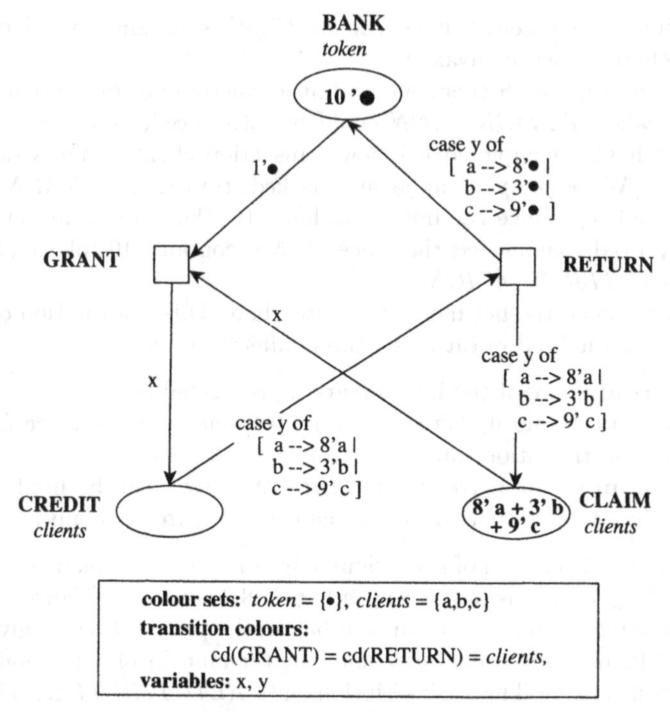

Fig. 8.12. Folding of the Banker's net from Figure 8.9 to a coloured net

This is represented by the notation */1* after the name of the place and is an abbreviation of an explicit notion given in the lower part of Figure 8.13 using so-called *complementary places*.

From this restriction it follows that data transmission has to preserve the order of messages. Therefore the net can be understood as a type of specification of the alternating-bit protocol. It is obvious from this specification that no data item is lost. Since this cannot be guaranteed by any real transmission media, the underlying layer (*physical layer*) is designed to assure this property. Modelling this layer means refining the sub-net consisting of the elements $\{A, s, B\}$. Usually it is assumed that the channel is capable of transmitting data in both directions, but only in one direction at a time. In addition, errors are detected and indicated by a special signal (*reliable half-duplex channel*). The channel is represented in Figure 8.14. Error events are modelled by occurrence of transitions g and k in place of transitions h and i respectively. The signal for error detection is F. The half-duplex property is not explicitly modelled, but will be fulfilled by the realisation of the alternating-bit protocol. Adding a place, as in Figure 2.10, the simultaneous occurrence of transitions h and i would be prevented. As usual, this detail is omitted also in our presentation. The nets in Figure 8.13 can be understood as formal specifications of the problem since correct transmission (i.e. no corruption, no duplication, preservation of message ordering) is ensured.

Combining the specification of the channel (Figure 8.14) with the protocol specification (Figure 8.13) we obtain the net in Figure 8.15. The realisation (due to [BS69]) is shown in Figure 8.16, which can be seen as a refinement of transitions A and B in the net of Figure 8.15. The behaviour of this net is briefly described as follows. Data items follow a path from the sending host (transition X) passing through a, c, h, e and b to the receiving host Y. By transition a a bit x, with the initial value $x = \mathit{1}$, is attached and will be removed by transition b, but only when the current value of the bit in the place s_9 is complementary. This is checked by the transition e via the arc labelled with the function expression $inv(x)$. When an error occurs transition g fires instead of h, producing a token F in the place s_{11}. Then a negative acknowledgement will cause a retransmission of the item. This is performed by the sequence of transitions n followed by i and q. After a correct transmission the next item will be sent with a complementary value $x = \mathit{0}$. When the acknowledgement is corrupted by a faulty channel, transition k occurs instead of i. This also causes a retransmission of the message by transition q. By this retransmission the message could be sent twice, which is detected since the bit has not alternated. Then the redundant message is deleted by transition m, which also initiates a new transmission of the acknowledgement.

As explained in the specification, channel errors are modelled by transitions g and k, resulting in the production of an error signal F. To satisfy the specification that a data item is eventually transmitted by the protocol, it is necessary to assume that after some finite number of steps, the channel is working correctly at least occasionally. This means that transition h has to behave *fairly* with respect to transition g. The same is assumed for i and k.

This example gives the modelling of a well-known protocol by a coloured Petri net. In addition it has been shown, how specifications of the protocol and the channel can be given by nets, and that the realisation can be understood as a net refinement. Thus the different layers of refined and unrefined nets correspond to the layers of abstraction of the protocol architecture.

At the end of this chapter, following an idea of [Obe81], the current example is used to show an unfolding of the coloured net into a place/transition net, which will contain at most one token in any place. This is of particular interest for this example since it reflects the nature of the alternating bit.

As a first step we unfold the circular structure of the coloured net of Figure 8.16 by omitting the controlling transitions q and m. To simplify the net, we also omit the data items d.

Most transitions of the coloured net from Figure 8.16 have two modes, one for the value 0 of the bit and one for the value 1. In the unfolding of Figure 8.17 the corresponding transitions bear the label 0 and 1 respectively, while the original name of the transition is associated with the dashed rectangle representing the refinement. Thus transition a is split into two transitions

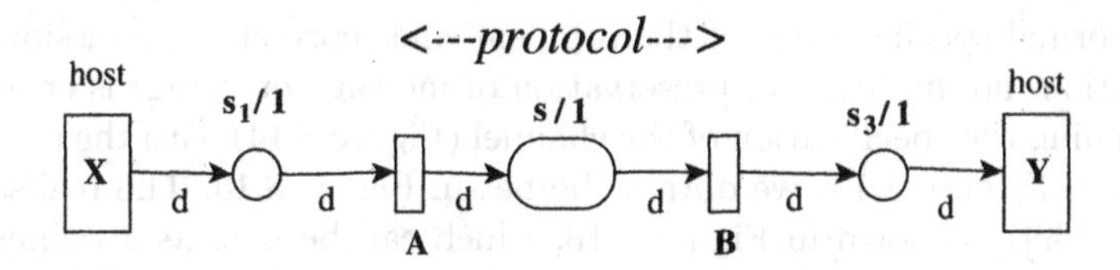

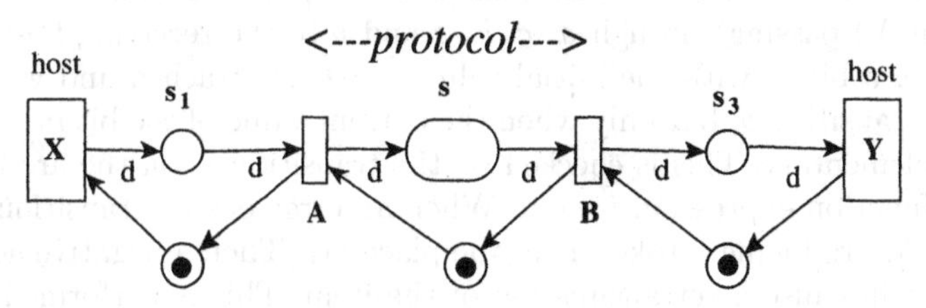

Fig. 8.13. Specification of the alternating-bit protocol

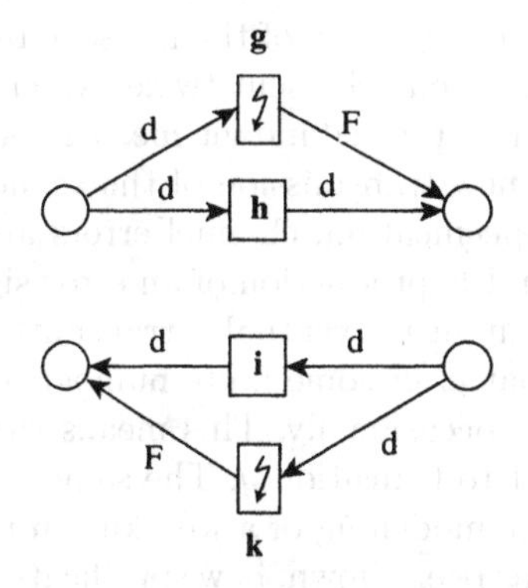

Fig. 8.14. Channel with error detection

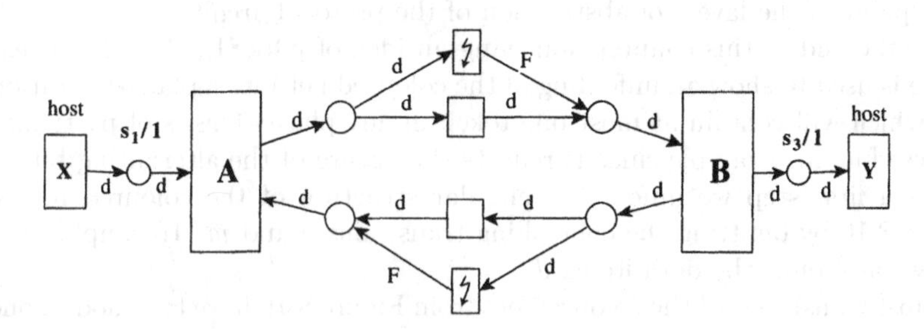

Fig. 8.15. Channel between hosts

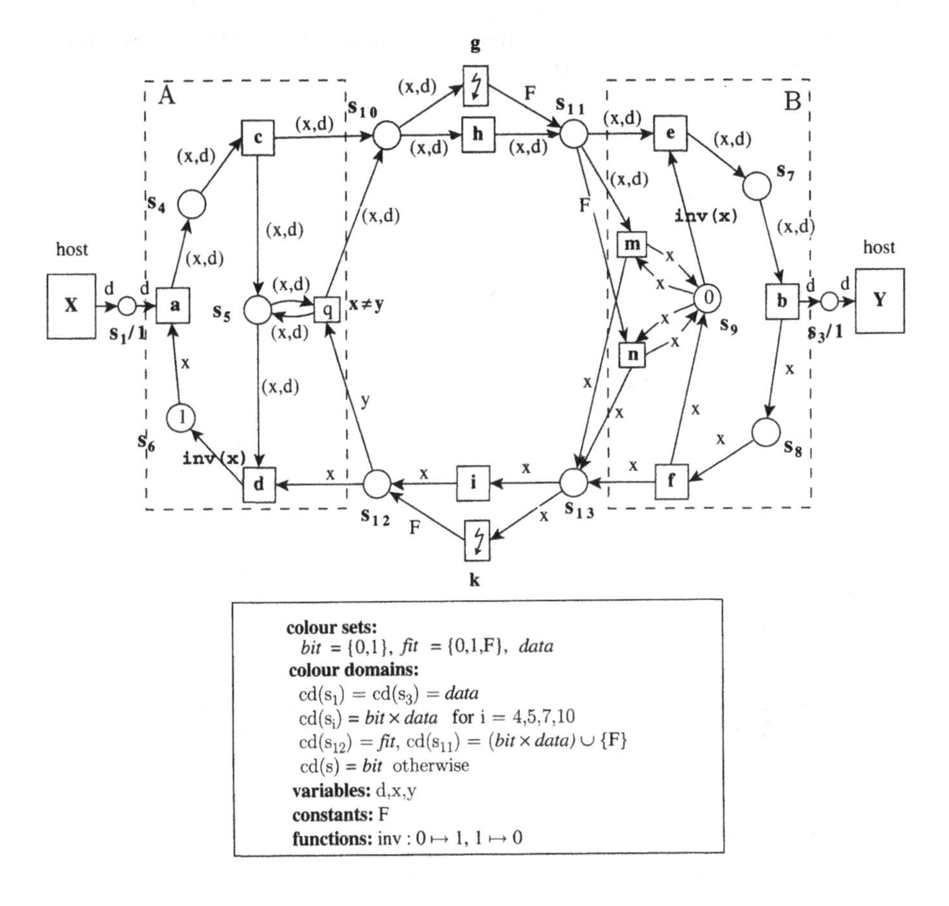

Fig. 8.16. Realisation of the alternating-bit protocol

labelled 0 and 1 that are referred to as $a(0)$ and $a(1)$. The same holds for places, e.g. s_4 is refined to $s_4(0)$ and $s_4(1)$. Hence in this net a number 0 or 1 in a place does not denote a token but rather a parameter of its name. The only token of Figure 8.17 is in the place $s_6(1)$ simulating the token 1 of the place s_6 in Figure 8.16.

The places s_{11} and s_{12} are exceptional as they have a colour set of three elements: 0,1,F. Therefore they are refined into three places, e.g. $\{s_{11}(0), s_{11}(1), s_{11}(\mathrm{F})\}$. The alternation of the bit by transition d is reflected by the crossing arcs from $d(0)$ and $d(1)$ to $s_6(1)$ and $s_6(0)$. The initial marking is represented by tokens and should not be confused with the labels 0 and 1 of the places.

In Figure 8.18 the refinements of the transitions q, m, and n are added. For the transition q a guard has to be satisfied. As the occurrence modes of q can be chosen as $\{(x = 0, y = 1), (x = 0, y = \mathrm{F}), (x = 1, y = 0), (x = 1, y = \mathrm{F})\}$, the refinement of q contains four transitions instead of two. (Recall that the domain of the variable y is the three-element set *fit*, whereas x can only have values from *bit*.)

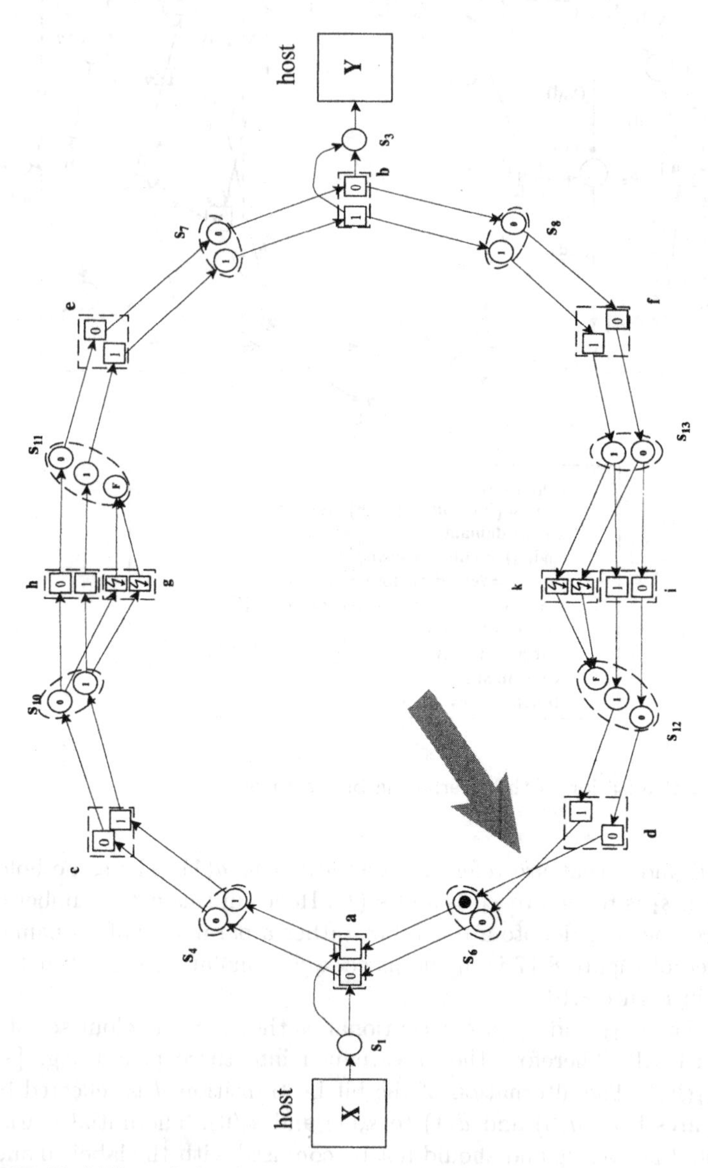

Fig. 8.17. Unfolding of the alternating-bit protocol – first step

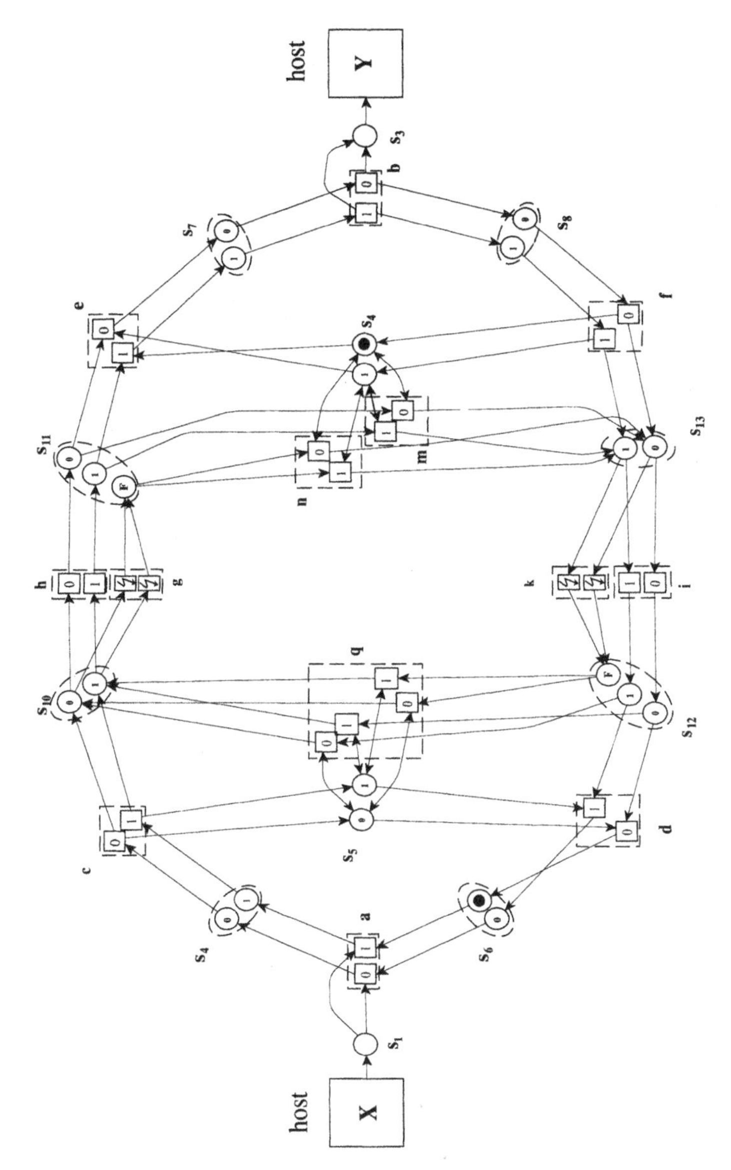

Fig. 8.18. Unfolding of the alternating-bit protocol – second step

9. Techniques*

In this chapter we give general principles of modelling with Petri nets. We will concentrate on the aspects that are specific to Petri nets. We shall discuss how the specific building blocks of Petri nets (places, transitions, arcs, and tokens) are used to model components and aspects of the problem.

A large part of this chapter is devoted to composition and decomposition of net models. A bottom-up modelling strategy starts by building models for simple subsystems and combining them into more complex ones until the desired model is obtained. The top-down approach decomposes the system to be modelled into subsystems, and decomposes these subsystems into smaller subsystems to the point where subsystems can simply be modelled as nets. Often the two approaches are combined. The gap between the system to be modelled and the building blocks of the modelling paradigm is narrowed by both decomposing the system and constructing some higher-level building blocks.

In the sections to come, we will discuss the use of Petri net building blocks for modelling. Then, we will consider the synthesis and decomposition of nets. We start with simple (place/transition) nets and then move on to extensions including colour, priority, and time.

9.1 Building Blocks

Petri nets consist of places (circles), transitions (squares or rectangles), directed arcs (arrows) and tokens (dots inside places).

Transitions are the active components of a Petri net. They are used to model various kinds of *actions*. Tokens are the volatile components and are used to model *objects*. Places represent the states that the objects can be in. Arcs represent the way in which objects are created or destroyed or change state because of the occurrence of an action.

* Author: M. Voorhoeve

A good way to start a model is to begin with an object class and to list its possible states. For each state, draw a place. Next, draw a transition for each possible state change, with an input arc to the old state and an output arc to the new state. An illustration is given in Figure 9.1.

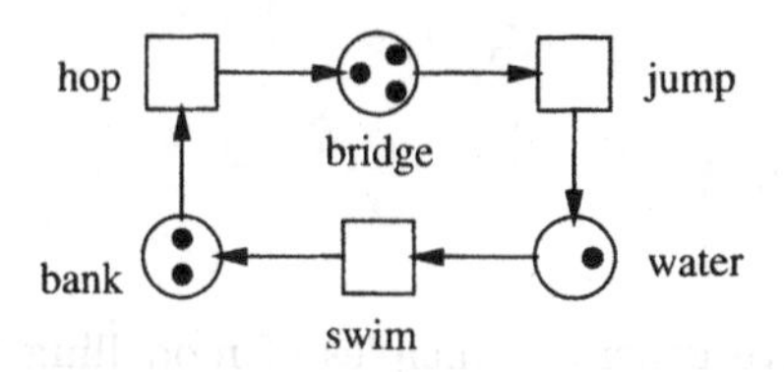

Fig. 9.1. Playful frogs

Here the tokens represent frogs that amuse themselves with jumping into a stream from a bridge, swimming to the bank and then hopping back to the bridge and starting all over again. Clearly, the frogs are the objects with three possible states and three actions that alter the state in a fixed order.

We can complicate the frog model by adding a beautiful girl who sometimes catches a frog that jumped from the bridge and kisses it. When the frog fails to become a prince, she disappointedly throws it into a nearby bush. The frog then hops back to the bridge to resume its play.

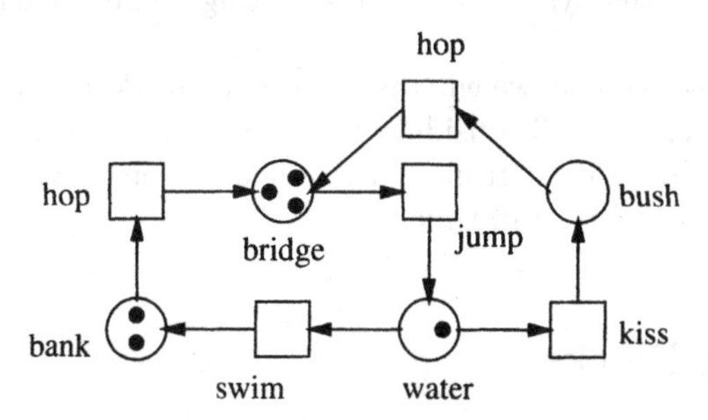

Fig. 9.2. Frogs with girl

The model is given in Figure 9.2. We add an extra state (`bush`) to the frog object and the actions (`kiss` and `hop`) leading to and from it, with a nondeterministic choice between the kissing and swimming action. Note that the same action can used in different states. This is modelled by different transitions with the same label.

In both models, no frog objects are created or destroyed, which is witnessed by the existence of a p-flow. Often *resource* objects have the same property. They can be available or occupied in several ways, but the number of resources stays the same. This way of modelling resources is used often for manufacturing applications (cf. Section 24.2).

Note that the same net (with one token in it) can be used to model the life cycle of a frog object. In Section 26.3, the models in Figures 26.4 and 26.5 are similar models that describe the life cycle (state-wise and action-wise) of objects in a certain class.

After modelling each object class, we can investigate how the various objects influence one another. In general, the interaction of objects must be modelled by communication of the sub-nets of each object class, which will be treated in the next section. Here, we give a simple example based on the frog fairy tale, which causes the addition of places and transitions. The girl from our story is really determined to marry a prince, so she sets herself to the task of kissing exactly one out of every three frogs that jump from the bridge.

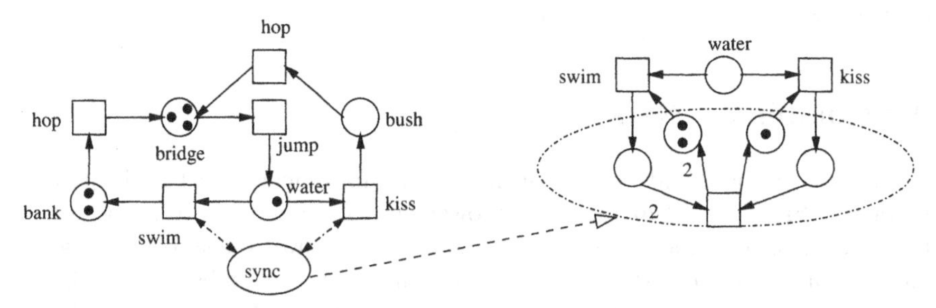

Fig. 9.3. Frogs with stubborn girl

The model is given in Figure 9.3. We start from Figure 9.2 and then restrict the firing of the swimming and kissing actions by adding extra places containing "permission" tokens. The first three frogs share two swimming permissions and one kissing permission. When these permissions have been used up, a blank transition fires that makes the same three permissions available.

The approach to modelling sketched above, first allowing all conflicts and then restricting them by adding extra places, is well established. The first example in Chapter 8 is very similar.

Note that even for this simple model, the need for a compositional or hierarchical approach becomes apparent. At the highest level, a "super-place" is modelled that restricts the firing of the conflicting transitions. At the lower level, this superplace is modelled in detail. A great deal of this chapter will be devoted to such techniques for combining or decomposing nets.

Also, the same model shows the need for blank or invisible components that are needed only to influence the behaviour of the visible objects. Workflow models (cf. Section 25.3) are oriented towards actions (tasks), so only blank tokens are used. It is even advisable to exclude any meaningful tokens (such as resources) from these models. The event-oriented modelling method in Section 10.2 follows the same principles.

The state-oriented method in Section 10.1 attaches much more importance to tokens and places. Temporal logic predicates are used to characterise the states of the nets to be modelled and the way in which these states develop from one another. Often, no importance is attached to the transition that fires, so it might as well be blank.

The object-oriented method in Section 10.3 uses both visible and invisible tokens, places, and transitions. An application of this method can be found in Section 26.3. The method uses a non-graphical specification language and generates Petri nets from specifications in that language. It is also the most bottom-up method.

The method that is used in Section 24.3 for modelling flexible manufacturing systems (FMS) aims at the direct top-down modelling of Petri nets and uses both state-oriented and event-oriented features. Blank tokens and transitions are kept to a minimum.

9.2 Combining Nets

Both the top-down and bottom-up modelling approaches are based upon a hierarchy. Smaller and hierarchically lower nets are combined to form larger (higher) nets, or higher nets are decomposed into lower ones. The nets will model some dynamic system, of which the components are subsystems. These subsystems will *communicate* to perform the functions that are required for the combined system.

There are two essentially different methods of communication: *asynchronous* and *synchronous* communication. Communication by electronic mail is asynchronous. Information is sent only one way, and the sending and receiving of the message does not necessarily occur at the same point in time. Since Petri nets use places as containers of information, it is by *place fusion* that asynchronous communication is modelled: the output place of one subsystem is fused with the input place of another subsystem. Similar to place fusion is *arc addition*: adding an input or output arc between a place in one subsystem and a transition in another.

Communication between people through the telephone is essentially synchronous. Both communicating parties have to be present during the communication, and information can be exchanged both ways. Synchronous communication among more than two parties is also possible (e.g. a meeting by telephone). Since transitions model activities in Petri nets, it is through *transition fusion* that synchronous communication is modelled: the activities in two (or more) subsystems have to be synchronised.

9.2.1 Place Fusion

Combining nets by means of place fusion is a simple and effective way to model (asynchronous) communication between sub-nets. In Figure 9.4, the left-hand sub-net can produce tokens for its place **smess**, and the right-hand sub-net can consume these tokens from its place **rmess**. By fusion of these places, communication between these sub-nets takes place.

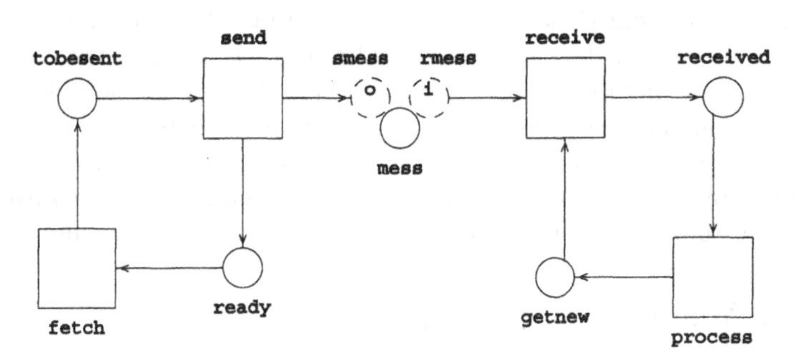

Fig. 9.4. Asynchronous communication by place fusion

This communication is *asynchronous*; the action that produces a token and the action that consumes that token cannot occur at the same time. Any number of actions may occur between the two and the consumption may even not occur at all.

When doing place fusion, it is good practice to consider one place as the main place that the others are fused with. That place will be the only one that can be initialised with tokens. The other places are mere fusion places or "pins" that are connected to the main place. This practice also resolves any naming problems: the name of the main place is of course retained. Pins can be divided into input pins, from which tokens are consumed, and output pins, into which tokens are produce. The combination of an input pin, output pin, and place is a *channel* for one-way token transfer. This is the situation depicted in Figure 9.4.

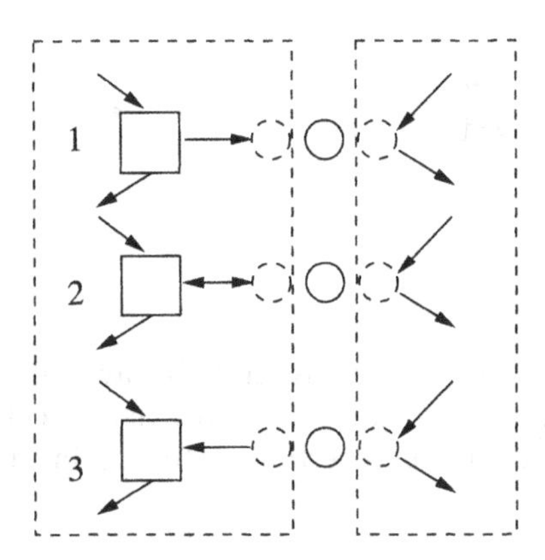

Fig. 9.5. More place fusion examples

Other methods for place fusion are depicted in Figure 9.5. The first situation resembles the channel concept; the difference is that the right-hand sub-net receives transfer tokens in a place that can also be filled internally. The second situation is common, too; here an action in the left-hand sub-net may or may not occur depending upon the state of the right-hand sub-net. The state of the latter does not change. The third situation is the most dangerous one, as both sub-nets may remove tokens from the fusion place.

Place fusion can also occur among three or more nets. The safe way of doing so is by allowing the connection of at most one of the sub-nets to the fusion place by input-only arcs. Two or more sub-nets *consuming* from the fusion place, such as (3) in Figure 9.5, is considered bad modelling practice.

9.2.2 Arc Addition

Arc addition is another way to communicate synchronously. *Input* arc addition is the addition of an arc from a place in one sub-net to a transition in another. By the extra arc, the firing of that transition is restricted since it needs an extra token. So, events of one sub-net become dependent upon the state in another sub-net.

The inverse *output* arc addition is also used. Input arc addition restricts the possible firings, whereas output arc addition extends them. Here, the state of one sub-net is modified by the occurrence of an event in the other sub-net, which is used to model message passing.

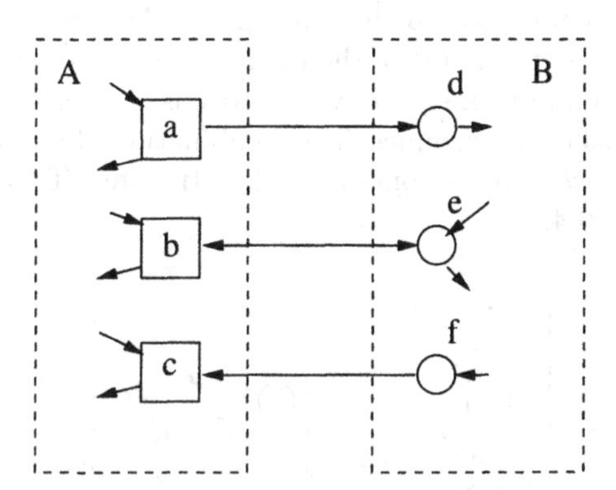

Fig. 9.6. Three ways of adding arcs

A third possibility is I/O arc addition. It behaves like input arc addition, since it restricts firings of the transition involved. It can be used to model events that are only possible in a certain state, e.g. in workflow models.

9.2.3 Transition Fusion

By fusion of transitions of sub-nets, *synchronous* communication between the sub-nets is modelled. When all the fusing transitions are enabled, the fused transition is enabled. The consumption of the fused transition consists of the sum of the consumptions of the fusing transitions and its production is the sum of the productions of the fusing transitions. In Figure 9.7, a transition fusion is depicted. The "send" transition of the left-hand sub-net is fused with the "receive" transition of the right-hand net, thus creating the "communicate" transition of the combined net.

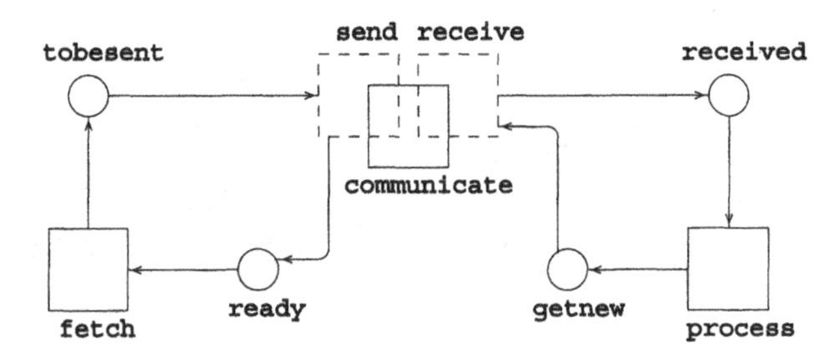

Fig. 9.7. Synchronous communication by transition fusion

Transition fusion is the most natural way to combine the nets that are created from modelling object classes. If we return to our frog example, the model in Figure 9.2 can be combined with the model of a girl object class; the girl can come and go to the stream and, while at the stream, can kiss a frog, but needs to wipe her mouth afterwards. The kissing of the girl and the being kissed of the frog are actions that can be synchronised by transition fusion, as shown in Figure 9.8.

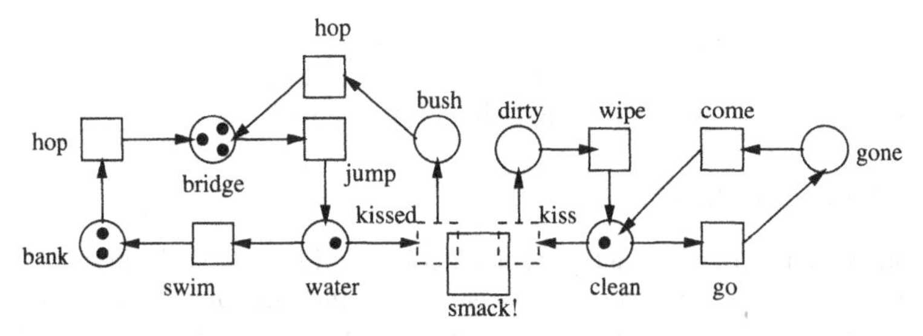

Fig. 9.8. Synchronisation of actions for two object classes

In Figure 9.9, various multiple-transition-fusion constructions are given. On the left-hand side, portions of the sub-nets are depicted together with the way that they are fused. On the right-hand side, the result is shown.

The top situation shows a "conjunctive" three-way fusion that can be used to model broadcasting. The other two situations show "disjunctive" fusions. In the second one, the transition t is fused with either u or v, resulting in transitions a and b. The arcs from t and to t are duplicated. The third example shows a two-by-two disjunctive fusion.

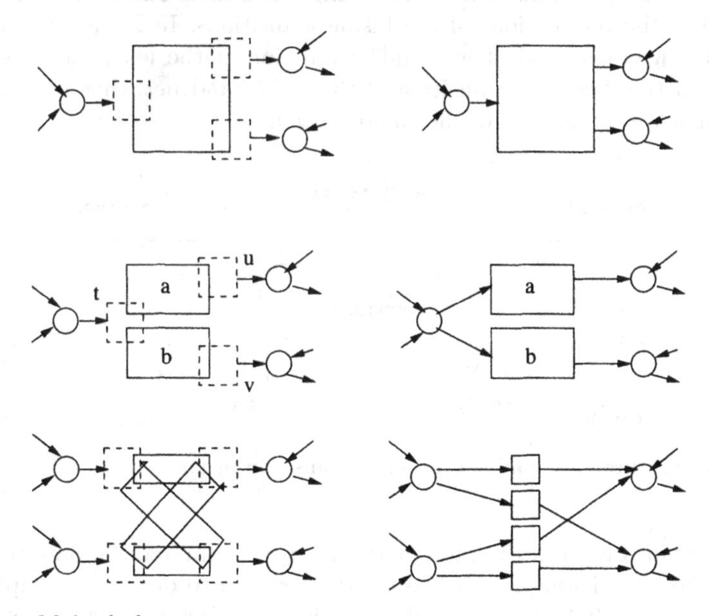

Fig. 9.9. Multiple fusions

9.3 High-Level Nets

In this section we shall discuss the extension of place/transition nets with colour, priority and time. We also very briefly consider fairness assumptions with respect to non-deterministic behaviour of nets.

9.3.1 Coloured Nets

In coloured nets, tokens exist in various colours. The set of all possible token colours may be finite or infinite. For systems engineering purposes, a *countably infinite* set of token colours is appropriate. A finite set of colours would be too restrictive and an uncountable one is not needed in practice and would preclude the computation of properties.

For any place in the net one prescribes a set of allowed colours. A transition that produces a token will not fire if it would produce a token of a colour that is not allowed in its destination place.

For every transition of a coloured net there is a relation between the colours of consumed and produced tokens. This relation can be described by means of pre- and post-conditions. If the transition fires, it consumes tokens that satisfy the pre-condition and it produces tokens that satisfy the post-condition. These pre- and post-conditions can be denoted by adding expressions to arcs and predicates to transitions.

The techniques of place fusion, transition fusion, and arc addition all carry over to coloured nets. Place fusion is the simplest technique; it requires only that the allowed colours of fused places be the same. With arc addition, an arc expression must be added too. Finally, transition fusion often requires that transitions be parametrised with a fusion relation depending on the transition parameters. We give a simple example taking the natural numbers as colours.

The nets in Figure 9.10 model an automatic teller machine (ATM) with the bank behind it. When the ATM is in the ready state, a client can ask for a certain amount of money. The ATM communicates this amount to the bank and waits for approval. When the approval arrives, the money is given to the client. Meanwhile, the requested amount is deducted from the client's account. If the account is deficient, approval will not be given. In the model here, this leads to a deadlock. In a complete model, a non-approval message would be sent from the bank to the ATM, leading to an error message from the ATM to the client.

The topmost net in Figure 9.10 contains a transition `in` that consumes tokens from the pin `amt_wanted` and the place `rdy`. The colour x of the token consumed from `amt_wanted` is copied to the pin `amt`. The fusion relation of `in` is thus that the token produced in `amt` must have the same colour as the token consumed from `amt_wanted`. The token consumed from `rdy` and the one produced for `w_appr` can have any colour. Note that in the combined net the place `w_appr` is superfluous. It is nevertheless good practice to include it in the ATM sub-net to ensure that a token from `ok` is consumed only if a previous token in `amt` has been produced.

We can see from the figure that the fusion relation of `chk` is that the token in `balance` must have a colour y exceeding the colour z from `amt`, that this colour must be copied to `ok` and that the new colour of the `balance` token must equal $y - z$. If $y \geq z$ does not hold, the `chk` transition will not fire.

The middle net has the same behaviour, but synchronous communication is used instead. Note that a place such as `w_appr` in the top net is no longer necessary. The fused transition c must have a relation that depends upon the colours of tokens consumed both at the bank (the balance) as well as at the ATM (the amount). To achieve this, fusion transitions have to be parametrised. Transition parameters consist of unique identifiers typed with colour sets. When transitions are fused, a relation between the parameters may be added. In the figure, the fusion transition a has a parameter x, and

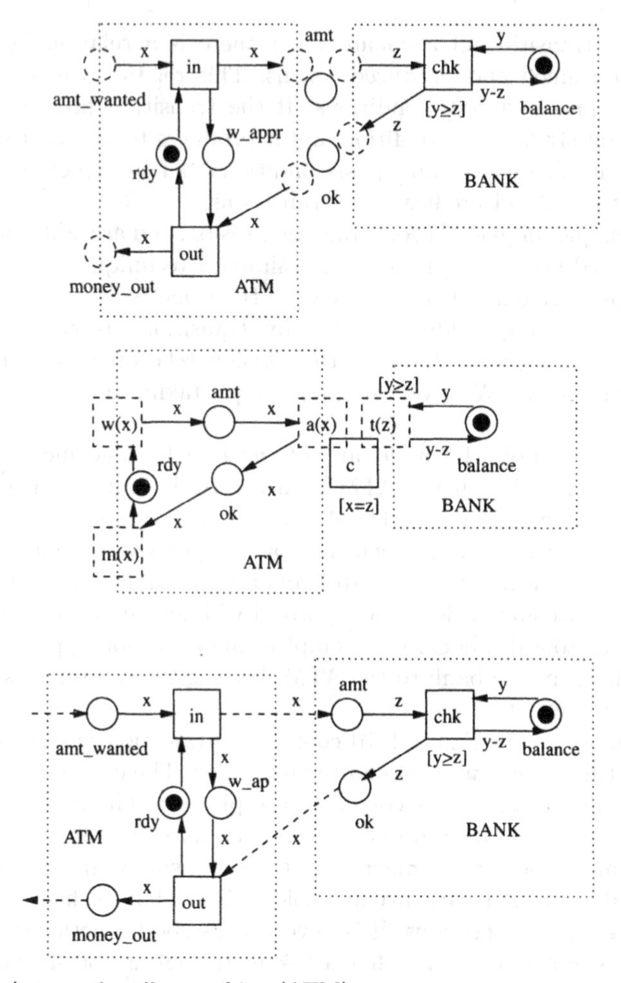

Fig. 9.10. Automatic teller machine (ATM)

t has parameter z. The fused transition c requires that the x parameter of a and the z parameter of t be equal.

In this way, the colour x of the token consumed at the ATM is transferred to the bank and the permission (which depends on y) is transferred back. The parameter y is present at the bank only and plays no role at the ATM. This is the reason that the fusion transitions are parametrised as shown.

The bottom net is the same, but using arc addition instead. For the subnet ATM to be meaningful, the place `w_ap` is coloured. With the addition of arcs comes an arc expression. Note that the parameters (x) of the interface transitions and the possible colours of the interface places have to be exported in order to allow syntactically correct arc additions. So, although arc addition looks very simple, there is more to it than meets the eye.

This example illustrates that transition fusion with coloured nets is less straightforward than place fusion. Arc addition is in between the two. Existing tools for modelling and simulating coloured nets, such as Design/CPN ([Jen92b]) and ExSpect ([HSV91, Bak96]), support only place fusion for this very reason.

9.3.2 Fairness, Priority, and Time

Nondeterminism is everywhere in Petri nets. In the case of a conflict, there is no preference about which firing will occur. Many analysis results are dependent upon rules that restrict the possible firing sequences which may occur in a given situation.

The fairness rules address the infinite firing sequences possible in a net system. A weak fairness rule states that every transition which is enabled will either cease to be enabled (because some conflicting transition fires) or will have fired after *finitely* many steps. A stronger fairness rule states that every transition that is enabled infinitely often will have fired after finitely many steps. In between is a rule stating that if n transitions are in conflict infinitely often, one of them will have fired after finitely many steps.

Priority directly addresses conflicting firings for which preferences can be indicated. A less-preferred firing will occur only if more-preferred firings cannot. In nets without colour, preferences are attached to transitions. In coloured nets, these preferences also depend on the bindings (the colours of the tokens to be consumed), so it is for example possible to indicate a preference for the largest token in some place. With parametrised transitions, the transition parameters can also be included in determining preferences.

The treatment of time in nets can be seen as a special case of priority. Tokens have a time stamp (e.g. their time of production) included in their colour. Any event (firing of a transition) may depend on these time stamps and the *occurrence time* of the event depends upon them. In the case of conflicting events, the one with the earliest occurrence time will happen. A real conflict remains only if the occurrence times are equal.

In tools such as ExSpect or Design/CPN, some specific choices have been made with respect to the possible occurrence times of events. Tokens in ExSpect have a time stamp that indicate their earliest possible time of consumption. The occurrence time of an event equals the maximum time stamp of the tokens consumed in it. The time stamp of input tokens cannot occur in the pre-condition of a transition, but only in the post-condition.

This means that the situation at the top of Figure 10.19 is hard to model in ExSpect. Here, a situation is sketched where a token in q can be consumed by transition a if a token is present in p or will arrive within 5 time units. If this does not happen, the token will be consumed by b. In Design/CPN, it is actually possible to test for time stamps in the pre-condition.

Another difference between timing in ExSpect and in Design/CPN is the handling of conflicts. The transition in the bottom net in Figure 10.19 might fire in two possible ways, consuming the tokens with time stamps 3 and 5 or with 4 and 5. Design/CPN has a kind of LIFO firing rule for timed nets allowing only the last firing (time stamps 4 and 5).

The above discussion shows that the modelling of time in dynamic systems is still at a preliminary stage and few true standards have emerged.

9.4 Decomposing Nets

In the previous sections we saw how more complex nets can be created from simpler ones for several kinds of nets. This can be used to directly model nets via the bottom-up strategy. For a top-down strategy we need to decompose the net into sub-nets (that are to be modelled later) and indicate how these sub-nets communicate.

The task of modelling the overall system then boils down to the modelling of individual subsystems that communicate adequately. The decomposition technique can be based on any combination of place fusion, transition fusion, and arc addition.

When decomposing nets by means of place fusion, the sub-nets can be compared to transitions. The difference between real transitions and these sub-nets is that any number of tokens can be consumed by them and any number produced. The relation between consumed and produced tokens may also depend upon the history of the sub-net, i.e. the events relating to the sub-net that have previously occurred.

When using colours, it is a good idea to define colour sets (data types) for the pins of the sub-nets. Some static type checking can then be used to detect modelling errors. In ExSpect and Design/CPN this is the standard modelling technique supported by the tool.

When decomposing nets by means of transition fusion, one can consider the sub-nets as abstract data types or objects. Communication with the sub-net takes place through fusion with some predefined transitions (methods). These methods may have parameters in the case of coloured nets.

It is perfectly natural to define sub-nets with both fusion places and fusion transitions. Such sub-nets can be compared to objects. Transitions are like methods and places are like attributes. The fusion places and transitions are the exported interface, whereas the other nodes belong to the implementation. One may then use both place and transition fusion to communicate. Another possibility is arc addition.

It is not difficult to show that place fusion, transition fusion, and arc addition are equivalent. By adding some extra components, it is possible to model one kind of communication (such as place fusion) by means of any other.

9.5 Conclusion

In this chapter, we have discussed the techniques of Petri net modelling. An important part of our discussion has been devoted to communication among sub-nets, in order to allow a divide-and-conquer modelling strategy. By a careful division into components and a wise selection of the ways that these components communicate, a modeller can concentrate on just a few aspects at a time.

The sketched techniques find their place within modelling methods, such as those described in the following chapters. However, techniques and even methods are merely aids to the modeller. What is most important is that he knows the purpose of the models to be constructed. He must restrain himself and not just model anything simply because he can.

10. Methods*

This chapter is devoted to the use of Petri nets within methods that support structured approaches for modelling systems and for validating and verifying them using the formal foundation provided. The three approaches presented each have their own way of Petri net modelling, aimed at verification and validation.

Section 10.1 is devoted to a state-oriented approach in which the starting point is the specification of temporal dependencies among the reachable states of a given system. Such an approach allows a problem-oriented description of the requirements for the system being built. This specification is transformed into a solution-oriented Petri net model by mapping the state invariants onto net invariants. This mapping is performed by following well-defined refinement steps. These steps allow a focus on the details needing formal verification.

Section 10.2 is devoted to an event-oriented approach. Such an approach starts from the interaction protocols that govern the flows of events among well-delimited subsystems of the whole system. First the subsystems are identified. Then, in a top-down and structured approach, it is shown how to model the protocol that governs the interactions among these subsystems. There are two levels at which the system can now be examined. At a very low level of granularity, all the events occurring in the system can be considered. Alternatively, the focus can be put on certain relevant events, abstracting from the others. These two levels correspond to two net models which can be shown to be equivalent by different kinds of bisimulation.

Section 10.3 is devoted to the presentation of trends in integration of nets and objects. Object orientation is now widely used for the structuring facilities it offers for building systems. However, object orientation in general does not handle the strong requirement for verification and validation which one faces when building systems. Different ways have been proposed to use Petri nets to help alleviate this shortcoming. Some of these approaches are presented in this section. Object orientation combined with nets can benefit from the two previous approaches. System designers can focus on the states

---

* Authors: R. Mackenthun (Section 1), M. Voorhoeve (Section 2), A. Diagne (Section 3)

of the objects during their life-cycle and their temporal dependencies. They can also concentrate on the events flowing among the objects of a system.

The following three sections show that nets are worth using because they enhance the modelling and design activities by verification and validation. They also enable different approaches (state-, event-, and object-oriented), each of them placing the focus on different aspects of the systems.

10.1 State-Oriented Modelling

The aim of this chapter is to give the reader an understanding of how to combine structured and/or intuitive approaches for system design with the formal techniques of Petri nets. This section gives a state-oriented view.

The method presented not only supports the design of a system, but also integrates aspects of verification into the development process.

This approach starts with a formal specification that will be refined in several steps. The specification is given in the temporal logic of UNITY presented in [CM88]. The intuitive UNITY-formalism is chosen since it is easy to understand, the expressive power is reasonably high, and it is based on interleaving semantics, the semantics mainly used in this book. In this section only a subset of the UNITY properties is used. A brief introduction to these properties is given in Section 10.1.1. The proof rules used in UNITY will not be introduced here. They are used to prove the correctness of the steps of the method, e.g. the construction of the proof graph (page 132). Such proofs are essential for the developer of the method but not for the user. For details of UNITY-proofs see Section 16.2.

In Section 10.1.2 it is shown how to transform the problem-oriented specification into a solution-oriented net model using Petri net techniques. The development of algorithms is partially a creative process. The method cannot replace the creative work, but it can give guidance to the creative developer.

In Section 10.1.3 the description of the modelling process is concluded by considering implementation details.

A simple mutual exclusion (ME) algorithm will be used to clarify the method. An advanced ME-problem will be solved as case study in Section 11.1.

The example used in this section has the following informal description: A fixed number of computers compete for a common resource. It is possible that the computers will never apply for the resource. The protocol should ensure mutual exclusion, and also ensure that a requesting computer will eventually get permission to use the resource.

To model the dynamic properties, the net model has to be extended as described in the following excursus.

Excursus: Firing Rules

In ordinary nets, transitions are not constrained to fire. However, many of system descriptions require dynamic properties such as *a transition will eventually fire*, i.e. after a finite number of steps the transition will occur. Additionally, in the case of a conflict it is often necessary to ensure that all conflicting transitions will eventually occur, if the conflict appears again and again.

These problems can be solved by adding special firing rules to transitions. In this section, three firing rules are distinguished. *Productive* transitions eventually occur in an infinite occurrence sequence if they are persistently enabled. This characteristic is also known as the *finite delay property* or as the *weak fair condition*. In a finite sequence, productive transitions must not be enabled in the last marking. *Fair* transitions (also known as *strongly fair*) eventually occur in an infinite occurrence sequence if they are enabled infinitely often. In a finite sequence, fair transitions must not be enabled in the last marking. All other transitions (called *normal*) are not constrained to fire.

In the following definition $\mathbb{N}^\omega$ is the set $\mathbb{N} \cup \{\omega\}$. The total order relation "$<$" of $\mathbb{N}$ is extended in such a way that all natural numbers are less than ω:

$$<_\omega = \{(a,b) \in \mathbb{N}^\omega \times \mathbb{N}^\omega | (a,b) \in < \vee (a \in \mathbb{N} \wedge b = \omega)\}$$

The length $|\mathbf{os}|$ of a finite occurrence sequence $\mathbf{os}$ is the number of transitions occurring in that sequence. The length of an infinite occurrence sequence is defined as ω. Note that "$<$" is used for both order relations, the original and the extended one, if the meaning can be deduced from the context. Additionally, $\mathbf{os}_{T,i}$ is the ith transition in an occurrence sequence $\mathbf{os}$ ($i \in \mathbb{N}$).

Definition 10.1.1. *Let $\langle \mathcal{N}, \mathbf{m_0} \rangle$ be a P/T net system and let* $\mathbf{fr}$ *be a mapping* $\mathbf{fr} : T \to \{prod, fair, normal\}$ *giving the firing rules of the transition.*

The set of **fr-conform** *occurrence sequences $Occ(\langle \mathcal{N}, \mathbf{m_0} \rangle)_{\mathbf{fr}}$ is the greatest subset of the set of finite and infinite occurrence sequences ($Occ(\langle \mathcal{N}, \mathbf{m_0} \rangle)$) with:*

$$\begin{aligned}
&\forall \mathbf{os} \in Occ(\langle \mathcal{N}, \mathbf{m_0} \rangle)_{\mathbf{fr}} \,.\; \forall t \in T : \mathbf{fr}(t) = prod \,.\; \forall k \in \mathbb{N} : k \leq |\mathbf{os}| \,. \\
&\qquad ((\forall j \in \mathbb{N} : k \leq j \leq |\mathbf{os}| \,.\; \mathbf{m}_j \xrightarrow{t}) \\
&\qquad \Rightarrow (\exists i \in \mathbb{N} : k+1 \leq i \leq |\mathbf{os}| \,.\, \mathbf{os}_{T,i-1} = t)) \\
\wedge\; &\forall \mathbf{os} \in Occ(\langle \mathcal{N}, \mathbf{m_0} \rangle)_{\mathbf{fr}} \,.\; \forall t \in T : \mathbf{fr}(t) = fair \,. \\
&\qquad ((\forall k \in \mathbb{N} : k \leq |\mathbf{os}| \,.\, \exists j \in \mathbb{N} : k \leq j \leq |\mathbf{os}| \,.\; \mathbf{m}_j \xrightarrow{t}) \\
&\qquad \Rightarrow (\forall k \in \mathbb{N} : k \leq |\mathbf{os}| \,.\, \exists i \in \mathbb{N} : k+1 \leq i \leq |\mathbf{os}| \,.\, \mathbf{os}_{T,i-1} = t))
\end{aligned}$$

For infinite sequences the consistency of the formal definition and the informal descriptions above is obvious. In the case of finite sequences no productive or fair transition can be enabled in the final marking. For productive transitions the implication would be falsified for $k = |os|$ since the first part of the implication would be true and the second part would be false, because the domain of variable i is empty. For fair transitions the first part of the implication would be true for all k and the second part would be false because for $k = |os|$ the domain of variable i is empty.

Figure 10.1 shows an example net that will be used to illustrate the definition. The firing rules of the transitions are: $\mathbf{fr}(a) = \mathit{normal}$, $\mathbf{fr}(b) = \mathit{prod}$, $\mathbf{fr}(c) = \mathit{fair}$, and $\mathbf{fr}(d) = \mathit{prod}$.

Table 10.1 gives the firing sequences as extended regular expressions, where ω is an infinite repetition. The corresponding occurrence sequences belong to the set of sequences given in the title of the column.

For instance, the occurrence sequence

$$\mathbf{os}_1 = \mathbf{m}_0, t_0, \mathbf{m}_1, t_1, \ldots, \mathbf{m}_5 = p, a, r, b, q, d, p, a, r, c, p$$

belongs to $Occ(\langle \mathcal{N}, \mathbf{m_0} \rangle)_{\mathbf{fr}}$ since the corresponding firing sequence is $abdac$. The length of the occurrence sequence is 5. To prove that $\mathbf{os}_1$ belongs to $Occ(\langle \mathcal{N}, \mathbf{m_0} \rangle)_{\mathbf{fr}}$ the first part of the conjunction in the definition must be true for transitions b and d. Since b and d are not enabled in $\mathbf{m}_5$, there exists no $k \leq 5$ such that b or d is persistently enabled from $\mathbf{m}_k$ to $\mathbf{m}_5$. Therefore, the implication is true for all $k \leq 5$. For the fair transition c the second part of the conjunction must be true. For $k = 5$ there exists no binding of variable j where transition c is enabled. Therefore the first part of the implication is false and the implication is true.

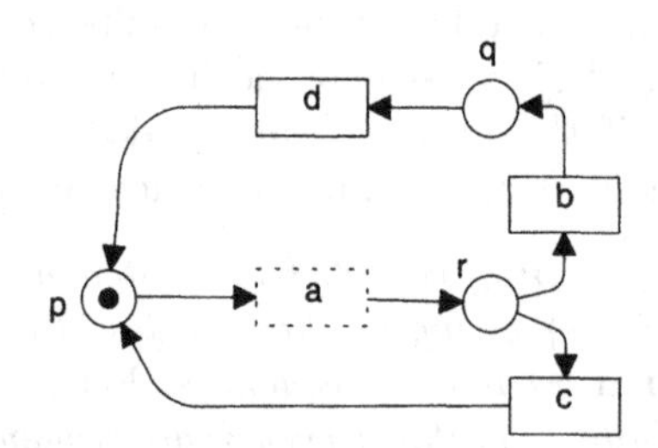

Fig. 10.1. Firing rules: simple example

Table 10.1. Firing sequences of the simple example

$Occ(\langle \mathcal{N}, \mathbf{m_0} \rangle)$	$Occ(\langle \mathcal{N}, \mathbf{m_0} \rangle)_{\mathbf{fr}}$	$Occ(\langle \mathcal{N}, \mathbf{m_0} \rangle) \setminus Occ(\langle \mathcal{N}, \mathbf{m_0} \rangle)_{\mathbf{fr}}$
$(a(bd+c))^\omega$	$((abd)^*ac)^\omega$	$(a(bd+c))^*(abd)^\omega$
$(a(bd+c))^*(\epsilon + a + ab)$	$(a(bd+c))^*$	$(a(bd+c))^*(a+ab)$

10.1.1 Specification

This section describes the specification of the algorithms as temporal logic expressions. The algorithm is embedded in an environment. The environment can be described in terms of temporal logic or as a Petri net model. The latter is possible since the environment has a fixed detailed behaviour from the beginning. Nothing will be changed in the environment of the model during the development process.

Environment. Here, the environment is given as a Petri net. In the context of ME-algorithms the environment gives an abstract behaviour of the competitors that use the ME-algorithm. The environment is called the *client* or *client unit*. The specification of the client is given in Figure 10.2. The behaviour of the client is modelled by the three states *not interested (cni)*, *interested (cint)*, and *in critical section (ccs)*. The first c in the name of a place means that the place belongs to a *client*. The transitions are *gets interested (cgi)*, *enters the critical section (cec)*, and *leaves the critical section (clc)*. Since a client is constrained to do the latter two actions to give control back to the ME-protocol after having obtained permission, the respective transitions are productive (solid border). The first action is optional since the client unit will not be forced to use its critical section. The corresponding transition is a normal one that is neither productive nor fair (dotted border). The above-named places and transitions model the so-called *state process* of the client. Initially, no clients are interested.

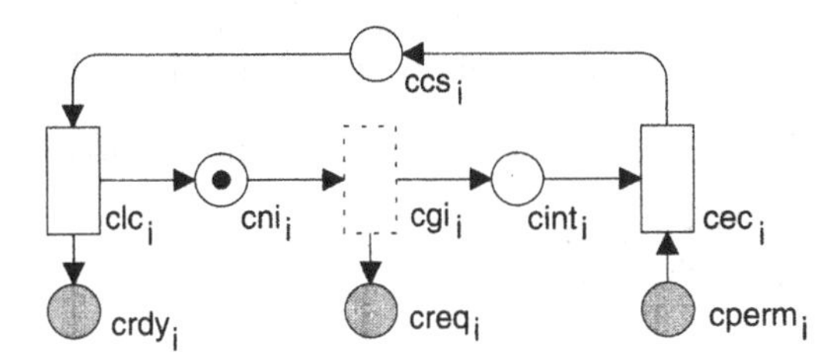

Fig. 10.2. The client unit

A client unit that gets interested sends a request to the protocol (*creq*). To enter the critical section it needs permission from the protocol (*cperm*). On leaving the critical section the client unit informs the protocol that it is ready (*crdy*).

All clients have the same structure. We use indices for the transitions and places to distinguish between them.

Protocol. In Petri nets states are usually modelled by markings. To specify the behaviour of a given part of a system, properties of single markings and temporal dependencies of different markings have to be expressed. Properties are conjunctions, disjunctions, or negations of statements about the number of tokens on the places in a marking $\mathbf{m}$. For instance $(\mathbf{m}(p) \geq 3) \wedge (\neg(\mathbf{m}(q) = 2))$ is a property of marking $\mathbf{m}$.The shorthand notation $p \geq 3 \wedge \neg(q = 2)$ will be used for this property. By default, p means $p = 1$. The notation is extended to sets of places P e.g. $P = 2$ means $\sum_{p \in P} \mathbf{m}(p) = 2$. $M_{[r]}$ is the set of markings where property r holds. Such markings are called r-markings.

When reasoning about dependencies of markings, it is an advantage to abstract from the transitions in the occurrence sequences of a net system $\langle \mathcal{N}, \mathbf{m_0} \rangle$. The resulting set is the set of marking sequences

$$\begin{aligned} MS(\langle \mathcal{N}, \mathbf{m_0} \rangle) := \; & \{\mathbf{m}_0, \ldots, \mathbf{m}_n \mid \forall i \in \{0, \ldots, n-1\} \,.\, \exists t \in T \,.\, \mathbf{m}_i \xrightarrow{t} \mathbf{m}_{i+1}\} \\ & \cup \{\mathbf{m}_0, \mathbf{m}_1, \ldots \mid \forall i \in \mathbb{N} \,.\, \exists t \in T \,.\, \mathbf{m}_i \xrightarrow{t} \mathbf{m}_{i+1}\}. \end{aligned}$$

The ith marking of a marking sequence $\mathbf{ms}$ is denoted as $\mathbf{ms}_i$.

UNITY introduces five basic temporal operators: UNLESS, IS STABLE, IS INVARIANT, ENSURES, and LEADS TO. In this section only the IS INVARIANT and the LEADS TO operators are used. The others would only be necessary to prove the correctness of the development steps presented. As mentioned before, these proofs are omitted here.

An invariant property r (r IS INVARIANT) of a net system $\langle \mathcal{N}, \mathbf{m_0} \rangle$ holds in the initial markings and all successor markings. Equivalent statements are *r holds in all reachable markings of $\langle \mathcal{N}, \mathbf{m_0} \rangle$* or *$r$ holds in all markings of all marking sequences of $\langle \mathcal{N}, \mathbf{m_0} \rangle$*. The meaning of an r LEADS TO s property (denoted as $r \mapsto s$) is that an r-marking is already an s-marking, or a successor marking is an s-marking. The s-marking will eventually be reached by the net system. For a given net system $\langle \mathcal{N}, \mathbf{m_0} \rangle$ and its set of marking sequences $MS(\langle \mathcal{N}, \mathbf{m_0} \rangle)$ the temporal properties are defined by

Definition 10.1.2. r IS INVARIANT *iff* $\mathrm{RS}(\langle \mathcal{N}, \mathbf{m_0} \rangle) \subseteq M_{[r]}$

Definition 10.1.3. $r \mapsto s$ *iff*
$\forall \mathbf{ms} \in MS(\langle \mathcal{N}, \mathbf{m_0} \rangle) \,.\, \forall i \in \mathbb{N} \,.\, (\mathbf{ms}_i \in M_{[r]}) \Rightarrow \exists j \in \mathbb{N} \,.\, (j \geq i) \wedge (\mathbf{ms}_j \in M_{[s]})$

Using these operators the ME-algorithm can be specified as follows:

$$\begin{array}{ll} \forall i \in \{0, \ldots, n-1\} \,. & creq_i \geq 1 \;\mapsto\; cperm_i \geq 1 \\ \forall i, j \in \{0, \ldots, n-1\} \wedge (i \neq j) \,. & ccs_i = 0 \vee ccs_j = 0 \text{ IS INVARIANT} \end{array}$$

The first property means that a requesting client i will eventually get permission and the second demands that the critical section be used under mutual exclusion.

10.1.2 Design

The design starts with the specification given above. The result will be an executable Petri net that includes several design ideas. The basic stages of the design process are:

1. Create the set of places used for the net model.
2. Design constraints describing the behaviour and the structure of the solution, which ensure at least the safety properties of the specification.
3. Add all transitions that do not violate the constraints.
4. Prove the dynamic properties.

Designing an algorithm is a creative process and needs some experience. Therefore, in some of the following steps there are no strict rules for how to proceed. Nevertheless some steps can be identified that will guide the user of the method.

There is no need to handle the steps strictly in the sequence in which they are presented here. In general, the normal design process will need several iterations.

Basic Method Restrictions. Until now the method has only been used in the area of control algorithms such as mutual-exclusion algorithms and election algorithms. This has influenced the models constructed by this method. The algorithms developed have certain restricting properties. These properties theoretically restrict the areas where a specification can be implemented. In practice most of the published control algorithms can be constructed using this method.

Some of the restrictions mentioned are fundamental to the methods. They are presented below, while some minor restrictions are described during the presentation of the steps.

- All places are bounded. Only P/T nets are used. In a more sophisticated version of the method coloured nets are used. In that version the number of tokens is also bounded, but the colours can be infinite sets ([Mac98]).
- During the design stage, transitions have no loops (forward and backward incidence on the same place), because the existence of loops cannot be concluded from invariants. Loops can be included at the implementation stage.
- Transitions that only consume tokens from or only produce tokens for places are not considered since they are not important for problems of control algorithms.

Step 0: Designing the Components and the Communication Structure. The first design step is the identification of components and their communication structure. The algorithm constructed in a later step has to ensure that all added transitions are either internal to a component or access the component and its interface (incoming or outgoing channels).

In this approach components of the system are units that communicate via unidirectional channels. This means that components can be distributed to different locations, since channel communication allows real distribution.

The result of this step is a set of identified components and a set of communication channels.

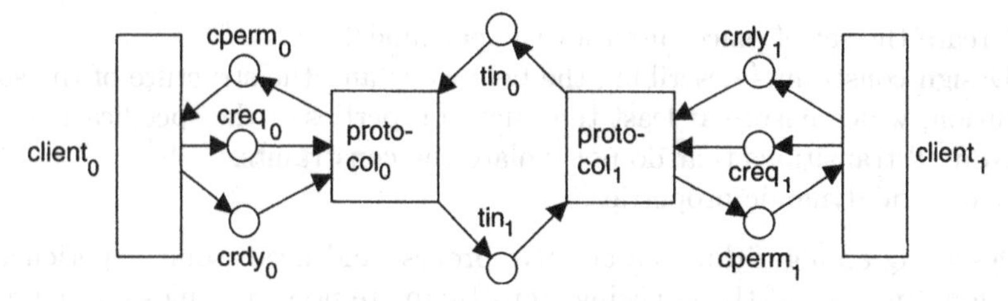

Fig. 10.3. The components and the communication structure

In the ME-example each client is a component. This design decision is made to separate the fixed parts in the client components from the changeable parts in the protocol. Furthermore, the solution will contain one local protocol component per competitor. The benefits are a modular structure and components that can be parametrised. This is especially interesting for symmetrical systems such as this one. Figure 10.3 gives the communication structure for a system with two competitors. A client unit and its protocol communicate via the channels *creq*, *cperm*, and *crdy*. Protocols communicate via the *tin*-channels (*token in*). Along the ring, the *tin*-channels go from competitors with lower numbers to competitors with higher numbers (modulo n).

Step 1: Designing Safety Properties with Place Invariants. In this method p-flows are used to implement the invariants of the specification. Using p-flows, linear invariants are expressed that are much better adapted to Petri nets than are the general invariants used in the specification. This step may entail the loss of some possible solutions, if specific invariants are not expressible as place invariants. Nevertheless, we get a wide range of solutions that are easy to implement.

The result of this step is a list of place invariants. Some invariants implement the invariants of the specification, others are used to integrate certain design ideas into the model.

Due to lack of space the design of place invariants will not be discussed in detail. Instead, the place invariants for the ME-example will be given and explained.

$$\forall i \in \{0,\ldots,n-1\}\,.\ \mathbf{m}(cni_i) + \mathbf{m}(cint_i) + \mathbf{m}(ccs_i) = 1 \quad (10.1)$$

$$\forall i \in \{0,\ldots,n-1\}\,.\ \mathbf{m}(ni_i) + \mathbf{m}(wt_i) + \mathbf{m}(cs_i) = 1 \quad (10.2)$$

$$\textstyle\sum_{i\in\{0,\ldots,n-1\}}(\mathbf{m}(cs_i) + \mathbf{m}(ut_i) + \mathbf{m}(tin_i)) = 1 \quad (10.3)$$

$$\forall i \in \{0,\ldots,n-1\}\,.\ \mathbf{m}(crdy_i) + \mathbf{m}(cperm_i) + \mathbf{m}(ccs_i) = \mathbf{m}(cs_i) \quad (10.4)$$

$$\forall i \in \{0,\ldots,n-1\}\,.\ \mathbf{m}(cni_i) + \mathbf{m}(creq_i) = \mathbf{m}(crdy_i) + \mathbf{m}(ni_i) \quad (10.5)$$

Places of the clients cannot be changed to develop refined models of the system, since the client model must not be changed. Therefore, in the local protocol a new state process is modelled which is strongly coupled with the state process of the client. Equation 10.1 describes the state processes of the clients. Equation 10.2 describes the state processes of the local protocols, where the abbreviations are as follows: ni_i for *not interested*, wt_i for *waiting for token*, and cs_i for *in critical section.*

The coupling of the two state processes is given by Equations 10.4 and 10.5: the client is only in its critical state *ccs* if the protocol is in its critical state *cs* (Equation 10.4), and the number of not-interested clients is given by the number of tokens on place *ni* of the protocol if there are no "update messages" in *crdy* or *creq* (Equation 10.5). Therefore place ni_i will contain a token when the current knowledge of the local protocol is that its client is not interested.

Typically in token algorithms, either the token is used by one of the competitors (cs_i, *in critical section*), or it is unused but still possessed by one of them (ut_i, *unused token*), or it is in one of the channels (tin_i, *token in*) (Equation 10.3). Equation 10.4 together with Equation 10.3 preserves the mutual-exclusion property of the specification.

In further design steps only those places are used that are introduced in the place invariants. This requirement might again reduce the number of possible solutions, but simplifies further steps. The set of places is

$$P = \bigcup_{i \in \{0,\dots,n-1\}} \{cni_i, cint_i, ccs_i, creq_i, crdy_i, cperm_i, cs_i, ni_i, wt_i, ut_i, tin_i\}.$$

From the invariants we can deduce the corresponding p-flows Φ, using the fact that the linear invariants are $\Phi \cdot \mathbf{m} = \Phi \cdot \mathbf{m}_0$. For details see Section 5.2.2.

The set of p-flows is $\Psi = \bigcup_{i \in \{0,\dots,n-1\}} \{\Psi_{1,i}, \Psi_{2,i}, \Psi_3, \Psi_{4,i}, \Psi_{5,i}\}$ with:

$$\begin{array}{lll l}
\Psi_{1,i}: & \forall p \in \{cni_i, cint_i, ccs_i\}. & \Psi_{1,i}[p] = 1, & \text{(inv. 10.1)} \\
 & \forall p \in P \setminus \{cni_i, cint_i, ccs_i\}. & \Psi_{1,i}[p] = 0 & \\
\Psi_{2,i}: & \forall p \in \{ni_i, wt_i, cs_i\}. & \Psi_{2,i}[p] = 1, & \text{(inv. 10.2)} \\
 & \forall p \in P \setminus \{ni_i, wt_i, cs_i\}. & \Psi_{2,i}[p] = 0 & \\
\Psi_3: & \forall p \in \bigcup_{i \in \{0,\dots,n-1\}} \{tin_i, ut_i, cs_i\}. & \Psi_3[p] = 1, & \text{(inv. 10.3)} \\
 & \forall p \in P \setminus \bigcup_{i \in \{0,\dots,n-1\}} \{tin_i, ut_i, cs_i\}. & \Psi_3[p] = 0 & \\
\Psi_{4,i}: & \forall p \in \{crdy_i, cperm_i, ccs_i\}. & \Psi_{4,i}[p] = 1, & \text{(inv. 10.4)} \\
 & \forall p \in \{cs_i\}. & \Psi_{4,i}[p] = -1, & \\
 & \forall p \in P \setminus \{crdy_i, cperm_i, ccs_i, cs_i\}. & \Psi_{4,i}[p] = 0 & \\
\Psi_{5,i}: & \forall p \in \{cni_i, creq_i\}. & \Psi_{5,i}[p] = 1, & \text{(inv. 10.5)} \\
 & \forall p \in \{crdy_i, ni_i\}. & \Psi_{5,i}[p] = -1, & \\
 & \forall p \in P \setminus \{cni_i, creq_i, crdy_i, ni_i\}. & \Psi_{5,i}[p] = 0 &
\end{array}$$

Step 2: Assigning the Places to the Components. Places are assigned to the components by the two functions δ_p and δ_c, giving the components where tokens are produced for places and consumed from places respectively. Places which are assigned to two components are *channels* between the components. Places assigned to one component are *internal places* of the component.

$$\forall i \in \{0,\ldots,n-1\} \,.\, \delta_p(p) = \begin{cases} client_i & \text{if } p \in \{cni_i, cint_i, ccs_i, \\ & \quad crdy_i, creq_i\} \\ protocol_i & \text{if } p \in \{cs_i, ni_i, wt_i, ut_i, \\ & \quad cperm_i, tin_{i\oplus 1}\} \end{cases}$$

$$\forall i \in \{0,\ldots,n-1\} \,.\, \delta_c(p) = \begin{cases} client_i & \text{if } p \in \{cni_i, cint_i, ccs_i, \\ & \quad cperm_i\} \\ protocol_i & \text{if } p \in \{cs_i, ni_i, wt_i, ut_i, \\ & \quad crdy_i, creq_i, tin_i\} \end{cases}$$

So, for instance, cni_i is internal to $client_i$, cs_i is internal to $protocol_i$, and tin_i is a channel from $protocol_{i\ominus 1}$ to $protocol_i$ ($\ominus$: minus modulo n, $\oplus$: plus modulo n).

Step 3: Designing Further Restrictions. Users should be given some further possibilities to describe the model in more detail, to reduce the complexity of later steps.

One aspect that has not been described yet is the direction of the transitions. The problem is explained in Figure 10.2 on page 123. If the client process had to be described, it could not be done only by Equation 10.1 on page 126, since there could have been a transition from place *ccs* to place *cint*. To describe the direction of the flow, transition sets of excluded transitions will be given. For instance, a transition from place *ni* directly to place *cs* should not be included in the model for design reasons.

This is achieved by giving a set T_{min} of "minimal" transitions. All transitions that have greater or equal forward and backward incidences are omitted.

For the ME-example T_{min} is chosen as

$$T_{min} = \bigcup_{i\in\{0,\ldots,n-1\}} \{t_{1,i}, t_{2,i}, t_{3,i}\}$$

where $\mathbf{Pre}[t_{1,i}, ni_i]$, $\mathbf{Post}[t_{1,i}, cs_i]$, $\mathbf{Pre}[t_{2,i}, wt_i]$, $\mathbf{Post}[t_{2,i}, ni_i]$, $\mathbf{Pre}[t_{3,i}, cs_i]$, and $\mathbf{Post}[t_{3,i}, wt_i]$ are all equal to 1, and all other entries in the incidence vectors are zero.

Step 4: Constructing an Executable Net Model. In this step the first executable net model will be constructed. The net model contains the places given above and all transitions that do not violate the given constraints.

The final set of added transitions is created in several steps. A first estimation is given by $T_0 = \{t_{\mathbf{a},\mathbf{b}} | \mathbf{a}, \mathbf{b} \in \mathbb{N}^{|P|}\}$. For all $t_{\mathbf{a},\mathbf{b}} \in T_0$ the vectors **Pre** and **Post** are given by $\mathbf{Pre}[\bullet, t_{\mathbf{a},\mathbf{b}}] = \mathbf{a}$ and $\mathbf{Post}[\bullet, t_{\mathbf{a},\mathbf{b}}] = \mathbf{b}$. This is the set of all transitions that have different incidences. To this point no constraints are taken into account.

T is the greatest subset of T_0 with the following restrictions for all $t \in T$:

1. The number of tokens that a transition consumes from or produces for a place p is less than or equal to the upper bound of the place $b(p)$ (basic restriction):

$$\forall p \in P \,.\, \mathbf{Pre}[p,t] \leq b(p) \vee \mathbf{Post}[p,t] \leq b(p)$$

2. No transition is only producing or only consuming tokens (basic restriction):

$$\mathbf{Pre}[\bullet,t] > \mathbf{0} \wedge \mathbf{Post}[\bullet,t] > \mathbf{0}$$

3. No transition consumes tokens from a place and produces tokens for the same place (no loops, basic restriction):

$$\forall p \in P \,.\, \mathbf{Pre}[p,t] = 0 \vee \mathbf{Post}[p,t] = 0$$

4. All transitions preserve the invariants (step 2):

$$\forall \Psi \in \Psi \,.\, \sum_{p \in P} \Psi[p] \cdot \mathbf{Pre}[p,t] + \Psi[p] \cdot \mathbf{Post}[p,t] = 0$$

5. Transitions only have an incidence on places of a single component and/or on places that represent channels to or from the component (steps 1 and 3):

$$\forall p_1, p_2 \in P \,.\, (\mathbf{Pre}[p_1,t] > 0 \wedge \mathbf{Post}[p_2,t] > 0) \Rightarrow (\delta_p(p_1) = \delta_c(p_2))$$

6. Transitions are restricted in their direction (step 4):

$$\forall t_{min} \in T_{min} \,.\, \mathbf{Pre}[\bullet,t] \not\geq \mathbf{Pre}[\bullet,t_{min}] \vee \mathbf{Post}[\bullet,t] \not\geq \mathbf{Post}[\bullet,t_{min}]$$

7. There exist no transitions in T of which the forward incidence is a linear combination of the forward incidences of the other transitions of T and of which the backward incidence is the same linear combination of the backward incidences of the other transitions. For all combinations of natural numbers $c_{\mathbf{a},\mathbf{b}}$ holds:

$$\begin{aligned} &\mathbf{Pre}[\bullet,t] = \textstyle\sum_{t_{\mathbf{a},\mathbf{b}} \in T \setminus t} c_{\mathbf{a},\mathbf{b}} \cdot \mathbf{Pre}[\bullet, t_{\mathbf{a},\mathbf{b}}] \\ \Rightarrow\ &\mathbf{Post}[\bullet,t] \neq \textstyle\sum_{t_{\mathbf{a},\mathbf{b}} \in T \setminus t} c_{\mathbf{a},\mathbf{b}} \cdot \mathbf{Post}[\bullet, t_{\mathbf{a},\mathbf{b}}] \end{aligned}$$

To skip the discussion of suitable firing rules for the moment, but to get a model that also satisfies LEADS TO properties, fair firing is assumed for all transitions. For a more detailed discussion see Section 10.1.3.

Finally the initial markings have to be added in such a way that they satisfy the invariants. Initially all local protocols have one token on place ni_i. One of the competitors has a token on place ut_i (symbolised by a grey token, e.g. in Figure 10.4). All other places of the protocols are empty.

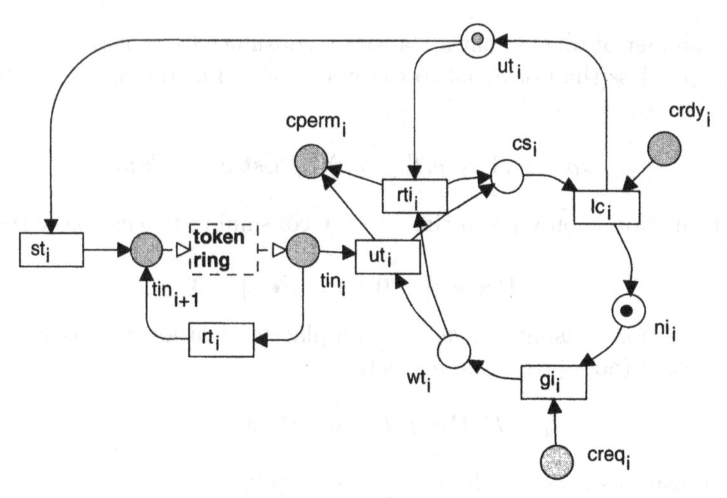

Fig. 10.4. The first executable solution

Step 5: Identifying the Meaning of the Transitions. Figure 10.4 shows the net that is constructed so far. The transitions have new names that are much more expressive than the ones used in the construction above.

The transitions are: *st* (*send token*), *rt* (*relay token*), *rti* (*reuse token immediately*), *gi* (*get interested*), *lc* (*leave critical section*), and *ut* (*use token*).

The dashed box named *token ring* in Figure 10.4 has no formal semantics and simply symbolises where the token leaves the local protocol towards the ring and where it comes back from the ring.

Step 6: Proving Dynamic Properties. The next step involves constructing a proof graph that is similar to the one used in [Wal95]. Initially it can be used to prove LEADS TO properties of a system.

Figure 10.5a shows a preliminary version of the proof graph for a system with two competing units. The nodes of the proof graph represent subsets of the reachable markings, e.g. $creq_1$ represents all markings m where $\mathbf{m}(creq_1) \geq 1$. Since all places are 1-safe, $creq_1$ for instance can be interpreted as $\mathbf{m}(creq_1) = 1$. Therefore, $creq_1$ represents all $creq_1$-markings

The property to be proven is $creq_1$ LEADS TO $cperm_1$. A proof graph is developed in two steps. First it must be investigated which markings are reached while executing a firing sequence that starts in a $creq_1$-marking and terminates in a $cperm$-marking.

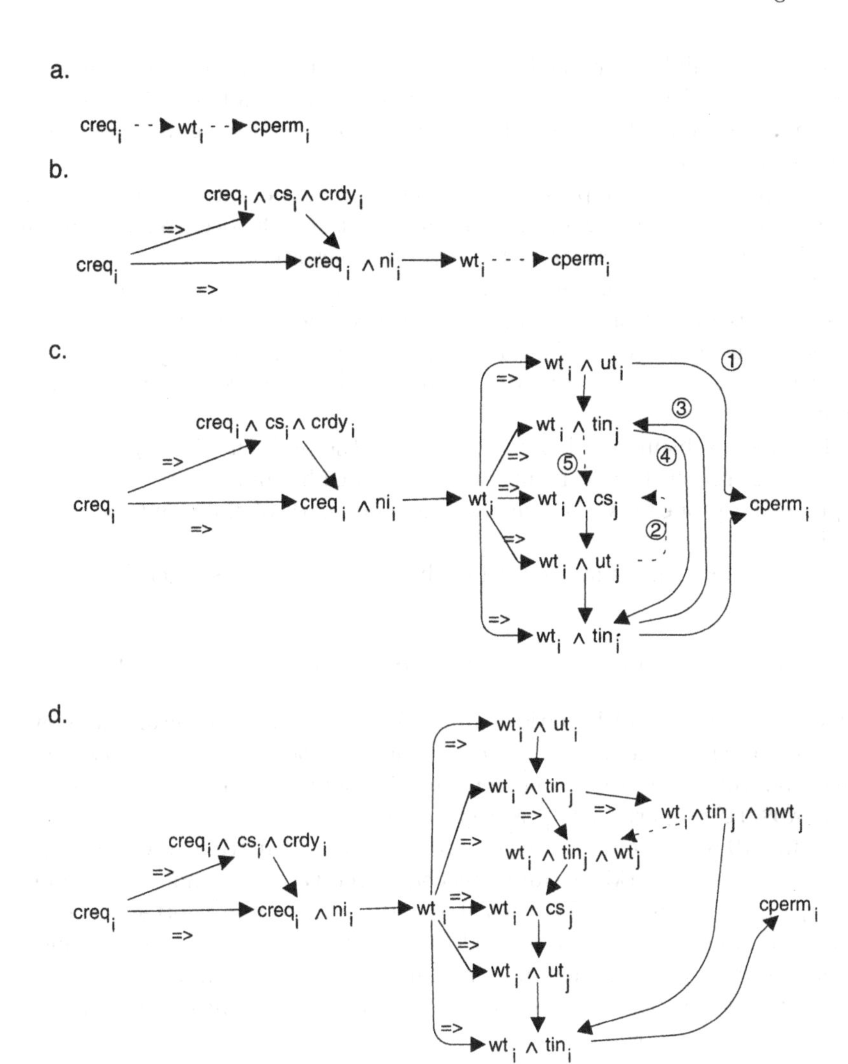

Fig. 10.5. Dynamic properties

Second, it will be examined if the transitions given in the proof graph between two sets of markings will eventually occur. In the following, transitions in the proof graph will be named *graph transitions* to distinguish them from Petri net transitions.

A graph transition represented by solid arcs will eventually occur if the current node is not left by another graph transition. The occurrence of graph transitions represented by dotted arcs is not guaranteed.

The goal is to find a refinement where the node $cperm_1$ can be reached by solid arcs only, from all nodes that are reached when firing the above mentioned firing sequences from $creq_1$-markings. If such proof graphs can be constructed for all required LEADS TO properties, then we have finished our proof.

The first problem that has to be solved is the problem of the granularity of the proof graph. How can subsets of markings be found that can be used to prove the property while keeping the complexity of the graph as low as possible?

The development is started using the partition that is given by the invariant:

$$\mathbf{m}(cni_1) + \mathbf{m}(creq_1) + \mathbf{m}(wt_1) + \mathbf{m}(cperm_1) + \mathbf{m}(ccs_1) = 1,$$

a linear combination of Equations 10.2, 10.4, and 10.5. This invariant shows that a marking is either a cni_1-, a $creq_1$-, a wt_1-, a $cperm_1$-, or a ccs_1-marking. This invariant or partition is chosen since it contains both the conditions of the LEADS TO property, $creq_1$ as well as $cperm_1$.

Figure 10.5a presents the proof graph for that partition. From a $creq_1$-marking only wt_1-markings can be reached directly, and from a wt_1-marking only $cperm_1$-markings can be reached directly. Both arcs are dotted, since the graph transitions are not guaranteed at that abstraction level. For instance, transition gi_1 is only enabled in a $creq_1$-marking if ni_1 contains a token too. So even the assumed strong fairness does not ensure the firing of the transition at this high-level view.

This can be changed by refining the sets of markings in the proof graph. In Figure 10.5b $creq_i$-markings are refined using invariant 10.2. The resulting nodes are $creq_i \wedge ni_i$, $creq_i \wedge wt_i$, and $creq_i \wedge cs_i$. From the invariants 10.4 and 10.5 it follows that the set of $creq_i \wedge wt_i$ is empty, and the set of $creq_i \wedge cs_i$-markings is equal to the set of $creq_i \wedge cs_i \wedge crdy_i$. Therefore a $creq_i$-marking is either a $creq_i \wedge ni_i$-marking or a $creq_i \wedge cs_i \wedge crdy_i$-marking. This fact is represented in the graph by the arcs annotated by a "$\Rightarrow$". The resulting proof graph proves the first part of the property ($creq_i \mapsto wt_i$).

To prove the part $wt_i \mapsto creq_i$ the wt_i-markings are partitioned into five subsets in Figure 10.5c. The partition is complete because a competitor cannot simultaneously wait for a token and use it (Equations 10.2 and 10.1) and because of Equation 10.3. The token is either unused, in a channel, or used by the competitors. This graph proves the required property.

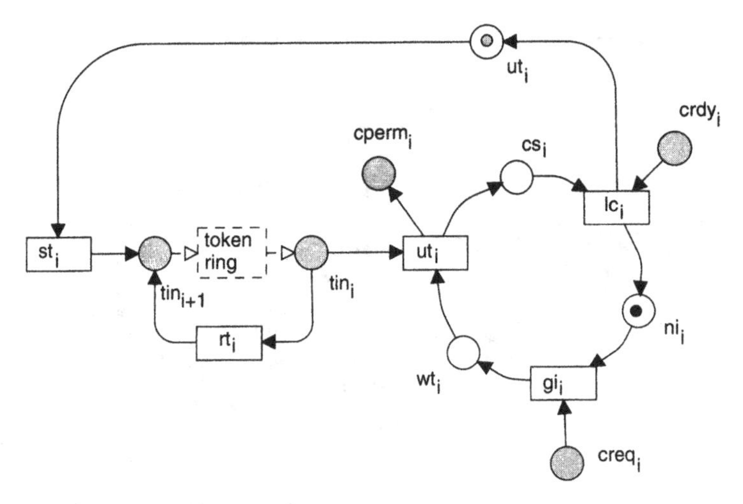

Fig. 10.6. An intermediate version

If the required property cannot be proved by further refinements, the solution ideas of earlier development steps and their consequences have to be changed.

Step 7: Removing Optional Transitions. If the property can be proved, then the proof graph is used for another task. There might be graph transitions that are optional since the LEADS TO property is also preserved without them. In this case we can remove the corresponding net transitions. By removing different sets of transitions, different (classes of) algorithms are produced. By removing the optional transition *rti* we get the net in Figure 10.6. The LEADS TO property is still valid, since only graph transitions 1 and 2 have been removed. Graph transition 3 cannot be removed since a removal of the corresponding net transition rt_1 would also remove graph transition 4 in the proof graph for $creq_2$ LEADS TO $cperm_2$. This problem is handled in the next subsection.

10.1.3 Implementation

The last step concerns the fairness of the transitions. Until now (strong) fairness has been assumed for protocol transitions. Fairness is not easy to implement and is in many cases unnecessarily strong. Fairness is only needed to leave a cycle of the proof graph. But even in such cases, strong fairness can often be replaced by the productive firing rule if some additional side-conditions are added that result in a removing of arcs from the proof graphs, thus breaking the cycle.

This aspect is explained using Figure 10.6. If place wt_i and place tin_i contain tokens, transition rti_i and transition ut_i are in conflict. There is a cycle in the proof graph representing the moving of the token along the ring

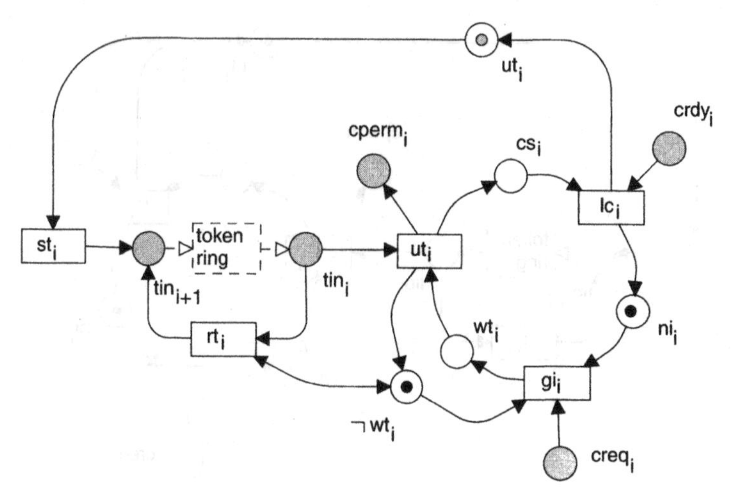

Fig. 10.7. The final version of the algorithm

(Figure 10.5c, graph transitions 3 and 4). So transition ut_i must fire fairly, since otherwise the token would be able to move around the ring forever. By adding a side-condition to transition rt_i (see Figure 10.7) the firing of ut_i is ensured and strong fairness can be replaced by the productive firing rule. The side-condition can be created as follows:

To leave the cycle, transition ut_i gets a higher priority than rt_i. This is implemented by restricting the enabling of rt_i to the markings where ut_i is not enabled, i.e. rt_i should only be enabled in $tin_i \wedge \neg wt_i$-markings. Therefore a place $\neg wt_i$ is constructed as the complementary place of wt_i and is added to transition rt_i as a side-condition.

Figure 10.5d shows the adapted proof graph which no longer has any cycles. This completes the construction of the algorithm.

10.1.4 Conclusion

The method presented here is well suited to the development of control algorithms. The aspect we focus on is the aspect of guiding a developer of the algorithm. The approach can also be used to classify algorithms in the same problem area as ME by giving the sequence of design decisions. Similar algorithms differ only in some later design decisions or at the implementation stage, whereas less closely related algorithms already differ in earlier design steps.

The presented version of the method uses only P/T nets. A sophisticated version would use coloured Petri nets. Such a version would really fulfil the requirements for a method in the area of control algorithms. It would preserve the main steps of this simpler version, refining some aspects of it. This simple method gives a good indication of how to use Petri nets to model a system in a state-oriented way, as well as how to use their formal semantics to include verification.

10.2 Event-Oriented Modelling

In this section we discuss Petri net modelling of discrete dynamic systems, based on the events within the modelled system.

We advocate a top-down and structured approach, starting with high-level modelling dividing the system into communicating subsystems, and ending with low-level modelling defining the *protocol* of a subsystem.

The high-level modelling uses the power and flexibility of nets, combined with a well-understood graphical representation, whereas the low-level modelling uses the conciseness of an algebraic notation.

From the complete net model one can derive the protocol, which is often too complicated to be of any value. However, after *abstracting* from internal communication, retaining only a few essential actions, the system often has to satisfy a simple protocol. The notion of *branching bisimilarity* can be used to verify that the complete system obeys this protocol.

Our approach to event-oriented modelling is subdivided into three parts, treating high-level modelling, protocol-oriented low-level modelling, and verification.

10.2.1 High-Level Modelling

High-level modelling aims to decompose the complete system into simpler subsystems. This decomposition can be made on physical and functional grounds. A system is physically decomposed into subsystems that can be observed to correspond to different entities in the real world (e.g. by their location), or it is functionally decomposed into subsystems that have different abilities. For instance, a human body can be physically decomposed into arms, legs, head, and torso, whereas a functional decomposition would feature muscles, blood circulation, nerves, and brain etc.

Subsystems are independent to a large extent; however they must be able to communicate with one another. Communication consists of token passing or full synchronisation. The former is modelled by place fusion and the latter by transition fusion. The net formalism allows both methods of communication, whereas algebraic methods such as CCS allow only synchronisation ([Mil89]).

When decomposing a system into subsystems, one should identify the subsystems, briefly describing their tasks, and one should describe the communication interface (fusion places and transitions) in each subsystem.

To illustrate the above concepts, we construct a model of a supermarket. The environment of the supermarket consists of customers, who enter the supermarket to buy goods, and suppliers who provide the goods to be bought.

So our initial model consists of three subsystems, called Consumer, Shop, and Supplier. The Shop system is the one that we are interested in, whereas the other two are modelled so as to simulate the shop's environment. This decomposition is physical as well as functional. It makes sense to model individual suppliers if the differences in their behaviour matter to us; however it is pointless to model individual customers.

The task of the Consumer subsystem is to generate customers visiting the supermarket with their various profiles, i.e. the goods which they need to buy as well as the goods which they are inclined to buy if they are attractively priced and/or displayed. The interface between the Shop and Consumer subsystems are "customer" tokens that enter and leave the shop via two places, which are connected to the subsystems via place fusion.

Likewise, it is natural to model two interface places between the Shop and Supplier subsystems, one containing the supply orders and one containing the deliveries. The resulting high-level model is displayed in Figure 10.8.

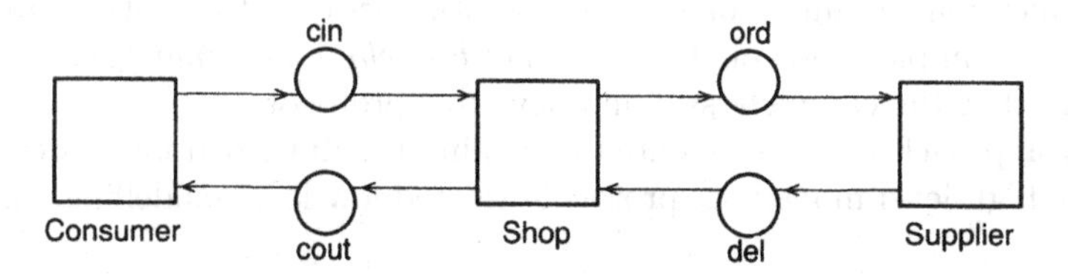

Fig. 10.8. Supermarket with environment

The Shop system may be decomposed functionally into subsystems Custs dealing with customers and Goods dealing with supplies and orders. The interface between Custs and Goods manifests itself by a sell action. At that moment the amount of goods on display is diminished and the amount of goods in the custody of the consumer increased. The sell action synchronises a load (`ld`) action of a product p on the customer side and a diminish (`dim`) action of the same product in the goods subsystem.

One way of modelling this event is through transition fusion. At the moment that a load action takes place in the Custs subsystem, a diminish action should take place in the Goods subsystem. Such load and diminish actions should be possible for every displayed product. Thus it is natural to model a single fusion transition that is parametrised with the type of product. The resulting decomposition is given in Figure 10.9.

Of course, a model with the same power would have been obtained by place fusion, making the set of displayed products available in the Custs subsystem. The advantages of transition fusion become apparent when e.g. a finance subsystem is added. Now a third action should be synchronised with the sell action, adding the value of the loaded product to the amount

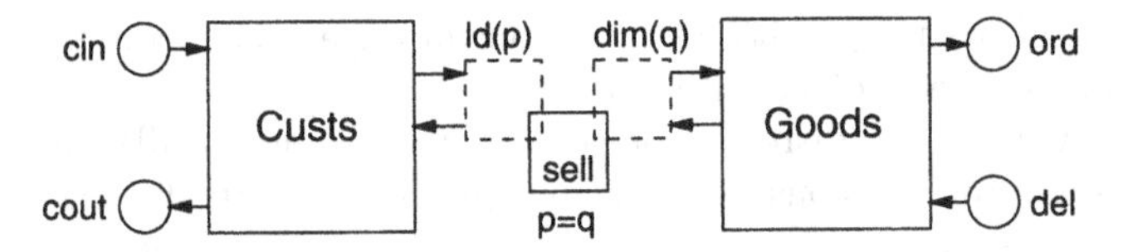

Fig. 10.9. Supermarket decomposition

of money to pay. A triple fusion is easily implemented, whereas a message-passing solution would necessitate the addition of extra fusion places.

10.2.2 Protocol Modelling

The high-level modelling should continue to the point where the subsystems can be understood as collections of independent objects. Each object has a *protocol* that describes the various *states* that an object may be in, and the *events* that the object may undergo in each state. An object has an initial state and may (or may not) terminate. This protocol can be described by an expression or "program" that can easily be converted into a net.

We consider here an example language for protocol modelling that can be used to supplement the high-level modelling described in the previous subsection. This language is a slight extension of the process algebra described in Section 16.5.3. The expressions describing protocols are built from elementary actions and some operators. The operators are related to the standard ones from programming, featuring *sequencing*, *choice*, and *iteration*. A fourth operator, the *free merge*, models the independent execution of its operands.

The basic constructs are similar to those in the languages PA ([BW90]) and PTNA ([BV95b]). The difference between our approach and process algebra is that the communication of subsystems is not modelled algebraically. The high-level modelling of the previous subsection is used instead.

Protocol Language Constructs. The simplest protocol is the one that cannot undergo any events at all from its initial state, which is denoted by the constant δ ("deadlock"). This protocol cannot terminate.

The second most simple protocol executes a single event and then terminates. An event consists of the production and consumption of tokens, combined with the firing of a transition. In our language, we assume that all places and transitions have a label, which need not be unique. Transitions and/or places having the same label cannot be distinguished by other subsystems that communicate or synchronise with them.

Consumptions from a place are denoted by decorating the place label with a question mark. Similarly, productions are denoted by decorating the place label with an exclamation mark. The firing of a transition is denoted by the undecorated transition label. The combination of these sub-events into one single event is denoted by a bar symbol. So $a?|b?|c|b!|b!$ denotes the single-event protocol where transition c fires, consuming a token from places a and b and producing two tokens into place b.

One may choose to disregard or *abstract from* some (or all) transition firings and some (or all) consumptions and productions. In this case, we have *observed* events that need not contain firings, productions, or consumptions. So $a?|b!$ denotes a consumption from a simultaneously with a production into b, where the transition firing (and maybe other consumptions and productions) have been abstracted from. There may even be events (and thus protocols) that are not directly observed at all. Such a *silent* event is denoted by τ. A silent event can be observed indirectly, since it may take an object to a new state.

From the simple one- (or zero-) step protocols above, more complex ones can be constructed by means of operators. Brackets can be used to indicate the order in which the operators are applied. The operator . denotes sequencing. If X and Y are protocols, then $X.Y$ is the protocol obtained by letting X, upon terminating, enable the initial actions of Y. For example $a?.b?$ is the protocol that first consumes a token from a and then terminates by consuming from b. This differs from $a?|b?$ where the tokens are consumed by the same transition. From the descriptions above, it is clear that $\delta.X = \delta$ for any protocol X.

The operator + denotes choice. If X and Y are protocols, then $X + Y$ is the protocol obtained by choosing between the two protocols. The sequencing operator has priority over the choice operator, which can be overruled by using brackets. So in $a?.(b?+c?.e!).d!$, the initial event is $a?$, after which either $b?$ or $c?$ can occur. After a c token is consumed, an e token is produced. Then, in both cases, $d!$ is the terminal event. A choice between X and δ equals X, since no events can occur in δ.

Note that the environment can influence a choice between $b?$ and $c?$ by providing a token for one place and not the other. This is called *external choice*. Of course, the environment can also provide both tokens, in which case the choice is made *internally*, i.e. by the object itself. In $a?.b?+a?.c?$ for example the choice is made when consuming the a token. This choice cannot be influenced by the environment, so it is always internal. This is an example of *nondeterminism*.

The free-merge operator is denoted $||$. If X and Y are protocols, then $X||Y$ is the protocol obtained by independently executing X and Y. Note that $X||Y$ terminates iff both X and Y terminate.

Last but not least is the possibility of splitting a protocol into subprotocols which are identified by names and defining these subprotocols separately. We shall use upper case letters for protocol names and lower case letters for events. For example a protocol P can be defined by the following equations:

$$\begin{aligned} P &= P_1.(P_2 + P_3) \\ P_1 &= a?|b? \\ P_2 &= c! \\ P_3 &= d!|e? \end{aligned}$$

By substitution, one derives that $P = a?|b?.(c! + d!|e?)$. This feature corresponds to a *procedure call* in programming. What makes it interesting is the possibility of *recursion*: defining equations containing the names of protocols to be defined. We restrict ourselves here to iteration. An iterative protocol P that can choose between protocol B or A followed by P itself (thus satisfying $P = A.P + B$) is represented as A^*B. A protocol P that iterates A forever (satisfying $P = A.P$) is represented as $A^*\delta$. This is correct, since P also satisfies $P = A.P + \delta$.

Construction of Nets for Protocols. We shall construct nets for each protocol expression. The places of protocol nets are divided into *internal* places and *fusion* places (pins). The internal places are called simply "places" in the what follows. They are abstracted from, so only consumptions from and productions into pins are observed. In the figures, pins are shaded.

A protocol net may possess a set of initial places, each marked initially with a single token, and a set of terminal places. The marking of every terminal place with a single token is called the *terminal marking*. The initial places have no incoming edges and the terminal places have no outgoing edges. All other places have both incoming and outgoing edges. Protocol nets are constructed in such a way that the net becomes dead as soon as the terminal marking is reached. This property can be proved easily by induction.

First we describe an auxiliary construction giving the *place product* with respect to two given disjoint sets of places in a net. An example of this construction is depicted in Figure 10.10.

Let N be a net and let $\{a_i \mid i \in I\}, \{b_j \mid j \in J\}$ be disjoint sets of places of N. The place product with respect to these two place sets is the net obtained by removing the places in the two sets and adding places $\{c_{i,j} \mid i \in I \wedge j \in J\}$. Every arc originally connected to a place a_k with $k \in I$ is replaced by arcs connected in the same way to $c_{k,j}$ for $j \in J$, and every arc originally connected to a place b_ℓ with $\ell \in J$ is replaced by arcs connected in the same way to $c_{i,\ell}$ for $i \in I$.

Note that if either set I or J is empty, the set of product places becomes empty too and the product net is obtained by removing places and arcs. Also note that the place product operation is commutative and associative.

The construction of a net for a given protocol is as follows. A deadlock net is totally empty. A single event is modelled by an initial and a terminal place, with a transition causing the event consuming from the initial place and producing into the terminal place.

To construct the net for $X.Y$, we juxtapose the nets for X and Y and apply the place product of X and Y with respect to the terminal places of

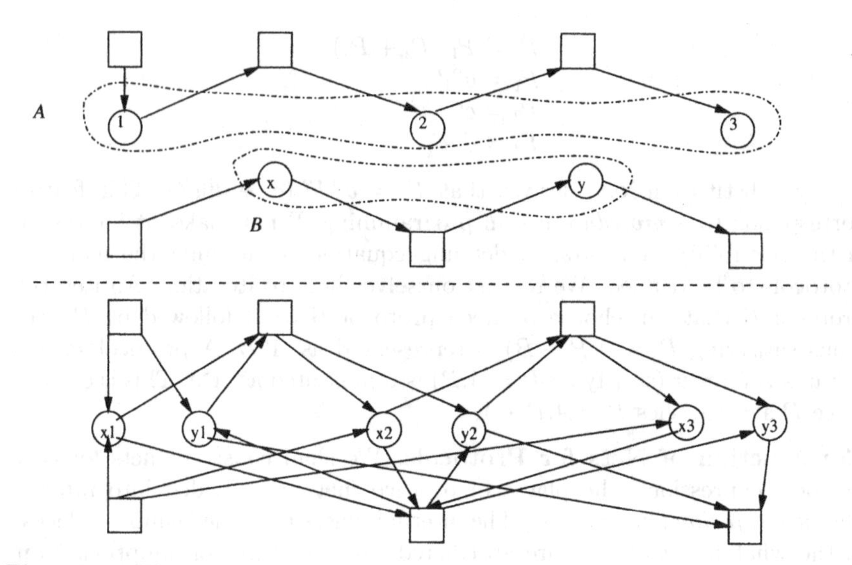

Fig. 10.10. Place product construction

X and the initial places of Y. Its initial places are the initial places of X and its terminal places are those of Y.

The net for $X + Y$ is constructed by juxtaposing the nets for X and Y and applying the place product with respect to the initial places of the two nets, making the new product places initial, and doing the same with respect to their terminal places, making the new places terminal.

The net $X||Y$ is constructed by juxtaposing the nets for X and Y. The initial places of the new net consist of the original initial places, and its terminal places consist of the original terminal places, provided both original nets possess them. Otherwise, all original terminal places and the arcs leading to them are deleted.

For iterative protocols X^*Y, we juxtapose two copies X, X' of the net for X and two copies Y, Y' of the net for Y. We take the place product of the initial places of X and Y as initial places. We take the place product of the terminal places of X, the initial and terminal places of X', and the initial places of Y' as intermediate places. We fuse the terminal places of Y and Y' and make these the terminal places of the protocol net.

In Figure 10.11 some examples of net construction are given. For the sake of clarity, no consumption or production of pin tokens is depicted, however these can be added easily.

Now that we can construct the protocol net for a single object, a net representing the behaviour of several objects with the same protocol is easily constructed. By marking the initial places with n tokens instead of one in the net corresponding to the protocol P, the behaviour of n objects with protocol

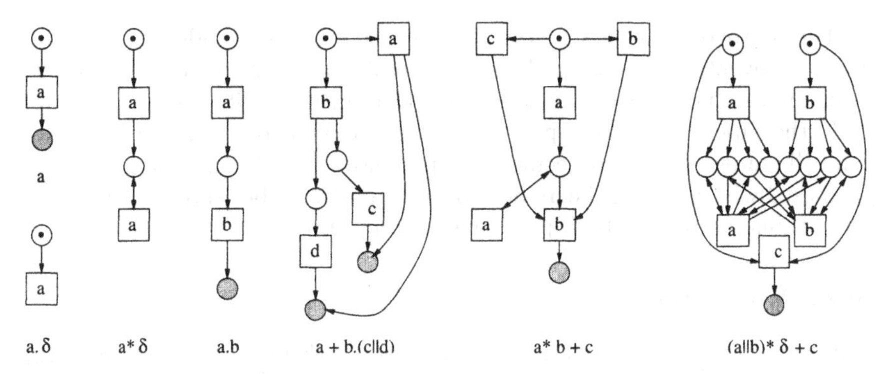

Fig. 10.11. Net constructions for protocols

P is modelled. By removing the initial and terminal places, the behaviour of an unbounded number of objects with protocol P is modelled.

As an example, we take the Custs subsystem of our supermarket. This supermarket has a bakery section where employees fetch, cut, and package fresh bread. Customers queue for their fresh bread. The checkout lanes constitute a second queue.

The protocol for a single customer consists of entering the supermarket (`ent`), loading products (`ld`), possibly interrupted by queuing for bread (`qb`) and being served (`sb`), finally queuing for checkout (`qc`), paying (`pay`) and leaving (`lv`). Formally this becomes the following expression, resulting in the net in Figure 10.12 for the Custs subsystem:

$$\texttt{cin?}|\texttt{ent.ld}^*((\texttt{qb.sb.ld})^*\texttt{qc}).\texttt{pay.lv}|\texttt{cout!}$$

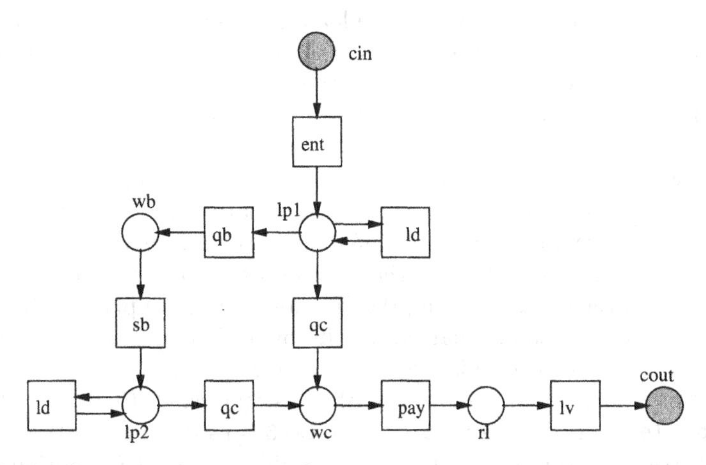

Fig. 10.12. Customers in supermarket

The protocol nets constructed above can be refined by adding colour. In this way, events that occurred in an object's past can be recorded in the object and influence decisions that are taken during its life-cycle.

In the supermarket example, the profile of the customer entering the system will influence which objects will be purchased, and the number of objects purchased will influence the amount of time needed to be served when checking out. Such data will be kept in the customer token.

10.2.3 Verification

The event-oriented construction of a net model sketched here is top-down, guided by intuition and heuristics. It is essential to perform verification and validation of the model in order to ensure that it has not diverged from what was intended. Validation can be performed by implementing the model with existing tools and simulating. This brings to light errors in the model i.e. behaviour that is not intended and must be corrected.

If the validation shows no errors, we want to verify their absence. The verification that we propose here proves the modelled net to be equivalent to a much simpler net after *abstracting* from events that are invisible to its environment.

Abstraction consists of declaring the non-essential events to be *silent*, which is done notationally by labelling them with the label τ. The net thus modified is then reduced modulo an equivalence relation that disregards τ-labelled actions as much as possible. The reduced net is often a simple protocol net that embodies all desired behaviour and nothing more. Equivalence notions that disregard silent events are surveyed in [PRS92].

The following example illustrates that reduction must be done with care. In a bank, clients can apply for loans, for which collateral is needed. The collateral is estimated, after which the loan file is sent to a responsible bank employee, who must follow the following procedure. If the estimated value of the collateral exceeds a percentage p of the loan, it is granted. If it is below a smaller percentage q, it is rejected. If it is in between, he may decide either way.

Abstracting from the estimate, we infer that the employee sometimes automatically grants the loan, sometimes automatically refuses it and sometimes decides there and then. This differs from the case where the employee is allowed to decide regardless of the collateral's value. This may not be of any concern to the client awaiting the decision (unless he plans to bribe the employee), but it certainly matters to the bank. The difference lies in the moment at which the choice is made, the *branching time*.

The "loan protocol" L can be described as follows, cf. Figure 10.13:
$L = \texttt{apply}.(\texttt{est1}.\texttt{grant} + \texttt{est2}.\texttt{refuse} + \texttt{est3}.(\texttt{grant} + \texttt{refuse}))$
When abstracting from the estimate, we obtain a protocol $\bar{L}$ described by
$\bar{L} = \texttt{apply}.(\tau.\texttt{grant} + \tau.\texttt{refuse} + \tau.(\texttt{grant} + \texttt{refuse}))$.
By simply omitting the τ's in the above expression, we would obtain

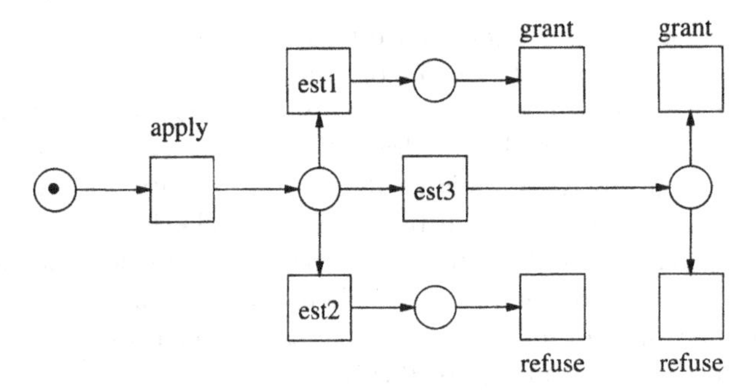

Fig. 10.13. Loan protocol net

$\bar{L} = \texttt{apply}.(\texttt{grant} + \texttt{refuse} + (\texttt{grant} + \texttt{refuse}))$,
which is further simplified to $\texttt{apply}.(\texttt{grant}+\texttt{refuse})$. Note that the above protocol expressions are nets as well. As we have seen above, the moments of choice are not adequately represented in this last expression.

In administrative procedures (and other protocols) there is a difference between a so-called *external choice* $a + b$, where the next action that is taken can be influenced from outside the organisation (e.g. by the arrival of a message or a timeout), and *internal choice* $\tau.a + \tau.b$, where the decision is taken within the organisation. This also illustrates the importance of the point at which choices are made.

In the what follows, we first look at nets without abstraction and consider *strong bisimilarity*, the simplest equivalence relation that respects moments of choice. We then define *branching bisimilarity* by showing how and when τ-labelled events can be eliminated. These equivalence notions are treated in [PRS92] and [GW96].

Strong Bisimilarity. As we have seen earlier, an event in a net corresponds to the firing of some transition, and the states of a net correspond to its marking (i.e. the distribution of tokens). Two nets are considered (strongly) *bisimilar* iff a correspondence between their states can be established such that in corresponding states every event in one net can be matched by a similar event in the other net leading to again corresponding states. This is a branching-time equivalence.

A formal definition of bisimilarity between two nets N, M is based on the existence of a relation R (called a *bisimulation*) between the reachable markings of nets N and M such that for any markings m, m' of M and n, n' of N such that nRm,

$$\begin{aligned} m \xrightarrow{e} m' &\Rightarrow \exists \bar{n} \,.\, n \xrightarrow{e} \bar{n} \wedge \bar{n} R m', \\ n \xrightarrow{e} n' &\Rightarrow \exists \bar{m} \,.\, m \xrightarrow{e} \bar{m} \wedge n' R \bar{m}. \end{aligned}$$

Whether two nets are bisimilar now depends on which events are possible in a given net. If one is only interested in the *functionality* of a system and not in its *efficiency*, it is a good choice to define single firings of transitions as events, thus obtaining *interleaving* bisimilarity.

If efficiency is important, all possible simultaneous firings of transitions can be taken as events (which may dramatically increase the number of possible events in a given state). This yields *step* bisimilarity. Figure 10.14 gives examples of interleaving and step bisimilar nets. The left-hand and middle nets are interleaving bisimilar but not step bisimilar, whereas the middle and right-hand nets are even step bisimilar. There are even finer notions of bisimilarity that take *causal* relations of events into consideration. These notions distinguish all three nets in Figure 10.14.

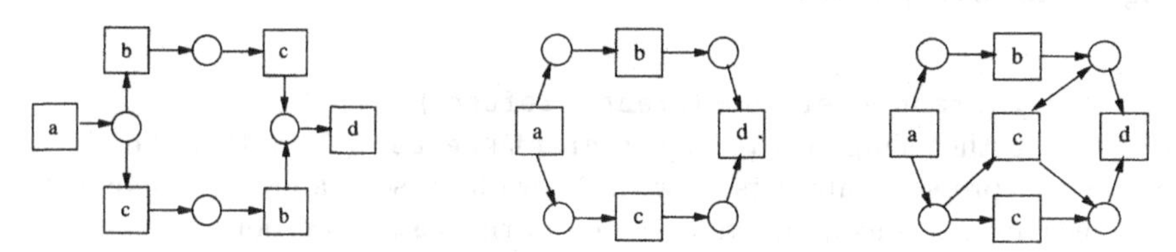

Fig. 10.14. Bisimilar nets

From the definition of bisimilarity, one may infer that protocols of bisimilar nets have exactly corresponding moments of choice. All the options of some net in a certain state are also open to a bisimilar net in a corresponding state and vice versa.

Branching Bisimilarity. Above, we defined "strong" bisimilarity that does not take the special nature of τ-labelled transitions into consideration. The label τ is considered as an ordinary label, so it is not possible to eliminate any silent event while staying in the same equivalence class.

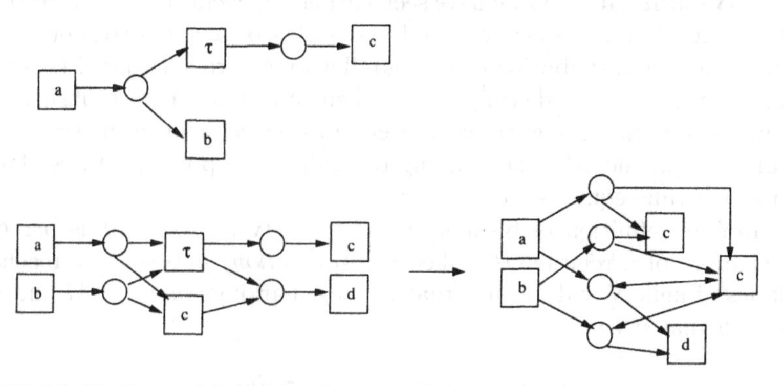

Fig. 10.15. Visible and invisible silent transitions

Branching bisimilarity is an equivalence relation containing strong bisimilarity, as it allows nets with silent events to be equivalent to nets without them. Given a net, one can determine silent transitions that are truly invisible, i.e. that do not make choices when firing. The upper net in Figure 10.15 contains a silent transition that is not invisible, since after it fires the event b can no longer occur. The lower net in the figure contains a silent transition that is invisible, since the possibility of an event c followed by d, which might have become impossible when the silent transition fired, in fact remains possible.

An invisible transition can be eliminated from a net modulo branching bisimilarity. In Figure 10.15, the lower net on the right-hand side is obtained by removing the invisible transition. Here, this involves taking the place product of the places that the invisible transition consumes from and those into which it produces. The two c transitions can then be merged modulo step bisimilarity, as in Figure 10.14.

Two nets are branching bisimilar iff they are strongly bisimilar after eliminating invisible transitions. A formal definition in terms of relations between markings can be found in e.g. [PRS92], and in terms of transition systems in [GW96].

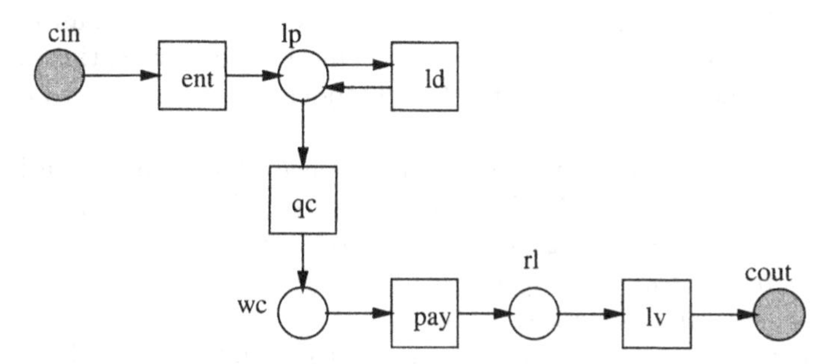

Fig. 10.16. Supermarket customers after abstraction

In general, it is easy to discover and eliminatie invisible transitions. For example, after abstracting from queuing and being served at the bakery, the supermarket net is branching bisimilar to the net shown in Figure 10.16. The behaviour of a customer becomes `cin?|ent.ld*qc.pay.lv|cout!` After further abstracting from e.g. the load action this becomes `cin?|ent.qc.pay.lv|cout!`. After abstracting from all actions, we obtain `cin?.cout!`. Note that branching bisimilarity presupposes *fair* iteration, ruling out the possibility that a customer will eternally load products without ever checking out.

10.2.4 Conclusion

The event-oriented modelling strategy described here uses top-down hierarchical modelling, until simple objects are obtained. For these objects, the protocol (or life-cycle) is described by a language that can be translated into a net. The net may be kept implicit, or can be constructed explicitly from the term describing it. One can verify the constructed net by concentrating on certain aspects, abstracting from the other ones, and using branching bisimilarity to prove equivalence to the expected behaviour with respect to that aspect.

10.3 Object-Oriented Modelling

The integration of Petri nets with object-oriented concepts is a rich research domain which has been tackled in many ways. It has had a considerable exposure at conferences devoted to Petri nets or object orientation, and some satellite workshops have been devoted to the combination.

Three approaches are used, namely giving a formal basis to an object-oriented language or methodology, extending Petri nets by the use of complex data types for tokens, and using object-oriented concepts directly in the Petri net formalism. Petri nets lack structuring facilities and this makes them less suitable for handling large distributed systems. Each of the three approaches aims to bridge that gap. Another immediate benefit is the enhancement of object-oriented methodologies with a formal method for verification and with prototyping aspects, since Petri net based models can be simulated and animated.

Association of objects with Petri nets hence allows one to satisfy the following imperative from Mellor and Shlaer (see [MS94]): *"The ability to execute the application analysis model is a sine qua non for any industrial-strength method, because analysts need to verify the behaviour of the model with both clients and domain experts."*

However, such an association is somewhat difficult because the intended modelling and structuring power of objects is often in conflict with the proving facilities of Petri nets. For instance, the use of complex data types for tokens and the support of concepts such as inheritance and polymorphism may cause a great loss of proving strength within formalisms. In this chapter, the correspondences between concepts from both approaches are summarised. Then for each of the main integration approaches, a description of selected work is given. Thereafter, a description of our method (the OF-Class/OF-CPN approach) is given. It aims to provide a good tradeoff between modelling/structuring needs on the one hand and verification facilities on the other.

10.3.1 Objects vs. Petri Nets

Object orientation in specification and programming is based on a set of structuring concepts such including (see [RBP+91]):

1. *Identity*: An object is an entity with a handle by which it can be uniquely identified and addressed. Identification is not possible in P/T nets but is easy in coloured nets.
2. *Classification*: This involves grouping intrinsically similar objects into classes. A class is therefore a pattern which describes the structure of each object of a collection. Classification can also be performed at the class level with the *meta-class* notion but the similarity is not forced to be intrinsic. Classification is achieved in nets by differentiating tokens which circulate in the same net structure (skeleton). The common aspects of objects are represented in the net structure and tokens carry specific features of objects.
3. *Modularity* and *encapsulation*: An object is an entity which owns data and performs operations to manipulate it. The many aspects of an object can be divided into a publicly accessible part (the *interface*) and a private one. Objects can thus be differentiated from one another by this *encapsulation principle*. In net semantics, the encapsulation principle can be achieved by discriminating between the interface and internal places and transitions.
4. *Interaction types*: These are mainly supported by some kind of message passing. An object can manipulate the data owned by another object by sending a message in order to invoke the appropriate method. The message passing can be synchronous or asynchronous.

To these basic concepts, other more elaborate ones such as inheritance (sub-typing and sub-classing), delegation, polymorphism, and dynamic binding are added to enhance genericity, reuse, and loose coupling. Inheritance and delegation allows an implementation to be shared and the subsequent aspect to be optimised. Dynamic binding and polymorphism shift method resolution to run-time. They enable polymorphic operations implemented by many methods, each one dedicated to carry out the operation in a specific context determined by the parameters provided.

Object orientation also introduces a multi-level abstraction on a system. It differentiates the internal implementation of objects from the interactions that can occur between them. Which operations an object performs is modelled by its interface and how these operations are performed is modelled in an encapsulated way to hide it from the environment. This allows modification of the encapsulated part provided it remains orthogonal to the interface.

There are many approaches to integrating Petri nets and object orientation, each focusing on the adaptation to Petri nets of some concepts from objects.

Modularity and encapsulation are naturally obtained by decomposing flat nets into sub-nets considered as well-defined entities. In each sub-net, one can consider parts of the net structure (generally state machines) as performing some given operation or method. Decomposition is used in a *top-down approach* (see Chapter 9). It allows one to handle large-scale nets and modularity is taken into account in the analysis of such nets. For instance, liveness and reachability can be analysed in an optimised way using modular nets ([CP92, NM94, Val94b]).

Identity and classification are achieved by colouring of tokens. The net structure models the class pattern and objects are modelled as tokens which need colour to be distinguished. For this part of the integration we face similar problems as when moving from P/T nets to coloured ones.

Interfacing and interactions are achieved in three ways as shown in Chapter 9:

- By fusion of shared bordering places which can model asynchronous data sharing;
- By fusion of shared bordering transitions which can model synchronous rendezvous communication;
- By arc addition which can model asynchronous message passing.

Interfacing by place fusion, transition fusion, and arc addition covers the different ways of synchronising between objects. These interfacing mechanisms enable a *bottom-up approach* (see Chapter 9).

Researchers have also investigated differentiated token types to support the *data-oriented view* of systems which is common in the object-oriented paradigm. For that purpose, tokens are sometimes modelled as nets or as algebraic data types.

10.3.2 Integration Approaches

This section presents what can be considered as the four main approaches in the integration of object-oriented concepts with Petri nets.

Multi-Level Abstraction. Firstly, some authors borrow the multi-level abstraction of object orientation and apply it directly to Petri nets. Objects – seen as entities with data and behaviour – are modelled as separate nets called *object nets* or algebraic data types. The whole system is modelled as a net, called the *system net*, where tokens are the *object nets* or algebraic data. Synchronisation is achieved by specialising the firing rule to the complex tokens (object nets). Although not using object concepts, the formalism proposed by Valk – Task/Flow EN systems – is a representative attempt to apply multi-level abstraction to Petri nets ([Val87b]). The model proposes a bi-level approach. Systems are considered to be finite sets of tasks partially ordered by precedence relations. A system is modelled by a *system net* with one token which is a net called an *object net*. The model has been extended in more recent versions to take into account more elaborate aspects of object orientation ([Val95]) as well as multi-level abstraction which allows one to model hierarchy. The enabling/firing rule of object nets expresses their autonomous behaviour. The enabling/firing rule of the system net expresses the interactions between objects and their combination into other objects (Figure 10.17).

In *OBJSA nets* Battiston and others [BDCM88] consider nets with objects as domains. The tokens flowing in the net are described in a dedicated object-oriented language. We will see below a language based on such principles. Transitions are labelled with method invocations on objects, while arcs describe the flow of objects in the system. This approach allows one to formalise the life-cycle of objects in a system. Dynamic creation and destruction of objects can be supported in this kind of model.

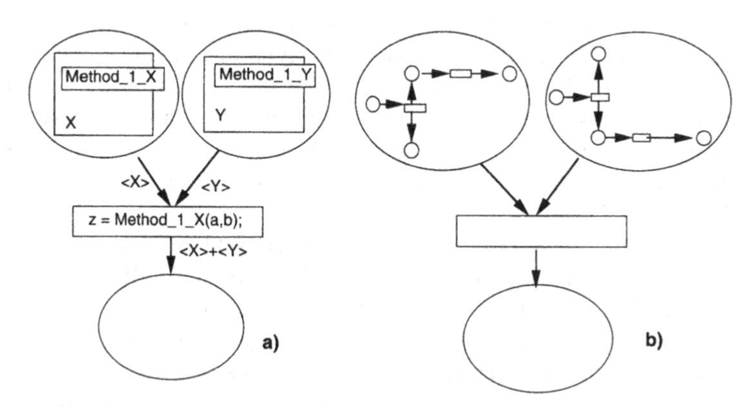

Fig. 10.17. Using multi-level abstractions in nets. In a) tokens are objects while in b) tokens are themselves nets.

The multi-level abstraction can be adapted to Petri nets in a recursive way ([Val95]), resulting in hierarchical description of systems. It allows us at level n ($n > 1$) to focus on interactions at level $n - 1$ whose tokens are nets modelling concurrent actions of the system.

Petri Nets as a Formal Basis for Object-Oriented Languages. Secondly, some authors are concerned with providing a formal basis for object-oriented languages. They mainly focus on enhancing algebraic data types using the control structure from Petri nets. Object life-cycles are modelled with nets. Tokens in places are objects that might be complex, and methods are attached to transitions. In the remainder of this section, the emphasis is on the integration approach which tries to give Petri net semantics to an object-oriented language. We present two object-oriented languages, CLOWN and CO-OPN.

CLOWN (CLass Orientation With Nets) is developed on top of OBJSA, a combination of algebraic Petri nets with the specification language OBJ which provides the modular features ([BCDC96]). CLOWN defines elementary classes with multiple inheritance. An elementary class in CLOWN has an interface which specifies a list of typed formal parameters expected from the partners (instances of other classes). It also defines instance variables and places to specify the correctness conditions on objects (accepted values for instance variables). Methods are described by giving the interacting partners and the pre- and post-conditions (residency in places and modifications on instance variables). Elementary classes can be combined into composite classes. Composition in CLOWN is based on method synchronisation (transition fusion, cf. Section 9.2.3).

A system in CLOWN is a set of classes (elementary or composite) for which an initial configuration is defined that is the set of live objects. Polymorphism is supported in order to enhance generic compositionality. CLOWN allows one to model concurrent systems with strong synchronisation between the components. It does not support dynamic instantiation. Methods of classes are specified in an axiomatic way by eventual synchronisation with a list of partners and pre- and post-conditions. Because of the synchronisation mechanism used for composition, objects can not exhibit internal concurrency. An example of a CLOWN specification is given in Figure 10.18 (adapted from [BCDC96]).

CO-OPN (Concurrent Object-Oriented Petri Net) relies on both algebraic specification and Petri nets ([BG91]). The first formalism enables us to describe the data structures and the functional aspects of systems while the second one supports dynamics and concurrency. Object orientation provides structuring aspects which these formalisms lack. It allows one to structure a system as a set of interacting components which can be organised hierarchically using multi-level abstraction. Interactions between objects can be synchronous or asynchronous and can be executed concurrently. Concurrency is supported both at the intra-object and the inter-object level. In its most recent versions, the language has achieved integration of the notions of class, inheritance, and sub-typing. CO-OPN is the only integration of nets and objects which takes into account the difference between sub-typing and sub-classing, the subject of a rich debate in the object-oriented research community. Methods are modelled by parametrised transitions with which other objects can synchronise.

In this type of approach, the contribution of Petri nets is to formalise the internal behaviour of objects and the interactions between them (see the graphic in Figure 10.18). An interface, in an object-oriented language, is a set of operation signatures without any additional structure. Petri nets allow some structuring of such interfaces by prescribing whether operations are performed sequentially or in parallel.

Nets with Object-Oriented Concepts. Thirdly, some authors investigate the use of object-oriented techniques more directly. These authors aim to support object-oriented concepts such as polymorphism, inheritance, and sub-typing/sub-classing directly in Petri nets. Nets modelling objects are interfaced by places or transitions to model the synchronisation semantics. Sibertin-Blanc [SB93] and others have first defined a coloured Petri net model

```
CLASS printer
INHERITS root
CONST id : PR-NAME; speed : NAT;
VAR owner : USERID; buffer:DOC; copies : NAT;
INTERFACE USER(name : USER-ID; ficle : DOC; qty : NAT);
PLACES ready : owner =/= none;

METHODS

load
WITH USER;
POST
owner <- USER.name;
buffer <- USER.file;
copies <- USER.qty;

print
PRE  copies > 0;
POST copies <- copies -1;

reset
WITH USER;
PRE  owner = USER.name and copies == 0;
POST  owner <- none;
NET
```

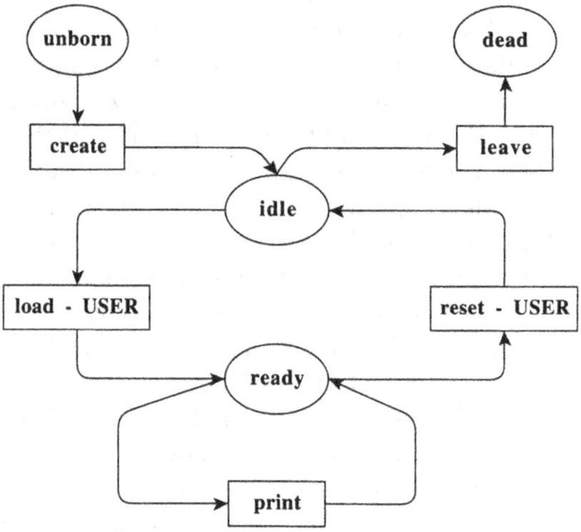

unborn
dead
create
leave
idle
load - USER
reset - USER
ready
print

Fig. 10.18. Example of a CLOWN class specification

to build complex distributed systems whose components interact according to the client-server paradigm. Requests and replies are based on an adaptation of message passing. Messages are issued by passing an appropriate token in an interface place or by firing an interface transition. The model has been extended in *Cooperative nets* to a full support of identification of objects, dynamic instantiation and binding, sub-typing, and inheritance ([SB94]). Tokens are described in an *ad hoc* data-type system which allows the sub-typing. Objects may pass the identity of other objects in messages (tokens). The late binding between objects can therefore be achieved in a satisfactory way.

In *Object Petri nets*, Lakos [Lak95] has achieved a full integration of basic and elaborate object-oriented concepts. The class hierarchy supports both token and sub-net types and allows us to take into account the multi-level abstraction of object orientation. This model is a full object-oriented version of the Petri net formalism.

The main drawback of this type of approach is that the authors emphasise structuring over proving. The complex types adopted for tokens prevent computation and interpretation of structural invariants for such formalisms. Models however can be simulated and animated.

Nets within an Object-Oriented Methodology. Finally, some authors integrate Petri nets as a formal basis into object-oriented methodologies for analysis and design, by giving Petri net semantics to object-oriented models. The analysis and design are performed in an object-oriented methodology whose steps are enhanced with Petri nets in order to formalise them.

In [Mol96], net-based semantics are provided for class diagrams which are commonly used to model structures of objects in a system. Static class diagrams can therefore be made executable and stand for prototypes of object specifications. The net-based semantics allows us to enhance object-oriented methodologies by validation, verification, and prototyping features.

Lakos and others have built a software engineering environment based on the semantics of object Petri nets and the methodology developed by Shlaer and Mellor [LK95]. They give correspondences between the object Petri net model and the four models supporting the Shlaer-Mellor methodology. The *Information Model* of the Shlaer-Mellor methodology models the static aspects of object orientation. It is mapped on a somewhat extended version of the object Petri net model. Associations are represented by adding classes to the model or attributes to the classes. The *State Model* of the methodology represents the life-cycle of instances of a given class. It is modelled in a state/transition way and therefore it is directly handled by Petri net concepts. The *Object Communication Model* shows the flow of events between objects in a given model. The hierarchical aspects of object Petri nets allow us to model each class *State Model* as a super place and to add transitions and arcs to model the event flow between them. Finally, the *Action Dataflow Diagram* of the methodology shows the actions executed when entering each state. It is considered to be more relevant to conceptual modelling and hence has a level of precision which does not match that of the three previous models. The authors do not discuss its correspondence with object Petri net models. This work enhanced the Shlaer-Mellor methodology with a prototyping level and most likely analysis and verification techniques adapted from coloured Petri nets such as place invariants will be developed in the near future.

10.3.3 A Multi-Formalism Approach Including Nets

This section presents another approach for the integration of nets for verification in a specification process. Some authors advocate the use of nets as a validation/verification tool to be used on demand. The specification is run in another environment which can be *ad hoc* or an object-oriented one. Transformation to nets is performed when verification/validation is needed. Such approaches are used in [DE96] and in [Lil96]. In [DE96], the authors have built a component model based on a specification language. The component model OF-Class (*object formalism class*) allows us to specify the components of a distributed system in order to enhance encapsulation of the components and explicit expression of interactions between them. Its main characteristics are highlighted in the discussion below. As well as this component model, a modular Petri net model called OF-CPN (*object formalism coloured Petri net*) is defined which is later in this section. Transforming OF-Class to OF-CPN (see below) is formal, fully automated, and supported by a tool. From the modular nets, a verification and validation method is developed, it too is presented below.

OF-CPN is a modular coloured Petri net model that has interfacing places and supports verification of systems modelled with OF-Class. The transformation from OF-Class to OF-CPN is fully automated and supported by a tool. Thus, the main objective here is to compute structural properties such as place invariants (interpreted as integrity constraints on the objects) and to prove behavioural properties by checking the model to detect liveness, deadlocking states, and all properties meaningful in the Petri net paradigm. Reactive distributed systems have some specific features. A *reactive system* is a system which maintains an interaction with its environment and which is not expected to terminate ([Rei92]). One therefore needs to ensure its safety and liveness. According to Lamport, safety means that *"something bad"* such as a deadlock or mutual exclusion violation does not occur, whereas liveness means that *"something good"* such as a resource access without starvation or response to a request does always occur ([Lam77]). For this purpose, there is a focus on formalising interactions.

The OF-Class Model. The OF-Class model does not aim to support complex object-oriented concepts such as inheritance and polymorphism. It exploits only the modularity and encapsulation which are intrinsic characteristics for components of a distributed system. The component metaphor adopted is the following: *a component in a distributed system is an entity able to manage resources, to satisfy the requests on them and to request other components when neccessary.*

Therefore, the main characteristics of the OF-Class component model are ([DE96]):

- Components of a distributed system are concrete entities which must have a life-cycle. They can have strong similarity and implement servicing policies such as competition or cooperation. We therefore need a two-level abstraction which corresponds to *classification/instantiation* in object orientation.
- Interactions between the components of a distributed system must be formalised. Each component defines exported services as interaction patterns

that the environment must respect when requesting such a service. An exported service is a set of operations with attached access semantics (synchronous or otherwise) and allowed sequencings. Each exported service defines a given class of clients for the component. A component must import services exported by others in order to use their operations. For a required service, one can have expectations of the results of the operations and specify some kind of *exception processing* when these expectations are not met.
- Components of a distributed system must be autonomous and active. They must be able to satisfy the requests from their environment. Moreover, they must be able to evaluate the state of managed resources and trigger some necessary processing on their own. A component can define some operations called *triggers* which can not be invoked by the environment. They correspond to asynchronous methods called at the creation of an instance of the component and their execution may be suspended depending upon a predicate on the state of local resources,
- in order to enable precise verification, the specification must be run at a sufficient level of detail. We do not need an axiomatic specification but a procedural one which describes precisely the two aspects of components of a distributed system: *which operations are performed* and *how they must be performed.*

For examples to illustrate the OF-Class model and its specification language, the reader is referred to Section 11.3 in Chapter 11 and to Section 26.3 in the Applications part of this book.

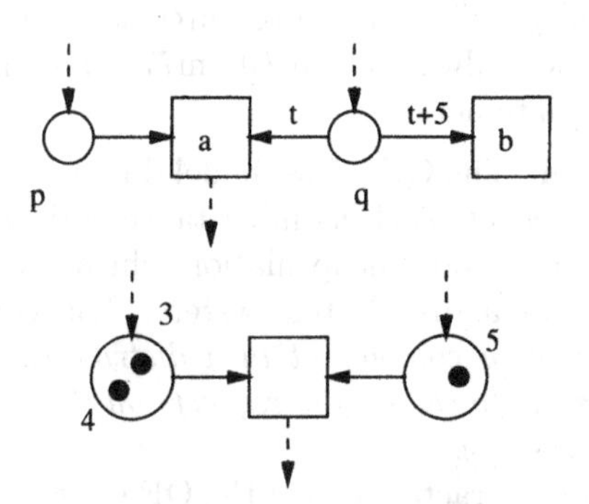

Fig. 10.19. Examples of timed nets

The OF-Class model allows a precise description of components of a distributed system, but it does not have any built-in facility for verification and validation. For that purpose, it is formally mapped onto a modular coloured Petri net, the OF-CPN model. The main interest in such an approach is the proof that nets can contribute in a multi-formalism environment for verification and validation purposes. Experts in an application domain do not need to create and manage nets for specification and verification. The main difficulty is to formally transform their specification into nets and to trace back the results of verification.

The OF-CPN Model.

Definition 10.3.1 (OF-CPN model). *An OF-CPN is a seven-tuple* $\langle \mathcal{N}, P_{\mathrm{acc}}, P_{\mathrm{res}}, P_{\mathrm{snd}}, P_{\mathrm{get}}, \Im_{\mathrm{acc-res}}, \Im_{\mathrm{snd-get}} \rangle$ *where:*

1. $\mathcal{N}$ *is a coloured Petri net system* $\langle P, T, \mathrm{Dom}, \mathbf{Pre}, \mathbf{Post}, \mathbf{guard}, \mathbf{m_0} \rangle$ *with:*
   a) P *is the set of places and* T *the set of transitions and* $P \cap T = \emptyset$,
   b) $\mathrm{Dom} : P \cup T \longrightarrow \Gamma^*$ *defines the colour domains for places and transitions,*
   c) $\mathbf{Pre}$ *and* $\mathbf{Post}$ *define respectively the backward and forward incidence colour functions:*
   $\mathbf{Pre}, \mathbf{Post} : P \times T \longrightarrow \mathrm{Bag}(Symb_{\mathrm{Dom}(P)})$,
   d) $\mathbf{guard}$ *defines the guards on transitions :*
   $\forall t \in T \,.\, \mathbf{guard}(t) : \mathrm{Bag}(Sym_{\mathrm{Dom}(t)}) \longrightarrow \mathcal{B} = \{\mathrm{True}, \mathrm{False}\}$,
   e) $\mathbf{m_0}$ *is a marking for* $\mathcal{N}$, *i.e.* $\forall p \in P \,.\, \mathbf{m_0}(p) \in \mathrm{Bag}(C_{\mathrm{Dom}(p)})$,
2. $P_{\mathrm{acc}} \subset P$ *is a set of places such that* $\forall p_{\mathrm{acc}} \in P_{\mathrm{acc}} \,.\, {}^{\bullet}p_{\mathrm{acc}} = \emptyset \wedge \mathbf{m_0}(p_{\mathrm{acc}}) = \langle\rangle$,
3. $P_{\mathrm{res}} \subset P$ *is a set of places such that* $\forall p_{\mathrm{res}} \in P_{\mathrm{res}} \,.\, p_{\mathrm{res}}{}^{\bullet} = \emptyset \wedge \mathbf{m_0}(p_{\mathrm{res}}) = \langle\rangle$,
4. $P_{\mathrm{snd}} \subset P$ *is a set of places such that* $\forall p_{\mathrm{snd}} \in P_{\mathrm{snd}} \,.\, p_{\mathrm{snd}}{}^{\bullet} = \emptyset \wedge \mathbf{m_0}(p_{\mathrm{snd}}) = \langle\rangle$,
5. $P_{\mathrm{get}} \subset P$ *is a set of places such that* $\forall p_{\mathrm{get}} \in P_{\mathrm{get}} \,.\, {}^{\bullet}p_{\mathrm{get}} = \emptyset \wedge \mathbf{m_0}(p_{\mathrm{get}}) = \langle\rangle$,
6. *the sets* P_{acc}, P_{res}, P_{snd}, *and* P_{get} *are pairwise disjoint,*
7. $\Im_{\mathrm{acc-res}} : P_{\mathrm{acc}} \longrightarrow P_{\mathrm{res}}$ *is a bijection such that:*
   a) $\forall (p_{\mathrm{acc}}, \Im_{\mathrm{acc-res}}(p_{\mathrm{acc}})) \in P_{\mathrm{acc}} \times P_{\mathrm{res}} \,.\, \forall t_n \in {}^{\bullet}(\Im_{\mathrm{acc-res}}(p_{\mathrm{acc}})) \,.$
   $\exists t_1, \ldots, t_{n-1} \in T \,.\, t_1 \in p_{\mathrm{acc}}{}^{\bullet} \wedge \forall i \in \{1, \ldots, n-1\} \,.\, t_i{}^{\bullet} \cap {}^{\bullet}t_{i+1} \neq \emptyset$,
   b) $\forall (p_{\mathrm{acc}}, \Im_{\mathrm{acc-res}}(p_{\mathrm{acc}})) \in P_{\mathrm{acc}} \times P_{\mathrm{res}} \,.\, \forall t_1 \in p_{\mathrm{acc}}{}^{\bullet} \,.$
   $\exists t_1, \ldots, t_n \in T \,.\, t_n \in (\Im_{\mathrm{acc-res}}(p_{\mathrm{acc}}))^{\bullet} \wedge \forall i \in \{1, \ldots, n-1\} \,.\, t_i{}^{\bullet} \cap {}^{\bullet}t_{i+1} \neq \emptyset$,
   c) $\forall (p_{\mathrm{acc}}, \Im_{\mathrm{acc-res}}(p_{\mathrm{acc}})) \in P_{\mathrm{acc}} \times P_{\mathrm{res}} \,.\, p_{\mathrm{acc}}{}^{\bullet} \cap {}^{\bullet}(\Im_{\mathrm{acc-res}}(p_{\mathrm{acc}})) = \emptyset$,
8. $\Im_{\mathrm{snd-get}} : P_{\mathrm{snd}} \longrightarrow P_{\mathrm{get}}$ *is a bijection such that:*
   a) $\forall (p_{\mathrm{snd}}, \Im_{\mathrm{snd-get}}(p_{\mathrm{snd}})) \in P_{\mathrm{snd}} \times P_{\mathrm{get}} \,.\, \forall t_n \in {}^{\bullet}(\Im_{\mathrm{snd-get}}(p_{\mathrm{snd}})) \,.$
   $\exists t_1, \ldots, t_{n-1} \in T \,.\, t_1 \in p_{\mathrm{snd}}{}^{\bullet} \wedge \forall i \in \{1, \ldots, n-1\} \,.\, t_i{}^{\bullet} \cap {}^{\bullet}t_{i+1} \neq \emptyset$,
   b) $\forall (p_{\mathrm{snd}}, \Im_{\mathrm{snd-get}}(p_{\mathrm{snd}})) \in P_{\mathrm{snd}} \times P_{\mathrm{get}} \,.\, \forall t_1 \in p_{\mathrm{snd}}{}^{\bullet} \,.$
   $\exists t_1, \ldots, t_n \in T \,.\, t_n \in (\Im_{\mathrm{snd-get}}(p_{\mathrm{snd}}))^{\bullet} \wedge \forall i \in \{1, \ldots, n-1\} \,.\, t_i{}^{\bullet} \cap {}^{\bullet}t_{i+1} \neq \emptyset$,
   c) $\forall (p_{\mathrm{snd}}, \Im_{\mathrm{snd-get}}(p_{\mathrm{snd}})) \in P_{\mathrm{snd}} \times P_{\mathrm{get}} \,.\, p_{\mathrm{snd}}{}^{\bullet} \cap {}^{\bullet}(\Im_{\mathrm{snd-get}}(p_{\mathrm{snd}})) = \emptyset$.

An OF-CPN is a modular coloured Petri net with interface places which can contain requests and responses (both incoming and outgoing). A source place in the interface (i.e. p such that $^\bullet p = \emptyset$) contains tokens coming from the environment. A sink place (i.e. p such that $p^\bullet = \emptyset$) in the interface contains tokens issued to the environment. The bijections $\Im_{\mathrm{acc-res}}$ and $\Im_{\mathrm{snd-get}}$ in Definition 10.3.1 establish the correspondence between the requests and the responses issued for them. These correspondences are important at the stage of verification to check the reliability of the components.

The places of an OF-CPN can be partitioned into interface places ($P_{\mathrm{acc}} \cup P_{\mathrm{res}} \cup P_{\mathrm{snd}} \cup P_{\mathrm{get}}$) and internal places (compare Section 9.2). The same holds for transitions. The interface transitions are the elements of ${P_{\mathrm{acc}}}^\bullet \cup {}^\bullet P_{\mathrm{res}} \cup {P_{\mathrm{snd}}}^\bullet \cup {}^\bullet P_{\mathrm{get}}$. These partitions are based on the *encapsulation principle* well-known in object orientation. They allow us to separate the observable behaviour (firing interface transitions) of a component from the internal behaviour (firing internal transitions).

The OF-CPN model is quite simple because it takes into account only the encapsulation and modularity concepts from object orientation. Now our modular net model is defined, let us consider the transformation from the component model OF-Class to this one.

Two interacting OF-CPNs O_1 and O_2 can be composed by merging the right interface places. If O_1 is the server and O_2 the client, the merging is defined by a morphism[1] as follows.

Definition 10.3.2 (Composition of OF-CPNs). *Two OF-CPNs O_1 and O_2 can be combined if there is a mapping $\zeta : P_{\mathrm{snd}}(O_1) \cup P_{\mathrm{get}}(O_1) \longrightarrow P_{\mathrm{acc}}(O_2) \cup P_{\mathrm{get}}(O_2)$ verifying:*

1. *$\zeta(P_{\mathrm{snd}}(O_1)) \subset P_{\mathrm{acc}}(O_2)$ and $\zeta(P_{\mathrm{get}}(O_1)) \subset P_{\mathrm{res}}(O_2)$,*
2. *$\forall p \in P_{\mathrm{snd}}(O_1)$, if $\zeta(p)$ is defined then $\zeta(\Im^{O_1}_{\mathrm{snd-get}}(p))$ is also defined and $\zeta(\Im^{O_1}_{\mathrm{snd-get}}(p)) = \Im^{O_2}_{\mathrm{acc-res}}(\zeta(p))$,*
3. *$\forall p \in P_{\mathrm{snd}}(O_1) \cup P_{\mathrm{get}}(O_1) . \mathrm{Dom}(\zeta(p)) = \mathrm{Dom}(p)$.*

Such a mapping ζ is also sometimes denoted $\zeta_{O_1 \to O_2}$ in the remainder of this chapter.

From OF-Class to OF-CPN. The transformation method is based on the one presented in [Hei92]. The basic principles are :

- Resources, variables, and parameters are mapped to places whose colour domains are determined by the type of the corresponding items.
- Elementary actions (e.g. assigning a value to a variable) are transformed into transitions while elaborate ones (e.g. loops and conditionals) correspond to sets of transitions with sequencing places (see below).
- Arcs are added to model the use of resources (resp. variables and parameters) by actions, and the arc-expressions model the effects of such uses.
- Special places called *sequencing places* are added to model the control flow in the component. If an action t_1 is sequentially followed by t_2, then there exists a *sequencing place* p such that $p \in {t_1}^\bullet \cap {}^\bullet t_2$.

[1] Each place is merged with its image by ϕ.

Verification and Validation. OF-CPN are modular coloured Petri nets which can be combined by place fusion according to offer/require service relations. This combination allows the system designer to build and validate/verify subsystems (sets of interacting components). Once a set of OF-CPNs is combined into a composite one, the interface places of the composite can be marked by tokens modelling requests from the environment. A composite (or even a simple) OF-CPN with such an abstraction of its environment is used for verification techniques such as reachability graph construction and structural invariants computation. The graph supports verification of safety and reliability properties.

Based on the (direct) executability of nets, models can be validated by simulation or animation. In this sense, validation means model execution and allows us to validate *a priori* some specific scenarios supplied by application-domain experts.

10.3.4 Conclusion

Integration of objects and Petri nets has been tackled in many ways. In all of them the two main achievements are:

- Enhancement of Petri nets by the addition of structuring facilities. The Petri net formalism is therefore more relevant for the modelling of large concurrent/distributed systems.
- Enhancement of object-oriented concepts by the formalisation of interactions and control structures. The object-oriented technology is therefore enhanced to support at least simulation and possibly verification for the modelling of large concurrent/distributed systems.

Petri nets provide a valid formalism for enhancing object-oriented concepts with formal semantics and/or validation and verification facilities. They have semantics able to support, at least partially, the elaborate constructions one can find in the software life-cycle (analysis, design, etc). Moreover, they provide validation, and verification by simulation or animation facilities which is *sine qua non* for any industrial-strength method ([MS94]).

11. Case Studies*

This chapter presents three case studies on developing a mutual exclusion (ME) algorithm for a ring architecture, one for each method described in the previous chapter. This section gives a brief introduction to the problem and a classification of the solutions within the ME-area.

The chosen algorithm does not work very efficiently. It is a compromise between two divergent requirements. On the one hand, the algorithm should be advanced enough to explain the features of the methods, but on the other hand it should be easily understood by readers who have no experience in using the methods. Improvements in efficiency would have caused overly complex structures.

The Problem. The access of n computers to a common resource (e.g. a printer) has to be organised. The printer can serve only one computer at a time. In addition to the mutual exclusion property it is required that a requesting computer must be eventually served, i.e. a job must be eventually printed.

An abstract representation of the problem is: n client units compete for a single (permission) token. The computer is represented in an abstract way by a client unit consisting of three states *client not interested* (doing some local work), *client interested* (waiting for the resource) and *client in its critical section*. The details will be given in each case study. In the following figures a client will be abstracted to one transition.

Token-based ME-algorithms contain a single token. A client unit has to possess the token if it wants to enter the critical section.

In a non-distributed system the problem can be solved as in Figure 11.1. There is one resource place (*res*) with a single token. The firing of a client transition represents the compound action of taking the token, using the critical section, and putting the token back into the resource place. If resource allocation is done by a fair strategy such as first-come-first-served, the model fulfils the specification. The place in the middle represents a kind of semaphore.

* Authors: R. Mackenthun (Section 1), M. Voorhoeve (Section 2), A. Diagne (Section 3)

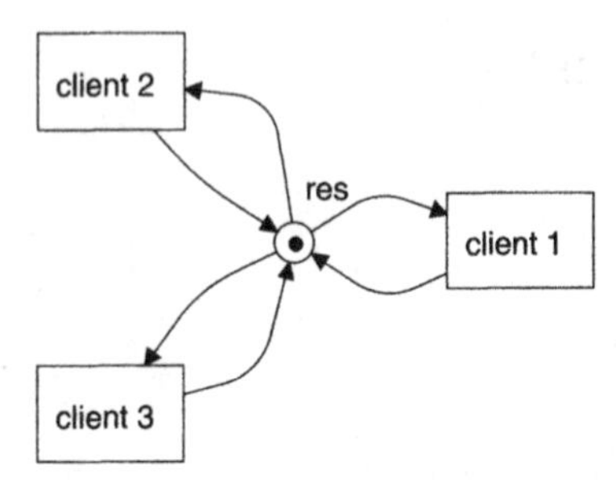

Fig. 11.1. A simple local solution

Since only one transition can fire at a time the mutual-exclusion property holds. The dynamic property is true if some fairness assumptions are added.

This first solution cannot be used for distributed systems. In distributed systems clients can communicate only via message-passing through channels. The important property of those channels is that messages can be sent only via a channel from one location and a message can be received only at one other location. If all three clients of Figure 11.1 are located at different sites, place *res* cannot be used as a channel. Figure 11.2 gives a solution where the clients can be distributed. It is a central-server model. The server is represented by the net in the dashed box in the middle of the figure. It decides which client is served next and sends the token to that client. The server and individual clients can each have their own location. All channels are between the server location and a client location. If the server allocates the resource by a fair strategy the dynamic properties are also fulfilled.

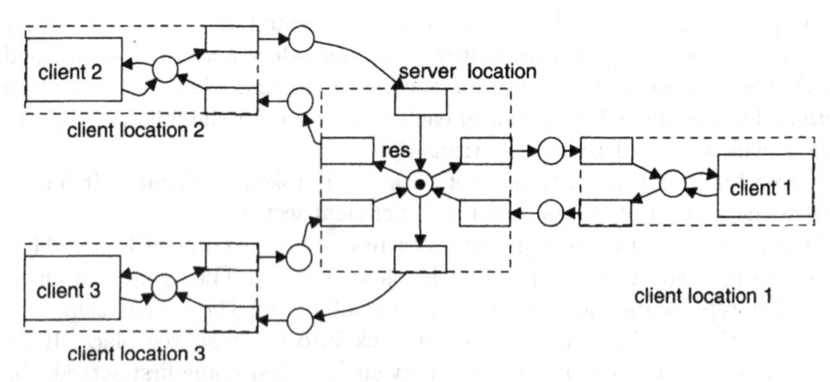

Fig. 11.2. The central-server solution

Even though this solution can be implemented in a distributed system it would not be called a distributed algorithm. In distributed algorithms all competing units have a similar behaviour or at least there is no unit that determines the behaviour of the whole system. Although no formal definition of that class of algorithms is presented here, it is obvious that the central-server model does not belong to this class since the server does control the behaviour of the whole system.

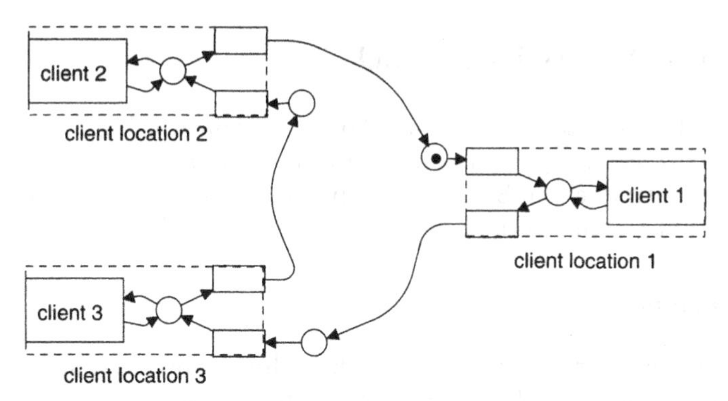

Fig. 11.3. The token-ring solution

An example for a distributed algorithm is the token-ring model in Figure 11.3. All units have the same structure. They are located on a logical ring. The token moves around the ring. If each client eventually gives the token to its neighbour and each requesting client eventually takes the token from the ring if it is coming around, the dynamic properties are ensured.

The Solution for the Case Studies. One could object that the described algorithm is a kind of distributed busy waiting. The token moves around the ring even if no client is interested. To avoid this behaviour a request mechanism is included.

The informal description of a possible solution is:

- A *functional unit* (FU) has used the token and still possesses it:
  - If the FU wants to re-enter the critical section it sends the request onto the ring.
  - If the FU gets a request it sends the token to its neighbour.
- An FU uses the token:
  - All incoming requests are delayed.
- An FU has no token and is not interested:
  - All incoming requests are relayed.
  - If the token arrives it is relayed.
  - If the FU wants to enter the critical section it sends a request.
- An FU has no token and is interested:
  - If the token arrives it is used.
  - All incoming requests are relayed.

This algorithm has sufficient complexity to be used for the case studies in this chapter. The reader should not be surprised that the solutions of all the case studies are slightly different. The representation of the algorithms is influenced by the method used to construct them. Nevertheless, the basic components are the same in all the case studies.

11.1 State-Oriented Approach

In Section 10.1 the state-oriented method is explained using a simple ME-example. This section presents an algorithm fulfilling the requirements as given in the introduction. Since all ME-problems share some aspects, some of the results from the method subsection will be reused.

11.1.1 Specification

The specification for the ME-problem has already been given in Section 10.1.1. All the results of that subsection can be reused.

11.1.2 Design

Step 0: Designing the Components and the Communication Structure. The structure of the components and the communication structure is shown for two competitors in Figure 11.4.

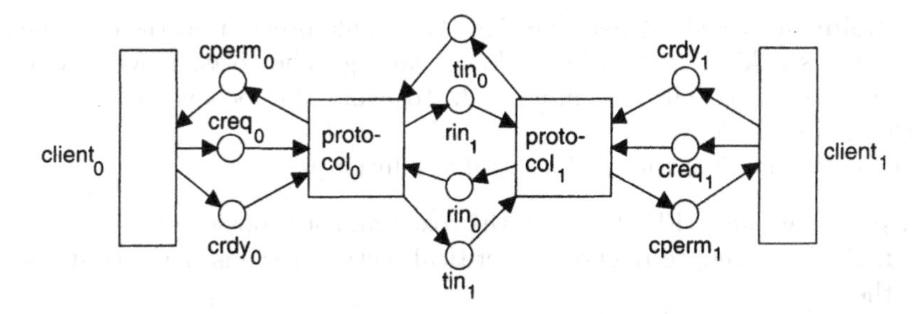

Fig. 11.4. The components and the communication structure

Only the channels for the requests (*rin*, *request in*) have been added to the structure in Figure 10.3. While the token channels go from protocols with lower numbers to protocols with higher numbers, the request channels go in the reverse direction.

Step 1: Designing Safety Properties with Place Invariants. The first five invariants are the same as in the method section:

$$\forall i \in \{0,\ldots,n-1\} .\ \mathbf{m}(cni_i) + \mathbf{m}(cint_i) + \mathbf{m}(ccs_i) = 1 \quad (11.1)$$

$$\forall i \in \{0,\ldots,n-1\} .\ \mathbf{m}(ni_i) + \mathbf{m}(wt_i) + \mathbf{m}(cs_i) = 1 \quad (11.2)$$

$$\sum_{i\in\{0,\dots,n-1\}}(\mathbf{m}(cs_i)+\mathbf{m}(ut_i)+\mathbf{m}(tin_i))=1 \quad (11.3)$$

$$\forall i\in\{0,\dots,n-1\}\,.\ \mathbf{m}(crdy_i)+\mathbf{m}(cperm_i)+\mathbf{m}(ccs_i)=\mathbf{m}(cs_i) \quad (11.4)$$

$$\forall i\in\{0,\dots,n-1\}\,.\ \mathbf{m}(cni_i)+\mathbf{m}(creq_i)=\mathbf{m}(crdy_i)+\mathbf{m}(ni_i) \quad (11.5)$$

The sixth invariant affects the request mechanism. To avoid unnecessary movement of the permission token, it should never leave a competitor if there is no request. This dynamic property can be presented as follows: On the ring there are as many request and permission tokens as competitors waiting for a token. If the permission token is possessed by a competitor then there are as many request tokens on the ring as waiting competitors. This ensures that there are always tokens moving as long as at least one competitor is waiting, and that no token is moving if no client is interested.

$$\sum_{i\in\{0,\dots,n-1\}}\mathbf{m}(wt_i)=\sum_{i\in\{1,\dots,4\}}(\mathbf{m}(rin_i)+\mathbf{m}(tin_i)) \quad (11.6)$$

The set of places is

$$P=\bigcup_{i\in\{0,\dots,n-1\}}\{cni_i, cint_i, ccs_i, creq_i, crdy_i, cperm_i, cs_i, ni_i, wt_i, ut_i, tin_i, rin_i\}.$$

The set of p-flows is $\Psi=\bigcup_{i\in\{0,\dots,n-1\}}\{\Psi_{1,i},\Psi_{2,i},\Psi_3,\Psi_{4,i},\Psi_{5,i},\Psi_6\}$ with

$$\begin{array}{lll}\Psi_{1,i}: & \forall p\in\{cni_i, cint_i, ccs_i\}\,. & \Psi_{1,i}[p]=1, \quad \text{(inv. 11.1)}\\ & \forall p\in P\setminus\{cni_i, cint_i, ccs_i\}\,. & \Psi_{1,i}[p]=0\end{array}$$

$$\begin{array}{lll}\Psi_{2,i}: & \forall p\in\{ni_i, wt_i, cs_i\}\,. & \Psi_{2,i}[p]=1, \quad \text{(inv. 11.2)}\\ & \forall p\in P\setminus\{ni_i, wt_i, cs_i\}\,. & \Psi_{2,i}[p]=0\end{array}$$

$$\begin{array}{lll}\Psi_3: & \forall p\in\bigcup_{i\in\{0,\dots,n-1\}}\{tin_i, ut_i, cs_i\}\,. & \Psi_3[p]=1, \quad \text{(inv. 11.3)}\\ & \forall p\in P\setminus\bigcup_{i\in\{0,\dots,n-1\}}\{tin_i, ut_i, cs_i\}\,. & \Psi_3[p]=0\end{array}$$

$$\begin{array}{lll}\Psi_{4,i}: & \forall p\in\{crdy_i, cperm_i, ccs_i\}\,. & \Psi_{4,i}[p]=1, \quad \text{(inv. 11.4)}\\ & \forall p\in\{cs_i\}\,. & \Psi_{4,i}[p]=-1,\\ & \forall p\in P\setminus\{crdy_i, cperm_i, ccs_i, cs_i\}\,. & \Psi_{4,i}[p]=0\end{array}$$

$$\begin{array}{lll}\Psi_{5,i}: & \forall p\in\{cni_i, creq_i\}\,. & \Psi_{5,i}[p]=1, \quad \text{(inv. 11.5)}\\ & \forall p\in\{crdy_i, ni_i\}\,. & \Psi_{5,i}[p]=-1,\\ & \forall p\in P\setminus\{cni_i, creq_i, crdy_i, ni_i\}\,. & \Psi_{5,i}[p]=0\end{array}$$

$$\begin{array}{lll}\Psi_6: & \forall p\in\bigcup_{i\in\{0,\dots,n-1\}}\{rin_i, tin_i\}\,. & \Psi_6[p]=1, \quad \text{(inv. 11.6)}\\ & \forall p\in\bigcup_{i\in\{0,\dots,n-1\}}\{wt_i\}\,. & \Psi_6[p]=-1,\\ & \forall p\in P\setminus\bigcup_{i\in\{0,\dots,n-1\}}\{rin_i, tin_i, wt_i\}\,. & \Psi_6[p]=0\end{array}$$

Step 2: Assigning the Places to the Components.

$$\forall i \in \{0,\ldots,n-1\}\,.\,\delta_p(p) = \begin{cases} client_i & \text{if } p \in \{cni_i, cint_i, ccs_i, crdy_i, creq_i\} \\ protocol_i & \text{if } p \in \{cs_i, ni_i, wt_i, ut_i, cperm_i, tin_{i\ominus 1}, rin_{i\oplus 1}\} \end{cases}$$

$$\forall i \in \{0,\ldots,n-1\}\,.\,\delta_c(p) = \begin{cases} client_i & \text{if } p \in \{cni_i, cint_i, ccs_i, cperm_i\} \\ protocol_i & \text{if } p \in \{cs_i, ni_i, wt_i, ut_i, crdy_i, creq_i, tin_i, rin_i\} \end{cases}$$

Step 3: Designing Further Restrictions. The set T_{min} is chosen as

$$T_{min} = \bigcup_{i \in \{0,\ldots,n-1\}} \{t_{1,i}, t_{2,i}\}$$

where $\mathbf{Pre}[t_{1,i}, wt_i]$, $\mathbf{Post}[t_{1,i}, ni_i]$, $\mathbf{Pre}[t_{2,i}, cs_i]$, and $\mathbf{Post}[t_{2,i}, wt_i]$ are all equal to 1, and all other entries in the incidence vectors are zero. So, the local protocol may not change its internal state directly from wt_i to ni_i or from cs_i to wt_i.

Step 4: Constructing an Executable Net Model. The result of the construction is shown in Figure 11.5 (already including Step 5).

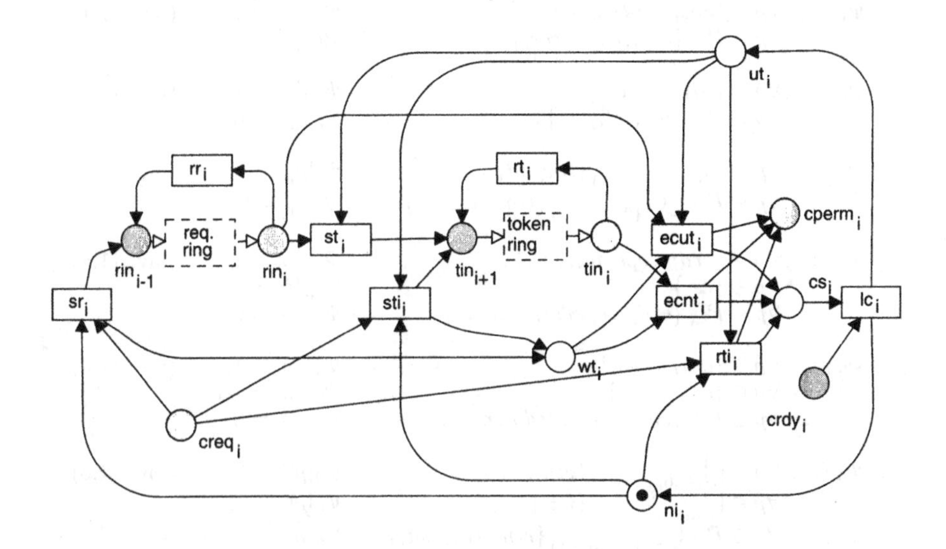

Fig. 11.5. Request token ring – first version

Step 5: Identifying the Meaning of the Transitions. The protocol includes the required solution. An interested competing unit sends a request(*sr* for *send request*). If it possesses the token, the token will be sent on an incoming request (*st* for *send token*). The critical section will be entered if a token arrives (*ecnt* for *enter critical section with a new token*). Not-interested competing units must be able to relay incoming tokens (*rt* for *relay token*) and units that do not possess the token must be able to relay incoming requests (*rr* for *relay request*).

Additionally there are some short cuts in the net. A token can be reused if it is still possessed by a competitor which gets interested again (*rti* for *reuse token immediately*). Since requests are used to activate the token, a unit that possesses the token can immediately send the token if it gets interested (*sti* for *send token immediately*). It can also enter the critical section if a request comes in (*ecut* for *enter critical section with a used token*). This transition can occur if a unit suddenly decides to reuse the token but has already sent a request.

Step 6: Proving Dynamic Properties. The dynamic property is proved by the graph in Figure 11.8. It contains some extended arcs, labelled by ($\Rightarrow$). The semantics is that the system being in one of the states represented by the starting node is already in one of the states represented by the ending nodes or it will eventually reach one of those states.

Step 7: Removing Optional Transitions. The proof graph shows that all short-cut transitions can be removed without violating the dynamic properties. The resulting proof graph is not shown here. The net is given in Figure 11.6.

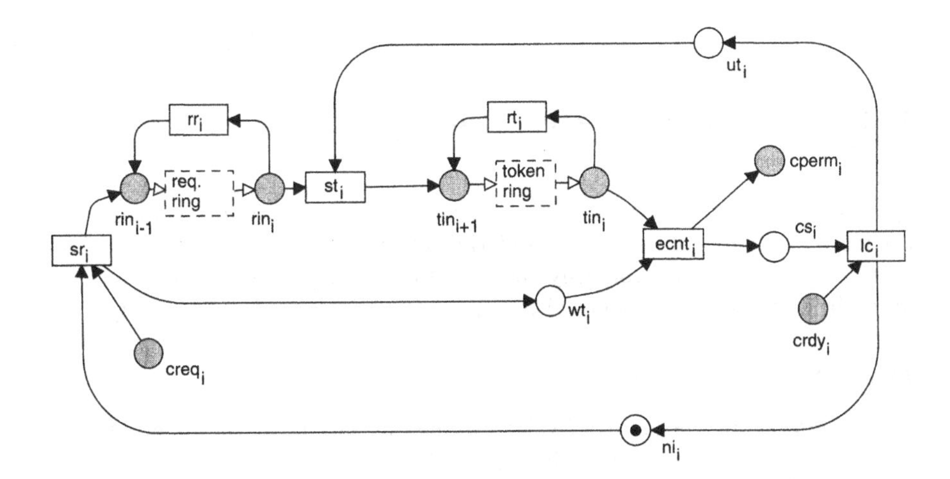

Fig. 11.6. Request token ring – intermediate version

11.1.3 Implementation

At the implementation stage, transitions *rr* and *rt* have to have a lower priority than the transitions *st* and *ecnt*. This is again achieved by adding loops (Figure 11.7). The proof graph for the final version is given in Figure 11.9.

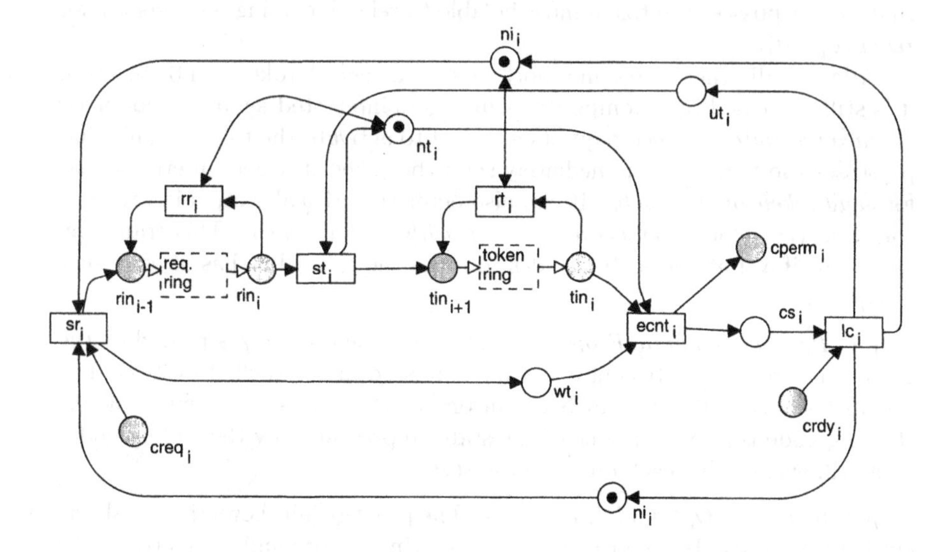

Fig. 11.7. Request token ring – final version

11.2 Event-Oriented Approach

In this section, we shall indicate how the request ring node can be modelled by an event-oriented approach. The interface between the nodes is modelled asynchronously by place fusion, and the interface between a node and its client is modelled synchronously by transition fusion. The section is terminated by verifying the models with respect to their desired functionality.

11.2.1 Modelling a Node

The modelling process starts by identifying the events in the external behaviour of a node. The events that are exchanged between the node and its client are the following firings of transitions:

`creq`	requesting enty into the critical section,
`cperm`	permitting entry into the critical section,
`crdy`	terminating the stay in the critical section.

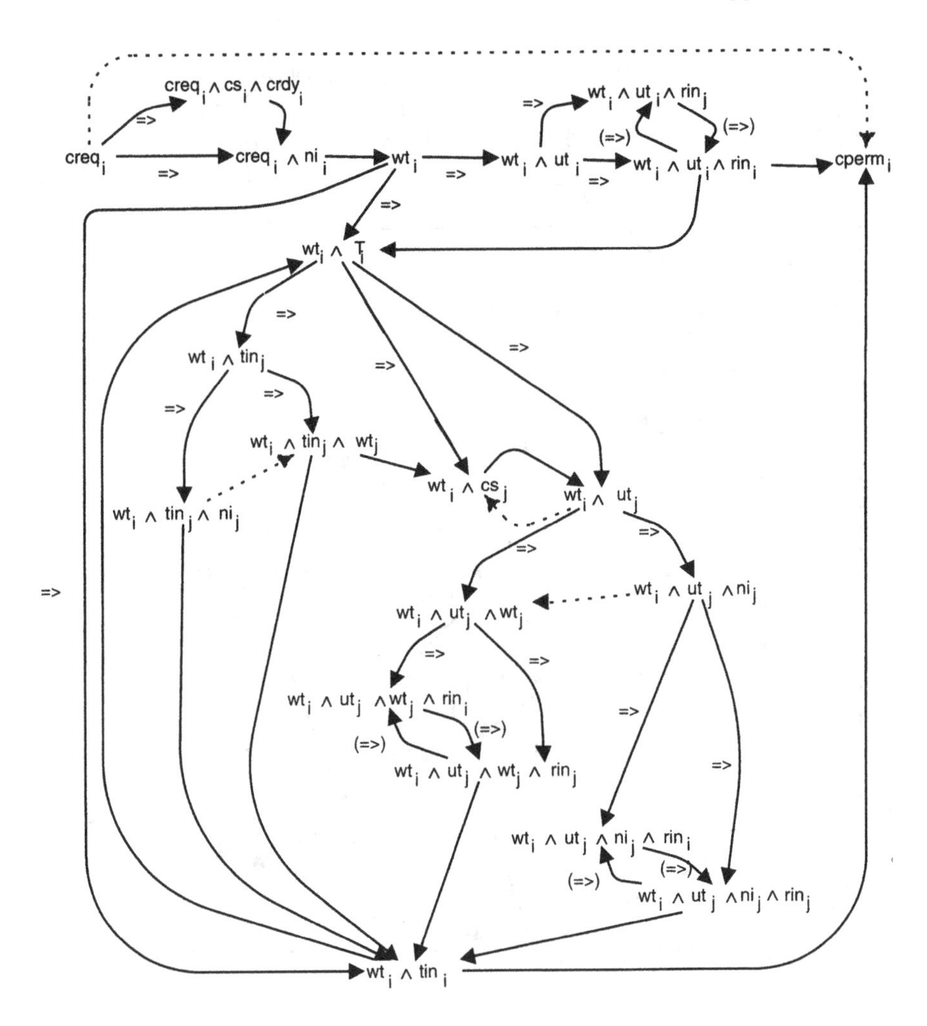

Fig. 11.8. Proof graph – first version

The protocol that the node and client must follow is that the client, when wanting to enter the critical section, issues a request and waits for permission. If the permission arrives, the critical section is entered and the actions therein are executed by the client. Upon termination of these actions, the critical section is left. A new request can then be issued.

The events of the node-node communication to ensure mutual exclusion are the following:

`treq?`	receiving request for token,
`treq!`	sending request for token,
`tok?`	receiving token,
`tok!`	sending token.

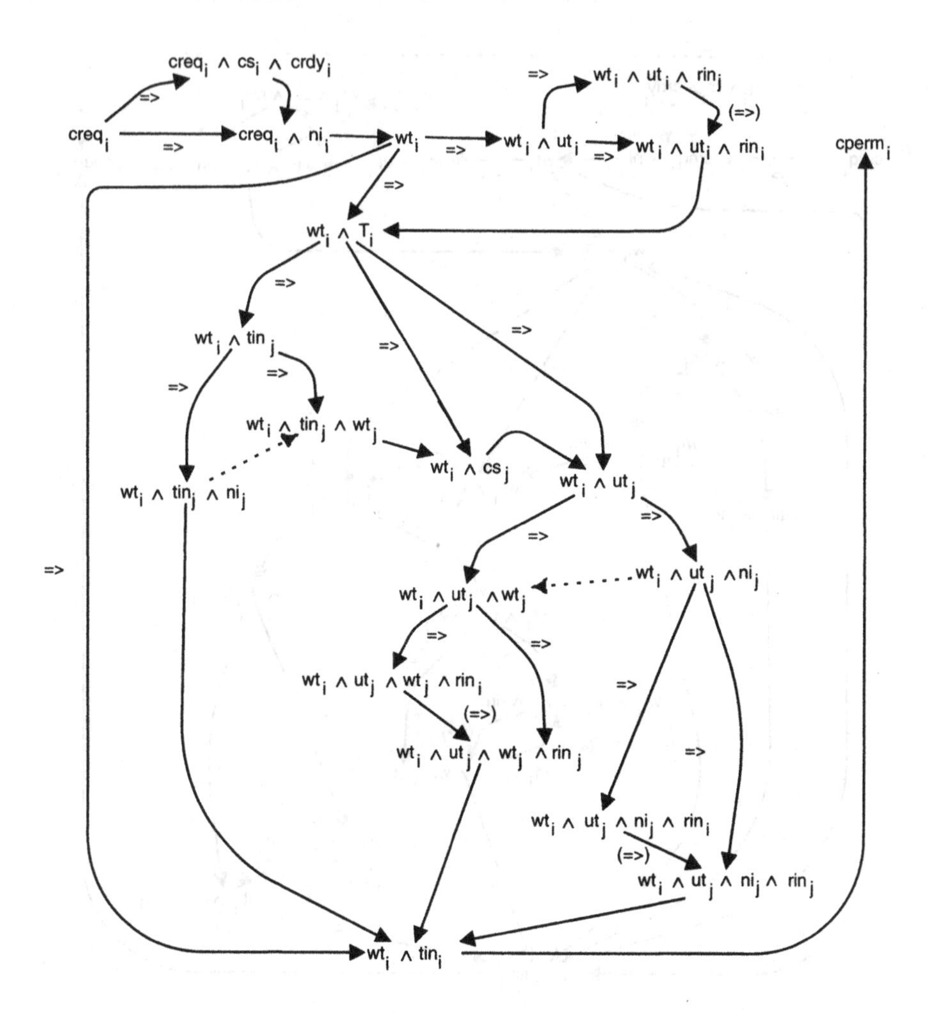

Fig. 11.9. Proof graph – final version

Let R be the protocol of a node that just has left the critical section, and thus possesses the token. Clearly, the `creq` and `treq!` events can be synchronised, as can the `cperm` and `tok?` events. Furthermore, after `crdy`, a `tok!` event must occur in order not to monopolise the token, and this `tok!` event can be synchronised with a `treq?` event. So the protocol R can be modelled by the expression

$$((\texttt{creq}|\texttt{treq!}||\texttt{treq?}|\texttt{tok!}).\texttt{tok?}|\texttt{cperm.crdy})^*\delta$$

By the algorithm for constructing a net from a given protocol we obtain the net of Figure 11.10. The net contains the following internal places:

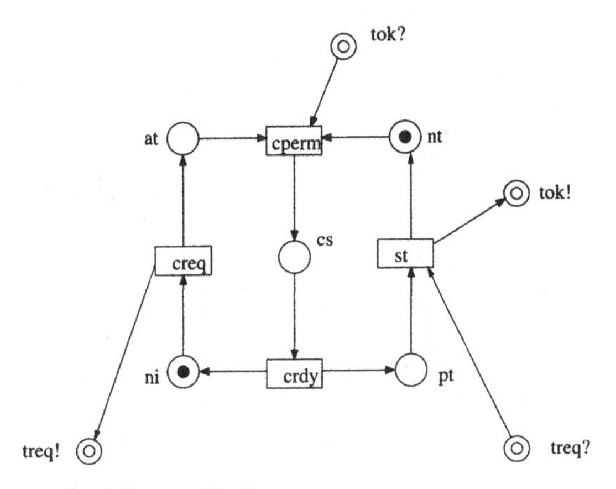

Fig. 11.10. Protocol without relaying

ni	not interested in entering critical section,
at	awaiting the token,
pt	having the token,
nt	not having the token,
cs	within critical section.

The above definition of R disregards the need to relay incoming requests and tokens. The relaying of tokens is performed when the token is not wanted by the node, so `ni` must be marked. The relaying of requests is performed when the token is no longer possessed by the node, so `nt` must be marked. This leads to additions to Figure 11.10, resulting in Figure 11.11.

The net contains the following internal transitions:

rt	relay token,
rr	relay request,
st	send token.

The ring is connected by means of place fusion, and each node is connected to its own client by transition fusion as indicated in Figure 11.12. The `treq!` pin of the k-th node and the `treq?` pin of the $(k+1)$-th node are connected to the place r^k. Similarly, the `tok!` pin of the k-th node and the `tok?` pin of the $(k+1)$-th node are connected to the place t^k. The given node is connected with its `treq!` pin to r^1, its `treq?` pin to r^n, its `tok!` pin to t^1 and its `tok?` pin to t^n. These places are initially not marked. We assume that initially one node has just left its critical section and thus is holding the token. The other nodes are in their rest state (not interested and without token).

The protocol of a client has been treated already. A client may either issue a request by synchronising with the `crq` action of his server or "fall asleep". He does not have to become interested, but if he does, he must synchronise with the permission and ready actions, completing the cycle.

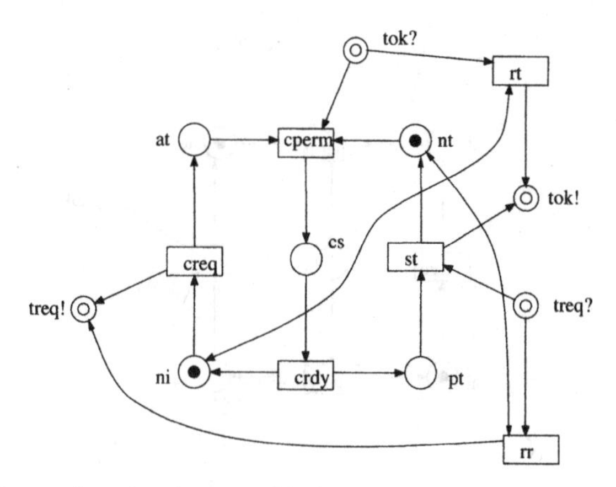

Fig. 11.11. Protocol with relaying added

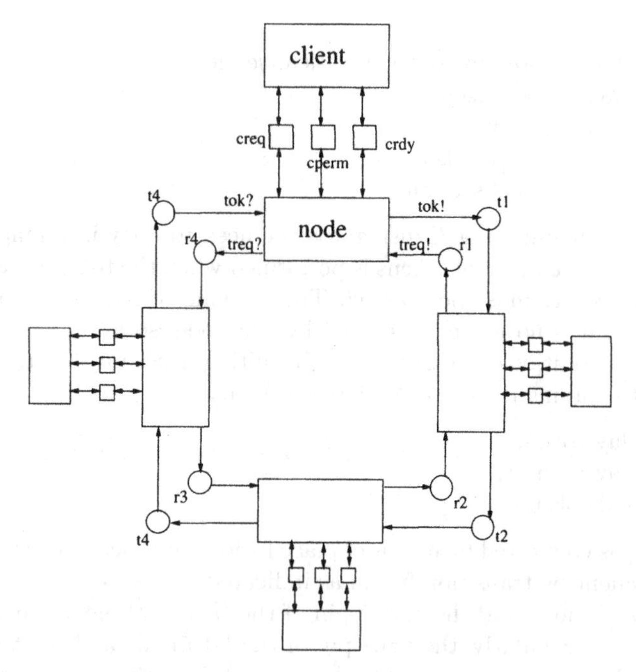

Fig. 11.12. Request token ring

The protocol expression thus becomes $(\mathbf{zzz}.\delta + \mathbf{crq.cp.crd})^*\delta$, giving the net in Figure 11.13.

11.2.2 Verification

The verification of the request ring protocol is done in two steps. First, we show that only one client can be in the critical section. This can be done

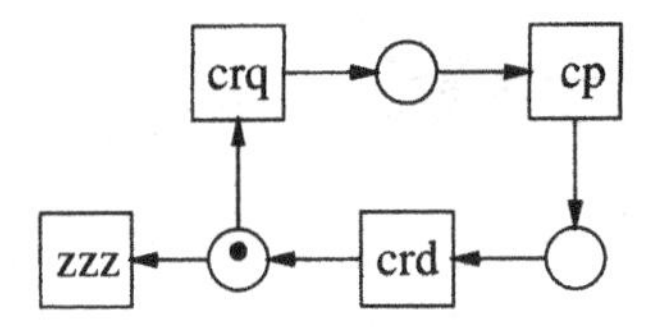

Fig. 11.13. Client protocol

by directly establishing a place invariant for the net; here we choose to first simplify the net modulo branching bisimilarity. In order to verify that in a ring of n nodes all clients are guaranteed to have access to the critical section, it is proved that a given node can accommodate the needs of its client.

In Figure 11.14, two neighbouring nodes from Figure 11.12 with their clients have been expanded with trivial simplifications. Our first abstraction step consists of abstracting from the relaying and token release events.

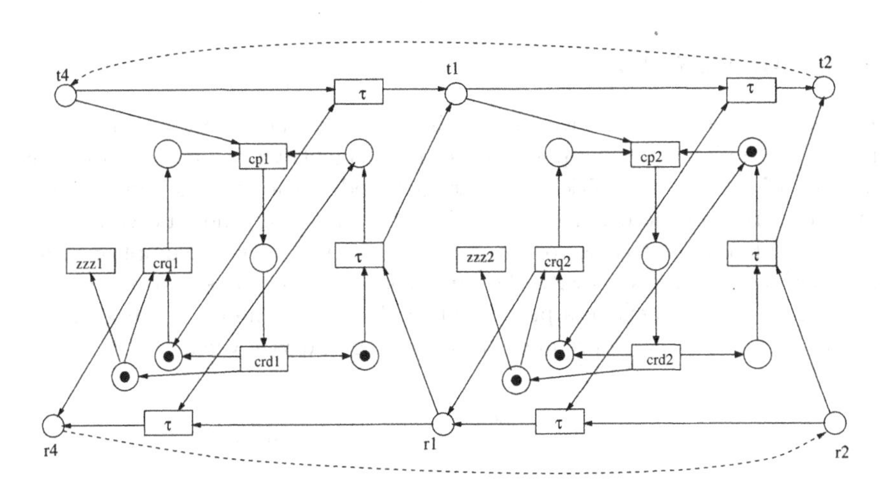

Fig. 11.14. First abstraction step

It turns out that the token release and the request relaying events are invisible. Removing them from the net, we obtain Figure 11.15. However, in doing so, an extra place `rp` (requests pending) has to be added. Its marking reflects the number of nodes that are interested in entering the critical section. Only if this place is marked will tokens be relayed. From the figure, it can be inferred that every `cp` event must be followed by a corresponding `crd` event before any new `cp` event can occur. Thus the critical section can be entered by at most one client at a time. This follows from a place invariant in both nets stating that there is only one "token" token in the ring.

Next, we abstract from all but one client's actions. It turns out that all silent events become invisible and the ring poses no restrictions upon the

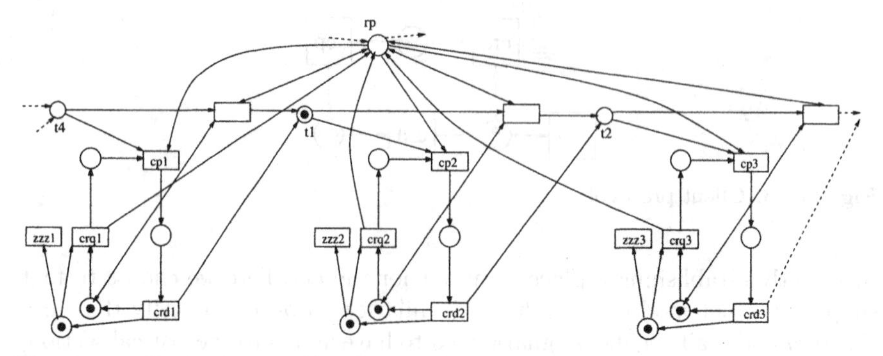

Fig. 11.15. Reduction modulo branching bisimilarity

client, since it becomes branching bisimilar to Figure 11.13. Thus every client is able to enter the critical section after a request.

11.2.3 Adding Colour

By adding colour to the above model it is possible to model any size of request ring. Each node in the ring possesses a unique identifying number n, ranging from 0 to $N-1$. The transitions are parametrised by the node identifier. These numbers are also the colour sets of the places. Node n sends its requests to node $n-1$ and its tokens to the node $n+1$. All other consumptions and productions are of tokens with the same colour as the transition parameter. Addition and subtraction is performed modulo N, so node $N-1$ sends its tokens to and receives its requests from node 0 and vice versa.

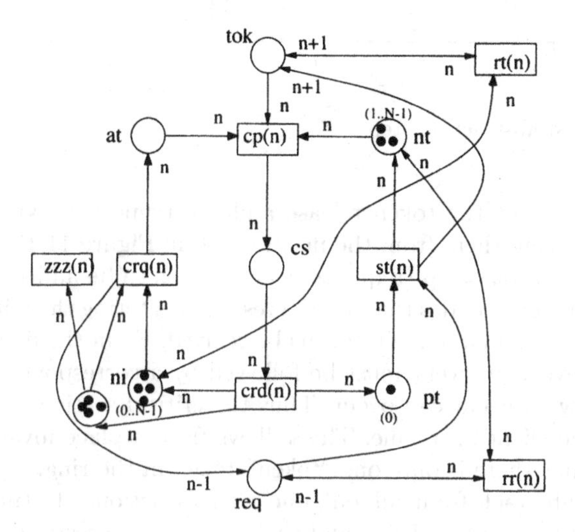

Fig. 11.16. Coloured request ring

The resulting coloured net is displayed in Figure 11.16. The state where node 0 possesses the token and no node is interested is depicted. Abstracting from the relaying and token release events and then removing invisible events gives Figure 11.17. The place **rp** is "uncoloured"; its marking indicates the number of waiting interested nodes. The properties of this simpler net can be studied.

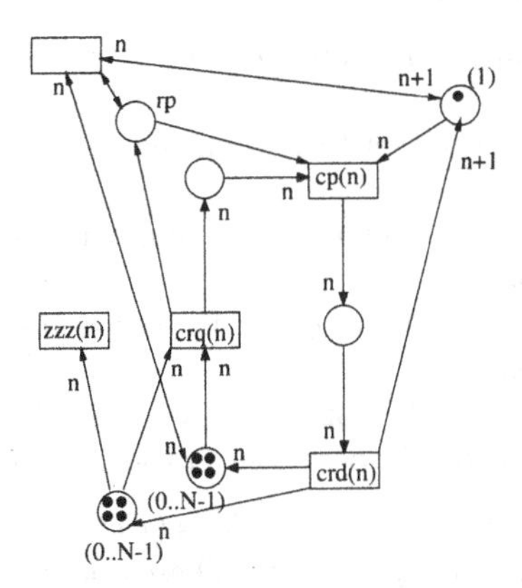

Fig. 11.17. Abstraction of Figure 11.16

11.3 Object-Oriented Approach

This section is dedicated to the specification of the mutual exclusion token ring in an object-oriented way. The object-oriented method focuses on how the events are implemented on each node rather than on the way they must happen (protocol-oriented approach) or their causal relations (temporal-logic-oriented approach). Events are implemented as modifications of the states of given nodes. These modifications influence the way later events are processed.

We therefore consider a class named *NodeCoordinator* whose instances relay tokens and requests and allow other objects located at the same node to enter and exit the critical section. An instance of the NodeCoordinator has a set of resources describing its current state. Events are processed according to the current state and can eventually modify it.

The NodeCoordinator is described using the OF-Class formalism described in Chapter 10. A Petri net is then synthesised from this description. This net is the entry point of the activities relevant to the Petri net formalism.

11.3.1 Structure of the NodeCoordinator

On each node, we have an instance of the NodeCoordinator OF-Class to coordinate the objects for entering and exiting the critical section. The NodeCoordinator instances communicate by means of signals sent on the corresponding signal interfaces. These signals are relayed or not according to the current state of the instance, e.g. an incoming request signal is relayed if the token is not present and the site is not willing to enter the critical section. If the site wants to enter the critical section, the incoming request is delayed. If the site is idle and holds the token, the request is discarded and the token is sent. The state of the node at a given moment is determined by its resources:

- *TokenPresent* is an integer resource set to 1 when the node holds the token and to 0 otherwise.
- *InCS* is an integer resource set to 1 when the node is in the critical section and to 0 otherwise.
- *CSWanted* is an integer resource for the number of local pending requests for the critical section.
- *PendingRequest* is an integer resource for the non-local pending requests for the critical section.
- *GoCS* is a semaphore for the control of the critical section access on each node. The first call to the P operation on the semaphore is blocking.

The NodeCoordinator has two operations (*EnterCS* and *ExitCS*) and two triggers (*RelayToken* and *RelayRequest*). These operations are called by local objects requesting or releasing the critical section. Triggers run automatically and forever to handle incoming tokens and requests.

The NodeCoordinator's main role is to send the token on demand when it is present, and to relay the requests and the token when the node is idle. Incoming requests are delayed when the node is in its critical section or waiting for the token. The NodeCoordinator also offers a service for other objects on the same node to enter and exit the critical section. When a request for the critical section occurs on a node, the NodeCoordinator sends a request and stops relaying incoming requests. It waits for the token in order to enter the critical section. When the node exits the critical section several cases are possible:

1. There is only one pending request on the node; it is discarded and subsequently the token is sent.
2. There are many pending requests on the node; they are all relayed except one for which the token is sent.
3. There is no pending request; the token is kept until a new request comes in; then the request is not relayed but the token is sent.

To ensure fairness, the protocol prevents a node from starving the others by using the token more than once each time it is present. A node, even when holding the token, has to send a request before each critical section access. If the token is present and all sites are idle, the request comes back after one loop and the token goes for a loop because requests are undistinguishable.

11.3.2 The NodeCoordinator in OF-Class Formalism

We now give a description of the NodeCoordinator in the OF-Class Formalism.

```
NodeCoordinator ISA OFClass
  INTERFACE
      EXPORTS {
        SERVICE AccessCS
            OPERATIONS { void : EnterCS(void),void : ExitCS(void) }
            AUTOMATA AccessCS is { EnterCS, ExitCS }
            INVOCATION-MODE synchronous }
      PROVIDES {
        SIGNALS {
           OutRequest TRANSMISSION synchronous,
              /* signal interface for incoming requests */
           OutToken TRANSMISSION synchronous
              /* signal interface for incoming token */}}
      EXPECTS {
        SIGNALS {
           InRequest FROM NodeCoordinator,
              /* signal interface for outgoing requests */
           InToken FROM NodeCoordinator
              /* signal interface for outgoing token */}}
  STRUCTURE
    RESOURCES {
      int : TokenPresent default false duplicated,
           /* 1 if the node holds the token */
      int : InCS default false duplicated,
           /* 1 if the node is in critical section */
      int : CSWanted default 0 duplicated,
           /* number of critical section requests */
      int : PendingRequest default 0 duplicated,
           /* number of pending requests to be relayed later */
      semaphore : GoCS init 0 duplicated
           /* permission semaphore for the critical section */ }
    OPERATIONS /* for critical section request and release on a node */
    {
      void : EnterCS(void)
      {
        CSWanted++; /* increasing the requests for a critical section */
        signal(outRequest); /* sending a request to other nodes */
        P(GoCS); /* waiting for a critical section permission */
        InCS := 1; /* the node is in critical section */
        return;
      }
      void : ExitCS(void)
      {
        CSWanted-; /* decreasing the requests for a critical section */
        InCS := 0; /* the node is no longer in critical section */
        if( PendingRequest != 0 ) then
          TokenPresent := 0;
          PendingRequest-; /* one pending request discarded */
          signal(outToken); /* sending the token for the discarded request */
          while(PendingRequest != 0) do
            signal(outRequest); PendingRequest-;
          done; /* relaying other pending requests if any */
        endif
        return;
      }
    }
    TRIGGERS /* for handling incoming token and requests automatically */
    {
      RelayToken(void)
      pre-condition true;
      {
        while(true) do
```

```
        signal(inToken); /* blocking to wait for the next incoming token */
        if (CSWanted != 0) then  /* a local critical section is wanted */
          TokenPresent := 1; /* marking the node as holding the token */
          V(GoCS); /* allowing a critical section access */
        else /* no critical section request on the node */
          signal(outToken); /* relay the token out */
        endif;
      done
    }
    RelayRequest(void)
    pre-condition true;
    {
      while(true) do
        signal(inRequest); /* blocking to wait for the next incoming token */
        if(TokenPresent == 1 and InCS == 0) then
          TokenPresent := false; signal(outToken);
        endif;  /* discard the request and send the token */
        if(TokenPresent == 1 and InCS == 1) then
          PendingRequest++;  /* delay the request */
        endif;
        if(TokenPresent == 0 and CSWanted != 0) then
          PendingRequest++;  /* delay the request */
        endif;
        if(TokenPresent == 0 and CSWanted == 0)
          signal(OutRequest);  /* relay the request */
        endif;
      done
    }
  }
ENDOFCLASS  /* The Node Coordinator is fully specified */
```

11.3.3 Net Synthesis from the NodeCoordinator Specification

From the description of the NodeCoordinator in OF-Class, we can synthesise a Petri net in order to verify the model. The net synthesis is based on the method we describe in Chapter 10. We give below a reduced version of the resulting net.

Here is the net modelling one NodeCoordinator instance. In order to have a model for one ring configuration, we use two or more such instances. The ring configuration is achieved as follows:

- The *OutRequest* place of each instance is merged with the *InRequest* place of the instance located at the downstream node;
- The *OutToken* place of each instance is merged with the *InToken* place of the instance located at the downstream node;
- One node holds the token (its *TokenPresent* place is marked 1).

11.3.4 Verification of Protocol Correctness

The Petri net modelling the NodeCoordinator has two interesting p-flows:

- The first one is: <x1>Idle + <x1>EnterCS + <x1>CS + <x1>ExitCS + <x1>GoIdle + <x1>Wait2 + <x1>Go + <x1>GoRelay + <x1>Relay = <x1>. This means that an object on a given node is always in one of these

```
CLASS
Process is 1 .. 10;
Flag is 0 .. 1;
Request is 1 .. 10;
DOMAIN
VAR
p in Process;
cw in Request;
tk in Flag;
cs in Flag;
pr in Request;
```

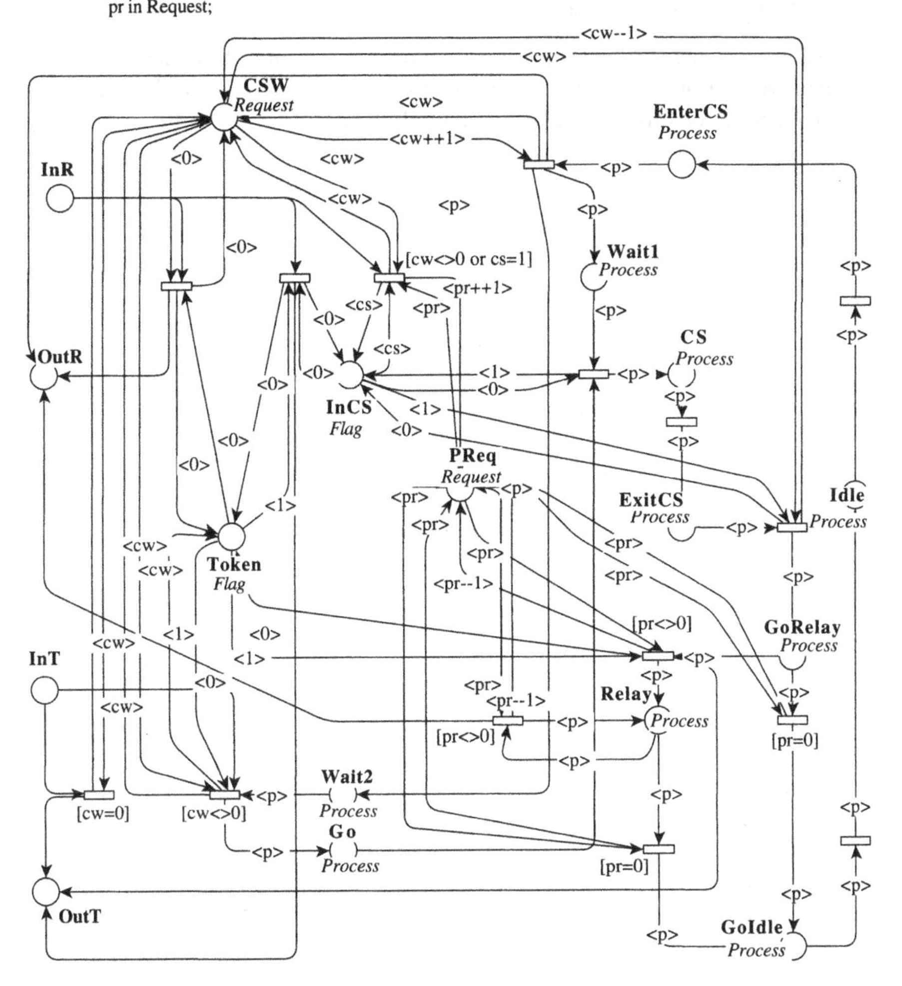

Fig. 11.18. The reduced version of the synthesised net

states. Hence the access to the critical section is correctly handled in this implementation of the protocol. This p-flow corresponds to a sub-net which is a state-machine; it shows that the process can have cyclic behaviour.

- The second one is: <x1>CS + <x1>Wait1 + <x1>ExitCS = <x1>CS + <x1>Wait2 + <x1>Go + <x1>ExitCS. This means that a request to the critical section access (Wait1) is either blocked waiting for the token (Wait2) or authorised (Go).

For a ring configuration with two nodes which we do not show here for sake of space, we compute a p-flow which means that the token is either present on a given node or circulating on the signal ports.

12. Conclusion

In the previous chapters, Petri net models have been presented for various systems. In retrospect, these systems (and the models for them) are somewhat similar. This is no coincidence, since Petri nets are well suited for modelling distributed systems, i.e. systems that consist of many largely independent subsystems that work towards a common goal.

The examples used were communications networks and logistic chains. It is possible to model a monolithic (non-distributed) system such as a computer program as a Petri net, but the net formalism will add little to the understanding of such systems.

Modelling a system as a Petri net requires the distributed nature of the net. In a full-fledged distributed system, each component has its own state, which is altered by the occurrence of events. These events can be either external, from the system's environment, or internal, from other components of the same system.

We have seen a modelling strategy based primarily on states and one based on events. These approaches are complemented by an object-orientation-based one, stressing both state components (attributes) and event components (methods). In [DNV90], it has been shown that all approaches have the same power, since it is possible to model states as future events, and to model the execution of an event as a state component.

In a large project, all three approaches can be used at different stages. Components are identified with the messages and protocols of communication between them. Next, one identifies local state components reflecting the protocol stages, together with predicates that one wants to make true or keep invariant. At the design and implementation stage, one identifies objects and their classes.

Methodologies that combine some or all of these approaches are still the topicof research. No doubt it will be important to have one theoretical framework for every "flavour" of modelling with Petri nets. However, it will largely be up to the modeller to choose an adequate strategy for the specific problem addressed.

Part III

Verification

13. Introduction: Issues in Verification*

The diversity of the verification methods developed for Petri nets and their extensions may be confusing for the engineer trying to choose appropriate techniques to solve his problem. This chapter aims to clarify the basis of such a choice by discussing some general issues involved in the design and application of a verification method:

- The net models that the method enables us to verify
- The kind of properties to be checked
- The families of methods
- The interplay of different methods.

On the one hand a net model with highly expressive power such as a coloured Petri net enables us to handle complex systems. On the other hand, this expressive power implies difficulties for the verification process (e.g. increasing complexity, semi-decidability, or restrictive types of properties). Independently, some high-level models enlarge the range of the results by introducing a parametrisation (e.g. abstract data types or variable cardinalities of domains).

The specification of the properties must address the following question: How should we define good behaviour of a net? Among the different answers, one can suggest:

- A family of properties expressing the behaviour of the net independently of its interpretation (e.g. liveness, boundedness)
- A language of properties adapted to dynamic systems and especially concurrent ones (e.g. linear time logic, branching time logic)
- Behaviour equivalence with another net modelling, for instance, a more abstract view of the system (e.g. bisimulation)
- The response to a sequence of tests (e.g. failure tests, acceptance tests).

The methods may be classified according to basic criteria. What types of nets are supported by the method? Does the method work at the structural level (i.e. the net description) or at the behavioural level (i.e. the reachability description)? Is the verification process entirely or partially automatic? What kinds of properties is the method able to check?

---

* Author: S. Haddad

Finally, to combine the different verification methods, it is necessary to understand what benefits one method can take from the results of another. Furthermore, this may also have an impact on the specification process: for instance, the system could be modelled with a very abstract net and exhaustive results, then refined while retaining as many results as possible.

The remainder of this chapter is organised as follows: In Section 13.1 we present a classification of net models and especially of coloured nets, then in Section 13.2 we discuss the types of properties one can check. In Section 13.3 we list and provide details about the criteria of each method, and we show how to combine them in Section 13.4. We conclude the chapter in Section 13.5 with an overview of the methods presented in this part of the book.

13.1 Classification of Nets

From the model of P/T nets, one can derive new models in different ways: restriction, extension, abbreviation, and parametrisation. In this section, we discuss the impact of these derivations on the verification methods.

13.1.1 Restriction of Nets

The most meaningful restrictions from the point of view of verification rely on the conflicts among transitions. For instance, the well-known model of free choice nets ([Bes87]) restricts the conflicts among transitions with the same input places. The impact of such a restriction is twofold: new algorithms with reduced complexity may be developed to check properties, and an equivalence between structural properties and behavioural properties may be established. A characteristic property which has been defined by Commoner states that any siphon must contain a marked trap (see Section 15.3 for an explanation). Thus a free-choice net is live if and only if it fulfils the Commoner property.

13.1.2 Extension of Nets

A net model is an extension of the P/T net model if its expressive power is strictly greater than that of the original model. The first extensions proposed for P/T nets aims to give more flexibility in the design process ([CDF91b, LC94]). The inhibitor arcs model the zero test; the transition priorities model, for instance, the interruption mechanism; and the flush arcs model, for instance, the crash of a machine, etc.

In most cases, the new model has the same expressive power as Turing machines, and thus the reachability problem – is one marking reachable from another one? – becomes undecidable ([Hac75]). However this drawback should not be overestimated:

- At first, many structural methods which produce results will remain unchanged. For instance, the computation of the flows is unaffected by the presence of inhibitor arcs since they do not induce movement of tokens.
- The hypothesis of bounded nets required for many state-based methods transforms the extension into an abbreviation. For instance, the inhibitor arcs can be modelled by the method of complementary places.
- Often, the modification brought to the verification methods is easy to develop and straightforward. For instance, the computing of the (extended) conflict sets among transitions handles the inhibitor arcs in an intuitive and natural way.

Nevertheless the extensions have a pernicious effect on the design process. Let us look at the two nets of Figure 13.1. These two nets model concurrent accesses to a file by readers and writers. The safety property is evenly ensured in both cases. However, the computation of the flows will give us this property directly in the first net, while it will give only information about the number of readers and writers in the second net. The key point is that the more the extensions are involved in the design process, the less the classical methods will give significant results. As a heuristic principle, one can state: "Use extensions only when necessary."

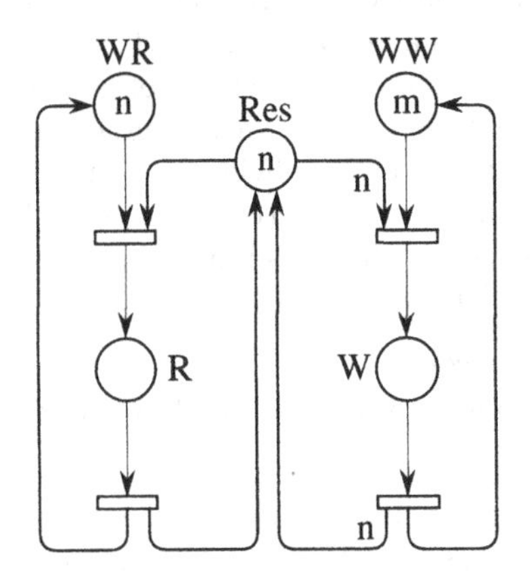

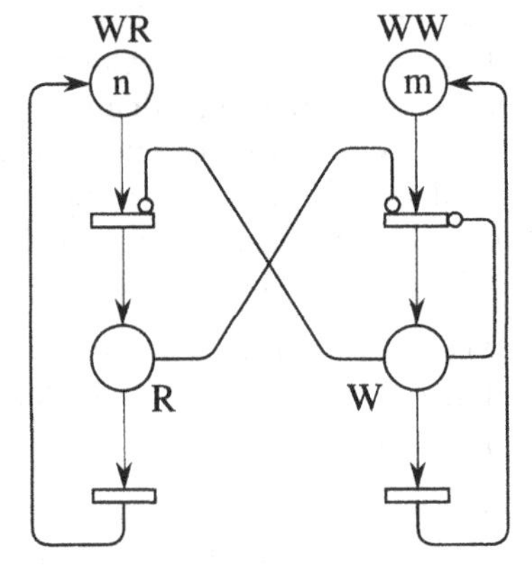

Fig. 13.1. Two Reader-Writer nets

13.1.3 Abbreviation of Nets

A net model is an *abbreviation* of a P/T-net model if:

1. There is a common semantics for the two models
2. For any net there is a semantically equivalent P/T net (generally larger).

A useful abbreviation of a P/T net is the model of coloured nets introduced by K. Jensen [Jen92b]. The main interesting features of this abbreviation are the information associated with tokens using colours and the ability to factorise activities with the help of firing instances of a transition. The unfolding is quite easy: any node (place or transition) is developed in a set of nodes indexed by the colours of its domain, and the arcs are defined according to the applications of colour functions.

Given this unfolding, it is not difficult to transform a verification method for P/T nets into a method for coloured nets:

1. Unfold the coloured net.
2. Apply the algorithm.
3. Interpret the results for the original coloured net.

However, such a transformation is unsatisfactory for two main reasons: the complexity of the algorithm depends on the size of the unfolded net, and it is sometimes difficult to interpret the results. So the main objective of a verification theory for coloured nets is to develop algorithms which do not require the unfolding of the nets.

In order to avoid the unfolding of nets, one is led to examine the syntax and properties of the colour functions. However, the general definition of coloured nets works at the semantic level. The easiest way to give a syntactical definition of a colour function is to represent it as an expression in which the constants denote bags of colours, the variables denote projections of colour domains, and operators denote operations on functions. Then syntactical conditions on an expression provide necessary and/or sufficient conditions on the denoted function. Let us take an example: in a net reduction (called a pre-agglomeration) a transition is required not to share its input places. In the coloured net, the equivalent condition is defined by:

1. The transition does not share its input places.
2. The colour functions which label the input arcs fulfil a condition called *quasi-injectivity.*

Rather than explain what quasi-injectivity is, let us say that there are numerous necessary or sufficient conditions for quasi-injectivity. In the example of Figure 13.2, we have depicted two coloured nets with their unfolded nets. The first unfolded net does not share the input places, and it can be detected directly in the expression of the coloured net since the expression is a tuple of all the variables of the transition. The second unfolded net shares the input places and it can also be detected directly in the expression of the coloured net since not all the variables of the transition appear in the expression.

Other important properties can be checked in expressions such as the algebraic structure for the flow computation or the symmetric structure with respect to a colour domain.

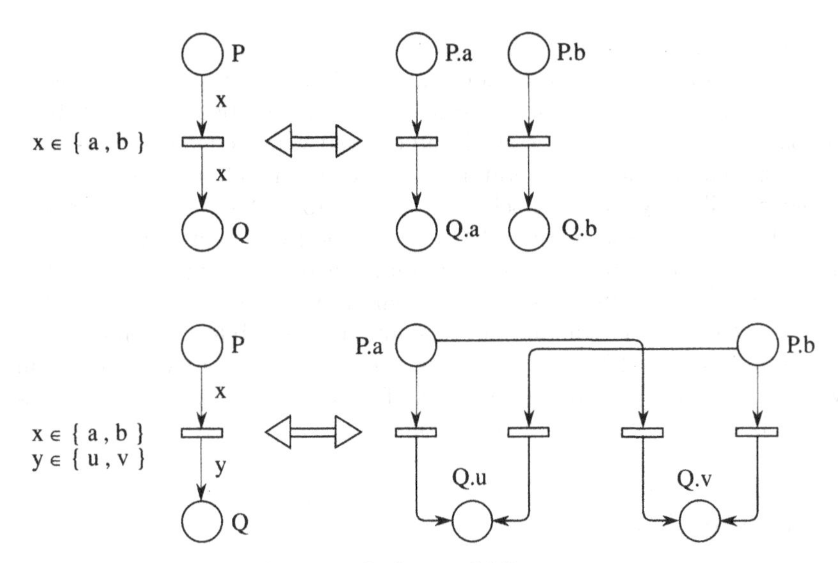

Fig. 13.2. Two coloured nets with their unfoldings

13.1.4 Parametrisation of Nets

A net model is a *parametrisation* of a P/T net if it denotes a family of P/T nets. Implicitly an unmarked P/T net is a parametrised Petri net ([CDF91a]) and we can already obtain results which do not depend on the initial marking. Nevertheless the parametrisation of nets is very interesting in the field of coloured nets since there are many ways to achieve it, among them:

- Abstract predicate/transition nets ([Gen88])
- Algebraic Petri nets ([Rei91])
- Well-formed Petri nets ([CDFH93]).

An abstract predicate/transition net is associated with first-order logic, and colour functions are expressions of this logic. Each interpretation of this logic provides a concrete predicate/transition net which is a syntactical denotation of a coloured Petri net. The main theoretical results of this parametrisation concern the existence of a normalised specification of such nets; this makes it possible to decide whether two nets are semantically equivalent.

An algebraic Petri net is associated with an abstract data type, and colour functions are expressions of this abstract data type. Each algebra which fulfils this abstract data type also gives a coloured Petri net. There are various results on algebraic Petri nets; for instance one can establish statements such as: If the net associated with an initial (or final) algebra has a certain property then a net associated with any algebra has that same property. The algebraic Petri nets can easily be integrated into a prototyping software environment, which is another advantage.

Well-formed Petri nets (WFNs) were introduced in order to develop efficient verification methods for coloured nets and parametrised coloured nets. The syntax of such nets relies on three basic constructions: the variables; some particular constants (the static subclasses) which denote colours with similar behaviour; and one operator, the successor function, which chooses the colour "following" a colour selected by a variable. Despite their restricted syntax, it has been shown that WFNs have the same expressive power as general coloured nets. The parametrisation is introduced by the cardinalities of colour domains. Reductions and flow computations exploit the parametrisation whereas the construction of the symbolic reachability graph operates on an unparametrised WFN. Numerous applications of the symbolic graph have been developed to obtain measures of performance (steady-state probabilities, bounds, tensorial decomposition, etc.).

13.2 Properties

The choice of properties for Petri nets raises the same problem as the choice of the Petri net model. Specifying a large set of properties forbids the development of efficient specialised algorithms whereas a restricted set of properties fails to express the various properties of protocols and/or distributed algorithms.

If one chooses to restrict the properties then these properties must be generic in the following sense: they must express the behaviour of the modelled system for a large range of interpretations. Let us see how such an interpretation is possible. We give below a non-exhaustive list of properties which, of course, do not cover all the general properties a net may have (see the discussion later on in this section):

- **Quasi-liveness.** "Every transition is fired at least one time" expresses a syntactically correct design in the sense that any activity or event must occur at least once in the net behaviour.
- **Deadlock-freeness.** "There is no dead marking" means that global deadlock never happens in the system.
- **Liveness.** "Every transition is fired at least once from any reachable marking" means that the system never loses its capacities.
- **Boundedness.** "There is a bound on the marking of any place" ensures that the system will reach some stationary behaviour eventually. Let us note that multiple stationary behaviours are possible.
- **Home state.** "There exists a marking which is reachable from any other marking" indicates that the system is able to re-initialise itself.
- **Unavoidable state.** "There exists a marking which can not be avoided indefinitely" indicates that the system must necessarily re-initialise itself.

Despite the generality of the previous properties, there will always be some features of behaviour that are not captured by a fixed set of properties.

For instance, "The firing of t_1 will eventually be followed by the firing of t_2" is a useful fairness property which is not mentioned above. Of course, one could include it, but there are many possible variants. Thus it is better to adopt a language of properties adapted to dynamical systems and especially concurrent ones. Among such languages, the temporal logic framework (e.g. linear time logic, branching time logic, etc.) has been widely used for Petri nets (see for instance [Bra90]). The reason for this development is twofold: most interesting properties of concurrency are expressed by simple formulas and the model checking associated with these logics can be easily transported to the reachability graph. In fact, by exploiting the structure of Petri nets, the complexity of model checking can be reduced; we will discuss this topic in Section 13.3.

The framework of temporal logic is interesting if one wants to verify a set of properties which characterises the desired behaviour of the modelled system. Nevertheless, starting from a global behaviour such as a set of services, specifying the correct formulae requires a great deal of work. Moreover the modeller is led to build more and more complex formulae in which the semantics becomes mysterious. In such cases, it is much simpler to specify the set of services with a Petri net and to compare the behaviour of the net modelling the services with the behaviour of the net modelling the protocol. However, it is necessary to define equivalence between nets. First, one has to distinguish between internal transitions (implementing the protocol) and external transitions (associated with the service interface). Then the projection of the protocol net language onto the external transitions should be equal to the language of the service net (language equivalence). However, the language equivalence does not capture the choices offered by a net upon reaching some state. Equivalence including language and choice can be defined by means of one of the numerous bisimulation definitions ([BDKP91]) which roughly say that whatever one Petri net can do (sequence and reached state) the other can simulate. The bisimulation has two interesting features :

- An efficient algorithm has been developed once the reachability graph of the Petri nets is built.
- For models such as process algebras, axiomatisation of equivalence is possible at the structural level.

Moreover, Petri nets refine the definition of equivalence by distinguishing true concurrency from interleaving concurrency (see Figure 13.3). Lastly (see Section 13.5) the Petri net model can be translated into a process algebra during the design of a system in order to facilitate rewriting techniques and equational reasoning. Again it should be noted that restrictions are required in order to avoid the undecidability of bisimulation for Petri nets ([Jan94]).

Another possibility (and the last we examine here) is the response of a Petri net, or more generally a transition system, to a sequence of tests ([Bri87]). A typical test application may be described as follows:

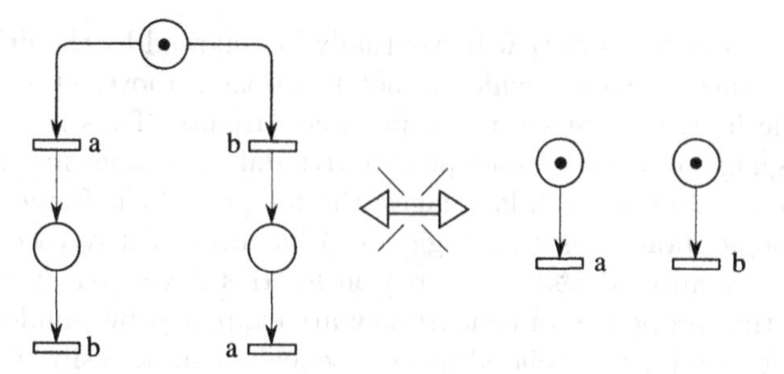

Fig. 13.3. Two nets which do not bisimulate concurrently

1. It starts with a specification (often a process algebra model).
2. Then it generates an intermediate object called a *success tree* which takes into account the sequence of transitions and the choice offered by the states.
3. This tree is transformed into a transition system called the canonical tester.
4. The synchronous product of the Petri net and the canonical tester is formed.
5. The observation of deadlocks in the product provides information on the failures of the implementation given by the Petri net.

13.3 Classification of Methods

There are different ways to discriminate between the methods: automatic verification versus manual verification, property verification versus property computation, specific Petri net methods versus general transition system methods. Let us develop each of these points.

One objective of formal models is computer-aided verification. At first sight, automatic verification may appear as highly desirable. However, there are some inherent limitations to automatic verification that the modeller should be aware of:

- There are numerous undecidable properties.
- Even for decidable properties, checking is so complex that it may become impractical.
- Automatic verification never takes into account the specifics of the modelled system.

Another advantage of manual verification is the insight it gives into the behaviour of the system. However, manual verification is prone to errors, and a sound (and sometimes complete) axiomatisation of proofs may help to develop correct proofs. The duality of the manual and automatic verification should be emphasised: for instance, the reduction or abstraction of nets may be done automatically whereas refinement of nets requires the participation of the designer. Yet these are two facets of the same theory.

The automatic method may check properties given by the modeller or simply generate valid properties of the model. Each method has its own drawbacks. Checking of properties is sometimes tricky: an inductive proof may not be obtained whereas it might be possible to find a stronger property which is inductive. Indeed, for a large class of transition systems, a property is true if and only if there is a stronger property which is inductive. On the other hand, the automatic generation of properties is generally limited in scope: a non-linear invariant will never be generated by the computation of the flows.

Different models may be employed when developing a system. Thus even if one models a system with Petri nets during the design phase, it is not neccessary that the verification be Petri-net-based. For instance, the compositional aspect of a model is not easily exploited by Petri net techniques. Translation into a process algebra may be fruitful in this particular case. Nevertheless there are some good reasons for usimg Petri net formalisms:

- The most important techniques from other models of parallelism have been adapted for Petri nets.
- Petri net verification is one of the most flexible because of the various methods.
- Some methods have no equivalents in other models (e.g. computation of the flows).
- Some other methods have equivalents but their application in Petri nets is easier (e.g. partial order methods).

However the most important criterion for verification techniques depends on which aspects of Petri nets are exploited. We will list these aspects before introducing the methods based on each:

- A Petri net is a graph and the token flows must follow the arcs of this graph; structural deadlocks are clearly based on this feature.
- A Petri net is a linear transformation of the vector of tokens and so linear algebra can be used (for instance, computation of the flows).
- A Petri net underlies event structures with causality and compatibility relations. Partial-order methods reduce the complexity of constructing the reachability graph.
- The colours of a domain often have the same behaviour. The symmetry methods also reduce the reachability graph, using equivalence relations.
- The application of logics is widely used in Petri nets: for instance a logic can encode semantics of Petri nets in such a way that one obtains properties by deduction, or one can build a graph of formulas where a formula naturally denotes a subset of reachable states.

Graph Theory. The examination of the graph structure leads to two different and complementary families of methods which are based either on the local structure or on the global structure. The local structure of a sub-net may make it possible to reason about its behaviour independently of the rest of the net. This is the key point of the reductions theory, in which the agglomeration of transitions corresponds to transforming a non-atomic sequence of transitions into an (atomic) transition ([Ber87, Had89, CMS87]). Even if they just simplify the net by, say, eliminating a transition, their impact is considerable. Indeed, in the reachability graph they eliminate all the intermediate states between the initial firing and the terminal firing of the sequence. Roughly speaking, an agglomeration divides by two the reachability space and thus n agglomerations have a reduction factor of 2^n. Analysing the global structure of the net can be done by restricting the class of Petri nets and developing polynomial algorithms for the standard properties (e.g. liveness). With no restrictions on the Petri nets, similar algorithms provide necessary or sufficient conditions for the standard properties.

Linear Algebra. Linear algebra techniques rely on the state-change equation, which claims that a reachable marking is given by the sum of the initial marking and the product of the incidence matrix with the occurrence vector of the firing sequence. Thus a weighting of the places which annuls the incidence matrix (i.e. a flow) is left invariant by any firing sequence. Similarly, a vector of transition occurrences which annuls the incidence matrix (i.e. a rhythm) keeps any marking invariant.

Thus there are two objectives for linear algebra techniques: computing a generative family of flows (respectively rhythms), and then applying the flows (respectively rhythms) to the analysis of the net. The computation of the flows is more or less easy depending on the constraints on the flows. For instance, the complexity of the computation of general flows is polynomial whereas unfortunately the computation of positive flows is not polynomial ([KJ87]). However, positive flows are often more useful than general flows and researchers have produced heuristics to decrease the average complexity ([CS89]). In P/T nets, such algorithms are now well known. The applications of flows and rhythms are numerous: they help to define reductions, they characterise a superset of the reachable set, they give bounds on maximal firing rates, they make it possible to compute synchronic distances between transitions, etc. Some of these applications are illustrated in Chapter 15.

State-based Methods. Before speaking about partial-order and colour analysis methods, we must point out that one common objective of these two methods is to reduce the complexity of the state-based methods. As these latter methods are, together with simulation, assuredly the most widely used ones, it is important to give an insight into the different ways to cope with the space complexity of the state graph. There are two approaches: manage the graph construction or build another graph.

An efficient management of the graph construction has an important advantage. It is independent of the structural model which generates the graph and thus can be applied to Petri nets, process algebras, etc. The two main methods of this type are the binary decision diagram (BDD) and on-the-fly verification.

Binary Decision Diagrams. Originally the BDD technique was defined to compress the representation of boolean expressions ([Ake78]). Any boolean expression is represented by a rooted acyclic graph where non-terminal nodes are variables of the expression with two successors (depending on the valuation of the variables), and there are two terminal nodes (true and false). In order to evaluate an expression one follows the graph from the root to a terminal node choosing a successor with respect to the chosen assignment. Since subexpressions occurring more than once in the expression are factorised, the gain may be very important.

The application of the BDD technique to graph reduction relies on the representation of a node by a bit vector, and the representation of the arc relation by an expression composed of variables denoting the bits of the vectors. It can be shown that the formula of modal logics can also be represented in this way, and lastly that the building of the graph and the property checking can be reduced to operations on BDDs. In a famous paper, [BCM$^+$90], this technique has been employed to encode graphs with 10^{20} states. A drawback of the method is that it is impossible to predict the compression factor even approximately.

On-the-Fly Verification. The on-the-fly technique is based on two ideas: state properties can be checked on each state independently, and in a finite state graph there is no infinite path with different states. Thus one does not build the entire graph but instead develops the elementary branches only. The only memory required is what is that needed for the longest elementary path of the graph ([Hol87]). In the worst case there is no gain, but on average the gain is important.

Moreover, the technique can be extended to check the properties of temporal logics ([JJ89]). There the trick is to dynamically develop the product of the state graph with an automaton (say for instance a Büchi automaton for the LTL formula) and check for particular states ([CVWY90]).

What is interesting in this method is its adaptation to the memory space of the machine. Indeed, one can add a cache of states which remembers a number of states that are not on the current path, thus reducing the development of the branch if a cache state is encountered. Another fruitful aspect of this method is that it can be combined with other reduction methods (for instance the partial-order method discussed below).

Partial-Order Methods. The partial-order methods rely on structural criteria to reduce the state graph, and are efficiently implemented on Petri nets. The two main methods – sleep set and stubborn set – associate a set of transitions to a state reached during the building, and use this set to restrict further developments of the graph. These sets of transitions are based on a basic structural (or possibly marking-dependent) relation between transitions. Two transitions are *independent* if their firings are not mutually exclusive. The independence property is structural if the pre-condition sets do not intersect, whereas it is marking-dependent if the bag sums of the pre-conditions do not exceed the current marking.

The sleep set method keeps track in a reached marking of independent transitions fired in other branches of the graph ([God90a]). The method ensures that if one fires a transition of this (sleep) set, one encounters an already reached marking. Thus the sleep set method "cuts" arcs on the reachability graph, but the number of states is left unchanged. Figure 13.4 illustrates such a process.

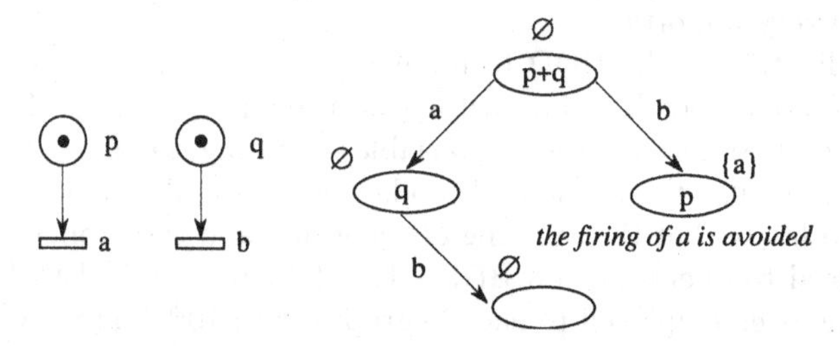

Fig. 13.4. Application of the sleep set method

Given a marking, a *stubborn set* of transitions is such that any sequence built with other transitions includes only independent transitions with respect to the stubborn set ([Val89]). Note that if the independency relation is marking-dependent then the independency must be fulfilled at the different firing markings. Then it can be shown that restricting the firing of an enabled transition in any stubborn set preserves the possibility of the other firing sequences. The building of the reduced graph is similar to that for the ordinary one except that:

- Once a state is examined, the algorithm computes a stubborn set of transitions including at least one enabled transition (if the marking is not a deadlock).
- The successors of the state are the ones reached by the enabled transitions of the stubborn set.

An interesting consequence is deadlock equivalence between the reduced graph and the original graph. Figure 13.5 illustrates such a process. Note that the initial stubborn set is {a,b,c} since starting from a one must include c and then b. Another possible stubborn set would have been {d,e}. The attentive reader will have noticed that an arc building would have been avoided by combining stubborn sets with sleep sets.

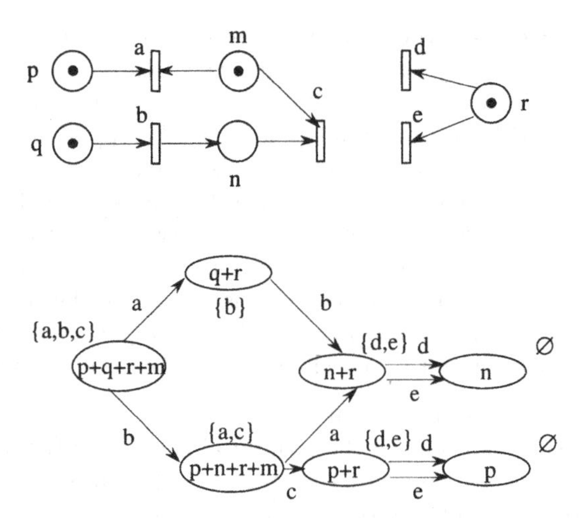

Fig. 13.5. Application of the stubborn set method

The stubborn set method requires more computations than the sleep set method since there is no incremental computation of the stubborn set and the computation includes disabled transitions. On the other hand, the reduction factor is often more important as here states are pruned. Nevertheless, the combination of the two methods is straightforward, and this improves the reduction factor ([GW91b]). What is more difficult to obtain is a large equivalence of properties between the reduced graph and the original one. Safety properties may be obtained if the property is taken into account during the building process. The handling of general liveness properties is not possible and one is restricted to the checking of special liveness properties ([Val93c]).

A third partial-order method is based on unfoldings of Petri nets. An unfolding of a Petri net is an acyclic Petri net in which places represent tokens of the markings of the original net, and transitions represent firings of the original net. One starts with the places corresponding to the initial marking and one develops the transitions associated with the firings of every initially enabled transition linking input places to the new transition and producing (and linking) output places; then one iterates the process. Of course if the net has an infinite sequence the unfolding would be infinite and thus this unfolding must be "cut". In order to produce finite unfoldings, different cut methods have been proposed ([McM92, ERW96]). The unfolded net is a very compact representation of the reachability set and thus safeness properties can be checked with a low space complexity (time complexity may also be reduced but not so significantly). Recently the method has been extended to support linear temporal logic verification ([CP96]). The principle is to build a graph of unfolded nets in which the relevant transitions for the property are always graph transitions.

Colour Structure Analysis. Colour structure analysis has many theoretical applications. Here we just mention three theoretical developments (which will be discussed in more detail in Chapters 15 and 16). The first important point is that a theoretical development may be applied to coloured nets and/or to parametrised coloured nets. As discussed before, parametrisation is better from the modeller's point of view but more difficult from the researcher's point of view. Moreover, there are two ways to obtain results for parametrised coloured nets: first develop a theory for unparametrised coloured nets and then adapt the conditions to include the parametrisation; or restrict the kind of parametrisation to develop a particular theory.

The reduction theory for coloured nets is based on the following approach:

1. Take a reduction for an ordinary Petri net.
2. Add coloured conditions to the structural conditions (i.e. conditions on the colour functions valuating the arcs); these coloured conditions are as weak as possible to ensure the structural conditions on the unfolded net for a set of reductions.
3. Check that there is a possible ordering of the set of reductions in the unfolded net.
4. Define a structural transformation similar to the original reduction with complementary coloured transformations; this transformation must correspond to the successive reductions of the unfolded net.

The parametrisation of the method is more or less straightforward since coloured conditions may be ensured by syntactical conditions on expressions (see the discussion above in Section 13.1)

The flow computation for coloured nets requires deeper analysis of the colour function structure. It appears that the cornerstone of the flow computation is the algebraic concept of generalised inverses. Colour functions are linear transformations on a set of bags and thus this algebraic concept is sound. Moreover, a elegant algorithm adapted from Gaussian elimination rules can be developed, provided that the successive generalised inverses may be computed. The space and time complexity are dramatically reduced and the flows are represented in a compact way which allows for natural interpretation.

Unfortunately, the parametrisation of this method is not possible. So researchers have used a different approach: colour expressions can be identified with polynomials. The idea is then to apply a Gaussian-like elimination to a ring of polynomials. The difficulty lies in the transformation (and the reciprocal transformation) from a colour function to a polynomial one. Some subclasses of well-formed nets have been successfully studied (regular nets, ordered nets) with this technique. Another way to obtain parametrised methods is not to require that the flow family be a generative family. Then simple methods can work on very general nets and give useful information (even if not complete).

The symbolic reachability graph of well-formed nets exploits the symmetry of colour functions with respect to the firing semantics. This symmetry leads to an equivalence relation between markings and transition firings. Once canonical representation of equivalence marking (and firing transitions) classes is defined, symbolic graph building is similar to ordinary graph building. Some studies show that the difference between the reduction factor of symmetrical methods and partial-order methods depends on the modelled system. These methods may again be combined. Another difference between symmetrical methods and partial-order methods is that very general properties may be checked on the symbolic reachability graph (indeed any formula of CTL*).

Logics. Logics is the support of reasoning about nets. Often some inductive rules or schemes are defined to derive properties. There are two ways to do so.

The first is an automatic (but semi-decidable) verification of a property. For instance, a safeness property must be true at the initial state and also true afterwards. Then one begins with the initial formula and derives successively stronger formulas using the firing rule until stability is obtained. This technique may be refined using a graph of formulas where a formula is an intensive representation of states.

The other direction is manual proofs using a proofs scheme. An example is given by a verification diagram which is in fact a directed acyclic graph of the formula used to prove safeness or fairness properties. Even if the proof schemes are very detailed, the verifier needs skill to obtain his proof.

13.4 Verification Process

The verification step is closely linked to the design process. Ideally, even the (formal) specification of the properties which should be satisfied by the system has to be checked. Indeed there are examples where the protocol meets the service requirements but the service is not correctly defined! Generally, the verification process is interleaved with the different steps of new designs. The reasons for these steps are multiple. Obviously, a new model is required if failures are detected. But if the design is incremental, then once a first step of verification is successful, the model is enriched with more detail before further checks.

The two main mechanisms for incremental design are refinement and composition. Here we focus on the consequences for the verification process. Sound refinement should be local so that the properties are retained. Nevertheless, it is clear to the skilled modeller that the hypothesis of locality must be formalised in order to preserve properties. The composition aims to combine results already obtained for the components. Much work has been done in this respect but considering the difficulties encountered by the theoretical research we have to impose some restrictions:

- A process must have restricted choice between synchronisation actions and local actions.
- The behaviour of a synchronisation must not – or only slightly – depend on previous synchronisations.

It should be noted that refinement and composition have their counterparts in the area of verification, namely, reduction and decomposition. These two aspects are very similar, however, there are particularities for the latter which are:

- The reduction (respectively decomposition) process should be automatic.
- The choice among reductions (respectively decomposition) is proof-oriented.
- If the congruence of reductions (respectively decomposition) rules is ensured then the order of application is irrelevant, and this is a great complexity reduction factor.

Here are some hints on how the different verification methods should be ordered. Starting from a Petri net, applying automatic structural methods has numerous advantages:

- It points out what was implicitly in the modelling. For instance, the process decomposition and message traffic are often described as flows of the net.
- It quickly discovers modelling bugs such as an unmarked structural deadlock.
- The established properties can dramatically simplify a deductive proof.
- Lastly it helps the modeller to choose the next verification methods. As a simple illustration, positive flows covering all the places of the net ensure the success of state-based methods and also give an upper bound on the size of the state space.

Once the structural methods have been fully exploited, the modeller can use the state-based methods. As said before, it may happen that the state space is too large to be generated. However, even in this case, the modeller has some alternatives:

- Classically, he can always simulate his net and the consequences are twofold: negative properties are shown and long runs without trouble help to develop confidence in the model.
- He can do on-the-fly verification which, although it takes much longer, can check all the important properties of the net. Here, the key point is that the complexity space is related to the longest simple path in the graph.
- He can generate a smaller object equivalent to the state graph with respect to some properties. Partial-order methods and symmetry methods typically produce such objects.

Alternatively, deduction methods avoid the space complexity problem. This does not mean that these methods should be used after the other methods have failed. Indeed, if the modeller has a clear idea of how his model fulfils its properties, he often develops a quick deductive proof of its correctness. Examples are numerous in the area distributed algorithms.

Lastly it should be pointed out that the deductive methods can be employed in the design process by means of system synthesis. Indeed, if the specification is a formula and the semantic models of the logic involved are the behaviour of the nets, then the system synthesis can be based on the satisfiability resolution. Using this resolution, one begins to produce a semantic model, then folds the semantic model in order to obtain a Petri net. The first step is often possible with modal logics (they verify the small model property), but the second step is technically and sometimes theoretically difficult. At the current time, this is an open field of research.

13.5 Overview

Chapter 14 describes state-based methods. It develops some of the techniques which we have presented above. First it shows how the computation of the state space may be managed efficiently. Then it introduces more precisely the partial-order methods, symmetries, and modular methods. There is more development on the use of symmetries including the implementation for well-formed coloured nets where this building can be completely formalised. This part ends with comparing these methods according to different criteria such as space and time reduction, property equivalence, and how these different methods may be combined.

The rest of the chapter takes into account the types of properties that can be checked and the impact on the graph building. An original technique of parametrised building is developed including the verification of temporal logics. Lastly, the problem of model checking is discussed as a whole.

In Chapter 15, structural methods are developed. Some accurate reduction rules are presented with special emphasis on the implicit place. The implicit place has a particular rôle since it simplifies the structure of the net and makes it possible to apply other techniques more efficiently. Moreover, implicit places have a strong connexion with positive flow computation as shown in the chapter. The linear algebraic techniques are then developed and the equivalence between behavioural properties and linear algebraic results is pointed out. Then siphons and traps are carefully studied since they are the cornerstone of necessary and sufficient conditions for liveness properties.

In the last part of the chapter, some syntactical subclasses are defined, showing what behavioural consequences can be established from the syntactical restrictions. The behavioural properties include fairness, liveness, deadlock-freeness, and the relation between reachable states and linear invariants.

Chapter 16 presents new techniques which cover an open field of research. The techniques presented in the beginning of that chapter are based on logics: rewriting logic, temporal logic, and linear logic. The relevance of linear logic is twofold: it provides Petri nets with an operational semantics, and a proof scheme for linear logic gives the proof of a property in the corresponding Petri net.

The last section is devoted to a technique that shows how to benefit from multi-formalisms. This technique starts from a specification of the system given in process-algebraic terms. Then it constructs a Petri net model of the system. The Petri net is simulated to show bad behaviour in order to reinforce confidence in the model. Finally the Petri net is transformed again into a process algebra so that the two process algebras (modelling the specification and the implementation) are equivalent. Emphasis is put on the design cycle rather than on technical aspects.

14. State-Space-Based Methods and Model Checking*

Previous chapters bring out the need for system designers to profit from automated verification techniques. Effectively, the known problem of state-space explosion makes it difficult to achieve an a priori understanding of the whole behaviour of any reasonable system. Verification techniques must exceed the capabilities of structural approaches which mainly allow us to check that the system is well-structured. One also wants to know the real semantics of the system to be designed.

The current chapter presents techniques which allow the automatic verification of a wide range of properties. It shows how some specific logics, namely temporal logics, can formally capture the required temporal aspects of properties. For instance, they can capture that all further executions after the sending of a message will contain the message receptions by the correct addresses.

Given the formal descriptions of both system and properties, the verification stage can be automated in a so-called "model checking procedure". Assuming that the system has a finite state space, the verification of a property consists in checking that none of the possible executions of the system state space may invalidate the ones induced by the temporal aspects of the property.

Model checking techniques offer a complete framework for verification purposes, however the related complexity depends mainly on the size of the state space, which grows exponentially in the number of represented objects. This drawback is mainly caused by concurrence and, in particular, the interleaving semantics used to represent any sequence of possible actions. To be applicable in an industrial context, model checking must be strongly associated with methods and techniques useful for reducing the problem, such as:

- Restrictions and abstractions which reduce the size of the represented systems
- Efficient state-space explorations which do not require the construction of the whole state space
- Limitations on the effects of the interleaving semantics

* Authors: C. Dutheillet (Sections 1 & 4), I. Vernier-Mounier (Sections 1 & 4), J.-M. Ilié (Sections 2 & 4), D. Poitrenaud (Sections 3 & 5)

- State space reductions by detection of symmetrical behaviours
- Techniques for coding the state space compactly.

For the first point, techniques such as Petri net reductions can be applied. In this chapter, we assume that the represented system is correctly reduced and concentrate our presentation on the other points. Hence, we aim to present automatic techniques which offer powerful memory reduction while retaining a capacity for verification.

The organisation is as follows: Section 14.1 recalls the principles of temporal logics and model checking, and Section 14.2 presents on-the-fly techniques which tend only to build the necessary portion of the state space with regards to the checking of a property. Section 14.3 discusses partial-order techniques which benefit from independence notions so as to avoid the representation of useless interleavings of state changes. Section 14.4 exploits the symmetries of system descriptions and introduces symbolic and parametrised approaches, and Section 14.5 demonstrates how to compact state space information. Section 14.6 contains our concluding remarks; in particular, we observe that the techniques presented can be used simultaneously, and give an overview of current research on model checking problems.

14.1 Properties, Temporal Logic, and Fairness

The verification of a system consists in checking whether it satisfies a set of properties that are derived from its specification. Properties of concurrent systems can be classified according to the type of behaviour they describe. Two important kinds of properties are the *liveness* and *safety* properties. A *liveness* property stipulates that "good things" do happen (eventually). They are properties that must be satisfied by all the executions. A *safety* property stipulates that "bad things" do not happen during the execution of a program. Actually, in [AS87] it has been shown that any property of executions can be decomposed into a safety property and a liveness property. The decomposition is such that the original property is the conjunction of the liveness property and the safety property.

For an automatic verification of a property of a system, one needs two things: a formal description of the system and a formal description of the property to be verified. In this section, we consider systems with a finite state space, i.e. that can be formally described by a bounded Petri net, and we focus on the description and verification of their properties. The verification methods actually depend on the class to which the property belongs.

The Petri net in Figure 14.1 models a mutual exclusion algorithm. Two processes try to access a critical section. Each process has three possible states: *Idle, Wait,* and *CS.* Transition *Req* is fired when a process in place *Idle* wants to enter the critical section. Once in place *Wait,* a process can enter the critical section (place *CS*) if and only if the other process is neither in place *CS* nor in place *Wait*. These conditions are modelled by the inhibitor arcs that link transition $Enter_i$ with places $Wait_j$ and CS_j. Inhibitor arcs are routinely used to ease the first steps of Petri net modelling, and stipulates that a place can inhibit the firing of a transition whenever it contains tokens. Finally, when a process leaves the critical section (transition *Free*) it goes back to place *Idle*.

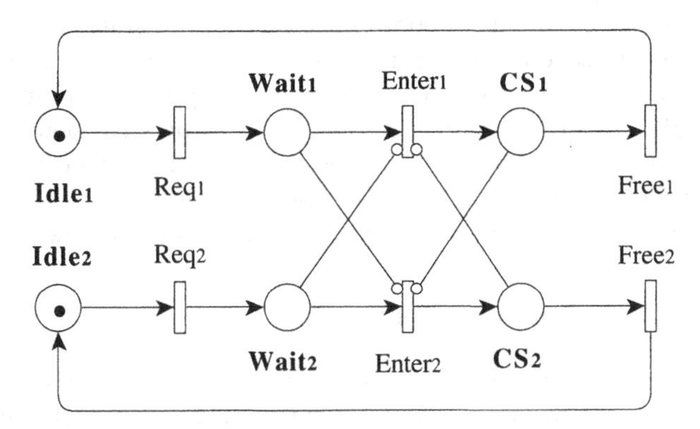

Fig. 14.1. Mutual exclusion algorithm

To verify that this mutual exclusion algorithm is deadlock-free and fair, we need to verify the following four properties :

P1 : The algorithm is deadlock-free.
P2 : At each state, there is at most one process in the critical section.
P3 : The algorithm is starvation free. Each process that asks for the critical section will obtain it.
P4 : The entries in the critical section are in the same order as requests.

The first two properties are *safety* properties. The "bad things" that do not happen are deadlocks and two processes simultaneously in the same place. The last two properties are *liveness* properties. The "good thing" that eventually happens when a process asks for a critical section is that the process will enter it according to the order of requests.

As we consider dynamic systems, the truth value of an assertion P on the behaviour of the system, such as "Process p1 enters the critical section", depends on the time it is pronounced. Hence, we need a language that takes into account the possibility for the truth value of a formula to change over time. This is the case for temporal logic, where two typical temporal operators are *sometimes* and *always*: *sometimes* P is true now if there is a future moment at which P becomes true, and *always* P is true now if P is true at all future moments.

Among the languages that can be used to formalise a property, an advantage of temporal logic is that the meaning of its operators is rather intuitive, while the range of properties that it can express is large. There exist several logics that correspond to different views of time. We consider here logics based on a discrete representation of time because the execution of Petri nets is performed with respect to discrete time. Actually, the different moments of an execution are the reachable states of the system. The reachability graph of a Petri net describes the set of all possible executions of the system, every execution being represented by a path in the graph. In this representation a state may have several successors. Since we identify states and moments, moments also have several successors. This view is known as branching-time.

We present in this section CTL*, which is a very general branching-time temporal logic, in the sense that it has a high expressive power. We also consider two restrictions to CTL*, namely CTL and LTL. Their interest lies in the fact that efficient algorithms have been developed to check the truth value of their formulae, while they retain a high expressive power.

14.1.1 The Temporal Logic CTL*

Let us come back to our example and try to see which operators would be useful to express the properties that we want to check. Since the system has several possible executions, we first need to specify if a property has to be true for one execution or for all possible executions. This can be done by means of two quantifiers, the existential quantifier $\Diamond$ meaning *there is a path from* and the universal quantifier $\Box$ meaning *for all paths from.* Property P1 requires that deadlock-freeness be true in every state of the system, i.e. *for all paths, always.* This is the same for Property P2. Property P3 relies on the dynamics of the system: *for all paths* starting from a state where a process asks for the critical section, *sometimes* this process enters the critical section. Property P4 imposes an order among the actions: if process 1 requests the critical section before process 2, then process 2 must not enter the critical section before process 1, which can also be expressed as "Process 2 enters the critical section" remains false *until* "Process 1 enters the critical section" is true. Hence, we can see that with a restricted set of operators, we can express a wide range of properties.

CTL* was introduced by Emerson and Halpern in [EH86]. Besides the operators presented above, the language includes an operator $\circ$, meaning "next time", making it possible to require that a property become true in the next state. As the notion of future state or future moment is not defined for a terminal state, a loop is added to the states with no successors. This is actually necessary because in the models presented below, only infinite executions are considered.

Syntax. The CTL* language is formed by the set of state formulae that are generated inductively by the set of rules below:

A state formula is:

- an atomic proposition;
- $\neg f_1, f_1 \wedge f_2$ where f_1, f_2 are state formulae;
- $\Box f, \Diamond f$ where f is a path formula.

Atomic propositions allow one to express elementary conditions on the values assigned to some variables in the system. When we consider place/transition nets, the atomic propositions are conditions on the number of tokens in places allowing one to test the markings.

A path formula is:

- a state formula;
- $\neg f_1, f_1 \wedge f_2$ where f_1, f_2 are path formulae;
- $\circ f_1, [f_1 \mathsf{U} f_2]$ where f_1, f_2 are path formulae.

The semantics of CTL* formulae is defined using the state graph.

Semantics. A structure for a CTL* formula (with a set AP of atomic propositions) is a triple $M = \langle S, R, \pi \rangle$ where:

S is a finite or enumerable set of states.
R is a total binary successor relation on S ($R \subseteq S \times S$) that gives the possible transitions between states.
π: $S \rightarrow 2^{AP}$ assigns truth values to the atomic propositions at each state.

For example, the truth value of the atomic proposition "$p \leq 2$" is true at each marking with at most two tokens in place p.

The structure $\langle S, R \rangle$ is the reachability graph of the model. A *path*, or an *execution*, is an infinite sequence of states $(s_0, s_1, \ldots)$ such that $\forall i \,.\, \langle s_i, s_{i+1} \rangle \in R$ (remember that each state has at least one successor, which is itself). If $x = (s_0, s_1, \ldots)$ is a path, we denote by x^i the *suffix path* $(s_i, s_{i+1}, \ldots)$.

For a structure M, a state $s_0 \in S$, and a path x we have:

$\langle M, s_0 \rangle \models p$	iff	$p \in \pi(s_0)$, for $p \in AP$
$\langle M, s_0 \rangle \models f_1 \wedge f_2$	iff	$\langle M, s_0 \rangle \models f_1$ and $\langle M, s_0 \rangle \models f_2$
$\langle M, s_0 \rangle \models \neg f$	iff	not $\langle M, s_0 \rangle \models f$
$\langle M, s_0 \rangle \models \Diamond f$	iff	for some path $x = (s_0, s_1, \ldots), \langle M, x \rangle \models f$
$\langle M, s_0 \rangle \models \Box f$	iff	for all paths $x = (s_0, s_1, \ldots), \langle M, x \rangle \models f$

$\langle M, x \rangle \models f$	iff	$\langle M, s_0 \rangle \models f$
$\langle M, x \rangle \models f_1 \wedge f_2$	iff	$\langle M, x \rangle \models f_1$ and $\langle M, x \rangle \models f_2$
$\langle M, x \rangle \models \neg f$	iff	not $\langle M, x \rangle \models f$
$\langle M, x \rangle \models [f_1 \mathsf{U} f_2]$	iff	$\exists i, i \geq 0.\ \langle M, x^i \rangle \models f_2$ and $\forall j \,.\, 0 \leq j < i, \langle M, x^j \rangle \models f_1$
$\langle M, x \rangle \models \circ f$	iff	$\langle M, s_1 \rangle \models f$

When we consider a net system, $\langle \mathcal{N}, \mathbf{m_0} \rangle$, $\langle \mathcal{N}, \mathbf{m_0} \rangle \models f$ is the notation for $\langle \mathrm{RG}(\mathcal{N}, \mathbf{m_0}), \mathbf{m_0} \rangle \models f$ where $\mathrm{RG}(\mathcal{N}, \mathbf{m_0})$ is the reachability graph of $\langle \mathcal{N}, \mathbf{m_0} \rangle$.

At this point, we can notice that the operators *sometimes* and *always* are not included in the language. They are merely notations and can be expressed by means of the operator *until*:

$$\langle M, x \rangle \models \mathrm{F} f \stackrel{\text{def}}{=} \langle M, x \rangle \models [\mathit{true} \,\mathsf{U}\, f]$$
$$\langle M, x \rangle \models \mathrm{G} f \stackrel{\text{def}}{=} \langle M, x \rangle \models \neg\, \mathrm{F} \neg f$$

In other words, $\mathrm{F}f$ states that f holds *sometimes* on the path, i.e. at least at one state. $\mathrm{G}f$ states that f holds at all the states of the path, i.e. *always*. We now apply the syntax of CTL* to express the properties that we want to check in our example. Let us consider property P3, which states that a process which is waiting for the critical section will eventually enter it.

Atomic propositions can express the fact theat a process is waiting for the critical section ($Wait_i$=1) or is in the critical section (CS_i=1). To simplify the formulae, we use the logical implication operator "⇒" with its usual meaning. Then we can find two different but equivalent ways of expressing property P3 with CTL*:

$$\langle \mathcal{N}, \mathbf{m_0} \rangle \models \bigwedge_{i \in \{1,2\}} \Box \mathrm{G}[Wait_i = 1 \Rightarrow \mathrm{F}(CS_i = 1)]$$
$$\langle \mathcal{N}, \mathbf{m_0} \rangle \models \bigwedge_{i \in \{1,2\}} \Box \mathrm{G}[Wait_i = 1 \Rightarrow \Box \mathrm{F}(CS_i = 1)]$$

The first expression corresponds to a view in which the different executions of the system are considered independently. In each execution, a state has a single successor; hence an execution of the system is seen as a linear sequence of states starting from the initial state. The property is satisfied if it is satisfied by all the executions. This view is known as *linear time*. In the second expression, every time a state in which a process is waiting for the critical section is encountered, all the executions initiated at this state are considered. Hence, a state may have several successors. This view is known as *branching time*.

From this, we notice not only that CTL* makes it possible to express a wide range of properties, but also that its grammar does not introduce too many constraints on the expression of a property. However, the drawback of the generality of CTL* is the complexity of the algorithms for verifying a formula. They are linear in the size of the model, i.e. the number of nodes of the reachability graph, but exponential in the size of the formula, i.e. the number of operators and atomic propositions.

Hence, various restrictions have been proposed; we present CTL (computation tree logic) which corresponds to the branching time view and LTL (linear time logic) which adopts the linear time approach.

The Temporal Logic CTL. The CTL logic was introduced by Clarke, Emerson, and Sistla in [CES86].

Syntax. The CTL language is given by:

$$f := p \mid \neg f \mid f \wedge f \mid \Box \circ f \mid \Diamond \circ f \mid \Box[f \mathsf{U} f] \mid \Diamond[f \mathsf{U} f]$$

where $p \in AP$ is an atomic proposition.

The semantics of CTL is the same as for CTL*. We illustrate the language by expressing the properties that we want to check in the mutual exclusion example.

To express property P1 with a temporal logic formula, we first have to express the possibility that a marking is deadlock free. This is true if it enables at least one transition. This property involves no temporal operator.

Let $\mathcal{N} = \langle P, T, \mathbf{Pre}, \mathbf{Post}, \mathbf{Inh} \rangle$ be a place/transition net with inhibitor arcs, then a reachable marking $\mathbf{m}$ is deadlock free if and only if it satisfies the formula:

$$f_{deadlockfree}(\mathcal{N}) = \bigvee_{t \in T} \bigwedge_{p \in P} (p \geq \mathbf{Pre}[p,t] \wedge p < \mathbf{Inh}[p,t])$$

When we consider the net of Figure 14.1 the formula $f_{deadlockfree}(\mathcal{N})$ is:

$$(Idle_1 \geq 1) \vee (Idle_2 \geq 1) \vee (CS_1 \geq 1) \vee (CS_2 \geq 1) \vee$$
$$((Wait_1 \geq 1) \wedge (CS_2 < 1)) \vee ((Wait_2 \geq 1) \wedge (CS_1 < 1))$$

Therefore, the whole net system is deadlock free if and only if all the reachable markings are. Hence, Property P1 is true for $\langle \mathcal{N}, \mathbf{m_0} \rangle$ iff

$$\langle \mathcal{N}, \mathbf{m_0} \rangle \models \Box \mathrm{G} f_{deadlockfree}(\mathcal{N})$$

Property P2 states that in every reachable marking, the two processes must not be simultaneously in the critical section. This last part of the property can be identified with the atomic proposition $(CS_1 + CS_2) \leq 1$. Therefore, Property P2 is true for $\langle \mathcal{N}, \mathbf{m_0} \rangle$ iff

$$\langle \mathcal{N}, \mathbf{m_0} \rangle \models \Box \mathrm{G}((CS_1 + CS_2) \leq 1)$$

For Property P4 to be true, the entries in the critical section must be in the same order as the requests. This means that if a process i has fired transition Req_i while process j was in place $Idle_j$, process i will be in the critical section before process j. With the "until" operator, we can say that place CS_j will stay empty until place CS_i is marked. We use the property that a process cannot be in several places at the same time. Therefore, for each marking, there is exactly one token in one of the places $Idle_i, Wait_i$ or CS_i. Therefore, Property P4 is true for $\langle \mathcal{N}, \mathbf{m_0} \rangle$ iff

$$\langle \mathcal{N}, \mathbf{m_0} \rangle \models \bigwedge_{i,j \in \{1,2\}, i \neq j} \Box \mathrm{G}[(Wait_i = 1 \wedge Idle_j = 1) \Rightarrow \Box[\neg(CS_j = 1) \mathsf{U} (CS_i = 1)]]$$

Verification. CTL formulae can be verified by means of rather intuitive algorithms that apply directly to the reachability graph. Algorithms 14.1, 14.2, 14.3, and 14.4 make it possible to check the basic CTL formulae. The first two algorithms are trivial. The idea of Algorithm 14.3 is to mark a state as soon as the value of the formula can be decided for this state. When the formula is true in a state, its value can also be decided for the predecessors of that state. This is done by means of procedure *Propagate.* A state can be unmarked after loop $\forall s\,.\,Verify\_\Diamond\mathsf{U}(s)$, meaning that the value of the formula has not yet been decided for it. This happens for states where f_1 is true but which have no (direct or indirect) successors for which f_2 is true. Hence the formula is false for these states. In Algorithm 14.4, the idea is that a state verifies the formula if f_2 is true in this state, or f_1 is true and all its successors verify the formula. This verification is performed in recursively. This time, the marking of states is used to stop the recursion, and states are marked as soon as they are considered. If a marked state is considered again, it has necessarily been reached through a path where f_1 is always true, otherwise the recursion would have been stopped before. If the value of the formula is not yet decided for this state, it means that the state is participating in the current recursion and thus it belongs to a cycle where f_1 is always true but f_2 is never true. Hence the formula is false in this state.

A complex formula is verified by repeatedly applying the algorithms on the simple formulae that comprise it. The overall complexity for checking a CTL formula is linear in the size of both the model and the formula.

Algorithm 14.1 $\Diamond \circ f$

for every state s **do**
  if $\exists s'$ successor of s such that $s' \models f$ **then**
    $s \models \Diamond \circ f$
  else
    $s \models \neg\Diamond \circ f$
  fi
od

Algorithm 14.2 $\Box \circ f$

for every state s **do**
  if $\forall s'$ successor of s, $s' \models f$ **then**
    $s \models \Box \circ f$
  else
    $s \models \neg\Box \circ f$
  fi
od

Algorithm 14.3 $\Diamond[f_1 \mathsf{U} f_2]$

```
Procedure Propagate_◇U(s)
if s is not marked then
    mark(s)
    if s ⊨ f2 or s ⊨ f1 then
        s ⊨ ◇[f1 U f2]
        ∀s' predecessor of s, Propagate_◇U(s')
    else
        s ⊨ ¬ ◇[f1 U f2]
    fi
fi

Procedure Verify_◇U(s)
if s is not marked then
    if s ⊨ f2 then
        mark(s)
        s ⊨ ◇[f1 U f2]
        ∀s' predecessor of s, Propagate_◇U(s')
    else
        if s ⊨ ¬ f1 then
            mark(s)
            s ⊨ ¬ ◇[f1 U f2]
        fi
    fi
fi

begin
    ∀s unmark(s)
    ∀s Verify_◇U(s)
    ∀s if s is not marked then
        s ⊨ ¬ ◇[f1 U f2]
    fi
end
```

The Temporal Logic LTL. The LTL logic was introduced by Pnueli in [Pnu81]. It is one of the most common versions of propositional linear-time temporal logic appearing in the computer science literature.

Syntax. The LTL language is given by

$$f := p \mid \neg f \mid f \wedge f \mid \circ f \mid [f \mathsf{U} f]$$

where $p \in AP$ is an atomic proposition.

All these formulae are path formulae. To be properties of a model, they have to be satisfied by all the possible executions of the model. The semantics of LTL formulae is defined for one execution. Hence, unlike CTL* and CTL, the successor function R in the structure for an LTL formula is a function $S \to S$ that for each state gives a unique next state.

As for CTL, we use properties P1, P2, and P4 to illustrate LTL.

Algorithm 14.4 $\square[f_1 \mathsf{U} f_2]$

```
Procedure Verify_□U(s)
if s is marked then
    if s ⊨ □[f1 U f2] then
        return(true)
    else
        return(false)
    fi
else
    mark(s)
    if s ⊨ f2 then
        s ⊨ □[f1 U f2]
        return(true)
    else
        if s ⊨ ¬ f1 then
            s ⊨ ¬ □[f1 U f2]
            return(false)
        else
            if ∀s' successor of s, Verify_□U(s') then
                s ⊨ □[f1 U f2]
                return(true)
            else
                s ⊨ ¬ □[f1 U f2]
                return(false)
            fi
        fi
    fi
fi

begin
    ∀s unmark(s)
    ∀s Verify_□U(s)
end
```

$$\text{P1} \quad \langle \mathcal{N}, \mathbf{m_0} \rangle \models \mathsf{G} f_{deadlockfree}(\mathcal{N})$$

$$\text{P2} \quad \langle \mathcal{N}, \mathbf{m_0} \rangle \models \mathsf{G}((CS_1 + CS_2) \leq 1)$$

$$\text{P4} \quad \langle \mathcal{N}, \mathbf{m_0} \rangle \models \bigwedge_{i,j \in \{1,2\}, i \neq j} \mathsf{G}[(Wait_i = 1 \wedge Idle_j = 1) \Rightarrow [\neg(CS_j = 1) \, \mathsf{U} \, (CS_i = 1)]]$$

Verification. The verification technique that we present for LTL is often referred to as the automata-theoretic approach. It is based on the fact that a property can also be characterised by the set of behaviours that satisfy it. This set of behaviours is described by a specific kind of automaton, namely a Büchi automaton (see [GPVW93] for instance). Such an automaton is finite, possibly contains several initial states, and has a set of particular states called the accepting states. A transition of this automaton is labelled with a set of atomic propositions. Informally, an infinite word is accepted by the Büchi automaton if it corresponds to an infinite execution of the automaton encountering infinitely often an accepting state. Figure 14.2 presents an example of such a Büchi automaton. State 1 is an initial state, and states 2 and 3 are accepting states. The language of the automaton, i.e. the set of accepted words, is $\mathit{True}^*(w_1 \wedge \neg cs_1)(\neg cs_1)^\infty + \mathit{True}^*(w_2 \wedge \neg cs_2)(\neg cs_2)^\infty$ where $*$ stands for finite and ∞ stands for infinite repetition. This automaton actually represents the formula:

$$\bigvee_{i,j \in \{1,2\}} \mathrm{F}[Wait_i = 1 \wedge \mathrm{G}(CS_i < 1)]$$

if w_i stands for $Wait_i = 1$ and $\neg cs_i$ stands for $CS_i < 1$.

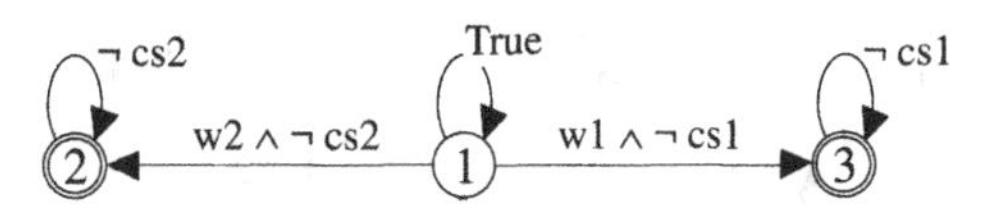

Fig. 14.2. Büchi automaton for Property ¬ P3

This temporal logic formula is the negation of Property P3. As a consequence, the automaton represents the set of behaviours for which Property P3 is false. In order to check whether our mutual exclusion model verifies Property P3, there are two possibilities: either check whether all the infinite behaviours represented by the reachability graph of the system are included in the set of behaviours that satisfy the property, or check whether there exists a behaviour in the reachability graph for which the negation of the property is true. In the first case, we prove that the property is true, and in the second we prove that it is false. The second approach is usually chosen because it is less expensive.

In this case, the verification is done by finding the intersection between the behaviours of the system and those represented by the Büchi automaton associated with the negation of the property. A way to compute the intersection is to construct the *synchronised product* of the reachability graph and the Büchi automaton. Informally, this product is an automaton in which the states are pairs (e_i, e_j) where e_i is a state of the reachability graph and e_j a state of the Büchi automaton. There is a transition between states (e_i, e_j) and (e_k, e_l) if and only if there is a transition between e_i and e_k in the reachability graph, there is a transition labelled by a between e_j and e_l in the Büchi automaton, and a is true in e_i. The initial states of the product graph are states (e_i, e_j) such that e_i is an initial state of the reachability graph and e_j is an initial state of the Büchi automaton. The accepting states of the product graph are states (e_i, e_j) such that e_j is an accepting state of the

Büchi automaton. We illustrate this construction on our example. The reachability graph of the mutual exclusion model is presented in Figure 14.3 and the synchronised product of this graph with the Büchi automaton associated with Property ¬P3 is given in Figure 14.4.

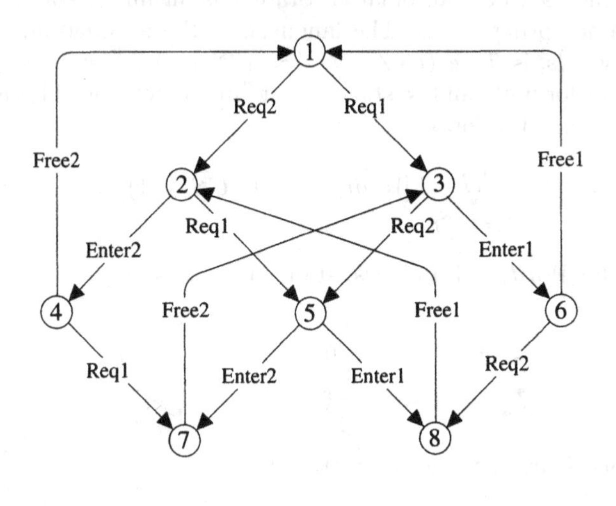

1 :	$Idle_1 + Idle_2$	5 :	$Wait_1 + Wait_2$
2 :	$Idle_1 + Wait_2$	6 :	$Idle_2 + CS_1$
3 :	$Idle_2 + Wait_1$	7 :	$Wait_1 + CS_2$
4 :	$Idle_1 + CS_2$	8 :	$Wait_2 + CS_1$

Fig. 14.3. Reachability graph for the mutual exclusion algorithm

The verification of a property by means of the resulting automaton completely depends on the class of this property. For safety properties, the resulting automaton has only one accepting state. Checking that the property is true reduces to verifying that this accepting state is not reachable from an initial state. For liveness properties, the resulting automaton should not accept infinite sequences that encounter infinitely often an accepting state. In other words, there must be no cycle reachable from the initial state that contains an accepting state.

In our example such a cycle, for instance $(82, 22, 52)$, exists. Hence, Property P3 is not true. Actually, this cycle shows the existence of an infinite sequence in which Process 2 infinitely waits for the critical section without entering it.

Linear time model checkers are exponential in the size of the formula but linear in the size of the model ([LP85, VW86]). However, advocates of linear time logics argue that the high complexity of the model checkers is acceptable for short formulae.

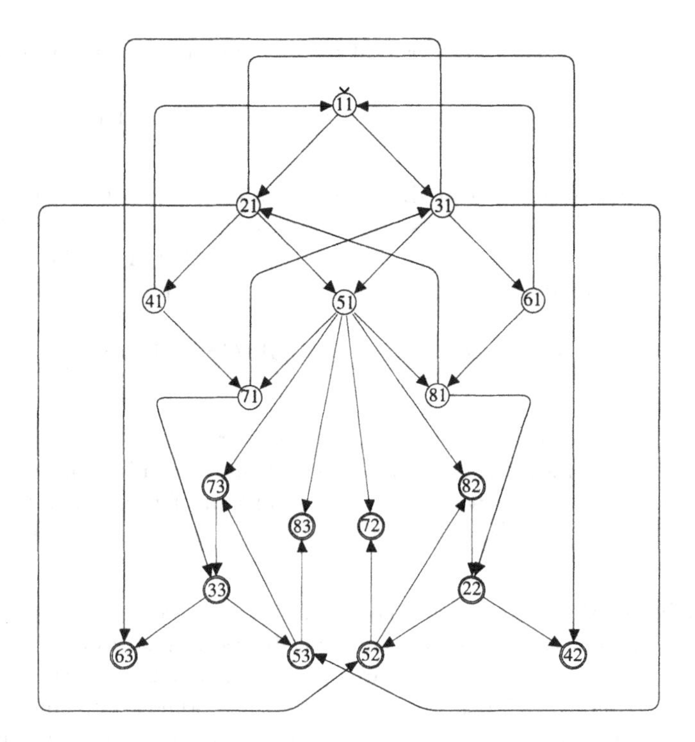

Fig. 14.4. Synchronised product of the RG and the Büchi automaton

Model Checking Under Fairness Assumptions. Very often, only some of the executions represented by the reachability graph are of interest when one wants to verify a formula. The executions that are eliminated are those which do not correspond to a possible behaviour of the system, for instance those where the scheduling of the tasks is not compatible with the scheduling algorithm executed by the system. In particular, the allocation of the processor is generally assumed to be fair, in the sense that a task that remains ready will eventually be executed. Hence, only *fair* executions should be considered when checking some property on the system.

Instead of trying to represent the fairness of the scheduler when modelling the behaviour of the system, it is more convenient to "filter" at the verification step the executions that should not be considered. The idea is to represent the fairness constraint by a temporal logic formula. This can be done easily with the linear time logic LTL and the verification of a formula f under fairness constraint f' will be done by checking that the model verifies the LTL formula $f' \Rightarrow f$.

Let us come back to our mutual exclusion example. We consider the case where the scheduler obeys the so-called *Strong Fairness* constraint, i.e. if an action is enabled infinitely often, then it is executed infinitely often. In the particular case of access to the critical section, this constraint can be written as:

$$\bigwedge_{i \in \{1,2\}} [\mathrm{GF} \mathit{enabled}(\mathit{Enter}_i)] \Rightarrow [\mathrm{GF}(CS_i = 1)]$$

where $\mathit{enabled}(\mathit{Enter}_i)$ is a predicate which is true each time Enter_i is enabled. In our example, it can also be expressed as $(Wait_1 \geq 1) \wedge (CS_2 < 1)$ for $i = 1$ and $(Wait_2 \geq 1) \wedge (CS_1 < 1)$ for $i = 2$. Hence, the starvation freeness in this strongly fair environment would be expressed as

$$\bigwedge_{i \in \{1,2\}} [\mathrm{GF} \mathit{enabled}(\mathit{Enter}_i)] \Rightarrow [\mathrm{GF}(CS_i = 1)] \Rightarrow P3$$

It is easy to check that this new property is true.

The example above uses the LTL formalism. Actually, fairness properties cannot be expressed with CTL. An extension of the language, the so-called Fair-CTL ([EL87]), has been proposed to take into account fairness constraints.

Conclusions. Many theoreticians and practitioners recognise the utility and appropriateness of temporal logic as a specification and verification tool for concurrent systems. However, they are divided into two groups: the advocates of linear time and the advocates of branching time. Some arguments for linear time logic can be found in [Lam80], whereas some arguments in favour of branching time logic are presented in [EH86]. In both cases, the arguments rely on the expressiveness and the complexity of formulae and verification algorithms.

Although the examples of the current section do not illustrate this fact, some properties can be expressed with a linear time logic and not with a branching time logic and vice versa. For instance, it is not possible to express the indeterminism of concurrent programs, i.e. possibilities of an execution (*possibility* properties), with a linear time logic. On the contrary, pure branching time logic is not suitable for the expression of fairness constraints.

We have seen that the complexity of verifying a formula is linear in the size of the model for both CTL and LTL. It is linear in the size of the formula for CTL but exponential for LTL. The complexity for verifying a CTL* formula is the same as for verifying an LTL formula, and Büchi automata are also often used in this case. Actually, Büchi automata are being used more and more even in the model checking of CTL formulae.

However, the complexity with respect to the formula is usually not the most important consideration since many interesting properties of a system can be expressed with short formulae. In contrast, the number of states of the model is often huge and it is very expensive to store the complete reachability graph in the memory of a computer. Hence, model checking techniques have been developed which either avoid storing the whole graph at the same time, or make it possible to check a property while constructing only a part of the graph. The next section of this chapter is devoted to the presentation of some of these techniques.

14.2 On-the-Fly Approaches

In Section 14.1.1, we have shown that automata-theoretic approaches offer a new understanding of model checking by reducing it to cycle detection. Cycle detection is a classical problem of graph theory which can be solved in linear time with respect to the size of the graph, and which mainly consists of computing the strongly connected components of a graph. Hence, one can state the validity of a formula by (1) building two automata: the state space of the system (i.e. the reachability graph) and the automaton which characterises the formula to be checked, (2) building the structure corresponding to the synchronised product of these automata, and (3) computing the strongly connected components over the result viewed as a graph. While this is a simple way to demonstrate the intuition of the model checking procedure, it is clearly not a reasonable way of implementation. In fact, one must face complexity problems induced by the size of the representations to be built, in particular:

- Because of combinatorial effects, formulae and state spaces can yield automata too large to be stored entirely in memory.
- The size of the synchronised product representation is strongly related to a product of nodes, thus can easily exceed the strict bound imposed by the size of the memory, even if the construction of the former automata succeed.
- The execution time required can be excessive since it is proportional to the size of the synchronised product representation.

Another way to proceed consists of checking the satisfaction of the property during the building of the synchronised product representation. Hence, one aims to finish the model checking as fast as possible, while reducing the need for memory. Various techniques based on this approach have been applied to check temporal logics and are known as *on-the-fly* model checking.

In this section, we concentrate on the easily implemented solutions that have been proposed for LTL properties. The starting point is a synchronisation between the state space of the system and a Büchi automaton which corresponds to the negation of the property to be checked. A synchronised product between a reachability graph and a Büchi automaton has been demonstrated in Section 14.1.1. In this context, the negation of the property is detected to be false (equivalently, the property is shown to be true) as soon as a reachable cycle that loops over an accepting state is found in the synchronised product. In this case, depth-first searches (DFS) can be used instead of computing strongly connected components.

An On-the-Fly Algorithm for Checking LTL Properties. In [CVWY92], the cycle search is performed using two kinds of *DFS*: DFS_1 is used to detect the states that are simultaneously reachable from the initial states and accepting from the Büchi automaton point of view, whereas the purpose of DFS_2 is to check whether any accepting state which is reachable from the initial state is also reachable from itself. There are two possible outcomes:

- The algorithm ends normally without finding any cycle.
- It aborts as soon as a cycle is detected, indicating that the property is false.

The two kinds of *DFS* can be nested in order to directly produce a sequence of states that invalidates the property, if such a sequence exists.

Because several reachable states may be accepting, the same path can be parsed several times (by distinct instances of DFS_2). Fortunately, the reachable accepting states can be ordered in such a way that the ones which have already been visited for a cycle detection need not be visited again in further searches. The parsing made by DFS_1 is used to order these states according to their most recent visits. This is based on the following key points:

- The *DFS* algorithm on a graph structure corresponds to the parsing of a tree provided every cycle is cut (i.e. a cycle is detected whenever the current state is already visited).
- The parsing of a tree can be performed from its root by a recursive procedure which is called for every possible successor of the current node.
- States which are accepting are detected during the backtracking process of DFS_1; hence the reachable accepting states which are the closest to the leaves are considered first.

Algorithm 14.5 follows this scheme (see also [CVWY92]). For the sake of simplicity, we consider that there is only one initial node in the synchronised product. The DFS_1 starts with a call to procedure *search*; moreover, as soon as a reachable accepting state is detected, a DFS_2 starts with a call to procedure *detectCycle*. Each *DFS* is associated with two kinds of data structures:

- An explicit stack is used to store the visited nodes of the current path ($Stack_1$ and $Stack_2$).
- An explicit heap is used to store the other visited nodes (H_1 and H_2).

The memory is assumed to be large enough that any node yet unvisited can be stored. The reader may refer to Section 14.5 (state-space caching and hashing compaction) in which this assumption is relaxed by the use of efficient data management techniques.

Example. Figure 14.5 depicts the behaviour of the algorithm on a synchronised product representation which is assumed to be developed on-the-fly.

Algorithm 14.5

```
    Initialise: Stack1 and Stack2 and H1 and H2 are emptied;
    enter s0 in H1;
    push s0 onto Stack1;
    search()
end
search() begin
    s = top of Stack1;
    for all s' = succ_formula,system(s) do begin
        if s' is NOT already in H1 ∪ H2 then
            enter s' in H1;
            push s' onto Stack1;
            search();
            fi
    od
    if s is accepting then
        transfer s from H1 to H2
        push s onto Stack2;
        detectCycle();
    fi
    pop s from Stack1;
end{search}
detectCycle() begin
    s = top of Stack2;
    for all s' = succ_formula,system(s) do begin
        if s' is bottom of Stack2 or in Stack1 then abortSearch(Stack1 · Stack2);
        if s' is NOT already in H2 then
            transfer s' from H1 to H2;
            push s' onto Stack2;
            detectCycle();
        fi
    od
    pop s from Stack2;
end{detectCycle}
```

1. The first part of DFS_1 develops a tree, the nodes of which are $0, 1, 2, 3, 4, 5$, and 6.
2. During the backtracking, node 3 is detected as an accepting node.
3. A DFS_2 is started from node 3, hence nodes 4, 5, and 6 are parsed again, with the partial result that no cycle is detected from node 3.
4. The second part of DFS_1 is developed, backtracking from node 3 and parsing nodes 7 and 8. One may note that nodes 6 and 0 are not visited again since they have been already visited by the first portion of DFS_1. Moreover, during the backtracking from node 8, node 7 is detected as the second accepting state to be dealt with.
5. So, a second DFS_2 is started from node 7, which parses nodes $8, 0, 1$, and 2 without parsing node 3 which was already visited during the first DFS_2. Finally, node 7 is reached again, causing a cycle to be detected and

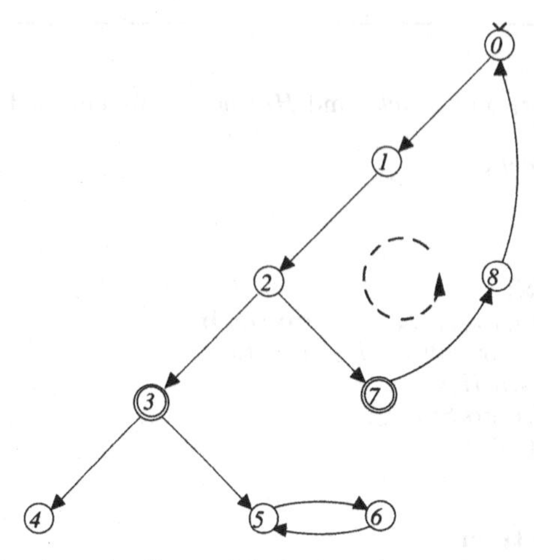

Fig. 14.5. Example of on-the-fly model checking

the search to be aborted. Actually, the stack of DFS_1 which currently stores nodes $0, 1, 2$, and 7 and that of DFS_2 which stores $8, 0, 1, 2$, and 7, yield the erroneous path.

Concluding Remarks. Model checking is a major foundation for checking temporal properties, and on-the-fly approaches contribute to the use of it. The algorithm presented deals efficiently with LTL formulae by considering their representations in Büchi automata. Moreover it is worth noting that the authors of [GPVW93] have presented an incremental translation of the LTL formula into a Büchi automaton which makes the on-the-fly approach possible.

Nowadays, LTL on-the-fly techniques have been integrated into the major tools of model checking, for instance SPIN [Hol97], PROD [VHHP95], and PEP [BG96]. It is worth noticing that extensions exist for CTL^* formulae, the truth values of which can also be captured within an automata-theoretic approach, by means of *weak alternating automata*.

Despite the efficiency of the on-the-fly algorithm, a large portion of the state space may be built and parsed; therefore, the next section will show how the Petri net formalism used to model the system can be exploited to improve model checking techniques. Also, some improvements will be obtained by efficient implementation (see Section 14.5).

14.3 Partial-Order-Based Approaches

Partial-order methods attack one of the main drawbacks of the standard interleaving semantics: transition firings are interleaved even if some of them can be executed in true concurrency. This may cause the representation to be exponentially larger than necessary. For instance, the state space of a system composed of n transitions that can be fired in true concurrency from some state contains $n!$ interleavings in order to represent all the firing sequences. Since the difference is only in the order of firings, the state-space representation can be reduced by defining some representative sequences for these interleavings. Mazurkiewicz in [Maz87] gave the first foundations of this new theory, by defining the notion of traces. A *trace* represents a set of firing sequences, such that two sequences in a class can be obtained from each other by successively permuting independent and adjacent firing occurrences. In practice, the independence relation between the system events is defined in opposition to dependence relations: e.g. two events may be in conflict (the firing of t_1 prevents that of t_2 or vice versa) or may be in causal relations (an occurrence of t_1 must be preceded by a firing of t_2).

The corresponding methods are called *partial-order methods* since dependence and independence relations feature the partial ordering of transition firings. Two rather distinct directions are followed: the first one proposes to reduce the reachability graph representation by removing the sequences which are redundant for the trace representation. The trace-based graph approach has been developed by several researchers, including Valmari [Val88, Val89, Val90b, Val91b, Val93c], Godefroid and Wolper [God90b, God94, GW91b, GW94], and Peled [Pel93, Pel94, KP92, PP94].

The second approach aims to represent the partial order of transition firings directly, by reusing the notions of concurrency and conflict from place/transition nets. This was initiated in [NPW80, Win87] by introducing a translation of any Petri net to a specific labelled Petri net, called a branching process. Such a translation is called an unfolding since any transition (and thus any place) may be represented several times within a process, according to possible firings of the transitions. In this representation, events which are independent are modelled by independent transitions, thus featuring any of their interleavings without representing them. Several works have developed this approach, demonstrating efficient algorithms for building such a structure and checking system properties on it: [McM95, Esp92a, Esp93, KKTT96, CP96].

Partial-order approaches, trace-based graphs, and net unfoldings are used to check a very large class of temporal properties. In the next sections, we recall their foundations and investigate related results as well as their algorithms of verification.

Figure 14.6 presents a simple net (and its reachability graph) which is used as an example to illustrate these different techniques.

14.3.1 Traces and Verification Issues

Within a state space, partial order is related to a basic relation of independence among certain transition firings, namely the diamond property, that is if one sequence $t_1.t_2$ is enabled from some state, then $t_2.t_1$ is also. Moreover

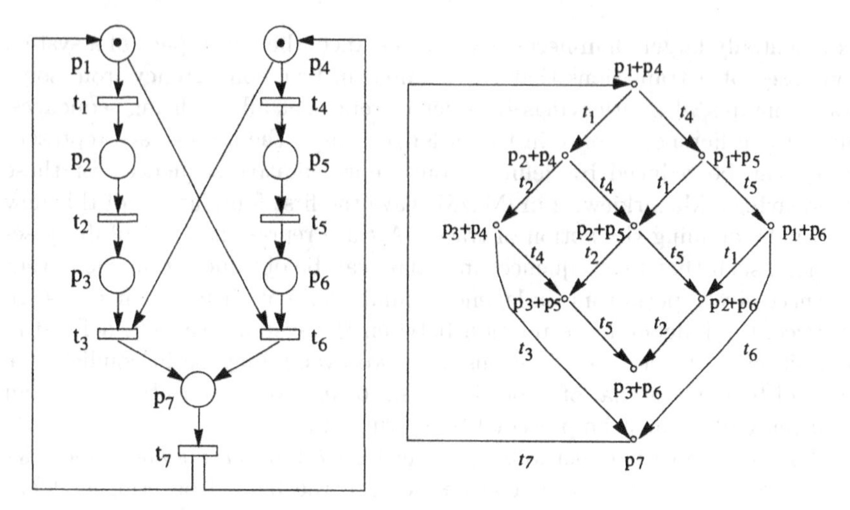

Fig. 14.6. A net and its reachability graph from state $p_1 + p_4$

both sequences reach the same target state. According to this independence relation, the scenarios of a system can be gathered in equivalence classes called traces in such a way that two sequences are equivalent if they can be obtained from each other by successively permuting adjacent independent transition firings. From the reachability graph depicted in Figure 14.6, one may note that (t_1, t_4) from state $p_1 + p_4$, (t_2, t_4) from state $p_2 + p_4$, (t_1, t_5) from state $p_1 + p_5$, and (t_2, t_5) from state $p_2 + p_5$ are pairs of independent transition firings. Therefore sequences from state $p_2 + p_4$ such as $t_2.t_4.t_5$, $t_4.t_2.t_5$, and $t_4.t_5.t_2$ can be regarded as equivalent and be represented by the same trace, e.g. $[t_2.t_4.t_5]$.

In a reachability graph, the number of interleavings can be reduced by focusing on representative sequences of traces. More precisely, a trace starting from the initial state of the reachability graph will be represented if each of its sequences corresponds to a prefix of another sequence obtained by a valid permutation of one path of the reduced graph. Observe that the prefix relation is introduced to obtain a more concise graph, for instance, sequence $t_4.t_2$ is a prefix of $t_4.t_2.t_5$, therefore it can be represented by trace $[t_2.t_4.t_5]$. Figures 14.7, 14.8, and 14.9 illustrate three reduced graphs with respect to the above definition (the corresponding technique will be detailed further in sections 14.3.2 and 14.3.3). In these figures, bold arcs represent the selected transition firings. The removed arcs are also drawn but with gray lines, to show the gain achieved by of each technique. These graphs differ according to their chosen representatives, but any of them covers all the traces obtained from the initial state of the system, that is all the traces contained in one the following three expressions: $[t_1.t_2.t_4.t_5]$, $[[[t_1.t_2.t_3 + t_4.t_5.t_6].t_7]^*.t_1.t_2.t_4.t_5]$, and $[[t_1.t_2.t_3 + t_4.t_5.t_6].t_7]^\infty$. The best case is obtained in Figure 14.9 since no more reduction can occur without loss of traces.

Another important point is that incorrect behaviours must not be introduced with respect to the properties to be verified. Hence, only reduced representations of the reachability graph are possible, thus avoiding the existence of incorrect states and incorrect sequences of transition firings. In the same way, sub-traces (in the sense of trace inclusion) are risky since one can conclude falsely that there is no causal relation in some path. For instance, without more information, the use of the graph in Figure 14.8 is confusing with respect to deadlock properties: sub-trace $t_1.t_4.t_5$ from the initial state leads to the representation of false deadlocks.

By using a valid trace-based graph, a large class of properties can be checked. A partial representation of traces is sometimes enough. In particular, the quasi-liveness property can be checked by saving only one trace per transition. Invariant properties are also simple since they can be checked by introducing a test transition, then non-quasi-liveness is required. For instance, the property which expresses that "places p_2 and p_5 are never marked simultaneously" can be tested by introducing a transition with p_2 and p_5 as input places and checking that the added transition is not quasi-live. General safety and liveness properties need more effort because one must take into account *observable* and *invisible* transitions with respect to a property to verify. Transitions are defined to be *observable* provided their firings could change the truth values of the atomic propositions comprising the formula, while the other transitions are called *invisible* because their effects are not observable in the proposed model checking. In the place/transition net context, a transition can be defined as observable whenever one of its input or output places is concerned with the atomic properties of the formula to verify. Actually, three key points must be considered, since temporal properties can be sensitive to the order of transition firings:

1. A sufficient set of traces must be preserved in order to cover all the interleavings of observable transition firings, thus allowing a comparison against sequences of the Büchi automaton of the formula.
2. A consequence of the first point is that reduction must involve invisible transitions only. This allows us to check the safety property and also liveness properties concerning infinite cycles over observable transition firings. However, a path which contains a loop over invisible transitions risks being lost. Such a path, called a *divergent sequence*, must be captured if one wants to check general liveness properties; hence there is a need to retain some loops of invisible transitions (but not necessarily all).
3. The third point is to define the kind of formulae that would benefit from trace-based graph techniques. Effectively, the satisfaction of point (1) suggests the building of the standard reachability graph as a trace-based graph in case where all transitions are observable. Of course, this would eliminate the benefit of using trace-based graph techniques. This is the case for state-based formulae containing the Next-time operator, therefore several works refer to LTL-$\circ$ where $\circ$ means the Next-time op-

erator ([WW96]). Such a logic belongs to a large family of logics that restrict the class of properties to stuttering-invariant properties. *Stuttering invariance* means that the truth value of a formula is not affected if some information needed for the verification is repeated through a finite sequence of states. Without ensuring stuttering invariance, all the transitions must be considered as observable (i.e. by definition, invisible transitions cause such repetition of information). In [PWW96], a fine algorithm is presented to decide stuttering invariance with regard to some property.

Of course, it would be irrelevant to detect (in)dependence relations of transition firings by building the reachability graph. In practice, trace-based graphs are built on-the-fly since it is possible to compute independence relations from the Petri net structure with respect to each visited marking. This computation is mainly based on an absence of firing conflicts among subsets of enabled transitions according to some state. Hence, it is possible to select only some of the enabled transitions from a state, provided their firings do not cause conflict with the other enabled ones. As a consequence, the enabled but unselected transitions at some state remain enabled in the next states until one decides to fire them. The graph obtained can thus be a reduced representation of the reachability graph, however, it is worth noting that it does not necessarily preserve important sequences allowing one to prove certain properties. Actually, two problems may appear whenever some transition is enabled but not selected:

- The first problem, called the *ignoring problem* of transition firings, is related to the enabled transitions at some state that are unselected and which remain unfired in all further reached states. This appears when some sets of transition firings are independent of others, whatever the state considered. A system composed of independent processes can be used as a straightforward example: a trace-based graph selecting only enabled transitions from only one independent process would leave aside the enabled transition firings of the other processes.
- The second problem, called *confusion cases*, is related to some state that remains unvisited although it is the beginning of sequences that are not represented in another way. Let us again consider the net of Figure 14.6 and the trace-based graph of Figure 14.7. Although transitions t_1 and t_4 are not in effective conflict from the initial state, both must be fired from this state. Effectively, if one of them is not fired, e.g. t_4, then every sequence containing the firing of t_6 is lost since state $p_1 + p_6$ is never reached. This is due to a structural conflict between t_1 and t_6 which is not effective in the initial state but which becomes effective after firing $t_4.t_5$. Thus, a dependence relation can be hidden behind causal relations.

We now discuss three kinds of solutions for building complete and valid reachability graph reductions which benefit from the trace-based approach.

The first one, persistent sets, is a direct application of the above presentation, while the sleep set technique is a complementary approach. Finally, the covering step technique allows one to build a particular graph which can also be viewed as a trace-based graph.

14.3.2 Persistent Set Searches

A set of enabled transitions called a *persistent set* is defined in every state so as to select each time a subset of transition firings which are independent from the other ones. In other words, if T is a persistent set in a state s and if one sequence $t.w$ is enabled from s such that t belongs to T and w is a sequence of transition firings taken out of T, then all the permutations mixing the firing of t with the transition firings of w are enabled and reach the same target state. As a brute force case, the set of all the transitions enabled from a state is a persistent set in that state.

With such a notion, the building of a trace-based graph is similar to the ordinary one except that:

- Once a state is examined, the algorithm computes a persistent set of transitions including at least one enabled transition (if the marking is not dead).
- The successors of the state are the ones reached by the enabled transitions of the persistent set. The persistent-set technique can be viewed as a linearisation of independent transition firings. Effectively, the enabled transitions which are not selected in a state remain enabled in the next states until they are fired. Consequently, the representation of the state space is reduced while preserving all deadlocks and the existence of infinite firing sequences. It is worth noting that the independence of persistent sets from the other transition firings would lead to the ignoring of some enabled transitions. Fortunately, this does not occur if the net is bounded and strongly connected ([Val89]). In other cases, the ignoring problem can be avoided by checking in each visited state the satisfaction of some additional constraints, called provisos. In [Val89], a first proviso is proposed based on the detection of every non-trivial terminal strongly connected component while building the visited state space. In case some enabled transition remains unselected within one of these, an additional persistent set is chosen so as to consider more enabled transitions. Another proviso, more simple, is proposed in [HGP92]. It consists of detecting states from which an enabled transition firing closes a circuit. In such a state, all the enabled transition firings are considered. The fact that there are no ignored enabled transitions yields a valid trace-based graph in which deadlocks as well as invariant state properties can be checked.

Several additional provisos have been proposed to enable model checking of (linear time) temporal properties. That of Valmari [Val90b] expresses the following two requirements: (1) from each visited state that is reached during the construction of the trace-based graph, if there exist invisible transitions then at least one is executed; (2) every cycle must contain at least one state where all enabled visible transitions are explored. Intuitively, the first requirement maintains cycles of invisible transitions in the trace-based graph, while the second ensures that visible transitions are not ignored. Several works have demonstrated how to build a trace-based graph ensuring requirement (2). In particular, another proviso can be used to force the firings of all the transitions enabled from the state currently treated during the graph construction each time this state is detected to close a cycle ([Val93c, Pel94]). Moreover, when model checking is performed by means of an on-the-fly synchronisation against the states of a Büchi automaton (see Section 14.2), it is possible to reduce the checking of the former two requirements: the first requirement must be ensured only when meeting accepting states, while the second is required in the case of non-accepting states ([Val93c]).

Based on the analysis of a Petri net structure, different algorithms have been proposed to efficiently compute persistent sets in each visited state, from the most simple one which limits the searches to conflict detections ([God94, Bau97]) to the most advanced one, the stubborn set technique ([Val89]). These techniques are presented next.

Stubborn Set Searches. A stubborn set is built at each state to select a subset of the enabled transition firings which are independent from the others. In order to avoid confusion cases, it can also contain disabled transitions for capturing conflict cases that might be effective in the next states. Hence, dependencies of transitions are analysed not only through conflicts but also through causality relations. Any selected transition must satisfy one of the following two conditions, depending on the result of its enabling test:

- If it is enabled, also select all the transitions which are in structural conflict with it.
- If it is disabled, then choose one of its deficient input places and also select all the input transitions of that place (*deficient* means that such a place does not have enough tokens and thus disables the transition).

This results in a fixpoint procedure since one may have to select some other transitions each time a transition is selected, depending on whether it is enabled or disabled. The worst case would yield a stubborn set containing all the enabled transitions of the state considered (no reduction appears). The stubborn set may also be a single enabled transition, without conflict against any other transition.

If the current state is not a deadlock state, then from the computation of a stubborn set, a persistent set is obtained which contains at least one enabled transition, and includes all the enabled transitions of the stubborn set. Figure 14.7 depicts a reduced graph obtained by applying the stubborn set technique at each visited state, with respect to the net of Figure 14.6. For instance, the stubborn set of the initial state is $\{t_1, t_6, t_5, t_4, t_3, t_2\}$ leading to

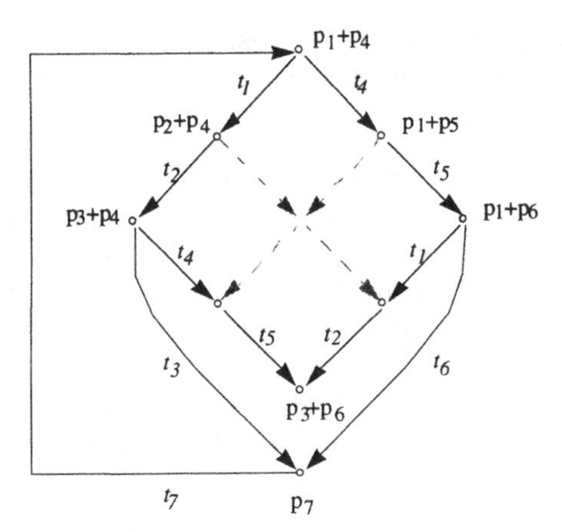

Fig. 14.7. Stubborn set method

the persistent set $\{t_1, t_4\}$. It can be built from transition t_1 which is enabled in the initial state; since t_1 is selected, t_6 is also due to conflict (t_1, t_6); then t_5 is selected because t_6 is disabled, p_6 is the input place of t_6, and p_6 is the output place of t_5; then t_4 is selected for similar reasons; then t_3 is selected since t_4 is enabled and in conflict with t_3; then t_2 is selected since t_3 is disabled; then t_1 is selected again since p_2 is disabled, which terminates the searching procedure.

Of course, the size of a stubborn set is very sensitive to the choice of the first enabled transition and the choice of an input place in the case of a disabled transition. There may exist several stubborn sets for a state. In order to ease the choice of a (small) stubborn set, Valmari observes that testing the inclusion of stubborn sets can be checked in linear time because it corresponds to a search of a minimal strongly connected component. The reader may refer to [Val89] for more details. For instance, in state $p_2 + p_4$ (i.e. after the first firing of t_1), there are two stubborn sets containing enabled transitions: $\{t_2\}$ is found if the analysis starts from transition t_2, whereas $\{t_4, t_3, t_2\}$ is found from transition t_4. The observation that $\{t_2\}$ is included in $\{t_4, t_3, t_2\}$ allows one to take $\{t_2\}$ as a persistent set in either case (note that t_2 is enabled). Thus, the selection of the enabled t_4 transition is postponed to further visited states.

The efficiency of a stubborn set implementation may depend on several parameters:

- The choice of an input place in the case of disabled transitions
- The choice of a persistent set since there may be several possible persistent sets for one state
- The choice of a procedure to detect and solve the ignoring problem of enabled transitions.

Conflict-Based Dependency Searches. A simpler technique for computing a persistent set appears in [GW91a]. It is based on the following two remarks:

1. A set of enabled transitions which is obtained by a fix point procedure on the conflict relation is a particular type of persistent set (in other words, starting from one enabled transition, this procedure recursively adds the transitions that are in structural conflict with the ones previously selected; in addition, it fails whenever a disabled transition risks being added).
2. It is always possible to avoid disabled transitions by using the following method: if the former construction fails, then return the entire set of enabled transitions.

Hence, all the transitions selected in the persistent set are enabled and have no conflict with those not selected. Often, this technique may yield larger persistent sets than those of the stubborn set technique, because the application of point (2) gives no reduction in a state. The former approach can however be efficient for some nets such as the one depicted in Figure 14.6. For this example, the conflict-based dependency approach obtains the same graph as that comuted by the stubborn set technique (see Figure 14.7). In fact, in this example the application of point (1) never fails, so point (2) is never applied.

For model checking, this technique is comparable to stubborn sets, and the same general solutions can be used. Complementary approaches have been developed by Bause. In [Bau97], the way to choose a convenient subset of the enabled transitions in a state is specified by means of a notion of transition priority (high and low), dynamically reconsidered from each state: low priority transitions cannot be fired if some higher priority transitions are enabled. Some interesting results are given which assume that the net is bounded and strongly connected. In particular, the dynamic priority rule ensures that liveness (in a Petri net sense) is preserved. Also, home states can be preserved given the following restriction: a high priority in a state may concern only a set of enabled transitions which are in (extended) equal conflict , one from another. In other words, these conflicting transitions are enabled and share the same input places, thus are either all enabled or all disabled. One may remark that a transition which has no conflict is a degenerate case of equal conflict. For instance, this is the case for transitions t_2 and t_5 in the net of Figure 14.6, and both are enabled from the state $p_2 + p_5$, therefore one may linearise their firings while preserving home state properties.

In [Bau97], it is also highlighted that, instead of an arbitrary policy, a careful selection of persistent sets can improve the graph reduction. The proposed algorithm in a sense mimics the firing of T-invariants to reduce the length of visiting cycles as much as possible. Experience shows that it should perform well when large parts of the net have an equal-conflict structure.

14.3.3 Sleep Set Searches

The sleep set technique, introduced by [God94], exploits the absence of effective conflict between some transition firings. It aims to avoid having some target states reached by several interleavings (due to the diamond property). In this technique, a specific set, called the *sleep set*, is associated with each state. Unlike stubborn sets, a sleep set of a state represents the enabled transitions that are not worth firing, because their own target states are known to be reached by another way. In this way, the sleep set technique proposes eliminating some arcs of the reachability graph.

Sleep sets are built on-the-fly during the graph construction, starting from an empty sleep set associated with the initial state. From an already computed state s and its sleep set, consider the firings of the enabled transitions successively except for those of the sleep set. Whenever the firing of one of these transitions, t, reaches a state not yet visited, define the sleep set for this state to be the one obtained when reaching s, augmented with all transitions considered in s before t, and purged of all transitions that are in effective conflict with t in s.

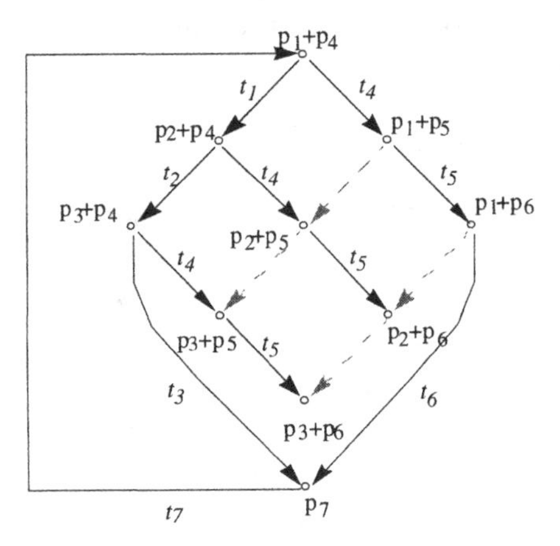

Fig. 14.8. Sleep set technique

The graph of Figure 14.8 is an example of reduction using the sleep set technique. According to the sleep set procedure, the sleep set of the initial state is empty, hence both transitions t_1 and t_4 are fired as usual. If t_1 is fired first, then state $p_2 + p_4$ is obtained first, and then $p_1 + p_5$ is yielded by the firing of t_4. The sleep set of $p_2 + p_4$ remains empty but that of $p_1 + p_5$ is $\{t_1\}$ due to the mutual independence between t_1 and t_4. From $p_2 + p_4$, all the enabled transitions are fired as usual since the associated sleep set is empty, causing in particular state $p_2 + p_5$ to be reached. From $p_1 + p_5$ and its sleep set $\{t_1\}$, only transition t_5 is fired although t_1 is enabled. So, state $p_2 + p_5$

is reached from the initial state but only by sequence $t_1.t_4$. More generally, one can observe that only state p_7 is reached twice while the other states are reached only once.

The main interest of the sleep set procedure comes from the fact that it strongly decreases the number of state matchings against sets of already visited markings. From a space point of view, the fact that all states are preserved indicates that very often the graph reduction is less important than with the stubborn set technique. The comparison of Figures 14.7 and 14.8 highlights such a case. The opposite case is also possible, in particular, sleep set reductions may exist because of an absence of effective conflict for some transition firings, while the persistent set procedure may yield the entire set of enabled transitions at every state.

The preservation of reachable markings allows one to verify all reachability properties, including state invariant ones. All traces are also represented since the removed arcs are useless, but they may be accompanied with sub-traces, such as $[t_1.t_4.t_5]$ with respect to trace $[t_1.t_4.t_5.t_2]$ from the initial state. Hence, some properties such as deadlocks can be checked only by assuming that the transitions of sleep sets are fired. More generally, the standard verification techniques based on the trace-based graph must be modified to take into account the sleep sets associated with the states.

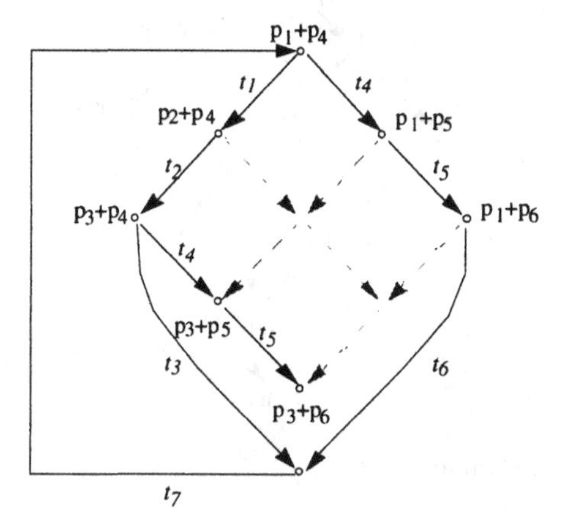

Fig. 14.9. Combining stubborn set and sleep set techniques

Sleep set techniques can also be used to reduce stubborn sets: a sleep set of transitions is removed from any considered stubborn set. Hence, valid trace-based graphs are obtained which are often reduced more than the ones of the standard stubborn set technique. This is highlighted by the graph of Figure 14.9 where the $p_2 + p_6$ intermediate state is no longer reached. Effectively, from the initial state, state p_1+p_6 is reached by the firing sequence

$t_4.t_5$ and $\{t_1\}$ is the selected sleep set of this state. Since $\{t_1, t_6\}$ is the persistent set of state $p_1 + p_6$, only t_6 is fired.

As with persistent sets, general properties can be checked since the proviso proposed by Godefroid to solve the ignoring problem for persistent set techniques remains valid. In fact, an improved version is used since a transition that appears in a sleep set is by definition not ignored.

14.3.4 Covering Step Graphs

The building of a covering step graph (CSG) has been proposed recently in [VAM96, VM97] in order to decrease the interleavings of transition firings. Basically, it takes advantage of independence of some enabled transitions to fire these transitions in a single step, i.e. simultaneously from a state. Hence a new structure, namely CSG, is obtained wherein nodes are taken among the reachable markings, and arcs may represent steps of simultaneous transition firings. For instance, four steps are represented from the initial marking of the CSG in Figure 14.10. The $p_3 + p_2$, $p_4 + p_2$, $p_1 + p_5$, and $p_1 + p_6$ intermediate states are void, causing deadlocks such as $p_3 + p_6$ to be found more rapidly.

By means of a linearisation of steps (i.e. fire the transitions of any step one by one), one can retrieve a valid trace-based graph such as that depicted in Figure 14.10. However, the construction of a CSG offers more since all the sequences of the reachability graph are covered. More precisely, every transition firing sequence in the reachability graph is represented by the prefix of a sequence, equivalent (in the sense of Mazurkiewicz) to a sequence of steps in the CSG. This can be seen by considering the graphs of Figure 14.10. For instance, sequence $t_3.t_2.t_5$ which exists from the initial marking of the reachability is covered by sequence $\{t_2, t_3\}.\{t_5\}$ in the associated CSG. In contrast, this sequence has no representation in the trace-based graph.

A CSG represents the firing sequences of the reachability graph, except that certain transitions are fired by steps according to some independence relations. In every state, the set of enabled transitions is partitioned into a set of transitions that must be fired on their own, and a set of mergeable transitions, i.e. transitions that can be included in steps. The CSG is constructed on-the-fly, once a definition is chosen for independence relations. A rather simple way to define such relations consists of testing in each state the absence of effective conflict with respect to the enabled transitions.

The algorithm for building a CSG is very similar to that for building a reachability graph, however the firing procedure from a state must follow three key points:

1. All the enabled transitions are fired.
2. An enabled transition firing is considered as mergeable if the transitions with which it is in direct conflict are also enabled.
3. There are as many steps as cases of merging with respect to the conflict relations among the mergeable transition firings.

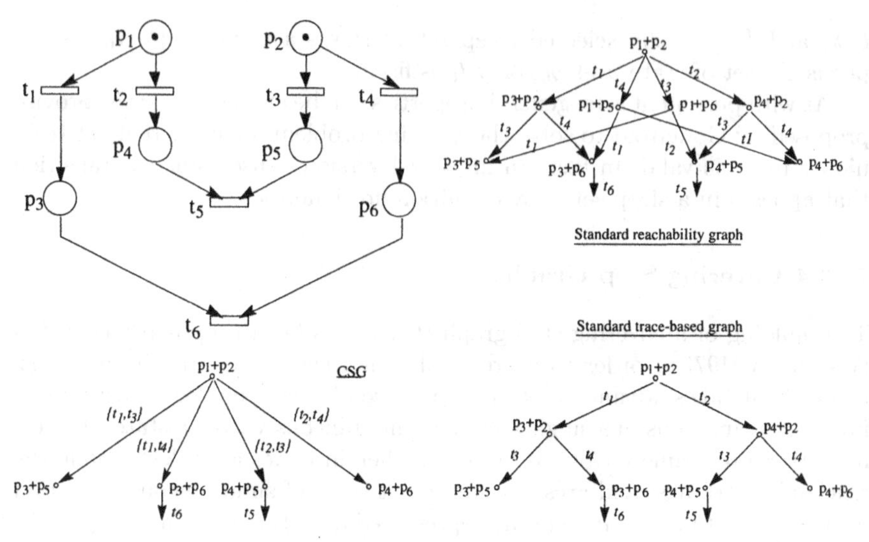

Fig. 14.10. A net, its reachability graph, a CSG, and a standard trace-based graph

Point (1) ensures that no starts of sequences are forgotten while point (2) avoids possible confusion cases because conflicts void merge operations. For instance, let us consider the initial marking of Figure 14.11. By point (1), transitions t_1, t_2, t_3, and t_4 are enabled and can be fired. Because of point (2), the firing of these transitions can be merged. Finally, point (3) ensures that every sequence is covered since every possible merge of independent transition firings is built. Since t_1 and t_2 are in conflict and so are t_3 and t_4, four possible merges occur from the initial state of the CSG of Figure 14.10: $\{t_1,t_3\}$, $\{t_1,t_4\}$, $\{t_2,t_3\}$, and $\{t_2,t_4\}$. Hence, the same transition firings can be repeated several times in steps.

It is worth noting that the efficiency of the CSG approach compared to other techniques such as persistent sets depends on the system considered. In the example of Figure 14.10, the factor of reduction is almost the same for the CSG as for the standard trace-based graph. Effectively, the combinatorial effect of step buildings is compensated by the fact that some (intermediate) states disappear in the CSG. In this example, states $p_3 + p_2$ and $p_4 + p_2$ are present in the trace-based graph while they disappear in the CSG. Moreover, from each of them, the firings of transitions t_3 and t_4 occur. In contrast the CSG of Figure 14.11, which corresponds to the net of Figure 14.6, highlights the opposite case since it contains only one possible step of two transition firings, and thus few reductions. The resulting CSG can be compared to the trace-based graphs of Figures 14.7 and 14.9 which are obtained by persistent set techniques.

Considerable reduction can appear with CSG when some processes in the system can flow independently but periodically reach the same places

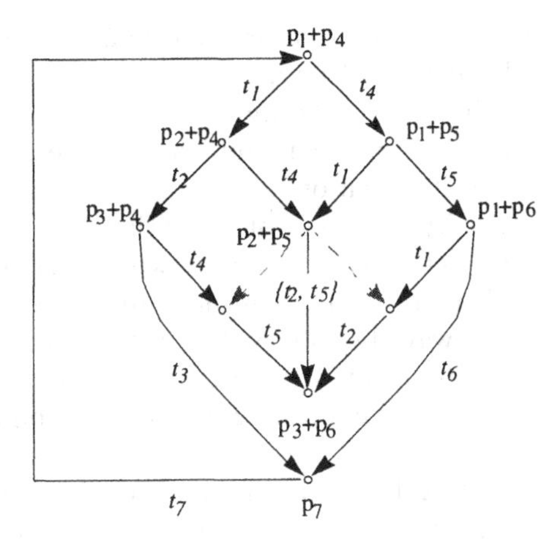

Fig. 14.11. A CSG for the net of Figure 14.6

(or equivalent ones). Examples, such as a model of a multi-copy distributed database, can be found in [VAM96]. In this example, a request broadcast from one site to the others causes the others to answer after some wait. Since there is no direct relation between the addressees, they can act synchronously in single steps while receiving a request, waiting, and answering. In contrast, persistent sets linearise the actions of the addressees. In this case, the more sites, the better the CSG representation with respect to standard trace-based graphs.

The nice property that all sequences are covered by the CSG implies that several path-based properties can be checked, such as deadlocks and above all the liveness of transitions (in a Petri net sense). In contrast with persistent set techniques, no proviso is needed. Moreover, the algorithm may easily be adapted to deal with (observable) transitions. In particular, model checking of the temporal logic property is enabled if every step is constrained to contain at most one observable transition firing. Hence, all the interleavings of observable transition firings are preserved.

14.3.5 Branching Process Techniques

Branching process techniques aim to obtain a direct representation of the partial order of system events, in terms of Petri nets. The basic method is called the unfolding computation. The second method, called the branching process graph computation, corresponds to a generalisation of the unfoldings to make the verification of temporal properties possible.

Branching Process and Unfoldings. The unfolding method was introduced by K.L. McMillan in [McM92] and various improvements exist: [Esp92a, Esp93] are concerned with the verification of properties;

[ERW96, Röm96] with implementation problems; and [KKTT96] with adaptation to more general classes of Petri nets.

As for the reachability graph, the aim is to characterise all the reachable markings and the events enabled from each of them. But rather than representing interleavings of transition firings, the basic idea is to model independent transition firings by independent transitions. Roughly speaking, an unfolding of a given Petri net yields another net wherein nodes are labelled by the elements of the original one; hence, a firing sequence or a reachable marking of an unfolding can always be interpreted in the context of the original net using the labels. The problem is to structurally constrain the unfolding in such a way that efficient procedures for its construction and analysis are defined.

The proposed solution exploits causality and conflict relations. Basically, one can define a labelled causal net, namely a process, for the representation of a Mazurkievicz trace ([Eng91]). A *process* of a net is defined as follows:

- It is acyclic.
- Each place has at most one input transition and at most one output transition.
- Each place (respectively transition) of the process is labelled by a place (respectively transition) of the original net.
- The environment of transitions of the original net must be preserved by defining a convenient set of input and output places for each transition of the process. Thus, with respect to any transition and to the corresponding transition in the original net, there is a one-to-one correspondence of labels of adjacent places.
- The initial marking of a process corresponds (via the labels) to the initial marking of the original net.

Because places have at most one input transition, any pair of process transitions are either in causal dependence (i.e. one must occur before the other) or independent (both can occur concurrently); hence, no conflict can appear.

Moreover, just as the initial marking of a process corresponds to that of the original net and the environment of the transitions is preserved, reachable markings and firing sequences of the process can be interpreted in the context of the original net.

Figure 14.12 presents two processes of the original net from Figure 14.6. The labels referencing nodes of the original net are in italic. The first process (Figure 14.12i) represents trace $[t_1.t_2.t_4.t_5]$ for which the e_1 and e_2 events are in causal dependence, as are e_3 and e_4. Nine reachable markings are represented: p_1+p_4, p_2+p_4, p_3+p_4, p_1+p_5, p_2+p_5, p_3+p_5, p_1+p_6, p_2+p_6, and p_3+p_6. In contrast to this process, that one of Figure 14.12ii is infinite. It describes the trace $[(t_1.t_2.t_3.t_7)^\infty]$.

In a process, any reachable marking is featured by a *cut*, which is a maximal (with respect to set inclusion) set of process places that can be marked

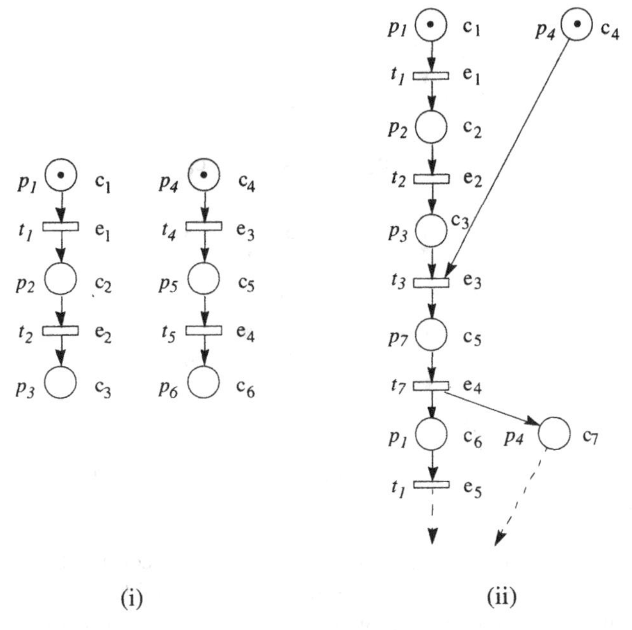

Fig. 14.12. One finite and one infinite process of the net of Figure 14.6

concurrently; in other words there is no causal relation between the places of a cut. For instance, the set $\{c_5\}$ is a cut of the process of Figure 14.12ii and corresponds to the reachable marking p_7 of the original net.

The notion of configuration in a process is the counterpart of cuts since it focuses on transitions instead of places. A *configuration* is a set of transitions downward closed with respect to the causal relation, so it characterises a prefix of a trace. For instance, set $\{e_1, e_2, e_3\}$ is a configuration of Figure 14.12i. With each transition of a process is associated a minimal configuration, that is the minimal set of transitions containing the given transition and forming a configuration. For instance, the minimal configuration of transition e_2 in the process of Figure 14.12i is the set $\{e_1, e_2\}$. Observe that there is a close relation between the notions of configuration and cut: to each configuration there corresponds a maximal cut ($\{c_3, c_4\}$ for $\{e_1, e_2\}$ in Figure 14.12i) and to each cut there corresponds a minimal configuration ($\{e_1\}$ for $\{c_2, c_4\}$ in Figure 14.12ii).

In [Eng91], Engelfriet has shown how a set of processes can be represented by a labelled occurrence net called a branching process. A *branching process* is a process in which conflicts are permitted. Hence, a branching process is also an acyclic net in which places can have at most one input transition but the number of output transitions is not constrained. As for a process, the environment of the transitions of the original net must be preserved. Figure 14.13 presents a branching process of the net from Figure 14.6.

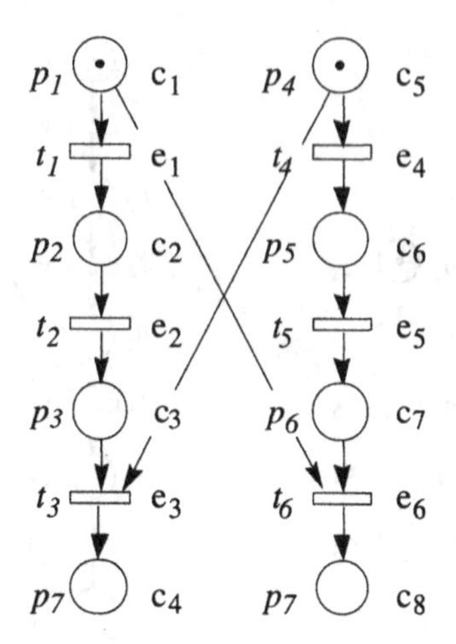

Fig. 14.13. A finite branching process of the net from Figure 14.6

The notions of configurations and cuts have to be redefined in the context of branching processes. To be a configuration, a set of transitions must not only be downward closed with respect to the causal relation of transitions but also must not contain any pair of conflicting transitions. For instance, set $\{e_1, e_2, e_4, e_5\}$ is a configuration of the process of Figure 14.13 but this is not the case for set $\{e_1, e_2, e_3, e_4\}$ because transitions e_3 and e_4 are in conflict. In a similar way, the set of places forming a cut must be concurrently reachable, i.e. there must be neither causal nor conflict relations between them. For instance, set $\{c_2, c_6\}$ is a cut whereas set $\{c_3, c_8\}$ is not a cut because both places are in indirect conflict (because of c_1).

Finally, an *unfolding* can be defined from the notion of branching process after introducing the notion of stability for a cut. A cut is *stable* with respect to its corresponding marking if all the transitions enabled from this marking are represented by output transitions of the cut. For instance, cut $\{c_3, c_5\}$ of the branching process of Figure 14.13 is stable because e_3 and e_4 are the output transitions of this cut and correspond to the enabled transitions (t_3 and t_4) from marking $p_3 + p_4$ in the original net. This is not the case for the cut $\{c_4\}$ because the transition t_7 is not represented in the branching process.

In this context, an unfolding of a net is a finite branching process such that each reachable marking of the original net is represented at least once by a stable cut. For instance, the branching process of Figure 14.14 is an unfolding of the net depicted in Figure 14.6. All the reachable markings of this net are represented and for each of them one can find a corresponding stable cut: $\{c_4\}$ for p_7 and $\{c_1, c_5\}$ for $p_1 + p_4$.

The main advantages of the unfolding are the following:

1. For a highly concurrent system, we can obtain a reduced representation of the reachable markings (compared to a classical approach).

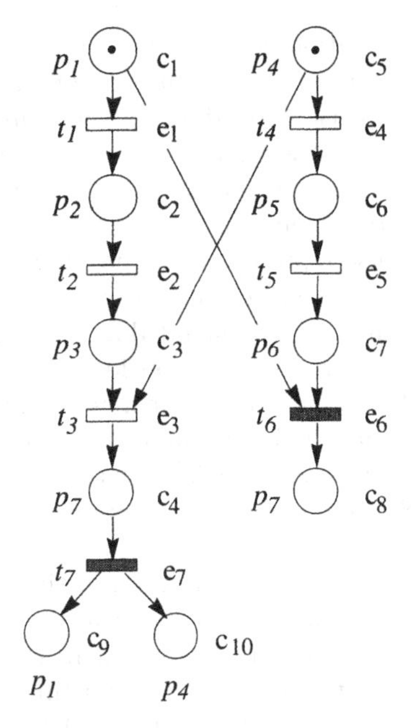

Fig. 14.14. A finite branching process (which is an unfolding) of the net from Figure 14.6

2. The acyclic nature of branching processes enables the specification of efficient procedures for the construction and verification of properties.

The remainder of this section is dedicated to the presentation of this second aspect.

The construction presented is restricted to safe nets. Its generalisation to k-bounded nets can be found in [KKTT96].

The construction of an unfolding starts with the generation of a place for each marked place in the initial marking of the original net. Adding a transition to the unfolding is done by selecting a set of concurrent places corresponding to the input set of an original transition. From such a transition, a place set corresponding to the output places of the original transition and the corresponding arcs can be generated. This process is done repeatedly.

This procedure leads to the construction of an infinite branching process if the original net is able to perform an infinite firing sequence. The termination of the procedure is achieved by the introduction of cutoff transitions. A cutoff transition is a transition of the branching process from which it is not necessary to continue the construction. In order to ensure the completeness and the stability of the construction, cutoff transitions have to be well chosen. Different definitions have been presented (see [McM92, Esp92a, KKTT96]). For all of them, a transition is a *cutoff* if there exists a cut of the branching process corresponding to the same reachable marking represented by the cut of the minimal configuration of the transition considered. Moreover, the configuration of the cut must not contain a cutoff transition and must be strictly inferior (in some sense depending on the definition) to that of the minimal configuration of the transition. The most simple decision procedure is that due to McMillan. The order relation on configurations is their size. For example, the cut of transition e_7 in Figure 14.14 corresponds to marking $p_1 + p_4$. This marking is also represented by the cut $\{c_1, c_5\}$. Event e_7 can be considered as a cutoff transition, because the minimal configuration of this cut is the empty set and thus strictly included in the minimal configuration of e_7. The output places of e_7 are then not considered in the next construction steps of the unfolding. The same reasoning cannot be applied to transition e_6 since the size of the minimal configuration of e_6 is equal to that of e_3. Hence another order has been defined in [Esp92a]. This order (and the one presented in [KKTT96]) is more elaborate and leads to more concise unfoldings in such situations.

It is worth noting that the implementation of the decision procedure for a set of concurrent places and configurations has a large influence on the efficiency of the whole algorithm. Interested readers are referred to [KKTT96, Röm96] for more details on this part of the algorithm.

Verification procedures on unfoldings take advantage of the particular structure of branching processes. This fact is highlighted by the detection of the presence of a deadlock.

The principle of the algorithm is to construct a configuration wherein there is a transition in (direct or otherwise) conflict for any cutoff transition of the branching process. Hence obviously some dead markings are reachable from the reachable marking corresponding to the cut of such a configuration.

The procedure is illustrated for the branching process of Figure 14.14. Initially, the configuration is set to the minimal configuration of a transition in conflict with a given cutoff transition. In our example, the configuration is set to the minimal configuration of e_4 (in conflict with the cutoff e_7). To complete the algorithm, a transition in conflict with the cutoff e_8 must be introduced into the configuration. Because e_1 satisfies these properties and because the union of its minimal configuration with the already constructed configuration also yields to a configuration (there is no transition in conflict), the algorithm leads to the construction of the configuration $\{e_1, e_4\}$. From the corresponding reachable marking $p_2 + p_5$, the dead marking $p_3 + p_6$ is reachable.

The verification of invariant properties is performed according to the same principle (i.e. construction of particular configurations). The related algorithms can be found in [Poi96]. Moreover, a model checking procedure dedicated to the verification of safety properties has been presented in [Esp92a, Esp93].

Branching Process Graph. The main objective of this technique is the construction of a graph which enables the verification of linear temporal logic formulas. From a formula, a branching process graph is constructed and classical model checking algorithms can be applied. The size of the graph is closely related to the number of observable transitions implied by the formula. In the worst case where all the transitions are observable, this technique leads to the construction of the complete reachability graph. This method is well adapted to the verification of stuttering-invariant formulae (LTL without the Next-time operator for instance).

A branching process graph is a graph wherein nodes are stable branching processes and arcs are labelled by transitions of the original net. A branching process of a node cannot contain any transition representing an observable transition. Then the firing of an observable transition is always represented by an arc of the graph, and a node represents adjacent reachable markings which can be reached from the initial marking of the node without firing an observable transition.

A branching process comprising a node can contain two particular kinds of transitions: cutoff transitions, as for unfolding, and external transitions which correspond to output arcs of the node. For a given arc, the initial marking of the target node must represent the same marking as the one represented by the cut associated with the external transition of the source node. Obviously, to be valid a branching process graph must contain a node for which the initial marking of the branching process corresponds to the initial marking of the original net and it must be stable and complete (i.e. all possible firings and reachable markings are represented in it).

Figure 14.15 presents a branching process graph of the net from Figure 14.6.

External transitions are represented as solid black rectangles and cutoff transitions are shown in grey. This graph has been constructed with respect to the observable transition set $\{t_5\}$.

The cut associated with the external transition e_5 of node 1 is $\{c_1, c_6\}$ and it corresponds to the marking $p_1 + p_6$ of the original net. The arc which is labelled t_5 corresponds to this external transition and leads to node 2, which has an initial marking again corresponding to marking $p_1 + p_6$.

This graph does not contain enough information so as to enable the desired model checking. Effectively, some infinite or dead sequences may not be represented in the graph. For example, the projection of the dead sequence $t_1.t_2.t_4.t_5$ on the observable transitions cannot be found. In fact, the graph has to be completed. In particular, any node which contains a dead marking must be marked as blocking. Similarly, it is possible that a node may represent a suffix of an infinite sequence, composed only of invisible transitions. This is because of the presence of cutoff transitions. For example, the infinite sequence $(t_1.t_2.t_3.t_7)^\infty$ is represented in node 1. Such nodes are the source of divergent sequences and have to be marked as divergent.

So, to be equivalent to the reachability graph as far as model checking is concerned, a node (denoted B) without a successor is added to the graph

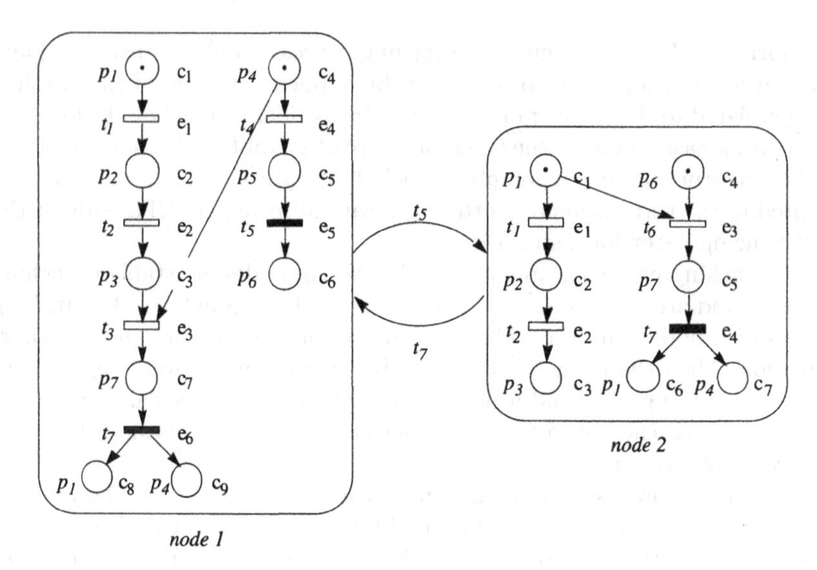

Fig. 14.15. A branching process graph with respect to $\{t_5\}$ as observable transition set

and all blocking nodes have this node as a successor (via an edge labelled by τ). Moreover, a loop is added on every divergent node, labelled by τ.

Figure 14.16 shows the equivalent graph derived from the branching process graph of Figure 14.15.

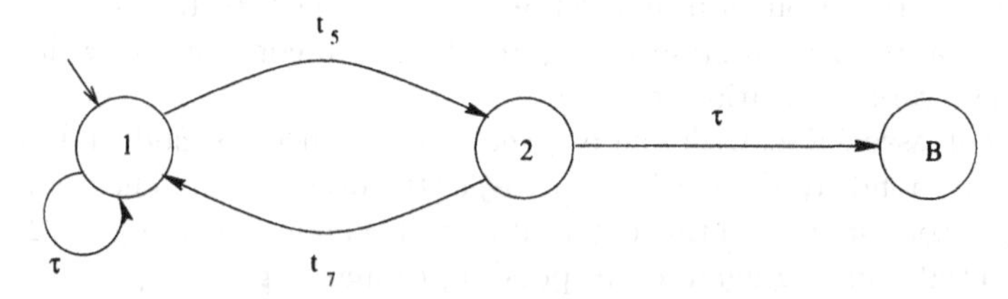

Fig. 14.16. Equivalent graph derived from the one of Figure 14.15

This graph can be used for verifying a temporal formula which requires the observation of transition t_5.

A node of a branching process graph is constructed as an unfolding except that observable transitions are taken into account. A transition labelled by an observable transition is always external. In order to limit the number of nodes of the graph, a transition for which the marking corresponding to the cut of its minimal configuration is also the initial marking of an other node can also be considered as an external transition. This is the case for transition e_4 of node 2 in Figure 14.15.

The detection of the blocking nodes is performed by applying the deadlock detection algorithm presented in Section 14.3.5 to each node. The detection of the divergence of a node is more difficult. A solution is to construct the branching process comprising the node using the inclusion set as the order for the definition of cutoff. Effectively, the presence of such a cutoff corresponds to the capability of the system to perform an infinite sequence.

Deadlock detection and the verification of invariant properties is performed as for unfolding, by applying the same algorithms on all the nodes of the graph. Moreover, the graph can be used to verify any linear temporal property (in this case, the observable transitions are the ones which can potentially modify the truth value of the atomic properties comprising the formula).

14.3.6 Conclusion

Partial-order techniques can be used without risk since they often reduce the size of the representation of the state space while preserving the ability to verify a large class of properties, from deadlocks to temporal logic formulae, e.g. LTL-$\circ$ ($\circ$ means the Next-time operator).

The partial-order techniques are based on dependency relations among system events, such as conflict and causality relations. We have presented three families of techniques which take advantage of Petri nets to handle these dependencies automatically:

1. The persistent sets which can be combined with sleep sets aim to linearise the independent transition firings within a reachability graph representation.
2. The covering step technique also deals with the reachability graph representation, but gathers some independent transition firings in steps.
3. The branching process techniques refines the Petri net representation to define the partial order of transition firings, thus with no direct expression of interleavings.

The comparison of these techniques is not easy because they do not necessarily offer the same reductions in all parts of the system state space.

As reachability graph techniques, the first two can be used efficiently to check properties on-the-fly, however a minimal graph is rarely obtained since sufficient conditions are applied to detect the partial order of transition firings. Moreover, depending on the chosen technique, sufficient conditions are applied during the construction to ensure that there is no loss of interesting traces with respect to some property to verify. In particular, persistent sets can be augmented to avoid ignoring problems or to maintain some of the divergent sequences. When using the covering step graph technique, the same transition can be fired several times in order to cover every prefix of sequences (i.e. a transition firing may appear in several steps). Here again, additional conditions are considered during the building of steps, not only to ensure the absence of confusion cases, but also to force the interleavings of transition firings that must be observed.

In contrast to the reachability graphs, the branching process technique offers a direct representation of the partial order of transition firings, however the building must be completed before starting the verification (i.e. it is not an on-the-fly procedure). The technique is especially interesting for two reasons: (1) the number of markings and transition firings of the standard reachability graph that refer to the same element of the branching process, and (2) the efficiency of the algorithms that run on the branching process structure. Here again, more effort is needed to maintain some of the divergent sequences. More precisely, a graph is built wherein each node is a branching process such that internal dead markings or divergent sequences are detected.

The partial-order techniques have already been implemented, in particular:

- In SPIN, a model checking is proposed using persistent sets and sleep sets ([Hol97]).
- In PROD, the model checking is based not only on persistent sets but also on other reduction methods such as the exploitation of symmetries and the use of BDD coding ([VHHP95]).
- In PEP, an unfolding construction is supplied ([Röm96]), from which safety properties can be checked.
- Moreover, a prototype of the branching process graph exists and is presented in [Poi96].

In fact, partial-order techniques can exceed the standard application domain of Petri nets. Several adaptations and extensions have been proposed in several languages according to possible representations of causality and conflicts. Models for SPIN are designed in PROMELA, a language used to specify communicating processes. Engineers may define possible access conflicts between data operators. Such conflicts can be taken into account to refine the dependency relations of transitions. In PROD, the specification is a coloured Petri net with C coloured functions which is unfolded into a standard Petri net so as to allow a verification process. In PEP, specifications are defined according to either a process algebra or its translation into a Petri net ([BG96]).

With such environments, both trace-based graph techniques and branching process techniques have demonstrated their ability to deal with large systems. However, given the difficulty of defining the best algorithm for a given system, more than one approach is generally proposed in software engineering platforms. In the PEP environment, branching process tools are accompanied by the ability to launch the PROD environment from the same description of the system. In the CPN-AMI environment, for which specifications are designed in terms of communicating objects or in coloured Petri nets, it is possible to launch PROD and PEP tools, as well as other tools which exploit techniques such as symmetries ([MT94]).

14.4 Symbolic and Parametrised Approaches

The idea common to the techniques that we present in this section is to reduce the size of the representation of the reachability graph by grouping states into classes. The reduction is always done in such a way that most important properties of the system can be checked on the reduced graph. However, the set of preserved properties depends on the reduction technique that is applied.

We start by presenting techniques where states are grouped according to some symmetry relation. A typical case where such relations occur is the mutual exclusion example presented in Section 14.1. There are two states, namely states 7 and 8, where one process is in the critical section and the other process is waiting for the resource. Since both processes potentially have the same behaviour, it does not matter which one is accessing the resource. This states are said to be symmetric and they can be grouped into a class.

The second class of techniques that we present are known as parametrised graphs. This approach is somehow orthogonal to the previous one. States are grouped according to the cardinalities of sets of processes that are in the same situation. A typical example would be a mutual exclusion with n processes, where one process accesses the critical resource and $i(i < n)$ processes are waiting for this resource. When leaving the critical section, the behaviour of the process may depend on whether there is someone waiting for the resource or not. However, it will usually be the same whether there is $1, 2$, or $(n-1)$ processes waiting for the resource. Hence, it would be interesting to group all these states into one single class. The interesting feature of such graphs is that they can be constructed independently of the value of n.

14.4.1 Symbolic Reachability Graph

The idea of the symbolic reachability graph (SRG) is to exploit the intrinsic symmetries of a system to obtain a compact representation of the reachable states. These symmetries occur when different components of a system have the same behaviour. It is often the case that such systems are represented with coloured Petri nets in which the equivalent components are identified by different colours. However, when dealing with a state of the system, the identities of the components may not be relevant. It is thus possible to define an equivalence relation between markings. Equivalent markings have the same distribution of tokens in places, but the colours of these tokens are different and correspond to the identities of similar components. The classes of markings thus obtained are the nodes of the SRG. Yet the problem remains of making it possible for the designer to exploit the symmetries of the system without having to define them in the model. By introducing a syntax in the definition of the colour functions, *well-formed nets* provide a modelling framework in which symmetries can be used automatically to reduce the size and complexity of the representation. In this section, we first give the definition of well-formed nets, then introduce the concepts of symbolic marking and symbolic firing that are used to build the SRG and we present some properties which can be checked directly on the SRG.

Well-Formed Nets. A *well-formed net* ([CDFH91, CDFH93]) is a coloured net where additional constraints are introduced in the definition of the different features. The goal is to ensure that a WN model will be somehow structured, and that it will be possible to use this structure to develop efficient analysis techniques.

The starting point in the definition of the WN colour syntax is the set of basic colour classes from which colour domains are constructed. A *basic colour class* is a non-empty, finite (possibly ordered) set of colours. It may be partitioned into several *static subclasses*: colours belonging to different static subclasses represent objects of the same type but with different behaviour. When a class is ordered, this order is cyclic, meaning that the successor of the last colour is the first colour of the class. A consequence is that if we want to represent a total order among colours (sites on a bus as opposed to sites on a ring in the cyclic case), we need to have one static subclass per colour of the class.

The colour domain of a place is defined by composition through the cartesian product operator of basic colour classes. An interpretation of this is that the information associated with tokens comprises one or more fields, and each field in turn has a type selected from the set of basic colour classes.

The colour domain of a transition defines the type of the parameters of the transition. Each parameter is associated with a variable on an arc connected to the transition, and has a type selected from a basic colour class. Restrictions on the possible bindings of the parameters can be defined by adding a *guard* to the transition. Hence, the colour domain of a transition is composed of two parts: the parameter type, and the guard defined as a Boolean expression of (a restricted set of) basic predicates on the parameters[1].

The arc functions are defined as weighted (and possibly guarded) sums of tuples. The elements composing the tuples are in turn weighted sums of *basic functions*, defined on basic colour classes and returning multisets of colours in the same class. The multiset returned by a tuple of basic functions is obtained by cartesian product composition of the multisets returned by the tuple elements. There are three types of basic functions: the projection function, the successor function, and the diffusion/synchronisation function.

[1] The basic predicates allowed are: $x = y, x = !y, d(x) = C_i^j, d(x) = d(y)$ where x and y are transition parameters of the same type, $!y$ denotes the successor of y, and $d(x)$ denotes the static subclass to which x belongs.

The syntax used for the projection function is x where x is one of the transition parameters. The syntax for the successor function is $!x$ where x is again one of the transition parameters. It applies only to ordered classes and returns the successor of the colour assigned to x in the transition binding. Finally, the syntax for the diffusion/synchronisation function is S_{C_i} (or $S_{C_i^j}$). It is a constant function that returns the set of colours of class C_i (or of static subclass $C_i^j \subset C_i$).

We illustrate different features of a well-formed net using the example in Figure 14.17. In this model, a set of processes (basic colour class $Proc = \{p_1, \dots, p_n\}$) may access a set of resources (basic colour class $Res = \{r_1, \dots, r_m\}$). To access a resource r, an idle process p sends a request (firing of t_1). We assume that a request is represented by the association of a process and the resource it tries to access. If several processes have requested the same resource, a selection must be done to determine which process will actually access it. This selection is performed by a (possibly repeated) firing of t_5. We do not care at this step what algorithm is used for the selection. However, the selection cannot be completed if there are still requests in transit, and it may take some time for all the processes to become aware of the request sent by p. This delay is represented by transition t_2. Hence, it is worth noticing that transition t_3, which models the actual access of a process to the resource, becomes enabled for $\langle p, r\rangle$ only once all the other requests for resource r have been discarded. In other words, there is no tuple $\langle -, r\rangle$ in RQ, nor in GS with the exception of $\langle p, r\rangle$, meaning that p is the only process whose request for r is still valid. Place PR is actually used to ensure that this condition is satisfied. It initially contains the set of all possible requests. Each time a process sends a request, the corresponding token is removed from PR. Each time a request is discarded, the corresponding token is added to PR. Note that t_3 can fire only if all the possible requests for r except $\langle p, r\rangle$ are in PR.

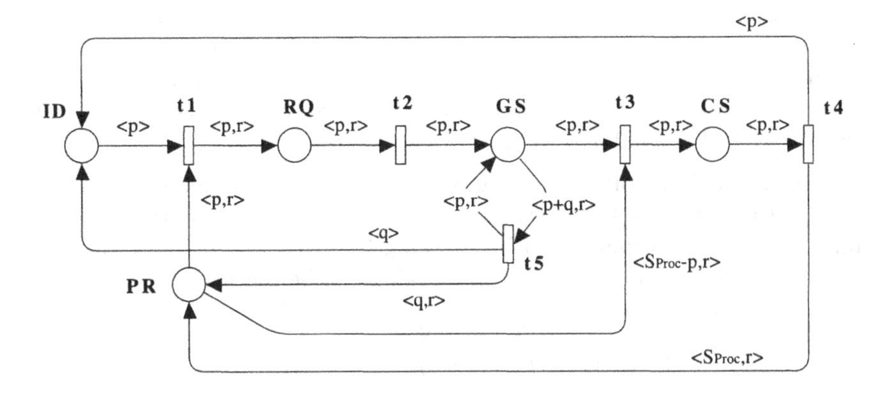

Fig. 14.17. WN model of a distributed critical section

For this model, we give the colour domain definition for some places and transitions. The colour domain of a transition is defined as a pair ⟨transition parameter type, guard⟩.

- $cd(ID) = Proc$
- $cd(PR) = Proc \times Res$
- $cd(t_5) = \langle\langle p,q,r\rangle \in Proc \times Proc \times Res, true\rangle$

The function $\langle S_{Proc} - p, r\rangle$ on the arc between PR and t_3, which can also be denoted by $\langle S_{Proc}, r\rangle - \langle p, r\rangle$, represents the sum of all tokens $\langle q, r\rangle$ for $q \in Proc$ except one token $\langle p, r\rangle$ whose actual identity is given by the binding of variables p and r.

The features of a well-formed net can be summarised in the following definition.

Definition 14.4.1 (Well-Formed Net). *A well-formed net (WN) is defined by a tuple* $\mathcal{N} = \langle P, T, \mathbf{Pre}, \mathbf{Post}, \mathbf{Inh}, \mathbf{pri}, \mathcal{C}, cd\rangle$ *where*

- *P and T are the sets of places and transitions.*
- $\mathbf{Pre}, \mathbf{Post}, \mathbf{Inh}$ *are arc functions.*
- $\mathbf{pri} : T \to \mathbb{N}$ *is a vector that associates with each transition t a priority. By default, we assume that $\forall t \in T, \mathbf{pri}[t] = 0$.*
- $\mathcal{C} = \{C_1, \ldots, C_n\}$ *is the set of basic colour classes. Basic colour classes are finite and disjoint, and every class C_i is possibly partitioned into static subclasses: $C_i = C_i^1 \cup \ldots \cup C_i^{n_i}$. An index h is defined such that colour classes with index i, $h < i \leq n$, are ordered, whereas those with index $i, 0 < i \leq h$, are not.*
- *cd is a mapping that associates with every place and transition of the net a colour domain defined by composition through the cartesian product operator of basic colour classes. The colour domain of a transition can be restricted by a guard.*

The use of $C_1, \ldots, C_n$ for denoting basic colour classes is a formal notation used in definitions and proofs. But most often, classes are denoted by more significant names, such as *Proc* for the class of processes in our example.

Since WNs are coloured nets, the dynamics of the model need not be redefined. The following section shows how WNs can be used to directly build a reduced reachability graph.

Symmetries, Symbolic Marking and Firing. We consider again the WN model of the distributed critical section, in the particular case of two processes and two resources. From an initial marking where all the processes are idle and the resources free, the following marking can be reached:

$$\mathbf{m}_1 = ID(p_3) + GS(\langle p_1 + p_2, r_1\rangle + \langle p_4, r_2\rangle) \\ + PR(\langle p_3 + p_4, r_1\rangle + \langle p_1 + p_2 + p_3, r_2\rangle)$$

In this state one process is idle, two have requested the same resource, and the last one has requested the other resource. It is easy to check that by taking an arbitrary permutation of the objects in basic colour classes *Proc* and *Res*, we obtain another legal state of the model with the same characteristics, for example, $\mathbf{m}_2 = ID(p_1) + GS(\langle p_2 + p_4, r_2\rangle + \langle p_3, r_1\rangle) + PR(\langle p_1 + p_3, r_2\rangle + \langle p_1 + p_2 + p_4, r_1\rangle)$. This transformation can be formalised with the following definitions.

Definition 14.4.2 (Colour Permutation). *Let h be such that $0 \leq h \leq n$ and let $\xi = \{s = \langle s_1, \ldots, s_n\rangle\}$ be a subgroup of the permutations on $C_1, \ldots, C_n$ such that:*

- *$\forall 0 < i \leq h, s_i$ is a permutation on C_i such that $\forall C_i^j, s_i(C_i^j) = C_i^j$;*
- *$\forall h < i \leq n, s_i$ is a rotation on C_i such that $\forall C_i^j, s_i(C_i^j) = C_i^j$. Note that this condition implies that if the number n_i of static subclasses of C_i is greater than 1 then the only allowed rotation s_i is the identity.*

Let $\langle c_1, \ldots, c_k\rangle$ be an element of a colour domain, i.e. a tuple of basic colours, and let $s \in \xi$. Then $s(\langle c_1, \ldots, c_k\rangle)$ is an element of the same colour domain defined by:

$$s(\langle c_1, \ldots, c_k\rangle) = \langle s_{(1)}(c_1), \ldots, s_{(k)}(c_k)\rangle$$

where $s_{(i)}$ is the permutation associated with the ith colour class in the colour domain.

In our example, by setting $C_1 = Proc$ and $C_2 = Res$, we have $s_1(p_1) = p_2, s_1(p_2) = p_4, s_1(p_3) = p_1, s_1(p_4) = p_3$, and $s_2(r_1) = r_2, s_2(r_2) = r_1$.

Definition 14.4.3 (Marking Permutation). *Let $\mathbf{m}$ be a marking and $s \in \xi$ a permutation. Then $s.\mathbf{m}$ is a marking defined by:*

$$\forall p \in P,\ \forall c \in cd(p),\ s.\mathbf{m}[p, s(c)] = \mathbf{m}[p, c]$$

A very important property of the model is that the firing property is preserved by applying a permutation both on the markings and the transition binding. In other words, for two markings $\mathbf{m}$ and $\mathbf{m}'$, for a transition t and a colour $c \in cd(t)$, if the firing of $\langle t, c\rangle$ leads from $\mathbf{m}$ to $\mathbf{m}'$ then the firing of $\langle t, s(c)\rangle$ leads from $s.\mathbf{m}$ to $s.\mathbf{m}'$, where s is any permutation in ξ.

Definition 14.4.4 (Symbolic Marking). *Let Eq be the equivalence relation defined by:*

$$\mathbf{m} Eq \mathbf{m}' \Leftrightarrow \exists s \in \xi, \mathbf{m}' = s.\mathbf{m}$$

An equivalence class of Eq is called a symbolic marking, denoted with $\widehat{\mathbf{m}}$.

We can now define a *well-formed system* $\mathcal{S} = \langle \mathcal{N}, \widehat{\mathbf{m}_0}\rangle$, which is a WN whose initial marking is a symbolic marking, i.e. a class of markings. If any permutation $s \in \xi$ leaves the initial marking invariant, the class is reduced to one element and the well-formed system is a net system. Otherwise the well-formed system defines a set of net systems.

Since a symbolic marking represents an equivalence class on the state space of the well-formed system, where the equivalence is in terms of possible basic colour permutations that yield the same behaviour, we can use it to derive directly and automatically a reduced representation of the reachability graph. The construction of this reduced graph, namely the *symbolic reachability graph*, requires that we define a unique representation for every symbolic marking, and a firing rule that applies directly to this representation. To achieve the first goal, we must choose an appropriate data structure to represent equivalence classes. From our example, it appears that for markings belonging to the same symbolic marking, what is relevant is not the actual identity of object but rather the distribution of objects in places. Our first abstraction thus consists of substituting object identifiers with variables. With each static subclass, we associate a set of variables. In the example, we can associate variables $\{x_1, \ldots, x_4\}$ with class *Proc* and $\{y_1, y_2\}$ with class *Res*. A possible representation of the symbolic marking would thus be

$$\begin{aligned}\widehat{\mathbf{m}} = ID(x_1) &+ GS(\langle x_2 + x_3, y_1\rangle + \langle x_4, y_2\rangle) \\ &+ PR(\langle x_1 + x_4, y_1\rangle + \langle x_1 + x_2 + x_3, y_2\rangle)\end{aligned}$$

To define the semantics of this representation, we need the definition of a *valid assignment.* An assignment of objects from a static subclass to the associated set of variables is said to be valid if the following three conditions are verified: 1) every variable is assigned an object; 2) the same object is not assigned to more than one variable; 3) if the class is ordered, adjacent objects are assigned to subsequently numbered variables. The symbolic marking $\widehat{\mathbf{m}}$ could thus represent the set of all ordinary markings that can be obtained from valid assignments of objects to the variables x_i and y_i. For instance, $\mathbf{m}_1$ and $\mathbf{m}_2$ are obtained from $\widehat{\mathbf{m}}$ by valid assignments.

Starting from this representation of a symbolic marking, it is rather natural to define a *symbolic firing rule* since the variables play the same role that objects play in ordinary markings. Hence from marking $\widehat{\mathbf{m}}$ it is possible to fire the *symbolic firing instance* $\langle t_5, \langle x_2, x_3, y_1\rangle\rangle$ from which the new symbolic marking $\widehat{\mathbf{m}}' = ID(x_1+x_3)+GS(\langle x_2, y_1\rangle+\langle x_4, y_2\rangle)+PR(\langle x_1+x_3+x_4, y_1\rangle+\langle x_1+x_2+x_3, y_2\rangle)$ is reached. The symbolic firing instance stands for all the ordinary instances that can be obtained by valid assignments of objects to variables. There is, however, a further step we can take to better exploit the grouping induced by the symbolic firing. Actually, by firing the symbolic instance $\langle t_5, \langle x_3, x_2, y_1\rangle\rangle$, we would have reached the same new symbolic marking (provided we use a unique representation for symbolic markings). It is possible to recognise in advance such a situation: all those variables that have the same *distribution of tokens in the places* can be used interchangeably in a transition instance. In our example, x_2 appears only in *GS* associated with y_1 and in *PR* associated with y_2. This is the same for x_3, hence interchanging them in the firing instance will not modify the symbolic marking, reached although we will obtain two different representations in which x_2 and x_3 are interchanged.

We now introduce the concept of *dynamic subclasses*, representing sets of objects that are not identified individually but that are known to be permutable with each other in any firing instance to produce markings that belong to the same equivalence class. A dynamic subclass is characterised by its cardinality (i.e. the number of different objects represented by the dynamic subclass) and by the static subclass to which the represented objects belong (i.e. we can group only variables belonging to the same static subclass). In case of ordered basic classes, only contiguous objects can be represented by the same dynamic subclass, and the ordering relation among objects is reflected by the ordering of the indexes of the dynamic subclasses. In marking $\widehat{\mathbf{m}}$, the two variables x_2 and x_3 have the same distribution of tokens in the places, so they can be grouped in the dynamic subclass Z^1_{Proc} of cardinality 2. To preserve the homogeneity of the representation, we create a dynamic subclass of cardinality 1 for all the other variables. The new symbolic marking representation can thus be written as:

$$\begin{aligned}\widehat{\mathbf{m}} = ID(Z^2_{Proc}) + GS(\langle Z^1_{Proc}, Z^1_{Res}\rangle + \langle Z^3_{Proc}, Z^2_{Res}\rangle) \\ + PR(\langle Z^2_{Proc} + Z^3_{Proc}, Z^1_{Res}\rangle + \langle Z^2_{Proc} + Z^1_{Proc}, Z^2_{Res}\rangle)\end{aligned}$$

where $|Z^1_{Proc}| = 2$ and all the other dynamic subclasses have cardinality 1.

The basic idea of the symbolic marking representation, i.e. using dynamic subclasses to represent sets of objects with the same marking, is not sufficient to ensure the uniqueness of the representation: we must add some constraints. Actually, we need to decide how to partition the static subclasses into dynamic subclasses of a given cardinality and how to properly name the dynamic subclasses. The detailed algorithm for computing a unique representation can be found in [CDFH93] and we will present here only the intuitive idea. The so-called *canonical* representation of a symbolic marking must be *minimal* and *ordered*. The minimality criterion requires that objects with the same marking be represented by a unique dynamic subclass. A representation of $\widehat{\mathbf{m}}$ with two dynamic subclasses Z^1_{Proc} and Z^4_{Proc}, each with cardinality 1, instead of Z^1_{Proc} with cardinality 2, would not be minimal. The ordering criterion consists of re-adjusting the dynamic subclass indexes in the minimal representation to obtain a minimum element in the lexicographic order defined on the markings. Without going into detail, it is clear that another possible representation for $\widehat{\mathbf{m}}$ could be $ID(Z^3_{Proc}) + GS(\langle Z^2_{Proc}, Z^1_{Res}\rangle + \langle Z^1_{Proc}, Z^2_{Res}\rangle) + PR(\langle Z^1_{Proc} + Z^3_{Proc}, Z^1_{Res}\rangle + \langle Z^2_{Proc} + Z^3_{Proc}, Z^2_{Res}\rangle)$ with $|Z^2_{Proc}| = 2$. The ordering algorithm defines in a unique way which representation to choose. Actually, this last representation is ordered.

In order to build the SRG directly starting from an initial marking $\widehat{\mathbf{m}_0}$ (i.e. without building the RG and then grouping markings into equivalence classes, which would be much easier but too costly), we now define a symbolic firing rule on the symbolic marking representations. In a symbolic firing instance, dynamic subclasses rather than objects are assigned to the transition parameters. This means that any object in the subclass can be assigned to the parameter. For every basic colour class C_i, we define a function λ_i, where $\lambda_i(x)$ gives the index of the dynamic subclass which is assigned to the xth occurrence of a C_i parameter in the transition colour domain. Coming back to the example of the firing of t_5 from $\widehat{\mathbf{m}}$, both p and q can be assigned Z^1_{Proc}. We thus have $\lambda_{Proc}(1) = 1$ and $\lambda_{Proc}(2) = 1$. When, for a transition, several parameters are assigned the same dynamic subclass, we also need to specify whether the parameters are instanced to the same object or to different objects of the dynamic subclass. Since Z^1_{Proc} represents two objects, we still have to specify whether or not p and q are assigned the same object. For this, we use a function μ_i. We set $\mu_i(x) = 1$ if the xth C_i parameter is the first to be assigned $Z_i^{\lambda_i(x)}$. If the yth parameter is the next to be assigned the same subclass, we set $\mu_i(y) = \mu_i(x)$ if it is assigned the same object as the xth parameter. We set $\mu_i(y) = 2$ if it is assigned another object, and so on: for every dynamic subclass, μ_i is incremented each time we select in the subclass an object that has not been instanced yet. Hence, $\mu_{Proc}(1) = 1$ and $\mu_{Proc}(2) = 1$ if p and q are assigned the same object in Z^1_{Proc}, whereas if they are assigned different objects, we get $\mu_{Proc}(1) = 1$ and $\mu_{Proc}(2) = 2$. If λ and μ are the collections of functions λ_i and μ_i respectively for each class C_i, a symbolic instance of t is defined by the pair $\langle \lambda, \mu \rangle$.

The notion of valid assignment can be extended to symbolic firings. Let us consider the symbolic firing $\widehat{\mathbf{m}} \xrightarrow{\langle t, \lambda, \mu \rangle}$. Given a valid assignment of $\widehat{\mathbf{m}}$, the jth parameter of t to be assigned in C_i is assigned an object belonging to the assignment of $Z_i^{\lambda_i(j)}$ for $\widehat{\mathbf{m}}$. For j and k such that $\lambda_i(j) = \lambda_i(k)$, if $\mu_i(j) = \mu_i(k)$, the jth and the kth parameters to be assigned in C_i are assigned the same object of $Z_i^{\lambda_i(j)}$, whereas they are assigned different objects if $\mu_i(j) \neq \mu_i(k)$. It is worth noticing that for one assignment of $\widehat{\mathbf{m}}$ there may be several assignments of $\langle \lambda, \mu \rangle$, which actually correspond to the different firings that are grouped within the symbolic firing.

Unfortunately, we cannot directly define a symbolic firing rule where dynamic subclasses play the same role that objects play in ordinary markings. In our example, t_5 is enabled from $\mathbf{m}_1$ if p and q are assigned p_1 and p_2 respectively. But it is not enabled from $\widehat{\mathbf{m}}$ if p and q are both assigned Z^1_{Proc}: the evaluation of the colour function would require two tokens $\langle Z^1_{Proc}, Z^1_{Res} \rangle$ in GS. We thus need to introduce the notion of *split symbolic marking*. The idea behind the splitting is to isolate in new dynamic subclasses the (arbitrarily chosen) objects that will be selected for the firing. Let us consider the case of a dynamic subclass Z_i^j of class C_i such that at least one object of Z_i^j is selected in the symbolic instance, i.e. $\exists x, \lambda_i(x) = j$. If C_i is ordered, Z_i^j is split into as many subclasses of cardinality 1 as the number of objects it represents. If not, a new subclass $Z_i^{j,k}$ of cardinality 1 is created for every pair (j, k) such that $\exists x, \langle \lambda_i(x), \mu_i(x) \rangle = \langle j, k \rangle$. The objects of Z_i^j that are not selected for the firing are put in a subclass $Z_i^{j,0}$ whose cardinality is the cardinality of Z_i^j minus the number of new subclasses. Z_i^j is then removed.

$$split(Z_i^j) = \begin{cases} \{Z_i^{j,k}, 0 < k \leq |Z_i^j|\} & \text{if } C_i \text{ is ordered} \\ \{Z_i^{j,0}\} \cup \{Z_i^{j,k}, \exists x : \langle \lambda_i(x), \mu_i(x) \rangle = \langle j, k \rangle\} & \text{otherwise} \end{cases}$$

For the symbolic firing we are considering in our example, where t_5 is instanced by two different processes and one resource, the split representation of $\widehat{\mathbf{m}}$ would be $ID(Z^2_{Proc}) + GS(\langle Z^{1,1}_{Proc} + Z^{1,2}_{Proc}, Z^{1,1}_{Res}\rangle + \langle Z^3_{Proc}, Z^2_{Res}\rangle) + PR(\langle Z^2_{Proc} + Z^3_{Proc}, Z^{1,1}_{Res}\rangle + \langle Z^2_{Proc} + Z^{1,1}_{Proc} + Z^{1,2}_{Proc}, Z^2_{Res}\rangle)$. In this representation, all the dynamic subclasses have cardinality 1. We do not have a subclass $Z^{1,0}_{Proc}$ because once two objects have been selected in Z^1_{Proc}, there are no objects left in this subclass. The same is true for $Z^{1,0}_{Res}$.

Now, using the split representation of a symbolic marking, dynamic subclasses can substitute objects into the transition enabling and firing. The evaluation of arc expressions and predicates does not change when dynamic subclasses of cardinality 1 replace objects in variable assignments. The symbolic firing is thus a three-step operation which is performed according to the following procedure:

Three-Step Symbolic Firing. The canonical representation of the symbolic marking $\widehat{\mathbf{m}}'$ obtained by firing $\langle t, \lambda, \mu\rangle$ in $\widehat{\mathbf{m}}$ (i.e. $\widehat{\mathbf{m}} \xrightarrow{\langle t,\lambda,\mu\rangle} \widehat{\mathbf{m}}'$) is computed in three steps:

1. Split $\widehat{\mathbf{m}}$ with respect to $\langle \lambda, \mu\rangle$. Let $\widehat{\mathbf{m}}_s$ be the representation of $\widehat{\mathbf{m}}$ obtained after the splitting.
2. Actually fire $\widehat{\mathbf{m}} \xrightarrow{\langle t,\lambda,\mu\rangle} \widehat{\mathbf{m}}'$. Obtain a (possibly) non-canonical representation $\widehat{\mathbf{m}}'_{nc}$ of $\widehat{\mathbf{m}}'$ by applying the incidence functions on $\widehat{\mathbf{m}}_s$.
3. Compute the canonical representation of $\widehat{\mathbf{m}}'$. Group dynamic subclasses of $\widehat{\mathbf{m}}'_{nc}$ to obtain a minimal representation, and order the subclasses to obtain the canonical representation of $\widehat{\mathbf{m}}'$.

Let us perform steps 2 and 3 on our example. By applying the incidence functions on the split marking, we obtain a new marking $\widehat{\mathbf{m}}'_{nc} = ID(Z^2_{Proc} + Z^{1,2}_{Proc}) + GS(\langle Z^{1,1}_{Proc}, Z^{1,1}_{Res}\rangle + \langle Z^3_{Proc}, Z^2_{Res}\rangle) + PR(\langle Z^2_{Proc} + Z^3_{Proc} + Z^{1,2}_{Proc}, Z^{1,1}_{Res}\rangle + \langle Z^2_{Proc} + Z^{1,1}_{Proc} + Z^{1,2}_{Proc}, Z^2_{Res}\rangle)$. In this marking, dynamic subclasses Z^2_{Proc} and $Z^{1,2}_{Proc}$ have the same distribution in places and can be grouped to form a single dynamic subclass of cardinality 2. After renaming the dynamic subclasses, we would obtain the following representation:

$$\begin{aligned} \widehat{\mathbf{m}}' = ID(Z^3_{Proc}) &+ GS(\langle Z^1_{Proc}, Z^2_{Res}\rangle + \langle Z^2_{Proc}, Z^1_{Res}\rangle) + \langle Z^2_{Proc}, Z^1_{Res}\rangle) \\ &+ PR(\langle Z^1_{Proc} + Z^3_{Proc}, Z^1_{Res}\rangle + \langle Z^2_{Proc} + Z^3_{Proc}, Z^2_{Res}\rangle) \end{aligned}$$

with $|Z^3_{Proc}| = 2$ and a cardinality of 1 for all the other dynamic subclasses. Although this representation looks very similar to the representation of $\widehat{\mathbf{m}}$, we can easily see that it cannot represent the same set of markings because the cardinalities of the dynamic subclasses are different. The space reduction obtained by using symbolic markings and firings is shown in Figure 14.18. The

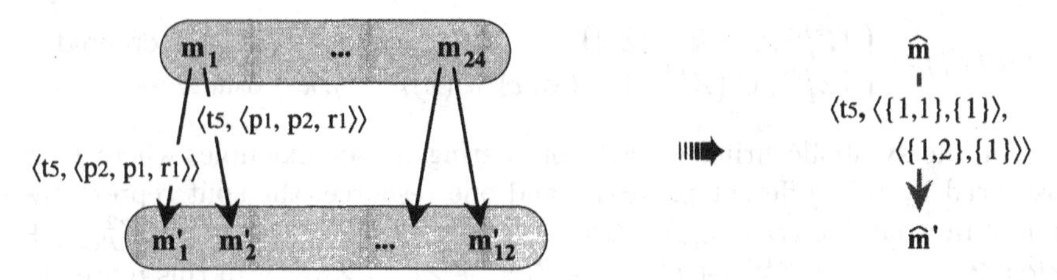

Fig. 14.18. Ordinary vs. symbolic firing

label associated with the symbolic firing of transition t_5 actually represents functions λ and μ in the following way: $\langle t_5, \langle \lambda_1, \lambda_2 \rangle, \langle \mu_1, \mu_2 \rangle \rangle$.

The algorithm for the symbolic reachability graph construction is then the same as for the ordinary reachability graph, except that we start from the canonical symbolic representation of the initial marking and we apply the symbolic firing rule instead of the ordinary firing rule. Of course, the construction is effective only in the case of a finite graph.

Properties of the Symbolic Reachability Graph. We present now some interesting properties of the symbolic reachability graph that can be exploited for a qualitative analysis of WN models. Other properties can be shown that are used for performance evaluation of WN models, which is beyond the scope of the present discussion. We assume here a symmetric initial marking, i.e. the well-formed system defines a single net system. In this case, we have the following set of equivalences:

1. $\mathrm{RS}(\mathcal{N}, \mathbf{m_0}) = \mathrm{SRS}(\mathcal{N}, \widehat{\mathbf{m_0}})$ (if we identify an equivalence class with the set of its elements).
2. $\mathrm{RS}(\mathcal{N}, \mathbf{m_0})$ is infinite $\Leftrightarrow$ $\mathrm{SRS}(\mathcal{N}, \widehat{\mathbf{m_0}})$ is infinite (remember that we consider only finite colour sets).
3. $\{\mathbf{m} \in \widehat{\mathbf{m}}\}$ is a home space for $\mathrm{RG}(\mathcal{N}, \mathbf{m_0})$ $\Leftrightarrow$ $\widehat{\mathbf{m}}$ is a home state for $\mathrm{SRG}(\mathcal{N}, \widehat{\mathbf{m_0}})$.
4. $\{\mathbf{m} \in \widehat{\mathbf{m}}\}$ is an unavoidable home space for $\mathrm{RG}(\mathcal{N}, \mathbf{m_0})$ $\Leftrightarrow$ $\widehat{\mathbf{m}}$ is an unavoidable home state for $\mathrm{SRG}(\mathcal{N}, \widehat{\mathbf{m_0}})$.
5. $\mathrm{RG}(\mathcal{N}, \mathbf{m_0})$ is deadlock-free $\Leftrightarrow$ $\mathrm{SRG}(\mathcal{N}, \widehat{\mathbf{m_0}})$ is deadlock-free.

Property 14.4.5 (Quasi-liveness). The following two propositions hold true:

i) $\mathbf{m}\xrightarrow{\langle t,c\rangle}$ in $\mathrm{RG}(\mathcal{N}, \mathbf{m_0})$ $\Longrightarrow$ $\widehat{\mathbf{m}}\xrightarrow{\langle t,\lambda,\mu\rangle}$ in $\mathrm{SRG}(\mathcal{N}, \widehat{\mathbf{m_0}})$ where $\widehat{\mathbf{m}}$ is the symbolic marking to which $\mathbf{m}$ belongs and $\langle \lambda, \mu \rangle$ is such that c is a valid assignment of $\langle \lambda, \mu \rangle$ in $\mathbf{m}$.
ii) $\widehat{\mathbf{m}}\xrightarrow{\langle t,\lambda,\mu\rangle}$ in $\mathrm{SRG}(\mathcal{N}, \widehat{\mathbf{m_0}})$ $\Longrightarrow$ $\forall \mathbf{m}$ a valid assignment of $\widehat{\mathbf{m}}$, $\forall c$ a valid assignment of $\langle \lambda, \mu \rangle$ in $\mathbf{m}$, $\mathbf{m}\xrightarrow{\langle t,c\rangle}$ in $\mathrm{RG}(\mathcal{N}, \mathbf{m_0})$.

Property 14.4.6 (Liveness). $\langle t, \lambda, \mu \rangle$ is quasi-live in $\mathrm{SRG}(\mathcal{N}, \widehat{\mathbf{m_0}})$ and $\widehat{\mathbf{m_0}}$ is a home state $\Longrightarrow$ $\langle t, c \rangle$ is live in $\mathrm{RG}(\mathcal{N}, \mathbf{m_0})$, where c is any valid assignment of $\langle \lambda, \mu \rangle$ in any valid assignment of $\widehat{\mathbf{m}}$.

Complexity of the Symbolic Approach. The efficiency of the symbolic approach for the construction of the reachability graph strongly depends on the intrinsic symmetry of the model: the more objects with equivalent behaviours, the more markings in an equivalence class, and so the higher the reduction ratio compared with the ordinary reachability graph. However, the value of the ratio is difficult to estimate because for a single model, the cardinality of equivalence classes is highly variable: it is one when all similar objects have the same marking, and the largest classes are obtained when all similar objects have different markings. There are examples ([CDFH93]) of "strongly symmetric" systems where the SRG is several orders of magnitude smaller than the RG.

The state-space reduction obtained with the SRG implies an increased computation cost but for reachability analysis, time is usually less critical than space. Compared with the RG construction, the most expensive step is the computation of a unique representative for each class of markings. It can be performed in a time of the same order of magnitude as the application of a symmetry. Hence, if the system is not symmetric, the time complexity for the SRG construction is the same order as that for the RG construction.

Symmetries and Model Checking. Emerson et al. are the first who have taken advantage of the symmetries of the system in order to improve the model checking of temporal formula.

The basic idea is that the verification process of a property is sometimes composed of a set of parsings over the state space, which are symmetric up to some permutation on atomic propositions. This implies that a graph of equivalence classes of markings could be used instead of the reachability graph.

At first glance, the symmetries existing within a verification process are reflected by permutations on atomic propositions which leave the expression of the property invariant. In formula $[true\, \mathsf{U}\, (p_1 \vee p_2)]$ (p_1 and p_2 are indexed atomic propositions), there are two such permutations: The identity, and permutation π such that $\pi(1) = 2$ and $\pi(2) = 1$. Because atomic propositions are defined from the colours of the system, one can derive easily the permutations of colours that can be considered when building the graph of equivalent markings. Such permutations must define symmetries with respect to both the state space and the formula to be checked.

It is worth noticing that the syntactical analysis of a formula does not necessarily yield all the symmetries that can be exploited. In the general case, this technique is efficient for very specific state properties containing propositional subformulae invariant by permutations. For instance, a formula such as $f = F \bigvee_{i \in I} p_i$, where i is an index of atomic propositions, allows all permutations on I.

In the context of *SRG*, the fact that all the colours of a static subclass can give the same behaviour in the system implies a useful property: each colour can be taken as a representative of any other colour of its static subclass. This has been exploited in [IA97] to follow some colours over the graph. A typical example is path formula P3 presented in Section 14.1: each process that asks for a critical section will obtain it. Roughly speaking, the solution consists of building a WN in which one of the processes is isolated from the others by the definitions of distinct static subclasses, then the formula is simplified in order to check if the property is verified for the isolated process alone.

Actually, the automata-theoretic approach can be used to find the admissible symmetries related to a formula. In the context of LTL properties, they are defined from groups of permutations of atomic properties which leave the Büchi automaton invariant. Nevertheless, one must be aware of the cost of such detection which is exponential in the general case. Therefore, researchers have investigated towards interesting subgroups/subsets of permutations ([AHI98]).

Extending the SRG to Partial Symmetries. The SRG approach can be applied efficiently only when the system exhibits a high degree of symmetry. However, even in this case, it often happens that the symmetry between objects is lost at some point in the behaviour of the system. For instance, let us consider again the example of the distributed critical section in Figure 14.17. We add the following information to the system: when several processes ask for the same resource, the process with the highest identity is selected for the access. This is done by adding to transition t_5 a guard $p < q$. Hence, we need to introduce an order among objects of the class *Proc*. However, the guard is meaningless if we use a circular order. Now, if we define a total order, the process with the highest identity should be isolated in a different static subclass. Actually, its behaviour is different from that of other processes in the sense that it has no successor. For a similar reason, the process with the highest identity among remaining processes should be isolated because its successor belongs to a subclass different from the successor of other processes. By an iterated reasoning, we conclude that there should be one static subclass per process. By the condition given in Definition 14.4.2, the only admissible rotation on class *Proc* is thus the identity function.

Such a situation is very penalising because two processes can never be grouped within the same dynamic subclass. Hence the grouping of markings into symbolic markings can be done only with respect to the resources. In this case, the SRG does not provide a drastic reduction compared with the reachability graph. Unfortunately, this drawback is common to all the techniques based on symmetries ([HJJJ84, CDFH91]). The system that we consider is however *partially symmetrical* in the sense that all processes have symmetrical behaviours, except when transition t_5 becomes enabled. It is usually the case that the symmetrical part of a system is much larger than the "asymmetrical" part. We present in this section a technique which extends the SRG by distinguishing the asymmetrical part from the remainder of the net. Compared with the SRG, the modifications that we propose for building an *extended symbolic reachability graph* (*ESRG*) are the following:

- Add to symbolic markings the relevant information in order to handle the firing of asymmetrical transitions. Information is added only when necessary.
- Redefine the symbolic firing rule for the asymmetrical part of the net.

The presentation of the method is organised into three steps: 1) the partitioning of transitions into a symmetrical and an asymmetrical part; 2) the representation of extended symbolic markings; 3) the building of the extended symbolic reachability graph. We will also present the major properties that can be checked directly on such graphs.

Partitioning of Transitions. The use of the operators "$<$", "$\leq$", "$>$", and "$\geq$" causes asymmetrical behaviour, since they need to distinguish among the objects of the tested classes. In terms of well-formed nets, such operators are expressed with membership tests, according to static subclasses. So the asymmetrical property of well-formed nets can be indicated by the specification of expressions, namely asymmetrical expressions, having membership tests. In the following, only one distinguished class is considered since the extension to several of these classes does not present any theoretical difficulty. Moreover, we assume that the distinguished class is partitioned into as many static subclasses as the number of objects in the class.

Definition 14.4.7 (Asymmetrical Variable). *Let C_d be the distinguished class. A variable X defined on C_d is said to be asymmetrical if and only if there exists a predicate function or a guard such that one of the following two conditions holds:*

i) The belonging of X to any static subclass of C_d is tested.
ii) X is in relation with an asymmetrical variable, by means of one of the following well-formed net operators: $=, \neq, !$.

In the following, such a predicate function or guard is said to be asymmetrical.

In WN, the instances of variables are local to transitions, therefore we use the term *asymmetrical variable with respect to a transition.* This means that the variable considered is used in an asymmetrical expression, either in predicate functions associated with the arcs adjacent to the transition or in the transition's guard. Such a transition is called an *asymmetrical transition.* A transition which is not asymmetrical is called a *symmetrical transition.*

Definition 14.4.8 (Asymmetrical and Symmetrical Transitions). *Let t be a transition of a well-formed net. Then t is said to be asymmetrical if and only if one of the following three conditions holds:*

i) There is a place p of P such that there is an asymmetrical predicate function in $\mathbf{Pre}[p,t]$ *or in* $\mathbf{Post}[p,t]$.
ii) The guard of t is asymmetrical.
iii) There is a place p of P such that there is a synchronisation/diffusion function in $\mathbf{Pre}[p,t]$ *or in* $\mathbf{Post}[p,t]$*, defined on the distinguished class.*

t is said to be symmetrical if and only if t is not asymmetrical.

Example of Transition Partition. In the net of Figure 14.17, C_1 is a distinguished class. Actually, p and q are defined on class C_1 and are asymmetrical with respect to the guard of t_5 and its "<" operator. So, t_5 is asymmetrical while the remaining transitions t_1, t_2, t_3, and t_4 are symmetrical.

Extended Symbolic Markings. An *extended symbolic marking* (ESM) may be viewed as a set of nodes: a *standard symbolic marking*, optionally associated with some *eventualities*. Such eventualities are the set of possible partial instances of the standard symbolic marking, with respect to the distinguished class. Hence, one has the ability to represent the behaviour of partially symmetrical systems.

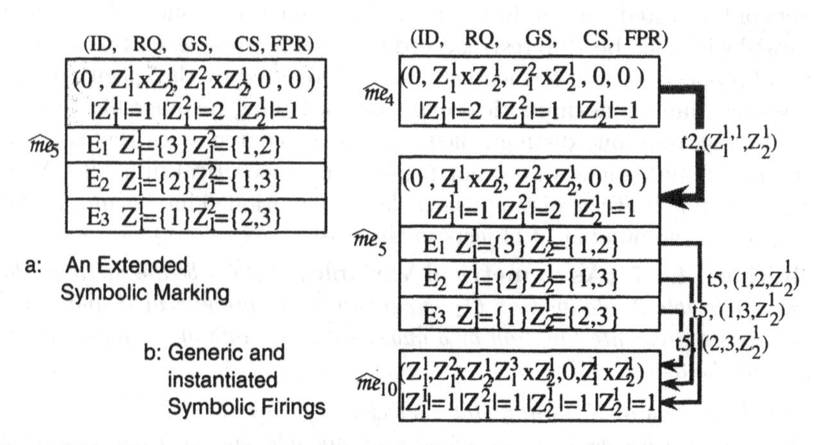

Fig. 14.19. Examples of extended symbolic markings and firings

Example. Figure 14.19a represents an example of an extended symbolic marking, which corresponds to the WN of Figure 14.17.

- The grey part corresponds to the classical representation of a standard symbolic marking. Class C_1 is split into two dynamic subclasses, Z_1^1 and Z_1^2, of cardinality 1 and 2 respectively, while class C_2 has one element represented by Z_2^1.
- The white part of the extended symbolic marking corresponds to the three associated eventualities: E_1, E_2, and E_3. The Z_1^1 and Z_1^2 dynamic subclasses are instantiated, since the distinguished class is C_1.

It is worth noting that eventualities are not markings, but partial instances of ESM. However, one of our aims is to represent them only when neccessary. In fact, the need to represent them is driven by the fact that two cases may occur with respect to an extended symbolic marking: (1) some asymmetrical transitions are enabled from at least one of the eventualities; (2) some of the eventualities are not reachable.

In order to decide on the construction of eventualities, we now define the notions of saturation and uniformity which are checked during the construction of each ESM:

Definition 14.4.9 (Saturation and Uniformity). *An* ESM *is said to be saturated if and only if all of its eventualities are reachable; it is said to be uniform if and only if the objects of the distinguished class have the same distribution in places.*

The following property highlights two cases for which the eventuality representation of ESMs can be useless. In this chapter, for reasons of clarity, the initial ESM is assumed to be uniform.

Property 14.4.10 (Reduction of Representation). An ESM can be represented by its underlying standard symbolic marking if one of the two following conditions holds:

i) The ESM is saturated and there is no asymmetrical transition enabled from it.
ii) The ESM is uniform.

Effectively, in the first case the whole set of eventualities of the ESM are reachable and enabled by the same symbolic instances of (symmetrical) transitions; hence, the underlying standard symbolic marking represents the ESM completely. In the second case, all the elements of the distinguished class are gathered in the same dynamic subclass, leading to the reduction of the set of eventualities to only one item.

Extended Symbolic Firing Rules. As for standard symbolic markings, the construction of an extended symbolic marking can be performed without computing the underlying reachable marking. The condition is that the extended symbolic firing rule takes an ESM into account, in order to build the resulting ESM representations directly. Our method to define a suitable symbolic firing rule is based on the idea that the static subclasses of the distinguished class must be introduced only to deal with asymmetrical transitions. Thus, we choose to build the standard symbolic marking of an ESM without considering the decomposition of the distinguished class into static subclasses; however, when necessary, another standard symbolic representation can be computed from it by considering any given eventuality. Hence, the enabling test can always be performed from a standard symbolic representation of ESM.

Three types of rules are defined (see [HITZ95] for a formal presentation):

i) The generic symmetrical firing occurs for symmetrical transitions, if the relevant extended symbolic marking is saturated. This case arises directly from the standard symbolic firing where the source and the target are standard symbolic markings.

ii) The instantiated symmetrical firing occurs also for symmetrical transitions, but in the absence of saturation. In this case, the source is a reachable eventuality, while the target is either an eventuality or a standard symbolic marking, depending on the ability to make the eventualities absent or not. Because of the symmetrical property of the transition, any reachable eventuality has the same ability of firings and reaches the same extended symbolic markings. Hence, the enabling test of transition is directly performed on the source standard symbolic marking, then the reached eventualities are deduced according to the values of the considered source eventualities.

iii) The (instantiated) asymmetrical firing occurs for asymmetrical transitions; here again, the source is a reachable eventuality, while the target is either an eventuality or a standard symbolic marking. Because of the asymmetrical property of the transition, the eventualities of an ESM may not have the same ability of firings (mixed existence of dead and live eventualities, target nodes may be different ...). To take the static subclasses into account, the dynamic subclasses of C_d must be refined first, which means defining one dynamic subclass for each colour of C_d. However, the asymmetrical firing stages are similar to those of the instantiated symmetrical firing.

Table 14.1 summarises the types of firings according to types of transitions and marking conditions.

Table 14.1. Use of Firing Types

type of transitions marking conditions	symmetrical	asymmetrical
saturated	generic symmetrical	(instantiated) asymmetrical
not saturated	instantiated symmetrical	(instantiated) asymmetrical

Example. Figure 14.19b presents examples of generic and instantiated firings. The extended symbolic markings $\widehat{\mathbf{me}}_4$, $\widehat{\mathbf{me}}_5$, and $\widehat{\mathbf{me}}_{10}$ are assumed to be reachable ESMs of the net of Figure 14.17. Note that $\widehat{\mathbf{me}}_4$ and $\widehat{\mathbf{me}}_5$ are assumed to be saturated, therefore their eventualities are not represented.

From $\widehat{\mathbf{me}}_4$, the $t2$ symmetrical transition is enabled. Since there is no asymmetrical transition enabled from $\widehat{\mathbf{me}}_4$, a generic symmetrical firing can occur by $t2$. Hence, the firing of $t2$ takes into account one item of Z_1^1 for variable p, isolated in the $Z_1^1, 1$ dynamic subclass, and the item of Z_2^1 for variable r. From $\widehat{\mathbf{me}}_5$, transition t_5 can be fired. Note that t_5 is asymmetrical, therefore the eventualities of $\widehat{\mathbf{me}}_5$ must be considered. Each of these eventualities is the source of a firing of t_5: from E_1: $p = \langle 1 \rangle$, $q = \langle 2 \rangle$, and $r = \langle Z_2^1 \rangle$; from $E2$: $p = \langle 1 \rangle$, $q = \langle 3 \rangle$, and $r = \langle Z_2^1 \rangle$; from E_3: $p = \langle 2 \rangle$, $q = \langle 3 \rangle$, and $r = \langle Z_2^1 \rangle$.

Extended Symbolic Reachability Graph. The definitions of extended symbolic markings and extended symbolic firing rules allow us to build a graph called the extended symbolic reachability graph (ESRG). An efficient algorithm is proposed in [HITZ95], which is an adaptation of the standard one. It is fundamentally based on the following two points:

- A canonical expression can be defined for any ESM since there is a canonical expression for the underlying symbolic marking. This allows easy comparisons of ESM*s*.
- Generic symmetrical firings must be privileged with respect to other kinds of firings. This prevents processing an instantiated symmetrical firing before having the ability to produce a generic symmetrical firing, covering it.

Moreover the following property, which is directly inherited from SRG theory, can be used to check saturation rapidly.

Property 14.4.11 (Propagation of Saturation). An ESM which is reached from a saturated ESM by means of a symmetrical transition firing is also saturated.

Example of ESRG. Figure 14.20 represents the ESRG for three processes sharing one resource. Two types of arcs must be distinguished: symbolic arcs (shown in bold) corresponds to generic symmetrical firings and link two underlying standard symbolic markings, while instantiated arcs (not in bold) correspond to the other firing types and link an eventuality to another node (eventuality or standard symbolic marking).

There are 11 ESMs in this graph whereas the corresponding symbolic reachability graph contains 30 standard symbolic markings. In this graph, all the extended symbolic markings are saturated and eventualities must be developed for only two ESMs. Effectively, each one is the target of a saturated symbolic node. However, the $\widehat{\mathbf{me}}_5, \widehat{\mathbf{me}}_6$, and $\widehat{\mathbf{me}}_7$ extended symbolic markings make the t_5 transition firable, therefore all the arcs are symbolic, except $\widehat{\mathbf{me}}_5$ to $\widehat{\mathbf{me}}_{10}, \widehat{\mathbf{me}}_6$ to $\widehat{\mathbf{me}}_7$, and $\widehat{\mathbf{me}}_7$ to $\widehat{\mathbf{me}}_8$, which are instantiated arcs. Moreover, one can note that $\widehat{\mathbf{me}}_0, \widehat{\mathbf{me}}_3$, and $\widehat{\mathbf{me}}_6$ are uniform. Since this is the case for $\widehat{\mathbf{me}}_6$, only the eventualities of $\widehat{\mathbf{me}}_5$ and $\widehat{\mathbf{me}}_7$ have to be represented.

Properties of the Extended Symbolic Reachability Graph. From a graph point of view, standard symbolic marking and eventualities of ESM must be considered as different nodes. Fortunately, the inclusion of eventualities according to some standard symbolic marking implies the existence of implicit arcs which can be taken into account together with explicit ones.

Notation: let t be a transition and let S and S' be two nodes of the ESRG. $S \xrightarrow{t,\hat{c}} S'$ represents an extended symbolic arc reaching S' from S, labelled by $(t, \hat{c})$. $S \xrightarrow{\phi} S'$ represents an extended symbolic path, ϕ, reaching S' from S.

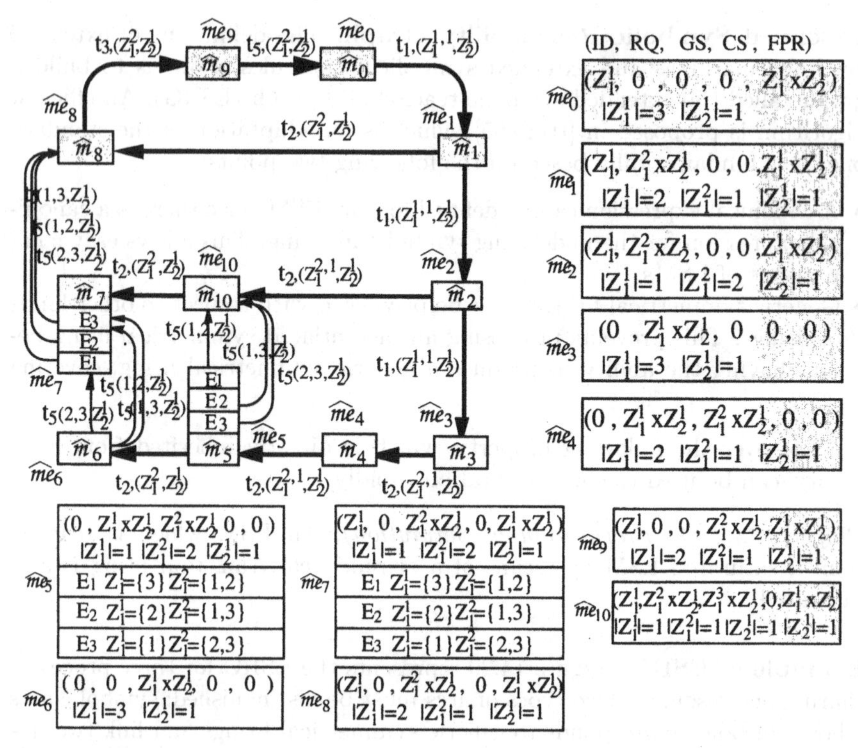

Fig. 14.20. ESRG Example

$M \in S$ means that M is an ordinary marking and it is represented by node S in the ESRG. $\widehat{\mathbf{me_o}}$ is the initial extended symbolic marking.

Property 14.4.12 (Preservation of Firing Sequences). Let $\mathbf{m}$ and $\mathbf{m}'$ be two markings of $\text{RG}(\mathcal{N}, \mathbf{m_0})$.
$\exists \sigma \mid \mathbf{m} \xrightarrow{\sigma} \mathbf{m}' \implies \exists S \xrightarrow{\Phi} S'$, with $\mathbf{m} \in S$ and $\mathbf{m}' \in S'$.

Property 14.4.13 (Relationship Between Arc and Ordinary Firing).
Let $S \xrightarrow{t,\hat{c}} S'$ be an arc of $\text{ESRG}(\mathcal{N}, \widehat{\mathbf{me_o}})$, then:
$\forall \mathbf{m} \in S, \exists \mathbf{m}' \in S', \exists c \in cd(t) \mid \mathbf{m} \xrightarrow{\langle t,c \rangle} \mathbf{m}'$.

The first property expresses the fact that any ordinary firing sequence is represented by an extended symbolic path. The second states that any extended symbolic arc represents at least one ordinary firing. As important consequences of the former properties, we can deduce that reachability and deadlock free problems can be directly checked on the ESRG. For the latter, the notion of dead marking must be re-expressed in order to take eventualities into account.

Property 14.4.14 (Reachability Equivalence). An ordinary marking is reachable from $\mathbf{m_0}$ if and only if it is represented by a node of $\text{ESRG}(\mathcal{N}, \widehat{\mathbf{me_o}})$.

Property 14.4.15 (Dead Marking). A marking $\mathbf{m}$ of $\mathrm{RG}(\mathcal{N}, \mathbf{m_0})$ is said to be dead if and only if there is no output arc from the eventuality or the standard symbolic marking which represents it in $\mathrm{ESRG}(\mathcal{N}, \widehat{\mathbf{me_o}})$.

However, in contrast to SRG, the knowledge of an extended symbolic firing sequence in an ESRG does not allow us to find the equivalent ordinary firing sequences. In fact, the ability to preserve firing sequences concerns transitions, but not their instances. This is because of our desire for conciseness in the representation of ESRG and our focusing on the preservation of the major property which is the reachability property. This leads us to define only a sufficient condition for detecting home spaces of markings, and to define liveness properties in terms of transitions, forgetting colours.

Property 14.4.16 (Home Space of Markings). Let $S \in \mathrm{ESRG}(\mathcal{N}, \widehat{\mathbf{me_o}})$ and let $M(S)$ denote the set of ordinary markings represented by S.
$M(S)$ is a home space if, for all S' of $\mathrm{ESRG}(\mathcal{N}, \widehat{\mathbf{me_o}})$, the following path belongs to $\mathrm{ESRG}(\mathcal{N}, \widehat{\mathbf{me_o}})$:
$(S' = S_1) \xrightarrow{\phi_g} S_m,\ S'_m \xrightarrow{\phi_i} (S_n = S)$
with ϕ_g i a path, the arcs of which correspond to generic symmetrical firings; and ϕ_i a path, any arc of which corresponds to an instantiated firing, either symmetrical or asymmetrical. Optionally, ϕ_g or ϕ_i may not exist.

Property 14.4.17 (Quasi-liveness). Let t be a transition. Then t is quasi-live if there is an arc the label of which contains t.

Property 14.4.18 (Liveness). Let t be a transition. Then t is live if the three following points hold: (1) t is quasi-live; (2) $\widehat{\mathbf{me_o}}$ is uniform; (3) $\widehat{\mathbf{me_o}}$ represents a home space of markings.

14.4.2 Symmetries in Nets

The notion of symmetry was first introduced for coloured Petri nets by Jensen et al. [HJJJ84], but it has since been extended to other classes of nets. Actually, the similarities of behaviours are related to a system and not to the model used to represent it. Hence symmetries can be defined for very general classes of nets ([Sta91, Sch95]), even if they have to be computed and cannot be obtained automatically as in the case for well-formed nets.

The general framework for computing symmetries in nets is the following. Let $\mathcal{N} = \langle P, T, F \rangle$ be a net (we use the algebraic representation given in Definition 2.2.1) where we denote by $\mathcal{I}$ the (arbitrary) set of possible inscriptions on places, transitions, or arcs and χ the function $P \cup T \cup F \to \mathcal{I}$ that assigns inscriptions to the elements of the net. Symmetries, as they have been defined in [Sta91], are bijections on the set $P \cup T$ that respect the node type and the arc relation. Usually, it is also required that they preserve the firing property. This can be done by defining an adequate equivalence relation on the arc inscriptions and requiring that a bijection maps an inscription onto an equivalent inscription. For models such as place/transition nets, the equivalence relation is trivial, namely the equality on natural numbers, but for more complex models it may be rather tricky to define. There is, to our knowledge, no automatic way to design such relations. Hence, for non-classical models, it relies on the skill of the designer.

The usual reason for using symmetries is to reduce the size of the reachability graph. However, the cost of testing whether or not a new marking $\mathbf{m}'$ should be included in this graph may be high. Actually, the reason for not including $\mathbf{m}'$ is that there is already a marking $\mathbf{m}$ in the graph such that $\mathbf{m}' Eq \mathbf{m}$, i.e. there exists a symmetry s such that $\mathbf{m}' = s.\mathbf{m}$. There are two strategies for deciding this property: either first compute and store the whole set of symmetries, and then test if there is one that satisfies the condition; or go through the set of already constructed markings, compare each with the new marking, and re-compute the set of symmetries at each comparison.

The first strategy is usually faster but expensive in storage space since the number of symmetries may be exponential with respect to the number of places in the net. Hence, taking into account that space is often the critical factor in the construction of the reachability graph, the second strategy should be preferred.

14.4.3 Parametrised Reachability Graph

The approach presented in this section also aims to build a reduced reachability graph by grouping states into classes. However, it strongly differs from the approach based on symmetries because:

- It applies to parametrised models, e.g. the number of processes involved in the distributed critical section is not known.
- The grouping of states depends on whether some condition is satisfied, e.g. there is at least one process waiting for the critical resource, and not on the number of processes that satisfy the condition as is the case for symmetries.

Actually, many parallel programs are described with a parameter, the number of processes, and are instantiated, i.e. the number of processes to be executed is fixed. It is usually impossible to extend to all the instantiated programs the results of the analysis of a particular one. With the classical property-verification algorithms, it is necessary to study all the instances of a parametrised program to be sure that the expected properties are verified whatever the value of the parameters. Of course, this is prohibitive since the number of possible values is infinite. Therefore new verification methods and algorithms are needed to study parametrised programs. It has been proved that the verification of properties without fixing the value of the parameters of the program is undecidable ([AK86, Suz88]). Therefore, the methods that are proposed can only solve part of the problem. We present model checking algorithms.

The model checking algorithms described in Section 14.1 of this chapter works on a representation of the state graph of the model. To build this graph, the number of each component of the parallel program must be known. Graph computation algorithms cannot be applied to the study of parametrised programs. Formal methods exist for checking temporal properties of parametrised programs when the number of processes is unknown. These all work on symmetrical parallel programs. Processes have symmetrical behaviour if they have identical possible executions that are independent of their identity ([Bou88]). Symmetric algorithms such as termination detection ([DS80]), mutual exclusion ([Dij65, Lam74]), or reaching agreement algorithms ([PSL80]), compose an important subclass of the parallel algorithms. They may be categorised according to two criteria ([Ray86]): the knowledge that a process has of its environment, and the influence of a process on the behaviour of the others. In the works presented, processes do not know about their identity and the number of other processes. They are however aware of the presence of other processes. Processes communicate through shared variables or directly. All the papers on parametrised programs assume the same hypothesis. The parametrised version of the example from Figure 14.1 in Section 14.1 respects this hypothesis. The number of processes that want to access the critical section is not defined. As the identity of processes is unknown, we cannot develop an algorithm which ensures that accesses are executed in the same order as requests. To make the parametrised approach more clear, we slightly modify the algorithm: a process enters the critical section only if no process is in it and *no other process is waiting for it.* The Petri net in Figure 14.21 represents the parametrised version of this algorithm. The value 2 on the inhibitor arc that links place *Wait* and transition *Enter* ensures that the process which enters the critical section is the only one that is waiting.

We present two classes of parametrised verification methods. The approach of the methods of the first class is to find an unparametrised program that satisfies the same properties as the instantiated programs. This particular program is called a representative program. Properties are verified on the representation of its behaviour. The methods differ in the choice of the representative programs and in the sets of properties that are preserved. The approach of the methods of the second class is to symbolically represent the behaviour of the parametrised program and to verify properties on this symbolic representation.

Representative Program. In [WL89, KM89] the representative program is supposed to be an "abstraction", with no parameter, of the behaviour of the parametrised program. The authors give the principle of the method. Its application depends on the model that is chosen to represent the programs. In [WL89], TCSP [Hoa85] is used while in [KM89] it is CCS [Mil89].

The same specification language is used to model the program and the expected properties. A property is viewed as a program whose executions are all

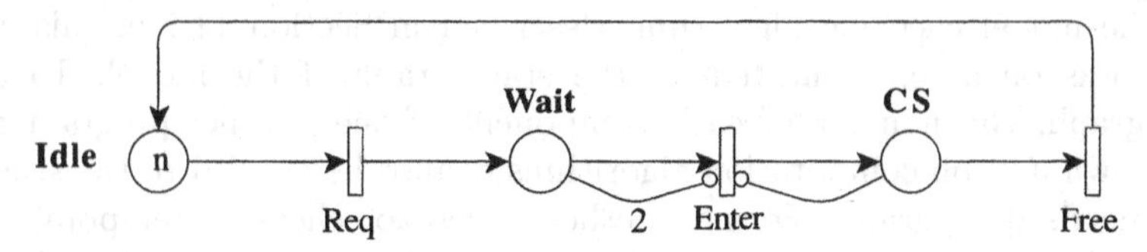

Fig. 14.21. Parametrised mutual exclusion algorithm

the possible ones that satisfy the property. The language may be compatible with the definition of an order between specifications. A specification $Spec_1$ is less than or equal to a specification $Spec_2$ means that all the possible behaviours of $Spec_1$ are behaviours of $Spec_2$. $Spec_2$ may have more behaviours than $Spec_1$. A specification satisfies a property if it is less than or equal to the program associated with the property.

The user must specify a program supposed to be *representative* one. The verification of the representativeness of the program and of the properties on this program is a succession of comparisons with respect to the difined order. If the representative program does not satisfy the property, we cannot draw conclusion about the instantiated programs. The $unsatisfaction$ may be due to a behaviour of the representative program that is not a behaviour of the instantiated ones. If the comparison can be performed by an algorithm, the verification is automatic once the representative program is defined. The existence of a representative program is undecidable.

In [RS93] the authors define sufficient conditions to ensure the existence of a representative program. Furthermore, the verification of the conditions gives this program if it exists. In [BSV94] the authors present a class of systems for which a representative program is the union of the behaviours of successive instances of the parametrised program. Their solution is to test the *representativeness* of the successive unions. If a representative program exists it will be found otherwise the verification will not stop.

The following methods are less general than the previous ones since they are restricted by the possible representative programs. These must be programs instantiated with n processes. The possible values of n depend on the methods.

In the methods proposed in [CG87, LSY94], the user has to provide the value of n. Algorithms are proposed to verify the equivalence between the representative program and the other instances. In [CG87] the user has to follow the verification algorithm, i.e. the algorithm will require some information from the user. In contrast, in [LSY94], once n is defined the user has nothing else to do. Since the language of temporal properties and the representation of the programs are not the same in the two methods, the verification procedures are different. In both cases the existence of a representative program is undecidable.

In the method proposed in [SG89, CGJ95] the representative program is the one instantiated with the smallest possible number of processes. The equivalence is defined with respect to a set of properties and not a logical language. If the set of properties is modified, the work has to be done again. If the parametrised program is specified for at least n_1 processes and if the instantiated programs have similar behaviour if they involve at least n_2 processes, either the equivalence can be proved only if $n_1 \geq n_2$. In compensation, the user has nothing to do during the verification of the equivalence. Two results are possible, the equivalence is proved or the verification algorithm fails, i.e. it detects that it cannot solve the problem and stops.

Symbolic Representation of Behaviour. In [GS92] the authors are interested in linear time temporal logic. The behaviour of the parametrised program is studied through the behaviour of one of the identical processes. The other processes are represented only by their communications with the studied process. The behaviour of the process studied is represented by an automaton. The algorithm defined in Section 14.1.1 is used to verify the properties. An algorithm that builds the automaton, from the specification of the parametrised program, is given. There are some restrictions on the properties that can be verified: they must not refer to the immediate successors of a state and they must concern the behaviour of a single process, i.e. they cannot refer to the states where the other processes are.

In [Ver96] a parametrised state graph that represents all the reachable states and executions of all the instantiated programs is defined. Its computation is automatic, and the specification of the program is a Petri net. The markings are parametrised. Two kinds of information are possible: the number of tokens in a place is known (the marking is given as usual by an integer), or the number of tokens in a place is unknown and it depends on the value of the parameter (the marking gives the smallest number of tokens that can be in the place). For example, the parametrised marking $(p \geq 2, q = 1, r \geq 1)$ represents the markings in which there are at least two tokens in place p, exactly one token in place q, at least one token in place r and no token in any other places . The number of tokens that are in the places corresponding to the states of the identical processes is a constant (the number of identical processes). In the previous example, if p, q, and r are states of processes, then the number of processes dispatched among the three places is a constant for each instantiated program. Therefore if we consider the program instantiated with 6 processes, the parametrised marking corresponds to the following set of markings $\{(p = 2, q = 1, r = 3), (p = 3, q = 1, r = 2), (p = 4, q = 1, r = 1)\}$. A parametrised marking represents an infinite set of markings since it is the union of sets of markings for an infinite number of instantiated programs. Two partial relations are defined for parametrised markings: an inclusion and a superiority relation. A parametrised marking pm is included in a parametrised marking pm' if and only if the set of markings represented by pm is included in the one represented by pm'. A parametrised marking pm is larger than or equal to pm' if and only if with each marking represented by pm' is associated a greater or equal marking represented by pm, and vice versa. The initial parametrised marking of the Petri net of Figure 14.21 is ($Idle \geq 1$), i.e. all processes – of which there exists at least one – are in place *Idle*.

In [Ver94a] the firing rule for building the symbolic graph is defined. A tree is first built. The computation of a branch ends when a parametrised marking included in one of its ancestors is computed. The possible executions from the new parametrised marking will be computed anyway. All the markings, represented by the same parametrised one, enable the same transitions. The set of enabled transitions is computed as usual with respect to the integer associated with each place. In one step, for each enabled transition, the set of successor markings is computed. Rules are defined to avoid the computation of infinite branches; these can be applied if certain precise conditions are satisfied. When the computation of an infinite branch is detected, the conditions are tested and if they are not satisfied the algorithm fails. It cannot compute a finite graph that represents the reachable markings and executions of all the instantiated programs.

The firing rule is divided into three steps:

1. The computation of the new values associated with each place. This is performed as usual. For each place, its associated integer is increased by the sum of the valuation of the arcs linking the fired transition to that place and decreased by the sum of the valuation of the arcs linking the transition to the fired transition.
2. The division of the parametrised marking obtained, if necessary. This is performed when the parametrised marking represents markings that do not enable the same transitions. A procedure to divide it into disjoint sets of markings is used.
3. For each parametrised marking obtained that is greater than one of its predecessors, we have to apply the procedure to avoid infinite branches. If the application conditions are not satisfied, the graph computation algorithm fails.

The parametrised reachability graph of the Petri net from Figure 14.21 is given in Figure 14.22. This graph has 13 vertices and 21 edges. The reachability graph of a program instantiated with n processes is $3 \times n$ edges and vertices. The parametrised graph is a symbolic representation of the reachable markings and executions of all the instantiated programs with 1, 2, ..., 1000, ... processes. The black vertex is the parametrised initial marking. A process can request entry into the critical section. The new parametrised marking is ($Idle \geq 0$, $Wait = 1$). It is divided into ($Idle = 0$, $Wait = 1$) and ($Idle \geq 1$, $Wait = 1$). The former represents a marking of the program instantiated with exactly one process, and enables transition *Enter*. It is reached when exactly one process is in place *Idle*. The firing of transition *Req* in these conditions is represented by the arcs labelled with *Req'*. The latter represents markings of the program instantiated with at least two processes and enables transitions *Req* and *Enter*. From this parametrised marking, when transition *Req* is fired the parametrised marking ($Idle \geq 1$, $Wait = 2$) is reached. It is greater than the previous one and if we continue, we will compute an infinite branch. We can observe that this branch will be composed of ($Idle \geq 1$, $Wait = 2$), ($Idle \geq 1$, $Wait = 3$), ... All these sets of markings can be represented by the parametrised marking ($Idle \geq 1$, $Wait \geq 2$). This modification does not change the set of reachable markings and executions computed. We could not apply this modification previously since the value of place *Wait* is compared to the value 2. We have to distinguish the markings with less than 2 processes in place *Wait* from the others. From ($Idle \geq 1$, $Wait \geq 2$), when transition *Req* is fired, the parametrised marking ($Idle \geq 1$, $Wait \geq 3$) is reached. It is included in ($Idle \geq 1$, $Wait \geq 2$). Therefore, we have a loop in the parametrised reachability graph.

If the number of tokens that each place contains is known exactly, the parametrised marking is a marking associated with an instantiated program. In the graph of Figure 14.22 this is the case for markings ($Wait = 1$) for the program with one process; and ($Wait = 2$) and ($Wait = 1, CS = 1$) for the program with two processes. If the number of tokens is not known exactly in at least one place, the parametrised marking represents markings of almost all the instantiated programs. If the program is instantiated with fewer processes than the sum of the integers associated with places that represent states of the identical processes, then the parametrised marking represents none of its reachable markings. The maximum of these sums is a bound below which the instantiated programs have particular behaviour. In the graph of Figure 14.22, this maximum is 4. The parametrised marking ($Wait \geq 3, CS = 1$) represents markings of programs instantiated with at least 4 processes.

Apart from the failure cases, this small example allows us to illustrate all the possible cases. The firing rule and the procedure to avoid infinite branches and failure cases are given in detail in [Ver96].

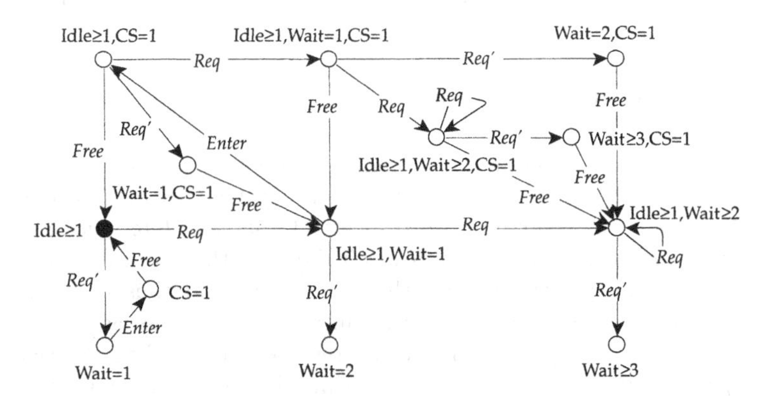

Fig. 14.22. Parametrised reachability graph

Once the graph is built, it can be used to check the satisfaction of properties. Properties verified by inspection of the set of reachable markings independently of their links are easily checked (i.e. deadlocks, state invariants, etc.). We consider the set of reachable parametrised markings that corresponds to the set of reachable markings of all the instantiated programs. We can check that the Petri net is not deadlock free when it is instantiated with at least two processes. When all the processes are in place *Wait* no more transitions are enabled (parametrised markings ($Wait = 2$) and ($Wait \geq 3$)). Model checking algorithms have been adapted for the parametrised symbolic graph. The main difficulty is that circuits of the symbolic graph may not represent "circuits" of the reachability graph of the instantiated programs. This is because of the procedure to avoid infinite branches. This is the case for the loop over parametrised markings ($Idle \geq 1$, $Wait \geq 2$) and ($Idle \geq 1$, $Wait \geq 2, CS = 1$). During the evaluation of a temporal property, it is necessary to distinguish "false" from "true" circuits. The verification algorithm for CTL formulas is presented in [Ver94b, Ver95]. Properties do not refer to the immediate successor of a state. The cases where the verification cannot be done in a finite time are detected and the algorithm fails on these cases.

Conclusion. Symmetry and parametrised approaches yield interesting solutions for the model checking problem. In particular, large systems having either a large or unknown number of similar components can be analysed. Several problems are still open; in particular, the extensions that have been proposed to deal with partial symmetry do not support the verification of the temporal property. Concerning parametrised approaches, none of the solutions presented can be used to study programs with several parameters. The main difficulty is that the satisfaction of some properties will be dependent on the relation between the values of the parameters. The fact that processes are not aware of their identity does not allow us to study problems such as the "philosophers' problem" where philosophers are on a ring and explicitly know the identity of their neighbours and the state where they are located.

14.5 Implementation Issues

This section presents three major implementation techniques which aim to restrain the amount of memory used during the construction of a reachability graph. The first two techniques are respectively named *state-space caching* and *hashing compaction*. They can be applied during the depth-first searches (DFS) that can be performed over the state space and thus in on-the-fly model checking (see Section 14.2 for an introduction). The last technique presented is based on a particular boolean representation called *binary decision diagrams* ([Bry92]). Currently, it is primarily used in different views of temporal logic (LTL, CTL*) in order to reach high data compression.

It is worth noting that the three techniques presented are not specific to Petri nets and have been used for a large variety of system design formalisms.

14.5.1 State-Space Caching

During the reachability graph exploration, the cost of detecting whether or not a marking is new is very high since each time it requires comparisons against all the already visited markings. A depth-first search, such as the ones defined for on-the-fly model checking, uses the following two structures:

- A stack of visited markings is used to store a current path while performing a model checking.
- A heap stores some markings which are already visited but are not in the current path, thus explorations from already visited states can be avoided.

Since memory is rather small with respect to any usual reachability graph, one may consider discarding the visited markings that are surely explored only once: such markings do not have to be kept in memory. As we saw in Section 14.3, stubborn set and sleep set techniques can serialise independent transition firings, so can limit the number of times a (target) marking is visited. Such techniques can be seen as a way to restrain the number of markings to be stored in memory. Unfortunately, it is not possible to decide which markings are visited only once in the general case. In other words, they cannot capture all cases.

Another way to limit the total amount of memory consists in accepting the loss of some information about the visited markings. When only a few nodes are reached several times, such an approach appears possible because the likelihood of reaching a visited marking again is low. It is worth noticing that the stack must not be limited, since this would violate a requirement of the model checking procedure.

The state-space caching technique proposed in [Hol85, JJ91, HGP92] follows this approach. It consists of using a fixed size cache to store not only the stack of the DFS but also as many other visited markings as possible. When the cache is full, all the markings which do not belong to the stack are eliminated. If the size of the cache is bigger than the maximal size of the stack during the search, the exploration will surely terminate. According to this limit, the larger is the cache size, the faster the search procedure, since this decreases the risk of having several parsings for the same paths. However, it is not necessary to consider the size of the graph; rather, a compromise is reached between a maximal size for the cache and the higher computing and searching times.

14.5.2 Hashing Compaction

The hashing compaction technique also aims to reduce the amount of memory to be used. Unlike state-space caching, the hashing compaction technique does not reduce the number of states kept in memory. Rather, it focuses on the reduction of the size of the stored markings, to the extent of representing each of them with a single bit (cf. compression techniques). This method was first presented by Holzmann in [Hol88]. He uses the idea that, at the price of possibly missing part of the state space, the amount of memory required by the searches could be substantially reduced. This method is fundamentally based on the use of hashing without collision detection.

In the basic algorithm, a static bit-array is used. The scheme is the following. Initially, the bit-array is set to off. When a new state is visited, its name is hashed to yield an index into the array; if the corresponding bit is on, then the state is considered to have already been visited; in the opposite case, the bit is set to on and the state is pushed onto the stack associated with the depth-first search. Because there is no detection of collision, the search can be partial and there is always a risk of missing a state. Hence, when this method is used, model checking appears more like an intelligent debugger than like a verifier. However, the author claims that in general, one can choose the size of the array to be large enough, and proposes hash function politics so that the number of collision remains arbitrarily small. For example, Holzmann recommends using a bit-array associated with two different hash functions. Hence, a state is considered to be visited whenever both the corresponding bits are set to on. A generalisation to several-bit arrays and to several hash functions can be found in [WL93]. Moreover, as proposed in [CVWY92], another way is to take a probabilistic point of view to diminish collision. The reader can find an efficient model checking procedure using hashing compaction in [CVWY92].

14.5.3 Boolean Manipulation

Other verification techniques are based on a work of Bryant [Bry86, Bry92] which proposes an operational data structure for managing sets of boolean functions efficiently in a very compact structure, namely *ordered reduced binary decision diagrams* (*ORBDDs*). This section addresses more specifically the model checking problem as introduced in [CES86] by Clarke et al. However, the reader may refer to other interesting works concerning the reachability analysis, [PRCB94], or structural properties, [GVC95].

After a short presentation of *ORBDD*, this section highlights the way to take advantage of this structure in order to perform a model checking of a (finite) place/transition net. A more detailed synthesis of *ORBDD* can be found in [Bry92].

Ordered Binary Decision Diagrams. Fundamentally, an *ORBDD* is a binary decision diagram (*BDD*), that is a rooted directed acyclic graph used to code any boolean function according to its variables. Such a graph ends with two leaf nodes (labelled 0 and 1) which encode the truth values of the boolean function. Moreover, any assignment of the encoded function corresponds to the definition of a path from the root to one of the leaf nodes.

The following coding procedure is followed according to the variables of a function: Each non-terminal node n is followed by exactly two successors (respectively $lo(n)$ and $hi(n)$) and is labelled by a boolean variable ($var(n)$). If the assignment of $var(n)$ is considered to be 1 then the successor is $hi(n)$, else it is $lo(n)$. Hence, a specific value of a function is given by considering for any variable a value and then a specific successor. For instance, both *BDD*s of Figure 14.23 encode the boolean function f defined by $f(a,b,c) = !ab + a!c$. As an illustration, the coding of $f(0,1,0)$ has been highlighted with bold edges. The fact that this assignment yields a *TRUE* value for the function is shown by the last edge, which reaches the 1 node. Observe that the other possible assignments for f are also represented, hence the function is completely described.

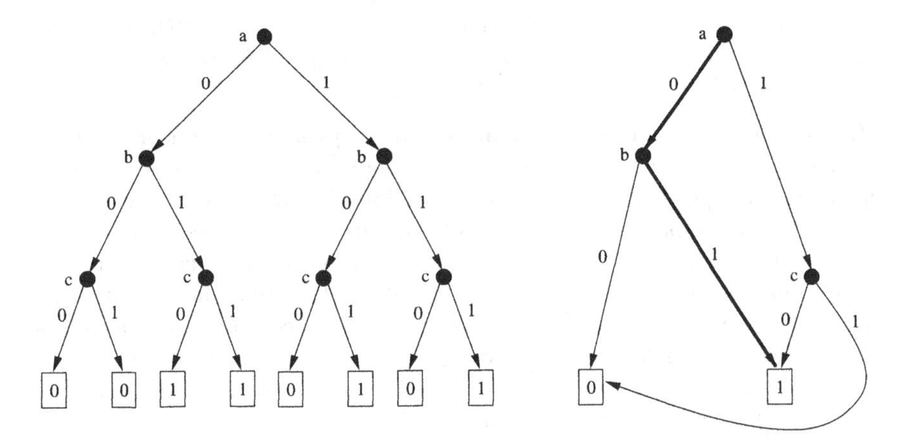

Fig. 14.23. An expanded *BDD* and the same reduced

Bryant has shown that *BDD* can always be reformulated in a canonical form, called *ordered reduced* BDD (*ORBDD*). Canonicity is useful for at least three reasons:

- Equivalence tests between functions are easily computable in terms of equality relations.
- With this canonical form, boolean binary operations can be calculated in polynomial time in the size of the *ORBDD*s involved.
- Moreover, some interesting problems such as satisfiability and tautology can be solved in constant time.

A *BDD* is said to be *ordered* when it is defined according to a total ordering of the variables, in other words, for any two non-terminal nodes, say u and v, the fact that v is a direct successor of u implies that inequality $var(u) < var(v)$ holds. For instance, both *BDD*s of Figure 14.23 are ordered according to the same order of variables: $a < b < c$.

Taking into account a given set of variables and an associated order, the canonical form derives from factorisations of isomorphic subgraphs. Two reduction rules are applied inductively:

- If a node n is such that $lo(n) = hi(n)$ then suppress n and redirect its incoming arcs to $hi(n)$.
- If two nodes n_1 and n_2 are such that $var(n_1) = var(n_2)$, $lo(n_1) = lo(n_2)$, and $hi(n_1) = hi(n_2)$ then suppress n_1 and redirect its incoming arcs to n_2.

For instance, the *BDD* depicted on the right-hand side of Figure 14.23 is ordered and reduced, while this is not the case for the other. Hence, only the right-hand *BDD* is in a canonical form.

Experiments show that the shape and size of the *ORBDD* depend on the variable ordering. The size can be exponential in the number of variables, however many useful boolean functions have a very compact *BDD* representation. A useful and efficient package implementing *ORBDD* operations has been developed by D.E. Long (see [BCL+94]).

State-Space Building Under Boolean Manipulations. The use of *BDD* coding to optimise the model checking problem has been intensively studied in the area of speed independent circuits. Mainly, it consists in a reformulation of the state-space information: not only are the states reconsidered but also the transition relation which symbolically represents the changes among states. One of the interesting features of Petri nets is that the transition relation is directly enhanced by the links between places and transitions. For higher specification languages which introduce objects more complex than Petri nets, different solutions exist which differ in the ways to code the transition relation. See [Enc95] for example.

In this section, we focus on the algorithm of Burch et al. [BCL+94] and apply it in the context of Petri nets. The required operations are the following:

- The existential quantification over boolean variables: $\exists_v$
- The substitution of a variable by another
- The classical logical operations: $\vee$, $\wedge$ and $\neg$

Also, the proposed solution uses the algorithm of Bryant which directly computes an *ORBDD* from a formula of the form $f|_{v=0}$ or $f|_{v=1}$ (i.e. f where variable v is set to either 0 or 1). This algorithm allows the computation of the *ORBDD* for $\exists_v[f]$ as $(f|_{v=0} \vee f|_{v=1})$ (i.e. evaluate to true if there exists a value of v such that f is verified). The substitution of a variable w by a variable v in a formula f is denoted by $f\langle v \leftarrow w\rangle$ and can be carried out by using the existential quantifier: $\exists_v[(v \Leftrightarrow w) \wedge f]$.

Let us now apply this approach to represent the state-space information of a place/transition net. For the sake of clarity, a safe place/transition net is considered because the markings of any such net can be simply coded by boolean variables. The generalisation to k-bounded nets (k being known) can be found in [PRCB94].

A marking of a net is coded by introducing as many variables as the number of places in the net. Any marked place with name p will be represented by a positive literal (boolean variable) of the same name, while empty places are represented in a similar way but with a complementary literal. The boolean representation of a safe marking is deduced from such notations by using logical conjunctions over such variable representations. In particular the case of the initial marking is $(S_0)(\mathcal{P})$:

$$(S_0)(\mathcal{P}) = (\bigwedge_{p \in P, m_0(p)=1} p) \wedge (\bigwedge_{p \in P, m_0(p)=0} \neg p)$$

Observe that disjunctions over such formulas can be used to represent a set of markings. Therefore, in the following, $(S)(\mathcal{P})$ will denote a set of markings coded with the variables of $\mathcal{P}$.

In the *BDD* approach, the building of the state space is performed inductively in the following way: from the set of already visited markings, a set of enabled transitions are concurrently fired and yield another set of computed markings. Then, the resulting markings are added to the markings previously visited to perform the next induction.

More precisely, two sets of boolean variables, denoted in the following $\mathcal{P}$ and $\mathcal{P}'$, are needed in order to model such two sets of markings. Moreover, a boolean function $f_t(\mathcal{P}, \mathcal{P}')$ is defined for each t transition of the net to express the firing conditions of t, from markings coded with $\mathcal{P}$ to those coded with $\mathcal{P}'$.

$$f_t(\mathcal{P}, \mathcal{P}') = (\bigwedge_{p \in {}^\bullet t} p) \wedge (\bigwedge_{p \in {}^\bullet t \setminus t^\bullet} \neg p') \wedge (\bigwedge_{p \in t^\bullet} p') \wedge (\bigwedge_{p \notin {}^\bullet t \cup t^\bullet} p \Leftrightarrow p')$$

So, pure input places are emptied while output places receive a token and other places are unchanged.

Hence, the global transition relation of a net N is defined as follows:

$$N(\mathcal{P}, \mathcal{P}') = \bigvee_{t \in T} f_t(\mathcal{P}, \mathcal{P}')$$

From this coding, one can compute the next set of visited firings, that is $F(S)$:

$$F(S) = S \cup \{s' \mid \exists s \in S, (s, s') \in N\}$$

Or, in terms of boolean expressions:

$$(F(S))(\mathcal{P}') = (S)(\mathcal{P}') \vee \exists_{p \in \mathcal{P}}[(S)(\mathcal{P}) \wedge N(\mathcal{P}, \mathcal{P}')]$$

Due to the $\exists_{p \in \mathcal{P}}$ operator, there remain only variables of $\mathcal{P}'$. Finally, an induction over $F(S)$ is easy to implement by applying a straightforward variable substitution to obtain the same expression but coded with variables of $\mathcal{P}$.

Thus, by noting

$$(F(S_i))(\mathcal{P}') = (S_{i+1})(\mathcal{P}')$$

one can compute successively S_0, $S_1 = F(S_0)$, $S_2 = F^2 = F(F(S_0))$, etc. Clearly, this sequence converges to the least fixed point of F, which exactly corresponds to all the reachable states. Of course, such a technique must be adapted to fire at each step only a subset of the enabled transitions, in case, for instance, of a synchronisation caused by a model checking procedure.

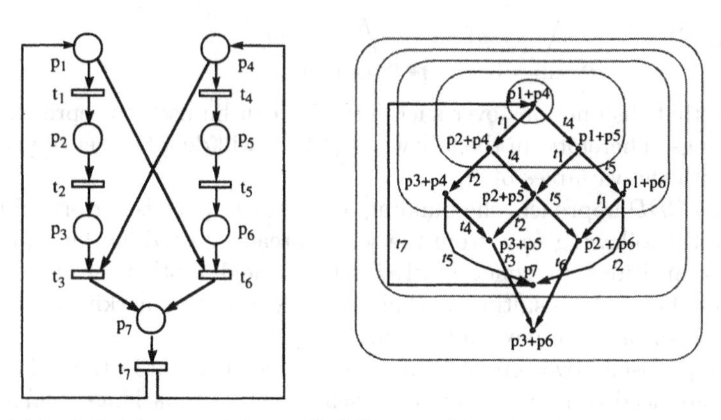

Fig. 14.24. A Petri net and its reachability graph

Let us evaluate the *BDD* method on the net of Figure 14.24 with $p_1 + p_4$ as the initial marking.

The corresponding transition relation can be expressed as follows:

$$\begin{aligned}
f_{t_1}(\mathcal{P},\mathcal{P}') = {} & (p_1) \wedge (\neg p_1') \wedge (p_2') \wedge \\
& [(p_3 \Leftrightarrow p_3') \wedge (p_4 \Leftrightarrow p_4') \wedge (p_5 \Leftrightarrow p_5') \wedge (p_6 \Leftrightarrow p_6') \wedge (p_7 \Leftrightarrow p_7')] \\
f_{t_2}(\mathcal{P},\mathcal{P}') = {} & (p_2) \wedge (\neg p_2') \wedge (p_3') \wedge \\
& [(p_1 \Leftrightarrow p_1') \wedge (p_4 \Leftrightarrow p_4') \wedge (p_5 \Leftrightarrow p_5') \wedge (p_6 \Leftrightarrow p_6') \wedge (p_7 \Leftrightarrow p_7')] \\
f_{t_3}(\mathcal{P},\mathcal{P}') = {} & [(p_3) \wedge (p_4)] \wedge [(\neg p_3') \wedge (\neg p_4')] \wedge (p_7') \wedge \\
& [(p_1 \Leftrightarrow p_1') \wedge (p_2 \Leftrightarrow p_2') \wedge (p_5 \Leftrightarrow p_5') \wedge (p_6 \Leftrightarrow p_6')] \\
f_{t_4}(\mathcal{P},\mathcal{P}') = {} & (p_4) \wedge (\neg p_4') \wedge (p_5') \wedge \\
& [(p_1 \Leftrightarrow p_1') \wedge (p_2 \Leftrightarrow p_2') \wedge (p_3 \Leftrightarrow p_3') \wedge (p_6 \Leftrightarrow p_6') \wedge (p_7 \Leftrightarrow p_7')] \\
f_{t_5}(\mathcal{P},\mathcal{P}') = {} & (p_5) \wedge (\neg p_5') \wedge (p_6') \wedge \\
& [(p_1 \Leftrightarrow p_1') \wedge (p_2 \Leftrightarrow p_2') \wedge (p_3 \Leftrightarrow p_3') \wedge (p_4 \Leftrightarrow p_4') \wedge (p_7 \Leftrightarrow p_7')] \\
f_{t_6}(\mathcal{P},\mathcal{P}') = {} & [(p_1) \wedge (p_6)] \wedge [(\neg p_1') \wedge (\neg p_6')] \wedge (p_7') \wedge \\
& [(p_2 \Leftrightarrow p_2') \wedge (p_3 \Leftrightarrow p_3') \wedge (p_4 \Leftrightarrow p_4') \wedge (p_5 \Leftrightarrow p_5')] \\
f_{t_7}(\mathcal{P},\mathcal{P}') = {} & (p_7) \wedge (\neg p_7') \wedge [(p_1') \wedge (p_4')] \wedge \\
& [(p_2 \Leftrightarrow p_2') \wedge (p_3 \Leftrightarrow p_3') \wedge (p_5 \Leftrightarrow p_5') \wedge (p_6 \Leftrightarrow p_6')]
\end{aligned}$$

The initial marking is coded by the predicate $(S_0)(\mathcal{P}) = p_1 \wedge p_4 \wedge \neg p_2 \wedge \neg p_3 \wedge \neg p_5 \wedge \neg p_6 \wedge \neg p_7$. The application of the predicate transformer F on

S_0 leads to the following set of states $\{p_1p_4, p_2p_4, p_1p_5\}$. The iteration of F is depicted in Figure 14.24 by concentric curves that become successively larger. It is worth noting that all the reachable markings are visited in four iterations.

14.5.4 Symbolic Model Checking

Temporal model checkers based on the *BDD* representation have been implemented to check real systems. *BDD* model checkers are called *symbolic model checkers* since the transition relations are described using variables. In fact, the classical algorithms of model checking are not changed, but predicate transformers must be introduced to match/synchronise the representation of the formula against the symbolic expression of the transition relation. More detail can be found in [BCL$^+$94].

14.5.5 Concluding Remarks on Implementation Issues

By means of techniques such as state-space caching, hashing compaction, or *ORBDD*s, the memory used for the building of state spaces can be drastically reduced. Such approaches are very promising since representations up to 10^{20} states have already been analysed using standard memory. Some systems that are unmanagable with standard approaches of model checking have been either verified or systematically debugged.

State-space caching is based on the depth-first search procedure, therefore it can be used to optimise on-the-fly model checking, specifically for linear temporal formulae (see Section 14.2). The only requirement is that the memory must be able to contain the maximal path of the search.

Hashing compaction corresponds to an abstraction of states, therefore one can miss some states in the case of imperfect hash functions or an insufficiently large hash table. By focusing on accepting states, it is demonstrated that some errors can be missed but the algorithms will never falsely claim that the system is incorrect ([CVWY92]). Therefore, model checking algorithms under hashing compaction should be viewed more as an intelligent validation tool able to perform several automated simulations with respect to a property, rather than as a verification tool.

For the sake of efficiency, state-space caching and hashing compaction techniques must be used in conjunction with any other compatible technique that can reduce the number of states matching, such as the sleep sets or stubborn sets. The SPIN tool ([Hol97]) of the AT&T Bell Laboratory implements such a combined approach and also allows the combination of state-space caching and hashing compaction in order to increase the scope of automated validation with respect to an LTL property.

The use of *ORBDD* is a rather different approach which represents the entire state space but factorises components of states in order to save a large amount of memory. Both LTL and CTL algorithms have been implemented, however the *ORBDD* approach appears to be better suited to CTL rather than LTL. Effectively, the firing mechanism is more efficient than the largest uniform progression throughout the state space is chosen, in other words, the cost to process firings from either one or several markings might be the same. Different tools implement verification using *ORBDD*s, among them PROD [VHHP95] and COSPAN [HHK96].

It is worth noticing that *BDD* approaches might not solve the combinatorial problem, so can fail to check some systems. Effectively, their efficiency is strongly related to the number of variables and above all to the selected choice of variable ordering. Hence, intermediate *BDD* representations yielded during the firing process can be very large, although the final representation could be sufficiently reduced. A factor of ten or even more is common. Finally, the use of two sets of variables to code the transition relation has a negative effect on performance when the system description becomes large.

Given this problem, approaches other than *BDD* are of interest ([CK97a, CVWY92]). Moreover, one may consider combining techniques such as *BDD*s with stubborn set construction and symmetry detection ([Tiu94, ABH+97]).

14.6 Synthesis and General Concluding Remarks

Throughout this chapter, we have shown that model checking is a general and totally automatic solution for verifying system properties. The price to pay is formalisations of both system and properties. In this context, temporal logic appears expressive enough for the modelling of a large class of properties while being sufficiently intuitive to be accepted by engineers.

The model checking problem has become a reality for systems engineering, given techniques which can reduce the state-space representation. The family of techniques that we have presented exploit orthogonal concepts which can be viewed as complementary approaches for reducing the size of the problem:

- Partial-order techniques provide efficient algorithms based on a system semantics which is much more economic and realistic than that of the interleaving of transition firings.
- Symmetry and parametrisation techniques cope with replications of objects in the system, and are particularly appropriate for checking fault tolerance problems, or protocol specifications of agent groups, etc.
- Efficient implementation techniques such as BDD compression and the on-the-fly approach provide a general framework for efficient model checking.

All these techniques can be used for example in the LTL model checker. However, in practice, it is very difficult to know which one yields the best rate of reduction. Honest comparisons, such as [CP96], compare systems models and demonstrate the relative efficiency of each technique. Therefore, several verification tools, such as PEP [BG96], PROD [VHHP95], SPIN [Hol97], and CPN-AMI [MT94], propose combining several approaches.

Currently, model checking is being extensively researched, and studies include:

- Extensions of the expressiveness of system models, by adding (real) time concepts ([DT98]), by considering unbounded or infinite colour domains, or by taking infinite state space into account ([CC99, MN95]).
- Extensions to more powerful logics which consider interesting new operators ([KMM+97]), including a time operator for dealing with real clocks ([DT98]) or a probabilistic operator for a better understanding of the truth value of a property.
- Improvements of state-space reduction techniques which relate to, for example, alternatives to BDDs ([NM94]), a better exploitation of symmetries ([AHI98]), or elaborate combinations of reduction techniques.
- Integration of model checkers within theorem provers in order to automate some stages in the verification of complex proofs ([Spr98]).

For a general (and not Petri-net-related) discussion of algorithms for state-space seach the reader is referred to [Zha99].

15. Structural Methods*

Verification methods based on the state space have a major drawback: the reachability graph must be computed in advance. The construction of this reachability graph is computationally a very hard problem. This is because the size of the state space may grow more than exponentially with respect to the size of the Petri net model (measured, for example, by the number of places). This problem is usually known as the *state-space explosion problem.* In [Val92] the reader can find a discussion of the size of the reachability graph obtained from a Petri net and the role of the concurrency in the state-space explosion problem.

Although these analysis techniques have the drawback mentioned above, for bounded net systems they are the more general methods and, in some cases, are the only way to verify a given property.

A successful way to cope with this problem has been so-called *structure theory.* The idea is to get useful information about the behaviour of the system from the structure of the net model (avoiding the construction of the reachability graph). Structure theory investigates the relationship between the behaviour of a net system and its structure, i.e. the linear algebraic and graph theoretic objects and properties associated with the net and the initial marking. The study of this relationship usually leads to a deep understanding of the system. The ultimate goals of structure theory are usually termed the *analysis problem*, i.e. the problem of alleviating state-space explosion by developing analysis methods that do not require the construction of the state space; and the *synthesis problem*, i.e. the problem of designing refinement and composition operators that are known to preserve the properties of interest. In this chapter we concentrate on the analysis problem within structure theory.

When general concurrent systems are considered, typical structural techniques give necessary or sufficient conditions on the properties studied. Nevertheless, the most satisfactory results are obtained when the scope is limited to restricted classes of systems and particular properties. The behaviour of general concurrent systems is of course richer, but sensible limitations lead to useful subclasses, able both to model certain practical systems and to give insight into the relationships between behaviour and structure in more general systems. The typical restrictions that are imposed aim to limit the interplay

* Authors: J. M. Colom, E. Teruel, M. Silva, and S. Haddad

between synchronisations and conflicts. On the one hand, these restrictions facilitate the analysis. On the other, some modelling capabilities are lost. The designer must find a compromise between modelling power and availability of powerful analysis tools, while one of the theoretician's goals is to obtain better results for increasingly larger subclasses.

In this chapter two intimately related families of structural analysis techniques will be considered:

- *Graph Theory.* Objects such as paths, circuits, handles, bridges, siphons, and traps, and their relationships, are investigated. Typically, only ordinary nets are considered, and the main results apply to specific properties, mainly boundedness, liveness, and reversibility ([BT87, CHEP71, ES91, Hac72, TV84]).
- *Linear Algebra and Convex Geometry.* These are techniques based on the state equation and/or the flows and semiflows. The semiflows can be used to prove properties such as boundedness, mutual exclusion, and liveness. More generally, the state equation can be used as a basic description of the system in order to prove or disprove the existence of markings or firing sequences fulfilling certain given conditions, eventually expressed as logic formulas ([Col89, Esp94, Sil85]). Typically, results for general P/T net systems are obtained ([Col89, CCS90, Lau87, Mem78, MR80, SC88, TCS93]), some of which may become especially powerful when applied to restricted subclasses together with graph-theory-based arguments ([Esp92b, Esp94, ES90, TS93]).

To facilitate the analysis, a large and complex system can be transformed (typically reduced) while preserving the properties to be analysed. Transformation rules preserve the behaviour and are often supported by structural arguments such as simple, efficient sufficient conditions. Net system reductions are presented on the next section with special emphasis in the implicit place concept.

15.1 Net System Reductions

In order to alleviate the state-space explosion problem, several techniques for obtaining *reduced state spaces* have been introduced. As an example we can cite the stubborn set method ([Val90c, Val92]). These techniques are used directly in the construction phase of the reachability graph for the original net model. In this section we review a different kind of reduction techniques named *net system reductions.* These reductions proceed by transforming the net structure and, sometimes, the initial marking.

From an operational point of view, the approach is based on the definition of a kit or catalogue of *reduction rules*, each one preserving a subset of properties (liveness, boundedness, reversibility, etc) for analysis. A reduction rule characterises a type of sub-net system (*locality principle*) which can be substituted by another (simpler) sub-net system.

The pre-conditions to be fulfilled have a *behavioural* and/or a *structural* formulation. Behavioural pre-conditions can be more powerful for a given initial marking, but their verification is usually much more complex. Thus the pre-conditions presented here are based on structural considerations and properties of the initial marking (i.e. the initial marking is considered as a parameter).

The design of a catalogue of reduction rules is based on a tradeoff between completeness (i.e. transformation capabilities) and usefulness (i.e. applicability).

Given a catalogue of reduction rules, analysis by reduction (the transformation procedure) is iterative in nature: Given the property (or properties) to be analysed, the subset of rules that preserve it (them) is applied until the system becomes irreducible. The irreducible system may be so simple that the property under study can be trivially checked (see Figure 15.2d). In other cases, the irreducible net is just "simpler" to analyse using another analysis technique (e.g. we can obtain a reduced state space in which it is possible to analyse the property that has been preserved in the reduction process). In other words, techniques for analysing net system models are complementary, not exclusive.

Reduction rules are transformation rules used for net analysis. When considered in the reverse sense they become expansion rules, used for net synthesis: i.e. for a stepwise refinement (or top-down) approach. Examples of this approach can be found in the context of the synthesis of live and bounded free-choice systems ([ES90]) or in the definition of subclasses of nets by the recursive application of classical expansion rules as in the case of macroplace/macrotransition systems ([DJS92]). Using this approach, with adequate expansion rules, the model will verify the specification by construction. This is interesting when compared with the more classical approach based on the iteration of the design and analysis phases until the specification is satisfied. The iterative process has two basic disadvantages:

1) The lack of general criteria for modifying (correcting) a model which does not meet the requirements
2) The operational difficulty inherent to the validation phase

Nevertheless, since no kit of reduction rules is complete (i.e. able to fully reduce any system), it is not possible to synthesise an arbitrary system by such stepwise refinements.

A very basic kit of reduction rules is presented. Additional details are given only for the rule of implicit places; these places are redundancies in the net system model: if an implicit place is removed, then (illusory) synchronisations disappear and other reduction rules can be applied.

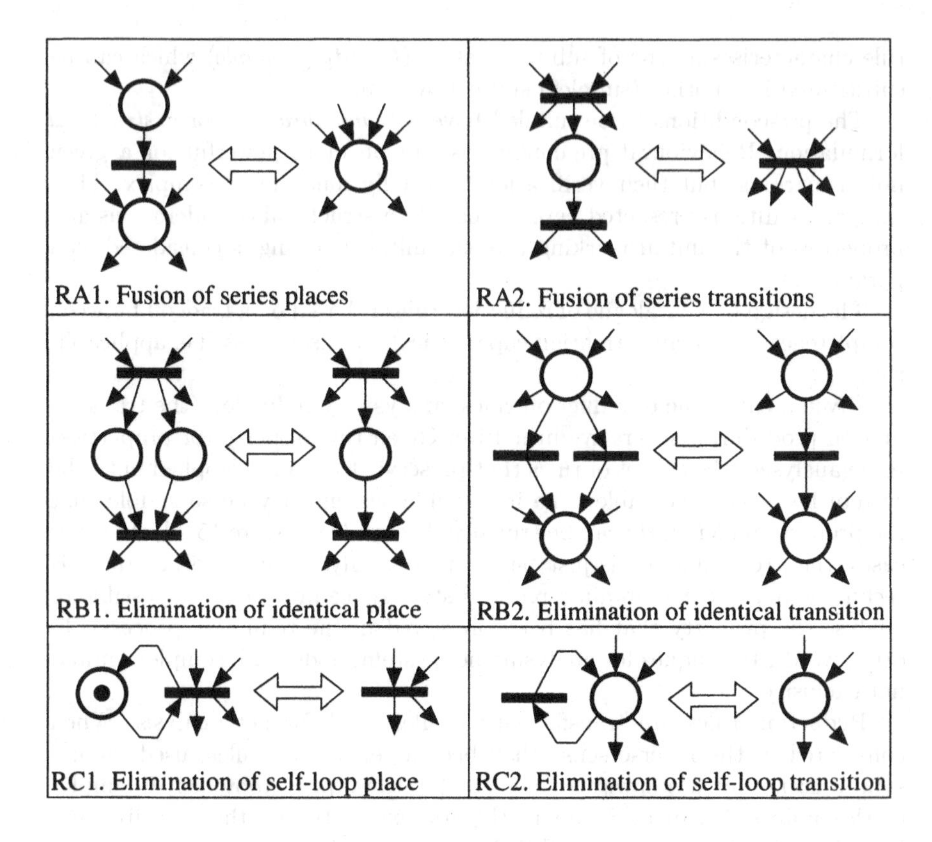

Fig. 15.1. A basic reduction kit

15.1.1 A Basic Kit of Reduction Rules

Figure 15.1 presents graphically the structural and marking conditions for a kit of very specific reduction rules. It is not difficult to observe that these rules preserve properties such as liveness and the bound of places (thus boundedness):

- *RA1* is a particular case of the *macroplace rule* ([Sil81]).
- *RA2* is a particular case of the *transition fusion rule* ([Ber87]).
- *RB1* and *RC1* are particular cases of the *implicit place rule* ([Sil85, SC88]), to be considered later in more detail. Observe that *RC1* can be trivially generalised by considering several self-loops in which the place always appears. Liveness, the bound of places, and reversibility are preserved. Moreover if the place contains several tokens, liveness, boundedness (in general not the bound of the net system), and reversibility are preserved.
- *RB2* and *RC2* are particular cases of *identical* and *identity transition rules* ([Ber87]) respectively.

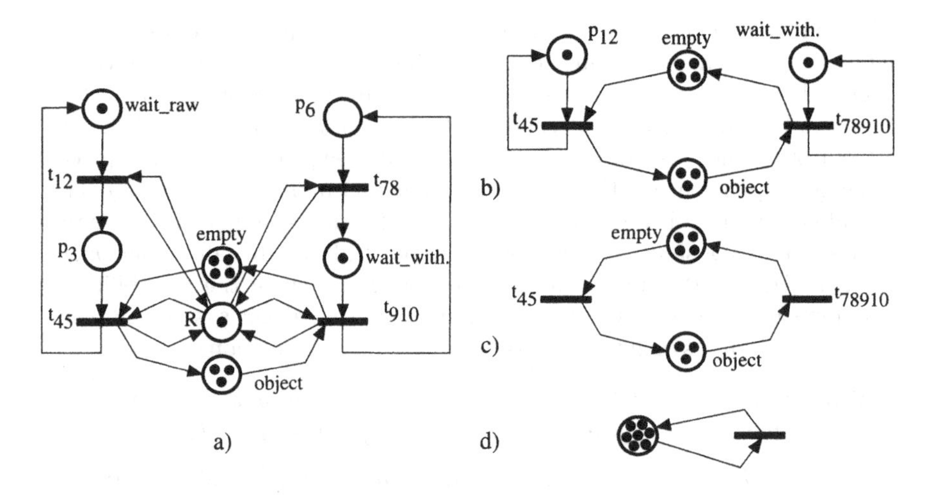

Fig. 15.2. The reduction process shows (see (d)) that the net system in Figure 5.8b) is live, 7-bounded, and reversible.

An interesting remark is the analogy between rules in the same row in Figure 15.1: Basically rules *RX2* are obtained from rules *RX1* by changing the role of places and transitions (*duality*) and reversing the arrows (important only for the *RA* rules).

Example 15.1.1. Let us now consider the net system in Figure 5.8b. The sub-net defined by $\mathrm{op}_1 - t_3 -$ wait_dep. verifies the pre-condition of rule *RA1*. Thus it can be reduced to a place, p_3 (Figure 15.2a). The same holds for $\mathrm{op}_2 - t_6 -$ wait_free which is reduced to p_6 (Figure 15.2a). The sub-nets $t_1 - \mathrm{load} - t_2$, $t_4 - \mathrm{deposit} - t_5$, $t_7 - \mathrm{unload} - t_8$, and $t_9 - \mathrm{withdrawal} - t_{10}$ can be reduced according to *RA2* (see t_{12}, t_{45}, t_{78}, and t_{910} in Figure 15.2a). Place R in Figure 15.2a is implicit (one of the trivial generalisations mentioned for *RC1*). Thus it can be removed, and wait_raw $- t_{12} - p_3$ and $t_{910} - p_6 - t_{78}$ can be reduced to p_{12} and t_{78910} respectively (see Figure 15.2b). Places p_{12} and wait_with. are implicit (*RC1*) in Figure 15.2b, thus the net system in Figure 15.2c is obtained. Playing the token game, a place (e.g. object) can became empty in Figure 15.2c and t_{45} – object – t_{78910} can be reduced (*RA2*) to a single transition (Figure 15.2d). Therefore, the original net system is live, 7-bounded, and reversible.

15.1.2 Implicit Places

A place in a net system is a constraint on the firing of its output transitions. If the removal of a place does not change the behaviour of the original net system, that place represents a redundancy in the system and can be removed. A place whose removal preserves the behaviour of the system is called an *implicit place.* Two notions of behaviour equivalence are used to define implicit places. The first considers that the two net systems have the same behaviour if they present the same fireable sequences. That is, this place can be removed without changing the *sequential observation* of the behaviour of the net system (i.e. the set of fireable sequences). Implicit places under this equivalence notion are called *sequential implicit places* (SIP). The second notion of equivalence requires that the two net systems must have the same sequences of steps. In this case the implicit places are called *concurrent implicit places* (CIP) and their removal does not change the possibility of simultaneous occurrences of transitions in the original net system. Implicit places model false synchronisations on their output transitions.

Definition 15.1.2. *Let $\mathcal{S} = \langle \mathcal{N}, \mathbf{m_0} \rangle$ be a net system and $\mathcal{S}' = \langle \mathcal{N}', \mathbf{m_0}' \rangle$ the net system resulting from removing place p from $\mathcal{S}$. The place p is a*

1. *Sequential implicit place (SIP) iff $\mathrm{L}(\mathcal{N}, \mathbf{m_0}) = \mathrm{L}(\mathcal{N}', \mathbf{m_0}')$, i.e. the removing of place p preserves all firing sequences of the original net.*
2. *Concurrent implicit place (CIP) iff $\mathrm{LS}(\mathcal{N}, \mathbf{m_0}) = \mathrm{LS}(\mathcal{N}', \mathbf{m_0}')$, i.e. the removing of place p preserves all sequences of steps of the original net.*

It is easy to see that if a place p is a CIP then it is also an SIP (since the preservation of the sequences of steps implies the preservation of the firing sequences). Nevertheless, the contrary is not true in general. Let us consider, for example, the net in Figure 5.2. The place p_6 is an SIP since its removal does not change the set of firing sequences (the reachability graphs of the original net system and the net system without place p_6 are the same). But the place p_6 is not a CIP because after its removal transitions b and c can occur simultaneously whereas in the original net system they are sequentialised (i.e. the steps are not preserved). In order for a SIP with self-loops to be a CIP, more tokens may be needed in its initial marking (in our example p_6 requires two tokens in the initial marking in order to be a CIP). In [Col89] it is proved that a self-loop-free SIP is also a CIP.

Let p be a CIP of the net system $\mathcal{S}$, and $\mathcal{S}'$ the net system $\mathcal{S}$ without the CIP p. Let σ be a fireable sequence of steps in $\mathcal{S}$ such that $\mathbf{m_0} \xrightarrow{\sigma} \mathbf{m}$. The sequence σ is also fireable in the net system $\mathcal{S}'$, i.e. $\mathbf{m_0}' \xrightarrow{\sigma} \mathbf{m}'$. This is because the removal of a CIP preserves the fireable sequences of steps of the net system. A trivial consequence of this is that the reached markings in $\mathcal{S}$ and $\mathcal{S}'$, firing the same sequence σ, are strongly related: $\forall q \in P \setminus \{p\}$, $\mathbf{m}[q] = \mathbf{m}'[q]$. Moreover, if $\mathbf{s}$ is a step enabled at $\mathbf{m}'$ then the following holds: $\mathbf{m}' \geq \mathbf{Pre}' \cdot \mathbf{s} \Longrightarrow \mathbf{m}[p] \geq \sum_{t \in (p^\bullet \cap \|\mathbf{s}\|)} \mathbf{s}[t] \cdot \mathbf{Pre}[p, t]$. If p is an SIP the previous property can be written in the following way: $\forall t \in p^\bullet$, $\mathbf{m}' \geq \mathbf{Pre}'[P', t] \Longrightarrow \mathbf{m}[p] \geq \mathbf{Pre}[p, t]$.

The elimination of a CIP or an SIP preserves deadlock-freeness, liveness, and marking mutual exclusion properties; but it does not preserve boundedness or reversibility. Moreover, the elimination of a CIP preserves the firing mutual exclusion property, but this is not true for SIPs.

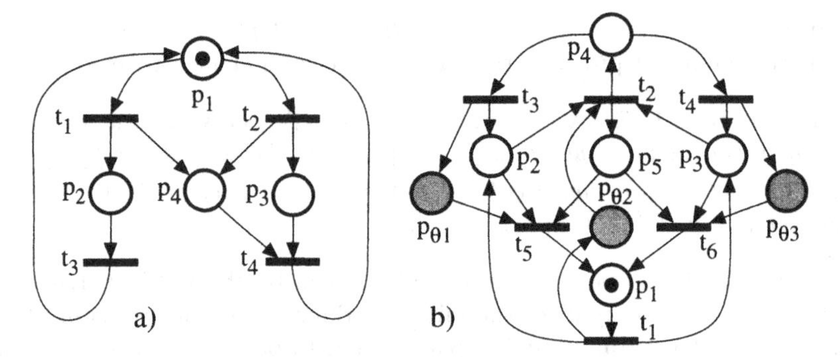

Fig. 15.3. a) Place p_4 is firing implicit but not marking implicit. After the removal of p_4 the "false" synchronisation in t_4 disappears. b) The places in the set $\{p_{\theta 1}, p_{\theta 2}, p_{\theta 3}\}$ (or $\{p_2, p_3, p_5\}$) are CIPs.

Example 15.1.3. The net system in Figure 15.3.a is unbounded (p_4 is the unique unbounded place) and non-reversible (also because of p_4). Place p_4 is a CIP. After the removal of p_4 the system becomes bounded and reversible! On the other hand, place p_6 in Figure 5.2 imposes firing mutual exclusion between b and c. Since p_6 is a SIP, the reduction rule does not preserve firing mutual exclusion. According to the definition, fireable sequences are preserved.

Sometimes it is practical to impose an additional condition on the definition of implicit places, requiring that their marking be redundant (computable) with respect to the markings of the other places in the net (i.e. a marking redundancy property). Let us consider the CIP $p_{\theta 1}$ of the net system $\mathcal{S}$ depicted in Figure 15.3b. This place is a CIP and its marking can be computed from the markings of places p_1, p_2, and p_5: $\forall \mathbf{m} \in \mathrm{RS}(\mathcal{S})$, $\mathbf{m}[p_{\theta 1}] = \mathbf{m}[p_1] + \mathbf{m}[p_2] + \mathbf{m}[p_5] - 1$. Such places will be called *marking implicit places.* Nevertheless, the marking of some implicit places cannot be computed only from the markings of the other places in the net. These places will be called *firing implicit places.* As an example consider the CIP p_4 in Figure 15.3a: $\forall \mathbf{m} \in \mathrm{RS}(\mathcal{S})$ such that $\mathbf{m_0} \xrightarrow{\sigma} \mathbf{m}$, $\mathbf{m}[p_4] = \mathbf{m}[p_3] + \boldsymbol{\sigma}[t_1]$). The classification of the implicit places into marking implicit places and firing implicit places can be applied to the two previously defined classes, CIP and SIP. Because of the additional condition, marking implicit places preserve the state space (i.e. the reachability graphs of the net system with and without p are isomorphic), therefore they also preserve boundedness and reversibility.

So far, implicit places have been presented in a behavioural setting. In order to do the verification we must resort to algorithms based on the reachability graph with its inherent limitations and high associated complexity. The structural formulation of the implicit place reduction rule requires the statement of a structure-based condition to be satisfied by the implicit place, and the characteristics of its initial marking. Places satisfying the structure-based condition will be called *structurally implicit places*; these are places that become implicit provided they are marked with enough tokens.

Definition 15.1.4. *Let $\mathcal{N}$ be a net. A place p of $\mathcal{N}$ is a structurally implicit place iff there exists a subset $I_p \subseteq P \setminus \{p\}$ such that $\mathbf{C}[p,T] \geq \sum_{q \in I_p} y_q \cdot \mathbf{C}[q,T]$, where y_q is a non-negative rational number (i.e. $\exists \mathbf{y} \geq 0$, $\mathbf{y}[p] = 0$ such that $\mathbf{y} \cdot \mathbf{C} \leq \mathbf{C}[p,T]$ and $I_p = ||\mathbf{y}||$.*

Obviously, the above structural condition can be checked in polynomial time. The next property gives the initial marking conditions which must be satisfied by a structurally implicit place if it is to be an SIP or a CIP. This condition is based on the solution of a linear programming problem (LPP 15.1 below). The linear program computes an upper bound on the minimal initial marking of a structurally implicit place which must be satisfied in order for the place to be an SIP or a CIP in the net system $\langle \mathcal{N}, \mathbf{m_0} \rangle$. Because LPPs are of polynomial time complexity ([NRKT89]), the evaluation of this condition also has this complexity.

Property 15.1.5. Let $\langle \mathcal{N}, \mathbf{m_0} \rangle$ be a net system. A structurally implicit place p of $\mathcal{N}$, with initial marking $\mathbf{m_0}[p]$, is an SIP (CIP) if $\mathbf{m_0}[p] \geq z$, where z is the optimal value of the LPP 15.1 with $\alpha = 1$ ($\alpha = \max\{\sum_{t \in p^\bullet} \mathbf{s}[t] | \mathbf{s} \in \mathrm{LS}(\mathcal{N}, \mathbf{m_0})\}$).

$$
\begin{aligned}
z = \min.\ & \mathbf{y} \cdot \mathbf{m_0} + \alpha \cdot \mu \\
\text{s.t.}\ & \mathbf{y} \cdot \mathbf{C} \leq \mathbf{C}[p,T] \\
& \mathbf{y} \cdot \mathbf{Pre}[P,t] + \mu \geq \mathbf{Pre}[p,t] \quad \forall t \in p^\bullet \\
& \mathbf{y} \geq 0, \mathbf{y}[p] = 0
\end{aligned}
\tag{15.1}
$$

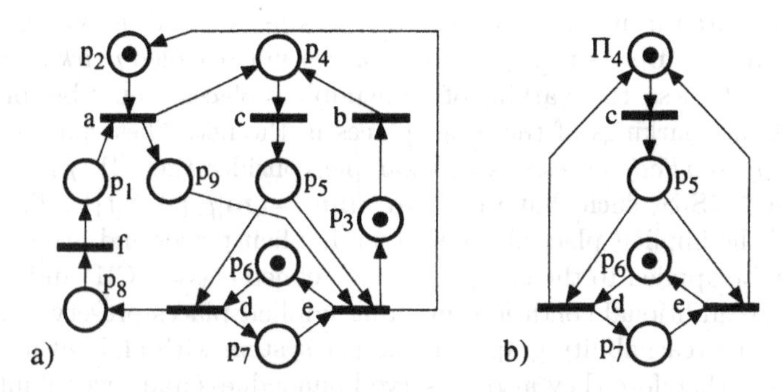

Fig. 15.4. Places p_9 and p_2 (or p_2 and p_7) are implicit.

If the optimal solution of the LPP 15.1, for a structurally implicit place p, verifies that $\mathbf{y} \cdot \mathbf{C} = \mathbf{C}[p,T]$, then p is a marking implicit place and the following holds: $\forall \mathbf{m} \in \mathrm{RS}(\mathcal{N}, \mathbf{m_0})$, $\mathbf{m}[p] = \mathbf{y} \cdot \mathbf{m} + \alpha \cdot \mu$.

Observe that a structurally implicit place can become implicit for any initial marking of places $P \setminus \{p\}$ if we have the freedom to select an adequate initial marking for it. This property is not true for CIPs (or SIPs) that are not structurally implicit places.

Example 15.1.6. Solving the LPP 15.1 for the place p_9 in Figure 15.4a with $\alpha = 1$ we obtain $z = 0$, for the optimal solution: $\mathbf{y} = [0, 0, 1, 1, 1, 0, 1, 0, 0]$ and $\mu = -1$. Moreover, $\mathbf{C}[p_9, T] = \mathbf{C}[p_3, T] + \mathbf{C}[p_4, T] + \mathbf{C}[p_5, T] + \mathbf{C}[p_7, T]$. Because $\mathbf{m_0}[p] \geq z = 0$, p_9 is an SIP (since p_9 is a self-loop-free place it is also a CIP) and can be removed. Since p_9 is a marking implicit place we can write: $\forall \mathbf{m} \in \mathrm{RS}(\mathcal{N}, \mathbf{m_0})$, $\mathbf{m}[p_9] = \mathbf{m}[p_3] + \mathbf{m}[p_4] + \mathbf{m}[p_5] + \mathbf{m}[p_7] - 1$.

Once p_9 is removed, a similar computation can be done for p_2 and p_2 is also shown to be a CIP. Figure 15.4b shows a reduced net system. It can be obtained by reducing $p_3 - b - p_4$ into a place (say p_{34}) (*RA1*) and finally $p_8 - f - p_1 - a - p_{34}$ into Π_4. Using the kit in Figure 15.1 together with the implicit place rule we can perform no more reductions: the net system is irreducible with respect to that kit of reduction rules. The rule *RA1* allows us to fuse Π_4 and p_5. The new place is implicit, so it can be removed. Then a cycle with $p_6 - d - p_7 - e - p_6$ remains. This can be reduced to a basic net, $p_6 - t_{de} - p_6$, with one token. Therefore the original net system is live and bounded. It is also reversible, but we cannot guarantee this because of the fusion of $p_3 - b - p_4$ into p_{34}.

15.2 Linear Algebraic Techniques

Analysis techniques based on linear algebra allow the verification of properties of a general net system. The key idea is simple, and has already been introduced: Let $\mathcal{S}$ be a net system with incidence matrix $\mathbf{C}$. If $\mathbf{m}$ is reachable from $\mathbf{m_0}$ by firing sequence σ, then $\mathbf{m} = \mathbf{m_0} + \mathbf{C} \cdot \boldsymbol{\sigma}$. Therefore the set of natural solutions $(\mathbf{m}, \boldsymbol{\sigma})$ of this state equation defines a linearisation of the reachability set $\mathrm{RS}(\mathcal{S})$, denoted $\mathrm{LRS}^{SE}(\mathcal{S})$. This set can be used to analyse properties such as marking and submarking reachability and coverability, firing concurrency, conflict situations, deadlock-freeness, mutual exclusion, k-boundedness, the existence of frozen tokens, and synchronic relations. To do so, the properties are expressed as formulas of a first order logic having linear inequalities as atoms, where the reachability or fireability conditions are relaxed by satisfiability of the state equation. These formulae are verified by checking the existence of solutions to systems of linear equations that are automatically obtained from the formulae ([Col89]). For instance, if $\forall \mathbf{m} \in \mathrm{RS}(\mathcal{S}) \,.\, \mathbf{m}[p] = 0 \vee \mathbf{m}[p'] = 0$; then places p and p' are in mutual exclusion. This is verified by checking the absence of (natural) solutions to $\{\mathbf{m} = \mathbf{m_0} + \mathbf{C} \cdot \boldsymbol{\sigma} \wedge \mathbf{m}[p] > 0 \wedge \mathbf{m}[p'] > 0\}$. *Integer linear programming problems* ([NW88]) in which the state equation is included in the set of constraints can be formulated for optimisation problems such as the computation of marking bounds, synchronic measures, etc. ([Col89, SC88]). This approach is a generalisation of classical reasoning using linear invariants ([Lau87, MR80]), and it bridges the domains of net theory and convex geometry, resulting in a unified framework for understanding and enhancing structural techniques ([Col89], see also Section 5.2.2).

Unfortunately, it usually leads only to semidecision algorithms (i.e. only necessary or only sufficient conditions) because in general $\mathrm{RS}(\mathcal{S}) \subset \mathrm{LRS}^{SE}(\mathcal{S})$. The undesirable solutions are called *spurious*.

Example 15.2.1 (Existence of spurious solutions and their consequences in the analysis). Let us consider the net system depicted in Figure 15.3b without the places $p_{\theta 1}$, $p_{\theta 2}$, and $p_{\theta 3}$. The corresponding net state equation has the following marking spurious solutions: $\mathbf{m}_1 = 2 \cdot p_4$, $\mathbf{m}_2 = 2 \cdot p_2$, $\mathbf{m}_3 = 2 \cdot p_3$, $\mathbf{m}_4 = 2 \cdot p_5$, $\mathbf{m}_5 = p_2 + p_4$, $\mathbf{m}_6 = p_3 + p_4$. The first four solutions allow us to conclude that p_2, p_3, p_4, and p_5 are 2-bounded, whereas they are actually 1-bounded (check it). The solutions $\mathbf{m}_2$, $\mathbf{m}_3$, and $\mathbf{m}_4$ are total deadlocks. Thus using the state equation we cannot conclude that this net system is deadlock-free.

Spurious solutions can be removed using certain structural techniques, consequently improving the quality of the linear description of the system ([CS90b]). For example, it is clear that by adding implicit places, a new system model with identical behaviour is obtained. For some net systems, if the implicit places are chosen carefully, the state equation of the new system may have no integer spurious solution preventing a conclusion about the bound of a place or the deadlock-freeness of the system.

Example 15.2.2 (Elimination of spurious solutions). The net system in Figure 15.3b is that considered in the previous example but containing three additional implicit places: $p_{\theta 1}$, $p_{\theta 2}$, and $p_{\theta 3}$. The above mentioned spurious solutions $\mathbf{m}_i, i = 1 \ldots 6$, are not solutions of the new state equation. Moreover, we can conclude now that the new net system and therefore also the original are 1-bounded and deadlock-free!

The algorithms based on linear algebra do decide in many situations, and they are relatively efficient, especially if the integrality of the variables is disregarded. This further relaxation may lower the quality, although in many cases it does not ([DE93, SC88]). Moreover, these techniques allow in an easy way an initial-marking *parametric* analysis (e.g. changing the number of customers, the size of resources, or the initial distribution of customers and/or resources). The application of these techniques to the analysis of boundedness and deadlock-freeness properties is illustrated in Sections 15.2.1 and 15.2.2 respectively.

In temporal logic terms, the approach outlined above is well suited for *safety* properties ("some bad thing never happens"), but not as well for *liveness* properties ("some good thing will eventually happen"). For instance, the formula expressing reversibility would be $\forall \mathbf{m} \in \mathrm{LRS}^{SE}(\mathcal{S}) : \exists \boldsymbol{\sigma}' \gneq 0 : \mathbf{m_0} = \mathbf{m} + \mathbf{C} \cdot \boldsymbol{\sigma}'$, but this is neither necessary nor sufficient for reversibility. The general approach for linearly verifying these liveness properties is based on the verification of safety properties that are necessary for them to hold, together with some inductive reasoning ([Joh88]). For instance, deadlock-freeness is necessary for transition liveness, and the existence of some *decreasing potential function* proves reversibility ([Sil89], see also Section 15.2.4).

Another important contribution of linear techniques to liveness analysis has been the derivation of *ad hoc* simple and efficient semidecision conditions. In Section 15.2.3, we present one of these conditions based on a rank upper bound of the incidence matrix, which was originally conceived when computing the *visit ratios* in certain subclasses of net models ([CCS91]).

The following subsections study marking bounds and boundedness, deadlock-freeness, structural liveness and liveness, and reversibility.

15.2.1 Bounds and Boundedness

The study of the bound of a place p, $\mathbf{b}(p)$, through linear algebraic techniques requires the linearisation of the reachability set in the definition of $\mathbf{b}(p)$ by means of the state equation of the net. In this subsection we assume that $\mathbf{m} \in \mathbb{R}^n$ and $\boldsymbol{\sigma} \in \mathbb{R}^m$. This linearisation of the definition of $\mathbf{b}(p)$ leads to a new quantity called the *structural bound* of p, $\mathbf{sb}(p)$:

$$\mathbf{sb}(p) = \sup\{\mathbf{m}(p) | \mathbf{m} = \mathbf{m_0} + \mathbf{C} \cdot \boldsymbol{\sigma} \geq 0, \boldsymbol{\sigma} \geq 0\} \tag{15.2}$$

Let $\mathbf{e_p}$ be the *characteristic vector* of p: $\mathbf{e_p}[q] := \text{if } q = p \text{ then } 1 \text{ else } 0$. The structural bound of p, $\mathbf{sb}(p)$, can be obtained as the optimal solution of the following linear programming problem (LPP):

$$\begin{aligned} \mathbf{sb}(p) = \ & \text{max. } \mathbf{e_p} \cdot \mathbf{m} \\ & \text{s.t. } \ \ \mathbf{m} = \mathbf{m_0} + \mathbf{C} \cdot \boldsymbol{\sigma} \geq 0 \\ & \qquad\ \ \boldsymbol{\sigma} \geq 0 \end{aligned} \tag{15.3}$$

Therefore $\mathbf{sb}(p)$ can be computed in polynomial time. In sparse-matrix problems (matrix $\mathbf{C}$ is usually sparse), good implementations of the classical *simplex method* lead to quasi-linear time complexities.

Because $\mathrm{RS}(\mathcal{S}) \subset \mathrm{LRS}^{SE}(\mathcal{S})$ in general, we have that $\mathbf{sb}(p) \geq \mathbf{b}(p)$ (recall example 15.2.1). Therefore, if we are investigating the k-boundedness of a place (i.e. $\mathbf{m}[p] \leq k$), we have a sufficient condition in polynomial time: if $\mathbf{sb}(p) \leq k$ then $\mathbf{b}(p) \leq k$ (i.e. p is k-bounded).

In what follows we argue from classical results in linear programming and convex geometry theories. We assume that the reader is aware of these theories (see, for example, [Mur83, NRKT89]); otherwise all the arguments needed are compiled and adapted in [SC88]. The important point here is to convey the idea that other theories are helpful for understanding in a general framework many sparse results on the behavior of net systems. The dual linear programming problem of 15.3 is the following (see any text on linear programming to check it):

$$\begin{aligned} \mathbf{sb}(p)' = \min. \ & \mathbf{y} \cdot \mathbf{m_0} \\ \text{s.t.} \ & \mathbf{y} \cdot \mathbf{C} \leq 0 \\ & \mathbf{y} \geq \mathbf{e_p} \end{aligned} \tag{15.4}$$

The LPP 15.3 always has a feasible solution ($\mathbf{m} = \mathbf{m_0}$, $\boldsymbol{\sigma} = 0$). Using duality and boundedness theorems from linear programming theory, both LPPs 15.3 and 15.4 are bounded (thus p is structurally bounded) and $\mathbf{sb}(p) = \mathbf{sb}(p)'$ iff there exists a *feasible solution* for the LPP 15.4: $\mathbf{y} \geq \mathbf{e_p}$ such that $\mathbf{y} \cdot \mathbf{C} \leq 0$.

The reader can easily check that LPP 15.4 makes, in polynomial time, an "implicit search" for the structural bound of p on a set of structural objects including all the p-semiflows. In this sense, we can say that analysis methods based on the state equation are more general than those based on linear invariants. That is, the dual LPPs of those based on the state equation consider not only the p-semiflows but also other structural objects with $\mathbf{y} \geq 0$ such that $\mathbf{y} \cdot \mathbf{C} \lneq 0$. On the other hand we must say that the computational effort needed when using the linear invariants is greater than that required when using the state equation, since the computation of the minimal p-semiflows (in some cases, an exponential number!) must be done prior to the study of the property.

From the above discussion and using the *alternatives theorem* (an algebraic form of the *Minkowski-Farkas lemma*) the following properties can be proved:

Property 15.2.3. The following three statements are equivalent:

1. p is structurally bounded, i.e. p is bounded for any $\mathbf{m_0}$.
2. There exists $\mathbf{y} \geq \mathbf{e_p}$ such that $\mathbf{y} \cdot \mathbf{C} \leq 0$ (place-based characterisation).
3. For all $\mathbf{x} \geq 0$ such that $\mathbf{C} \cdot \mathbf{x} \geq 0$, $\mathbf{C}[p, T] \cdot \mathbf{x} = 0$ (transition-based characterisation).

Property 15.2.4. The following three statements are equivalent:

1. $\mathcal{N}$ is structurally bounded, i.e. $\mathcal{N}$ is bounded for any $\mathbf{m_0}$.
2. There exists $\mathbf{y} \geq \mathbf{1}$ such that $\mathbf{y} \cdot \mathbf{C} \leq 0$ (place-based characterisation).
3. For all $\mathbf{x} \geq 0$ such that $\mathbf{C} \cdot \mathbf{x} \geq 0$, $\mathbf{C} \cdot \mathbf{x} = 0$; i.e. $\nexists\ \mathbf{x} \geq 0$ s.t. $\mathbf{C} \cdot \mathbf{x} \gneq 0$ (transition-based characterisation).

15.2.2 Deadlock-Freeness and Liveness

Deadlock-freeness concerns the existence of some activity from any reachable state of the system. It is a necessary condition for liveness, although in general not sufficient. When no part of the system can evolve, it is said that the system has reached a state of *total deadlock* (or *deadlock* for short). In net system terms, a deadlock corresponds to a marking from which no transition is fireable. In order to study deadlock-freeness with linear algebraic techniques, the property must be expressed as a formula of a first-order logic having linear inequalities as atoms, in which the reachability or fireability conditions are relaxed by satisfiability of the state equation. The formula to express that a marking is a deadlock consists of a condition for every transition indicating that it is disabled at such a marking. This condition consists of several inequalities, one per input place of the transition (indicating that the marking of that place is less than the corresponding weight) linked by the "$\vee$" operator (because a lack of tokens in a single input place disables the transition). We give below a basic general sufficient condition for deadlock-freeness based on the absence of solutions satisfying simultaneously the net state equation and the formula expressing the total deadlock condition mentioned above.

Proposition 15.2.5. *Let $\langle \mathcal{N}, \mathbf{m_0} \rangle$ be a net system. If there exists no solution $(\mathbf{m}, \boldsymbol{\sigma})$ for the system*

$$\begin{aligned} &\mathbf{m} = \mathbf{m_0} + \mathbf{C} \cdot \boldsymbol{\sigma} \\ &\mathbf{m} \geq 0, \boldsymbol{\sigma} \geq 0 \\ &\bigvee_{p \in {}^\bullet t} \mathbf{m}[p] < \mathbf{Pre}[p, t]; \forall t \in T \end{aligned} \tag{15.5}$$

then $\langle \mathcal{N}, \mathbf{m_0} \rangle$ is deadlock-free.

Obviously, the deadlock conditions are non-linear, because they are expressed using the "$\vee$" operator. However, we can express the above condition by means of a set of linear systems as follows. Let $\alpha : T \to P$ be a mapping that assigns to each transition one of its input places. If there exists no α such that the system

$$\begin{aligned} &\mathbf{m} = \mathbf{m_0} + \mathbf{C} \cdot \boldsymbol{\sigma} \\ &\mathbf{m} \geq 0, \boldsymbol{\sigma} \geq 0 \\ &\mathbf{m}[\alpha(t)] < \mathbf{Pre}[\alpha(t), t]; \forall t \in T \end{aligned} \tag{15.6}$$

has a solution, then $\langle \mathcal{N}, \mathbf{m_0} \rangle$ is deadlock-free. The problem is that we have to check it for *every* mapping α of input places to transitions so we have to check $\prod_{t \in T} |{}^\bullet t|$ systems of linear inequalities. If every transition has exactly one input place (e.g. state machines) then only one system needs to be checked, but in general the number might be large. Nevertheless it is possible to reduce the number of systems to be checked, while preserving the set of *integer solutions*. For this purpose, the work [TCS93] presents four simplification rules of the deadlock condition using information obtained from the net system, and a simple net transformation leading to an equivalent net with respect to the deadlock-freeness property in which the enabling conditions of transitions can be expressed linearly. As a result, deadlock-freeness of a wide variety of net systems can be proved by verifying the absence of solutions to a single system of linear inequalities. Even more, in some subclasses it is known that there are no spurious solutions which are deadlocks, so the method decides on deadlock-freeness ([TS93]). The following example presents the deadlock-freeness analysis of the net system in Figure 5.8b using this technique.

Example 15.2.6 (Deadlock-freeness analysis and simplification rules). Let us consider the net system in Figure 5.8b. The direct application of the method described in proposition 15.2.5 requires us to check $\prod_{t \in T} | {}^{\bullet}t| = 36$ linear systems of the form presented in Equation 15.6. Nevertheless, we show below that we can reduce the deadlock-freeness analysis on this net to the checking of a unique linear system by applying the simplification rules presented in [TCS93]. Solving the LPP 15.3 for the places of the net system we obtain the following: $\mathbf{sb}(\mathrm{p}) = 1$ for all $p \in P \setminus \{\mathrm{empty}, \mathrm{object}\}$; and $\mathbf{sb}(\mathrm{empty}) = \mathbf{sb}(\mathrm{object}) = 7$ (the same can be obtained from the linear invariants in Equations 5.1–5.4). The transitions t_1, t_4, t_7, and t_9 are those presenting complex conditions giving rise to a large number of linear systems. The simplification of these conditions is as follows:

a) The non-fireability condition of t_1 is $(\mathbf{m}[\mathrm{wait\_raw}] = 0) \vee (\mathbf{m}[\mathrm{R}] = 0)$. Taking into account the fact that $\mathbf{sb}(\mathrm{wait\_raw}) = \mathbf{sb}(\mathrm{R}) = 1$, we can apply a particularisation of rule 3 in [TCS93] to replace the complex condition by a unique linear inequality: Let t be a transition such that each input place verifies that its structural bound is equal to the weight of the output arc joining it to t. The non-fireability condition for transition t at a marking $\mathbf{m}$ is $\sum_{p \in {}^{\bullet}t} \mathbf{m}[p] \leq \sum_{p \in P} \mathbf{Pre}[p, t] - 1$. That is, the number of tokens in the input places of t is less than needed. Therefore, for the transition t_1 this linear condition is: $\mathbf{m}[\mathrm{wait\_raw}] + \mathbf{m}[\mathrm{R}] \leq 1$.
b) The non-fireability condition of t_7 is $(\mathbf{m}[\mathrm{wait\_free}] = 0) \vee (\mathbf{m}[\mathrm{R}] = 0)$. In a similar way to the case of transition t_1, we replace this condition by $\mathbf{m}[\mathrm{wait\_free}] + \mathbf{m}[\mathrm{R}] \leq 1$, since $\mathbf{sb}(\mathrm{wait\_free}) = \mathbf{sb}(\mathrm{R}) = 1$ and rule 3 in [TCS93] can be applied.
c) The non-fireability condition of t_4 is $(\mathbf{m}[\mathrm{wait\_dep.}] = 0) \vee (\mathbf{m}[\mathrm{R}] = 0) \vee (\mathbf{m}[\mathrm{empty}] = 0)$. Since $\mathbf{sb}(\mathrm{wait\_dep.}) = \mathbf{sb}(\mathrm{R}) = 1$ and $\mathbf{sb}(\mathrm{empty}) = 7$ (i.e. only one input place of t_7 has an $\mathbf{sb}$ greater than the weight of the arc), rule 4 of [TCS93] can be applied. Then, the complex condition is replaced by the following linear condition:

$$\mathbf{sb}(\mathrm{empty}) \cdot (\mathbf{m}[\mathrm{wait\_dep.}] + \mathbf{m}[\mathrm{R}]) + \mathbf{m}[\mathrm{empty}] \leq \\ \mathbf{sb}(\mathrm{empty}) \cdot (\mathbf{Pre}[\mathrm{wait\_dep.}, T] + \mathbf{Pre}[\mathrm{R}, T]) + \mathbf{Pre}[\mathrm{empty}, T] - 1$$

i.e. $7(\mathbf{m}[\mathrm{wait\_dep.}] + \mathbf{m}[\mathrm{R}]) + \mathbf{m}[\mathrm{empty}] \leq 14$.
d) The non-fireability condition of t_9 can be reduced to the following linear condition similarly to the case of transition t_4: $7(\mathbf{m}[\mathrm{wait\_with.}] + \mathbf{m}[\mathrm{R}]) + \mathbf{m}[\mathrm{object}] \leq 14$.

Applying the previously stated simplifications, the deadlock-freeness analysis for the net system in Figure 5.8b is reduced to verifying that there exists no solution $(\mathbf{m}, \boldsymbol{\sigma})$ for the following single linear system (the reader can check that the system has no solutions):

$$\begin{array}{ll}
\mathbf{m} = \mathbf{m_0} + \mathbf{C} \cdot \boldsymbol{\sigma} & \\
\mathbf{m} \geq 0, \boldsymbol{\sigma} \geq 0 & \\
\mathbf{m}[\text{wait_raw}] + \mathbf{m}[\text{R}] \leq 1; & \text{for } t_1 \\
\mathbf{m}[\text{load}] = 0; & \text{for } t_2 \\
\mathbf{m}[\text{op}_1] = 0; & \text{for } t_3 \\
7(\mathbf{m}[\text{wait_dep.}] + \mathbf{m}[\text{R}]) + \mathbf{m}[\text{empty}] \leq 14; & \text{for } t_4 \\
\mathbf{m}[\text{deposit}] = 0; & \text{for } t_5 \\
\mathbf{m}[\text{op}_2] = 0; & \text{for } t_6 \\
\mathbf{m}[\text{wait_free}] + \mathbf{m}[\text{R}] \leq 1; & \text{for } t_7 \\
\mathbf{m}[\text{unload}] = 0; & \text{for } t_8 \\
7(\mathbf{m}[\text{wait_with.}] + \mathbf{m}[\text{R}]) + \mathbf{m}[\text{object}] \leq 14; & \text{for } t_9 \\
\mathbf{m}[\text{withdrawal}] = 0; & \text{for } t_{10}
\end{array} \tag{15.7}$$

Linear invariants may also be used to prove *deadlock-freeness*. Using the linear invariants in Equations (5.1–5.4), we shall prove that our net system in Figure 5.8.b is deadlock-free.

If there exists a deadlock, no transition can be fired. Let us try to construct a marking in which no transition is fireable. When a unique input place of a transition exists, that place must be unmarked. So $\mathbf{m}[\text{load}] = \mathbf{m}[\text{op}_1] = \mathbf{m}[\text{deposit}] = \mathbf{m}[\text{op}_2] = \mathbf{m}[\text{unload}] = \mathbf{m}[\text{withdrawal}] = 0$, and the linear invariants in Equations (5.1–5.4) reduce to:

$$\mathbf{m}[\text{wait\_raw}] + \mathbf{m}[\text{wait\_dep.}] = 1 \tag{15.8}$$
$$\mathbf{m}[\text{wait\_free}] + \mathbf{m}[\text{wait\_with.}] = 1 \tag{15.9}$$
$$\mathbf{m}[\text{empty}] + \mathbf{m}[\text{object}] = 7 \tag{15.10}$$
$$\mathbf{m}[\text{R}] = 1 \tag{15.11}$$

Since R should always be marked at the present stage, to prevent the firing of t_1 and t_7 places wait_raw and wait_free should be unmarked. The linear invariants are reduced once more, leading to:

$$\mathbf{m}[\text{wait\_dep.}] = 1 \tag{15.12}$$
$$\mathbf{m}[\text{wait\_with.}] = 1 \tag{15.13}$$
$$\mathbf{m}[\text{empty}] + \mathbf{m}[\text{object}] = 7 \tag{15.14}$$
$$\mathbf{m}[\text{R}] = 1 \tag{15.15}$$

Since $\mathbf{m}[\text{wait\_dep.}] = \mathbf{m}[\text{wait\_with.}] = 1$, to avoid the firing of t_4 and t_9 $\mathbf{m}[\text{empty}] + \mathbf{m}[\text{object}] = 0$ is needed. This contradicts Equation (15.14), so the net system is deadlock-free. A more compact, algorithmic presentation of the above deadlock-freeness proof is:

if $\mathbf{m}$[load] + $\mathbf{m}$[op$_1$] + $\mathbf{m}$[deposit] + $\mathbf{m}$[op$_2$] + $\mathbf{m}$[unload] + $\mathbf{m}$[withdrawal] ≥ 1
  then one of t_2,t_3,t_5,t_6,t_8, or t_{10} is fireable
  else if $\mathbf{m}$[wait_raw] + $\mathbf{m}$[wait_free] ≥ 1
    then one of t_1 or t_7 is fireable
    else one of t_4 or t_9 is fireable

As a final remark, we want to point out that liveness can be proved for the net system in Figure 5.8b. Liveness implies deadlock-freeness, but the reverse is not true in general. Nevertheless, if the net is consistent and has only one minimal t-semiflow, as in the example where the unique minimal t-semiflow is **1**; then any infinite behaviour must contain all transitions with relative firings given by the t-semiflow. Thus deadlock-freeness implies, in this case, liveness.

15.2.3 Structural Liveness and Liveness

A necessary condition for a transition t to be live in a system $\langle \mathcal{N}, \mathbf{m_0} \rangle$ is its eventual infinite fireability, i.e. the existence of a firing repetitive sequence σ_R containing t: $\exists \sigma_R \in \mathrm{L}(\mathcal{N}, \mathbf{m_0})$ such that $\mathbf{m_0} \xrightarrow{\sigma_R} \mathbf{m} \geq \mathbf{m_0}$ and $\boldsymbol{\sigma}_R[t] > 0$.

Using the state equation as a linearisation of the reachability set, an *upper bound* on the number of times t can be fired in $\langle \mathcal{N}, \mathbf{m_0} \rangle$ is given by the following LPP ($\mathbf{e_t}[u]$:= if $u = t$ then 1 else 0):

$$\begin{aligned} \mathbf{sr}(t) = \ & \max.\ \mathbf{e_t} \cdot \boldsymbol{\sigma} \\ & \text{s.t.} \quad \mathbf{m} = \mathbf{m_0} + \mathbf{C} \cdot \boldsymbol{\sigma} \geq 0 \\ & \qquad\quad \boldsymbol{\sigma} \geq 0 \end{aligned} \tag{15.16}$$

The dual of the LPP 15.16 is:

$$\begin{aligned} \mathbf{sr}(t)' = \ & \min.\ \mathbf{y} \cdot \mathbf{m_0} \\ & \text{s.t.} \quad \mathbf{y} \cdot \mathbf{C} \leq -\mathbf{e_t} \\ & \qquad\quad \mathbf{y} \geq 0 \end{aligned} \tag{15.17}$$

We are interested in characterising the cases where $\mathbf{sr}(t)$ goes to infinity. The LPP 15.16 has $\mathbf{m} = \mathbf{m_0}$ and $\boldsymbol{\sigma} = 0$ as a feasible solution. Using first duality and unboundedness theorems from linear programming and later the alternatives theorem, the following properties can be stated:

Property 15.2.7. The following three statements are equivalent:

1. t is structurally repetitive (i.e. there exists a "large enough" $\mathbf{m_0}$ such that t can be fired infinitely often).
2. There does not exist $\mathbf{y} \geq 0$ such that $\mathbf{y} \cdot \mathbf{C} \leq -\mathbf{e_t}$ (place-based perspective).
3. There exists $\mathbf{x} \geq \mathbf{e_t}$ such that $\mathbf{C} \cdot \mathbf{x} \geq 0$ (transition-based perspective).

Property 15.2.8. The following three statements are equivalent:

1. $\mathcal{N}$ is structurally repetitive (i.e. all transitions are structurally repetitive).
2. There does not exist $\mathbf{y} \geq 0$ such that $\mathbf{y} \cdot \mathbf{C} \lneq 0$
3. There exists $\mathbf{x} \geq \mathbf{1}$ such that $\mathbf{C} \cdot \mathbf{x} \geq 0$

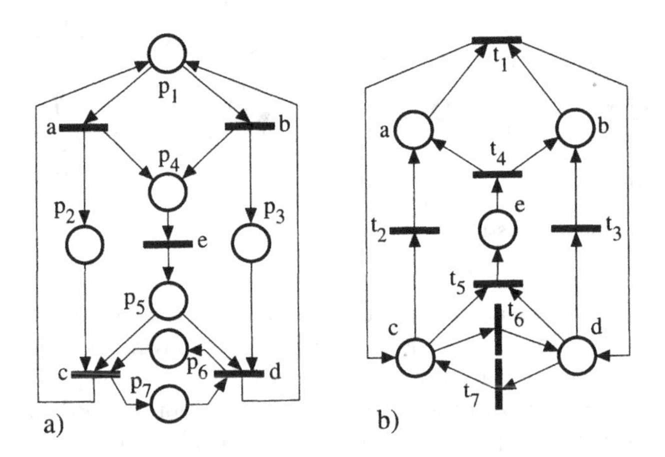

Fig. 15.5. Two conservative and consistent, structurally non-live nets:
a) rank($\mathbf{C}$) = 4, |EQS| = 3, thus $\mathcal{N}$ is not structurally live;
b) rank($\mathbf{C}$) = 4, |EQS| = 4, |CCS| = 3, thus no answer.

Additionally, the following classical results can be stated ([MR80, Bra83, Sil85]):

Property 15.2.9. Let $\mathcal{N}$ be a net and $\mathbf{C}$ its incidence matrix.

1. **if** $\mathcal{N}$ is structurally live **then** $\mathcal{N}$ is structurally repetitive.
2. **if** $\mathcal{N}$ is structurally live and structurally bounded **then** $\mathcal{N}$ is conservative ($\exists \mathbf{y} \geq \mathbf{1}$ such that $\mathbf{y} \cdot \mathbf{C} = 0$) and consistent ($\exists \mathbf{x} \geq \mathbf{1}$ such that $\mathbf{C} \cdot \mathbf{x} = 0$).
3. **if** $\mathcal{N}$ is connected, consistent, and conservative **then** it is strongly connected.
4. **if** $\mathcal{N}$ is live and bounded **then** $\mathcal{N}$ is strongly connected and consistent.

The net structures in Figure 15.5 are consistent and conservative, but there does not exist a live marking for them. A more careful analysis allows us to improve the above result with a *rank condition* on the incidence matrix $\mathbf{C}$ of $\mathcal{N}$. Before the introduction of this improved result we need to introduce certain structural objects related to conflicts.

Conflicts in sequential systems are clearly situations in which two actions are enabled so one must be chosen to occur. For instance, Figure 15.6 (a) shows a conflict between t and t'. The situation becomes more complicated in the case of concurrent systems, where the fact that two transitions are

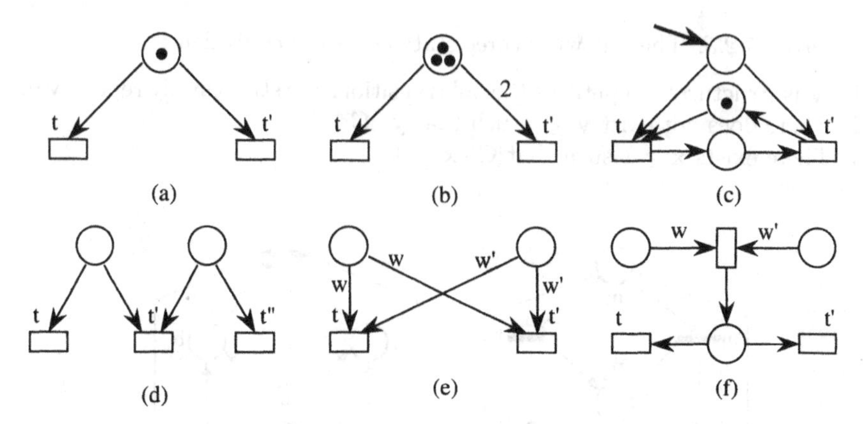

Fig. 15.6. Conflicts and structural conflicts

enabled does not necessarily imply that we must choose one. Sometimes the "sequential" definition — there is a conflict when two transitions are enabled and the occurrence of one disables the other — is suitable, namely in 1-bounded systems. But in other cases a new definition is needed. Consider now the marking that puts two tokens into the place of Figure 15.6 (a). The occurrence of t does not disable t' and vice versa, but the firing of one decreases the enabling degree of the other: i.e. each token must decide which way to go. Formally, *there is a conflict situation when the enabling vector is not an enabled step.* In Figure 15.6 (b), the occurrence of t or t' does not disable the other, but the firing of t' decreases the enabling degree of t from three to one. The enabling vector is $3t + t'$, whereas the (maximal) enabled steps are $3t$ and $t + t'$. By the way, this example shows that *conflict does not imply the absence of concurrency*: t and t' are involved in a conflict, but they could occur concurrently, as in $t + t'$.

The basic net construct used to model conflicts is a place with more than one output transition, i.e. a distributor place. In fact, distributor places are needed to model conflicts, but the converse is not true. Because of the regulation circuit in Figure 15.6 (c), t and t' are never in effective conflict although they share an input place. The output transitions of a distributor place are said to be in *structural conflict relation* ($\langle t_i, t_j \rangle \in \mathrm{SC}$) when ${}^\bullet t_i \cap {}^\bullet t_j \neq \emptyset$. This relation is reflexive and symmetric, but not transitive. Its transitive closure is named the *coupled conflict relation*, and it partitions the transitions of a net into *coupled conflict sets* (CCS(t) denotes the coupled conflict set containing t). In Figure 15.6 (d), t and t'' are not in structural conflict relation but they are in coupled conflict relation, through t'.

In the literature, structural conflicts are often simply called "conflicts", but we prefer to add the adjective "structural" to better distinguish from the behavioural, thus dynamical, notion of (effective) conflict, which depends on the marking. As we have noted, a structural conflict makes possible the existence of an effective conflict, but does not guarantee it, e.g. Figure 15.6 (d), except in the case of *equal conflicts*, where all the transitions in structural conflict have the same pre-condition. Transitions t and t' are said to be in *equal conflict relation*, $\langle t, t' \rangle \in EQ$, when $t = t'$ or $\mathbf{Pre}[P, t] = \mathbf{Pre}[P, t'] \neq \mathbf{0}$. This equivalence relation partitions the transitions into *equal conflict sets.* The equal conflict set containing t is denoted by EQS(t). Figure 15.6 (e) shows an equal conflict set.

Relating the rank of the incidence matrix to the number of (coupled or equal) structural conflicts in a net improves the previous conditions on structural liveness:

Property 15.2.10. Let $\mathcal{N}$ be a net and $\mathbf{C}$ its incidence matrix.

1. **if** $\mathcal{N}$ is live and bounded **then** $\mathcal{N}$ is strongly connected, consistent, and $\text{rank}(\mathbf{C}) \leq |\text{SEQS}| - 1$.
2. **if** $\mathcal{N}$ is conservative, consistent, and $\text{rank}(\mathbf{C}) = |\text{SCCS}| - 1$, **then** $\mathcal{N}$ is structurally live and structurally bounded.

The condition in property 15.2.10.1 has been proved to be sufficient for some subclasses of nets ([Des92, TS94, TS96]). Observe that even for structurally bounded nets, we do not have a complete characterisation of structural liveness. Since $|\text{SCCS}| \leq |\text{SEQS}|$, there is still a range of nets which satisfy neither the necessary nor the sufficient condition for being structurally live and structurally bounded! The added rank condition allows us to state that the net in Figure 15.5a is structurally non-live. Nevertheless, nothing can be said about structural liveness of the net in Figure 15.5b.

Property 15.2.10 is purely structural (i.e. the initial marking is not considered at all). Nevertheless, it is clear that an overly small initial marking (e.g. the empty marking) makes any net structure non-live. A less trivial lower bound for the initial marking based on marking linear invariants is based on fireability of every transition. If $t \in T$ is fireable at least once, for any p-semiflow $\mathbf{y}$, then $\mathbf{y} \cdot \mathbf{m_0} \geq \mathbf{y} \cdot \mathbf{Pre}[P, t]$. Therefore:

Property 15.2.11. If $\langle \mathcal{N}, \mathbf{m_0} \rangle$ is a live system, then $\forall \mathbf{y} \geq 0$ such that $\mathbf{y} \cdot \mathbf{C} = 0$, $\mathbf{y} \cdot \mathbf{m_0} \geq \max_{t \in T}(\mathbf{y} \cdot \mathbf{Pre}[P, t]) \geq 1$.

Unfortunately no characterisation of liveness exists in linear algebraic terms. The net system in Figure 5.1.b with a token in p_5 is consistent, conservative, fulfils the rank condition, and has all p-semiflows marked, but it is non-live.

15.2.4 Reversibility and Liveness

Let us use now a *Liapunov-stability-like* technique to prove that the net system in Figure 5.8b is reversible. This serves to illustrate the use of marking linear invariants, and inductive reasoning for the analysis of liveness properties.

As a preliminary consideration that simplifies the remainder of the proof, the following simple property will be used: Let $\langle \mathcal{N}, \mathbf{m}_1 \rangle$ be a reversible system and $\mathbf{m_0}$ reachable from $\mathbf{m}_1$ (i.e. $\exists \sigma \in L(\mathcal{N}, \mathbf{m}_1)$ such that $\mathbf{m}_1 \xrightarrow{\sigma} \mathbf{m_0}$). Then $\langle \mathcal{N}, \mathbf{m_0} \rangle$ is reversible.

Assume $\mathbf{m}_1$ is like $\mathbf{m_0}$ (Figure 5.8b), but with: $\mathbf{m}_1[\text{wait\_raw}] = \mathbf{m}_1[\text{empty}] = 0$, $\mathbf{m}_1[\text{wait\_dep.}] = 1$ and $\mathbf{m}_1[\text{object}] = 7$.

Let us prove first that $\langle \mathcal{N}, \mathbf{m}_1 \rangle$ is reversible. Let $\mathbf{w}$ be a non-negative place weighted such that $\mathbf{w}[p_i] = 0$ iff p_i is marked in $\mathbf{m}_1$. Therefore, $\mathbf{w}[\text{wait\_dep.}] = \mathbf{w}[\text{R}] = \mathbf{w}[\text{object}] = \mathbf{w}[\text{wait\_with.}] = 0$, and $\mathbf{w}[p_j] > 0$ for all the other places. The function $\mathbf{v}(\mathbf{m}) = \mathbf{w} \cdot \mathbf{m}$ has the following properties: $\mathbf{v}(\mathbf{m}) \geq 0$ and $\mathbf{v}(\mathbf{m}_1) = 0$

For the system in Figure 5.8b a stronger property holds: $\mathbf{v}(\mathbf{m}) = 0 \iff \mathbf{m} = \mathbf{m}_1$. This can be clearly seen because $\mathbf{w} \cdot \mathbf{m} = 0 \iff \mathbf{m}[\text{wait\_raw}] = \mathbf{m}[\text{load}] = \mathbf{m}[\text{op}_1] = \mathbf{m}[\text{deposit}] = \mathbf{m}[\text{empty}] = \mathbf{m}[\text{op}_2] = \mathbf{m}[\text{wait\_free}] = \mathbf{m}[\text{unload}] = \mathbf{m}[\text{withdrawal}] = 0$. Furthermore, it is easy to check the following: $\mathbf{m}_1$ is the present marking $\iff t_9$ is the unique fireable transition.

If there exists (warning: in Liapunov-stability criteria the universal quantifier is used!) a finite firing sequence (i.e. a finite trajectory) per reachable marking $\mathbf{m}_i$ such that $\mathbf{m}_i \xrightarrow{\sigma_k} \mathbf{m}_{i+1}$ and $\mathbf{v}(\mathbf{m}_i) > \mathbf{v}(\mathbf{m}_{i+1})$, then in a finite number of transition firings $\mathbf{v}(\mathbf{m}) = 0$ is reached. Because $\mathbf{v}(\mathbf{m}) = 0 \iff \mathbf{m} = \mathbf{m}_1$, a proof that $\mathbf{m}_1$ is reachable from any marking has been obtained (i.e. $\langle \mathcal{N}, \mathbf{m}_1 \rangle$ is reversible).

Premultiplying the net state equation by $\mathbf{w}$ we obtain the following condition: **if** $\sigma_k = t_j$ **then** $[\mathbf{w} \cdot \mathbf{m}_{i+1} < \mathbf{w} \cdot \mathbf{m}_i] \iff \mathbf{w} \cdot \mathbf{C}[P, t_j] < 0$

Now, removing from Figure 5.8 the places marked at $\mathbf{m}_1$ (i.e. wait_dep., R, object, wait_with.) and fireable transitions (i.e. t_9) an acyclic net is obtained, so there exists a $\mathbf{w}$ such that $\mathbf{w} \cdot \mathbf{C}[P, t_j] < 0$, $\forall j \neq 9$.

For example, taking as weights the levels in the acyclic graph we have:

$$\mathbf{w}[\text{op}_1] = \mathbf{w}[\text{unload}] = 1 \tag{15.18}$$

$$\mathbf{w}[\text{load}] = \mathbf{w}[\text{wait\_free}] = 2 \tag{15.19}$$

$$\mathbf{w}[\text{wait\_raw}] = \mathbf{w}[\text{op}_2] = 3 \tag{15.20}$$

$$\mathbf{w}[\text{deposit}] = \mathbf{w}[\text{withdrawal}] = 4 \tag{15.21}$$

$$\mathbf{w}[\text{empty}] = 5 \tag{15.22}$$

and $\mathbf{w} \cdot \mathbf{C} = [-1, -1, -1, -1, -1, -1, -1, -1, +4, -1]$. In other words, the firing of any transition, except t_9, decreases $\mathbf{v}(\mathbf{m}) = \mathbf{w} \cdot \mathbf{m}$.

Using the algorithmic deadlock-freeness explanation from previous sections, the reversibility of $\langle \mathcal{N}, \mathbf{m}_1 \rangle$ is proved (observe that the p-invariants in Equations (5.1–5.4) persist for $\mathbf{m}_1$):

if $\mathbf{m}[\text{load}] + \mathbf{m}[\text{op}_1] + \mathbf{m}[\text{deposit}] + \mathbf{m}[\text{op}_2] + \mathbf{m}[\text{unload}] + \mathbf{m}[\text{withdrawal}] \geq 1$
then $\mathbf{v}(\mathbf{m})$ can decrease firing t_2,t_3,t_5,t_6,t_8, or t_{10}
else if $\mathbf{m}[\text{wait\_raw}] + \mathbf{m}[\text{wait\_free}] \geq 1$
then $\mathbf{v}(\mathbf{m})$ can decrease firing t_1 or t_7
else $\mathbf{v}(\mathbf{m})$ can decrease firing t_4, or t_9 is unique fireable transition (iff $\mathbf{m}_1$ is the present marking)

Because $\mathbf{m_0}$ is reachable from $\mathbf{m}_1$, e.g. by firing $\sigma = (t_9t_{10}t_6t_7t_8)^5t_4t_5$, $\langle\mathcal{N}, \mathbf{m_0}\rangle$ is a reversible system.

Once again liveness of the system in Figure 5.8b can be proved, because the complete sequence $\sigma = t_1t_2t_3t_4t_5t_9t_{10}t_6t_7t_8$ (i.e. containing all transitions) can be fired. Since the system is reversible, no transition loses the possibility of firing (i.e. all transitions are live).

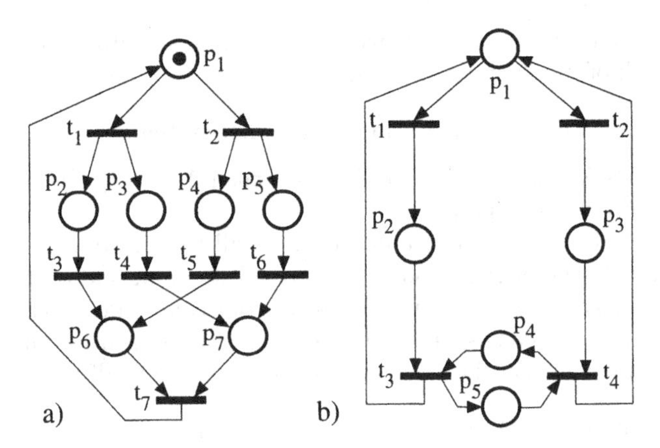

Fig. 15.7. Two consistent and conservative free-choice nets: a) structurally live, rank(**C**) = 5, |EQS| = 5; b) structurally non-live, rank(**C**) = 3, |EQS| = 2.

15.3 Siphons and Traps

By means of graph-theory-based reasoning it is possible to characterise many properties of net subclasses. *Siphons* (also called *structural deadlocks*, or often simply *deadlocks*) and *traps* are easily recognisable subsets of places that generate very particular sub-nets.

Definition 15.3.1. *Let $\mathcal{N} = \langle P, T, F \rangle$ be an ordinary net.*

1. *A* siphon *is a subset of places, Σ, such that the set of its input transitions is contained in the set of its output transitions: $\Sigma \subseteq P$ is a siphon $\iff {}^\bullet\Sigma \subseteq \Sigma^\bullet$.*
2. *A* trap *is a subset of places, θ, such that the set of its output transitions is contained in the set of its input transitions: $\theta \subseteq P$ is a trap $\iff \theta^\bullet \subseteq {}^\bullet\theta$.*

$\Sigma = \{p_1, p_2, p_4, p_5, p_6\}$ is a siphon for the net in Figure 15.7a: ${}^\bullet\Sigma = \{t_7, t_1, t_2, t_3, t_5\}$, while $\Sigma^\bullet = {}^\bullet\Sigma \cup \{t_6\}$. Σ contains a trap, $\theta = \Sigma \setminus \{p_5\}$. In fact θ is also a siphon (it is minimal: no further siphons can be obtained by removing places).

Siphons and traps are reverse concepts: A subset of places of a net $\mathcal{N}$ is a siphon iff it is a trap on the reverse net, $\mathcal{N}^{-1}$ (i.e. that obtained by reversing the arcs and its flow relation, F).

The following property "explains" the names of structural deadlocks or siphons (think of "soda siphons") and traps.

Property 15.3.2. Let $\langle \mathcal{N}, \mathbf{m_0} \rangle$ be an ordinary net system.

1. If $\mathbf{m} \in \mathrm{RS}(\mathcal{N}, \mathbf{m_0})$ is a deadlock state, then $\Sigma = \{p | \mathbf{m}[p] = 0\}$ is an unmarked (empty) siphon.
2. If a siphon is (or becomes) unmarked, it will remain unmarked for any possible net system evolution. Therefore all its input and output transitions are dead. So the system is non-live (but can be deadlock-free).
3. If a trap is (or becomes) marked, it will remain marked for any possible net system evolution (i.e. at least one token is "trapped").

If a trap is not marked at $\mathbf{m_0}$ and the system is live, $\mathbf{m_0}$ will not be recoverable from those markings in which the trap is marked. Thus:

Corollary 15.3.3. *If a live net system is reversible, then* $\mathbf{m_0}$ *marks all traps.*

Remark 15.3.4. For live and bounded free-choice systems a stronger property holds: Marking all traps is a necessary and sufficient condition for reversibility ([BCDE90]). The net system in Figure 15.7a is reversible. Nevertheless, if $\mathbf{m_0} = [0, 1, 0, 0, 1, 0, 0]$, the new system is live and bounded but not reversible: The trap $\theta = \{p_1, p_3, p_4, p_6, p_7\}$ is not marked at $\mathbf{m_0}$.

A siphon which contains a marked trap will never become unmarked. So this more elaborate property can be helpful for some liveness characterisations.

Definition 15.3.5. *Let $\mathcal{N}$ be an ordinary net. The system $\langle \mathcal{N}, \mathbf{m_0} \rangle$ has the* marked-siphon-trap property, *MST-property, if each siphon contains a marked trap at* $\mathbf{m_0}$.

A siphon (trap) is *minimal* if it does not contain another siphon (trap). Thus, siphons in the above statement can be constrained to be minimal without any loss of generality.

The MST-property guarantees that all siphons will be marked. Thus no dead marking can be reached, according to property 15.3.2.1. Therefore:

Property 15.3.6. If $\langle\mathcal{N}, \mathbf{m_0}\rangle$ has the MST-property, the system is deadlock-free.

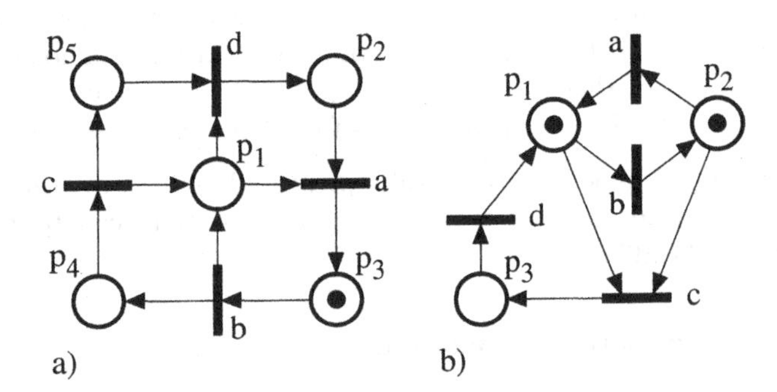

Fig. 15.8. For the two nets, the MST-property does not hold, but: a) the simple net is live and bounded; b) the non-simple net is non-live (although deadlock-free) and bounded.

Figure 15.8 presents some limitations of the MST-property for liveness characterisation.

Remark 15.3.7. The MST-property is sufficient for liveness in simple net systems, and necessary and sufficient for free-choice net systems. As a corollary, the *liveness monotonicity* result is true for the case of live free-choice systems: If $\langle\mathcal{N}, \mathbf{m_0}\rangle$ is a live free-choice system, then for all $\mathbf{m_0}' \geq \mathbf{m_0}$, $\langle\mathcal{N}, \mathbf{m_0}'\rangle$ is also live. The previous result does not apply to simple net systems. The system in Figure 5.1b is simple, and $\Sigma = \{p_1, p_2, p_7\}$ is a siphon ($^\bullet\Sigma = \{t_3, t_4, t_1\}$, $\Sigma^\bullet = {}^\bullet\Sigma \cup \{t_2\}$) that does not contain any trap. If we assume $\mathbf{m_0}[p_5] = 1$, t_2 can be fired and Σ becomes empty, leading to non-liveness.

15.4 Analysis of Net Subclasses

In this section we briefly overview some of the analytical results for certain subclasses that we define in Section 15.4.1. We organise the material around properties instead of describing the results for each subclass, which would lead to redundancy. (Of course, properties of large subclasses, such as EQ systems, are inherited by their subclasses, such as FC or DF systems.)

Our intention is to show how the restrictions imposed in the definitions of subclasses, at the price of a loss of some modelling capabilities, facilitate the analysis. The designer must find a compromise between modelling power and the availability of powerful analysis tools, while one of the theoretician's goals is to obtain better results for increasingly larger subclasses.

The general idea behind the structure theory of net subclasses is to investigate properties that every net system in the subclass possesses, instead of analysing each particular system. These general properties are useful in two ways:

- The designer knows that the system (if it belongs to an appropriate subclass) behaves "well" (e.g. liveness monotonicity, existence of home states).
- General analysis methods become more applicable or more conclusive (e.g. model checking for FC, liveness analysis for all the subclasses considered).

The technical development of the results presented, and many other details that are beyond the scope of this succinct presentation, can be found in [DE95, RTS96, TCS97, TS96].

15.4.1 Some Syntactical Subclasses

Historically, subclasses of ordinary nets have received special attention because powerful results were obtained for them early on. In this presentation some of them appear as subclasses of their weighted generalisations for the sake of conciseness. Regarding the modelling power, clearly some subclasses have less than others if the former are properly included in the latter. Also, the weighted generalisations have more modelling power than their ordinary counterparts since, in general, the ordinary implementations of weights do not preserve the (topological) class membership.

Join-Free and State Machines. A P/T net $\mathcal{N}$ is *join-free (JF)* when no transition is a join, i.e. $|{}^\bullet t| \leq 1$ for every t. With these nets, proper synchronisations cannot be modelled. $\mathcal{N}$ is a *weighted P-net* when every transition has one input and one output place, i.e. $|{}^\bullet t| = |t^\bullet| = 1$ for every t. An ordinary weighted P-net is a P-net or *state machine (SM)*, so called because when marked *with only one token* each place represents a possible global state of the (sequential) system. With more than one token concurrency appears: an SM with k tokens represents k instances of the same sequential process evolving in parallel. Given an adequate stochastic interpretation, strongly connected SMs correspond to closed Jackson queuing networks.

Distributor-Free and Marked Graphs. A P/T net $\mathcal{N}$ is *distributor-free (DF)* when no place is a distributor, i.e. $|p^\bullet| \leq 1$ for every p. With these nets, conflicts cannot be modelled. They are also called *structurally persistent* because the structure enforces persistency, that is, the property that a transition can be disabled only by its own firing. $\mathcal{N}$ is a *weighted T-net* when every place has one input and one output transition, i.e. $|{}^\bullet p| = |p^\bullet| = 1$ for every p. An ordinary weighted T-net is a T-net or *marked graph (MG)*, the name due to a representation as a graph in which the nodes are the transitions and the arcs joining them are marked (that is, places have been eliminated). For example, MGs can model activity ordering systems, generalising PERT graphs, job-shop systems with fixed production routing and machine sequencing, flow lines, and Kanban systems. For instance, the net in Figure 15.9 (a) is an MG. Given an adequate stochastic interpretation, strongly connected MGs correspond to fork/join queuing networks with blocking.

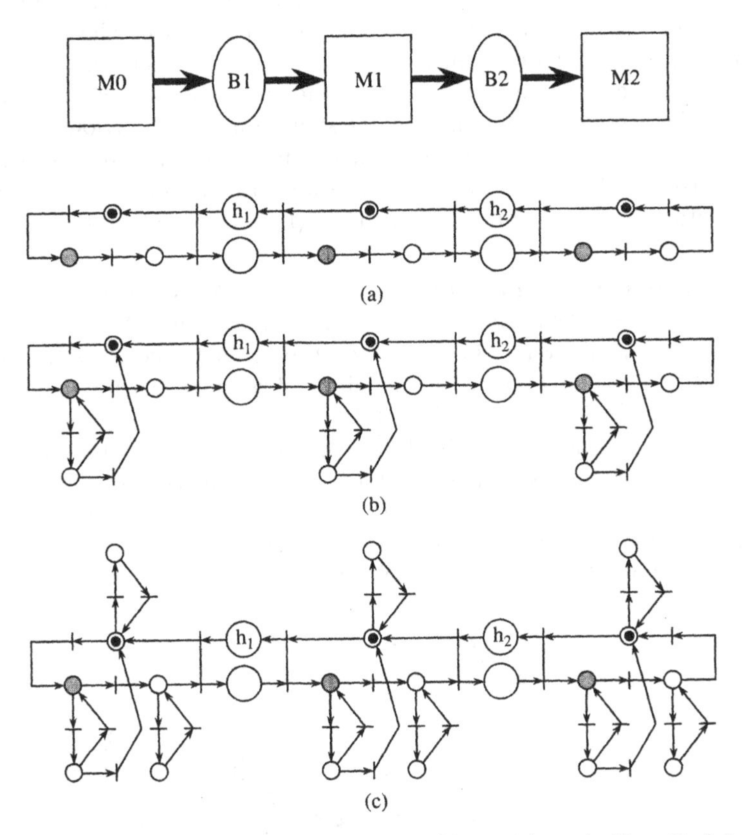

Fig. 15.9. Modelling a flow line with three machines and two buffers. Each buffer is modelled with two places, for the parts and "holes" respectively (the latter initially marked with h_i holes). Each machine is modelled with a state machine, initially idle, where the "working-state" is shaded; they follow a blocking-after-service policy (they start their work even if there are no holes in the output buffer, so they might stay blocked before unloading). The different models consider: (a) reliable machines, (b) machines with operation-dependent failures (may fail only when working), and (c) machines with time-dependent failures (may fail at any time). Scrapping (the part is discarded) is possible in the case of unreliable machines.

Equal Conflict and Free Choice. A P/T net $\mathcal{N}$ is *equal conflict (EQ)* when every pair of transitions in structural conflict are in equal conflict, i.e. they have the same pre-incidence function: ${}^\bullet t \cap {}^\bullet t' \neq \emptyset$ implies $\mathbf{Pre}[P,t] = \mathbf{Pre}[P,t']$. An ordinary EQ net is an *(extended) free-choice net (FC)*. Free-choice nets play a central role in the theory of net systems because there are powerful results for their analysis and synthesis, while they allow the modelling of systems including both conflicts and synchronisations. It is often said that FCs can be seen as MGs enriched with SM-like conflicts or, equivalently, SMs enriched with MG-like synchronisations. However, they cannot model mutex semaphores or resource sharing, for instance. The net in Figure 15.9 (b) is FC. The fundamental property of EQ systems is that whenever a marking enables some transition t, then it enables every transition in $\text{EQS}(t) = \text{CCS}(t)$. It can be said that the structural and behavioural notions of conflict coincide. It is also said that conflicts and synchronisations are neatly separated, because it is easy to transform the net so that no output of a distributor place is a join: in Figure 15.6, (f) is the result of transforming (e).

Asymmetric Choice, or Simple. A P/T net $\mathcal{N}$ is *asymmetric choice (AC)*, sometimes called *simple*, when it is ordinary and $p^\bullet \cap p'^\bullet \neq \emptyset$ implies $p^\bullet \subseteq p'^\bullet$ or vice versa. In these nets, the conflict relation is transitive. They generalise FC, and allow the modelling of resource sharing to a certain extent. The net in Figure 15.9 (c) is AC.

The above subclasses are defined through a global constraint on the topology. The relations between them are illustrated in the graph of Figure 15.10, where a directed arrow connecting two subclasses indicates that the source properly includes the destination, and the constructs depicted illustrate the typical situations that distinguish each subclass.

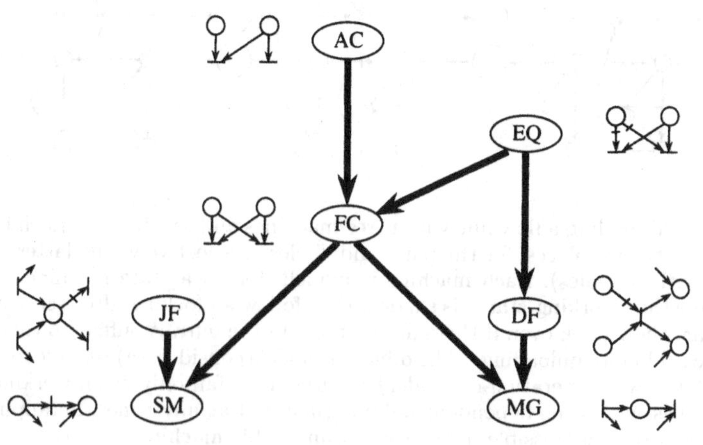

Fig. 15.10. Relations between some basic syntactical subclasses

Modular Subclasses. Subclasses can also be defined in a modular way, by giving some modules and how to interconnect them. Very often the modules are monomarked SMs, representing sequential systems which run in parallel communicating in a restricted fashion. A few examples follow.

Superposed automata systems (SA) are composed by monomarked SMs synchronised by transition merging, that is, via rendezvous. They lead to general — although structured — bounded systems models. For instance, all the nets in Figure 15.9 are SA (if the capacities of the buffers are one).

Systems of buffer-cooperating functional entities are modules (depending on the kind of modules we obtain different subclasses) synchronised by message-passing through buffers in a restricted fashion. A P/T system $\mathcal{S}$ is in this class when:

- $P = B \uplus \biguplus_i P_i$, $T = \biguplus_i T_i$. The net systems $\mathcal{S}_i$ generated by P_i and T_i are the *functional entities* or modules, and the places of B are the *buffers*.
- For every $b \in B$, there exists i such that $b^\bullet \in T_i$, that is, buffers are output private. Moreover if $t, t' \in T_i$ are in EQ relation in $\mathcal{N}_i$, then $\mathbf{Pre}[b,t] = \mathbf{Pre}[b,t']$, that is, buffers do not modify the EQ relations of the modules. These restrictions on buffers prevent competition.

If the modules are monomarked SMs we obtain *deterministically synchronised sequential processes (DSSPs)*. If they are EQ systems we obtain systems of cooperating EQ systems. These can be buffer-interconnected again, leading to a hierarchical class of systems, recursively defined, which is called $\{SC\}^*EQS$, standing for systems of cooperating systems of cooperating EQ systems. They allow the modelling of hierarchically coupled cooperating systems. The net systems in Figure 15.9 (a) and 15.9 (b) can be seen as (rather trivial) examples of buffer-cooperating systems, in which the places modelling the buffers are precisely the buffers, while each machine is modelled by an SM.

Systems of Simple Sequential Processes with Resources (S^3PRs) are SMs synchronised by restricted resource sharing. The restrictions require that there be a place in each SM which is contained in every cycle and does not use any resource (an "unavoidable idle state"), and that every other place use one (possibly shared) resource. They allow the modelling of rather general flexible manufacturing systems, or similar systems where resource sharing is essential.

15.4.2 Fairness and Monopolies

In some systems, *impartiality* (or *global fairness*, i.e. every transition appears infinitely often in infinite sequences) can be achieved *locally* (every solution of a (local) conflict that is effective infinitely often is taken infinitely often):

Theorem 15.4.1. *Let $\mathcal{S}$ be a bounded strongly connected EQ system or DSSP. A sequence $\sigma \in \mathrm{L}(\mathcal{S})$ is globally fair iff it is locally fair.*

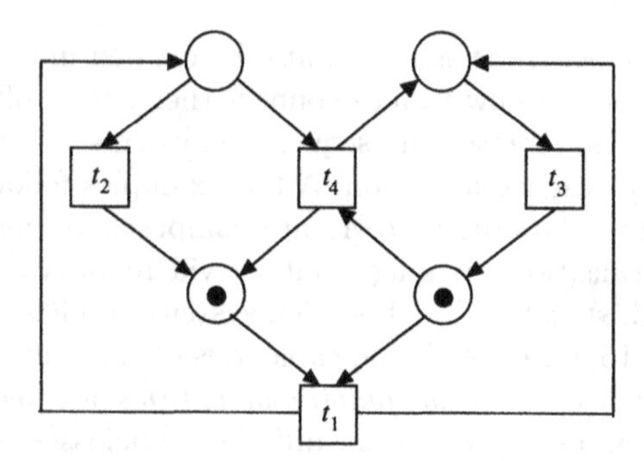

Fig. 15.11. A net system where local fairness does not guarantee impartiality, and which can exhibit monopoly situations.

This property is not true in general. Take for instance the net system in Figure 15.11. The sequence $\sigma = \{t_1\, t_2\, t_3\}^\omega$ is locally fair (actually, during the occurrence of σ no conflict is effective at all), but not globally fair since t_4 never occurs. Conversely, the sequence $\sigma = \{t_1\, t_3\, t_4\, t_3\, t_1\, t_2\, t_3\}^\omega$ is globally fair but not locally fair since whenever t_2 and t_4 are in conflict, t_4 wins.

The equivalence of local and global fairness has two important consequences. The first one is the equivalence of liveness and deadlock-freeness, which facilitates the analysis of liveness because it suffices to check the weaker property of deadlock-freeness:

Theorem 15.4.2. *Let $\mathcal{S}$ be a bounded strongly connected EQ system or DSSP. Then $\mathcal{S}$ is live iff it is deadlock-free.*

The second consequence is relevant for the eventual interpretation of the model. Assume, for instance, that the system in Figure 15.11 is interpreted so that transitions occur after a deterministic delay equal to their index. Then the system repeats the occurrence of $t_1\, t_2\, t_3$, never giving a chance to t_4, although it was live in the autonomous model: the interpretation has destroyed liveness, leading to a *monopoly* situation (the "resources" needed by t_4 are "monopolised" by t_2).

This can never happen to a bounded strongly connected EQ system or DSSP, assuming that the interpretation allows progress (i.e. a transition that is continuously enabled eventually occurs). By imposing a fair conflict resolution policy, which can be done in a distributed fashion provided structurally conflicting transitions are allocated together, it is guaranteed that no action in the system becomes permanently disabled if the autonomous model was live.

15.4.3 Confluence and Directedness

Persistent systems, which include structurally persistent ones (DFs), possess a strong *confluence* property: whenever from a given marking we reach two different markings by firing two distinct sequences, then we can complete both sequences, each with the firings remaining with respect to the other, reaching the same marking in both cases ([LR78]). Confluence is closely related to determinacy ([KM66]): interpreting sequences as executions and transition occurrences as operations, if from a given point two different executions may occur, then depending on operation times or other external matters, each operation in one execution will eventually occur in the other (assuming progress), possibly in a different order and with a different timing.

Moreover, confluence facilitates checking liveness (non-termination) of persistent systems: it suffices to find a repeatable sequence that contains every transition. This is because such a repeatable sequence allows us to construct a sequence greater than any given sequence σ fireable from the initial marking, and this proves that σ can be continued to enable the repeatable sequence.

For systems which are not persistent, the presence of effective conflicts may destroy confluence. *Directedness* is a weaker property which states that a common successor of arbitrary reachable markings always exists; this property holds for some subclasses:

Theorem 15.4.3. *Let $\mathcal{S}$ be a live EQ system or DSSP. Let $\mathbf{m_a}, \mathbf{m_b} \in \mathrm{RS}(\mathcal{S})$. Then $\mathrm{RS}(\mathcal{N}, \mathbf{m_a}) \cap \mathrm{RS}(\mathcal{N}, \mathbf{m_b}) \neq \emptyset$.*

Informally, directedness means that the effect of a particular resolution of a conflict is not "irreversible": there is a point where the evolution merges with that which would have been if another decision had been made. The existence of *home states*, i.e. states that can ultimately be reached after any evolution, follows from directedness and boundedness:

Theorem 15.4.4. *Live and bounded EQ systems or DSSPs have home states.*

The system in Figure 5.3.b is an example of a live and 1-bounded system without home states.

This is an important property for many reasons:

- The system is known to have states to return to, which is often required in reactive systems. Choosing one such state as the initial one makes the system *reversible*, i.e. $\mathbf{m_0}$ can always be recovered.
- Model checking is greatly simplified, since there is only one terminal strongly connected component in the reachability graph.
- Under a Markovian interpretation (e.g. as in *generalised stochastic Petri nets*, [ABC+95]), *ergodicity* of the marking process is guaranteed; otherwise, simulation or computation of steady-state performance indices could be meaningless.

15.4.4 Reachability and the State Equation

As was discussed in Section 15.2, reachable markings are solutions to the state equation but, in general, the converse is not true: some solutions of the state equation may be spurious. This limits the use of the state equation as a convenient algebraic representation of the state space.

Fortunately, stronger relations between reachable markings and solutions to the state equation are available for some subclasses:

Theorem 15.4.5. *Let $\mathcal{S}$ be a P/T system with reachability set* RS *and linearised reachability set w.r.t. the state equation* LRS^{SE}.

1. *If $\mathcal{S}$ is a live weighted T-system, or a live and consistent source private DSSP, then* $\text{RS} = \text{LRS}^{\text{SE}}$. *Moreover, if it is a live MG, then the integrality constraints can be disregarded.*
2. *If $\mathcal{S}$ is a bounded, live, and reversible DF system, then* $\mathbf{m} \in \text{RS}$ *iff* $\mathbf{m} \in \text{LRS}^{\text{SE}}$ *and the unique minimal t-semiflow of the net is fireable at* $\mathbf{m}$.
3. *If $\mathcal{S}$ is a live, bounded, and reversible FC system, then* $\mathbf{m} \in \text{RS}$ *iff* $\mathbf{m} \in \text{LRS}^{\text{SE}}$ *(integrality constraints can be disregarded) and every trap is marked at* $\mathbf{m}$.
4. *If $\mathcal{S}$ is a live EQ system or a live and consistent DSSP, and* $\mathbf{m_a}, \mathbf{m_b} \in \text{LRS}^{\text{SE}}$, *then* $\text{RS}(\mathcal{N}, \mathbf{m_a}) \cap \text{RS}(\mathcal{N}, \mathbf{m_b}) \neq \emptyset$.

We can take advantage of the above statements in a diversity of situations. For instance, the reachability characterisation for live MGs allows us to analyse some of their properties through linear programming. Even the last, and weakest, statement in the above theorem — a directedness result at the level of the linearised reachability graph — can be very helpful. Since, in particular, it implies that there are no spurious deadlocks in live EQ systems, or live and consistent DSSPs, the deadlock-freeness analysis technique presented in Section 15.2.2 — which in these cases requires a single equation system — allows us to decide the property.

Figure 5.4.6 shows an example of a live and 1-bounded system with spurious deadlocks.

15.4.5 Analysis of Liveness and Boundedness

One of the properties which supports the claim that "good" behaviour should be easier to achieve in some subclasses than in general systems is liveness monotonicity with respect to. the initial marking. This means that liveness, provided that the net is "syntactically" correct as we shall specify later, is a matter of having enough tokens in the buffers (customers, resources, initial data, etc.). In contrast, in general systems the addition of tokens may well cause deadlocks because of poorly managed competition. For instance, in the net system of Figure 5.1.b adding a token (in p_5) to the initial marking destroys liveness.

Theorem 15.4.6. *Let $\langle \mathcal{N}, \mathbf{m_0} \rangle$ be a live EQ system or DSSP. The EQ system or DSSP $\langle \mathcal{N}, \mathbf{m_0} + \Delta\mathbf{m_0} \rangle$, where $\Delta\mathbf{m_0} \geq \mathbf{0}$, is also live.*

Often, a net system is required to be live and bounded. As we saw in Section 5.2.1 the verification of liveness can be difficult, so we want to avoid it when possible. In some cases we are able to decide using structural methods alone; in other cases we can characterise the nets that can be lively and boundedly marked, so the costly enumeration analysis must be used only when there is a chance of success.

Theorem 15.4.7. *Let $\mathcal{N}$ be an EQ or DSSP net. A marking $\mathbf{m_0}$ exists such that $\langle \mathcal{N}, \mathbf{m_0} \rangle$ is a live and bounded EQ system or DSSP iff $\mathcal{N}$ is strongly connected, conservative (or consistent), and* $\mathrm{rank}(\mathbf{C}) = |\mathrm{SEQS}| - 1$. *Moreover, in EQ systems, liveness of the whole system is equivalent to liveness of each P-component (the P-sub-nets generated by the minimal p-semiflows).*

Particular cases of the above result are well known in net theory. For instance, in the ordinary case, the P-components of an FC net are strongly connected SMs, which are live iff they are marked, so the liveness criterion can be stated as "there are no unmarked p-semiflows". In the case of MGs, which are always consistent and have $\mathrm{rank}(\mathbf{C}) = |\mathrm{SEQS}| - 1 = |T| - 1$, the existence of a live and bounded marking is equivalent to strong connectedness. Since their P-components are their circuits, liveness can be checked by removing the marked places and verifying that the remaining net is acyclic.

15.5 Invariants and Reductions for Coloured Petri Nets

15.5.1 Invariants

As in ordinary Petri nets, one of the main aspects of the structural verification of CPNs is the generation of invariants. But before developing techniques for such a problem, some points must be clarified:

- How can we express an invariant of CPNs and especially a linear invariant?
- How can a family of (linear) invariants be characterised as a generative family of invariants?
- How can we build a (generative) family of (linear) invariants?

The aim of this section is to answer these three questions concisely.

Presentation of Linear Invariants. The choice of an adequate definition of invariants should meet the following requirements. Firstly, an invariant of a high-level Petri net must be a high-level invariant. For instance, if we model processes by colours, an invariant should express properties of the behaviour of one particular process but also of the behaviour of any process, or else of the behaviour of any process except a particular one etc. On the other hand, the definition should enable mathematical developments leading to efficient algorithms for computing invariants.

One can try to use the definition of ordinary invariants, i.e. a invariant is a weighted sum of the marking of the places left invariant by the firing of any transition. However such a definition involves two hidden extensions:

- What can the weights on place markings be?
- There are multiple ways to fire a transition (as many as the size of the colour domain of the transition).

The essential point in the definition below is that the weights are colour functions. The interpretation is that applying the colour function to the place marking corresponds to extracting the relevant part of the information contained in this marking for a given invariant. In order to be mathematically sound, this function must have as its domain the colour domain of the place and as its codomain a common domain for the weights of the same invariant. This codomain may be viewed as the interpretation domain of the invariant, and this requirement ensures that the weighted sum of the marking places is well-defined.

Definition 15.5.1. *A linear invariant* $\mathbf{v}$ *of a coloured Petri net* $\mathcal{N}$ *is defined by:*

- *$cd(\mathbf{v})$ the colour domain of the invariant, and*
- *$\forall p \in P$, $\mathbf{v}(p)$ a function from $\mathbb{Z}^{cd(p)}$ to $\mathbb{Z}^{cd(\mathbf{v})}$,*

such that: $\forall \mathbf{m}$ reachable marking, $\sum_{p \in P} \mathbf{v}(p)(\mathbf{m}(p)) = \sum_{p \in P} \mathbf{v}(p)(\mathbf{m_0}(p))$

The net of Figure 15.12 models a database with multiple copies. The access grant of the database is centralised and subject to mutual exclusion.

The database is shared by a set of sites represented by the colour domain *Sites*. In order to modify the database, an idle site (in place *idle*) must get the grant (a neutral token in place *mutex*); and once it has modified the file, it sends messages to the other sites. The entire action is modelled with transition t_1 and the content of the message is not modelled. Then the other sites update their own databases (transition t_3) and send an acknowledgment (transition t_4). Once the active site has received all the acknowledgments, it releases the grant (transition t_2).

In order to simplify the net, accessing and modifying the database is modelled with a single transition (indivisible step) while the updating of the other sites is modelled with a place (divisible step).

Initially there is one token per site in place *idle* and a neutral token in place *mutex*.

Let us give an initial example of a linear invariant:

$$\begin{aligned} cd(\mathbf{v}) &= Sites \\ \mathbf{v} &= \langle x \rangle.idle + \langle x \rangle.wait + \langle x \rangle.update \end{aligned}$$

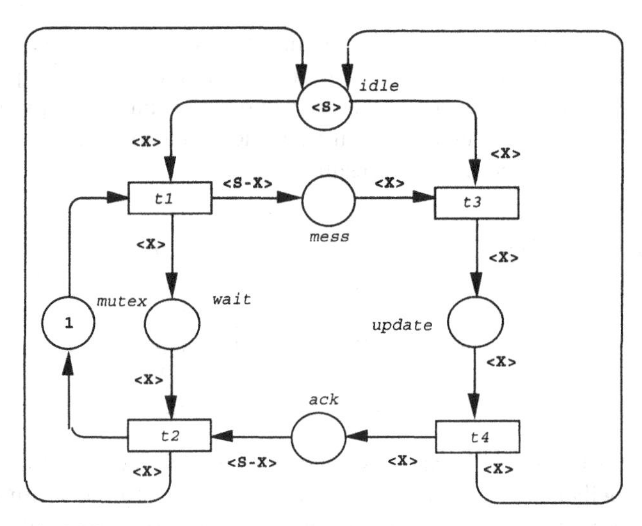

Fig. 15.12. A CPN model of a replicated database

This linear invariant describes the behaviour of any site: either the site is idle, or it waits for the acknowledgments, or it updates its database.

Computation of Linear Invariants. Now we focus on the computation of a family of linear invariants. Ideally, we want a generative family of invariants, i.e. any linear invariant should be obtained by a linear combination of the invariants of the family. Keeping in mind the definition of a linear invariant of a well-formed net, we allow the coefficients of the combination to be functions with identical requirements on the domain and codomain. It should be mentioned that this is the only way to control the size of a generative family and moreover (as we shall see in what follows) to obtain significant flows, directly as items of the family or by linear combination.

We will not give an algorithm in this subsection but instead we will use the previous example to show how to compute a generative family of invariants. Then we will interpret our family of invariants.

The cornerstone of all the algorithms is handling the incidence matrix in a way similar to Gaussian elimination. However, the elimination rules must be applied under conditions which ensure that no linear invariant will be "forgotten".

We start with the incidence matrix **C** of our example:

$$
\begin{array}{ccccccl}
 & \text{t1} & \text{t2} & \text{t3} & \text{t4} & & \\
\mathbf{C} = & \left(\begin{array}{l} \langle x\rangle \\ -\langle x\rangle \\ \langle s-x\rangle \\ \text{-1} \\ 0 \\ 0 \end{array}\right. &
\begin{array}{l} -\langle x\rangle \\ \langle x\rangle \\ 0 \\ 1 \\ 0 \\ -\langle s-x\rangle \end{array} &
\begin{array}{l} 0 \\ -\langle x\rangle \\ -\langle x\rangle \\ 0 \\ \langle x\rangle \\ 0 \end{array} &
\left.\begin{array}{l} 0 \\ \langle x\rangle \\ 0 \\ 0 \\ -\langle x\rangle \\ \langle x\rangle \end{array}\right) &
\begin{array}{l} \langle x\rangle.\mathit{wait} \\ \langle x\rangle.\mathit{idle} \\ \langle x\rangle.\mathit{mess} \\ 1.\mathit{mutex} \\ \langle x\rangle.\mathit{update} \\ \langle x\rangle.\mathit{ack} \end{array}
\end{array}
$$

On the right, the initial family of invariants is shown, i.e. each place weighted by the identity function of its colour domain. Now we proceed by a standard rule: adding to a line another one which has been premultiplied by a function. The only restriction on this rule is the consistency of domains and codomains of functions. The result is:

$$\begin{array}{cccc} t1 & t2 & t3 & t4 \end{array}$$
$$\begin{pmatrix} \langle x\rangle & -\langle x\rangle & 0 & 0 \\ 0 & 0 & -\langle x\rangle & \langle x\rangle \\ 0 & \langle s-x\rangle & -\langle x\rangle & 0 \\ 0 & 0 & 0 & 0 \\ 0 & 0 & \langle x\rangle & -\langle x\rangle \\ 0 & -\langle s-x\rangle & 0 & \langle x\rangle \end{pmatrix} \begin{array}{l} \langle x\rangle.wait \\ \langle x\rangle.idle + \langle x\rangle.wait \\ \langle x\rangle.mess - \langle s-x\rangle.wait \\ 1.mutex + 1.wait \\ \langle x\rangle.update \\ \langle x\rangle.ack \end{array}$$

We apply to this new matrix a second rule that eliminates a line for which one of the coefficients is the only non-null coefficient of its column (here the first line); we require that this coefficient be an injective mapping:

$$\begin{array}{cccc} t1 & t2 & t3 & t4 \end{array}$$
$$\begin{pmatrix} 0 & 0 & -\langle x\rangle & \langle x\rangle \\ 0 & \langle s-x\rangle & -\langle x\rangle & 0 \\ 0 & 0 & 0 & 0 \\ 0 & 0 & \langle x\rangle & -\langle x\rangle \\ 0 & -\langle s-x\rangle & 0 & \langle x\rangle \end{pmatrix} \begin{array}{l} \langle x\rangle.idle + \langle x\rangle.wait \\ \langle x\rangle.mess - \langle s-x\rangle.wait \\ 1.mutex + 1.wait \\ \langle x\rangle.update \\ \langle x\rangle.ack \end{array}$$

The third rule we apply is identical to a Gaussian elimination rule, i.e. delete a null column (here $t1$):

$$\begin{array}{ccc} t2 & t3 & t4 \end{array}$$
$$\begin{pmatrix} 0 & -\langle x\rangle & \langle x\rangle \\ \langle s-x\rangle & -\langle x\rangle & 0 \\ 0 & 0 & 0 \\ 0 & \langle x\rangle & -\langle x\rangle \\ -\langle s-x\rangle & 0 & \langle x\rangle \end{pmatrix} \begin{array}{l} \langle x\rangle.idle + \langle x\rangle.wait \\ \langle x\rangle.mess - \langle s-x\rangle.wait \\ 1.mutex + 1.wait \\ \langle x\rangle.update \\ \langle x\rangle.ack \end{array}$$

We can iterate this process to eliminate the column $t2$, using the property that $\langle s-x\rangle$ is injective:

$$\begin{array}{ccc} t2 & t3 & t4 \end{array}$$
$$\begin{pmatrix} 0 & -\langle x\rangle & \langle x\rangle \\ 0 & -\langle x\rangle & \langle x\rangle \\ 0 & 0 & 0 \\ 0 & \langle x\rangle & -\langle x\rangle \end{pmatrix} \begin{array}{l} \langle x\rangle.idle + \langle x\rangle.wait \\ \langle x\rangle.mess - \langle s-x\rangle.wait + \langle x\rangle.ack \\ 1.mutex + 1.wait \\ \langle x\rangle.update \end{array}$$

The elimination of $t3$ ends the algorithm since it simultaneously eliminates $t4$. We obtain the following family of invariants:

- The state of any site – either the site is idle, or it waits for the acknowledgments, or it updates its database:

$$\langle x \rangle.wait + \langle x \rangle.idle + \langle x \rangle.update$$

- The state of the database – either the grant is present or a site is waiting to complete its transaction:

$$1.wait + 1.mutex$$

- The synchronisation between sites – if a site is waiting then either another site has a message for the current transaction, or it updates its copy, or it has sent its acknowledgment:

$$\langle x \rangle.mess + \langle x \rangle.ack + \langle x \rangle.update - \langle s - x \rangle.wait$$

Some other significant linear invariants may be obtained by combining the previous invariants. For instance, the following invariant says that either the grant is present and all the sites are idle, or the grant is absent and the idle sites are exactly those which will receive a message or have sent their acknowledgment:

$$\langle x \rangle.idle - \langle s \rangle.mutex - \langle x \rangle.mess - \langle x \rangle.ack$$

Additional Remarks. The computation of linear invariants as we have described in the previous section can not be straightforwardly extended to general coloured nets. We list below the three problems one must address in order to provide a general algorithm:

- How to combine lines in order to cancel items of the matrix?
- How to ensure that the last coefficient of a column is injective?
- How to handle the previous two operations in a parametrised way, i.e. independently of the size of the colour domains (in our example, the number of sites)?

A general algorithm for coloured nets has been proposed in [Cou91]. The key point of the algorithm is the intensive use of generalised semi-inverses. Its only restriction is that the size of the colour domains is fixed.

On the other hand, with restrictions to subclasses of coloured nets, it is possible to obtain algorithms which handle parameters ([HG86, HC88, MV87]). These algorithms transform the incidence matrix of functions into a set of matrices with coefficients taken from a ring of polynomials. The variables correspond to the parameters of the net. At this point, it is enough to apply a Gaussian-like elimination on these matrices. Each vector solution is finally transformed (in an inverse way) to a linear invariant.

15.5.2 Reductions

A reduction of a net is defined by some conditions of application and a method of transformation such that the reduced net has the same behaviour as the original one with respect to generic properties, if the original net satisfies the conditions. Reduction theory has been mainly developed by Gérard Berthelot [Ber87] who has proposed ten reductions covering a wide area of applications. Introducing a new reduction is interesting if:

- It covers a common behavioural situation (as will be described in the next subsection);
- The application conditions can be checked in a efficient way (e.g. by examination of the net structure or by linear invariant computation).

A generalisation to high-level nets has been proposed by different authors ([CMS87, Gen88, Had89]). We will follow the last reference to introduce some reductions. Then we will illustrate them on the example of database management and finally we will discuss a methodology for defining new reductions.

Some Reductions. Two reductions for CPNs (more information may be found in the previous reference) are enough for our example. The style of definition will be informal: we will give the general interpretation of the reduction, the definition of application conditions, and the method of transformation, with for each item a corresponding interpretation. The set of properties preserved is essentially the same for both reductions and includes liveness, boundedness, home state existence, unavoidable state existence, etc.

IMPLICIT PLACE SIMPLIFICATION. This reduction deletes a place which never on its own forbids the firing of a transition. Such a place occurs in two ways: either it is a redundant place which explicitly model information implicitly contained in the other places, or the place was not originally implicit but other reductions have transformed it.

The existence of an implicit place is ensured by a particular invariant. Such a reduction illustrates an indirect use of invariants for model verification.

Definition 15.5.2. *Let $\mathcal{N}$ be a coloured Petri net. A place p is implicit iff:*

1. *There exists a linear invariant* $\mathbf{v}$ *such that:*
   - $cd(\mathbf{v}) = cd(p)$
   - $\mathbf{v}(p)$ *is the identity function of* $\mathrm{Bag}(cd(p))$
   - $\forall p' \neq p, \forall c \in cd(p'), -\mathbf{v}(p')(c) \in \mathrm{Bag}(cd(p'))$
2. $\forall t \in T, \forall c \in cd(t), \sum_{p' \in P} \mathbf{v}(p')(\mathbf{Pre}(p', t)(c)) \geq \sum_{p' \in P} \mathbf{v}(p')(\mathbf{m_0}(p'))$

Condition *2* ensures that in the initial marking, p will not on its own forbid the firing of any transition due to the positivity constraints included in condition *1*. The fact that $\mathbf{v}$ is a linear invariant ensures that condition *2* is also true for all reachable markings.

The transformation deletes the place and the bordering arcs.

Definition 15.5.3. *The reduced net* $\mathcal{N}_r = \langle P', T', \mathbf{Pre}', \mathbf{Post}', \mathcal{C}', cd' \rangle$ *with initial marking* $\mathbf{m_0}'$ *obtained from the net* $\mathcal{N}$ *the simplification of the implicit place* p *is defined by:*

- $P' = P - \{p\}$
- $T' = T$
- $\mathcal{C}' = \mathcal{C}$
- $\forall t \in T', \forall p' \in P', cd'(t) = cd(t)$ *and* $cd'(p') = cd(p')$
- $\forall t \in T', \forall p' \in P', \mathbf{Pre}'(p', t) = \mathbf{Pre}(p', t)$ *and* $\mathbf{Post}'(p', t) = \mathbf{Post}(p', t)$
- $\forall p' \in P', \mathbf{m_0}'(p') = \mathbf{m_0}(p')$

POST-AGGLOMERATION OF TRANSITIONS. The principle of post-agglomeration is the following. Suppose we are given H, a set of transitions which represent global actions and which lead to an intermediate state represented by a token in the place p. Suppose that the way to leave this intermediate state is to fire a transition f which represents a local transition (i.e. without synchronisation). Then the firing of any transition of H could be immediately followed by the firing of f without modifying the global behaviour of the net. Hence, the place p may be deleted and the firings of transition f included in the firing of any transition of H. However, one must define carefully the new items of the **Pre** and **Post** matrices. For simplicity, we present here a restricted version of this reduction.

We begin by introducing the concept of a safe colour function; this is required by the subsequent definition. A safe function produces (or consumes depending on the arc) at most one token per colour of the place. In a model, the functions are usually safe functions.

Definition 15.5.4. *A colour function from* Bag(E) *to* Bag(F) *is safe iff:*

$$\forall c \in E, \forall c' \in F, f(c)(c') \leq 1$$

Definition 15.5.5. *Let* $\mathcal{N}$ *be a coloured Petri net,* p *a place which has for output the transition* f *and for input the set* H *of transitions with* $f \notin H$. *One can post-agglomerate* f *with* H *iff:*

1. $\forall h \in H, \mathbf{Post}(p, h)$ *is a safe function.*
2. $cd(f) = cd(p)$ *and* $\mathbf{Pre}(p, f)$ *is the identity function.*
3. *Initially* p *is unmarked.*
4. *Any firing of* f *produces tokens.*
5. *The only input place of* f *is* p.

Condition *1* ensures that for a given colour, a transition of H produces one token per firing. Condition *2* ensures that such a token can be consumed by the firing of f with the same colour. Condition *3* could be overcome by substituting $\mathbf{m_0}$ by a set of initial markings after emptying the place p. Nevertheless the place p is seldom marked. Condition *4* is necessary for preserving boundedness equivalence. The last condition together with condition *2* is crucial since it ensures the immediate firing of f once p is marked.

The transformation deletes the place p and the bordering arcs. Moreover, any function on output arcs of a transition of H is obtained as the sum of the previous function and the combined effect of the output of this transition followed by adequate firings of f (this explains the composition of functions).

Definition 15.5.6. *The reduced net* $\mathcal{N}_r = \langle P', T', \mathbf{Pre}', \mathbf{Post}', \mathcal{C}', cd' \rangle$ *with initial marking* $\mathbf{m_0}'$ *obtained from the net* $\mathcal{N}$ *by post-agglomeration of the set of transitions* H *and the transition* f *is defined by:*

- $P' = P - \{p\}$
- $T' = T - \{f\}$
- $\mathcal{C}' = \mathcal{C}$
- $\forall t \in T', \forall p' \in P', cd(t) = cd(t)$ *and* $cd(p') = cd(p)$
- $\forall t \in T' - H, \forall p' \in P', \mathbf{Pre}'(p', t) = \mathbf{Pre}(p', t)$ *and* $\mathbf{Post}'(p', t) = \mathbf{Post}(p', t)$
- $\forall h \in H, \forall p' \in P', \mathbf{Pre}'(p', h) = \mathbf{Pre}(p', h)$ *and* $\mathbf{Post}'(p', h) = \mathbf{Post}(p', h) + \mathbf{Post}(p', f) \circ \mathbf{Post}(p, h)$
- $\forall p' \in P', \mathbf{m_0}'(p') = \mathbf{m_0}(p')$

Application to the Example. Figure 15.13 shows the reduction process for the model of database management. We could have reduced the initial net to a single transition using reductions other than those presented here. Nevertheless the final net presented in the figure is small enough for us to analyse its behaviour. The net language is the set of prefixes of $(\bigcup_{c \in Sites} t1(c).t2(c))^*$. Its interpretation is straightforward: the behaviour of the net is an infinite sequence of database modifications done in mutual exclusion.

Let us give some explanations of the applications of the reductions:

- The flow associated with the implicit place *idle* is that obtained by the combination of the generative family in the Section 15.5.1.
- The flow associated with the implicit place *wait* is:

$$1.wait - \frac{1}{n-1}.mess - \frac{1}{n-1}.update - \frac{1}{n-1}.ack$$

  where n is the number of sites.
- During the post-agglomeration around *mess*, the value on the arc from $t1$ to *ack* is obtained by composition of the functions $\langle s - x \rangle$ and $\langle x \rangle$.

Methodology for Obtaining High-level Reductions. Here we give the method that one can use to define sound new reductions for CPNs starting from a reduction of Petri nets. This methodology includes two steps: the specification and the validation of a new reduction.

SPECIFICATION OF A HIGH-LEVEL REDUCTION. As for the ordinary reduction, one must specify application conditions and transformation rules. During this specification, the conditions must be decomposed into two kinds:

- Structural conditions as close as possible to those for the ordinary reduction.

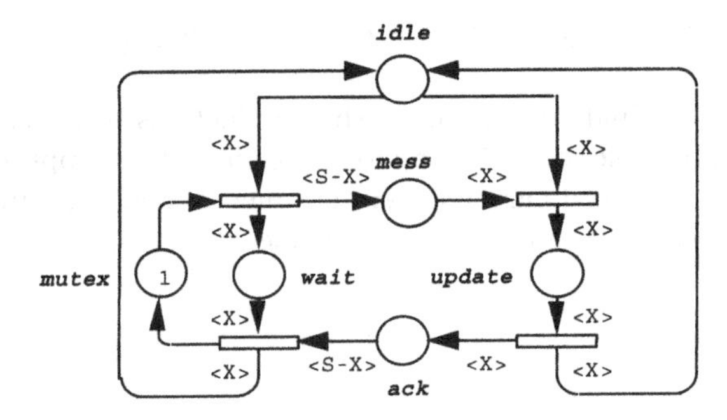

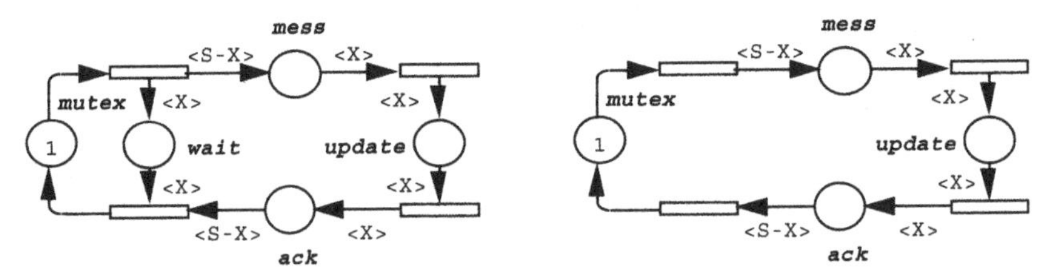

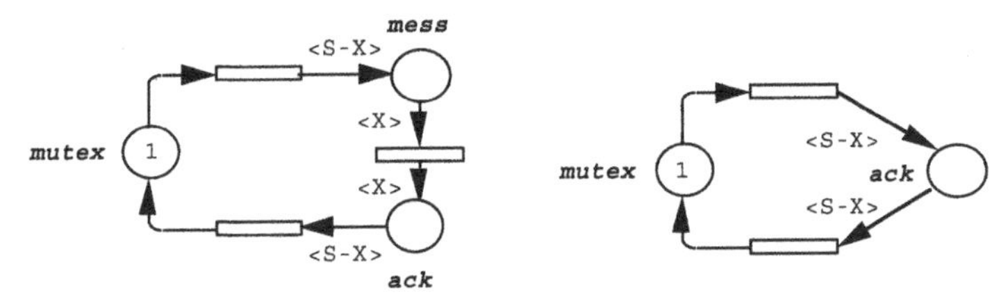

Fig. 15.13. The reduction of the CPN model of a replicated database

- Functional conditions that must be as weak as possible in order to obtain a good structure for the unfolded Petri net corresponding to the CPN.

The transformation rule must follow these two principles:

- It must not increase the size of the colour domain (since in this case there is a hidden extension of the net).
- It enables only significant operations for the colour functions (such as for instance, composition or inverse) in order to keep the new functions manageable.

VALIDATION OF A HIGH-LEVEL REDUCTION. Once the reduction has been defined, the proof that it is sound must be done in the following way:

Unfolding: Show that in the unfolded net a set of ordinary reductions fulfils its conditions.

Reduction sequence: Find an ordering of these reductions such that a reduction is still applicable after the previous ones have been applied.

Folding: Show that the ordinary reduced net may be folded to give the reduced CPN given by the transformation rule.

16. Deductive and Process-Algebra-Based Methods*

Five different approaches based on logical reasoning and process algebraic method swill be presented in this chapter. The sections describe different methods that may seem unrelated at first sight but have certain connections that make their presentation worthwhile. Nevertheless the sections of this chapter are self-contained so there is no preferred order of reading.

The first section entitled "A Rewriting Semantics for Algebraic Nets" gives an informal introduction to algebraic specifications, rewriting, and a generalisation of algebraic Petri nets which is useful in practice. It also describes the close connection between Petri nets and rewriting techniques, and conveys a unified view from the perspective of rewriting logic. Algebraic nets are a form of high-level nets combining the advantages of both algebraic specifications and ordinary Petri nets. An algebraic net should be regarded as a specification that has coloured nets as models. In contrast with standard presentations of algebraic nets, the emphasis of this section is on executable models and strategy-guided net execution. To reflect the requirements of practical modelling problems, we employ an expressive specification language, namely membership equational logic, that unifies and generalises existing approaches such as many-sorted and order-sorted algebra.

In the second section assertional reasoning for Petri nets is introduced. The aim is to provide a logic for the specification and verification of coloured Petri nets that is intuitive enough to be used for manual informal reasoning, but also complex enough to allow for a rigorous formal treatment and its use in connection with computer-aided interactive theorem proving. To this end we employ a generalisation of the elegant UNITY approach introduced by Chandy and Misra in [CM88]; this generalisation deals with arbitrary labelled transition systems instead of UNITY programs. In the context of this book we will restrict our attention to the special case where the labelled transition system is given in terms of a coloured net. Even though UNITY logic contains a fragment of linear time temporal logic, the former is better regarded as an extension of assertional reasoning with a few simple temporal operators. After presenting the basic kinds of assertions and a number of proof rules in a way that avoids known complications and is inspired by Misra's new

* Authors: M.-O. Stehr (Sections 1 & 2), B. Farwer (Sections 3 & 4), T. Basten (Section 5)

approach to UNITY, the use of these rules in the context of coloured Petri nets is demonstrated by applying them to a simple verification example.

In Section 16.3 we examine a special temporal logic for representing the enablement of computations. Using this logic one can, for instance, express the fact that two computations cannot be executed concurrently by the statement $\Box\neg E(\alpha \parallel \beta)$, where $\alpha \parallel \beta$ denotes the parallel composition of the computations α and β and $E(x)$ is a predicate stating the enablement of x, i.e. the formula could be paraphrased: "It will always hold that the parallel enablement of both α and β is not true at any instance of time."

The approach described in Section 16.3 is based on the static structure of nets. It enriches reductions (cf. Chapter 15) with a temporal logic through which preservation and reflection of different classes of properties is shown for reductions defined by morphisms in a category of marked P/T nets. The morphisms define a notion of simulation that allows us to prove properties of complex nets by showing that they can be simulated by some net for which that same property has already been proved.

This approach is followed by a section on the connection between linear logic and Petri nets which in some sense represents the foundation for the rewriting semantics presented in the first section. Apart from the specification of net properties and a basic result connecting reachability with derivability within the fragment of linear logic presented here, we show the possibility of introducing a new concept of nondeterminism into Petri net theory. This concept of nondeterministic transition is motivated by yet another fragment of linear logic incorporating only one additional connective.

Finally, Section 16.5 presents a method utilizing an ACP-like process algebra for the specification and verification of nets. In this approach, the desired properties are specified by algebraic expressions which are then used to construct a suitable net model. Algebraic methods can be used to check that this model satisfies the initial specification by using algebraic methods. In this sense the approach gives rise to a unifying perspective of the two disciplines of process algebra and Petri nets.

The presentation of these advanced concepts requires a formal – sometimes technical – exposition which is essential only for those who want to directly apply the methods. Readers interested in the basic ideas should not be put off by these technicalities but should simply skip the passages containing the more formal reasoning. We believe the remaining parts will still give some insight into the approaches discussed in this chapter.

16.1 A Rewriting Semantics for Algebraic Nets

Since the introduction of predicate/transition nets in [GL79] and [GL81], a variety of further general high-level net models have been developed, including coloured nets ([Jen81] and [Jen92b]) and relation nets ([Rei85b]). Such high-level nets reflect the practical need to deal with flow and transformation of individual data objects rather than indistinguishable tokens. The general idea is to extend place/transition nets with inscriptions in a language capable of expressing operations and/or predicates on data objects. Employing a programming language for this purpose as favoured in [Jen92b] allows *executable net models*, whereas a specification language yields system models which are more abstract but in general not executable. A minimal requirement for *executability* is that enablement can be effectively checked given a transition t together with an occurrence mode β, and that the effect of firing the transition element (t, β) can be computed. Execution of a net usually involves some *strategy* to select or to find transition elements which are enabled and to initiate their firing if this is desired.

Algebraic specification languages combine the advantages of specification and programming languages. The theory of algebraic data types and specifications or more generally universal algebra is well-studied (see e.g. [MG85], [EM85], or [Wec92]) and is appealing because of its simple model theory and its operational semantics. The latter can be given in terms of term rewriting systems which have themselves been the subject of intensive research (see e.g. [HO80], [DJ90], and [Klo92] for a survey).

In addition to the high-level net models mentioned above, a number of different approaches combining Petri nets and algebraic specifications exist (see e.g. [Vau85], [BCC+86], [Vau87], [RV87], and [Rei91]). In contrast to these references the following informal introduction to algebraic specifications and Petri nets will put particular emphasis on the operational point of view. Our ultimate goal is to obtain an executable formal representation of a system which can be useful for exploration, validation, or prototyping purposes. In order to achieve this with minimum effort we show that a deep connection between term rewriting and algebraic Petri nets can be exploited which is beyond the use of term rewriting as a means to execute algebraic specifications. For this purpose, algebraic nets will be equipped with a rewriting semantics using the framework of rewriting logic introduced in [Mes92] as a unified model of concurrency.

As far as we know the consideration of executability in the context of algebraic nets is new. Current research in the context of algebraic nets seems to be more oriented toward specification and verification without taking into account a unifying view such as the one we suggest here. Another reason discouraging execution of algebraic nets may be that the standard many-sorted algebraic specifications are too restrictive to be useful in practice, and yet another factor may be the lack of tools, in particular efficient tools. This situation is quite different from that for coloured Petri nets ([Jen92b]), where support for execution is considered to be the main feature of a Petri net tool. To make the algebraic approach more useful in practice we propose to replace many-sorted algebra by the more general membership equational logic ([Mes98b, BJM97]), a sublogic of rewriting logic. As a tool to execute and analyse such nets we use the Maude rewriting engine ([CDE+99, CELM96]) which implements an executable sublanguage of rewriting logic.

The objective of the present section is to give a brief introduction to algebraic specification and rewriting techniques, and to introduce a general form of algebraic net specifications based on membership equational logic together with a translation into rewrite specifications. Instead of a formal treatment, which would require dealing with many technicalities, we try to convey the main ideas using a running example, namely, a distributed network algorithm, which can be modelled as an algebraic net and executed in a controlled way using the Maude language ([CDE+99, CELM96]). It should however be mentioned that the techniques presented here are of a general nature and do not depend on any particular implementation.

16.1.1 Algebraic Specifications

Algebraic specification languages exist in many flavours. All of them define sorts of data objects and operations on these objects, specifying their abstract behaviour using an equation-based language.

Many-sorted equational specifications are a simple and well-studied class of algebraic specifications. They are given by a many-sorted signature Σ and a possibly empty set of equations E. A *many-sorted signature* defines *sorts* together with *sorted constant symbols* and *sorted function symbols* which determine the set of *sorted terms*.

For the purpose of introducing the syntax and semantics of algebraic specifications we employ specification fragments which will be partly reused in Section 16.1.3, where the main example, an algebraic Petri net specification, is discussed. The following specification declares constant symbols `a`,`b`,`c`,`d`,`e` of sort `Id` intended to represent identifiers. It also declares an operation `idPair` intended as a constructor for pairs of identifiers. Finally, two operations `fst` and `snd` are declared which are intended to represent the first and the second projection of a pair, as expressed by the two equations.

```
sorts Id IdPair .

ops a b c d e : -> Id .
op idPair : Id Id -> IdPair .

vars x x' y y' : Id .

op fst : IdPair -> Id .
eq fst(idPair(x,y)) = x .

op snd : IdPair -> Id .
eq snd(idPair(x,y)) = y .
```

The syntax we use is typical of algebraic specification languages.[1] The keywords **sort** and **sorts** introduce the sorts used in the specification. **op** and **ops** declare constant and function symbols together with their arity (the sorts of their arguments, if any) and coarity (the sort of their result). **var** and **vars** declare sorted variables. **eq** introduces a new equation which may involve some of the variables that have been declared.

Above we discussed the intended meaning of a specification. This should be made more precise. A specification has a simple algebraic semantics called the *initial algebra semantics*:[2] Each sort is interpreted as the set of *ground terms*, i.e. terms without variables, of this sort with the condition that two elements of a sort are *identified*, i.e. considered equal, iff they are E*-equivalent*, i.e. if they can be proved to be equal using equations E which are part of the specification.[3] The initial algebra formalises the idea that: (1) a sort contains only elements that can be built using the constant and function symbols of the signature, and (2) elements which are not enforced to be equal by the specification are distinct in the semantics.[4]

Under certain conditions a specification is *executable*, i.e. it can be equipped with an operational semantics based on *reduction*, the replacement of a subterm by another one which is known to be equal and simpler in a certain sense.[5] For this purpose we require that all variables used on the right-hand side appear already on the left-hand side. Applying an equation $u = v$ to a given term w means finding a subterm of w that matches u and replacing that subterm by the term determined by v under the given match. In this way, equations are viewed as directed left-to-right *reduction rules* in the operational semantics and it is assumed that the corresponding *reduction relation* between terms is confluent and terminating. *Confluence* means that if a term t can be reduced to different terms u and v then both of them can be further reduced to a common term t'. *Termination* means that every reduction sequence is finite. These two properties ensure that each term reduces, by successively applying reduction rules in an arbitrary way, to a unique *normal form* representing the *value* of the original term. An algebraic specification is *executable* iff the equations are of the form explained above and the re-

[1] It is actually the syntax of OBJ3 [GWM+92] which has also been employed in the more recent language Maude [CELM96].

[2] See [MG85] for different views of the initial algebra semantics. There are other possible semantics including the loose algebra semantics which interprets a specification as a class of algebras satisfying all equations. This is used when a specification is designated as a theory, but it will not be used in this introduction.

[3] The mathematically appropriate way to construct this initial algebra is a quotient algebra construction ([EM85]). Here we leave the interpretation function implicit by using ground terms both on the syntactic and semantic level (although with different notions of equality).

[4] These two properties are also known under the slogan "no junk and no confusion". See [EM85] for details.

[5] Simplicity can be made mathematically precise using the concept of reduction ordering.

duction relation is confluent and terminating. Consequently, an executable algebraic specification can be seen as a first-order functional program, where every term has a unique value.

The following example is an executable specification of lists over the sort of identifiers already introduced. `emptyIdList` and `idList` are the only constructors. `emptyIdList` represents the empty list and `idList(x,l)` adds a new head element `x` to a list `l`. The function `inIdList(x,l)` checks if `l` contains an element `x`.

```
sort IdList .

vars l l' : IdList .

op emptyIdList : -> IdList .
op idList : Id IdList -> IdList .

op inIdList : Id IdList -> Bool .
eq inIdList(x,emptyIdList) = false .
eq inIdList(x,idList(x',l')) =
   if x == x' then true else inIdList(x,l') fi .
```

We assume a predefined sort `Bool` with constant symbols `true` and `false` as well as a construct `if_then_else_fi` with the obvious meaning.

From a theoretical point of view every total computable function can be specified by an executable equational algebraic specification ([BT80]). Nevertheless it has been found that many-sorted equational algebra is not expressive enough for practical purposes. This has led to several extensions, e.g. conditional equations, subsorts, and overloading. *Order-sorted algebra* is a generalisation of many-sorted algebra which contains all these features (see [GD94] for a survey) and has been implemented in the language OBJ3. To execute our examples we have adopted a more recent development, namely *membership equational logic* (introduced in [Mes98b, BJM97]), which simplifies and generalises order-sorted algebra considerably and is the logic of Maude's equational sublanguage ([CDE+99, CELM96]). However, in the following we will use only its order-sorted fragment.

With some modifications of the list specification we can obtain a specification of finite sets: First, we declare `Id` to be a subsort of a new sort `IdSet` intended to represent sets of identifiers. This allows us to conceive every single identifier as a singleton set.[6] Secondly, we generalise our constructor `idSet`, casting it into a more symmetric form as shown below.

```
sort IdSet .
subsort Id < IdSet .
```

[6] An alternative is to define an explicit coercion function `singleIdSet` from `Id` to `IdSet`.

```
vars s s' s'' : IdSet .

op emptyIdSet : -> IdSet .
op idSet : IdSet IdSet -> IdSet
   [assoc comm] .

eq idSet(s,s) = s .
eq idSet(emptyIdSet,s) = s .

op inIdSet : Id IdSet -> Bool .
eq inIdSet(x,emptyIdSet) = false .
eq inIdSet(x,idSet(x',s')) =
   if x == x' then true else inIdSet(x,s') fi .
```

`emptyIdSet` represents the empty set and `idSet` represents set union. The annotations in square brackets state that `idSet` is an associative and commutative operator. The first two equations express idempotence of `idSet` and the fact that `emptyIdSet` is an identity element of `idSet`. In analogy to `inIdList` the function `inIdSet` checks containment of an element in a set.

For the algebraic semantics the annotations `assoc` and `comm` are tantamount to imposing the following equations:

```
idSet(idSet(s,s'),s'') = idSet(s,idSet(s',s''))
           idSet(s,s') = idSet(s',s)
```

However from an operational point of view we have to avoid nonterminating reduction sequences which occur if the second equation is seen as a reduction rule. A well-studied solution is to employ the technique of *reduction modulo structural equations*, where the objects to be reduced are not single terms but (possibly infinite) equivalence classes of terms. For this purpose the set E of equations is partitioned into two classes E_S and E_R: The class of *structural equations* E_S determines the equivalence classes, and the class of *reduction rules* E_R determines a set of confluent and terminating rules on these equivalence classes. Syntactically, certain kinds of structural equations can be introduced together with their associated operator by annotations in square brackets, whereas reduction rules are introduced by the keyword `eq`. One way to implement the operational semantics is to work with terms again, now interpreted as representations of equivalence classes. To perform reduction modulo structural equations automatically, we have to ensure that a matching algorithm modulo the structural equations E_S is available. This is indeed the case for the equations in our example, stating associativity and commutativity. The use of a matching algorithm for E_S allows us to identify E_S-equivalent terms on a conceptual level. Hence our specification of finite sets is indeed executable.

In the context of Petri nets, finite multisets play a central role. Using the fact that finite multisets can be seen as free commutative monoids we obtain the following algebraic specification of a *finite multiset sort* **idBag** over **Id**:

```
sort IdBag .

subsort Id < IdBag .

op emptyIdBag : -> IdBag .
op idBag : IdBag IdBag -> IdBag
   [assoc comm id: emptyIdBag] .
```

emptyIdBag represents the empty multiset and **idBag** represents multiset union. The annotations in square brackets specify that **idBag** is an associative, commutative operator with **emptyIdBag** as an identity element. Algebraically these annotations are equivalent to the following equations:

```
idBag(idBag(s,s'),s'') = idBag(s,idBag(s',s''))
         idBag(s,s') = idBag(s',s)
     idBag(emptyIdBag,s) = s
     idBag(s,emptyIdBag) = s
```

Operationally, these are structural equations and a matching algorithm modulo these equations is used for the evaluation of multiset terms.

For an associative operator **f** such as **idBag** we use a convenient syntax `f(x1,x2,x3)` abbreviating `f(f(f(x1,x2),x3))` which is also employed for more arguments. For instance, `idBag(a,b,b,a,d)` denotes a multiset of identifiers which has also been written as $2\mathtt{a}+2\mathtt{b}+\mathtt{d}$ in the context of coloured nets.

16.1.2 Rewriting Specifications

An algebraic specification is not equipped with an explicit notion of change. Its initial semantics is an algebra which is a static entity fully determined by the specification. A Petri net on the other hand describes state transitions which may occur in a structured state space. In order to capture state changes in an algebraic framework we employ a generalisation of algebraic specifications called rewrite theories or *rewrite specifications*. The underlying theory of rewriting logic has been developed in [Mes92].

A rewrite specification consists of an algebraic specification (Σ, E) and rewrite rules R. Ground terms (again, we assume that identifications have been performed according to equations in E) are interpreted as states, and rewrite rules are schemes describing possible state transitions. A rewrite rule of the form $u \to v$ states that a subterm matching u, which describes a part of a system's state, can have a local state transition by replacing such a part by the term obtained by applying the given match to v. As an example consider the following specification of a binary nondeterministic choice operator.

```
op choice : Id Id -> Id .

rl [left] : choice(x,y) => x .
rl [right] : choice(x,y) => y .
```

A rewrite rule is introduced by `rl` followed by a label in square brackets. The above specification gives two possible successor states for a state of the form `choice(x,y)`, namely `x` and `y`. For instance, `choice(a,c)` may evolve to `a` or to `c`.

An interesting application of rewrite specifications is to provide a rewriting semantics of place/transition nets using the slogan "Petri nets are monoids" advocated in [MM90].[7]

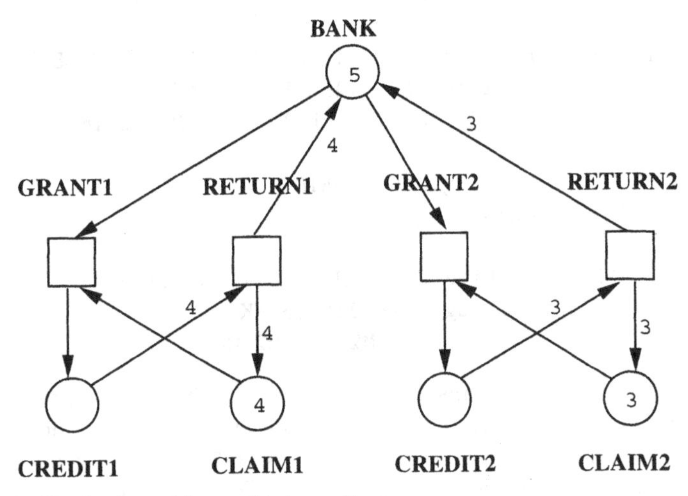

Fig. 16.1. Banker's problem with two clients

The instance of the banker's problem (cf. Section 8.2) depicted as a place/transition net in Figure 16.1 can be immediately translated into the following rewriting specification. We presuppose a specification of the boolean conjunction and, and of an equality predicate ==.[8]

```
sorts Token Marking .
subsort Token < Marking .
```

[7] It is noteworthy that the categorical semantics presented in that work and also the relation between place/transition nets and linear logic explained in [MOM91] inspired the development of rewriting logic.

[8] In Maude, and is declared in a predefined module and the equality predicate == is provided in a built-in way by simplifying both terms in the equality and comparing their identity under confluence and termination assumptions.

```
op emptyMarking : -> Marking .
op marking : Marking Marking -> Marking
   [assoc comm id: emptyMarking] .

var m : Marking .

op BANK : -> Token .
op CREDIT1 : -> Token .
op CREDIT2 : -> Token .
op CLAIM1 : -> Token .
op CLAIM2 : -> Token .

rl [GRANT1]  : marking(BANK,CLAIM1) => CREDIT1 .

rl [RETURN1] : marking(CREDIT1,CREDIT1,CREDIT1,CREDIT1) =>
               marking(BANK,BANK,BANK,BANK,
                       CLAIM1,CLAIM1,CLAIM1,CLAIM1) .

rl [GRANT2]  : marking(BANK,CLAIM2) =>
               CREDIT2 .

rl [RETURN2] : marking(CREDIT2,CREDIT2,CREDIT2) =>
               marking(BANK,BANK,BANK,
                       CLAIM2,CLAIM2,CLAIM2) .
```

Here we applied the translation of place/transition nets into rewriting logic given in [Mes92]:[9] A marking is represented as an element of sort `Marking` which is a multiset sort over the sort `Token` of tokens. `marking` is the corresponding multiset union operator. For each place there is a constant, called *token constructor*, representing a single token residing in that place. For each transition there is a rule, called *transition rule*, stating that its pre-set marking may be replaced by its post-set marking.

The operational semantics extends the operational semantics of the underlying algebraic specification: Again, we will identify only those terms which are forced to be equal by structural equation E_S. A rewriting specification is *executable* iff the underlying algebraic specification is executable[10] and it satisfies a property called *coherence* ([Vir94]) formulated between reduction rules and rewrite rules: Coherence means that if a term u rewrites in a single step to v and u reduces to u', then u' can be further reduced to a term u''

[9] This also corresponds closely to a translation of place/transition nets into linear logic which will be explained in Section 16.4.

[10] In order-sorted specifications and in membership equational specifications executability does not only require confluence and termination but also that equations are sort-decreasing (cf. [CDE+99]). However, for the examples we use in the present section there is no need to elaborate on this.

that rewrites to a term v' which is E-equivalent to v in a single step. Coherence ensures that rewrite steps are not in conflict with reduction steps. This allows us to use a technique which may be called *rewriting modulo a reduction system*: A term is reduced to normal form before a rewrite rule is applied. By coherence, reduction to normal form does not destroy the applicability of rewrite rules. It is easy to verify that the specification above is indeed coherent and therefore executable.

Although rewrite rules describe the dynamics of a concurrent system, equations are usually an important part of a rewrite specification: Equations can be used to specify the functional part of a system description which is preferably executable by viewing these equations as reduction rules. Structural equations on the other hand can be used for expressing symmetries of data or state representations. The equations for associativity and commutativity and the identity laws which have been used to represent a (distributed) marking are of this kind.

As demonstrated by the example above, there is an important difference between the reduction rules in executable algebraic specifications and rewrite rules in executable rewrite specifications: The state transition relation induced by rewrite rules is in general neither terminating nor confluent, although there may be situations where this is the case.[11] In order to execute a rewrite specification the user typically supplies a strategy which successively tries to instantiate selected rewrite rules and initiate rewriting steps.

From a more abstract point of view a strategy is very similar to a theorem-proving tactic which is designed to establish the existence of certain executions. In applications such as net execution the choice of a strategy will be guided by the need to explore the behaviour of the system under certain conditions.

An efficiently executable sublanguage of rewriting logic that supports rewriting modulo all combinations of associativity, commutativity, and (left and right) identity laws has been realised in the Maude language [CELM96]. Although details about specifying strategies are beyond the scope of this introduction, it is interesting to mention that Maude favours the use of a strategy language based on reflection, i.e. on the capability of representing a specification as an object in the language itself. Indeed, its reflective capabilities provide a very flexible way to specify rewrite strategies. In Maude a rewrite strategy operates on the meta-level of the specification to be executed, i.e. it considers the rewrite specification as a first-class object and controls its execution. Indeed, strategies are typically formulated in a user-definable and extensible strategy language which is itself defined in rewriting logic. In this way a strategy becomes again a rewriting specification and we do not have to resort to an external strategy language as is usually the case in tactic-based theorem provers. Details about strategies and the reflection

[11] Only terminating systems with a unique final state can be described by terminating and confluent rewrite rules. So this generalisation is a practical necessity.

mechanism of Maude can be found in [Cla98, CDE+99]. A quite different approach to rewrite strategies has been implemented in the ELAN language [BKK97], another implementation of rewriting logic.

16.1.3 Algebraic Nets

Instead of giving a precise formal definition of algebraic nets, which would also require introducing technicalities of algebraic specification, we try to convey the main idea common to different variants of algebraic nets. Also we restrict our attention to the initial algebra semantics introduced earlier.[12] From the technical point of view our definition is closer to the one in [KR96] and [KV98] generalising [Rei91] by so-called *flexible arcs* which transport variable multisets of tokens.[13]

An *algebraic net specification* consists of an algebraic specification Σ and an inscribed net (P, T, F) such that the following conditions are satisfied:

1. Every place $p \in P$ is inscribed by a *place sort* $C(p)$, meaning that each token residing in this place must be an element of the place sort. We assume that the specification has a finite multiset sort, denoted by $\text{Bag}(C(p))$, over each place sort $C(p)$ such that we can represent the tokens residing in a place as a term of its multiset sort.
2. We assume a set of sorted *variables* which can be used in the entire net. Each transition t uses a subset of these variables, called *local variables* $V(t)$. Moreover each transition is inscribed by a *guard* $G(t)$, i.e. a term of sort `Bool` over the local variables $V(t)$.
3. Each arc $(p,t) \in F$ or $(t,p) \in F$ carries an *arc inscription* which is a multiset term, denoted $W(p,t)$ or $W(t,p)$, specifying the multiset of tokens which is transported by the arc in the definition of transition occurrence below. This term can involve only local variables of t. Moreover, we require that $W(p,t)$ and $W(t,p)$ are terms over the multiset sort $\text{Bag}(C(p))$.
4. There is a distinguished *initial marking* $\mathbf{m_0}$, where a *marking* $\mathbf{m}$ assigns to each place $p \in P$ a ground term of the multiset place sort $\text{Bag}(C(p))$.

A transition element (t, β) consists of a transition t together with a *binding* β of its local variables to ground terms of appropriate sorts. The *preset* $\mathbf{Pre}[\bullet, t](\beta)$ of a transition element (t, β) is the marking which assigns to each place $p \in P$ the multiset obtained by interpreting the inscription $W(p,t)$ of the arc (p,t) under the binding β. Here $W(p,t)$ is defined as the empty multiset if $(p,t) \notin F$. The *postset* $\mathbf{Post}[\bullet, t](\beta)$ is defined correspondingly using inscriptions $W(t,p)$.

---

[12] As explained in [Rei91] algebraic nets can represent a whole class of systems under the loose algebra semantics. We do not exploit this possibility here.

[13] A difference is that we use membership equational logic for the underlying algebraic specification instead of many-sorted algebra.

A transition element (t, β) may *occur* at a marking $\mathbf{m}_1$ *yielding* a marking $\mathbf{m}_2$ iff the guard $G(t)$ is true under the binding β and there is a marking $\mathbf{m}$ such that for all $p \in P$ we have

$$\mathbf{m}_1(p) = \text{Union}_p(\mathbf{m}(p), \mathbf{Pre}[p,t](\beta)) \text{ and}$$
$$\mathbf{m}_2(p) = \text{Union}_p(\mathbf{m}(p), \mathbf{Post}[p,t](\beta)),$$

where $\text{Union}_{C(p)}$ is the multiset union operation defined for the multiset sort $\text{Bag}(C(p))$ in the algebraic specification.

Writing the occurrence rule in this unusual way makes it evident that the occurrence of a transition element replaces its pre-set by its post-set, whereas the remainder of the marking, here denoted by $\mathbf{m}$, is not involved in this process. This is a key observation which will be exploited for the rewriting semantics.

As an example we will model a distributed network algorithm. It is an algorithm which solves the *gossiping problem*: Every agent in the network has a piece of information which should be communicated to every other agent in the network. For simplicity we abstract from the exchange of information. Instead we are only interested in the fact that a *synchronised state* will be reached where every agent knows that it has heard about every other agent. The algorithm we use is a slight modification of the algorithm GOSSIP in [Tel91] which also appeared in [Tel94]. It is a different presentation of the re-synchronisation algorithm proposed by Finn in [Fin79]. The algorithm is appropriate for any non-trivial, directed, strongly connected network assuming that agents (i.e. the nodes of the network) are equipped with unique identities.

For identifiers and sets of identifiers we will use the specifications of `Id` and `IdSet` already defined. The network is represented as a directed graph. We represent such a graph as a finite multiset of identifier pairs avoiding multiple occurrences of pairs (for uniformity reasons we prefer to view a set as a multiset here).

```
sort IdPairBag .

subsort IdPair < IdPairBag .

op emptyIdPairBag : -> IdPairBag .
op idPairBag : IdPairBag IdPairBag -> IdPairBag
   [assoc comm id: emptyIdPairBag] .

vars p p' : IdPair .
vars g g' : IdPairBag .

op inIdPairBag : IdPair IdPairBag -> Bool .
eq inIdPairBag(p,emptyIdPairBag) = false .
eq inIdPairBag(p,IdPairBag(p',g)) =
```

```
      if p == p' then true else inIdPairBag(p,g) fi .

   op network : -> IdPairBag .
   eq network = idPairBag(idPair(a,b),idPair(b,c),idPair(c,d),
                          idPair(d,a),idPair(b,d),idPair(d,b)).
```

The auxiliary function `incoming` defined next is used with `network` as the first argument. `incoming(network,x)` denotes the multiset of headers of messages to be received by `x` from its *input neighbours*, i.e. those agents for which the network has a channel to `x`.

```
   op incoming : IdPairBag Id -> IdPairBag .
   eq incoming(emptyIdPairBag,y') = emptyIdPairBag .
   eq incoming(idPairBag(idPair(x,y),g),y') =
      if y == y'
      then idPairBag(idPair(y,x),incoming(g,y'))
      else incoming(g,y')
      fi .
```

For the net inscriptions we need two further variables and additional sorts `Knowledge` and `Message` with the only constructors being `knowledge` and `message` respectively. `knowledge(x,s)` represents the fact that agent `x` knows the set `s`. `message(x,y,ok,ho)` represents a message directed to `x`, sent out by `y`, with contents `ok` and `ho`.

```
   var ho ho' ok ok' : IdSet .

   sorts Message Knowledge .

   op knowledge : Id IdSet -> Knowledge .
   op message : Id Id IdSet IdSet -> Message .
```

Finally, in addition to `IdBag`, `IdPairBag` we add further obligatory multiset sorts, namely `MessageBag` and `KnowledgeBag`, which allow us to specify markings, in particular the initial one.[14]

```
   sorts MessageBag KnowledgeBag .

   subsort Message < MessageBag .
   subsort Knowledge < KnowledgeBag .

   op emptyMessageBag : -> MessageBag .
```

[14] The concept of multiset is a typical candidate for a parametrised specification (cf. [GWM+92]). However, our example is simple enough to get along without parametrisation.

```
op messageBag : MessageBag MessageBag -> MessageBag
   [assoc comm id: emptyMessageBag] .

op emptyKnowledgeBag : -> KnowledgeBag .
op knowledgeBag : KnowledgeBag KnowledgeBag -> KnowledgeBag
   [assoc comm id: emptyKnowledgeBag] .
```

The inscribed net of the gossip algorithm is shown in Figure 16.2. Names of places and transitions are written in bold font. Sorts associated with places are written in sans serif. Guards are the terms of sort `Bool` specified close to their corresponding transitions. In this example only **SEND** has an explicit guard. If not specified the guard is simply `true`.

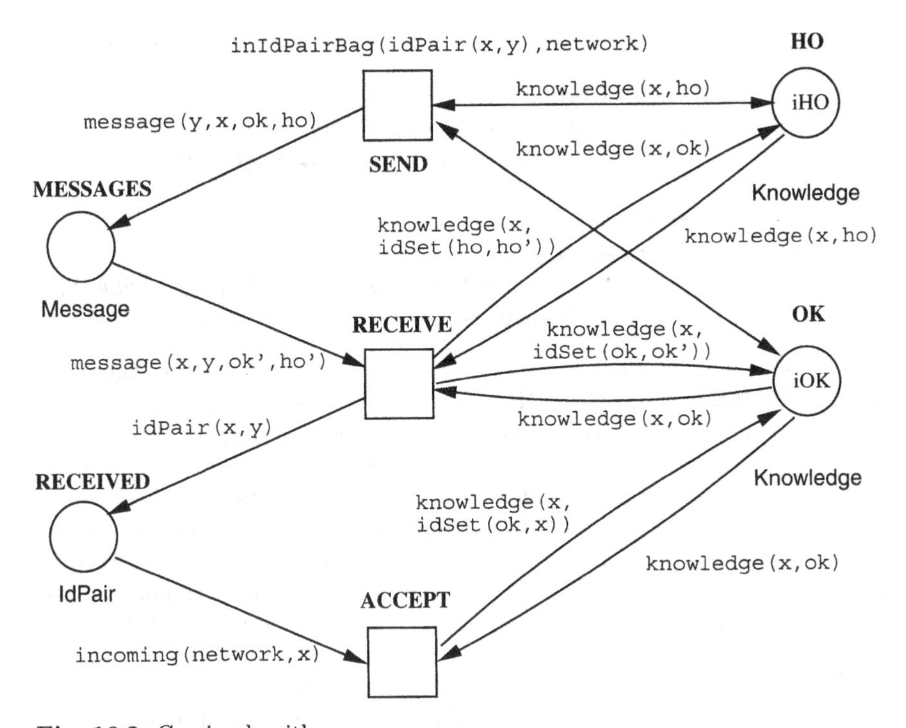

Fig. 16.2. Gossip algorithm

The places **HO** and **OK** contain the local knowledge of each agent about other agents. A token `knowledge(x,ho)` in **HO** represents the fact that agent `x` has directly or indirectly heard of the agents in the set `ho`. A token `knowledge(x,ok)` represents the fact that agent `x` has heard of all neighbours of agents in the set `ok`. For each agent `x` each of these places will contain precisely one token of the form `knowledge(x,...)`. So we can write $\mathbf{HO}_\mathtt{x}$ and

$\mathbf{OK_x}$ to denote the unique sets ho and ok, respectively. In the initial marking an agent has heard only of itself, i.e. $\mathbf{HO_x}$ contains only x and $\mathbf{OK_x}$ is empty.

That the properties explained above are indeed maintained during the execution of the algorithm is not completely obvious. One technique to prove such invariants is using the UNITY style temporal logic as presented in the next section. In the present section, however, the focus is on validation of the algorithm by net execution, not on verification.

For the network `network` we obtain the following initial marking $\mathbf{m_0}$ which is also depicted in Figure 16.2 by multiset terms within places:

$$\mathbf{m_0}(\mathbf{MESSAGES}) = \texttt{emptyMessageBag},$$
$$\mathbf{m_0}(\mathbf{RECEIVED}) = \texttt{emptyIdPairBag},$$
$$\mathbf{m_0}(\mathbf{HO}) = \texttt{iHO},\ \mathbf{m_0}(\mathbf{OK}) = \texttt{iOK}.$$

Here iHO and iOK are defined by extending the specification as follows:

```
op iHO : -> KnowledgeBag .
eq iHO = knowledgeBag(knowledge(a,a),knowledge(b,b),
                       knowledge(c,c),knowledge(d,d)) .

op iOK : -> KnowledgeBag .
eq iOK = knowledgeBag(
           knowledge(a,emptyIdSet),knowledge(b,emptyIdSet),
           knowledge(c,emptyIdSet),knowledge(d,emptyIdSet)).
```

Each single agent x can locally check if it has heard of all other agents by checking if $\mathbf{HO_x}$ is a subset of $\mathbf{OK_x}$. According to the explanation above this inclusion means that if agent x has heard of agent y, then x has also heard of all neighbours of y. Using connectivity of the network and the fact that x has heard of x itself this implies that x has heard of all agents in the network.

Under an appropriate fairness assumption (see below) it can be proved that the algorithm will eventually reach a synchronised state where $\mathbf{HO_x} = \mathbf{OK_x}$ holds for all agents x.[15] To keep the example as simple as possible and to emphasise the main idea, neither local nor global termination detection has been built into the net model. Also for simplicity, there is no mechanism to prevent the agents from sending superfluous messages.

The net contains a place **MESSAGES**, which serves as a message pool, and another place **RECEIVED**, which is locally used by agents to remember which messages have been received. The message pool makes the Petri net structure independent of the actual network used. The only place where the concrete network has to be mentioned is in the equation defining the constant `network`.[16]

[15] Notice that this means not only reachability of a synchronised state but also that such a state will be reached eventually.

[16] This way of modelling a network has been employed in [Rei98] in a number of interesting case studies on modelling and verification using algebraic nets.

Three transitions are sufficient to provide a natural net model of the algorithm. Each transition can be interpreted as an action that an agent can perform. Notice that all transitions are accessing and possibly modifying the local knowledge of an agent via arcs to **HO** and **OK**. We will explain the role of each of these transitions briefly:

1. Sending a message from an agent `x` to an agent `y` is modelled by the transition **SEND**, which deposits a message `message(y,x,ok,ho)` in the message pool. The message carries the current state of the agent's knowledge $\mathbf{OK}_x$ and $\mathbf{HO}_x$ in `ok` and `ho` respectively. The fact that such a message can only be sent over an existing channel from `x` to `y` is reflected by the boolean condition `inIdSet(idPair(x,y),network)`.
2. Receipt of a message by an agent `x` is modelled by transition **RECEIVE**, which takes a message `message(x,y,ok',ho')` from the message pool. At the same time the agent updates its knowledge. This is expressed by the subterms `idSet(ho,ho')` and `idSet(ok,ok')` which state that new knowledge (i.e. `ok'` and `ho'`) is simply added by set union, i.e. without removing previous knowledge. The fact that `x` has received a message from `y` is recorded in the place **RECEIVED**.
3. As soon as an agent `x` has received messages from all its input neighbours the transition **ACCEPT** is enabled. This is expressed by the arc inscribed by `incoming(network,x)`. The occurrence of **ACCEPT** updates the agent's knowledge: The fact that `x` has heard of all its neighbours is added to the local knowledge $\mathbf{OK}_x$ via the arc `knowledge(x,idSet(ok,x))`.

Observe that the transition **SEND** is enabled continuously for every channel directed from `x` to a neighbour `y`. For instance, there is a possible execution sequence sending the same message again and again. To ensure that a synchronised state is reached eventually we assume an appropriate form of *weak group fairness*, i.e. we admit only occurrence sequences satisfying the condition, that if a weakly fair group of transition elements is continuously enabled it will occur eventually. Here a group of transition elements is just a set of transition elements and it is said to be enabled or to occur if one of its elements is enabled or occurs respectively. In our example we specify that for each transition and for fixed bindings for `x`, `y`, `ho'`, and `ok'` the set of transition elements obtained by the different bindings for `ho` and `ok` is a weakly fair group. The idea behind this choice is that `ho` and `ok` are determined by a local variable of the agent, and fairness should abstract from the state of local variables. The importance of weak group fairness in the context of coloured Petri nets has already been observed in [MV91, Mac91]. It will be further discussed and incorporated in a UNITY-style temporal logic in the next section.

The fairness requirement ensures that **SEND** occurs eventually and repeatedly under all possible bindings for x and y. It also ensures that

RECEIVE and **ACCEPT** will eventually occur after becoming enabled, regardless of possibly concurrent changes to the local variables **HO** and **OK**. Observe, however, that it is still possible for an unbounded number of messages to accumulate in place **MESSAGES** and also in place **RECEIVED**.[17]

For our concrete network `network` of agents the net will eventually reach a marking satisfying the stable synchronisation condition $\mathbf{HO}_\mathtt{x} = \mathbf{OK}_\mathtt{x}$ for all agents `x`. It is a marking **m** with

$$\mathbf{m}(\mathbf{HO}) = \mathbf{m}(\mathbf{OK}) = \mathtt{knowledgeBag}(\mathtt{knowledge}(\mathtt{a}, \mathtt{idSet}(\mathtt{a}, \mathtt{b}, \mathtt{c}, \mathtt{d})), \mathtt{knowledge}(\mathtt{b}, \mathtt{idSet}(\mathtt{a}, \mathtt{b}, \mathtt{c}, \mathtt{d})), \mathtt{knowledge}(\mathtt{c}, \mathtt{idSet}(\mathtt{a}, \mathtt{b}, \mathtt{c}, \mathtt{d})), \mathtt{knowledge}(\mathtt{d}, \mathtt{idSet}(\mathtt{a}, \mathtt{b}, \mathtt{c}, \mathtt{d})))$$

16.1.4 Rewriting Semantics

We have already discussed a rewriting semantics for the place/transition net of the banker's problem. Using the distributed algorithm of the previous section we will demonstrate that a rewriting semantics can also be provided for algebraic net specifications. Our semantics is designed to cope with flexible arcs, such as the one used in the distributed algorithm between the place **RECEIVED** and the transition **ACCEPT**.

The rewriting semantics will be given as a rewriting specification extending the specification of the algebraic net. For each place we add a *token constructor*, representing the fact that a particular token resides in that place. A difference with respect to the place/transition net rewriting semantics is that tokens carry information, which is reflected in the fact that token constructors are functions rather than constants. So the token constructor can be seen as a function *tagging* an object with information about the place to which it belongs.

```
sort Token Marking .
subsort Token < Marking .

op emptyMarking : -> Marking .
op marking : Marking Marking -> Marking
   [assoc comm id: emptyMarking] .

op MESSAGES : Message -> Token .
op RECEIVED : IdPair -> Token .
op HO : Knowledge -> Token .
op OK : Knowledge -> Token .
```

[17] In practice a more restricted execution strategy (see below) could avoid this unbounded accumulation of messages, but the correctness of the algorithm does not depend on the assumption of such a strategy.

Each token, i.e. each element of sort Token, can be seen as a singleton marking because of the subsort declaration above. A marking is represented as an element of Marking, a multiset sort over Token. A token MESSAGES(message(x,y,ok,ho)) represents the fact that a message(x,y,ok,ho) resides in the place **MESSAGES**, and a corresponding interpretation holds for the other token constructors RECEIVED, HO, and OK.

Also, the translation of transitions into rewrite rules can be done in full analogy with the place/transition net translation: Each transition is represented as a rewrite rule, also called *transition rule*, replacing its pre-set marking by its post-set marking. If the transition has a guard, that guard becomes a condition of the rewrite rule, and the keyword crl is used to introduce such a conditional rule.

```
crl [SEND]:    marking(HO(knowledge(x,ho)),
                       OK(knowledge(x,ok))) =>
               marking(HO(knowledge(x,ho)),
                       OK(knowledge(x,ok)),
                       MESSAGES(message(y,x,ok,ho)))
               if inIdPairBag(idPair(x,y),network) .

rl [RECEIVE]:  marking(HO(knowledge(x,ho)),
                       OK(knowledge(x,ok)),
                       MESSAGES(message(x,y,ok',ho'))) =>
               marking(HO(knowledge(x,idSet(ho,ho'))),
                       OK(knowledge(x,idSet(ok,ok'))),
                       RECEIVED(idPair(x,y))) .
```

When formulating the transition rule for **ACCEPT** we are faced with the problem of how to translate the flexible arc between **RECEIVED** and **ACCEPT** appropriately. Of course we would like to express that the multiset idPairs(x,in(network,x)) is removed from place **RECEIVED**, but this presupposes an interpretation of places, as containers of objects, which is different from our current one, where tokens are tagged objects "mixed up in a soup together with other tokens".

The mathematically most elegant solution which has also the advantage of preserving the concurrent nature of Petri nets is the linear extension of RECEIVED to multisets. For this purpose we *generalise* the token constructor RECEIVED which has already been declared above and we add two equations expressing the linearity of RECEIVED, which will also be called *place linearity equations*:

```
op RECEIVED : IdPairBag -> Marking .
eq RECEIVED(emptyIdPairBag) = emptyMarking .
eq RECEIVED(idPairBag(b,b')) =
     marking(RECEIVED(b),RECEIVED(b')) .
```

Using these equations the transition rule for **ACCEPT** can be formulated naturally and in complete analogy with the rules above:

```
rl [ACCEPT]: marking(OK(knowledge(x,ok)),
                     RECEIVED(incoming(network,x))) =>
             OK(knowledge(x,idSet(ok,x))) .
```

According to our initial explanation a place can be seen as the tag of a token indicating the place in which the token resides. This is what we call the *tagged-token view*. The place linearity equations suggest a complementary view which is encountered more often in the context of Petri nets: A place is simply a container of objects. We will call this the *place-as-container view*. The place linearity equations express our intention to consider both views as equivalent. Place linearity equations are needed as soon we exploit the place-as-container view: This is not only the case if we have places with flexible arcs but also if we want to specify the initial contents of places by finite multisets. So for the rewriting semantics we extend all token constructors to multisets and we add place linearity equations such as the one given for **RECEIVED** for each of them.

In order to obtain an executable rewrite specification we first have to make the algebraic specification executable. There is a possibility of non-termination due to the second equation for `RECEIVED`: `idPairBag(b,b')` matches the term `emptyIdPairBag` with `b` and `b'` bound to `emptyIdPairBag`. This is caused by the fact that `emptyIdPairBag` is an identity element of the constructor `idPairBag`. An easy solution is to add a condition restricting the variables `b` and `b'` to non-empty multisets. After this modification the equations, viewed as reduction rules, are terminating and confluent. Hence, our algebraic specification is executable. Unfortunately, this is not yet true for the rewrite specification, because it is not coherent: In a state where the rule `ACCEPT` is applicable it may happen that the place linearity equations for `RECEIVED` are applied as reduction rules so that the rewrite rule `ACCEPT` loses its applicability. However, we can carry out a trivial semantics-preserving translation yielding a rule

```
var m : Marking .

crl [ACCEPT]: marking(OK(knowledge(x,ok)),m) =>
              OK(knowledge(x,idSet(ok,x)))
              if m == RECEIVED(incoming(network,x)) .
```

Now the rewrite specification is coherent and can be executed using a suitable strategy. Remember that we have assumed weak fairness in order to guarantee that the algorithm reaches a synchronised state. So every strategy which produces execution sequences satisfying weak fairness will be useful for validating the algorithm. For instance, the simple strategy which cycles through all transitions and their bindings for `x` and `y` again and again will finally lead to a synchronised state. However, with this strategy many superfluous messages are produced by the transition **SEND**. As a variation of this strategy we could give **ACCEPT** priority over **RECEIVE**, and **RECEIVE** over **SEND**. In this way messages tend to be produced on demand. The strategy language that we used to specify such strategies is itself defined in rewriting logic and has constructs similar to languages used to formulate tactics in theorem provers.

Last but not least, it is worth pointing out that there is a more systematic way to obtain an executable rewriting semantics of an algebraic net specification that is already executable in a suitable sense: Since the place linearity equations express different ways of looking at the same marking of an algebraic net, it would be reasonable to assign them to the class of structural equations expressing symmetries of the state representation instead of using them as reduction rules. To execute such a specification directly we would need a matching algorithm which implements matching modulo associativity, commutativity, identity, and linearity.[18] However, generalising an idea suggested in [Mes98a], matching modulo associativity, commutativity, identity, and linearity can be simulated by matching modulo associativity, commutativity, and identity using a simple semantics-preserving translation that uses the place linearity equations as reduction rules (just as we have used them in our example) and makes essential use of subsorts and overloading. Unfortunately, further details about this translation are beyond the scope of this introduction.

16.1.5 Final Remarks

The distance between Petri nets and algebraic specification and rewriting techniques is smaller than it may appear at first sight. It has been demonstrated that rewriting techniques can be used for the controlled execution of algebraic Petri nets in a very natural way. More work is necessary to develop specialised and general purpose strategies for efficient execution of Petri nets. We think that a framework based on algebraic specification and rewriting provides a convenient basis for this.

Despite all these possibilities one should keep in mind that net execution may be useful for locating flaws in the specification or design phase but it cannot replace the verification of a system with infinitely many possible executions. In some cases however, e.g. if the state space is finite, general strategies performing certain analysis tasks, such as reachability and deadlock analysis or, more generally, temporal logic model checking, can be used for automatic verification of Petri nets. Again a rewriting-based language provides a convenient environment for developing such analysis strategies.

---

[18] A particular advantage of the Maude engine is that its modular design favours the extension of its set of matching algorithms.

To verify general systems we propose another application of rewriting semantics: It is a good candidate for providing a symbolic representation of algebraic Petri nets and also of their processes in theorem-proving environments which usually rely on or support rewriting techniques. From our point of view, Petri net execution is a very special case of automated theorem proving. An execution engine generates a constructive proof that a particular finite execution is possible in a given system. Hence in the long term we propose that an integrated tool for verification and execution of Petri nets should be embedded into a general purpose tactic-based theorem proving environment which allows reasoning about Petri nets in an expressive language with partial automation. From this point of view net execution strategies appear as very special case of theorem proving tactics. As a logical framework, i.e. as a language that is convenient and expressive enough to specify deductive systems of different kinds, rewriting logic is an interesting candidate to be used as a basis for building theorem proving tools and formal analysis environments as it has been shown in [Cla98, Dur99, SM99].

In general, rewriting semantics allows us to exploit existing languages and techniques in order to build tools for the execution and verification of algebraic net models. It facilitates integration of Petri nets with other paradigms which can also be given a rewriting semantics, e.g. object-oriented concurrent programming ([Mes93]). But it can also be useful for theoretical purposes: For instance, rewriting semantics allows us to specialise the elegant categorical semantics of rewriting logic, yielding a concurrency semantics for algebraic Petri nets and extensions that can be expressed within the framework of rewriting logic.

16.2 Assertional Reasoning

Formal and informal reasoning about concurrent and nondeterministic systems is often carried out directly at the level of executions. This type of operational reasoning is prone to errors, since a concurrent system typically admits a huge variety of possible executions. In informal proofs sometimes certain representative executions are chosen, but again it is difficult to make sure that all essential behaviours are covered. Assertional reasoning, on the other hand, deals with logical statements that hold for all executions of programs. Typically, this method is based on a logic, and a corresponding deductive system is used for the reasoning process.

A logic well suited for this purpose is UNITY logic. It is a temporal logic which has already been employed in Section 10.1 on state-oriented modelling to illustrate a design methodology for distributed algorithms and systems using P/T nets. The objective of the present section is to adapt UNITY logic to support verification of coloured nets. The idea of combining UNITY logic and coloured nets is not new and has already been investigated in [MV91] and [Mac91]. Also the approach to verifying algebraic nets presented in [WWV+97] continuing earlier work on elementary net systems ([DGK+92, Kin95b]) is inspired by UNITY logic, although a partial-order semantics is chosen in these references instead of the interleaving semantics that we will consider below. A recent approach which uses separate operators for both of these semantics is presented in [Rei98].

UNITY logic was introduced as part of the UNITY methodology proposed by Chandy and Misra in [CM88] for reasoning about parallel programs written in the UNITY programming language. UNITY logic and derivatives have since been employed in other contexts (see, for instance, [Sha93] in addition to the references above) and even today the UNITY approach still attracts many researchers. One reason is certainly the elegance and simplicity that is achieved in UNITY while retaining a level of expressiveness that is sufficiently high for many purposes. The maturity of UNITY logic for several applications has been demonstrated by numerous case studies (see e.g. [CM88], [Sta93], and [CK97b]). Furthermore, its simplicity has led to the existence of rigorous formalisations of the logic within general proof assistants which support the interactive development of proofs ([APP94, HC96, Ste98b, Pau99]). Continuing the work started in [Ste98b] the author has recently developed an embedding of a generalisation of UNITY, based on arbitrary labelled transition systems[19] and incorporating a notion of group fairness, into the *calculus of inductive constructions* ([BBC+99]), a rich type theory with dependent types that contains higher-order logic. All the proof rules have been not only mathematically but also formally verified in this general setting using the COQ proof assistant ([BBC+99]). The result is the core of a verified temporal logic library that can be instantiated for specific system models such as coloured Petri nets and developed further into a concrete model for verification. In the present section we do not emphasise the formal aspects, but merely present the use of this approach in the context of coloured Petri nets by means of informal set theory.

In contrast to [CM88] and [Ste98b], a more semantic presentation is chosen here by defining the main operators, namely INVARIANT and LEADS TO, directly in operational terms. Instead of introducing a closed language of temporal logic, the temporal operators above are just abbreviations involving the reachability relation and occurrence sequences of a given net system. All other temporal operators are introduced as auxiliary notions intended to guide the activity of proving assertions involving INVARIANT and LEADS TO.

[19] A *labelled transition system* consists of a set of states S, a set of events E, and a transition relation $\rightarrow \subseteq S \times E \times S$. The labelled transition relation between markings of P/T nets and coloured nets gives rise to a labelled transition system in the obvious way.

Proof rules are formulated as theorems. Since we do not propose a closed system of proof rules completeness is not an issue here.[20]

After fixing the central notions of state predicate and state function, the presentation starts with elementary safety and liveness operators CO and TRANSIENT and defines the operators UNLESS and ENSURES in terms of these. Finally, the more complex liveness operator LEADS TO is introduced on top of ENSURES.

The operators ENSURES and TRANSIENT were introduced in [Cha94] and proposed as a basis for a new approach to UNITY in [Mis95]. We follow this new approach, but a more important aspect and major generalisation over the original UNITY logic (possible because of our choice of labelled transition systems as system models) is that we adopt a fairness notion that takes enabledness into account, in contrast to the unconditional fairness of UNITY. Furthermore, we consider fair groups of events instead of single events giving rise to the notion of group fairness which is more adequate for coloured nets. Together with the definitions of temporal operators, simple but useful proof rules are given. Most of these rules are known from [CM88], but it is remarkable that they are still valid in the more general setting of labelled transition systems with group fairness. Finally, a simple example is given to demonstrate the basic ideas of assertional reasoning about coloured nets.

Although a UNITY style temporal logic can be developed in the general framework of labelled transition systems, we assume throughout this section on assertional reasoning that we are interested in the dynamic behaviour of a coloured net $\mathcal{N} = (P, T, \mathbf{Pre}, \mathbf{Post}, C, cd)$ with initial marking $\mathbf{m_0}$. The states of the system are the markings, and the transition relation between markings is defined by the usual firing rule in coloured nets. As an execution semantics of such a net $\mathcal{N}$ we will first consider the set of *free occurrence sequences* $\mathit{FreeOcc}(\mathcal{N}) = \{\mathbf{m}_1\ (t_1, \beta_1)\ \mathbf{m}_2\ (t_2, \beta_2)\ \ldots \mid \mathbf{m}_i \xrightarrow{t_i, \beta_i} \mathbf{m}_{i+1}\}$, which are finite or infinite alternating sequences of markings $\mathbf{m}_i$ and transition elements (t_i, β_i) with $\beta_i \in cd(t_i)$ (β_i is an occurrence mode of t_i). Notice that we do not require them to start in the initial marking $\mathbf{m_0}$. Instead the initial condition will be part of the temporal logic specification of a system. Later we will restrict the semantics to admissible occurrence sequences $\mathit{AdmOcc}(\mathcal{N}, \mathcal{WF}) \subseteq \mathit{FreeOcc}(\mathcal{N})$ satisfying a particular weak fairness requirement $\mathcal{WF}$. According to the usual semantics of coloured nets each transition $t \in T$ will be conceived as the set $\{(t, \beta) \mid \beta \in cd(t)\}$ of transition elements obtained by unfolding the transition. The coloured net obtained from $\mathcal{N}$ by unfolding all transitions will be denoted by $\tilde{\mathcal{N}}$. $\tilde{T}$ denotes the set of transitions of $\tilde{\mathcal{N}}$, i.e. the set of all transition elements of $\mathcal{N}$. The coloured net $\mathcal{N}$ can be seen as a structured version of $\tilde{\mathcal{N}}$. Of course, nets can be struc-

[20] Actually, there are (relative) completeness results for the original UNITY logic ([Pac92a, Pac92b, Kna94, Pae95, HC96]) which can be seen as an important special case of the approach presented here.

tured according to different criteria. Structuring is a means to make complex net models comprehensible without changing their behaviour.

16.2.1 State Predicates and Functions

State predicates and state functions are introduced in order to argue about predicates and functions evolving in time.

Definition 16.2.1. *A state function f is a function from the set of markings into some domain. A state predicate p is a particular state function into the boolean domain* $\{\text{false}, \text{true}\}$. $p(\mathbf{m})$ *abbreviates* $p(\mathbf{m}) = \text{true}$. *In this case we also say that* $\mathbf{m}$ *satisfies p. A family of state predicates P over I associates a state predicate P_i with every $i \in I$. Subsequently, p, q, r and P, Q, R are reserved for state predicates and families of state predicates respectively. Logical connectives, quantifiers, and operators are naturally lifted to the level of state predicates and state functions. Additionally the everywhere operator $[p]$ is available on state predicates: $[p]$ holds iff $p(\mathbf{m})$ is true for all markings* $\mathbf{m}$.

As an example let $p1$ and $p2$ be places of colour $\mathbb{N}$. Each place p can be conceived as a state function by setting $p(\mathbf{m}) = \mathbf{m}[p]$. Also $p1 + p2$ can be seen as a state function defined by $(p1 + p2)(\mathbf{m}) = p1(\mathbf{m}) + p2(\mathbf{m})$, where $+$ denotes multiset union. Moreover, $p1 = \emptyset \wedge |p1 + p2| = 1$ is a state predicate with $(p1 = \emptyset \wedge |p1 + p2| = 1)(\mathbf{m}) \Leftrightarrow p1(\mathbf{m}) = \emptyset \wedge |p1(\mathbf{m}) + p2(\mathbf{m})| = 1$.

16.2.2 Basic Assertions

The definitions of temporal operators will be based on two kinds of basic assertions, inspired by Hoare triples. Originally, Hoare triples of the form $\{p\}\ s\ \{q\}$ with state predicates p and q were used as assertions about a program (statement) s ([Hoa69]). In the partial correctness interpretation, such a Hoare triple states that if the execution of s is initiated in a state satisfying p then either the execution of s does not terminate or it terminates in a state that satisfies q. In the total correctness interpretation such a Hoare triple also states that s terminates in every state satisfying p. Subsequently, we will adopt a similar notation for assertions about transitions of coloured nets.

For a transition t, an occurrence mode $\beta \in cd(t)$, state predicates p and q, and a subset $\tilde{T}' \subseteq \tilde{T}$ of net transitions we define the following kinds of basic state predicates and assertions:

Definition 16.2.2 (Basic state predicates).

1. *Enabled*$(\tilde{T}')(\mathbf{m})$ *holds iff there is a transition element $(t, \beta) \in \tilde{T}'$ such that t is enabled under β at marking* $\mathbf{m}$.

2. $\mathcal{SI}(\tilde{T}')(\mathbf{m})$[21] *holds iff* $\mathbf{m} \in \mathrm{RS}(\tilde{\mathcal{N}}', \mathbf{m_0})$, *where* $\mathrm{RS}(\tilde{\mathcal{N}}', \mathbf{m_0})$ *denotes the set of markings reachable in* $\tilde{\mathcal{N}}'$ *from* $\mathbf{m_0}$.

Definition 16.2.3 (Basic assertions).

1. $\{p\}\ \tilde{T}'\ \{q\}$ *holds iff for each* $\mathbf{m}$ *satisfying* p *and for each* $(t, \beta) \in \tilde{T}'$ *the following is true:* $\mathbf{m} \xrightarrow{t,\beta} \mathbf{m}'$ *implies* $\mathbf{m}'$ *satisfies* q.
2. $\langle p \rangle\ \tilde{T}'\ \langle q \rangle$ *holds iff for each* $\mathbf{m}$ *satisfying* p *there is a* $(t, \beta) \in \tilde{T}'$ *enabled at* $\mathbf{m}$ *and for each* $(t, \beta) \in \tilde{T}'$ *the following is true:* $\mathbf{m} \xrightarrow{t,\beta} \mathbf{m}'$ *implies* $\mathbf{m}'$ *satisfies* q.

Instead of a singleton set of transition elements $\{(t, \beta)\}$ we often write just (t, β). Observe that if (t, β) is not enabled at p the assertion $\{p\}\ (t, \beta)\ \{q\}$ holds trivially, but the assertion $\langle p \rangle\ (t, \beta)\ \langle q \rangle$ is not satisfied. We separated these two kinds of assertions, since temporal safety assertions can be defined using triples of the first kind, whereas definitions of liveness assertions will also involve triples of the second kind, since enabledness of certain transitions is obviously needed to make progress.

Notice that, as in Hoare's original definition, all conceivable states are considered in the preceding definition and not only reachable ones. The justification is that the set of reachable markings is usually complicated and unknown. More importantly, the set of reachable markings is not robust under composition of nets (cf. [Kin95a]). So for compositional reasoning quantification over all markings is more appropriate, since it depends on fewer assumptions about the system behaviour.

It is easy to see that the following proof rules are sound and complete for basic assertions $\{p\}\ \tilde{T}'\ \{q\}$ and $\langle p \rangle\ \tilde{T}'\ \langle q \rangle$ respectively.

Theorem 16.2.4 (Proof rules for basic assertions).

$$\frac{\forall \mathbf{m} \,.\, \forall (t, \beta) \in \tilde{T}' \,.\, p(\mathbf{m} + \mathbf{Pre}[\bullet, t](\beta)) \Rightarrow q(\mathbf{m} + \mathbf{Post}[\bullet, t](\beta))}{\{p\}\ \tilde{T}'\ \{q\}}$$

$$\frac{[p \Rightarrow \mathit{Enabled}(\tilde{T}')] \qquad \{p\}\ \tilde{T}'\ \{q\}}{\langle p \rangle\ \tilde{T}'\ \langle q \rangle}$$

16.2.3 Safety Assertions

Once basic state predicates and basic assertions have been defined, UNITY logic can be built on top of these concepts. As a first step, safety assertions are defined. Later, liveness assertions are added. The presentation of UNITY

[21] Strictly speaking, it is not necessary to take this state predicate as a basic notion, since it can be defined as the strongest invariant using the notion of invariant to be introduced later.

logic chosen here is based on elementary operators CO and TRANSIENT instead of UNLESS and ENSURES respectively. This is only a minor change compared to the original presentation of UNITY in [CM88], but it facilitates the understanding of the operators.

Another noteworthy point is that all operators are relative to a subset of transition elements $\tilde{T}' \subseteq \tilde{T}$ specifying a particular view of the net. By $\tilde{\mathcal{N}}'$ we denote the net obtained from $\tilde{\mathcal{N}}$ by removing all transitions not contained in $\tilde{T}'$. Temporarily adopting a certain view focuses on the changes performed only by transition elements contained in that view. Firing of other transitions is excluded under this view. Views can be a useful structuring mechanism for managing large models and their verification. At this point one might think of $\tilde{T}'$ as the set of all transitions $\tilde{T}$ representing a full view of the net. Later, proper subsets of transitions will be used to indicate that certain properties hold for partial views of the net. Since components can be conceived as special views, this notation will also be suited for compositional reasoning.

The assertion p CO q, which is introduced next, captures the fact that if p holds at some point then q holds at the next state. p UNLESS q means: p holds at least until q holds. This includes the possibility that q will never hold after p and p will remain true forever. In other words, if we want to leave a state satisfying p but not q then the only possibility is to move directly into a state still satisfying p or into a state satisfying q. The assertion p STABLE requires that once p becomes true it remains true forever. If p is inductively invariant (p IND. INVARIANT) it is additionally required that p holds initially.[22] In contrast to these assertions, which are defined using quantification over *all* markings, p INVARIANT means that p holds for all *reachable* markings.

Definition 16.2.5 (Safety assertions).

1. p CO q IN $\tilde{T}'$ *iff* $\{p\}\ \tilde{T}'\ \{q\}$.
2. p UNLESS q IN $\tilde{T}'$ *iff* $p \wedge \neg q$ CO $p \vee q$ IN $\tilde{T}'$.
3. p STABLE IN $\tilde{T}'$ *iff* p CO p IN $\tilde{T}'$.
4. p IND. INVARIANT IN $\tilde{T}'$ *iff* $p(\mathbf{m_0})$ *and* p STABLE IN $\tilde{T}'$.
5. p INVARIANT IN $\tilde{T}'$ *iff for all* $\mathbf{m} \in \mathrm{RS}(\tilde{\mathcal{N}}', \mathbf{m_0})$ *we have* $p(\mathbf{m})$.

Notice that p INVARIANT IN $\tilde{T}'$ can also be written as $[\mathcal{SI}(\tilde{T}') \Rightarrow p]$ and hence it is definable in terms of our basic concepts.[23]

[22] The separation between invariant and inductive properties has already appeared in [Kel76] where P/T nets with inscriptions are used as models for parallel programs and, in contrast to the original UNITY approach, we will maintain this distinction.

[23] Indeed the state predicate $\mathcal{SI}(\tilde{T}')$ can be used to express assertions relativised to reachable states, which are important for avoiding unnecessarily strong high-level specifications. As explained in [Ste98b] there is no need to introduce new temporal operators to express relativised assertions (cf. [San91]). However, we will not make use of relativised assertions in this introductory presentation.

The operators CO and UNLESS as well as the properties of being stable and an inductive invariant can be directly verified for a given coloured net using the proof rules for basic assertions and the definitions above. Some obvious proof rules for CO are given in the following.

Theorem 16.2.6 (Proof rules for CO).

false CO p

p CO true

$$\frac{[p \Rightarrow p'] \quad p' \text{ CO } q}{p \text{ CO } q} \qquad \text{(strengthening)}$$

$$\frac{p \text{ CO } q \quad [q \Rightarrow q']}{p \text{ CO } q'} \qquad \text{(weakening)}$$

$$\frac{p \text{ CO } q \quad p' \text{ CO } q'}{p \wedge p' \text{ CO } q \wedge q'} \qquad \text{(conjunction)}$$

$$\frac{p \text{ CO } q \quad p' \text{ CO } q'}{p \vee p' \text{ CO } q \vee q'} \qquad \text{(disjunction)}$$

$$\frac{\forall i \in I : P_i \text{ CO } Q_i}{(\exists i \in I : P_i) \text{ CO } (\exists i \in I : Q_i)} \qquad \text{(general disjunction)}$$

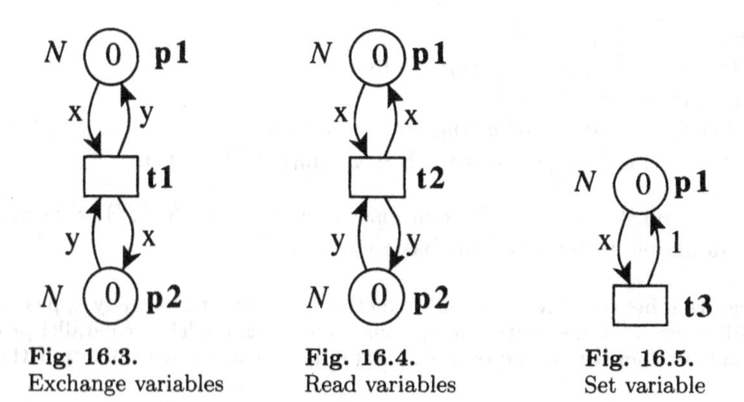

Fig. 16.3. Exchange variables

Fig. 16.4. Read variables

Fig. 16.5. Set variable

To illustrate the distinction between invariants and inductive invariants consider the coloured net in Figure 16.3. The only colour domain we use here is the set of natural numbers N. Define state predicates $p :\Leftrightarrow (p1 = \{0\}) \wedge (p2 = \{0\})$ and $q :\Leftrightarrow (p1 = \{0\})$. Both of them are invariants, since they hold in every reachable marking. p is even an inductive invariant, since it holds initially and $\{p\}\ t1\ \{p\}$ is true. However, q is not an inductive invariant, as $\{q\}\ t1\ \{q\}$ does not hold. Notice that $t1$ abbreviates $\{(t1, \beta) \mid \beta \in cd(t1)\}$, i.e. the unfolding of the transition $t1$ in the given coloured net.

In contrast to inductive invariants, invariants are defined using the set of reachable markings and it is not immediately clear how they may be verified. Fortunately, all of them can be proved using inductive invariants as the following simple theorem states; it appeared in [Kin95a] in the more general setting of transition systems.

All of the following assertions will refer to the same arbitrary but fixed view $T' \subseteq T$ without explicitly mentioning it.

Theorem 16.2.7 (Invariance theorems).

1. *If* p' INVARIANT *and* $p' \Rightarrow p$ *then* p INVARIANT.
2. *If* p IND. INVARIANT *then* p INVARIANT.
3. p INVARIANT *iff there is some* p' *such that* $p' \Rightarrow p$ *and* p' IND. INVARIANT.

Proof. The first statement is obvious. For the second, assume an inductive invariant is given. To prove that it is also an invariant, that is, it holds for all reachable markings, induction over the length of occurrence sequences starting from the initial marking succeeds. To prove the third statement, consider $\mathcal{SI}(\tilde{T}')$, the state predicate that holds exactly for all reachable markings of $\mathcal{N}'$. Clearly, $\mathcal{SI}(\tilde{T}')$ is the strongest invariant in the sense that every other invariant is implied by $\mathcal{SI}(\tilde{T}')$. Moreover, $\mathcal{SI}(\tilde{T}')$ is an inductive invariant since $\mathcal{SI}(\tilde{T}')(\mathbf{m_0})$ holds and $\mathcal{SI}(\tilde{T}')$ STABLE, since every successor of a reachable marking is again reachable. Now, for every invariant p there is an inductive invariant p', namely the strongest invariant $\mathcal{SI}(\tilde{T}')$, with $p' \Rightarrow p$. The converse of 3 is proved by 1 and 2.

The first part of the theorem states that invariants are closed under implication, a property which is in general not true for inductive invariants,[24] as the previous example demonstrates. The second statement of the theorem is that inductive invariants are invariants whereas the converse does not hold, which is again shown by the example. The last part of the theorem can be interpreted as a completeness result for proving invariants: Every invariant can be derived using an appropriate inductive invariant.

[24] Assuming the UNITY substitution axiom, which allows us to substitute UNITY invariants by true and vice versa, in our context would force the class of inductive invariants to be closed under implication giving rise to a contradiction. So we do not assume this axiom, maintaining a clear separation between the class of invariants which is closed under implication and the class of inductive invariants which is in general not closed under implication. Indeed, it is known that the substitution axiom is unsound ([San91]) if non-reachable states are also considered in the definition of the operators as is done here in order to allow for compositional reasoning.

Next we turn our attention to proof rules involving the operator UNLESS. The weakening, conjunction, and disjunction rules are particularly useful. The simple conjunction and disjunction rules are easy consequences.

Theorem 16.2.8 (Proof rules for UNLESS).

false UNLESS p

p UNLESS p (reflexivity)

p UNLESS $\neg p$ (antireflexivity)

$$\frac{p \text{ UNLESS } q \quad [q \Rightarrow r]}{p \text{ UNLESS } r} \qquad \text{(weakening)}$$

$$\frac{p \text{ UNLESS } q \quad p' \text{ UNLESS } q'}{(p \wedge p') \text{ UNLESS } (p \wedge q') \vee (p' \wedge q) \vee (q \wedge q')} \qquad \text{(conjunction)}$$

$$\frac{p \text{ UNLESS } q \quad p' \text{ UNLESS } q'}{(p \vee p') \text{ UNLESS } (\neg p \wedge q') \vee (\neg p' \wedge q) \vee (q \wedge q')} \qquad \text{(disjunction)}$$

$$\frac{p \text{ UNLESS } q \quad p' \text{ UNLESS } q'}{(p \wedge p') \text{ UNLESS } (q \vee q')} \qquad \text{(simple conjunction)}$$

$$\frac{p \text{ UNLESS } q \quad p' \text{ UNLESS } q'}{(p \vee p') \text{ UNLESS } (q \vee q')} \qquad \text{(simple disjunction)}$$

Further useful rules are given for the conjunction of stable and inductive invariant predicates and their conjunction with CO and UNLESS assertions.

Theorem 16.2.9 (Proof rules for STABLE and INVARIANT).

$$\frac{p \text{ STABLE} \quad q \text{ STABLE}}{p \wedge q \text{ STABLE}} \qquad \text{(STABLE conjunction)}$$

$$\frac{p \text{ IND. INVARIANT} \quad q \text{ IND. INVARIANT}}{p \wedge q \text{ IND. INVARIANT}} \qquad \text{(INVARIANT conjunction)}$$

$$\frac{p \text{ CO } q \quad r \text{ STABLE}}{p \wedge r \text{ CO } q \wedge r} \qquad \text{(CO/STABLE conjunction)}$$

$$\frac{p \text{ UNLESS } q \quad r \text{ STABLE}}{p \wedge r \text{ UNLESS } q \wedge r} \qquad \text{(UNLESS/STABLE conjunction)}$$

16.2.4 Liveness Assertions

Whereas safety assertions express that something must not happen, liveness assertions state that something will happen eventually (cf. [Lam77]).[25] We will use a special class of liveness assertions stating that if some condition holds at some point then some (other) condition will eventually be reached. In contrast to safety assertions, (non-trivial) liveness assertions can only be proved if certain fairness assumptions about the occurrence of transitions are made. Without such assumptions there is no need for a transition to occur even if it is enabled, and the system is not forced to make any progress. Unfortunately, the UNITY notion of unconditional fairness[26] is not appropriate for coloured nets. Instead we will adopt a generalisation of weak fairness, explained below.

Let us first recapitulate the requirement of weak fairness (also called productivity) for transitions of P/T nets introduced in Section 10.1. A transition t which is designated as *weakly fair* behaves in the following way: If t is permanently enabled then t will occur eventually.[27] Weak fairness can be easily lifted to a transition t of a coloured net as follows: If t is permanently enabled under some occurrence mode β then t will occur eventually under this occurrence mode β. This definition of weak fairness in coloured nets is motivated by the view of a coloured net transition as an abbreviation for a set of transition elements obtained by unfolding. Unfortunately, weak fairness in coloured nets is not sufficient to guarantee progress in the case where a transition is permanently enabled but in different occurrence modes. A typical situation is that a transition needs to access the value of a variable (modelled as a place with a single coloured token) which is permanently modified by other transitions. Under the assumption of weak fairness, the transition may not succeed in accessing the variable. This immediately justifies the use of a more flexible notion of fairness that has already been informally introduced in the previous section and will be made more precise below.

As in the linear time temporal logic presented in [Fra86] we employ a more general notion of weak fairness, which is called *weak group fairness* in what follows. Instead of a set of weakly fair transition elements a set of weakly fair groups (i.e. sets) of transition elements can be specified.[28] We want to ensure that if a weakly fair group is permanently enabled then some transition of the group will eventually occur. This will be made more precise below.

---

[25] This should not be confused with liveness in Petri nets. A formal definition of safety and liveness properties is given in [AS85].

[26] Every UNITY statement is selected infinitely often regardless of its guard.

[27] Weak fairness is treated in [Fra86] and [MP92]. In the latter reference it is called justice.

[28] Weak group fairness can also capture the original notion of fairness in UNITY when UNITY programs are modelled as coloured Petri nets. This is explained in [MV91] and [Mac91] where group fairness is called group productivity.

At the beginning of this section we introduced $FreeOcc(\mathcal{N})$ as the set of free occurrence sequences which do not need to satisfy any fairness assumptions. Now let $\mathcal{WF}$ be a *weak fairness specification*, i.e. a set of subsets of $\tilde{T}$. Each $\tilde{T}' \in \mathcal{WF}$ is called a *weakly fair group*. A set $\tilde{T}' \subseteq \tilde{T}$ is said to be enabled at $\mathbf{m}$ iff some $(t, \beta) \in \tilde{T}'$ is enabled at $\mathbf{m}$. A free occurrence sequence $\mathbf{m}_1\ (t_1, \beta_1)\ \mathbf{m}_2\ (t_2, \beta_2)\ \ldots$ is said to be *weakly fair with respect to a transition group* $\tilde{T}' \subseteq \tilde{T}$ iff for each index i such that $\tilde{T}'$ is enabled at all $\mathbf{m}_k$ with indices $k \geq i$, there is an index j with $j \geq i$ such that $(t_j, \beta_j) \in \tilde{T}'$. In particular, this excludes the possibility that $\tilde{T}'$ is enabled in the last marking of a finite occurrence sequence. For the remainder of this section we will consider only the subset of *admissible occurrence sequences* $AdmOcc(\mathcal{N}, \mathcal{WF}) \subseteq FreeOcc(\mathcal{N})$ which satisfy the weak fairness specification $\mathcal{WF}$, i.e. they are weakly fair with respect to all groups in $\mathcal{WF}$. For technical reasons, mainly to obtain an elegant definition of TRANSIENT, we assume that $\mathcal{WF}$ always contains the empty group $\emptyset$.

The reason that we do not require all conceivable (groups of) transition elements to be weakly fair is that for certain applications, e.g. for modelling external requests, transitions should not be forced to occur even if they are permanently enabled.[29]

The assertion p TRANSIENT means that p cannot hold permanently, since it will eventually be falsified due to a *single* weakly fair group of transitions. On top of TRANSIENT the assertion p ENSURES q is defined, requiring that p holds until q holds and that p and $\neg q$ cannot hold permanently because of a single weakly fair group of transitions which falsifies the condition p and $\neg q$. This implies that p holds until q holds and q will eventually become true.

Definition 16.2.10 (TRANSIENT and ENSURES assertions).

1. p TRANSIENT IN $\tilde{T}'$ *iff there is a weakly fair group* $\tilde{T}'' \in \mathcal{WF}$ *such that* $\tilde{T}'' \subseteq \tilde{T}'$ *and* $\langle p \rangle\ \tilde{T}''\ \langle \neg p \rangle$.
2. p ENSURES q IN $\tilde{T}'$ *iff* p UNLESS q IN $\tilde{T}'$ *and* $p \wedge \neg q$ TRANSIENT IN $\tilde{T}'$.

Theorem 16.2.11 (Proof rules for TRANSIENT).

false TRANSIENT

$$\frac{p \text{ TRANSIENT}}{p \wedge q \text{ TRANSIENT}} \qquad \text{(strengthening)}$$

Many of the following rules are similar to those for UNLESS. Notice, however, that a rule closely corresponding to the (simple) disjunction rule for UNLESS does not hold for ENSURES.

[29] The idea that different fairness requirements should be distinguished on the basis of individual transitions is also present in [MP92], [DGK+92], and [Rei98].

Theorem 16.2.12 (Proof rules for ENSURES).

false ENSURES p

p ENSURES p (reflexivity)

$$\frac{[p \Rightarrow q]}{p \text{ ENSURES } q} \quad \text{(implication)}$$

$$\frac{p \text{ ENSURES } q \quad [q \Rightarrow r]}{p \text{ ENSURES } r} \quad \text{(weakening)}$$

$$\frac{p \text{ ENSURES false}}{[\neg p]} \quad \text{(impossibility)}$$

$$\frac{p \text{ UNLESS } q \quad p' \text{ ENSURES } q'}{p \wedge p' \text{ ENSURES } (p \wedge q') \vee (p' \wedge q) \vee (q \wedge q')} \quad \text{(UNLESS/ENSURES conj.)}$$

$$\frac{p \text{ UNLESS } q \quad p' \text{ ENSURES } q'}{(p \wedge p') \text{ ENSURES } (q \vee q')} \quad \text{(UNLESS/ENSURES simple conjunction)}$$

$$\frac{p \text{ ENSURES } q}{p \vee r \text{ ENSURES } q \vee r} \quad \text{(simple disjunction)}$$

$$\frac{p \text{ ENSURES } q \quad r \text{ STABLE}}{p \wedge r \text{ ENSURES } q \wedge r} \quad \text{(ENSURES/STABLE conjunction)}$$

Subsequently, LEADS TO is introduced as the main operator for expressing liveness assertions. The meaning of p LEADS TO q is that if p holds at some point q will eventually hold. In contrast to the assertion p ENSURES q, p does not necessarily hold until q holds but may become false in the meantime.

Definition 16.2.13 (LEADS TO assertions).
p LEADS TO q IN $\tilde{T}'$ *iff for every occurrence sequence* $\mathbf{m}_1$ (t_1, β_1) $\mathbf{m}_2$ (t_2, β_2) $\ldots \in$ *AdmOcc*$(\tilde{\mathcal{N}}', \mathcal{WF})$ *the following is true: If* $p(\mathbf{m}_i)$ *holds for some index* i *then there is an index* $j \geq i$ *such that* $q(\mathbf{m}_j)$.

In the following list many rules similar to those of UNLESS and ENSURES can be found. Note that in contrast to UNLESS and ENSURES, weakening *and* strengthening rules are available. In addition to these and the first three rules the cancellation and induction rules are also frequently used.[30]

[30] In [CM88] LEADS TO is defined inductively as the smallest operator satisfying the first three rules below. For the purpose of this introductory presentation we have chosen a more intuitive operational definition of LEADS TO similar to [Lam77]

Theorem 16.2.14 (Proof rules for LEADS TO).

$$\frac{p \text{ ENSURES } q}{p \text{ LEADS TO } q} \qquad \text{(basis)}$$

$$\frac{p \text{ LEADS TO } q \quad q \text{ LEADS TO } r}{p \text{ LEADS TO } r} \qquad \text{(transitivity)}$$

$$\frac{\forall i \in I : P_i \text{ LEADS TO } q}{(\exists i \in I : P_i) \text{ LEADS TO } q} \qquad \text{(disjunction)}$$

$$\text{false LEADS TO } p$$

$$p \text{ LEADS TO } p \qquad \text{(reflexivity)}$$

$$\frac{[p \Rightarrow q]}{p \text{ LEADS TO } q} \qquad \text{(implication)}$$

$$\frac{p \text{ LEADS TO false}}{[\neg p]} \qquad \text{(impossibility)}$$

$$\frac{[p \Rightarrow p'] \quad p' \text{ LEADS TO } q}{p \text{ LEADS TO } q} \qquad \text{(strengthening)}$$

$$\frac{p \text{ LEADS TO } q \quad [q \Rightarrow q']}{p \text{ LEADS TO } q'} \qquad \text{(weakening)}$$

$$\frac{\forall i \in I : P_i \text{ LEADS TO } Q_i}{(\exists i \in I : P_i) \text{ LEADS TO } (\exists i \in I : Q_i)} \qquad \text{(general disjunction)}$$

$$\frac{p \text{ LEADS TO } q \vee q' \quad q' \text{ LEADS TO } r}{p \text{ LEADS TO } q \vee r} \qquad \text{(cancellation)}$$

$$\frac{p \text{ LEADS TO } q \quad r \text{ STABLE}}{p \wedge r \text{ LEADS TO } q \wedge r} \qquad \text{(LEADS TO/STABLE conjunction)}$$

$$\frac{p \text{ LEADS TO } q \quad r \text{ UNLESS } r'}{p \wedge r \text{ LEADS TO } (q \wedge r) \vee r'} \qquad \text{(progress-safety-progress)}$$

but with the essential difference that execution sequences not starting with the initial marking/state are also considered.

For a finite set I:

$$\frac{\forall i \in I : P_i \text{ LEADS TO } Q_i \lor r \quad \forall i \in I : Q_i \text{ UNLESS } r}{(\forall i \in I : P_i) \text{ LEADS TO } (\forall i \in I : Q_i) \lor r} \qquad \text{(completion)}$$

For a state function f with domain X and a well-founded strict partial order $(\prec)$ *on X:*

$$\frac{\forall x \in X : (p \land f = x) \text{ LEADS TO } (p \land f \prec x) \lor q}{p \text{ LEADS TO } q} \qquad \text{(induction)}$$

The rule of induction can be applied to every well-founded partial order (i.e. a partial order where all descending chains are finite) such as the usual order on natural numbers or the inclusion order on finite multisets. The induction rule can be motivated as follows: If p LEADS TO $p \lor q$ then there is a potential repetition of states satisfying p. If f is an appropriate variant function which is known to decrease for each round (the period between two states satisfying p) then this repetition cannot continue forever (due to well-foundedness). Hence, the condition q has to be reached eventually.

16.2.5 Elementary Compositionality

Given two nets with disjoint sets of transitions, their composition is simply the union of these nets. This definition allows shared places which are intended to be used for (asynchronous) communication among the components. Since the components are (essentially) determined by their sets of transitions, the components can also be conceived as particular views on the composed net. Instead of introducing a formal composition on nets it is more convenient to assume that the composed net is already given. Then different views of this single net are considered by specifying subsets of transitions. The notion of composition employed here is related to place fusion (see Section 10.2) since different components given by disjoint sets of transitions can exchange information only via shared places.[31]

Recall that CO, UNLESS, and ENSURES have been defined using all markings and not only the reachable ones. This is indeed necessary to make them compositional as the following example suggests: Clearly, the reachability graphs (containing markings reachable from the initial one) of the nets in Figures 16.3 and 16.4 are identical. However, it should also be clear that under composition of each one with the net in Figure 16.5 the behaviour is completely different. The reason for this phenomenon is simply that an essential part of the system structure remains hidden if the reachability relation and the temporal operators are restricted to the set of reachable markings.[32]

[31] Another notion of composition, called superposition in UNITY (also contained in [CM88]), is related to transition fusion with certain restrictions. This is, however, beyond the scope of this presentation.

[32] An in-depth analysis of this issue is contained in [Kin95a].

The following theorems will be useful for combining properties derived for component views to obtain properties for the joint view. They can be easily extended to unions of more than two views. Here we assume $\tilde{T}', \tilde{T}'' \subseteq \tilde{T}$.

Theorem 16.2.15 (Union theorems).

$$\frac{p \text{ CO } q \text{ IN } \tilde{T}' \quad p \text{ CO } q \text{ IN } \tilde{T}''}{p \text{ CO } q \text{ IN } \tilde{T}' \cup \tilde{T}''} \qquad \text{(CO union)}$$

$$\frac{p \text{ UNLESS } q \text{ IN } \tilde{T}' \quad p \text{ UNLESS } q \text{ IN } \tilde{T}''}{p \text{ UNLESS } q \text{ IN } \tilde{T}' \cup \tilde{T}''} \qquad \text{(UNLESS union)}$$

$$\frac{p \text{ TRANSIENT IN } \tilde{T}'}{p \text{ TRANSIENT IN } \tilde{T}' \cup \tilde{T}''} \qquad \text{(TRANSIENT union)}$$

$$\frac{p \text{ ENSURES } q \text{ IN } \tilde{T}' \quad p \text{ UNLESS } q \text{ IN } \tilde{T}''}{p \text{ ENSURES } q \text{ IN } \tilde{T}' \cup \tilde{T}''} \qquad \text{(ENSURES/UNLESS union)}$$

$$\frac{p \text{ STABLE IN } \tilde{T}' \quad p \text{ STABLE IN } \tilde{T}''}{p \text{ STABLE IN } \tilde{T}' \cup \tilde{T}''} \qquad \text{(STABLE union)}$$

$$\frac{p \text{ IND. INVARIANT IN } \tilde{T}' \quad p \text{ STABLE IN } \tilde{T}''}{p \text{ IND. INVARIANT IN } \tilde{T}' \cup \tilde{T}''} \qquad \text{(INVARIANT/STABLE union)}$$

$$\frac{p \text{ UNLESS } q \text{ IN } \tilde{T}' \quad p \text{ STABLE IN } \tilde{T}''}{p \text{ UNLESS } q \text{ IN } \tilde{T}' \cup \tilde{T}''} \qquad \text{(UNLESS/STABLE union)}$$

$$\frac{p \text{ ENSURES } q \text{ IN } \tilde{T}' \quad p \text{ STABLE IN } \tilde{T}''}{p \text{ ENSURES } q \text{ IN } \tilde{T}' \cup \tilde{T}''} \qquad \text{(ENSURES/STABLE union)}$$

Unfortunately, a rule similar to those above does not hold for LEADS TO. Compositional reasoning about liveness can be carried out at the level of ENSURES and, finally, LEADS TO properties can be derived on the basis of the composed net. An interesting possibility offered by UNITY logic that is not addressed in this introductory presentation is the possibility conditional assertions which were introduced in [CM88] to enhance compositional reasoning. A conditional assertion is of the form $A \Rightarrow B$ where A and B are conjunctions of safety or liveness assertions. The meaning is that the assertions in B hold under the assumption of A. Conditional assertions are useful for stating specifications in a rely/guarantee style where the correct behaviour B of a component relies on certain guarantees A maintained by the environment. For particular approaches to compositionality where A contains only safety properties we refer the reader to [HC96] and [CK97b].

16.2.6 A Simple Example

The coloured net $\mathcal{N}$ depicted in Figure 16.6 computes the square of a natural number by successive addition of odd numbers.[33] The only colour set used here is the set of natural numbers denoted by N. Moreover, all three transitions are assumed to be weakly fair, i.e. each transition element forms a singleton weakly fair group; formally we assume $\{(t, \beta)\} \in \mathcal{WF}$ for each $(t, \beta) \in \tilde{T}$. Without this requirement the net is not obliged to show any progress.[34] Initially, the input value iv is stored in place pm and the initial marking is given by

$$\mathbf{m_0}(p1) = \{0\} \wedge \mathbf{m_0}(p2) = \emptyset \wedge \mathbf{m_0}(p3) = \emptyset \wedge$$
$$\mathbf{m_0}(pm) = \{iv\} \wedge \mathbf{m_0}(pi) = \{1\} \wedge \mathbf{m_0}(podd) = \{1\}.$$

After termination the result is expected in place $p3$.

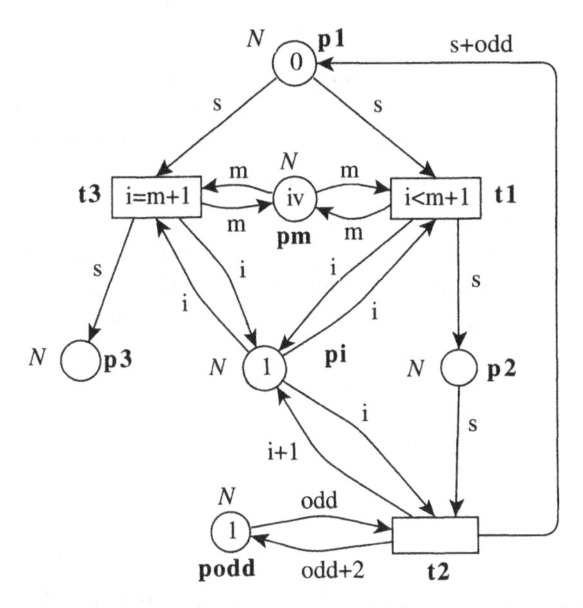

Fig. 16.6. A coloured net calculating the square of a natural number

[33] It is taken from [Val93a] where it was used as an example to demonstrate verification of non-linear place invariants in coloured nets.

[34] In contrast to the gossip algorithm, discussed in the previous section, which presupposed group fairness, the simple example we are going to verify does not have any variables that are accessed concurrently. Hence weak fairness is sufficient here. Further examples, in particular a verification of the main properties of the gossip algorithm using the approach we present here can be found in [Ste98a].

The following proof will be carried out for an arbitrary but fixed input value iv. According to the informal specification given above, the initial condition $IC(\mathbf{m})$ is defined by $\mathbf{m} = \mathbf{m_0}$ and the termination condition TC is $p3 \neq \emptyset$. Once this condition holds the result of the computation, i.e. the square of the input value, should be contained in $p3$.

In order to convey an elementary idea of compositional reasoning the net can be partitioned into two components corresponding to the views $t1 \cup t3$ and $t2$. The first component can be seen as a controller which checks for termination and uses the second component as a server to perform addition operations if necessary. As an abbreviation those assertions which are not explicitly restricted to a particular view of the net always refer to the full view $t1 \cup t2 \cup t3$.

The verification task can be split into two parts, the verification of partial and of total correctness. Partial correctness is a safety assertion: Once the termination condition is reached the result is correct. This can easily be expressed as an invariant assertion:

$$TC \Rightarrow \Sigma p3 = iv^2 \text{ INVARIANT} \tag{16.1}$$

Total correctness is stronger than partial correctness since it requires an additional liveness assertion: Eventually the termination condition is reached. This can be naturally expressed using the LEADS TO operator:

$$IC \text{ LEADS TO } TC \tag{16.2}$$

Both of these assertions will be proved below.

The following state predicates $I_1, \ldots, I_6$ are inductive invariants as can be easily verified by applying Definition 16.2.5: Each of them holds for the initial marking and is preserved by every occurrence of a transition.

$$\begin{array}{lll} I_1 & :\Leftrightarrow & |p1| + |p2| + |p3| = 1 \\ I_2 & :\Leftrightarrow & |pi| = 1 \\ I_3 & :\Leftrightarrow & |pm| = 1 \\ I_4 & :\Leftrightarrow & |podd| = 1 \\ I_5 & :\Leftrightarrow & \Sigma pm = iv \\ I_6 & :\Leftrightarrow & \Sigma podd = 2\Sigma pi - 1 \end{array}$$

For a multiset A of integers, $|A|$ is the cardinality and ΣA denotes the sum of all elements of A taking multiple occurrences into account. In the present example we deal only with singleton multisets $A = \{x\}$ so ΣA is used only to obtain the contents of A, that is x.

Instead of proving invariants directly for the full net they can be established in a compositional way. For instance, after verifying that I_1 is an inductive invariant in $t1 \cup t3$ and I_1 STABLE in $t2$ by inspection, the INVARIANT/STABLE union rule can be used to obtain that I_1 is an inductive invariant in $t1 \cup t2 \cup t3$. Of course, also a different partition into views may be used e.g. the finer one consisting of three views $t1$, $t2$, and $t3$.

We now prove that $\Sigma p1 + \Sigma p2 + \Sigma p3 = (\Sigma pi - 1)^2$ is an invariant. Clearly, it is not an inductive invariant because its preservation under firing of t_2 depends on the contents of *podd*. The following state predicate I_7 strengthens this equation by the restriction on *podd* given by I_4 and I_6. Now, it can be verified by inspection of the net that I_7 is indeed an inductive invariant.

$$I_7 :\Leftrightarrow I_4 \wedge I_6 \wedge \Sigma p1 + \Sigma p2 + \Sigma p3 = (\Sigma pi - 1)^2$$

We will carry out this proof in detail: Since I_4 and I_6 are already known to be inductive invariants they are preserved by every occurrence of a transition. So we can concentrate on $\Sigma p1 + \Sigma p2 + \Sigma p3 = (\Sigma pi - 1)^2$. I_7 is clearly an inductive invariant in $t1 \cup t3$ since it holds initially and the occurrence of $t1$ and $t3$ preserves I_4, I_6, and $\Sigma p1 + \Sigma p2 + \Sigma p3$ as well as Σpi. To conclude that I_7 is an inductive invariant in $t1 \cup t2 \cup t3$ it is sufficient to prove that I_7 STABLE in $t2$ and to apply the INVARIANT/STABLE union rule. Stability of I_7 can be checked as follows: Assume $\mathbf{m}'$ is the successor marking of $\mathbf{m}$ under $t2$ and $I_4(\mathbf{m})$, $I_6(\mathbf{m})$, and $(\Sigma p1 + \Sigma p2 + \Sigma p3)(\mathbf{m}) = ((\Sigma pi - 1)^2)(\mathbf{m})$ holds. We have $(\Sigma p1 + \Sigma p2 + \Sigma p3)(\mathbf{m}') = (\Sigma p1 + \Sigma p2 + \Sigma p3)(\mathbf{m}) + (\Sigma podd)(\mathbf{m})$ and $(\Sigma pi)(\mathbf{m}') = (\Sigma pi)(\mathbf{m}) + 1$. The goal is $(\Sigma p1 + \Sigma p2 + \Sigma p3)(\mathbf{m}') = ((\Sigma pi - 1)^2)(\mathbf{m}')$. But this follows from the previous two equations and $(\Sigma p1 + \Sigma p2 + \Sigma p3)(\mathbf{m}) + (\Sigma podd)(\mathbf{m}) = ((\Sigma pi)^2)(\mathbf{m})$, which is proved next. Exploiting the assumptions $I_6(\mathbf{m})$ and $(\Sigma p1 + \Sigma p2 + \Sigma p3)(\mathbf{m}) = ((\Sigma pi - 1)^2)(\mathbf{m})$ the last equation is implied by $((\Sigma pi - 1)^2)(\mathbf{m}) + (2\Sigma pi - 1)(\mathbf{m}) = ((\Sigma pi)^2)(\mathbf{m})$ which is obviously true.

Two further inductive invariants I_8 and I_9 capture the obvious relation between *pi* and *pm* immediately after firing $t1$ and $t3$. They can also be obtained directly from the net.

$$I_8 :\Leftrightarrow I_1 \wedge I_2 \wedge I_3 \wedge (p2 \neq \emptyset \Rightarrow \Sigma pi < \Sigma pm + 1)$$
$$I_9 :\Leftrightarrow I_1 \wedge I_2 \wedge I_3 \wedge (p3 \neq \emptyset \Rightarrow \Sigma pi = \Sigma pm + 1)$$

To prove that $\Sigma pi \leq \Sigma pm + 1$ is an invariant we again have to resort to a stronger inductive invariant, which can be obtained by conjunction with I_8.

$$I_{10} :\Leftrightarrow I_8 \wedge \Sigma pi \leq \Sigma pm + 1$$

At this stage partial correctness can already be verified. For this purpose we have to establish assertion 16.1 but again this cannot be done directly since $TC \Rightarrow \Sigma p3 = iv^2$ is not an inductive invariant. However, by the inductive conjunction rule, the following I_{11} is an inductive invariant.

$$I_{11} :\Leftrightarrow I_5 \wedge I_7 \wedge I_9$$

Now the desired invariant 16.1 establishing partial correctness is implied by I_{11}.

In the remainder of this example our objective is to prove the assertion 16.2 which is additionally necessary for total correctness. For this purpose it is convenient to collect $I_1, \ldots, I_{10}$ into a single inductive invariant using the conjunction rule for inductive invariants.

$$I :\Leftrightarrow I_1 \wedge \ldots \wedge I_{10}$$
$$I \text{ IND. INVARIANT}$$

It can also be seen from the net that the following ENSURES properties hold. Remember that to establish an ENSURES assertion, the corresponding UNLESS assertion is verified first and then a transition has to be found which realises the progress (the TRANSIENT part of ENSURES). On the other hand, the following assertions are quite intuitive on the basis of the given net. In 16.3 the progress is witnessed by $t1$ and $t3$ whereas in 16.5 it is witnessed by $t2$.
For all m, i:

$$\begin{array}{c} p1 \neq \emptyset \wedge p2 = \emptyset \wedge p3 = \emptyset \wedge pm = \{m\} \wedge pi = \{i\} \wedge i \leq m+1 \\ \text{ENSURES} \\ (p1 = \emptyset \wedge p2 = \emptyset \wedge p3 \neq \emptyset \wedge pm = \{m\} \wedge pi = \{i\} \wedge i = m+1) \vee \\ (p1 = \emptyset \wedge p2 \neq \emptyset \wedge p3 = \emptyset \wedge pm = \{m\} \wedge pi = \{i\} \wedge i < m+1) \\ \text{IN } t1 \cup t3. \end{array} \tag{16.3}$$

For all m, i:

$$\begin{array}{c} p1 \neq \emptyset \wedge p2 = \emptyset \wedge p3 = \emptyset \wedge pm = \{m\} \wedge pi = \{i\} \wedge i \leq m+1 \\ \text{STABLE IN } t2. \end{array} \tag{16.4}$$

For all i:

$$\begin{array}{c} p1 = \emptyset \wedge p2 \neq \emptyset \wedge pi = \{i\} \wedge podd \neq \emptyset \\ \text{ENSURES} \\ p1 \neq \emptyset \wedge p2 = \emptyset \wedge pi = \{i+1\} \wedge podd \neq \emptyset \\ \text{IN } t2 \end{array} \tag{16.5}$$

For all i:

$$\begin{array}{c} p1 = \emptyset \wedge p2 \neq \emptyset \wedge pi = \{i\} \wedge podd \neq \emptyset \\ \text{STABLE IN } t1 \cup t3 \end{array} \tag{16.6}$$

Applying the ENSURES union rule to 16.3 and 16.4 and also to 16.5 and 16.6 we obtain two ENSURES assertions for the joint view $t1 \cup t2 \cup t3$.
For all m, i:

$$\begin{array}{c} p1 \neq \emptyset \wedge p2 = \emptyset \wedge p3 = \emptyset \wedge pm = \{m\} \wedge pi = \{i\} \wedge i \leq m+1 \\ \text{ENSURES} \\ (p1 = \emptyset \wedge p2 = \emptyset \wedge p3 \neq \emptyset \wedge pm = \{m\} \wedge pi = \{i\} \wedge i = m+1) \vee \\ (p1 = \emptyset \wedge p2 \neq \emptyset \wedge p3 = \emptyset \wedge pm = \{m\} \wedge pi = \{i\} \wedge i < m+1) \\ \text{IN } t1 \cup t2 \cup t3. \end{array} \tag{16.7}$$

For all i:

$$p1 = \emptyset \wedge p2 \neq \emptyset \wedge pi = \{i\} \wedge podd \neq \emptyset \\ \text{ENSURES} \\ p1 \neq \emptyset \wedge p2 = \emptyset \wedge pi = \{i+1\} \wedge podd \neq \emptyset \\ \text{IN } t1 \cup t2 \cup t3 \tag{16.8}$$

Since we have I STABLE, 16.7 and 16.8 can be modified using the STABLE conjunction rule for ENSURES and the results can be converted into LEADS TO assertions by means of the basis rule for LEADS TO.

For all m, i:

$$I \wedge p1 \neq \emptyset \wedge p2 = \emptyset \wedge p3 = \emptyset \wedge pm = \{m\} \wedge pi = \{i\} \wedge i \leq m+1 \\ \text{LEADS TO} \\ (I \wedge p1 = \emptyset \wedge p2 = \emptyset \wedge p3 \neq \emptyset \wedge pm = \{m\} \wedge pi = \{i\} \wedge i = m+1) \vee \\ (I \wedge p1 = \emptyset \wedge p2 \neq \emptyset \wedge p3 = \emptyset \wedge pm = \{m\} \wedge pi = \{i\} \wedge i < m+1) \tag{16.9}$$

For all i:

$$I \wedge p1 = \emptyset \wedge p2 \neq \emptyset \wedge pi = \{i\} \wedge podd \neq \emptyset \\ \text{LEADS TO} \\ I \wedge p1 \neq \emptyset \wedge p2 = \emptyset \wedge pi = \{i+1\} \wedge podd \neq \emptyset \tag{16.10}$$

The condition $podd \neq \emptyset$ can be removed since it is already implied by I.

For all i:

$$I \wedge p1 = \emptyset \wedge p2 \neq \emptyset \wedge pi = \{i\} \\ \text{LEADS TO} \\ I \wedge p1 \neq \emptyset \wedge p2 = \emptyset \wedge pi = \{i+1\} \tag{16.11}$$

Each LEADS TO assertion in the following family can be derived from an assertion of the previous family by the strengthening rule for LEADS TO:

For all m, i:

$$I \wedge p1 = \emptyset \wedge p2 \neq \emptyset \wedge p3 = \emptyset \wedge pm = \{m\} \wedge pi = \{i\} \wedge i < m+1 \\ \text{LEADS TO} \\ I \wedge p1 \neq \emptyset \wedge p2 = \emptyset \wedge pi = \{i+1\} \tag{16.12}$$

Now the LEADS TO cancellation rule can be applied to 16.9 and the previous assertion.

For all m, i:

$$I \wedge p1 \neq \emptyset \wedge p2 = \emptyset \wedge p3 = \emptyset \wedge pm = \{m\} \wedge pi = \{i\} \wedge i \leq m+1 \\ \text{LEADS TO} \\ (I \wedge p1 = \emptyset \wedge p2 = \emptyset \wedge p3 \neq \emptyset \wedge pm = \{m\} \wedge pi = \{i\} \wedge i = m+1) \vee \\ (I \wedge p1 \neq \emptyset \wedge p2 = \emptyset \wedge pi = \{i+1\}) \tag{16.13}$$

The following family of LEADS TO assertions follows from the previous one. Actually, the case where $\Sigma pm + 1 - \Sigma pi = v$ holds is the only non-trivial one. Notice that $\Sigma pm = iv$ by I_5. Moreover, we have $\Sigma pi = i$ on the left-hand

side and $\Sigma pi = i + 1$ on the right-hand side. The reason why $(<_{\mathbb{N}})$, i.e. the restriction of $(<)$ to natural numbers, can be used instead of $(<)$ is that $I_1 0$ which is part of I guarantees that $\Sigma pm + 1 - \Sigma pi \in \mathbb{N}$.
For all m, i, v:

$$\begin{array}{c} I \wedge p1 \neq \emptyset \wedge p2 = \emptyset \wedge p3 = \emptyset \wedge pm = \{m\} \wedge pi = \{i\} \wedge \\ i \leq m + 1 \wedge \Sigma pm + 1 - \Sigma pi = v \\ \text{LEADS TO} \\ (I \wedge p1 = \emptyset \wedge p2 = \emptyset \wedge p3 \neq \emptyset \wedge pm = \{m\} \wedge pi = \{i\} \wedge i = \\ m + 1) \vee \\ (I \wedge p1 \neq \emptyset \wedge p2 = \emptyset \wedge pi = \{i + 1\} \wedge \Sigma pm + 1 - \Sigma pi <_{\mathbb{N}} v) \end{array} \tag{16.14}$$

Applying the weakening rule for LEADS TO with $pm = \{m\} \wedge pi = \{i\} \wedge i = m + 1 \Rightarrow \Sigma pi = \Sigma pm + 1$ and the fact that $I \wedge p1 \neq \emptyset$ implies $p3 = \emptyset$ yields:
For all m, i, v:

$$\begin{array}{c} I \wedge p1 \neq \emptyset \wedge p2 = \emptyset \wedge p3 = \emptyset \wedge pm = \{m\} \wedge pi = \{i\} \wedge \\ i \leq m + 1 \wedge \Sigma pm + 1 - \Sigma pi = v \\ \text{LEADS TO} \\ (I \wedge p1 = \emptyset \wedge p2 = \emptyset \wedge p3 \neq \emptyset \wedge \Sigma pi = \Sigma pm + 1) \vee \\ (I \wedge p1 \neq \emptyset \wedge p2 = \emptyset \wedge p3 = \emptyset \wedge \Sigma pm + 1 - \Sigma pi <_{\mathbb{N}} v) \end{array} \tag{16.15}$$

Now observe that the condition $i \leq m + 1$ can be removed since it is implied by I (more precisely I_{10}) in presence of $pm = \{m\}$ and $pi = \{i\}$. Furthermore, m and i occur only on the left-hand side of LEADS TO. So the general disjunction rule for LEADS TO can be applied. The only assumption about pi and pm that remains is that both are singletons. This condition, however, can be removed since it is already implied by I.
For all v:

$$\begin{array}{c} I \wedge p1 \neq \emptyset \wedge p2 = \emptyset \wedge p3 = \emptyset \wedge \Sigma pm + 1 - \Sigma pi = v \\ \text{LEADS TO} \\ (I \wedge p1 = \emptyset \wedge p2 = \emptyset \wedge p3 \neq \emptyset \wedge \Sigma pi = \Sigma pm + 1) \vee \\ (I \wedge p1 \neq \emptyset \wedge p2 = \emptyset \wedge p3 = \emptyset \wedge \Sigma pm + 1 - \Sigma pi <_{\mathbb{N}} v) \end{array} \tag{16.16}$$

Clearly, $(<_{\mathbb{N}})$ is a well-founded order such that the induction rule for LEADS TO applies. We obtain:

$$\begin{array}{c} I \wedge p1 \neq \emptyset \wedge p2 = \emptyset \wedge p3 = \emptyset \\ \text{LEADS TO} \\ I \wedge p1 = \emptyset \wedge p2 = \emptyset \wedge p3 \neq \emptyset \wedge \Sigma pi = \Sigma pm + 1 \end{array} \tag{16.17}$$

The left-hand side of this assertion is implied by the initial condition IC. Hence, LEADS TO strengthening and weakening yields

$$IC \text{ LEADS TO } p3 \neq \emptyset \tag{16.18}$$

which is just assertion 16.2, completing the proof of total correctness.

16.2.7 Extensions of the Logic

The state-based part of UNITY logic (i.e. the fragment without Hoare-style assertions, TRANSIENT, ENSURES, and the concept of views) can be conceived as a fragment of linear time temporal logic with next-time operator $\circ$. For instance, it can be embedded into the linear time temporal logic LTL (cf. Section 14.1.1, [Pnu81], [MP92]) by interpreting:

$$
\begin{array}{rll}
p \text{ CO } q & \text{as} & \Box(p \Rightarrow \circ\, q), \\
p \text{ UNLESS } q & \text{as} & \Box(p \wedge \neg q \Rightarrow \circ\,(p \vee q)), \\
p \text{ STABLE} & \text{as} & \Box(p \Rightarrow \circ\, p), \\
p \text{ IND. INVARIANT} & \text{as} & (IC \Rightarrow p) \wedge \Box(p \Rightarrow \circ\, p), \\
p \text{ INVARIANT} & \text{as} & IC \Rightarrow \Box p, \\
p \text{ LEADS TO } q & \text{as} & \Box(p \Rightarrow \Diamond q).
\end{array}
$$

Here IC is the initial condition which is satisfied only for the initial marking of the coloured net. Recall that only p INVARIANT and p IND. INVARIANT depend on the initial condition. Of course, to be consistent with the execution semantics which we use in this section we have to consider the set of free admissible occurrence sequences as the LTL semantics. If we consider only occurrence sequences starting at the initial marking it is impossible to express UNITY logic assertions, since they can express properties at non-reachable states (at least in the reinterpretation of UNITY that we have adopted here).

From this more general point of view, the main UNITY operators are just an abbreviation for frequently used formulae of linear time temporal logic. If further operators turn out to be necessary they can easily be defined. This supports our view that the logic presented here should not be seen as a closed system but should be extended if necessary for certain applications.

Generalising the original presentation of UNITY logic, we incorporated the standard notion of weak fairness and more generally weak fairness for groups of transition elements by modifying the original definition of TRANSIENT. That weak and strong fairness are central issues for the specification and verification of concurrent systems has already been observed in [LPS81] and should also be clear from the example in Section 10.1, where UNITY logic was used as a specification language for state-oriented modelling with P/T nets. In that chapter weak and strong fairness for certain transitions was assumed to ensure important liveness assertions needed to meet the specification.[35]

Of course, it is not possible to express the occurrence of transitions directly in a (primary) state-based temporal logics such as UNITY or MP logic. However, depending on the expressibility of the logic it may be possible to approximate by temporal formulae the effect of weak and strong fairness on the state. For instance, UNITY logic can express weak fairness assertions of

[35] Strong fairness is treated in [Fra86] and also in [MP92] where it is called compassion.

the form $(\Diamond\Box p) \Rightarrow (\Box\Diamond q)$ for atomic formulae p and q since this MP logic formula is equivalent to $\Box\Diamond(\neg p \vee q)$ which is in turn equivalent to the UNITY assertion true LEADS TO $(\neg p \vee q)$ (see [BT95]). On the other hand, UNITY logic is not expressive enough to capture strong fairness assertions of the form $(\Box\Diamond p) \Rightarrow (\Box\Diamond q)$.

Instead of viewing fairness requirements as assumptions expressed in the temporal logic itself, the UNITY approach encapsulates them in the definition of the temporal operators TRANSIENT and ENSURES. Just as IND. INVARIANT is an auxiliary operator for deriving INVARIANT assertions, we regard TRANSIENT and ENSURES as auxiliary operators for deriving LEADS TO assertions. It is possible to adjust their definitions to add further kinds of fairness specifications. For instance, as described in [Rao95], it is possible to integrate strong fairness for transitions by redefining ENSURES appropriately. An alternative that we prefer, however, is to preserve the definitions of TRANSIENT and ENSURES and their compositional properties, and to add a proof rule for strong (group) fairness at the level of LEADS TO. Although details are beyond the scope of this introduction, we would like to point out that such a rule can make use of a strong (group) fairness specification $\mathcal{SF}$ in analogy to the weak (group) fairness specification $\mathcal{WF}$ we are already using, and it can be justified using a more restricted execution semantics $AdmOcc(\mathcal{N}, \mathcal{WF}, \mathcal{SF})$ that takes $\mathcal{WF}$ and $\mathcal{SF}$ into account.

16.2.8 Combination with Other Methods

Of course, results obtained by other methods, such as model checking or structural methods, can be incorporated into the reasoning process to the extent that they can be expressed within the logic.

In particular, techniques for determining S-invariants are well suited for supporting the verification process. For instance, in the example above the (inductive) invariants $I_1, \ldots, I_6$ are actually linear S-invariant equations. To verify this notice that the functions $|p|$ and Σp are linear mappings from multisets of $\mathbb{N}$ into $\mathbb{N}$. I_6, for example, is equivalent to the equation $W(\mathbf{m}) = 1$ with linear weight function $W(\mathbf{m}) = 2\Sigma pi(\mathbf{m}) - \Sigma podd(\mathbf{m})$ mapping markings to natural numbers. To verify that $W(\mathbf{m}) = 1$ is indeed an S-invariant equation it remains for us to check it for the initial marking and to verify the *flow balance condition* $W(\mathbf{Pre}[\bullet, t](\beta)) = W(\mathbf{Post}[\bullet, t](\beta))$ for every transition t under every binding $\beta \in cd(t)$.

In contrast with $I_1, \ldots, I_6$, the invariant $\Sigma p1 + \Sigma p2 + \Sigma p3 = (\Sigma pi - 1)^2$ contained in the (inductive) invariant I_7 is clearly non-linear. Nevertheless it is possible to treat this equation as an extended S-invariant using an approach developed in [Val93a] and [Val93b]. We will demonstrate the basic idea below.

The invariant in question is captured by the equation $W(\mathbf{m}) = 0$ using the non-linear weight function $W(\mathbf{m}) = (\Sigma pi(\mathbf{m}) - 1)^2 - \Sigma p1(\mathbf{m}) - \Sigma p2(\mathbf{m}) - \Sigma p3(\mathbf{m})$. However, this function is almost linear, in the sense that it is linear in all non-clearing places, that is $p1$, $p2$, and $p3$. A *clearing place* is a place that is always empty after some transition of its post-set fires. In [Val93a] and [Val93b] it is proved that the usual flow balance condition is sufficient to establish the equation. Again, the flow balance condition is $W(\mathbf{Pre}[\bullet, t](\beta)) = W(\mathbf{Post}[\bullet, t](\beta))$. Whereas this equation is obvious for $t1$ and $t3$ the case of $t2$ is more intricate: If β is the binding given by (s, i, odd) we have $W(\mathbf{Pre}[\bullet, t2](\beta)) = (i-1)^2 - s = i^2 - (2i-1) - s$ and $W(\mathbf{Post}[\bullet, t2](\beta)) = i^2 - (s + odd) = i^2 - odd - s$. Flow balance can now be established using the inductive invariants I_2, I_4, and I_6 which ensure that $odd = 2i - 1$. Notice that we have additionally used I_2 which is not present in I_7. So we have actually proved the stronger assertion that $I_2 \wedge I_4 \wedge I_6 \wedge \Sigma p1 + \Sigma p2 + \Sigma p3 = (\Sigma pi - 1)^2$ is an inductive invariant.

16.2.9 Final Remarks

Even in our simple example it has become evident that verification of concurrent systems requires general mathematical results and techniques in addition to pure temporal reasoning. For instance, we implicitly employed straightforward logical rules and also simple results about natural numbers. The framework we used is informal set theory, and the temporal logic is embedded into set theory by its set-theoretic semantics. If we are interested in formalising correctness proofs to support verification via interactive computer-aided theorem proving, there are interesting alternatives to set theory, namely higher-order logics and type theories, e.g. those formalisms employed in [APP94], [HC96], and [Ste98b]. In any case it should be clear that isolated proof systems supporting only temporal reasoning cannot cope with the complex verification problems which we encounter in practice. From this point of view a combination of different methods seems to be inevitable, and remains an important research challenge which goes far beyond net theory.

16.3 A Logic of Enablement

In this section we discuss a temporal logic language $\mathcal{T}$ that allows us to argue about the enablement of transitions or so-called computations. In addition to the standard temporal operators of future necessity and eventuality, we need to define conditional modalities for which the standard semantics has to be extended along the lines of PTL or the propositional modal mu-calculus (see for example [Sti96, Bra92]). The semantics given to $\mathcal{T}$ will be essentially event-based. Our arguments will use a special predicate using conditional modalities that represents the enablement of a transition/computation which we call $E(t)$ for transition t for instance.

We study reductions from complex nets to much simpler test nets and apply some preservation and reflection results for the reduction morphisms. Assume that some property of a simple net $\mathcal{N}'$ has already been proved. Then if we can show that the application of a reduction ρ on a more complex net $\mathcal{N}$ leads exactly to $\mathcal{N}'$ such that the same property holds for $\mathcal{N}$, we speak of the *reflection* of properties of net behaviours. Conversely we say a property φ is *preserved* by a morphism ρ from net $\mathcal{N}$ to net $\mathcal{N}'$ if it holds for $\mathcal{N}'$ whenever it can be shown to hold for $\mathcal{N}$. This can be seen as a formalisation of concepts informally proposed by Olderog in [Old91].

16.3.1 Morphisms, Reductions, and Simulation

In this section we will define the basic notions that will be needed to study relations between nets. We start by setting up a categorical framework, then we define morphisms, and finally convey our understanding of simulation between nets.

A Category of Nets. In the literature there exist many categorical representations of Petri nets and net dynamics. Since we are not interested in the mathematical theory of categories in the present context, we will only outline for the more mathematically inclined reader the fundamental category in which we will be working.

A category with markings of Petri nets as objects and computations as morphisms is called a *behaviour category* in [BG92]. Other authors have defined categories with behaviour categories as objects, e.g. [MM88, DMM89]. We will instead follow Brown and Gurr's [BG94] notion of a category of marked nets and reductions.

MNet$^+$ is a category with marked multi-loop-free Petri nets as objects and reverse reductions of nets as morphisms. A *multi-loop* is a transition that outputs more than one token to any one of its input places, or conversely consumes more than one token from an output place. Formally we define:

Definition 16.3.1. *The category* **MNet**$^+$ *has net systems (i.e. marked P/T nets)* $\mathcal{S} = \langle \mathcal{N}, \mathbf{m} \rangle$ *consisting of a multi-loop-free Petri net* $\mathcal{N} = (P, T, \mathbf{Pre}, \mathbf{Post})$ *and a marking* $\mathbf{m} : P \to \mathbb{N}^{|P|}$ *as objects, and morphisms* $\langle f, F \rangle$ *such that* $f : T \to T'^{+}$ *and* $F : P' \to P$ *are partial functions. Furthermore we require that the left diagram holds for* $\leq$ *and the right for* $\geq$:

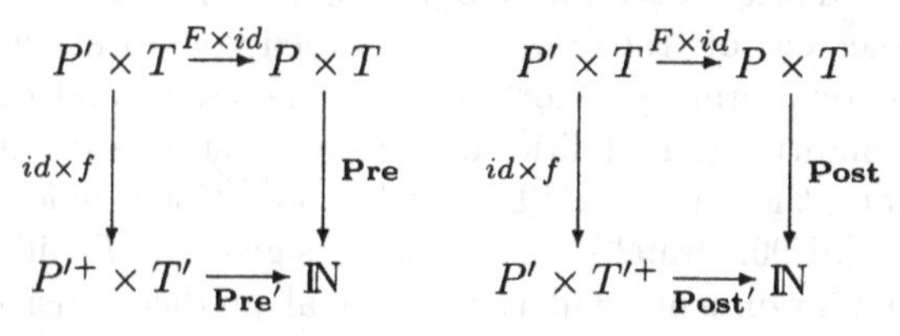

Put in standard set-theoretic terms we require $\forall t \in T . \forall p' \in P' . (\mathbf{Pre}[Fp', t] \geq \mathbf{Pre}'[p', ft]) \wedge (\mathbf{Post}[Fp', t] \leq \mathbf{Post}'[p', ft])$, *i.e. the image net uses at most as many tokens from, and produces at least as many tokens to, the corresponding places as the original net.*

Composition is defined component-wise in the morphisms.

We will allow an isolated place in our net systems that is labelled by $\star$. This will be needed to abstract from certain places of the original net. Isolated places will contain no tokens and will formally be written as a net system $\bot = \langle \{\star\}, \{\star\}, \mathbf{0}, \mathbf{0}, \mathbf{0} \rangle$, such that the pre- and post-condition vectors as well as the marking vector are zero vectors.

Remark 16.3.2. Note that the definition of $\mathbf{MNet}^+$ is based solely on the structure of the nets. There will be no need to construct any form of the reachability graph to apply its morphisms, hence avoiding the state-space explosion problem encountered in many traditional approaches to model checking of nets.

The net theoretic interpretation of categorical products and coproducts is not needed to understand the rest of the section. We remark though that the product represents choice and the coproduct is interpreted as the parallel composition of nets from $\mathbf{MNet}^+$.

Notion of Simulation. We first define a relation $F^+ \subseteq \mathbf{P} \times \mathbf{P}'$ on multisets of places based on functions $F : P' \to P$, such that $(\mathbf{m}, \mathbf{m}') \in F^+$ iff $\forall p' \in P' . \mathbf{m}[F(p')] \leq \mathbf{m}'[p']$ holds. The following notion of simulation is easily verified to hold for all morphisms in $\mathbf{MNet}^+$ and also every simulation is a morphism in our category.

Definition 16.3.3 (Simulation). *A morphism $\langle f, F \rangle$ is a* simulation *between two net systems $\mathcal{S} = \langle \mathcal{N}, \mathbf{m} \rangle$ and $\mathcal{S}' = \langle \mathcal{N}', \mathbf{m}' \rangle$ iff $(\mathbf{m}, \mathbf{m}') \in F^+$ and*

$$\forall (\mathbf{m}_1, \mathbf{m}_1') \in F^+ . \left(\mathbf{m}_1 \xrightarrow{t} \mathbf{m}_2 \quad \Rightarrow \quad (\mathbf{m}_1' \xrightarrow{f(t)} \mathbf{m}_2' \wedge (\mathbf{m}_2, \mathbf{m}_2') \in F^+) \right) .$$

There are other kinds of reductions defined on nets that serve a similar task, namely that of proving properties of nets. Most of these reductions are designed to simplify the net by some reduction rules that preserve properties such as boundedness or liveness, such that it is trivial to show that the simplified net has that property. The rules for such reductions are in general much more intricate and their application is very limited compared with the morphisms studied in this approach. For a more detailed introduction to such reductions, see Section 15.1.

Computations. We write computations, i.e. sequential and parallel composition of transitions, as the product and sum respectively of the individual transitions involved.

Definition 16.3.4 (Finite Computation, Computation Sequence). *A* finite computation *of a net system $\langle P, T, \mathbf{Pre}, \mathbf{Post}, \mathbf{m} \rangle$ is defined recursively as:*

1. *A single transition $t \in T$ and the "non-action" ι (identity step) are finite computations.*

2. *Given two finite computations u and v their parallel composition $u \| v$ is a finite computation.*
3. *Given two finite computations u and v their sequential composition $u \cdot v$ is a finite computation.*
4. *Only those expressions formed in a finite number of steps from 1.–3. are finite computations.*

A computation sequence $\gamma = \gamma_0, \gamma_1, \ldots$ *is a finite or infinite succession of finite computations γ_i. Since any finite computation sequence can be viewed as an infinite one that has a suffix of infinitely many identity steps, we will use only infinite computation sequences.*

Remark 16.3.5. Any finite computation sequence $\gamma = \gamma_0, \gamma_1, \ldots, \gamma_n$ can be rewritten as a finite computation $\gamma_1 \cdot \gamma_2 \cdot \cdots \cdot \gamma_n$. Infinite computations in which all computation steps are single transitions are called *executions* of a net, i.e infinite sequential processes.

Let $\mathcal{C}_{\mathcal{N},\mathbf{m}}$ denote the set of all finite computations of the net $\mathcal{N}$ with initial marking $\mathbf{m}$.

16.3.2 A Temporal Logic for Nets

In the following we define a temporal dynamic logic that will be used to describe properties of marked Petri nets. The logic presented here will be given an event-based semantics and the basic property of enablement of a single transition or of a computation will be the starting point for describing further properties of net dynamics.

Syntax. The logical language $\mathcal{T}$ used is given by

$$\Phi := true \mid \neg\Phi \mid \Phi \wedge \Phi \mid \Diamond\Phi \mid \langle t\rangle\Phi \mid \forall x\,.\,\Phi$$

where t is any closed term and x is a variable. As usual for any formulae α and β of $\mathcal{T}$ we can define the following composite formulae:

$$false := \neg true, \qquad \alpha \vee \beta := \neg(\neg\alpha \wedge \neg\beta), \qquad \exists x\,.\,\alpha := \neg\forall x\,.\,\neg\alpha,$$

$$[t]\alpha := \neg\langle t\rangle\neg\alpha, \qquad \Box\alpha := \neg\Diamond\neg\alpha,$$

$$\alpha \Rightarrow \beta := \neg\alpha \vee \beta, \qquad \alpha \Leftrightarrow \beta := \alpha \Rightarrow \beta \wedge \beta \Rightarrow \alpha.$$

Semantics. An *interpretation* θ is a partial function from the logical constants to computations of a net system $\langle\mathcal{N}, \mathbf{m}\rangle$. Following the definition of the denotation for formulae from $\mathcal{T}$ it is possible to specify a range of net properties:

$$
\begin{aligned}
\|true\|_\theta &= \mathcal{C}_{\mathcal{N},\mathbf{m}} \\
\|false\|_\theta &= \emptyset \\
\|\neg\Phi\|_\theta &= \{\gamma \mid \gamma \notin \|\Phi\|_\theta\} \\
\|\Phi \vee \Psi\|_\theta &= \|\Phi\|_\theta \cup \|\Psi\|_\theta \\
\|\Phi \wedge \Psi\|_\theta &= \|\Phi\|_\theta \cap \|\Psi\|_\theta \\
\|[t]\Phi\|_\theta &= \{\gamma \mid \forall k \in \mathbb{N} \,.\, \gamma_0, \ldots, \gamma_k = \theta(t) \Rightarrow \gamma_{k+1} \in \|\Phi\|_\theta\} \\
\|\langle t\rangle\Phi\|_\theta &= \{\gamma \mid \exists k \in \mathbb{N} \,.\, \gamma_0, \ldots, \gamma_k = \theta(t) \wedge \gamma_{k+1} \in \|\Phi\|_\theta\} \\
\|\Box\Phi\|_\theta &= \{\gamma \mid \forall k \in \mathbb{N} \,.\, \gamma_{k+1} \in \|\Phi\|_\theta\} \\
\|\Diamond\Phi\|_\theta &= \{\gamma \mid \exists k \in \mathbb{N} \,.\, \gamma_{k+1} \in \|\Phi\|_\theta\} \\
\|\forall\alpha.\Phi\|_\theta &= \{\gamma \mid \forall x \in \mathrm{dom}(\theta) \,.\, \gamma \in \|\Phi[\alpha/x]_\theta\} \\
\|\exists\alpha \,.\, \Phi\|_\theta &= \{\gamma \mid \exists x \in \mathrm{dom}(\theta) \,.\, \gamma \in \|\Phi[\alpha/x]_\theta\}
\end{aligned}
$$

We say a marked net $\langle\mathcal{N}, \mathbf{m}\rangle$ *models* a formula Φ iff there is an interpretation θ under which Φ is satisfied. In this case we write $\langle\mathcal{N}, \mathbf{m}\rangle \models_\theta \Phi$.

The satisfaction relation is defined as follows:

$$
\begin{array}{lll}
\langle\mathcal{N}, \mathbf{m}\rangle \models_\theta true & & \\
\langle\mathcal{N}, \mathbf{m}\rangle \models_\theta \neg\Phi & \text{iff} & \langle\mathcal{N}, \mathbf{m}\rangle \not\models_\theta \Phi \\
\langle\mathcal{N}, \mathbf{m}\rangle \models_\theta \Phi \wedge \Psi & \text{iff} & \langle\mathcal{N}, \mathbf{m}\rangle \models_\theta \Phi \quad \text{and} \quad \langle\mathcal{N}, \mathbf{m}\rangle \models_\theta \Psi \\
\langle\mathcal{N}, \mathbf{m}\rangle \models_\theta [t]\Phi & \text{iff} & \forall \mathbf{m}' \,.\, \mathbf{m} \xrightarrow{\theta(t)} \mathbf{m}' \quad \Rightarrow \quad \langle\mathcal{N}, \mathbf{m}\rangle \models_\theta \Phi \\
\langle\mathcal{N}, \mathbf{m}\rangle \models_\theta \Box\Phi & \text{iff} & \forall a.\forall \mathbf{m}' \,.\, \mathbf{m} \xrightarrow{a} \mathbf{m}' \quad \Rightarrow \quad \langle\mathcal{N}, \mathbf{m}\rangle \models_\theta \Phi \\
\langle\mathcal{N}, \mathbf{m}\rangle \models_\theta \forall x \,.\, \Phi & \text{iff} & \forall \alpha \in \mathrm{dom}(\theta) \,.\, \langle\mathcal{N}, \mathbf{m}\rangle \models_\theta \Phi[\alpha/x]
\end{array}
$$

The Enablement Predicate. Since our reasoning about net properties is based on the enablement of computations, it is convenient to define an enablement predicate $E(t)$ as $\langle t\rangle true$. It is easily shown that $\langle\mathcal{N}, \mathbf{m}\rangle \models_\theta E(t)$ whenever the interpretation of the term t is enabled in the net system $\mathcal{S} = \langle\mathcal{N}, \mathbf{m}\rangle$. Conversely $\langle\mathcal{N}, \mathbf{m}\rangle \models_\theta \langle t\rangle false$ whenever $\theta(t)$ is not enabled, and so $\langle\mathcal{N}, \mathbf{m}\rangle \models_\theta \neg E(t)$ holds.

Some properties that can be stated in $\mathcal{T}$ are:

formula	property
$E(t)$	enablement of $\theta(t)$
$\exists x \,.\, \neg E(x)$	existence of some non-enabled computation
$\forall x \,.\, \neg E(x)$	deadlock
$\Box\exists x \,.\, E(x)$	"there is always some computation that is enabled"
$\exists x \,.\, \Box E(x)$	"there exists a transition that is always enabled"

In practice many properties can be described by state formulae from $\mathcal{T}$, i.e. formulae of the form $true \mid E(t) \mid \alpha \wedge \alpha \mid \neg\alpha$. Among these properties are (for state formulae Φ and Ψ):

1. Safety properties $\Box\Phi$, such as mutual exclusion described by $(\Box\neg E(t\|t') \;=\; \neg\Diamond E(t\|t'))$
2. Progress properties $\Diamond\Box\Phi \vee \Box\Diamond\Psi$ such as the (strong) fairness described by $\Box(\Box\Diamond E(t_0) \Rightarrow \Box\Diamond E(t_1))$
3. Persistence properties $\Diamond\Box\Phi$

4. Termination properties $\Diamond\Phi$
5. Recurrence properties $\Box\Diamond\Phi$.

Many of these properties can be shown to be invariant against the application of simulations as defined above. To do this we need to define what we mean by invariance in this setting. We have made a distinction between safety and progress properties here, becaus we will see that different preservation results apply to the formulae of the example above.

Preservation and Reflection of Properties. We are interested in both directions of the application of morphisms to marked Petri nets. When looking at the reduction aspect, i.e. the transformation from a more complex net to a simpler one, we will use the *preservation* of properties for our arguments. When considering the abstraction aspect we are interested in *reflection* of properties. We start by formalising these concepts:

Definition 16.3.6 (Preservation). *The satisfaction of a property Φ of a marked net $\langle \mathcal{N}, \mathbf{m} \rangle$ is said to be* preserved *by a morphism $\langle f, F \rangle$ if the following holds for the image net $\langle \mathcal{N}', \mathbf{m}' \rangle$:*

$$\langle \mathcal{N}, \mathbf{m} \rangle \models_\theta \Phi \quad \Rightarrow \quad \langle \mathcal{N}', \mathbf{m}' \rangle \models_{f\theta} \Phi.$$

Φ-computations are preserved by a morphism $\langle f, F \rangle$ iff

$$\forall \gamma . \gamma \in \|\Phi\|_\theta \quad \Rightarrow \quad f\gamma \in \|\Phi\|_{f\theta}.$$

Definition 16.3.7 (Reflection). *The satisfaction of a property Φ of a marked net's image $\langle \mathcal{N}', \mathbf{m}' \rangle$ is said to be* reflected *by a morphism $\langle f, F \rangle$ if the following holds for the source net $\langle \mathcal{N}, \mathbf{m} \rangle$:*

$$\langle \mathcal{N}', \mathbf{m}' \rangle \models_{f\theta} \Phi \quad \Rightarrow \quad \langle \mathcal{N}, \mathbf{m} \rangle \models_\theta \Phi.$$

Φ-computations are reflected by a morphism $\langle f, F \rangle$ iff

$$\forall \gamma . f\gamma \in \|\Phi\|_{f\theta} \quad \Rightarrow \quad \gamma \in \|\Phi\|_\theta.$$

We will often omit the the term "satisfaction" and simply talk about the preservation or reflection of a formula. We will also omit mention of a particular morphism if the result holds for every morphism in the category.

Theorem 16.3.8. *The following preservation results for formulae from $\mathcal{T}$ can be obtained:*

1. *A quantifier-free formula Φ is preserved if it does not contain an instance of the connective $[t]$.*
2. *true-computations are preserved and true is preserved.*
3. *$E(t)$ and $E(t)$-computations are preserved.*
4. *If $\langle f, F \rangle$ preserves Φ- and Ψ-computations then also $\Phi \wedge \Psi$-, $\Phi \vee \Psi$-, $\Diamond$-, $\langle t \rangle\Phi$-, $\forall x . \Phi$-, and $\exists x . \Phi$-computations are preserved.*

5. *If* $\langle f, F\rangle$ *preserves* Φ*-computations and* f *is injective, then* $\langle f, F\rangle$ *also preserves* $[t]\Phi$*-computations.*

Theorem 16.3.9. *The following reflection results for formulae from* $\mathcal{T}$ *can be obtained:*

1. *A quantifier-free formula* Φ *is reflected if it does not contain an instance of the connective* $\langle t\rangle$.
2. Φ*-computations are reflected iff* $\neg\Phi$*-computations are reflected.*
3. $\langle f, F\rangle$ *reflects* Φ *if* $\langle f, F\rangle$ *reflects* Φ*-computations.*
4. $\neg E(t)$ *is reflected.*
5. *If* $\langle f, F\rangle$ *reflects* Φ*- and* Ψ*-computations then* $\langle f, F\rangle$ *also reflects* $\Phi \wedge \Psi$*-,* $\Box\Phi$*-,* $[t]\Phi$*-, and* $\forall x \,.\, \Phi$*-computations.*
6. *If* $\langle f, F\rangle$ *reflects* Φ*-computations and* f *is injective, then* $\langle f, F\rangle$ *also reflects* $[t]\Phi$*-computations.*

For proofs of these results the reader is referred to [BG94].

Example 16.3.10. Property 1 (safety) from page 365 is reflected, but property 2 (progress) is not reflected by all morphisms.

16.3.3 The Concept of a Test Net

There are several simple nets whose properties have been extensively studied in the past or which can be efficiently checked using traditional model checking techniques. Using these nets, we can apply our category of morphisms to show that the same properties hold for more complex nets. Therefore we call the simple nets *test nets* as proposed by C. Brown [BG94]. The basic idea is that if it is easy to show that a complex net is simulated by a test net (or vice versa) and we know that the simulation relation depends on a reduction morphism that has been shown to preserve certain properties, then it is also easy to deduce these properties for the net itself (by simply applying the morphism in question).

By applying some morphism from our category we obtain a net $\mathcal{N}'$ from $\mathcal{N}$. The morphisms will be designed to reduce the complexity of the net $\mathcal{N}$, to which a preservation theorem is to be applied, such that $\mathcal{N}'$ will in general have fewer states or transitions. If applying a reflection theorem the converse will be the case. If a net is "sufficiently" simple to show the satisfaction of some formula it will be called a *test net.*

Minimality. We are interested mainly in morphisms between nets, such that the behaviour of one net is an image of another net's behaviour under a specific morphism. For reasons of efficiency it is natural to consider only morphisms that do not allow any superfluous components. Such morphisms are called *minimal.*

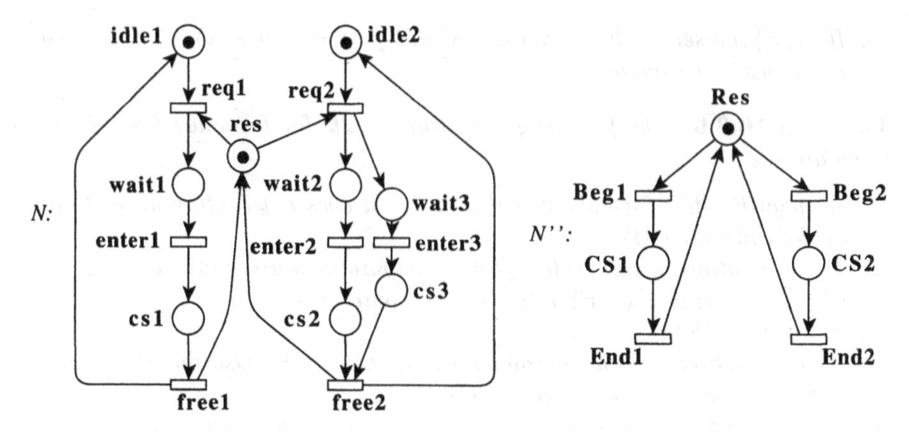

Fig. 16.7. Nets for mutual exclusion algorithm with semaphore

Definition 16.3.11 (Minimality). *A morphism* $\langle f, F\rangle$ *is said to be* minimal *if for every computation sequence* γ' *of* $\langle \mathcal{N}', \mathbf{m}'\rangle$, *there exists a unique inverse image with respect to* f *such that* $f(\gamma) = \gamma'$. *In other words, if we have* $\mathcal{C}_{\mathcal{N}',\mathbf{m}'} \subseteq f(\mathcal{C}_{\mathcal{N},\mathbf{m}})$ *then* $\langle f, F\rangle$ *is minimal.*

A formula $\Phi \in \mathcal{T}$ is said to be *minimally preserved* if all minimal morphisms $\langle f, F\rangle$ preserve Φ. Minimal reflection is defined analogously.

Theorem 16.3.12. *If* $\langle f, F\rangle$ *is a minimal morphism between two net systems the following preservation results may be obtained for formulae from* $\mathcal{T}$*:*

1. *If* $\langle f, F\rangle$ *preserves* Φ*-computations then* $\langle f, F\rangle$ *also preserves* Φ.
2. *If* $\langle f, F\rangle$ *preserves* Φ*-computations then* $\langle f, F\rangle$ *preserves* $[t]\Phi$.

Corollary 16.3.13. *If* Φ *is minimally preserved then* $[t]\Phi$ *is minimally preserved. If* Φ *is minimally reflected then* $\langle t\rangle\Phi$ *is minimally reflected.*

Example 16.3.14. Property 2 (progress) from Section 16.3.2 (page 365) is minimally reflected by morphisms $\langle f, F\rangle$ where f is an injection. This progress property is also preserved.

16.3.4 Example: Mutex

Let us now turn to one of the previous example properties of nets, the mutual exclusion property. Consider the net N from Figure 16.7. We would like to show that the places cs_1 and cs_2 cannot both hold a token at the same time and thus the transitions $enter_1$ and $enter_2$ are never concurrently enabled. This would be formally expressed by the enduring holding of the negation of the concurrent enablement of the transitions: $\Box\neg E(enter_1 \| enter_2)$.

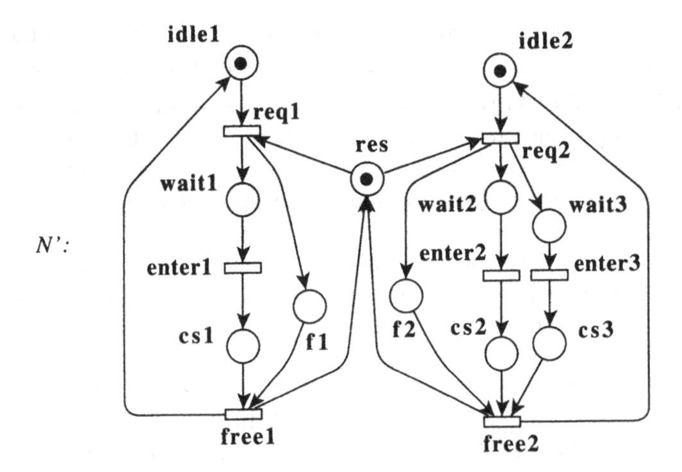

Fig. 16.8. Net modification N' for showing mutual exclusion in N

Consider the two propositional variables α and β. Let the interpretation θ assign the following computations to the variables: $\theta(\alpha) = free_1 \| free_2$ and $\theta(\beta) = enter_1 \| enter_2 \| enter_3$.

We take for granted that the transitions End_1 and End_2 of the test net $\mathcal{N}''$ have the mutual exclusion property. Let $\mathbf{m}_{Res}$ denote the marking of $\mathcal{N}''$ in which the only token in the net occupies place Res. Then the following holds:

$$\langle \mathbf{m}_{Res}, \mathcal{N}'' \rangle \models_{f\theta} \Box \neg E(\alpha) \tag{16.19}$$

As we know from Theorem 16.3.9, $\neg E(t)$-properties are reflected, thus we have to find a morphism from N to N'' (actually to the coproduct of N'' with $\bot$, since we will need the isolated place $\star$). But this is not possible directly, so we have to consider a modest extension N' of N as depicted in Figure 16.8. It is easily seen that N' is behaviourally equivalent to N, since it adds only two redundant places.

Now specify the morphism $\langle f, F \rangle$ by $f(req_i) = Beg_i$, $f(free_i) = End_i$, $f(enter_j) = \star$, and $F(Res) = res$, $F(CS_1) = cs_1$, and $F(CS_2) = cs_2 \| cs_3$ for $i \in \{1,2\}, j \in \{1,2,3\}$. It is easy to check that $\langle f, F \rangle$ is indeed a morphism in $\mathbf{MNet}^+$ and that $\mathcal{N}$ is multi-loop-free. The reasoning for the ME-property of $\mathcal{N}$ goes as follows:

From Theorem 16.3.9 it follows that $\langle \mathbf{m}_0, \mathcal{N}' \rangle \models_\theta \Box \neg E(\alpha)$, where $\mathbf{m}_0$ denotes the marking of $\mathcal{N}'$ in which places res, $idle_1$, and $idle_2$ contain one token each. We also have $\langle \mathbf{m}_0, \mathcal{N}' \rangle \models_\theta [\beta] E(\alpha)$, i.e. whenever $enter_1, enter_2$, and $enter_3$ fire simultaneously, $free_1$ and $free_2$ are both enabled. We have just shown that this cannot be the case, so we deduce that no computation of $\langle \mathbf{m}_0, \mathcal{N}' \rangle$ can ever enable $\theta(\beta)$, thus $\langle \mathbf{m}_0, \mathcal{N}' \rangle \models_\theta \Box \neg E(\beta)$ holds, meaning that in no reachable marking are places cs_1 and cs_2 simultaneously marked.

In a similar fashion we could also show that cs_1 and cs_3 cannot hold a token simultaneously, but cs_2 and cs_3 can be simultaneously marked.

Modularity of the Approach. The application of morphisms allows large parts of a complex net to be made invisible in a simulation of some aspect of the net. By successively concentrating on parts of the net and simulating them on suitable test nets we have a modular approach for verifying different system properties.

It is then possible to collect a basic set of test nets to deal with different kinds of situations such as the mutual exclusion problem shown in the example. Other test nets could be specified to deal with a variety of properties including those mentioned above.

For instance, the simple net consisting of a single marked place that acts as a side condition to one transition for each branch of the choice situation in the initial marking of N can serve as a test net to show that the net N from Figure 16.7 is deadlock-free. Absence of a deadlock can then be shown to be a preserved property of this net. The construction of an appropriate morphism is left to the reader as an exercise.

Requirements and Complexity Issues. The complexity of the basic check of whether a given morphism belongs to the category $\mathbf{MNet}^+$ is linear in the size of the net. However, as in the case for other techniques of model checking, this approach also has some drawbacks:

1. The morphisms used to verify a certain property have to be designed manually. A fully automatic method does not seem possible.
2. Properties of the net have to be expressed in temporal logic. The semantics of the temporal logic used by our approach allows only event-based properties to be expressed directly; arguments about markings and P-invariants have to be given indirectly by statements over the transitions enabled by the markings in question.
3. The question of which logical calculus is best suited is still open. In future work it would be desirable to construct a compositional proof system along the lines of [Win90] to exploit the modularity of this approach and the additional structure in the category $\mathbf{MNet}^+$ (i.e. the existence of products and coproducts to represent choice and parallel composition of processes).

16.4 Linear Logic and Petri Nets

Linear logic has been shown to be well suited for describing Petri nets and their dynamics, however few attempts have been made to develop linear logic for analysing nets and proving properties beyond the reachability of certain markings.

Before going into the details of the Petri net representation we will briefly and informally characterise the connectives of linear logic, the main idea of which is to split the conjunction (respectively disjunction) into two different versions, one that is resource sensitive and another that behaves more or less like the classical connective. In order to maintain the power of classical calculi one has to introduce modalities, the so-called exponentials. The linear constants and connectives that will be used in this section are:

$\otimes$ (*times, tensor*) is the multiplicative resource-sensitive version of classical conjunction, e.g. $A \otimes B$ in linear logic means that both resources A and B are present at the same time, whereas $A \otimes A$ means that two instances of the same resource are present.

$\parr$ (*par*) is the multiplicative disjunction, $A \parr B$ means: if not A then B.

$\multimap$ (*entails*) is the multiplicative implication, such that $A \multimap B$ means that we get one new instance of resource B while consuming exactly one instance of resource A.

The additive conjunction & (*with*) expresses a kind of deterministic choice, e.g. in a situation where both resources are offered to you but you cannot grab both of them, $A \& B$ is the representation of your choice between A and B. The additive disjunction $\oplus$ (*plus*) on the other hand represents nondeterminism or a choice on the systems side, i.e. given $A \oplus B$ it is at the system resource managers discretion to give you either A or B. You can only be sure not to leave empty-handed.

Linear *negation* $(\cdot)^{\perp}$ could be called a dept in monetary terms. In general $A^{\perp}$ is an input slot for using up one instance of resource A.

The exponential ! (*of course*) is the storage operator, also called the operator of reusability, which makes a resource arbitrarily available.

The multiplicative constants are 1 and $\perp$, where 1 is the unit of the multiplicative conjunction, meaning truth only in isolation, and $\perp$ is the unit of *par* representing a placeholder for nothingness. The units of the additive conjunction and disjunction are $\top$ and 0, representing true truth and falsity respectively (i.e. truth or falsity in any context).

16.4.1 Basic Relationship

We will give a short survey of the work done by C. Brown [Bro89], N. Martí-Oliet and J. Meseguer [MOM91], U. Engberg and G. Winskel [EW90]. We use the notion of a marked net, i.e. that of an instantaneous description of a net. Only formulae from propositional linear logic that contain the tensor product $\otimes$, linear implication $\multimap$, the additive connective of choice &, the storage-operator ! and the constant 1 will be used in this section.

Terminology. We denote sequents by $\Gamma \Rightarrow \Delta$, where Γ and Δ are multisets of formulae, and $\Rightarrow$ is a metasymbol that has the meaning of entailment in the calculus. The multisets are usually written in list notation, omitting any superfluous braces or parentheses, i.e. a sequent will often be written as $A_1, \ldots, A_n \Rightarrow B_1, \ldots, B_m$. The semantics given to such a sequent in classical terms is $A_1 \wedge \cdots \wedge A_n \Rightarrow B_1 \vee \cdots \vee B_m$. For linear logic the multiplicative fragment is used to give semantics to such a sequent, i.e. $A_1 \otimes \cdots \otimes A_n \Rightarrow B_1 \parr \cdots \parr B_m$ which is equivalent to the occasionally used one-sided sequent $\Rightarrow A_1^{\perp} \parr \cdots \parr A_n^{\perp} \parr B_1 \parr \cdots \parr B_m$ or $\Rightarrow A_1^{\perp}, \ldots, A_n^{\perp}, B_1, \ldots, B_m$ for short.

Table 16.1. Inference rules for $\mathcal{L}_{Petri}$

$$A \Rightarrow A (\text{Identity}) \qquad \Rightarrow 1 (1)$$

$$\frac{\Gamma \Rightarrow A \quad \Delta, A \Rightarrow B}{\Gamma, \Delta \Rightarrow B}(\text{Cut}) \qquad \frac{\Gamma, A, B, \Delta \Rightarrow C}{\Gamma, B, A, \Delta \Rightarrow C}(\text{Exchange})$$

$$\frac{\Gamma, A, B \Rightarrow C}{\Gamma, A \otimes B \Rightarrow C}(\otimes \text{L}) \qquad \frac{\Gamma \Rightarrow A \quad \Delta \Rightarrow B}{\Gamma, \Delta \Rightarrow A \otimes B}(\otimes \text{R})$$

$$\frac{\Gamma \Rightarrow A \quad \Gamma \Rightarrow B}{\Gamma \Rightarrow A \& B}(\&\text{R}) \qquad \frac{\Gamma, A \Rightarrow C}{\Gamma, A \& B \Rightarrow C}(\&\text{L1}) \qquad \frac{\Gamma, B \Rightarrow C}{\Gamma, A \& B \Rightarrow C}(\&\text{L2})$$

$$\frac{\Gamma \Rightarrow A \quad \Delta, B \Rightarrow C}{\Delta, \Gamma, A \multimap B \Rightarrow C}(\multimap \text{L})$$

$$\frac{\Gamma, !A, !A \Rightarrow B}{\Gamma, !A \Rightarrow B}(\text{Contraction}) \qquad \frac{\Gamma, A \Rightarrow B}{\Gamma, !A \Rightarrow B}(\text{Dereliction})$$

$$\frac{\Gamma \Rightarrow B}{\Gamma, !A \Rightarrow B}(\text{Weakening})$$

Sequent Calculus for Linear Logic. Only the inference rules from the fragment of linear logic shown in Table 16.1 will be needed. We use the two-sided version of the sequent calculus rules here, since it appeals more naturally to our intuition.

Remark 16.4.1. The calculus given above is a fragment of the full intuitionistic linear logic calculus and thus gives an interleaving semantics for Petri nets. It is nevertheless possible to give a true concurrency semantics to Petri nets by using a multi-conclusion fragment of linear logic.

Let us take a look at an example of a simple Petri net from which we will derive a linear logic representation. In the remainder of this section we will use the following abbreviation:

$$a^n := \underbrace{a \otimes \cdots \otimes a}_{n}$$

Example 16.4.2. Starting from the Petri net system shown in Figure 16.9 we can obtain a natural set of formulae from the fragment $\mathcal{L}_{Petri}$ of linear

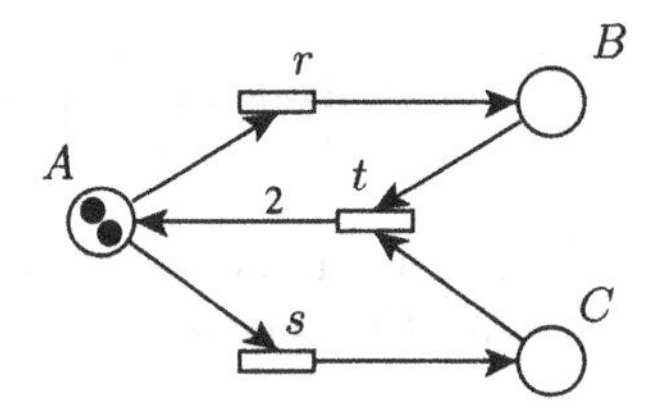

Fig. 16.9. Petri net for Example 16.4.2

logic to describe it. We start by describing the current marking of the net by introducing one resource of type p_i for each token in place p_i and writing these as a tensor product. In our net the current marking is therefore A^2 indicating the presence of two tokens in place A.

We shall now construct subformulae for each transition of the net:

t is represented by the formula $B \otimes C \multimap A^2$, meaning that firing transition t consumes one token each from places B and C, while producing two tokens in place A.

For transitions r and s we have: $A \multimap B$ and $A \multimap C$ respectively.

As usual every transition may fire as often as desired, as long as the pre-conditions are satisfied. Therefore we have to precede each formula that represents a transition by the storage operator !. Hence the complete description of transition t would be: $!(B \otimes C \multimap A)$, making t available *ad infinitum.* By putting all subformulae together we arrive at an instantaneous description of the net system:

$$A^2 \otimes !(A \multimap B) \otimes !(A \multimap C) \otimes !(B \otimes C \multimap A^2)$$

In the manner outlined in the previous example it is possible to construct for every net system $\mathcal{S}(\mathbf{m}) = \langle \mathcal{N}, \mathbf{m} \rangle = \langle P, T, F, W, \mathbf{m} \rangle$, its canonical formula $\Psi_{\mathcal{S}(\mathbf{m})}$ by forming the tensor product of the following formulae:

- For a transition t with non-empty pre-conditions ${}^\bullet t$ and non-empty post-conditions $t^\bullet$, construct

$$!\left(\bigotimes_{p \in {}^\bullet t} p^{W(p,t)} \multimap \bigotimes_{q \in t^\bullet} q^{W(t,q)} \right).$$

- In the special cases where a transition t has no pre-conditions (i.e. a *source transition*) construct for each such transition the formula

$$!\left(1 \multimap \bigotimes_{q \in t^\bullet} q^{W(t,q)} \right) \text{ or equivalently } !\left(\bigotimes_{q \in t^\bullet} q^{W(t,q)} \right).$$

- For all transitions t without any post-conditions (i.e. *sink transitions*) construct the linear logic formula

$$!\left(\bigotimes_{p\in {}^\bullet t} p^{W(p,t)} \multimap \bot\right) \quad \text{or equivalently} \quad !\left((\bigotimes_{p\in {}^\bullet t} p^{W(p,t)})^{\bot}\right).$$

- Construct for the current marking $\mathbf{m}$ and all places $p \in P$ with $\mathbf{m}[p] = n$, $n \geq 1$ the formulae p^n. Thus for the complete marking

$$\bigotimes_{p\in P, \mathbf{m}[p]\geq 1} p^{\mathbf{m}[p]}.$$

Having now a representation of a Petri net in linear logic we can state the following soundness and completeness theorem.

Theorem 16.4.3. *A marking* $\mathbf{m}'$ *is reachable in* $\mathcal{S}(\mathbf{m})$ *iff for the corresponding canonical formulae the sequent* $\Psi_{\mathcal{S}(\mathbf{m})} \Rightarrow \Psi_{\mathcal{S}(\mathbf{m}')}$ *is provable in* $\mathcal{L}_{Petri}$.

Proof. See [Bro89] for a detailed proof using induction on the number of steps made in the derivation, especially looking at the last rule used for the *only-if*-branch. The *if* branch is straightforward.

16.4.2 Specification of Net Properties

The view of linear logic presented in this section is based on the work of Engberg and Winskel in [EW90]. Places are henceforth represented as atomic propositions in the logical calculus, such that the well-formed-formulae are constructed in the following manner:

$$A := \top \mid 0 \mid 1 \mid a \mid A \otimes A \mid A \multimap A \mid A \& A \mid A \oplus A$$

An atomic proposition a is interpreted as the *downwards closure* of the associated marking $\mathbf{a}$, i.e. the set of markings from which $\mathbf{a}$ is reachable. In other words: An assertion is represented by the set of requirements sufficient to establish it.

The denotation of a formula for a given net $\mathcal{N}$ is then defined as follows, where $\mathcal{M}$ is the set of all markings of the net in question:

$$\begin{aligned}
\|\top\|_{\mathcal{N}} &= \mathcal{M} \\
\|0\|_{\mathcal{N}} &= \emptyset \\
\|1\|_{\mathcal{N}} &= \{\mathbf{m} \mid \mathbf{m} \to \mathbf{0}\} \\
\|a\|_{\mathcal{N}} &= \{\mathbf{m} \mid \mathbf{m} \to \mathbf{a}\} \\
\|A \otimes B\|_{\mathcal{N}} &= \{\mathbf{m} \mid \exists \mathbf{m}_A \in \|A\|_{\mathcal{N}}, \mathbf{m}_B \in \|B\|_{\mathcal{N}} : \mathbf{m} \to \mathbf{m}_A + \mathbf{m}_B\} \\
\|A \multimap B\|_{\mathcal{N}} &= \{\mathbf{m} \mid \exists \mathbf{m}_A \in \|A\|_{\mathcal{N}} : \mathbf{m} + \mathbf{m}_A \in \|B\|_{\mathcal{N}}\} \\
\|A \& B\|_{\mathcal{N}} &= \|A\|_{\mathcal{N}} \cap \|B\|_{\mathcal{N}} \\
\|A \oplus B\|_{\mathcal{N}} &= \|A\|_{\mathcal{N}} \cup \|B\|_{\mathcal{N}}
\end{aligned}$$

From $q_1 \otimes q_2 := \{\mathbf{m} \mid \exists \mathbf{m}_1 \in q_1, \mathbf{m}_2 \in q_2 \,.\, \mathbf{m} \to \mathbf{m}_1 + \mathbf{m}_2\}$ it is easily seen that this semantics also treats two formulae connected by $\otimes$ as simultaneously available resources. The connective & is interpreted as a choice between two possible resulting markings. From the interpretation of 1 it follows that for any A with respect to a net $\mathcal{N}$ we have $\models_{\mathcal{N}} A$ if and only if $\mathbf{0} \in \|A\|_{\mathcal{N}}$ holds.

The following theorem is easily shown:

$$\text{If} \quad \mathbf{m} \in \|A\|_{\mathcal{N}} \quad \text{and} \quad \mathbf{m}' \to \mathbf{m} \quad \text{then} \quad \mathbf{m}' \in \|A\|_{\mathcal{N}} \quad \text{holds.}$$

In other words: If $\mathbf{m}$ is a marking sufficient for A, then every marking $\mathbf{m}'$ from which $\mathbf{m}$ is reachable is also sufficient for A. It now becomes obvious that in our semantics any marking from which $\mathbf{a}$ is reachable is sufficient for $\mathbf{a}$. The semantics gives the linear logic connectives the expected meaning, e.g. $A \oplus B$ is the denotation of the union of sufficient conditions for A and B, whereas the cutset of markings sufficient for A and B is denoted by $A \& B$.

Theorem 16.4.4. *Let* $\mathbf{m}$ *and* $\mathbf{m}'$ *be arbitrary markings. Then the following holds:*

$$\mathbf{m} \to \mathbf{m}' \quad \textit{iff} \quad \models \mathbf{m} \multimap \mathbf{m}'$$

Given the fact that $(\mathbf{m} \otimes \top) = \downarrow \{\mathbf{m}' \mid \mathbf{m}' \geq \mathbf{m}\}$, where $\downarrow$ denotes the canonical extension of the downward closure of a marking, we can express some properties of the net from Example 16.4.2:

- From the initial marking A^2 a marking is reachable in which place B holds a token: $\models A^2 \multimap B \otimes \top$
- From a marking in which place A holds a token, there exists some firing sequence such that in the resulting marking place B holds a token: $\models A \otimes \top \multimap B \otimes \top$
- From a marking in which place A holds a token, it is possible to reach a marking in which either place B or place C contains a token, but we cannot tell which: $\models A \otimes \top \multimap (B \oplus C) \otimes \top$
- From a marking in which place A holds a token, it is possible that both places B and C become marked, but not necessarily at the same time: $\models A \otimes \top \multimap (B \& C) \otimes \top$
- From the initial marking A^2 every marking is reachable that contains at least one marked place from the set $P' = \{A, B\}$, but again we do not know which: $\models A^2 \multimap (\bigoplus_{a \in P'} a) \otimes \top$

The most interesting part of this characterisation of a net's property lies in the meaning of the classical-like conjunction in the antecedent, which in this case represents the deterministic choice between two possible resulting markings.

16.4.3 Linear Logic for Representation of Coloured Nets

There is an obvious way in which coloured Petri nets are representable within the same fragment of linear logic used in the preceding sections. The encoding used here is easily arrived at by a standard unfolding of a coloured net. One problem arises when considering infinite colour domains: In the unfolding there will be infinitely many transitions for each transition of the coloured net that has an incoming or outgoing arc labelled by a variable of an infinite colour domain. In this case we would have to use infinitary linear logic formulae as considered in [Far96]. If we restrict ourselves to finite colour domains the canonical linear logic formula is constructed as in the following example.

Example 16.4.5. Consider the transition depicted below and assume the multiset marking $\mathbf{m}[A] = \{1, 2, 4\}, \mathbf{m}[B] = \{4\}, \mathbf{m}[C] = \{\}, \mathbf{m}[D] = \{\}$

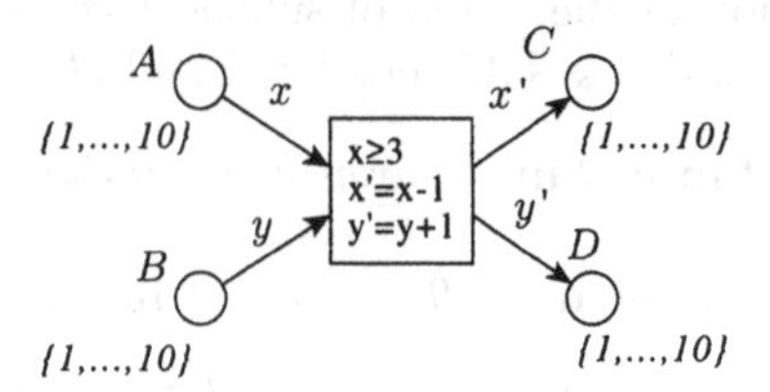

Fig. 16.10. Example of a coloured transition

This excerpt from a coloured Petri net can be represented by

$$A_1 \otimes A_2 \otimes A_4 \otimes B_4 \otimes \bigotimes_{\substack{x \in \{1,\dots,10\} \\ y \in \{1,\dots,10\} \\ x' \in \{1,\dots,10\} \\ y' \in \{1,\dots,10\} \\ (x \geq 3) \wedge (x' = x-1) \wedge (y' = y+1)}} !(A_x \otimes B_y \multimap C_{x-1} \otimes D_{y+1})$$

It is easily seen from the example above that the guard expression is not incorporated directly into the logic, but is used as a set-theoretic expression in the construction. In general a coloured transition t with exactly one variable per incoming and outgoing edge inscription is translated into a formula by means of the following construction:

- For each token with value x that is located in place S, one copy of S_x is in the tensor product representing the current marking.
- A transition with pre-set $\{p_1, \dots, p_n\}$ – the input places having colour domain $c_1, \dots, c_n$ respectively – and post-set $\{q_1, \dots, q_m\}$ – the output places having colour domain $d_1, \dots, d_m$ respectively – and incoming edges with inscriptions $x_1, \dots, x_n$, outgoing arc inscriptions $y_1, \dots, y_m$, and guard $g(x_1, \dots, x_n, y_1, \dots, y_m)$ is represented by:

$$\bigotimes_{\substack{x_i \in c_i, i \in \{1,\dots,n\} \wedge \\ y_j \in d_j, j \in \{1,\dots,m\} \wedge \\ g(x_1,\dots,x_n,y_1,\dots,y_m)}} !\left(\bigotimes_{i \in \{1,\dots,n\}} p_{i_{x_i}} \multimap \bigotimes_{j \in \{1,\dots,m\}} q_{j_{y_j}} \right)$$

Again the guard is used to restrict the possible bindings for the incoming variables in a set-theoretic way. This construction furthermore makes clear the need for an infinitary version of linear logic in the case of infinite colour domains, such as the set of natural numbers.

16.4.4 The Principle of Backward Reasoning

So far only the positive multiplicative fragment has been used for specifying Petri nets, but it is possible to give some negated formulae a meaning within the realm of net theory. Specifically, negative formulae can be interpreted as questions asked about the hypothetical enablement of transitions, or the possibility of reaching certain markings.

If, for instance, we have a transition with input places A, B and output place C, we can state that in order to fire the transition and obtain a token in place C we need tokens in both places A and B. We could simply try to negate both the premise and the conclusion of the transition's representation in linear logic, hence arriving at $C^\perp \multimap A^\perp ⅋ B^\perp$. The disjunctive character of the consequence is slightly misleading though, because in order to fire the transition and reach a marking in which C holds a token, we have to put one token each in places A and B, i.e. a debt in both places would better be expressed by the tensor product of A and B.

It would probably make more sense to consider the formula $C^\perp \otimes A \multimap B^\perp$ (and $C^\perp \otimes B \multimap A^\perp$) since it is clear that we would readily be willing to provide a token in place B if we knew that there was already a token in place A such that the transition may then fire. A theory using this approach has yet to be established.

In any case, having fixed a linear logic calculus we can use results from the computation of invariants for a given net, and combine them with deductions in our formal logic to reason about possible markings or restrictions on possible markings (e.g. some mutual exclusion property). This method however has to be applied manually.

16.4.5 Nondeterministic Transitions

Allowing a yet larger class of formulae including a very restricted use of the additive connective $\oplus$ we can take into consideration derivations in the $(!, \oplus)$-Horn fragment of linear logic containing only $(!, \oplus)$-Horn sequents, i.e. only sequents of the following kind are allowed:

$$A_1 \otimes \cdots \otimes A_k, !\Gamma \Rightarrow B_1 \otimes \cdots \otimes B_l$$

where the A_i and B_j are positive literals and Γ is a multiset of formulae of either of the two kinds (the C_i and D_j are also assumed to be positive literals):

- $C_1 \otimes \cdots \otimes C_n \multimap D_1 \otimes \cdots \otimes D_m$
- $C_1 \otimes \cdots \otimes C_p \multimap ((D_{1,1} \otimes \cdots \otimes D_{1,q_1}) \oplus \cdots \oplus (D_{r,1} \otimes \cdots \otimes D_{r,q_o}))$

The former kind of formula is exactly the one used to represent transitions in ordinary Petri nets. The latter can be used to represent nondeterministic transitions, i.e. transitions that have a set of sets of post-conditions. In nets without capacity restrictions on places the enablement of a transition is defined for nondeterministic transitions in exactly the same way as in ordinary nets. The difference in firing such a transition is that there are many possibilities for the post-set, one of which is chosen by the transition. This is viewed as internal nondeterminism of the net. An example of a nondeterministic transition is given in Figure 16.11 where the different post-sets are marked by inscriptions [i] and [j] on the outgoing arcs.

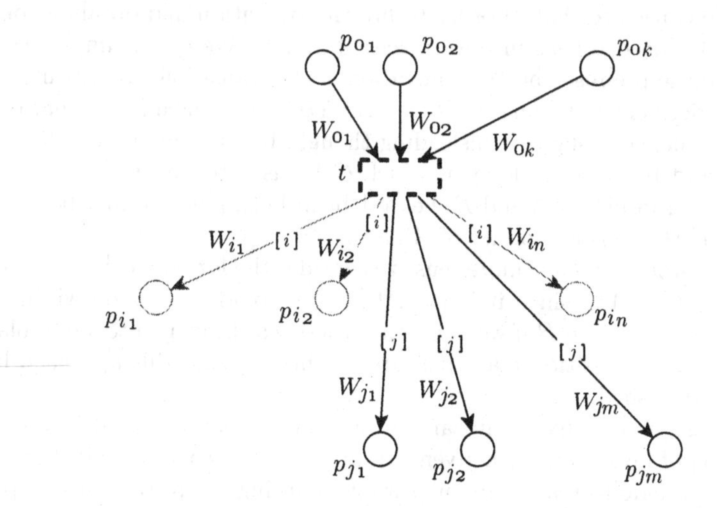

Fig. 16.11. Example of a nondeterministic transition

In [Kan94] this kind of transition has been considered and the undecidability of the reachability problem for nondeterministic Petri nets is proved by reduction to vector games (a variation of vector addition systems).

Theorem 16.4.6. *The problem of whether there exists a firing sequence in a nondeterministic Petri net that takes the initial marking* $\mathbf{m}_0$ *precisely to the marking* $\mathbf{m}$ *is undecidable.*

This result is used in [Far98] and [Far99b] to show that reachability in the system net is undecidable for generalised elementary object systems and for linear logic Petri nets[36]. These are Petri net systems that consist of an

[36] Linear logic Petri nets (LLPN) – defined in [Far98], [Far99b], and [Far99a] – are Petri nets with linear logic formulae as tokens.

elementary net system called the system net, a P/T net called the object net, and an interaction relation between transitions of both nets. The tokens of the system net are defined to be the object net processes restricted by the interaction relation – which requires synchronisation of certain pairs of transitions – and the firing rule (cf. [Val96b, Val96a, Val98, Val95]). Object systems generalise and extend the model of task/flow systems defined earlier in [Val91a]. We give an outline of these constructions in the following paragraphs.

Generalised EOS. The main idea of object systems is the use of Petri net systems as tokens in a Petri net. This is an extension of the task systems defined in [Val91a]. For this reason a distinction is made between the so-called system net and the object nets. We focus on elementary object systems where both the system net and the object nets are elementary Petri net systems.

Definition 16.4.7 (Elementary Object System (EOS)). *An* elementary object system *is a tuple* $EOS = (SN, \widehat{ON}, Rho, type, \widehat{\mathbf{M}})$ *where*

- $SN = (P, T, W)$ *is a net (i.e. an EN system without initial marking), called the* system net *or* environment net *of EOS,*
- $\widehat{ON} = \{ON_1, \ldots, ON_n\}$ $(n \geq 1)$ *is a finite set of EN systems, called the* object nets *or* token nets *of EOS, denoted by* $ON_i = (B_i, E_i, F_i, \mathbf{m}_{0i})$,
- $Rho = (\rho, \sigma)$ *is the* interaction relation, *consisting of a* system/object interaction relation $\rho \subseteq T \times \boldsymbol{E}$ *where* $\boldsymbol{E} := \bigcup\{E_i | 1 \leq i \leq n\}$, *and a symmetric* object/object interaction relation $\sigma \subseteq (\boldsymbol{E} \times \boldsymbol{E}) \setminus id_E$,
- $type : W \longrightarrow 2^{\{1,\ldots,n\}} \cup \mathbb{N}$ *is the* arc type function, *and*
- $\widehat{\mathbf{M}}$ *is a marking as defined in Definition 16.4.8.*

Definition 16.4.8. *The set* $\boldsymbol{Obj} := \{(ON_i, \mathbf{m}_i) | 1 \leq i \leq n, \mathbf{m}_i \in R(ON_i)\}$ *is the* set of objects *of the EOS. An* object-marking *(O-marking) is a mapping* $\widehat{\mathbf{M}} : P \longrightarrow 2^{\mathbf{Obj}} \cup \mathbb{N}$ *such that* $\widehat{\mathbf{M}}(p) \cap \mathbf{Obj} \neq \emptyset \Rightarrow \widehat{\mathbf{M}}(p) \cap \mathbb{N} = \emptyset$ *for all* $p \in P$.

The preceding definitions from [Val98] do not cover the occurrence rule of object systems. We will briefly and informally summarise the main impact of the occurrence rule: The tokens in object systems may exhibit a dynamic behaviour since they are themselves Petri nets. In addition they can synchronise with one another or with the system net. Synchronisation with other object nets is not in principle restricted to the case where they occupy the same place. The occurrence rule for object systems is constructed to allow the distributed parallel execution of an object net in the presence of a strict fork-and-join structure, i.e. partially executed object nets may only be joined if their processes are compatible in the sense that they can all be extended to a valid object net process. For a detailed discussion of the occurrence rule the reader is referred to [Val98].

We generalise elementary object systems (EOSs) so that the object net may be an arbitrary P/T net. The system net still has to be an elementary net system.

Definition 16.4.9 (Generalised EOS).
A generalised elementary object system is an object system that satisfies the following conditions:

1. *The system net* $\langle P_S, T_S, F_S, W_S, \mathbf{m}_S \rangle$ *is an elementary net system.*
2. *The object net* $\langle P_O, T_O, F_O, W_O, \mathbf{m}_O \rangle$ *is an ordinary P/T net system.*
3. *The interaction relation* $\rho \subseteq T_S \times T_O$ *is an arbitrary relation between system and object transitions.*

Enablement of transitions is defined as in elementary object systems.

Theorem 16.4.10. *The reachability problem for generalised elementary object systems relative to the system net occurrence sequences is undecidable.*

Proof. We show that nondeterministic (sometimes called generalised) Petri nets are reducible to generalised elementary object systems such that the admissible processes of the nondeterministic net are exactly the admissible system net processes of the object system.

Note that for any transition t of a nondeterministic Petri net its post-set $t^\bullet$ is a set, i.e. $t^\bullet \subseteq 2^P$, where 2^P denotes the powerset of P. If $\text{card}(t^\bullet) > 1$ then t is a nondeterministic transition, otherwise t is an ordinary transition.

Nondeterministic Petri nets have an undecidable reachability problem as proved by Kanovich in [Kan94]. Thus reachability in GEOSs is also undecidable for any given system net process.

The simulating GEOS will be built of a system net which consists of only one place connected to transitions named exactly as all those of the nondeterministic Petri net. The firing of these system net transitions will be restricted by the interaction relation that allows for a choice of transitions in case the original transition was nondeterministic. The nondeterministic transition from Figure 16.11 would be simulated by the GEOS from Figure 16.12. The interaction relation is indicated by $\langle 1 \rangle$, meaning that the system net transition must interact with one of the object net transitions marked by the same number. Both partial constructions must then be applied to all remaining transitions of the nondeterministic net.

We will now formalise the aforementioned reduction:

To construct an equivalent generalised elementary object system $OS = (SN, ON, \rho)$ for a given nondeterministic Petri net system $\mathcal{S}(\mathbf{m}) = \langle \mathcal{N}, \mathbf{m} \rangle$ with sets of places P and transitions T, flow relation F and weight function W,[37] we have to construct a system net SN, an object net ON, and an appropriate interaction relation ρ. We assume without loss of generality that for all $t \in T$ the set of post-conditions is $t^\bullet = \{X_0, \ldots, X_{n_t}\}$, where $X_i \in 2^P$ with associated weight functions $W_i(t, p)$ for all $p \in X_i$, $i \in \{0, \ldots, n_t\}$ and $n_t := \text{card}(t^\bullet)$.

[37] Note that the weight function is deterministic for incoming arcs but is composed of several possibilities for the nondeterministic outgoing branches!

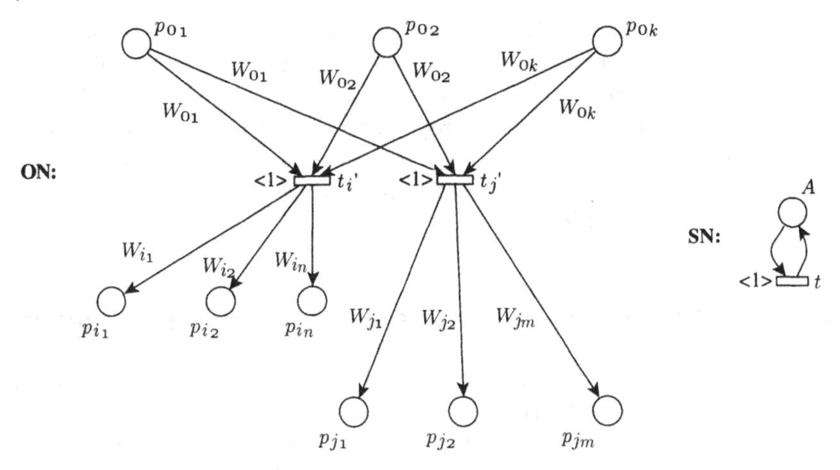

Fig. 16.12. Example of a nondeterministic transition simulated by a GEOS

1. $SN = \langle \{A\}, T, F_S, W_S, \mathbf{m}_S \rangle$, with $\mathbf{m}_S[A] = 1$, and $\forall t \in T.W_S(A,t) = 1 \wedge W_S(t,A) = 1 \wedge (A,t) \in F_S \wedge (t,A) \in F_S$
2. $ON = \langle P, T_O, F_O, W_O, \mathbf{m} \rangle$, with $T_O = \{t_i' \mid t \in T \wedge i \in \{1,\ldots,n_t\}\}$, $\forall t \in T\,.\forall i \in \{0,..,n_t\}\,.\,(p,t) \in F \Rightarrow (p,t_i) \in F_O \wedge W_O(p,t_i) = W(p,t)$ and $\forall t \in T.\forall i \in \{0,..,n_t\}.\forall p \in X_i.(t_i,p) \in F_O \wedge W_O(t_i,p) = W_i(t,p)$
3. $\rho = \{(t,t_i') \mid t \in T \wedge i \in \{1,\ldots,n_t\}\}$

It is left to the reader to show formally that the set of system net occurrence sequences of OS determined by the set of SN-processes for OS, is equal to the set of occurrence sequences of $\mathcal{S}$, and that the resulting markings of N and ON correspond to each other with respect to the simulation relation defined above.

Clearly for each possible post-set of any nondeterministic transition in $\mathcal{S}$ there exists exactly one transition in the object net of the corresponding object system OS. On the other hand, there are no other transitions to choose from unless there is also a choice of the nondeterministic transition to fire in $\mathcal{S}$. The reverse direction is also true, i.e. whenever some system net transition is enabled according to the firing rule of object nets, the corresponding nondeterministic transition is also enabled if we transfer the object net marking to $\mathcal{S}$.

16.4.6 Bibliographic Remarks

Introductions to linear logic, apart from Girard's original paper [Gir87], are available in [Sce93], [Tro93], [Tro92], and [Wad93]. For a thorough treatment of decidability and complexity issues of the various fragments of linear logic we refer the reader to [LSS92]. Some remarks on the possibility of backward reasoning can be found in [GPC95]. A detailed account of linear logic as a semantic framework for object Petri nets can be found in [Far99a]. The topic of dynamic modifications of the Petri net structure is tackled in [Far99c] and [Far00a]. A discussion of issues related to value semantics and reference semantics for object Petri nets has recently appeared in [Val00], while a comparison with the object Petri net formalism of Lomazova, called *nested Petri nets*, can be found in [Far00b].

16.5 Verifying Petri Net Models Using Process Algebra

In this section, we describe an example of the combined application of two formal methods, namely Petri nets and process algebra ([BW90]), to design and verify a distributed system. First, we briefly explain the engineering method that is the basis for combining the two methods. The key aspect is that it integrates system design with system verification. Secondly, we explain the net model and the process algebra used. Thirdly, we apply the integrated method to the development of a simple production unit. Finally, we explain what extensions of the method are possible and what still needs to be done to make the method useful in practice. An extensive treatment of the material presented in this section can be found in [Bas98, Chapter 3].

16.5.1 Method

Petri nets and process algebra are both formal methods that focus on the dynamic behaviour of systems. In distributed systems, the order in which communications between different parts of the system occur is often crucial. Does a indeed always happen before b? Can the system receive c while it is waiting for d? We are less interested in questions related to the data structures being used in the implementation of the system, and the correctness of computations local to some specific part of the system. This does not mean that these issues are less important. However, other methods, such as assertional reasoning based on predicate logic, are more useful for those purposes. As a consequence, the method presented in this section is particularly useful for the design of distributed systems in which the communication protocol between the various parts of the system plays an important role. It is not well suited for data-oriented applications.

However, if Petri nets and process algebra both focus on dynamic system behaviour, what then is the use of combining them? The answer is simple: They complement each other very well. Petri nets have an easy-to-understand graphical representation and are well suited for describing the dynamic behaviour of a system including the states in which the system can be. Hence, Petri nets are very useful for purposes of system validation and simulation. Process algebra, on the other hand, is a compositional, purely symbolic formalism, designed to compare the dynamic behaviour of different systems.

It is most often used in verification. By applying term rewriting techniques and equational reasoning, it can be verified whether an implementation satisfies a given specification, where both specification and implementation are algebraic terms.

Based on the above observations, we arrive at the following engineering method consisting of four separate activities. Each activity is explained in more detail below.

1. Give a specification of the system in process-algebraic terms;
2. Construct a Petri net model of the system;
3. Use simulation to validate the specification and to test whether the Petri net conforms to the specification;
4. Verify the correctness of the net with respect to its specification.

Algebraic Specification. In the early stages of design, it is useful to specify in a concise way the order in which the actions in a system can occur. Process algebra is well suited for that purpose. Assume, for example, that a system is initialised by executing either an a or a b, after which it performs a c and d in parallel. In process-algebraic terms, this is denoted as follows: $(a+b)\cdot(c \parallel d)$, where $+$ denotes choice, $\cdot$ denotes sequential composition, and $\parallel$ denotes parallel composition.

Petri Net Model. Once one has a clear understanding of the dynamic behaviour of a system, it is usually not very difficult to construct a coloured Petri net whose behaviour conforms to the algebraic specification of the system. The reason for starting with an algebraic specification instead of a Petri net model, is that the former is much more concise and easier to change than the latter. Also, in a coloured Petri net model, many details must be filled in which are not addressed in the specification step, such as data types, timing of events, etc. At this stage, there is not yet a formal relation between the algebraic specification and the Petri net model.

Simulation. The Petri net model of a system can be simulated using tools such as ExSpect [Bak97] or Design/CPN [JCHH91]. The specification and the net model can be validated, and corrected if necessary. Simulation is an excellent means to discover the more obvious mistakes in both the algebraic specification and the coloured net model. However, using simulation alone, it is usually impossible to guarantee that all possible errors are discovered.

Verification. In the verification phase, the Petri net model is formally translated into an algebraic expression using the theory of [BV95a, BV95b]. By means of term rewriting techniques and equational reasoning, it is verified that the behaviour of the net conforms to its algebraic specification. This step is useful for discovering the more subtle mistakes in a specification and the corresponding Petri net.

The presentation of the four activities suggests an order. In an ideal situation, they are indeed applied in the order presented. However, needless to say, in practice they are seldom clearly separated. Systems engineering is a complex process where one often works on several of the four activities simultaneously. If one discovers an error during one step, it may be necessary to redo some of the work in other steps. Also, if one follows a top-down design strategy, it is possible to first design and verify a system at a high level of abstraction before adding more detail. When adding detail, it is of course recommended to repeat the design cycle.

16.5.2 Hierarchical Place/Transition Nets

To illustrate the method of the previous subsection, we use a hierarchical variant of ordinary P/T nets. Since this section focuses on the verification part of the method, which is done in process algebra, a formal definition of hierarchical P/T nets is omitted. Instead, they are explained using the example of Figure 16.13.

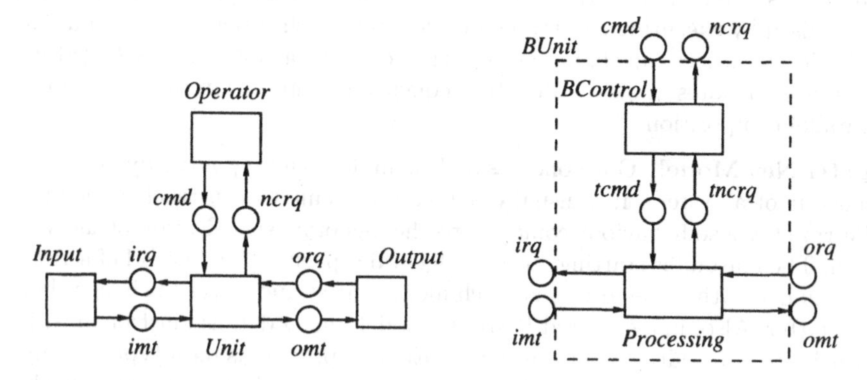

Fig. 16.13. A basic production unit and its environment

The left-hand-side of Figure 16.13 shows a hierarchical P/T net modelling a production unit in its environment. A hierarchical P/T net may contain places, transitions, and *sub-nets*. The net in the example consists of four sub-nets connected via places. It does not have any transitions. Sub-nets may be instantiated with other hierarchical P/T nets. For example, sub-net *Unit* is instantiated with the hierarchical net *BUnit*, which models a basic unit and is depicted on the right-hand-side of Figure 16.13. The net *BUnit* consists itself of two sub-nets. The dashed box divides *BUnit* into an internal and an external part, yielding a so-called *open* net. The external part or *interface* may consist only of places, which are called *pins*. The other places of the net are the *internal* places. When using an open net in a higher-level net, its pins must be mapped onto places. In the example of Figure 16.13, the mapping is simply the identity mapping. The absence of a dashed box means that a net is *closed*.

Since transitions are not always visible in a hierarchical P/T net, it is difficult to describe the behaviour of such a net in terms of transition firings. Instead, the behaviour of hierarchical P/T nets can be expressed nicely in terms of (simultaneous) consumption and production of tokens. For example, the *Operator* sub-net in Figure 16.13 can send a command to the unit by putting a token in place *cmd*. The unit may receive the command by consuming the token, after which it can request input by putting a token in place *irq*.

Based on this observation, it is possible to give concrete form to the goal of the specification step of our method. At this step, we must specify the order in which consumption and production of tokens may occur. The possible behaviour given above for the interaction between a unit and its environment, for example, could act as a (partial) specification for a unit. This automatically yields the goal of the verification step. We must verify whether the observed behaviour of an implementation of a net satisfies its specification. For example, considering net *BUnit* in Figure 16.13, we might see the following behaviour. Commands received through *cmd* are transferred to *Processing* via *tcmd*. After receiving the command, *Processing* issues an input request by putting a token in place *irq*. If we abstract from the command transfer between *BControl* and *Processing*, then the behaviour of *BUnit* conforms to the behaviour specified for the interaction between the unit and its environment. Thus, we may conclude that *BUnit* satisfies (this part of) its specification.

Other details of the example in Figure 16.13 are not yet important. They are further explained in Section 16.5.4, where the method of the previous subsection is applied to the development of a production unit. In the next subsection, we introduce a process algebra which is designed to reason about the token game as described above.

16.5.3 A Brief Introduction to Process Algebra

This section briefly introduces a process algebra in the style of ACP[BW90] which can be used to reason about hierarchical P/T nets. The algebra is taken from [BV95b].

An ACP-like process algebra consists of a *signature* and a set of *axioms*. The signature defines the *sorts* of the algebra and its *functions* including the constants. The functions of an algebraic theory can be used to build *terms* representing *processes*. In our case, processes can be thought of as P/T nets. The axioms state which process terms are considered equal.

The algebraic theory for reasoning about P/T nets is parametrised with a universe of constants U, which are identifiers of places in a P/T net. Other sorts in the signature of the theory are the sort of atomic actions A, the sort of actions AC, and the sort of processes P; each atomic action is an action and each action is a process ($A \subseteq AC \subseteq P$).

An atomic action is either a consumption of a token from a place a in U, denoted as $a?$, or a production of a token in a, denoted $a!$. An action consists of an arbitrary number of simultaneous consumptions and/or productions. A function $\_ \mid \_ : AC \times AC \to AC$, called the *synchronous merge*, is used to construct actions. For example, $a? \mid a? \mid b!$ denotes the consumption of two tokens from place a and the production of one token in place b. Sort AC contains one special action τ which denotes the unobservable or silent action. The introduction of this action is useful for hiding the internal behaviour of open P/T nets.

The most simple processes in our theory are the actions and a special process δ, called *deadlock* or *inaction*. Other processes can be constructed by means of the following operators, which are all functions of type $P \times P \to P$: $\_ + \_$, denoting choice, $\_ \cdot \_$, denoting sequential composition, $\_ ^* \_$, called the binary Kleene star, denoting iteration, $\_ \| \_$, called the merge, denoting parallel composition, and $\_ \lfloor\!\lfloor \_$, called the left merge, also denoting parallel composition but with the restriction that the first action is executed by the left operand. Axioms for these operators, which may clarify their intuitive meaning, can be found in Table 16.2. The operators which have not yet been mentioned are explained below. The binding precedence of operators is as follows. Unary operators bind stronger than binary operators. Sequential composition and the Kleene star bind stronger than all other binary operators. Choice binds weaker than all other operators. In Table 16.2, a is a place label in U; d is either an action in AC or δ; e, f, and g are actions in AC, and x, y; and z are processes in P.

Some of the axioms for the above-mentioned operators may need an explanation. Axiom $A4$ states the right distributivity of sequential composition over choice. The fact that the left-distributivity axiom is absent implies that processes which have different moments of choice are considered to be different. That is, our theory is a so-called *branching-time* theory. Axioms $A6$ and $A7$ show that δ is indeed a natural representation of inaction. Axioms $M1$ through $M4$ define parallel composition using the left merge as an auxiliary operator. It follows from these axioms that the process algebra presented in this section is an *interleaving* theory. Axioms $ASC1$ and $ASC2$ are the so-called axioms of standard concurrency, defining some desirable properties of parallel composition. $BKS1$ through $BKS3$ axiomatise the binary Kleene star. Axiom $BKS1$ shows that process $x ^* y$ means zero or more repetitions of process x followed by a single execution of y. Axiom AT says that only the visible part of an action is observed. Axioms $B1$ and $B2$ state that silent actions can be removed provided the moments of choice remain the same.

The operators introduced so far can be used in the specification step of a design process to specify the dynamic behaviour of the system under development. In general, such a specification takes the form of one or more terms containing only the above-mentioned operators. The two operators that have not been mentioned so far are used in verifying the behaviour of a P/T net against its specification. The most important one is the *causal state operator* $\lambda^{\_}_{\_}\_$.

Table 16.2. Basic axioms for reasoning about equality of processes

$x + y = y + x$	$A1$	$e \mid f = f \mid e$	$S1$
$(x + y) + z = x + (y + z)$	$A2$	$(e \mid f) \mid g = e \mid (f \mid g)$	$S2$
$x + x = x$	$A3$		
$(x + y) \cdot z = x \cdot z + y \cdot z$	$A4$	$x \parallel y = x \lfloor\!\lfloor y + y \lfloor\!\lfloor x$	$M1$
$(x \cdot y) \cdot z = x \cdot (y \cdot z)$	$A5$	$d \lfloor\!\lfloor x = d \cdot x$	$M2$
$x + \delta = x$	$A6$	$d \cdot x \lfloor\!\lfloor y = d \cdot (x \parallel y)$	$M3$
$\delta \cdot x = \delta$	$A7$	$(x + y) \lfloor\!\lfloor z = x \lfloor\!\lfloor z + y \lfloor\!\lfloor z$	$M4$
$\lambda_{\mathbf{s}}^{I}(\delta) = \delta$	$CSO1$	$(x \lfloor\!\lfloor y) \lfloor\!\lfloor z = x \lfloor\!\lfloor (y \parallel z)$	$ASC1$
$\mathbf{c}e \restriction I \subseteq \mathbf{s} \ \Rightarrow \ \lambda_{\mathbf{s}}^{I}(e) = e$	$CSO2$	$(x \parallel y) \parallel z = x \parallel (y \parallel z)$	$ASC2$
$\mathbf{c}e \restriction I \not\subseteq \mathbf{s} \ \Rightarrow \ \lambda_{\mathbf{s}}^{I}(e) = \delta$	$CSO3$		
$\lambda_{\mathbf{s}}^{I}(e \cdot x) =$		$x \,^{*}\, y = x \cdot (x \,^{*}\, y) + y$	$BKS1$
$\quad \lambda_{\mathbf{s}}^{I}(e) \cdot \lambda_{\mathbf{s}-\mathbf{c}e+\mathbf{p}e\restriction I}^{I}(x)$	$CSO4$	$x \,^{*}\, (y \cdot z) = (x \,^{*}\, y) \cdot z$	$BKS2$
$\lambda_{\mathbf{s}}^{I}(x + y) = \lambda_{\mathbf{s}}^{I}(x) + \lambda_{\mathbf{s}}^{I}(y)$	$CSO5$	$x \,^{*}\, (y \cdot ((x + y) \,^{*}\, z) + z) = (x + y) \,^{*}\, z$	$BKS3$
$e \mid \tau = e$	AT	$x \cdot \tau = x$	$B1$
		$x \cdot (\tau \cdot (y + z) + y) = x \cdot (y + z)$	$B2$
$a \in I \ \Rightarrow \ \tau_I(a?) = \tau$	$TAC1$	$a \in I \ \Rightarrow \ \tau_I(a!) = \tau$	$TAP1$
$a \notin I \ \Rightarrow \ \tau_I(a?) = a?$	$TAC2$	$a \notin I \ \Rightarrow \ \tau_I(a!) = a!$	$TAP2$
$\tau_I(\delta) = \delta$	TAD	$\tau_I(e \mid f) = \tau_I(e) \mid \tau_I(f)$	$TA1$
$\tau_I(\tau) = \tau$	TAT	$\tau_I(x + y) = \tau_I(x) + \tau_I(y)$	$TA2$
$\tau_I(x \parallel y) = \tau_I(x) \parallel \tau_I(y)$	$TAM1$	$\tau_I(x \cdot y) = \tau_I(x) \cdot \tau_I(y)$	$TA3$
$\tau_I(x \lfloor\!\lfloor y) = \tau_I(x) \lfloor\!\lfloor \tau_I(y)$	$TAM2$	$\tau_I(x \,^{*}\, y) = \tau_I(x) \,^{*}\, \tau_I(y)$	$TA4$

The intuitive meaning of $\lambda_{\mathbf{s}}^{I}(x)$, where I is a set of place identifiers, $\mathbf{s}$ a bag of place identifiers, and x a process, is a P/T net x with internal places I and marking $\mathbf{s}$. The operator is axiomatised by five axioms, $CSO1$ through $CSO5$. The auxiliary functions $\mathbf{c}$ and $\mathbf{p}$, yielding for each action in AC the bag of its consumed and produced tokens respectively, are defined as follows. For all $a \in U$ and $e \in AC$, $\mathbf{c}\tau = \emptyset$, $\mathbf{c}a? = a$, $\mathbf{c}a! = \emptyset$, $\mathbf{c}(a? \mid e) = a + \mathbf{c}e$, and $\mathbf{c}(a! \mid e) = \mathbf{c}e$; $\mathbf{p}\tau = \emptyset$, $\mathbf{p}a? = \emptyset$, $\mathbf{p}a! = a$, $\mathbf{p}(a? \mid e) = \mathbf{p}e$, and $\mathbf{p}(a! \mid e) = a + \mathbf{p}e$. The notation $\mathbf{x} \restriction D$ means that a bag $\mathbf{x}$ is restricted to some domain D. Axiom $CSO2$ can now be read as follows. It states that an action e may occur provided all the tokens it consumes from *internal* places are available in the marking $\mathbf{s}$. Axiom $CSO3$ says that if not enough tokens are available, the action cannot be executed, resulting in a deadlock. The reason for not considering pins in the marking is that we want to determine the behaviour of a P/T net under the assumption that the environment is responsible for producing tokens into and consuming tokens from pins. Axiom $CSO4$ states that the result of executing an action e is that consumed tokens are removed from the marking, whereas tokens produced in internal places are added to it. Axioms $CSO1$ and $CSO5$ should be clear without further explanation.

The last operator in the theory is the operator τ_I. It renames consumptions from and productions into places from $I \subseteq U$ to the silent action τ. It is mainly used to hide the internal behaviour of open P/T nets.

The axioms of Table 16.2 are the basis of every verification. However, they are not always sufficient. To conclude this introduction to process algebra, one more axiom is given. The *Recursive Specification Principle* for the binary Kleene star (*RSP**) is a conditional axiom which gives for iterative processes a solution in terms of the binary Kleene star. The requirement "x guarded in $y \cdot x$" means that y cannot terminate successfully without executing at least one visible action. A formal definition of guardedness is omitted here and can be found in [Bas98, BV95a].

$$x, y, z : P$$

$$\frac{x = y \cdot x + z, \qquad x \text{ guarded in } y \cdot x}{x = y \mathbin{^*} z} \qquad RSP^*$$

The idea of verifying the behaviour of a P/T net against a specification is now straightforward. First, an algebraic term is constructed from the net, representing its observable behaviour. Secondly, the algebraic framework presented in this subsection is used to prove that the observable behaviour is equal to the specification. The production-unit example in the next section will clarify this approach. The formal definition of the algebraic semantics of hierarchical P/T nets based on the process algebra of this subsection can be found in [BV95a, BV95b]. In [Bas98, Chapter 3], an extensive treatment of algebraic semantics for P/T nets is given.

16.5.4 The Production Unit

In this subsection, we apply the method introduced in the previous sections to the development of a production unit. This case study is a simplified version of a problem that originated from Dutch industry.

Informal System Requirements. The informal description of the production unit and its required behaviour is as follows. The most simple version of the unit can perform a single operation on a single piece of unprocessed material. First, it must receive a command. Then it requests material, performs the operation specified in the command, and waits for an output request. Upon receiving an output request, it delivers the processed material. Typically, the processing of material takes the most time in the whole system. While the unit is processing material, it must be ready to receive the next command. In this way, the delay between two operations in a unit is minimised.

It must be possible to combine several basic units into a more complex unit by connecting them in series. For this purpose, a *generic* controller must be developed that controls *two* units. These two units are not necessarily basic. They can be more complex units built from basic units and controllers. In this way, it must be possible to construct arbitrarily long series of basic units, using only two components, namely the basic unit and the generic controller. Therefore, it is essential that the interface behaviour of a series of basic units is similar to the behaviour of a single basic unit.

The last requirement in particular is rather vague, but, as is often the case in practice, that is the way the problem was phrased. The case study is a simplified version of the real case, because we assume that no errors can occur during the processing of the material. In reality, this is not true. In the concluding remarks, we briefly return to this point.

Below, we discuss the design and verification of the basic unit. At the end of this subsection, we say a few words about the generic control component and the more complex units.

Algebraic Specification. First, we start with the algebraic specification of a basic unit. From the informal description above, we derive the following actions that a unit can perform:

cmd? : receive a command	*ncrq*! : send a new-command request
irq! : request input material	*orq*? : receive an output request
imt? : receive input material	*omt*! : deliver output material

The behaviour of a basic unit can now be specified as follows:

$$BUnit = cmd? \cdot ((irq! \cdot imt? \cdot (ncrq! \cdot cmd? \parallel orq? \mid omt!))^* \delta) \qquad (16.20)$$

Note that it can be derived from the axioms in Table 16.2 that for any process x, $x^*\delta = x \cdot (x^*\delta)$. Hence, $x^*\delta$ denotes a non-terminating repetition of process x.

The above specification deserves a few words of explanation. It is not difficult to see that *BUnit* first receives a command and then enters a non-terminating repetition. Each cycle of the repetition starts with an input request followed by the receipt of material. Upon receiving material, the unit starts its processing task, which does not result in a visible action. It continues by sending a new-command request followed by the receipt of the next command, and *in parallel* it may deliver the output material, provided of course that the processing is completed. After it has delivered the output *and* received a new command, it starts with the next cycle of its non-terminating loop. In the above specification, the unit refuses to receive an output request if the output material is not available. Only if both the request and the material are available, is the request processed and the material delivered. A slightly different, but equally correct, possibility is to let the unit first receive the output request and then deliver the material.

The next step is to construct a net model of the basic unit. Although in this section, we use P/T nets, the actual case study was done with the tool ExSpect which uses coloured Petri nets. However, as mentioned earlier, we omit all the details about data structures and timing, and focus instead on the hierarchical construction and the communication structure.

Petri Net Model. We have already seen the Petri net model of the basic unit and its environment in Figure 16.13. The environment consists of an input system, an output system, and an operator. The environment is simply given for the sake of completeness. We do not go into detail about any of these three sub-nets. At the interface of the basic unit, we recognise the places we could expect after reading the list of actions a basic unit can perform. As mentioned, the implementation of the basic unit is divided into two separate subsystems, a control system, *BControl*, and a processing system, *Processing*. The control system transfers commands and new-command requests between the environment and the processing system.

The next two steps are simulation and verification. However, for this purpose, we need to have at least a specification of the systems *BControl* and *Processing*. Given such specifications, it is in principle possible to simulate the processing unit without detailing the net models of *BControl* and *Processing*. However, current tools such as ExSpect and Design/CPN do not allow simulation of such incomplete net models. Therefore, we proceed as follows. First, we give an algebraic specification of the systems *BControl* and *Processing*. Then we verify that the unit behaves correctly, provided the subnets for *BControl* and *Processing* satisfy their specification. Next, net models for *BControl* and *Processing* are given. Finally, the complete model of the basic unit is simulated and verified.

Algebraic Specification. The specification for *BUnit* shows that the receipt of a command and the sending of a new-command request alternate. Given the informal requirement that *BControl* transfers commands and new-command requests between the environment and the processing system, the following specification for *BControl* makes sense:

$$\mathit{BControl} = (\mathit{cmd}? \mid \mathit{tcmd}! \cdot \mathit{tncrq}? \mid \mathit{ncrq}!) * \delta \tag{16.21}$$

For the processing part, we arrive at the following specification. Actions are processed simultaneously as much as possible. Upon receiving a command, an input request is sent; upon receiving material, a new command is requested; and upon receiving an output request, the output material is delivered.

$$\mathit{Processing} = (\mathit{tcmd}? \mid \mathit{irq}! \cdot \mathit{imt}? \mid \mathit{tncrq}! \cdot \mathit{orq}? \mid \mathit{omt}!) * \delta \tag{16.22}$$

Note that in terms which are equal up to associativity, brackets are often omitted. It may be surprising that neither *BControl* nor *Processing* contains explicit parallelism. However, if we have a close look at the specifications, we see that *Processing* produces a token in place *tncrq* at the same time that it receives input material. Hence, the transfer of the new-command request to the environment by system *BControl* can occur in parallel with the delivery of output material by system *Processing*. This is exactly the desired result. The informal argument of this paragraph is confirmed by the verification in the next step of the design cycle.

Verification. Given the specifications for *BControl* and *Processing*, we can now verify the behaviour of the basic unit. The formal algebraic semantics for the behaviour of the Petri net *BUnit* is the following algebraic term (see [BV95a, BV95b]):

$$\tau_I \circ \lambda_\emptyset^I(BControl \parallel Processing) \ ,$$

where I is the set of internal places $\{tcmd, tncrq\}$. The above term can be interpreted as follows. The *unrestricted* behaviour of system *BUnit* is simply the parallel composition of the behaviour of all its components. The causal state operator $\lambda_\emptyset^I$ restricts this behaviour to all possible firing sequences allowed by the initial marking, which in the above case is the empty bag of tokens. The abstraction operator τ_I hides consumptions from and productions into internal places. By means of equational reasoning, it can be shown that the behaviour of the basic unit satisfies its specification. That is,

$$\tau_I \circ \lambda_\emptyset^I(BControl \parallel Processing) = BUnit \ .$$

The details of the verification are as follows. First, we introduce some abbreviations.

$tc = cmd? \mid tcmd!$	transfer command
$tn = tncrq? \mid ncrq!$	transfer new-command request
$pc = tcmd? \mid irq!$	process command
$ri = imt? \mid tncrq!$	receive input
$po = orq? \mid omt!$	produce output

The first part of the verification considers the complete behaviour of the basic unit without hiding internal details. Only after we have derived a simple expression for $\lambda_\emptyset^I(BControl \parallel Processing)$ will its internal behaviour be hidden.

$$
\begin{aligned}
& \lambda_\emptyset^I(BControl \parallel Processing) \\
= \ & \{ \text{ Specifications } BControl \text{ and } Processing;\ BKS1, A6 \text{ (Both } 2\times) \ \} \\
& \lambda_\emptyset^I(tc \cdot tn \cdot BControl \parallel pc \cdot ri \cdot po \cdot Processing) \\
= \ & \{ \ M1 \ \} \\
& \lambda_\emptyset^I(tc \cdot tn \cdot BControl \ ⌊\!⌊ \ pc \cdot ri \cdot po \cdot Processing \\
& \qquad + pc \cdot ri \cdot po \cdot Processing \ ⌊\!⌊ \ tc \cdot tn \cdot BControl) \\
= \ & \{ \ M3, A6, BKS1 \text{ (All } 2\times) \ \} \\
& \lambda_\emptyset^I(tc \cdot (tn \cdot BControl \parallel Processing) \\
& \qquad + pc \cdot (ri \cdot po \cdot Processing \parallel BControl)) \\
= \ & \{ \ CSO5, CSO4 \ (2\times), CSO2, CSO3 \ \} \\
& tc \cdot \lambda_{tcmd}^I(tn \cdot BControl \parallel Processing) \\
& \qquad + \delta \cdot \lambda_\emptyset^I(ri \cdot po \cdot Processing \parallel BControl) \\
= \ & \{ \ A7, A6 \ \} \\
& tc \cdot \lambda_{tcmd}^I(tn \cdot BControl \parallel Processing)
\end{aligned}
$$

Summarising, the above derivation yields

$$\lambda_\emptyset^I(BControl \parallel Processing) = tc \cdot \lambda_{tcmd}^I(tn \cdot BControl \parallel Processing) \ .$$

In a similar way, it is possible to derive the following results.

$$\begin{aligned}
&\lambda^I_{tcmd}(tn \cdot BControl \parallel Processing) = \\
&\qquad pc \cdot \lambda^I_\emptyset(tn \cdot BControl \parallel ri \cdot po \cdot Processing) \\
&\lambda^I_\emptyset(tn \cdot BControl \parallel ri \cdot po \cdot Processing) = \\
&\qquad ri \cdot \lambda^I_{tncrq}(tn \cdot BControl \parallel po \cdot Processing) \\
&\lambda^I_{tncrq}(tn \cdot BControl \parallel po \cdot Processing) = \\
&\qquad tn \cdot \lambda^I_\emptyset(BControl \parallel po \cdot Processing) \\
&\qquad + po \cdot \lambda^I_{tncrq}(tn \cdot BControl \parallel Processing) \\
&\lambda^I_\emptyset(BControl \parallel po \cdot Processing) = \\
&\qquad tc \cdot \lambda^I_{tcmd}(tn \cdot BControl \parallel po \cdot Processing) \\
&\qquad + po \cdot \lambda^I_\emptyset(BControl \parallel Processing) \\
&\lambda^I_{tcmd}(tn \cdot BControl \parallel po \cdot Processing) = \\
&\qquad po \cdot \lambda^I_{tcmd}(tn \cdot BControl \parallel Processing) \\
&\lambda^I_{tncrq}(tn \cdot BControl \parallel Processing) = tn \cdot \lambda^I_\emptyset(BControl \parallel Processing)
\end{aligned}$$

The above equations can be combined to derive a result for the expression $\lambda^I_{tcmd}(tn \cdot BControl \parallel Processing)$. The reason for considering this expression and not any of the others, is that the specification of *BUnit* tells us that we may expect $\lambda^I_{tcmd}(tn \cdot BControl \parallel Processing)$ to be the beginning of an iteration.

$$\begin{aligned}
&\lambda^I_{tcmd}(tn \cdot BControl \parallel Processing) \\
=&\quad \{ \text{ Repeated substitution } \} \\
&pc \cdot ri \cdot (\, tn \cdot (\, tc \cdot po \cdot \lambda^I_{tcmd}(tn \cdot BControl \parallel Processing) \\
&\qquad\qquad +po \cdot tc \cdot \lambda^I_{tcmd}(tn \cdot BControl \parallel Processing) \\
&\qquad\qquad) \\
&\qquad +po \cdot tn \cdot tc \cdot \lambda^I_{tcmd}(tn \cdot BControl \parallel Processing) \\
&\qquad) \\
=&\quad \{ \, A4 \; (2\times) \, \} \\
&pc \cdot ri \cdot (tn \cdot (tc \cdot po + po \cdot tc) + po \cdot tn \cdot tc) \\
&\qquad \cdot \lambda^I_{tcmd}(tn \cdot BControl \parallel Processing) \\
=&\quad \{ \text{ Derivation below } \} \\
&pc \cdot ri \cdot (tn \cdot tc \parallel po) \cdot \lambda^I_{tcmd}(tn \cdot BControl \parallel Processing)
\end{aligned}$$

$$\begin{aligned}
&tn \cdot (tc \cdot po + po \cdot tc) + po \cdot tn \cdot tc \\
=&\quad \{ \, M2 \; (3\times) \, \} \\
&tn \cdot (tc \mathbin{⌊\!⌊} po + po \mathbin{⌊\!⌊} tc) + po \mathbin{⌊\!⌊} tn \cdot tc \\
=&\quad \{ \, M1 \, \} \\
&tn \cdot (tc \parallel po) + po \mathbin{⌊\!⌊} tn \cdot tc \\
=&\quad \{ \, M3 \, \} \\
&tn \cdot tc \mathbin{⌊\!⌊} po + po \mathbin{⌊\!⌊} tn \cdot tc \\
=&\quad \{ \, M1 \, \} \\
&tn \cdot tc \parallel po
\end{aligned}$$

Axioms *A6* and *RSP** yield

$$\lambda^I_{tcmd}(tn \cdot BControl \parallel Processing) = (pc \cdot ri \cdot (tn \cdot tc \parallel po))^* \delta \ .$$

Note that the guardedness requirement for *RSP** is fulfilled. Combining the above result with the result derived for $\lambda^I_\emptyset(BControl \parallel Processing)$ yields

$$\lambda^I_\emptyset(BControl \parallel Processing) = tc \cdot ((pc \cdot ri \cdot (tn \cdot tc \parallel po))^* \delta) \ ,$$

which already looks very similar to the specification of the basic unit. In the second part of the verification, the internal behaviour in the above result is hidden. Recall that I is the set of internal places $\{tcmd, tncrq\}$.

$$\begin{aligned}
& \tau_I \circ \lambda^I_\emptyset(BControl \parallel Processing) \\
= \quad & \{ \text{ Previous result; substitution of abbreviations } \} \\
& \tau_I(cmd? \mid tcmd! \cdot ((tcmd? \mid irq! \cdot imt? \mid tncrq! \cdot \\
& \qquad\qquad (tncrq? \mid ncrq! \cdot cmd? \mid tcmd! \parallel orq? \mid omt!) \\
& \qquad\quad)^* \delta)) \\
= \quad & \{ \text{ Axioms for the abstraction operator } \tau_I \} \\
& cmd? \mid \tau \cdot ((\tau \mid irq! \cdot imt? \mid \tau \cdot (\tau \mid ncrq! \cdot cmd? \mid \tau \parallel orq? \mid omt!))^* \delta) \\
= \quad & \{ AT \} \\
& cmd? \cdot ((irq! \cdot imt? \cdot (ncrq! \cdot cmd? \parallel orq? \mid omt!))^* \delta)
\end{aligned}$$

Hence,

$$\begin{aligned}
& \tau_I \circ \lambda^I_\emptyset(BControl \parallel Processing) \\
= \ & cmd? \cdot ((irq! \cdot imt? \cdot (ncrq! \cdot cmd? \parallel orq? \mid omt!))^* \delta) \\
= \ & BUnit \ ,
\end{aligned}$$

which is the desired result. We have formally *proved* that the behaviour of the P/T net *BUnit* is indeed as given by its algebraic specification of Equation 16.20, provided of course that the sub-nets *BControl* and *Processing* are implemented according to their specifications of Equations 16.21 and 16.22. So the next step is to give the net models for those two systems.

Petri Net Model. Figure 16.14 shows the Petri net models of *BControl* and *Processing*. System *BControl* keeps track of the state of the unit. A unit is either *ready* to receive a command or *busy* processing. System *Processing* can receive a command if it is *empty.* Upon receiving a command, it sends an input request and enters a state called *waiting for material.* Next, it receives *raw material.* Processing turns raw material into *processed material.* Finally, the unit delivers the processed material, returning to the empty state.

Simulation and Verification. At this point, the P/T net model for the basic unit is complete, which means that it can be simulated using tools such as ExSpect and Design/CPN. Since the net of the production unit is very simple, the simulation results are not very surprising. Therefore, we do not go into details.

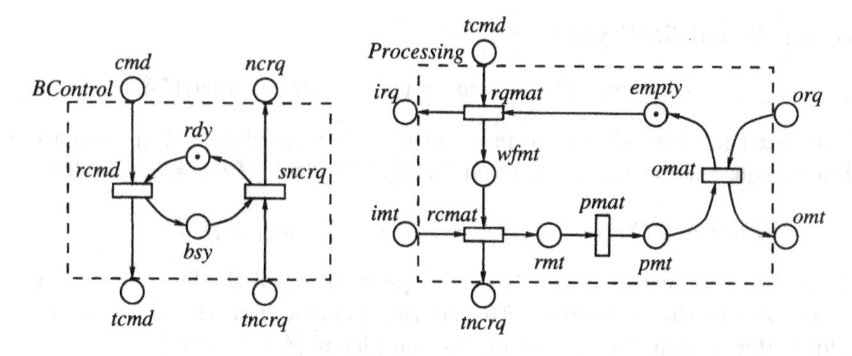

Fig. 16.14. The subsystems of the basic unit

The final step in the design cycle of the production unit is the verification of the systems *BControl* and *Processing*. Details are left to the reader. We simply state the results. Transition names are used as abbreviations for the synchronous merge of their consumptions and productions. The algebraic semantics for *BControl* is as follows.

$$\tau_I \circ \lambda^I_{rdy}(rcmd \parallel sncrq) \ ,$$

where I is the set of internal places $\{rdy, bsy\}$. The semantics for *Processing* is the following term.

$$\tau_J \circ \lambda^J_{empty}(rqmat \parallel rcmat \parallel pmat \parallel omat) \ ,$$

where J is the set of places $\{empty, wfmt, rmt, pmt\}$. It is straightforward to verify that these two terms are equal to the specifications of *BControl* and *Processing* given in Equations 16.21 and 16.22. The verifications follow the pattern of the verification of *BUnit*. However, the verification of *Processing* is very tedious when only the basic axioms of Table 16.2 plus RSP^* can be used. The so-called *expansion theorem*, which is a generalisation of $M1$ to an arbitrary number of processes in parallel, greatly simplifies the calculations. The theorem can be found in [BV95a, BV95b].

As a consequence of the previous verification steps, we may conclude that the complete hierarchical net model of the basic production unit, given in Figures 16.13 and 16.14, satisfies its specification.

Properties of a Basic Unit. At this point, it is time to return to the requirement that the behaviour of a series of basic units must be similar to the behaviour of a single basic unit. We ask ourselves: What are the fundamental properties in the dynamic behaviour of a basic unit? The most obvious one is related to the interface between the unit and the operator. As we have seen already, the receipt of a command and the sending of a new-command request alternate. We can prove this property formally. Hiding all actions in specification *BUnit* of Equation 16.20 except *cmd*? and *ncrq*! yields the following behaviour:

$$(cmd? \cdot ncrq!) * \delta \ , \tag{16.23}$$

which shows that the basic unit of Figures 16.13 and 16.14 indeed alternates commands and new-command requests. The calculations are as follows. Let I be the set $\{irq, imt, orq, omt\}$.

$$\begin{aligned}
& \tau_I(BUnit) \\
= & \quad \{ \text{ Equation 16.20 } \} \\
& \tau_I(cmd? \cdot ((irq! \cdot imt? \cdot (ncrq! \cdot cmd? \parallel orq? \mid omt!)) * \delta)) \\
= & \quad \{ \text{ Axioms of the abstraction operator } \tau_I \} \\
& cmd? \cdot ((\tau \cdot \tau \cdot (ncrq! \cdot cmd? \parallel \tau \mid \tau)) * \delta) \\
= & \quad \{ B1, AT \} \\
& cmd? \cdot ((\tau \cdot (ncrq! \cdot cmd? \parallel \tau)) * \delta) \\
= & \quad \{ M1, M3, M2 \} \\
& cmd? \cdot ((\tau \cdot (ncrq! \cdot (cmd? \parallel \tau) + \tau \cdot ncrq! \cdot cmd?)) * \delta) \\
= & \quad \{ M1, M2\ (2\times), B1 \} \\
& cmd? \cdot ((\tau \cdot (ncrq! \cdot (cmd? + \tau \cdot cmd?) + \tau \cdot ncrq! \cdot cmd?)) * \delta) \\
= & \quad \{ A6, B2, A6 \} \\
& cmd? \cdot ((\tau \cdot (ncrq! \cdot cmd? + \tau \cdot ncrq! \cdot cmd?)) * \delta) \\
= & \quad \{ A6, B2, A6 \} \\
& cmd? \cdot ((\tau \cdot ncrq! \cdot cmd?) * \delta)
\end{aligned}$$

This result is almost the result we are looking for. The final step using RSP^* is as follows.

$$\begin{aligned}
& cmd? \cdot ((\tau \cdot ncrq! \cdot cmd?) * \delta) \\
= & \quad \{ BKS1, A6 \} \\
& cmd? \cdot ((\tau \cdot ncrq! \cdot cmd?) \cdot ((\tau \cdot ncrq! \cdot cmd?) * \delta)) \\
= & \quad \{ A5\,(3\times), B1 \} \\
& (cmd? \cdot ncrq!) \cdot (cmd? \cdot ((\tau \cdot ncrq! \cdot cmd?) * \delta))
\end{aligned}$$

Hence, $A6$ and RSP^* yield

$$cmd? \cdot ((\tau \cdot ncrq! \cdot cmd?) * \delta) = (cmd? \cdot ncrq!) * \delta \ ,$$

which is the desired result.

Another characterising property is that the unit behaves as a one-place buffer if we look at the stream of material. Hiding the appropriate actions yields:

$$\tau \cdot ((imt? \cdot omt!) * \delta) \ , \tag{16.24}$$

which states that the unit after initialisation, expressed by the silent action τ, enters a cycle in which it alternates the input of material and the output of material. The details of the proof are left to the reader.

A final property is that the unit can accept two commands before it must produce output. That is, it behaves as a two-place buffer with respect to commands. Hiding the appropriate actions yields the following behaviour:

$$cmd? \cdot ((cmd? \parallel omt!) * \delta) \ , \tag{16.25}$$

which is indeed the behaviour of a two-place buffer. This last property confirms that the unit satisfies the requirement that it must be able to receive a new command while it is still busy processing.

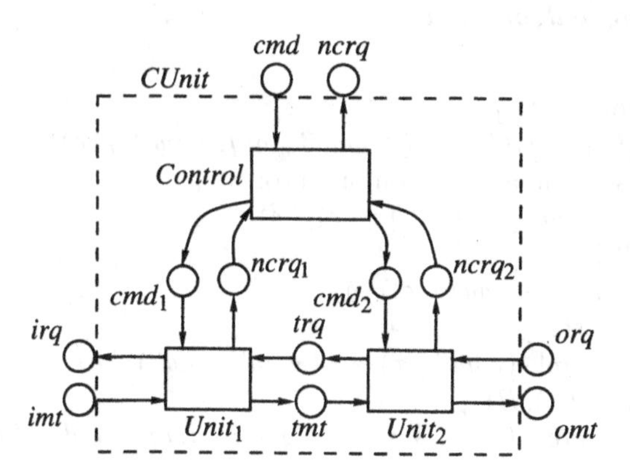

Fig. 16.15. Two units in series

Complex Units. So, what happens if we combine basic units into more complex units? As mentioned, it must be possible to construct arbitrarily long series of basic units, using only two components, namely the basic unit and a generic controller for two units. We do not discuss the entire design trajectory. Figure 16.15 simply shows the high-level net of a complex unit consisting of two units in series. The basic idea is that the control component receives a command from the environment and forwards this to the two units. Commands may be forwarded only if the receiving unit is ready to accept them. In its most simple form, the complex unit consists of two basic units. However, by instantiating $Unit_1$ and $Unit_2$ with complex units themselves, it is possible to construct arbitrarily long series of basic units.

The question is what properties a complex unit must satisfy. Obviously, in order to be compatible with a single basic unit, it must alternate commands with new-command requests. That is, if we calculate the behaviour of system *CUnit* and hide the appropriate actions, it must satisfy requirement 16.23. Furthermore, looking at the stream of material, it is not difficult to see that a complex unit consisting of N basic units should behave as an N-place buffer. Each basic unit can contain one piece of material. Finally, in order to exploit the parallelism in the basic units, a unit consisting of N basic units should be able to receive $N+1$ commands before having to give its first output. These last two properties are generalisations of properties 16.24 and 16.25 derived above for the basic unit.

We conclude this discussion with the results we have obtained for complex units thus far. We have calculated the complete behaviour of a unit consisting of two basic units in series. We also have verified the above three properties for such a unit. The verifications are not difficult for someone experienced with the theory. However, they are very tedious and take many pages. Verifying the properties for a series of three basic units is not recommended without the support of tools. For more details on the production-unit case study, the reader is referred to [Bas98, Section 3.8].

16.5.5 Concluding Remarks

In this section, we have discussed an engineering method combining Petri nets with process algebra. The example of the production unit shows the possibilities of the approach.

To date, the algebraic verifications do not incorporate data; they focus on the order of consumption and production of tokens. An important extension of the method is the inclusion of data in the style of μCRL[GP94]. Such an extension makes it possible to handle pre-conditions, or guards, in coloured nets, which is important since pre-conditions are used to steer the flow of tokens in a net. Furthermore, the induction principles of μCRL make it possible to derive properties for coloured nets with parameters. For example, assume that the production unit of Figure 16.15 is instantiated with one unit of N basic units in series and another unit of M basic units in series. Using the induction sprinciples of μCRL, it is possible to prove the buffer properties of the unit for arbitrary N and M instead of specific instances of N and M. Although such a parametrised proof is more difficult than a proof for specific instances of the parameters, in the long run it can save a lot of work.

Another useful extension of the method is the following. As mentioned, the case study discussed in this section is simplified in the sense that units always operate correctly. However, in reality, processing errors may occur. If such an error occurs, a unit must be reset, after which processing may resume. This means that error-handling facilities must be added to the specifications and the net models. It would be nice if some of the verifications discussed in this section could be reused in verifying the correctness of the extended production units. The theory presented in [AB97a, BA99] can be used for this purpose. However, it needs to be adapted to the framework presented here.

Finally, the method discussed in this section is only useful in practice with proper tool support. It is simply impossible to do any real-world verifications completely by hand. First, it is necessary to develop tools supporting algebraic reasoning in general. Current tools are usually built for a specific process algebra and, hence, not useful for the process algebra used in our method. In a later phase, process-algebra tools and Petri net tools must be combined to form one integrated design environment. An initial study of the possibility of providing support for ACP-style algebraic reasoning in the general proof environment PVS [SRI00] can be found in [BH99].

17. Conclusion

The diversity of the verification methods developed for Petri nets and their extensions may confuse the engineer trying to choose the best approach for his problem. This part of the book aimed to clarify the basics of such a choice by discussing some general issues involved in the design and application of a verification method:

- the net models that the method enables one to verify;
- the types of properties one wants to check;
- the families of methods;
- the interplay of different methods.

On the one hand a net model with a high expressive power such as a coloured Petri net enables us to handle complex systems. On the other hand, this expressive power implies difficulties for the verification process (e.g. complexity increase, semi-decidability, restrictive types of properties).

Each type of analysis method has been presented in a different chapter of this part.

Chapter 14 described the state-based methods. We showed that the computation of the state space may be managed efficiently. Then we introduced more precisely partial-order methods, symmetries, and modular methods. There was an emphasis on the use of symmetries including implementation for well-formed coloured nets where this construction can be completely formalised. This part ended with a comparison of these methods under different criteria such as space and time reduction and property equivalence, and showed how these different methods can be combined.

The subsequent sections of the chapter extended the previous work by taking into account the types of properties that one can check and the impact on the graph construction. An original technique of parametrised construction developed including the verification of temporal logics. Finally, the problem of model checking was discussed in general terms.

In Chapter 15, structural methods were developed. Accurate reduction rules were presented with emphasis on the implicit place. The implicit place has a particular role since it simplifies the structure of the net and enables us to apply other techniques more efficiently. Moreover, implicit places have a strong connection with positive flow computations as shown in the chapter.

Linear algebraic techniques were then developed, and the equivalence between behavioural properties and linear algebraic results was shown. Then siphons and traps were carefully studied since they are the cornerstone of necessary and sufficient conditions for liveness properties.

In the last part of the chapter, some syntactical subclasses were defined showing the behavioural consequences which can be established from the syntactical restrictions. The behavioural properties include fairness, liveness, deadlock-freeness, and the relation between reachable states and linear invariants.

Chapter 16 presented new techniques from an open research field. The first technique was based on a sublanguage of rewriting logic. It combined ideas from term rewriting, logics, and algebraic specification in an executable form. The second technique used assertional reasoning in UNITY for proving properties of Petri nets. Other modal logics were shown to capture the notion of transition enablement, while yet another technique was based on linear logic. The attraction of linear logic is twofold: it gives an operational semantics to Petri nets, and a proof scheme for linear logic gives the proof of a property in the corresponding Petri net. The fifth technique helped us to understand how to benefit from multi-formalisms. This technique started from a specification of the system given in process-algebraic terms. Then it constructed a Petri net model of the system. The Petri net was simulated to exhibit bad behaviour in order to reinforce trust in the model. Finally the Petri net was transformed again into a process algebra such that the two process algebras (modelling the specification and the implementation) were equivalent. Emphasis was put on the design cycle rather than on technical aspects.

Part IV

Validation and Execution

18. Introduction

Validation is one of the central tasks of system development. The modelled system has to match the expectations of the user/client/customer. A variety of possibilities is available. The models can be executed, simulated, animated, inspected, tested, debugged, observed, controlled, etc. This incomplete list can easily be enlarged. However, in this book the emphasis is on the validation of Petri net models for systems engineering applications. This allows us to concentrate on the major areas of Petri net validation, namely prototyping, net execution, and code generation. Petri net concepts can be applied in some of the most important areas of systems engineering: concurrency and distribution.

Software engineering is a good candidate for presenting systems engineering validation concepts. Many aspects of software can be expressed by Petri nets in a natural and adequate way. Because of the operational semantics and the ability to express the dynamic behaviour of systems (models) by the token game, prototyping based on Petri nets is an approach which covers all the tasks from initial idea to final product. In this part direct execution and indirect execution (by code generation) for the purpose of validation of Petri net models will be discussed extensively. Without tools there would be no efficient handling of large and complex distributed systems, and therefore the relevant tools for Petri-net-based systems engineering are presented.

Chapter 19 introduces validation within the systems engineering area and relates it to prototyping and tools. Chapter 20 introduces different types of distributed execution concepts and shows their relevance. Chapter 21 covers the (automatic) generation of code for distributed software systems from a Petri net. Conclusions follow in Chapter 22.

19. Systems Engineering and Validation*

The objective of this chapter is to describe the importance of validation for systems engineering approaches, to introduce the notion of validation, to explain prototyping, and to sketch the relevant tools in these areas in relation to Petri nets.

19.1 Software Life-Cycle and Validation

A major problem during the specification process is determining the user-formulated requirements which a system has to fulfil. Using verification to adress this problem was discussed in Part III. The contribution of validation is explained here. Due to the complexity of systems, a complete verification or complete validation is in most cases not possible. The goal of verification is to prove that certain properties hold or do not hold. To do this the description of the properties is checked against the specification. The goal of validation is to confirm that the specification meets the original requirements of the user. Therefore, validation and verification complement each other.

The definition of validation within the software engineering process varies greatly and depends to a large extent on the specific software engineering approach. Depending on the goals and concepts used, different means can be used to validate specifications, models, or systems.

The most well-known approach is the waterfall model and its modifications ([Boe81]). There, a validation is done at the different phases of the development process. However, nothing is said about how this task is to be performed. Other models such as the spiral model (see [Boe86]) are more precise but still leave several questions open to the modeller. The V-model (see [Som96]) explicitly encourages modellers to verify and validate the current results. At the different levels of the V-Model the validation process is determined. But how this is done remains open. Other approaches do not consider validation to be a central task in building software or systems. A topic which is closely related to validation is testing. There are test approaches (see [Som96]) which cover a wide range of the aspects of validation. However, they mostly work as post mortem approaches. Test cases can be generated

* Authors: D. Moldt (Sections 1, 2 & 4), F. Kordon (Section 3)

during the validation process or they can be generated prior to verification tasks.

Here *validation* is seen as the process of checking whether the system or model behaves as the user expects it to. Therefore, there are no formal means of validation in the general case. However, to support the tasks of engineers during validation, different means are available from software engineering, such as inspection, observation, and/or testing of simulated, animated, or executed models. In the phase of building the models all variants of validation can already be applied. The graphical structure of net models allows easy communication about the planned system. In the initial modelling the emphasis is laid on inspection, i.e. reading the diagrams. The further the process proceeds, the greater the need to understand the behaviour of the model. Here the advantages of the operational semantics and the token game become of interest. The net models can usually be executed by a simulator or animator (see Chapter 20). An additional step is to transform them into code and then to execute them (see Chapter 21). In addition, available verification techniques and methods can be used (see Part III). By combining verification and validation techniques we can improve confidence in the validity of the model.

An important software engineering approach with respect to validation is prototyping. Prototyping includes the tasks of modelling and validation. However, prototyping is used in many different ways. Therefore in Section 19.3 the notion is discussed in more detail.

It is important to notice that different approaches imply a different set of validation techniques. Not all possible validation techniques are used in every project. A description of validation in general follows in the next section.

19.2 Validation

The term validation is often confused with verification. The definition of verification is given above (see Part III). In general, verification is the task of checking that a model matches a given specification. A model is correct if it fulfils all the properties given in the specification.

Validation is the task of checking whether a model or system fulfils the expectations of the user, customer, or client. This includes the behaviour, the functionality, and the structure. While the static structure can be checked by inspection of a model, behaviour and functionality are best checked by execution. It is important to notice that involving the user is a central issue in this context. This adds an informal component to the validation process. Nevertheless, formal techniques such as simulation or code generation can be applied.

In the context of validation, the meeting of requirements can be viewed narrowly or broadly. With the narrow view, semi-formal approaches are mainly used. The broad view aims to allow the modeller/user/engineer to feel satisfied with the specification/system model so far. This extends the narrow view with formal approaches. These formal approaches include the use of previously verified sub-specifications to model a certain compound action, as is explained in Chapter 21.

Engineers emphasise the parts of a system which are critical and therefore under stronger and more precise investigation. Appropriate approaches are chosen according to the needs of the engineer. For example, the proof that a certain part of a model is live, independently of the rest of the model, is often important for an engineer. If the requirements have been precisely described, then verification techniques are normally used. If the requirements have not been fixed in an exact manner, then validation in the narrower sense is the only possible way to check the fulfilment of the requirements, since there is nothing against which to verify the specification.

There is a close connection with refinement and replacement processes, which will be described in more detail in Section 19.3. Both activities are central to the process of prototyping. Refinement covers the integration of models (or model parts), where one or more models are a more detailed version of the refined one. Replacement can be taken literally: (some part of) the whole model is replaced by one or several parts. This can even lead to a more abstract model.

How can specifications or models be validated? There are different solutions when using nets:

1. Observation and inspection of static net models – in this case the nets are not executed
2. Simulation and observation, covering inspection, performance, accounting, testing, debugging, and diagnosis
3. Animation
4. Code generation and code execution.

Here again the advantages of nets are the formal basis, the operational semantics, and the graphical interface. In addition to the verification opportunities, the direct execution of nets allows us to animate and simulate nets with little effort. Whereas simulation is directly related to the net formalism, animation is closely related to the application context. The same is true for most other interface-related topics which are not discussed here. Code generation from nets allows us to execute them in their final (and sometimes different) execution environment. Furthermore, different tools provide different observation, testing, and debugging possibilities. All this allows and supports prototyping approaches.

Because we use the narrow view of validation, verification techniques as presented in Part III are not considered here. The predominant aspect of validation comes from the operational semantics of nets which can be used for the execution of a specification. For this execution several cases have to be distinguished. The first is the direct execution of nets by a simulation tool. The second is indirect execution, which is achieved by code generation followed by the execution of the generated code. Indirect execution can be done within the environment used for the creation of the net, or a different environment. The tools can either execute the net directly or compile it before execution. This will be considered in more detail in the following chapters.

It is important to notice that there is always some risk. For verification, there is the possibility of a false proof, or that the wrong property was proved, or that incorrect assumptions about the environment have been made. (The latter two problems should be detected by validation.) For validation, we must take into account the fact that the tests performed usually do not cover all cases. Therefore, there remains a degree of uncertainty about the behaviour. However, this is often the case when building complex systems. For critical parts of a system, validation in the broader sense applies, because there are proofs of properties of the system. It is important to test even proved parts, as illustrated by the explosion of the Ariane rocket in June 1996. There, an approved system was transferred to a new platform without sufficient testing (see [Lio96]). The system fulfilled the specification, however, this specification did not describe the desired behaviour. The assumptions about the behaviour of the environment were wrong, hence the Ariane 5 showed a higher performance than the Ariane 4 in the tests but not in practice. An important question is how to make the risk of a failure or malfunction as small as possible. The techniques discussed make an important contribution but the way in which the engineer proceeds is also important. Therefore, in Section 19.3 prototyping as an approach is sketched.

19.3 Prototyping as an Approach

Since the beginning of the '90s, prototyping has become a popular and fashionable technique. However, this term has many definitions. In this section, we briefly recall the origin of this technique and some widely accepted definitions of prototyping. Finally, we propose a new and richer vision.

19.3.1 The Original Problems

There are three main issues that prototyping can address:

(1) Initial requirements: Software engineers must correctly understand the requirements of the client. A prototype which behaves similarly to the intended system can help to confirm that what the engineer proposes is what the client wants.
(2) Experimentation: It is worthwhile to check that an implementation strategy is correct before making a large investment. If it is a dead end, the strategy can be abandoned. Otherwise, the experimentation validates it and may provide some extra information about the global investment required for the complete design and implementation.

(3) Closing the Gap: There is a gap between the implementation requirements (the detailed solution that is proposed to the client by the software engineer) and the implementation itself. Ideally the implementation should be closely related to the requirements.

These three issues lead to different techniques. In (1), it is obvious that the prototype will be discarded as soon as there is an agreement between the engineer and the client. Here the main benefit of prototyping is that it clarifies the problem and its potential solution. In (2) the prototype may be either discarded or retained, depending on the care that was put into its design and implementation. In (3), the prototype is a product that has to be reliable and prototyping is then generally reduced to code generation.

It is obvious that if the prototyping process is well established and uses tools, firms can save time and money on both development ((1) and sometimes (2)) and maintenance aspects ((3) and sometimes (2)):

- Technique (1) reduces the misunderstandings of initial requirements, which are the source of 70% of bugs ([Hol96]).
- Technique (2) prevents major design mistakes. The throw-away product that is obtained allows experimentation with an implementation strategy before using it in the final product.
- Technique (3) minimises the difference between the specification and the corresponding code.

Prototyping is also called *Rapid Prototyping* when it is supported by tools because the *time to market* of a product may be greatly reduced. Rapid Prototyping has already been exploited ([GHT91, Bur93]), even for real-time applications ([Cas91, LSB93]). The results are satisfactory for large projects because the required tools are very expensive to develop.

19.3.2 Prototyping Taxonomy

There are many ways to interpret the three issues discussed in the previous section. If we emphasise aspects (1) and (2) only, the prototype does not need to be retained but the results should be carefully documented to produce detailed requirements for the final system. If we emphasise aspect (1), then prototyping aims to produce a final product. According to a similar interpretation, [Hal91] and [AH93] propose the following classification: *throw-away*, *incremental*, or *evolutionary*.

In the throw-away approach (also called experimental prototyping) only a subset of the requirements is generally implemented. This provides information about design choices and/or development costs. A throw-away prototype may be:

- Simulated: Usually, it cannot be executed outside the simulation environment for which it is designed ([Bur90, PG92]). Most prototyping environments allow integration of external procedures into the prototype to provide a good evaluation.

- Executed: It is able to run outside the environment in which it is designed. This approach is mainly used for larger or long-term projects ([Cas91]). Many executed prototypes are implemented within very sophisticated environments such as Smalltalk ([GR89]).

In the incremental approach, new parts of a system are successively added to a system kernel ([HI88]). The designed software architecture guides all the development phases. This approach is very similar to the traditional development process. The main problem is that this technique relies on the software architecture; if it is not robust, reliable, or pertinent, the prototype cannot be fully operational.

The main objective in the evolutionary approach is to preserve the flexibility of both the system architecture and functions. The prototype is improved by successive refinement of the previous version ([BKMZ84, Von90]). This approach is also called *waterfall prototyping* ([Flo84]).

Both incremental and evolutionary approaches produce a version 1.0 of the final system.

19.3.3 Key Issues in Prototyping

Throw-away prototyping is now a well-accepted technique in both the software and hardware fields. In the former field the goal is to evaluate requirements. In the latter domain, the goal is to implement a system using "soft" techniques (PLA in [PTVdB90], simulation in [LP90], etc.) in order to test it intensively. The production cost of any hardware system explains the interest of the hardware community in prototyping.

Incremental and evolutionary prototyping are not yet as popular as throw-away prototyping because, most of the time, the implementation operation is so expensive that the prototype replaces the application specification. Then, modification, extensions, and maintenance are performed at the prototype level and the *model*, if any, becomes increasingly out of date. A solution to this problem is to automate the process. This is outlined for both software ([Luq89]) and hardware ([DC90]) systems. A quick and low-cost procedure allows the system designer to work at the specification level instead of the program level. This may be valuable when the system is complex (distributed systems for example).

The notion of a model is very important. A model is a description of the system to be prototyped. Such a description may be general or detailed. Of course, the accuracy of computed results depends on the level of detail in the description.

In throw-away approaches, the expected information is a better understanding of what the client wants; so the construction of the model is not important. Models may even be expressed using high-level programming languages.

In incremental and evolutionary approaches, the construction of the model must be done carefully because code is generated (or deduced) from the model. So the use[1] of a formal representation such as Petri nets is valuable because properties can be computed from the model. The most important point is that properties can be deduced from the model in order to help a designer increase his knowledge of the system at a semantic level instead of a syntactic level.

With respect to these considerations, most of the studies outline the following points:

- It is rather difficult to determine when a model becomes a prototype ([Hal91, EG92, DK96]). A good illustration is that some approaches rely on very sophisticated programming languages: the model is the prototype.
- The quality of the results depends on the formalism used to design the specification of the system ([Luq92]). For this reason, formal specification techniques are valuable because they enable the computation of properties.
- While there is no universal formalism which fits every problem, the choice of an input formalism is not easy ([MGK89]). Particularities of an application domain must be considered.

The problem arises mainly because the boundary between model and prototype is not really clear. Hence there is little difference between modelling (specifying the design) and implementing (introducing operational aspects). When the design and operational issues are considered simultaneously, errors due to the misunderstanding of the application logic may be introduced. Such errors are difficult to detect.

19.3.4 Extended Definition of Prototyping

Prototyping should not be reduced to code generation. It also involves operations such as modelling, evaluation of the model, and evaluation of the prototype. It is thus more than a technique, it is a strategy. Figure 19.1 illustrates the prototyping approach. The main characteristic is that there are two distinct entities: the model, which is defined using a set of adapted representations, and the prototype.

If a system designer works on model version N, he may evaluate this model and refine it using the information provided by the environment. The approach may be formal or not. However, he cannot get complete information about the system execution: some aspects can only be studied when the prototype runs in its final execution environment.

[1] By "use", we mean either the direct description of the system by means of Petri nets, or the description of a system by means of a representation that is related to Petri nets, such as the one presented in Section 10.3.3.

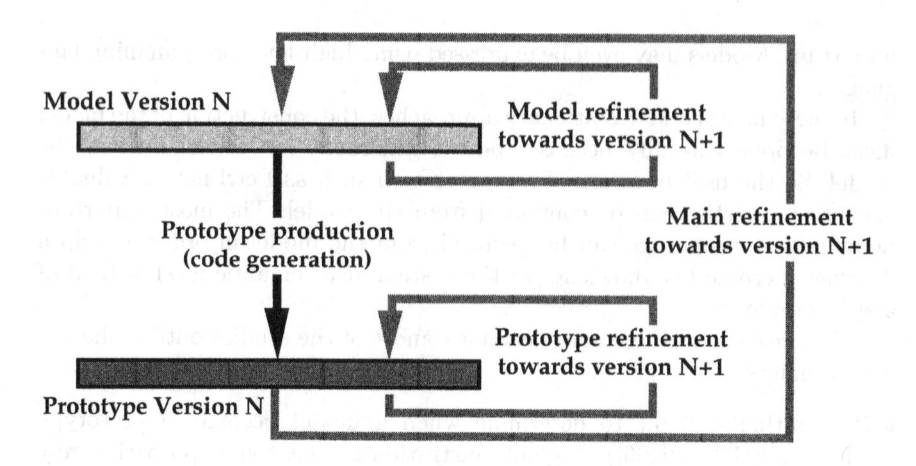

Fig. 19.1. Prototyping as an approach

Code generation provides him with a prototype that is an exact image of the model and that is ready to run in the final execution environment. This prototype may be evaluated and modified (in distributed systems, discrete process location may be investigated) and, if it is not satisfactory, the information provided can be reused to refine model version N into a version N+1 that should correct the problems.

This extended definition of prototyping may be seen as an extension of evolutionary prototyping that emphasises the differences between a *model* and a *prototype*. It takes advantages of operations (validation, evaluation, ...) that are enabled at both levels. This is especially valuable when the model level relies on a formal modelling technique (Petri nets). At this stage, we do not further define the words "model" and "prototype".

19.4 Tools

Petri nets can be applied to areas as diverse as software architecture, distributed system design, protocol verification, etc. In each of these areas Petri net tools may be used to clarify the structure of the system and to analyse or verify its properties. In general, Petri nets will be used mainly in the analysis and system specification phase.

There are two different classes of tools that employ Petri nets. First of all are tools that are built for a specific application area and that use Petri nets as part of their modelling or execution formalism. Examples are process centred software engineering environments (MELMAC, Process Weaver) and work flow management systems (COSA). Their task is to support a specific application area using some Petri net variant. Often the Petri net itself is hidden to a large extent from the end-user.

The second class consists of general-purpose Petri net tools. They are independent of a specific application area and provide the user with the means to construct, analyse, and simulate specific classes of Petri nets. These tools are usually stand-alone tools that do not interface directly to application-specific tools.

In this section we will concentrate on general-purpose Petri net tools. In general, such a Petri net tool consists of at least some of the following components:

- Editor
- Simulator
- Analyser
- Code generator
- Animator
- Repository.

Editor. The editor is used to create, modify, and store a Petri net model. An editor may be textual or graphical, or a combination of both. A graphical editor allows for the direct drawing of the Petri net. It supports the usual functions of a drawing package. In the case of hierarchical nets a navigation mechanism, eventually supported by multiple windows, may be needed to view all the components.

The editor should support the syntax of the underlying Petri net variant of the tool. Some editors support one or more very specific Petri net variants (syntax), while others allow the user to supply his own net variant. Editors that are targeted towards a specific Petri net variant will usually have a more user-friendly interface.

Petri net editors are often tightly coupled to the rest of the tool and are in most cases bound to a specific platform.

Simulator. The simulator executes the Petri net that has been constructed e.g. via the editor. Simulation can be in batch or interactive mode. In the latter mode, the user may interact with the execution of the Petri net, for example by inspecting, adding, or removing tokens. By setting breakpoints or going back and forth in the execution graph, the user has control over the simulation. In this way he can debug the model to detect incorrect-functions.

The simulation can be performed with or without animation. Without animation the report of the simulator will usually concentrate on numerical and statistical results.

Existing simulators are mostly bound to a specific editor or animator, and it is not easy to incorporate them in another Petri net environment or in another application.

The efficiency of most simulators will decrease rapidly for (very) large nets. In some cases it is possible to use the simulator as the end-product, e.g. to drive a real-time system. In such a case the net should usually be rather small, in order to meet timing requirements.

Analyser. The analyser derives static or dynamic properties of a Petri net model. Examples of such properties are S- and T-invariants. Another important area is the analysis of the reachability graph, to obtain information on e.g. liveness, fairness, and boundedness of the net. In order to avoid a state explosion, tools should support the operation on sub-nets, reduced nets or net generated graphs. Some properties can be calculated without computing the whole reachability graph, others may need the whole graph (e.g. model checking).

For stochastic Petri nets it may be possible to use Markov analysis to analyse the time behaviour. For timed Petri nets, time analysis is able to calculate, for example, bounds on arrival times of tokens in output places.

As is the case for the simulators, the analyser components are mostly coupled very tightly to a specific tool environment and are not available as subroutine packages.

Code Generator. A code generator generates executable code for a specific software/hardware platform. In some sense the simulator also generates code. However, a simulator usually interprets the Petri net, whereas we refer here to compilable code. An important problem is transforming the inherently parallel Petri net to sequential code.

At the moment, available code generators offer the possibility of generating code (or code skeletons) for small systems. It is not easy to target the code to a specific platform.

Animator. In the animator the user observes and (sometimes) may interact with the external behaviour of the net. The animation that is provided may be realistic or non-realistic. In the first case, the user sees a realistic scene (for example a factory) that the Petri net is supposed to describe. In the latter case, the user sees the tokens flowing through the Petri net.

Tools that offer animation mostly concentrate on non-realistic animation and show the state of the net as it evolves in time. Realistic animation may be obtained by attaching code to transitions or places. Especially in combination with graphical interfaces, newer products provide animation that expresses Petri net aspects in a way that is intuitive for users since it refers to application-dependent representations.

Repository. The Petri net models that are constructed by the editor, and in some cases the results of simulation runs and analysis sessions, are stored in a repository. In this sense a Petri net tool is just another CASE tool that manipulates system models.

Most available Petri net tools store their models in a proprietary data format, directly in the operating file system. Some tools make available the syntax of these files so that it is possible to interface to other applications. If they use a standard database system, this repository can also be used to query and update the model directly without the help of the Petri net tool itself. Version and configuration management of the Petri net models is relatively easy for an open repository.

In summary, existing Petri net tools offer adequate means to design, execute, animate, debug, and analyse Petri net models. When comparing Petri net tools to ordinary CASE tools, one observes that most Petri net tools are rather closed and are not easy to couple to other Petri net tools or to embed in standard software development environments. However, there is some ongoing work to agree on a common basis for an exchange format which is based on XML (extensible markup language).

An overview of existing Petri net tools can be found on the Internet at the World of Petri Nets homepage (http://www.daimi.au.dk/PetriNets/).

20. Net Execution*

For some years, there has been a growing consensus on the distinction between requirement analysis and software design.

- The requirement analysis phase of the software development life-cycle is concerned both with the elicitation of requirements from a customer, and their formalisation into a specification of the system behaviour.
- The design phase aims to produce an architecture suitable for code generation and to assign pieces of behaviour to components of the architecture.

This distinction also fits that between the verification of the specification and the validation of the software. Observation plays an important part in the two phases of the software life-cycle, as well as in dynamic software validation using simulation.

During the requirement analysis, the behaviour of the target system must be extensively considered before deciding that a particular course of events is an allowed behaviour. So the elicitation of requirements from a customer stresses observation and interactive execution under the control of the engineer ([DDBD94]).

During the design phase, on the contrary, the emphasis is put on the reverse side of observation, namely hiding or encapsulation. In order to define a module, or a component, one has to hide the internal behaviour of the piece of software. Furthermore, defining an equivalence relation on behaviours, for instance observational equivalence, allows the freedom to embed a component into a system. However, the designer must rely on a formal model for components plugging and for unwinding a compound system behaviour. Without such a model, simulation and prototyping techniques would be worthless.

This chapter is concerned with net execution in this setting; it focuses on software simulation and prototyping. Let us note that "net execution" may be misleading because a Petri net model is not precise enough to enforce a particular course of execution and because the execution semantics of such model has to be changed:

- Even if some transition is enabled, being idle is always an allowed behaviour of a net; the process driving the behaviour must be distinguished from the constraints modelled by the net.

* Author: P. Barril

- A sequence of transitions formalises an execution trace well enough for most purposes. However, some highly concurrent systems require a step semantics, that is, must consider the synchronous firing of a multiset of transitions.
- To model concurrence or asynchrony, one behaviour of the system may correspond to a set of execution traces.

In order to deal with these multiple execution semantics, we follow the execution paradigm of [Tau88] for Petri net execution. Every transition is attached to a step process divided into two sections:

- A *black box section* where the process stays for an unknown period of time, driven by its own concerns and which is left by requesting permission to fire a multiset of transitions.
- A *management section* where the step process tries to get permission to fire the transitions, and initiates the marking update if permission is granted before going back into the black box section. A process which is granted permission is committed to firing and cannot change its mind; it must initiate the moves of tokens before going back into the black box section or issuing other requests.

This process-driven approach separates the control of a behaviour from its correctness with regard to the firing rule. Computation and communications between the black box sections can be used to search for a particular behaviour or to enforce a higher-order form of control using causal or synchronic properties. We will say that computations in the black boxes are running at the semantic level. Because the intended uses of the Petri net engines are simulation and prototyping of distributed systems, we do not assume a central clock. A step which is granted permission is committed to firing, but takes some time to update the marking. Furthermore, the occurrence of a step is mirrored by a piece of code which performs the corresponding tasks in the black box sections and which also takes an unspecified amount of time to complete. In this asynchronous setting, our notion of step is not the usual one: the transitions are concurrently enabled, but their occurrences can overlap in time, without being required to do so.

We are concerned only with conservative strategies for allowed behaviours. This means that, assuming a way to suspend execution, the system stops at a reachable marking, where all tokens are causally accounted for by a sequence of steps in the Petri net model. A device for stopping execution is a snapshot algorithm running at the semantic level, rejecting any new requests, but allowing each committed transition to complete.

More formally, we adopt the semantics which unfolds a P/T net into a *prime event structure* generated by occurrences of transitions. Following [Sas95], tokens are just grains of causality corresponding to the events that move them into their places, modulo symmetries making two tokens in the same place indistinguishable. So the mapping from a commitment event in the simulation model to an occurrence event in the formal model is harmless if we allow a transition in the simulation to loop, continually requesting permission until tokens promised by committed transitions are delivered. Using the algebraic event structure in this way defines enabling without reference to a virtual global marking. In order to define disabling, we need to duplicate each generator of the formal event model: an occurrence maps to a commitment event and a completion event. Some algorithms in the following sections test only for enabling conditions and solve conflicts: they perform only input co-ordination and do not try to synchronise moving tokens or care about completion events. On the contrary, when simulating atomicity of a step occurrence or initiating a snapshot we do care about completion events: both require output co-ordination and quite different control techniques. The observation of the simulated system as well as its prototyping requires us to compute a partial or global virtual marking using a snapshot.

We sketch several algorithms and discuss them with regard to a distributed object software platform target. The main quality criteria are the efficient use of distribution and the feasibility for the target platform. The most important properties are obviously *correctness*, *deadlock-freeness*, and *productivity*. Beware, these properties are properties of an algorithm, not properties of the Petri net model! Furthermore one has to check that these properties are well defined with regard to our "waiting for promised tokens" semantics.

- *Correctness* is the fact that only enabled steps are permitted to occur.
- *Deadlock-freeness* means that no infinitely long delay is introduced by the management of the algorithm. For instance, conflicting transitions may easily result in such a deadlock without some care. Deadlock-freeness of the execution engine implies that the execution of a deadlock-free net (cf. Section 5.1) is infinite.
- *Productivity* means that a request is rejected only if between the announcement of the request and the rejection, there was a moment when the step was not enabled. This is a *weak fairness property* (no step is permanently enabled without occurring), also known as a *finite delay property* (cf. Sections 9.3.2 and 10.1).

Designing and engineering algorithms to ensure productivity is difficult: weaker properties are also of interest. First notice that productivity and correctness require solving conflicts and answering positively with permission first, delaying rejections until the system has evolved. So the kind of finite delay considered will exclude *livelocks*, that is, loops in the management system which do not eventually grant a permission when at least one of the candidates is enabled.

Less formal properties of a simulation platform are related to its user interface and its software engineering. For instance, a particular user might sometimes obtain an out-of-date view of the system state. At other times, he might observe the actual state of a suspended execution. The *honesty* of the user interface requires that the user view of the system has actually happened during execution, and also that the difference between an out-of-date and a real-time view is clearly shown on the screen. The *scalability* of the simulation platform with regard to the number of sites involved is also an issue: *deployment* and *maintenance* of a large-scale distributed environment require well-defined procedures.

Furthermore, a simulation of a particular Petri net model needs an installation, that is, an initial placement of components (places, transitions) and a set-up of remote communications between them. A large part of a simulation platform supports this set-up phase. We nevertheless deal only with the execution phase in the following sections.

20.1 Centralised Control

We are looking for a distributed control of the management subsystem because we want to use the P/T net model for prototyping or simulating a distributed system. We feel that a centralised control may easily overlook some race condition in the target distributed system. Yet, most simulation environments for high-level Petri nets (HLPNs) use a centralised management because finding enabled transitions is already a difficult and time-consuming task.

The enabling rule for HLPNs requires binding transition parameters to token attributes. For instance, the occurrence rule for coloured PNs defines an occurrence mode binding variables to colours (cf. Section 4.3). A first idea is to place some syntactic restrictions on the form of parameters and attributes in order to use symbolic pattern matching instead of brute-force exhaustive checking of alternative bindings. The *well-formed Petri nets* ([CDFH91]) achieve this for coloured nets because they allow symbolic operations on the colour function associated with an edge. The model incurs no modelling power loss: every coloured net can be transformed into a well-formed net ([Dut92]), and the semantics of folding an ordinary Petri net is preserved. The enabling tests can be further optimised ([IR93]) using token dependency.

A second idea is that partial bindings of transition parameters to token attributes are in fact resources. In this approach, the management system updates a working memory of partial bindings each time a transition occurs, instead of starting the enabling test from scratch. The KRON interpreter for HLPNs ([BMMV93]) uses the phenomenon of "*temporal redundancy*" to justify the idea. KRON means Knowledge Representation Oriented Net. It relies on the similarities between HLPNs and a rule-based production system, more precisely, the OPS5 expert system shell. The core of the system is an adaptation of the RETE matching algorithm which takes advantage of temporal redundancy by recording a graph structure storing both the state of a partial match and an evaluation strategy for rule pre-conditions. The

KRON matching algorithm returns the set of enabled transitions, and conflict resolution relies on policies defined in external libraries. The HLPN is intended to model the dynamic of knowledge revision through time. Knowledge update is usually a slow process and the number of tokens involved in a transition is small in relation to the total number of tokens in the marking. Using temporal redundancy is particularly efficient in this setting. The system described in [VB90] was developed earlier, but pushes the idea further, for a special case of P/T nets using objects (in fact, OPS5 entities) as tokens. First, the sorting network of the matching algorithm takes into account the partition of the working memory according to places. Furthermore, tokens of the Petri net and elements of the working memory are the same: they are tuples of objects satisfying relations. These relations are satisfied as long as the token remains in the place and a tuple may only be split during transition firing, but not during substitution for variables in matching. Thus the use of bindings as resources is strict. The system goal is the implementation of a real-time system using rule-based production systems: it uses moreover a compilation transform to encode run-time control information into tokens.

This last example shows that testing for enabling and solving conflict is already difficult for centralised implementations of HLPNs, so we limit ourselves to ordinary Petri nets in the following sections about distributed implementations. Furthermore, according to the prototyping framework described earlier, compilation of control is not considered and the management uses a strict interpreter approach to net execution. All control information about execution, for instance the distinction between *representing* and *synchronisation places* ([CSV86]) or the concept of *triggering places* ([Val87a]), is handled at the semantic level. In fact, one of the intended uses of the prototyping environment is to help the engineer to redesign the Petri net model to take into account control of the simulated system in terms of such concepts.

Remark 20.1.1. The analysis of conflicts leads to a mild improvement and a management distributed over locksets. By definition, all input places of a transition are in the same lockset and two transitions with distinct input locksets are independent. This definition of locksets deals with conflicts between transitions and cannot be used with a step semantics. Furthermore the structure of the net can be very unfavourable.

20.2 Distribution of Control over Places

We discuss now the distribution of management over places: the place objects or processes are responsible for checking pre-conditions and solving conflicts.

Putting a priority on resources is a well-known technique for solving conflicts. In our setting, the resources are located in places: priorities are defined by a total ordering over places. Thus this first algorithm will be called "polling

---

Algorithm 20.1 (Polling places in a fixed order)

Token reservations are queued by the place process which is responsible for denying overly large requests and granting the others in a first-in-first-out order:

- A step wanting to fire notifies its input places by sending the message `reserve(step, tokencount)` which is queued by the receiving place process.
- If there are enough tokens, the tokens required by the first step in the queue are marked as reserved and the step is notified by a `result(true)` message.
- Places are polled sequentially in a fixed order: a step waits for a `result(_)` notification before reserving tokens in its next input place or cancelling its reservations in the previous ones.
- The removal of reservations on failure is the responsibility of step processes: if a reservation in an input place is denied by a `result(false)` message, the step cancels its previous reservations by sending a `release(tokenCount)` message. The reserved tokens return in the free token pool.
- The effective firing is also under the control of steps: `take(tokencount)` removes reserved tokens from input places while `put(tokencount)` updates the output places marking.
- The place process is a sequential loop:
  - First process one message (`reserve`, `release`, `take`, and `put`);
  - Traverse the queue and deny reservations by a `result(false)` message for transitions requiring more than the actual marking `tokenCount`;
  - Check the first reservation and grant it if there are enough free tokens, that is, if `requiredToken` $\leq$ `tokenCount` $-$ `reservedCount`.

---

places in a fixed order". Another feature is the use of token reservations in a way similar to a two-step commitment protocol for distributed transactions.

Algorithm 20.1 is clearly correct. It is also deadlock-free and productive independently of any communication model. Assuming no message loss and a finite delay property for transition firing, for firing control by step, and for cancellation of reservations, we shall prove: $\forall s \in \mathit{Step} \,.\, \forall p \in \mathit{Place} \,.\, H(t,\ p)$ where $H(t,\ p)$ denotes the property "the step process s requiring `reserve(s, tokenCount)` from place p will get a `result(_)` answer in a finite time".

The proof uses the linear ordering $\{p_j,\ 1 \leq j \leq n\}$ of places and the inductive hypothesis $\forall s \in \mathit{Step} \,.\, \forall p_j,\ i < j \leq n \,.\, H(t,\ p_j)$. This hypothesis is void for $i = n$. Assume a reservation from s is waiting in the queue of place p_i, it will be answered when it reaches the front position of the queue and `reservedCount` decreases to zero. By the induction hypothesis, transitions owning reserved tokens will eventually send `take(tokenCount)` or cancel their reservation, in both cases decreasing `reservedCount`, so the place will eventually answer the front reservation and remove it. □

The implementation of the algorithm using distributed objects such as the *Java remote method invocation* (RMI) platform is straightforward ([AG96]). Java RMI is a lightweight client-server platform, without a sophisticated object request broker. Yet this remote object approach offers several benefits over a simpler *remote procedure call* (RPC). First, server applications, which instantiate and export remote objects, bind them to a name into a registry server. This name is a string in an URL-like syntax. A client gets a remote reference to an object by connecting to a registry and looking for the name of the object. Secondly, objects of the proper type can be exported as arguments of a remote call, allowing servers to call-back clients. These two features simplify the bootstrap of sophisticated peer-to-peer communication patterns. Finally, remote references are first-class objects: the RMI system maintains their identity, they can be used as look-up keys in a dictionary and are automatically garbage-collected. Notice however that RMI handles *remote objects* and not *remote classes*. The remote instantiation of new objects relies on creational *design patterns* called *factory methods* or *factories* ([GHJV95]).

```
import java.rmi.*;

public interface PRunTimeIntf extends PInitIntf, Remote {

/*  step is a local or remote reference to the calling step
    in order to delay answering by a result(boolean) message.
    tokenCount is the weight of the receiver place in the
    minimal pre condition of the step.
*/

    public void reserve(SRunTimeIntf step, int tokenCount)
                                    throws RemoteException;
    public void release(int tokenCount) throws RemoteException;
    public void take(int tokenCount) throws RemoteException;
    public void put(int tokenCount) throws RemoteException;
}
```

Fig. 20.1. Place interface for polling

A client interacts with servers only through abstract interfaces used to type remote references. After designing the abstract interface for remote interaction (for instance, `PRuntimeIntf` for places, Figure 20.1), one writes a concrete class for implementation (for instance, `PImpl` for places) and runs the RMI compiler which generates a stub and a skeleton class. The stub (client side) and skeleton (server side) are dynamically loaded to forward remote invocation using the RMI transport layer. Therefore, they must be located in the proper search path. Then you must start the RMI registry on each server site, then start the server applications. For our purpose, the servers are factories, constructing a new place or a new transition on the server, and returning a remote reference to the client. A client application invokes the server for placement of places and transitions. Then it uses their remote references to link and initialise the net and its initial marking (using the `PInitIntf` and `TInitIntf` remote interfaces).

```
import java.rmi.*;
import java.rmi.server.UnicastRemoteObject;
import com.objectspace.jgl.*

public class PImpl implements PInitIntf, PRuntimeIntf,
                    UnicastRemoteObject {
    private String label;
    private int tokenCount, reservedCount;
    private Deque queue;
    class Reservation {
        SRunTimeIntf step;
        int required;
        };

    public void init() {
    ...
    }

    private synchronised void clean() {
        for ( DequeIterator i = queue.begin();
                    !i.equals(queue.end()); i.advance() ) {
            (Reservation r = i.get());
            if r.required > tokenCount {
                r.step.result(false);
                queue.remove(i);
            }
        }
    }

    private synchronised void check() {
        Reservation r = queue.front();
        if r.required <= tokenCount - reservedCount {
            r.step.result(true);
            reservedCount = reservedCount + r.required;
            queue.remove(1);
        }
    }
}
```

Fig. 20.2. Queue management for polling

The run-time interface for the `Place` class (Figure 20.1) uses a reference to the calling step in order to delay answering reservations. The corresponding methods in the implementation are declared with the **`synchronised`** modifier in order to enforce the sequential loop structure of the algorithm: **`synchronised`** methods must acquire the receiver object lock before running and so their activations are mutually exclusive. The queue uses class `Deque` from the *Objectspace Java Generic Library*; its management is shown in Figure 20.2. The class `Step` just initialises itself by computing its pre-conditions from the constituent transitions before running the reservation protocol.

The sequential polling for reservation is a communication bottleneck in the previous algorithm. A transition willing to occur should instead concurrently reserve tokens in its input places. The next algorithm achieves this, and relies on a distributed election game between transitions to solve conflicts. It focuses on conflict resolution, not on concurrent firing, and its execution semantics is limited to transitions (as opposed to steps). Because a winning transition secures unique access to its input places, the execution is interleaved between each lockset.

Algorithm 20.2 (Distributed election game)

The token reservation queues are under the management of transitions, and conflict resolution uses a local round-robin priority list of transitions for each place.

- Each place manages a local list defining a total priority order on outgoing transitions.
- Each transition manages two subsets of its input places:
  - *enqueuedPlaces* is the set of places where the transition is waiting for lower priority transitions to cancel their earlier reservations, but may be removed from the queue if another transition wins the reservation.
  - *reservedPlaces* is the set of places where the transition is at the front of the queue.
- A reservation is denied if there are not enough tokens or if a higher priority transition is already waiting.
- In the event of such a denial, a transition must cancel all its previous reservations.
- A transition wins the game when all its input places are reserved, that is, when it holds the front position in each queue.
- An input place receiving a `take` message retrieves the winning transition which is at the front of its queue and removes it. Then the place changes its priority by moving the transition to the end of the local priority list. The place finally notifies the transitions requiring too many tokens and removes them from the queue.
- A transition has to check or be notified of changes in all its enqueuedPlaces:
  - Cancellations of competing transitions move the transition to a better position;
  - Success of a competing transition may force the transition to cancel its reservations.

The implementation written in OCCAM on parallel processors is given in [Tau88]. In the distributed-object setting one has to handle proper termination of the algorithm: the late delivery of cancellation messages to unused places after a winning event in a round may cause undue denial of lower priority transitions in the next round. The proper termination of a round guarantees that all the queues are emptied before the process driving execution starts the next round of the game. Java uses the so-called "synchronous message-sending protocol", so each message is acknowledged before the current thread of control ([Lea96]) sends the next one, but improper multithreading can abruptly dispose of a thread and spoil the implementation.

This distributed election algorithm is due to Vinkowski [Vin81] who has shown its non-productivity but proves a weaker property, strong enough to exclude livelocks. This *weak productivity* property states that if each member of a set of processes is looping asking permission to occur, and if the set contains a process which is enabled, then one process of the set (not necessarily the enabled one) will eventually get permission to fire. We will sketch the proof assuming that the local priority list of each place manages all transitions. Although the local priority list of the implementation is restricted to transitions which have an output edge from the place, any extension to a total order on transitions in the initial state leads to a local priority list which is total, because only transitions corresponding to output edges can change their priority by firing. Under this assumption, we must prove that if a non-empty set of transitions is competing and if there are enough tokens, the election game can neither deadlock nor reach a livelock.

- In a deadlock situation, each transition either holds a queued reservation for each of its input places, or has been forced to cancel its reservations and is absent from the queue for each of its input places. Because one transition is enabled, there is at least one transition enqueued in each of its input places. Now there is no winner, so there is a transition with a lower priority in a better position in the queue for some input place, and in a deadlock situation, this transition is enqueued in each of its input places. So assuming a deadlock, one can construct an infinite sequence of transitions with strictly decreasing priority, which is impossible.
- In a livelock situation, there is an infinite sequence of moves but no winner. Because only winning moves change the marking and the local priority lists, both the marking and the priorities remain the same forever, and the set of possible situations of the game is finite. Because there are enough tokens, only a transition of higher priority already waiting in a queue can force a transition to cancel its reservations. So there is at least one transition t that is not forced to cancel its reservations. Let $S(t)$ be the set of places where t holds a reservation. Similarly, among the competing transitions which need none of the place of $S(t)$, there is one that is not forced to cancel its reservations. This construction ends up with a partition of the set of places involved such that each class is held by a transition which is not obliged to cancel and such that every other transition needs a place from the set. That implies only a finite number of moves is possible, contradicting the livelock assumption. □

We will not describe the possible implementations. The game protocols for a transition can run in a different thread for each input place. The sharing of places and their queues imply some synchronisation between these threads. But this amounts to a further distribution of control over edges, which is the subject of the next section.

20.3 Distribution of Control over Edges

Because place processes can communicate only with one (step or transition) process at a time, they are bottlenecks in the previous algorithms. So a place process is further divided into a *place centre* and several *place parts* managing a pair of incoming-outgoing edges.

Assume first that all places have non-empty pre- and post-sets. A *place part* of p is any $\langle t_{\text{in}}, p, t_{\text{out}} \rangle$ for $t_{\text{in}} \in {}^{\bullet}p$ and $t_{\text{out}} \in p^{\bullet}$. We shall say that $\langle t_{\text{in}}, p, t_{\text{out}} \rangle$ is an *input part* of t_{out} and an *output part* of t_{in}. The place parts can store tokens: tokens in $\langle t_{\text{in}}, p, t_{\text{out}} \rangle$ are reserved for t_{out} consumption and the actual marking for a place is the sum of the token counts in the place centre and every place part. Places with empty pre- or post-sets are dealt with by special edges $\langle \bullet, p, t_{\text{out}} \rangle$ and $\langle t_{\text{in}}, p, \bullet \rangle$ whose meaning is that tokens of the part are borrowed from or delivered to the place centre.

The idea of the third algorithm is to divide management into two phases:

- If there are enough tokens in each input part of a transition willing to occur, the transition occurs without further notice.
- Otherwise, tokens are gathered from the place part into the place centre and the management reduces to a distribution of control over place (for instance, polling places in a fixed order, for this particular place).

[Tau88] gives an OCCAM implementation of this algorithm. In our distributed-object setting, the implementation relies on multithreading. *Java threads* are *lightweight processes* competing for execution on a particular Java virtual machine (VM). One use of multithreading is to increase server responsiveness by handling client connections in separate threads. Notice that programming a connection protocol in a separate thread is conceptually clearer. Furthermore, a thread may be suspended instead of waiting for input-output or for the response from a remote server. Thus multithreading leads to important optimisations. On the other hand, objects are shared by concurrent threads and an unexpected change of the active thread may well leave an object in an inconsistent state, which raises the issue of *thread-safe programming*. The Java thread model provides methods for accessing the scheduler, but the scheduler behaviour depends on the execution platform and VM implementation. So the best way to write thread-safe portable code is not to try to interfere with the VM scheduling of threads, but to use more abstract control structures such as *object-monitors*.

Object-monitors manage locks on objects. More precisely, they guard a body of code, which is usually a `synchronised` method call, by a mutual-exclusion semaphore associated with the object. The thread which owns the lock has an exclusive access: other threads trying to access the object or to acquire the lock are queued. These monitors are hidden from the programmer and supported at the language-level through the `synchronised` modifier. They are a variant of re-entrant Hoare's monitors ([Hoa74]): the thread owning the lock of an object can issue a locking request on the same object without deadlock. Hoare's monitors provide a synchronisation device through so-called *condition variables* representing waiting queues of processes. The process owning the lock may enter a queue or awaken other processes by signalling condition variables. Such monitors have proved themselves useful for object-oriented simulation of discrete-event systems ([Béz87]). The synchronisation devices of Java monitors are simpler because in an asynchronous environment, it is more convenient to let each woken thread check for itself the proper condition and then either proceed or go back to sleep. Thus Java monitors provide a unique *waitset* instead of several condition variables. The point of using monitors is that well-designed shared objects define monitor invariants. These assertions involving both the state of the shared object and the threads suspended in its waitset are valuable for validation.

The distribution of control over edges uses several threads of controls sharing access to a place. The partition into place centre and place parts is built into the control structure synchronising the threads. The first thread runs a variant of the polling reservation algorithm. Other threads run at the semantic level and execute the firing sequence built into place parts. The part marking is stored as a temporary variable in the thread's working memory. The main control thread is responsible for both synchronisation and redistribution of tokens. Tokens in place parts are marked as reserved for the thread running the reservation algorithm, so that the actual marking of the place is an invariant.

The Java language permits the execution of a block of code as a critical section under the control of the monitor associated with any local object. But when trying to lock a remote reference, the Java compiler generates code which will lock its proxy (its stub, using RMI terminology), instead of the intended remote object. So the only way to lock a remote object is through the remote invocation of a synchronised method. Coordination using the monitor's synchronisation device usually involves several monitors, and may be tricky in the RMI distributed setting. Yet this technique appears to be very valuable for output coordination.

20.4 Multithreading and Synchronisation

While control threads are programming entities confined to a particular site and virtual machine, their proper management is the key for designing and synchronising concurrent activities and processes involving several processing sites. Again, this management is not an operational scheduling matter, but relies on more abstract control entities: for instance, *monitors* or *object-channel.*

Object-channels implement point-to-point synchronous communication between sequential processes of the *CSP model* ([Hoa85]). Their Java implementation uses monitors and can be extended to asynchronous (buffered) or multicasting (multiple readers) channels. The communication protocol between place and transition for the distributed game algorithm, for instance, is conveniently implemented by an object-channel. In fact, most of the OCCAM language (including the *alternative construct*) was implemented in Java ([HBVB97]). Thus, several algorithms from [Tau88] for net execution can be simulated in Java. Notice however that dynamic instantiation of new channels, at run-time, is a powerful feature which is not possible with OCCAM.

Let us illustrate the use of monitors for output synchronisation in our prototyping framework. The idea is to generalise monitors so that they simultaneously lock a group of distributed objects thus creating a critical coordination space, before unlocking all objects in the group. We call such a structure a *causal box*, because all the messages exept the ones involved in the coordinated threads inside the box must occur before the box is created or after its dissolution. Creation and dissolution of a causal box is quite similar to a two-phase commitment protocol: first acquire the lock of each target object in the group and wait until each one is locked, then start the activities and wait until each activity inside the box terminates, finally unlock each target object in the group.

The atomicity of transition occurrences can be simulated according to this scheme: the movement of tokens takes place inside a causal box grouping all output places of the firing transition. In the same way, making a partial snapshot of a group of places is implemented by putting them into a box before reading their marking. In both these examples, the activities inside the box are compiled. However, dynamic invocation of messages using *reflectivity* and an object-channel to control operations inside the box is possible. Such techniques are useful for debugging a multithreaded execution algorithm at the management level. They are also useful for inspecting the control running at the semantic level and driving a particular simulation.

Thus multithreading achieves a better distribution of control in two ways: through the scheduling of finer structures and through the restriction of the scope of a control structure. The definition of time-out for message acknowledgement is not a basic feature of the RMI system, but can be simulated using threads. Multithreading introduces the possibility of building fault detectors based on time-out, a first step towards fault tolerance and simulation of large-scale distributed systems.

20.5 Asynchrony

It does not seem possible to display a consistent global real-time view of the system to the user of the simulation environment without losing all the benefits of the asynchronous approach. The multithreading techniques above are intended to handle and to take advantage of asynchrony. Furthermore, we argue that asynchrony is an important feature of the kind of distributed system we want to simulate, and that synchronous simulations of asynchronous systems are unfaithful. Let us make clear that a synchronous approach to distributed simulation is possible, and indeed exists.

Our solution is to display most of the time a partial consistent view of the system which is usually out of date. The user of the simulation can ask for a partial snapshot to get information on the actual state of the system. The implementation relies on multicasting transition firing events using the Java event system. The user interface (UI) on a particular site focuses on a subset of places and updates a local copy of their marking: it subscribes to the transitions events for all transitions which are incident to the set of places of interest. Event multicasting is not causally ordered: events from different sites may be delivered in a different order to different clients, and worse, causally related events from different sites may be delivered in the wrong order. Therefore a client UI uses its local view of the marking to compute a local set of enabled transitions as in Section 20.1. An occurrence event is queued until it becomes enabled and the local set of enabled transitions is updated when handling or dequeuing a locally enabled occurrence event. Assuming no loss of events, the local view is faithfully following the actual distributed execution of the net. However the execution semantics has changed and no longer takes steps into account.

The set of places of interest may be changed, which requires a snapshot to get a consistent local copy of their marking. Besides the marking of places, the user may be interested in the set of enabled transitions or, in our framework, the set of transitions wishing to occur. This introduces new complications into the event set. Furthermore, this requires new presentations on the screen. The display of token moves is not enough to convey all these pieces of information. One has to design a full-fledged user interface, including alternative presentations and interaction with the user.

A distributed simulation platform is suitable for team prototyping. For instance, two users can set up two disconnected Petri nets and simulate fusion of transitions by a step. Such a feature, and more generally, using the semantic level to control the execution of the net, requires the synchronisation of clients on different sites. We rely on logical clocks (cf. [Lam78]) which are vectors counting the transition occurrences. While the topic of *causal multicast* (cf. [MR93]) is quite different, there are striking similarities between the ISIS causal broadcast protocol ([BSS91]) and message delivery in our framework, which uses both a vector clock and a local enablement for occurrence events. A better understanding of causality in Petri nets is however required to support a higher order control of the execution as well as team prototyping.

The last issue concerns the reproducibility of an execution. One user may be interested in the replay of a particular execution, in order to inspect it more closely, using different tools of the interface. While the algorithms in the management section are deterministic, their execution depends on the delivery order of remote messages. These delivery delays are not predictable, because of variations both of the network and machine loads. So, restarting the simulation on the same marked net may well lead to a different course of execution. Registering a global trace is inefficient. For HLPNs, the trace must register the bindings of parameters to token attributes and can be very large. On the contrary, replaying a trace is made easier by the process-driven control of our framework. Ideas along these lines, based on the "*Instant Replay Algorithm*" may be found in [Pru98].

20.6 Conclusion

We have given the outline of a distributed implementation of Petri nets in a simulation framework. We have also given a sketch of the proofs of several algorithms: The development of such a simulation tool must rely on formal methods. In order to apply it to the design and prototyping of distributed systems, its specification must be carefully laid out and the implementation must be correct. Notice that the formal specification of properties related to *fairness* or *productivity* is not obvious for HLPNs.

The implementations of the algorithms were briefly described. In addition to this programming point of view, we want to stress that the software engineering of the simulation platform is of equal importance. On the one hand, the multithreaded-object Java technology is evolving rapidly, most notably in the domain of user interfaces. On the other hand, virtual machine implementations, just-in-time compilers and the operating systems themselves are evolving independently. In practice, these layers of software platforms may not be bug free. A preventive approach to software maintenance is required in order to build a simulation platform above them.

Another kind of evolution is related to the incremental addition of features to the simulation environment in order to satisfy new requirements of the engineers using it. One can guess that presentation, user interfaces, and team development will introduce new requirements. In another direction, the simulation of large-scale distributed systems requires dealing with faults and fault tolerance. A careful software design may succeed and sufficiently simplify the programming task in such an intricate setting. However, a more architectural approach must complement the brute-force programming approach we take here.

21. Code Generation*

Petri nets as an initial specification introduce the possibility of verification of many properties of the system to be built. Petri nets also introduce the possibility of obtaining many interesting properties for the implementation which do not appear in the initial specification. For example, the fact that a net can be partitioned into a set of disjoint state machines can be used to propose a distributed implementation. The advantage of deducing the implementation from its formal specification is that it avoids (or reduces) the inherent errors associated with the coding task.

Code Generation Purposes

Code generation shares with *net execution* the "animation goal" of a system specification. However, the result is quite different. In code generation, the produced code aims to be *standalone*: it can be run outside the environment that produced it. In contrast, net execution remains closely related to the simulator that animates the specification[1].

This chapter deals with *code generation* from Petri nets. By code generation, we mean "production of programs (hardware or software) that implement a Petri net specification, i.e. that have the same behaviour". Simulation techniques were discussed in Chapter 20.

Thus, a *prototype* is the result of code generation. We also use the word *prototyping* to refer to the methods (or methodologies) that aim to produce "good" code from a specification. A brief classification of these techniques (which usually combine modelling, validation, and code generation) was presented in Chapter 19.

Whatever the input specification or the application domain, a prototype is software/hardware suitable for many purposes such as:

- Evaluation of the final system: the produced software corresponds to a "pre-release" useful for checking *a posteriori* some constraints on the system.

* Authors: W. El Kaim, F. Kordon

[1] In many simulation environments, it is possible to "connect" the model to "external pieces of code" executed according to the model (e.g. the corresponding transition fires). However, we cannot consider such an execution as being "independent" of the software environment that produced it.

The idea is to uncover information that might not have been outlined in the formal specification such as the validation of time constraints or the interface with the execution environment of the application.
- Evaluation of implementation strategies: this may be relevant for evaluating performance of a distributed system (modelled with Petri nets) according to a given task allocation.

It is interesting if the produced code leads to the final application (i.e. it respects some performance[2] and/or other quality criteria[3]). However, this should not be the main goal of code generation, even if we emphasise performance of the prototype. The main point is the evaluation of something that is no longer a system specification (or model) and not yet the final application (but close). Of course, a fast, automatic, and low-cost code generation procedure is required: if many tracks are to be investigated, a low-cost procedure is essential.

This chapter considers distributed software systems. However, most of the techniques described should be applicable to other types of systems. They are not restricted to a particular type of programming language. However, they rely strongly on message passing which is a commonly used technique in distributed systems.

Structure of this Chapter

Prototyping from Petri nets has been investigated since the late '70s. Three classes of studies are presented in Section 21.1.1. It appears that the most important point relies on an interpretation of the model semantics. We give some clues in Section 21.1.2 and propose a particular technique for interpreting a Petri net specification in order to partition it into a set of concurrent programs. This technique has been implemented in a code generator from coloured Petri nets (Section 21.2).

This partitioning technique relies on *prototype objects* corresponding to pre-defined (Petri-net-like) patterns to be implemented. Section 21.3.1 discusses the implementation of such prototype objects. For better reuse and evaluation of the produced code, interfaces with existing pieces of code may be necessary. So some "linking" techniques are investigated in Section 21.3.2.

The performance of a distributed system relies strongly on the allocation of its components. The prototype is more flexible if code generation is distinct from software component allocation (tasks, resources, ...). The prototype self-configures during its initialisation, and its architecture may evolve during execution time (process migration). It is easier when component allocation comes from a configuration file because the file can be modified without

[2] Please note that generation of efficient code raises "compilation-like" problems. This is now widely accepted as an industrial issue.

[3] Such criteria may be readability of the code (for later modification), capability of connection to the execution environment, ...

having to go through the entire code generation (and compilation) procedure. We show in Sections 21.3.3 and 21.3.4 how the information obtained from the Petri net and its properties can be used to get interesting clues about a "good" allocation strategy.

Petri nets are very suitable for modelling distributed systems. However, it is difficult to handle large-scale systems. Moreover, the nets may not contain all the information required for code generation (for example, about high-level objects that lead to complex submodels but a simple implementation). So Section 21.4 addresses code generation for both the association with another high-level (but not formal) model, and the encapsulation of Petri nets.

21.1 Petri Net Approaches to Code Generation

Code generation from Petri nets has been investigated for a long time. Studies propose producing hardware or software implementations of Petri net models (we focus in this section on the latter). Section 21.1.1 summarises the studies of Petri net code generation and Section 21.1.2 focuses on the major problems addressed by these works.

21.1.1 State of the Art

Many studies, summarised in [CSV86, Kor94], have been dedicated to the prototyping of parallel systems specified by means of Petri nets. Characteristics of these studies are presented in Table 21.1. These works represent three approaches to code generation from Petri nets: the *centralised approach*, the *totally decentralised approach*, and the *hybrid approach*.

Centralised Approach. The first approach aims to implement a centralised "token player" investigating each transition in the model and checking its fireability. Very efficient evaluation filters have been studied to reduce the number of scanned transitions, to avoid the bottleneck that appears when the net grows ([CSV86, MKM86]). However, this approach does not preserve the parallelism of the model because the "token player" is a sequential program. This strategy was quickly discarded as a way to generate software systems.

However, the centralised approach is still considered in flexible manufacturing systems where the token player is considered as a *scheduler*. In this case, a task represents an atomic action processing input data into output data. Such actions are naturally associated with the firing of transitions.

The scheduler is then centralised but it can send the tasks to be executed to discrete processors. Such a scheduler is difficult to distribute because it needs to handle the global state of the net. In this context, the most efficient techniques of centralised Petri net simulation are still considered to be satisfactory implementation techniques ([BC94]).

The scheduling of such tasks may also be computed and then statically implemented in a program. Such a strategy addresses scheduling problems. Such approaches are studied in [LLRKS93] for non-scheduling problems but they do not rely on Petri nets. Petri nets are used in [HM95] but in the case of cyclic scheduling.

Table 21.1. Summary of work on code generation from Petri nets

Ref	PN class	Implementation		Techniques		Comments
		Hard	Soft	Execution	Approach	
[SV82] (1982)	Binary PN	mono processor	Assembler	Interpreted	Centralised	-Use of a "scanner" to investigate transition. -Evaluation of discrete techniques.
[NHS83] (1983)	P/T nets + inhibitor arcs	mono processor	PL/1 and PL/S	Compiled	Centralised	-Selective investigation of transitions. -Use of an intermediary language.
[VCBA83] (1983)	Binary PN	mono processor	Assembler	Interpreted	Centralised (extension for distribution)	-Selective investigation of transitions.
[Thu85] (1985)	A form of coloured PN	mono processor	Pascal with tasks	Interpreted	Centralised	-First study for coloured Petri nets.
[BM86] (1986)	PROT nets (a form of coloured nets)	mono & multi processor	Ada83	Compiled	Partially distributed	-Two types of processes, one for places and the other for transitions.
[MKM86] (1986)	Control-Nets	mono processor	PL	Interpreted	Centralised	-Selective investigation of transitions based on both input and output places.
[Hau87] (1987)	P/T nets	mono & multi processor	C under Unix	Compiled	Distributed	-Each node in the net is implemented by a process.
[Tau88] (1988)	P/T nets	multi processor (transputers)	Occam	Compiled	Distributed	-Each node in the net is implemented by a process.
[Bré90] (1990)	P/T nets	multi processor (transputers)	Occam	Compiled	Hybrid	-Manual partitioning of the net into communicating processes.
[KEC90] (1990)	P/T nets	mono & multi processor	Ada83	Compiled	Hybrid	-Manual partitioning of the net into communicating processes.
[Pal91] (1991)	P/T nets	mono processor	Ada83	Compiled	Centralised	-Description of an approach in the design of complex systems that involves both Petri nets and HOOD.
[KE91, KP91, Kor92] (1991/92)	coloured PN	mono & multi processor	Ada83	Compiled	Distributed	-Automatic partitioning into processes (deduced from the structure of the coloured net), coloured invariants.
[EKK94] (1994)	coloured PN	multi processor	Ada83	Compiled	Distributed	-Proposal for a process allocation on a hardware architecture.
[LK95] (1995)	coloured PN + hierarchy	mono processor	C++	Compiled	Distributed	-Code generation from object-oriented Petri nets.
[Zak96] (1996)	coloured PN + time	multi processor	Ada83 + libraries	Interpreted	Distributed	-Integration of pieces of external C code.
[Hul97] (1997)	Algebraic PN	multi processor	Ada95	Interpreted & compiled	Distributed	-Hierarchy management in input Petri nets. -Progressive substitution of generated code by hand written code.

Totally Decentralised Approach and Hybrid Approach. Some work considered a totally distributed execution ([Hau87, Tau88]). Each place and each transition was implemented by means of a process. This approach preserves the parallelism of the model, but becomes inefficient when the net grows or when it has large sets of colours. Since the overhead introduced by conflict management (by means of messages, semaphores, ...) ruins performance, a thorough investigation for efficient code[4] was necessary. The resulting code's drawback is that some Petri nets cannot be processed. However, when the Petri net is structured "properly"[5], transformation into an equivalent, but decomposable, net is possible.

A natural extension of the hybrid approach is to enhance a Petri net in order to find high-level information about its structuring. Some work dedicated to code generation (tools are being implemented) can be found in [LK95] and [BFR92, Hul97]. Some related work more likely dedicated to high-level Petri nets and modelling also relies on this orientation: [SB94, Bas95, BCAP95, Val98, Lak96]. We can also mention [Zak96] as an interesting work that deals with Petri nets plus some timing constraints. In this work, the generated code is a token player linked to external C/C++ functions, stored in a library, that manage time constraints.

Code generation is also used to speed up simulation/animation of Petri nets, as in [PG92]. In that case, the generated code is centralised. The produced code then communicates with a front-end displaying graphical actions.

21.1.2 Parallel Interpretation of a Petri Net

Evaluation of the parallelism in a Petri net model is important since both centralised and decentralised approaches have poor performance when the model becomes very large.

A solution is to split the net into a set of components that can be separately generated (and then concurrently executed). Sequential state machines ([Hac74]) are interesting candidates because they can be implemented by a sequential program (a process). Here are two accepted partitioning approaches:

- Techniques based on BDD (Binary Decision Diagrams) [McM93]. The idea is to directly exploit information deduced from the state graph generated by the Petri net ([GVC95]).
- Techniques based on place invariants. The idea is to use information deduced from the net structure. We will focus on this approach.

A state machine usually corresponds to a place invariant in a Petri net model (in contrast, a place invariant is not automatically a sequential state machine). So, for the hybrid approach presented above, it is useful to define a way to select "good" place invariants which lead to state machines. In

[4] Prototypes containing up to 500 Ada processes were experimented with by the authors on a single workstation.

[5] A "proper" structuring may be achieved by sub-net composition using the techniques mentioned in Chapter 9.

[KP91, Kor92], such properties are defined for P/T invariants. Extensions that use coloured invariants to deduce state machines are investigated in [BP93, BP94]. In this chapter, we define a *PN-process* to be a sequential state machine computed from the Petri net model[6].

The model is then interpreted as a set of PN-processes *plus* communication mechanisms. Two possible communication mechanisms are widely accepted in Petri nets:

- Places (asynchronous communication): this corresponds to the composition of sub-nets by means of *channel places* ([Sou89]).
- Transitions (synchronisation): this corresponds to a transition fusion between discrete PN-processes[7] ([BC93, BCE94]).

Colours do not raise any particular problem. Studies such as [CSV86, Kor92, Bré93] propose managing them as data structures computed and manipulated by programs generated from the Petri net model. When the net is split into state machines (hybrid approach), management should be defined according to the following rules:

- If coloured tokens are located in communication (or channel) places, they are treated like incoming or outgoing messages.
- If coloured tokens are located in state places in a sequential state machine, they are considered as local variables.
- Management of colours in shared transitions is similar to that for channel places. Data is exchanged between the PN-processes involved and a server that manages the corresponding synchronisation.

Deduction of a generic architecture that is relevant for most programming languages (Figure 21.1) is now possible. The prototype potentially contains the following components:

- PN-processes, each being associated with a computed sequential state machine
- Channel places manager (potentially distributed)
- Synchronisation manager (potentially distributed)
- Colour manager which takes care of both type constraints and type compositions (potentially distributed)

---

[6] A sequential state machine is quite different from a task as introduced in the definition of the centralised approach (Section 21.1.1). Tasks always produce a result in a finite time while state machines may not (they may contain loops in which inputs are consumed and/or outputs are produced).

[7] Please note that the semantics of a synchronisation is that of a *multi-rendezvous*. A multi-rendezvous is an extension of the "classical" rendezvous mechanism ([And91]) that is commonly used in CSP [Hoa85] and Ada [DoD83]. However, N processes can synchronise on a multi-rendezvous. The rendezvous is a particular case in which $N = 2$.

- Prototype manager (potentially distributed) which manages initialisation, handles low-level communications (between PN-processes and the channel manager, between PN-processes and the synchronisation manager), and detects termination of the generated system (usually when all state machine tokens have reached a state place without successors).

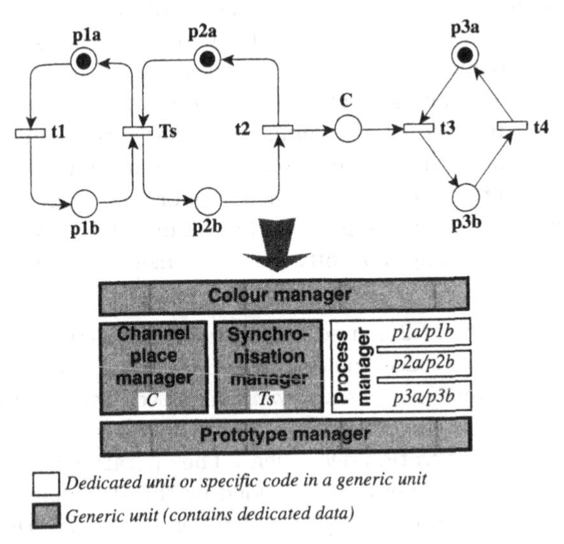

Fig. 21.1. An example of a Petri net with the corresponding prototype structure

Figure 21.1 shows a sample model in which three PN-processes can be detected. Each PN-process should be implemented as a specific module that describes its sequential behaviour. The channel place C and the synchronisation Ts are managed by dedicated modules. All components rely on the prototype manager and use primitives offered by the colour manager (a library deduced from the colour classes in the model and offering basic colour management functions such as successor or predecessor).

Managers of PN-processes *p1a/p1b*, *p2a/p2b*, and *p3a/p3b* contain *local* transitions (i.e. those that do not correspond to a synchronisation). In *p2a/p2b*, transition *t2* is local and the corresponding code must take place in the PN-process. In contrast, Ts corresponds to a communication. It is implemented by a server; in the program corresponding to *p2a/p2b*, it is implemented as a communication with this server. A typical algorithm for local and synchronisation transitions is provided in Section 21.3.1.

Please note that the only specific pieces of the produced code are related to PN-processes. The code implementing the managers is almost independent of the input Petri net: only the required data structures are specific to the net. In the example of Figure 21.1, descriptions for C, Ts, and the colour domain should be generated.

21.2 A Petri Net Partitioning Algorithm

The partitioning algorithm we present is the one proposed in [KP91, Kor92]. It relies on non-coloured place invariants. Extensions that deal with coloured invariants can be found in [Pey93, BP94]. The algorithm has four steps:

- Computation of the *structural model* (defined below in Section 21.2.1).
- Computation and selection of positive place invariants (called P-semiflows in this section) for the structural model. These P-semiflows will be used in the next step to deduce sequential state machines.
- Evaluation of partitioning properties in order to eliminate useless P-semiflows and to find the set of combinations that corresponds to a possible partition of the model into communication state machines.
- Computation of *prototype objects* which support code generation, according to the generic architecture presented in Figure 21.1. A prototype object is an entity corresponding to a software component of the generic architecture.

The model of Figure 21.2 is used to illustrate the partitioning procedure. It models the increment or decrement of a set of bank accounts (unit by unit) according to a list of queries. It is *a priori* composed with two PN-processes. The operation generator (covered by places *s3*/*s4*) generates either an operation increment and an operation decrement. The operation manager (covered by places *s1*/*s2*) takes an operation provided by the operation generator and applies it to the current user. It has one channel place *value* that effectively represents a shared variable containing all pairs <user, current bank amount> and one synchronisation (*getQ*).

21.2.1 Transformation into a Structural Model

The first step of the algorithm is to extract a *structural model* called $\mathcal{N}_s$. $\mathcal{N}_s$ is a P/T net structurally equivalent to $\mathcal{N}$. The transformation procedure from $\mathcal{N}$ to $\mathcal{N}_s$ is:

1. Each place in $\mathcal{N}$ becomes a place in $\mathcal{N}_s$. Colour domains (if any) are discarded. Only the initial number of tokens remains as the initial marking of each place in $\mathcal{N}_s$.
2. Each transition of $\mathcal{N}$ becomes a transition of $\mathcal{N}_s$. Guards (if any) are discarded.
3. Each arc of $\mathcal{N}$ becomes an arc of $\mathcal{N}_s$. Only the required number of tokens remain in the arc valuation of $\mathcal{N}_s$.

Remark: When $\mathcal{N}$ is a P/T net, $\mathcal{N} = \mathcal{N}_s$ because the structural model describes the net structure without any information about variables and colours (which are implemented by means of data structures).

Figure 21.3 shows the net obtained from that of Figure 21.2. Note that there are three non-coloured tokens in places *s2* and *value*, while place *s3* contains only one token.

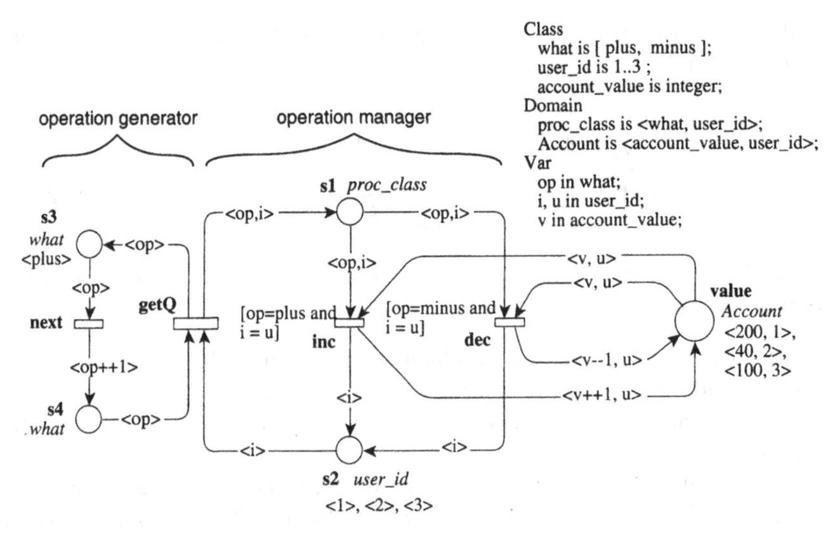

Fig. 21.2. Example to illustrate the partitioning algorithm. Colour classes, domains, and variable declarations are given in the upper right of the net. Object names are outlined using **bold** characters, colour domains of places are given using *italic* characters.

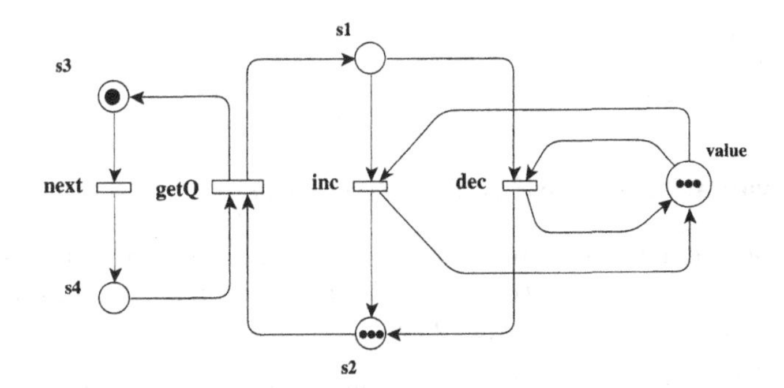

Fig. 21.3. Structural model of the Petri net presented in Figure 21.2

21.2.2 Computation and Selection of Positive Place Invariants

Next, a generative family of P-semiflows on $\mathcal{N}_s$, denoted $\mathcal{F}$, has to be computed. The algorithm[8] will not be detailed here. To compute P-semiflows for the examples presented in this chapter, we have used the GreatSPN tool [Chi91].

We already indicated that invariants of $\mathcal{F}$ do not necessarily represent PN-processes. We must first compute $\mathcal{F}_p$, a subset of $\mathcal{F}$ that contains only

[8] The most frequent implementation of this calculus relies on the Farkas algorithm ([CS90a]).

"good" candidates for PN-processes. Elements of $\mathcal{F}_p$ should satisfy the following property[9]:

$$\forall f \in \mathcal{F}_p. \forall p \in \|\mathbf{f}\|. \forall (t,t') \in p^\bullet \times {}^\bullet p.$$
$$\mathbf{Pre}[p,t] = \mathbf{Post}[t',p] = 1 \quad \wedge \sum_{p \in \|\mathbf{f}\|} \mathbf{m}[p] \geq 1 \tag{21.1}$$

Condition 21.1 means that the elements of $\mathcal{F}_p$ should satisfy the following conditions:

1. For each place covered by an element of $\mathcal{F}_p$, the valuation of both the input and output arcs should be equal to one (the program counter of a process, and therefore a PN-process, must remain constant throughout its execution.
2. There is at least one token in the set of places covered by $\mathcal{F}_p$ (the total number of tokens represents the total number of PN-process instances in the prototype). If this condition is not satisfied, then there is no potential instantiation of this PN-process and the prototype cannot run.

The structural model of Figure 21.3 has three P-semiflows:

1. sf_1: $value = 3$
2. sf_2: $s4 + s3 = 1$
3. sf_3: $s1 + s2 = 3$

21.2.3 Evaluation of Partitioning Properties

To partition the net into state machines, it is now necessary to find all possible combinations $\mathcal{F}_d \subset \mathcal{F}_p$ that satisfy both condition 21.2 and 21.3. These conditions select subsets of $\mathcal{F}_p$ in which

- all transitions (instructions) take place in a state machine (condition 21.2);
- no elements of $\mathcal{F}_d$ have a place in common (condition 21.3):

$$\bigcup_{i=1}^{|\mathcal{F}_d|} [\bigcup_{j \in \|\mathbf{f_i}\|} (p_j{}^\bullet \cup {}^\bullet p_j)] = T \tag{21.2}$$

$$\forall f_1, f_2 \in \mathcal{F}_d. \quad \|\mathbf{f_1}\| \cap \|\mathbf{f_2}\| = \emptyset \tag{21.3}$$

Each $\mathcal{F}_d$ then corresponds to a possible partition of the net into a set of communicating PN-processes.

In the example, there are two possible subsets $\mathcal{F}_d$ that satisfy the above conditions:

---

[9] For a place invariant $f = \sum_{i=1}^{|P|} a_i.p_i$, $\|\mathbf{f}\|$ corresponds to all the places p_i for which $a_i > 0$.

1. $\mathcal{F}_{d_1} = \{sf_2, sf_3\}$
2. $\mathcal{F}_{d_2} = \{sf_1, sf_2\}$

The partition $\mathcal{F}_{d_1}$ corresponds to the one that was *a priori* expected. $\mathcal{F}_{d_2}$ describes another possible partition of the net. They are both discrete views of the system (Figure 21.4). Partitioning according to $\mathcal{F}_{d_2}$ relies on asynchronous communication only (*s1* and *s2* are considered as channel places). The choice of a partition has to be done by the designer. He uses criteria such as the number of PN-processes found, the type of communication between PN-processes, the semantic meaning of processes, etc.

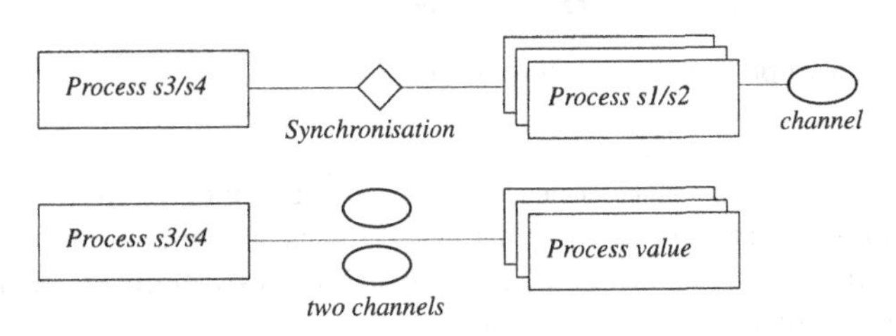

Fig. 21.4. The two views introduced by partitioning $\mathcal{F}_{d_1}$ and $\mathcal{F}_{d_2}$

21.2.4 Computation of Prototype Objects

When a given partition has been selected, prototype objects may be generated. These are the ones previously identified (PN-processes, channels, and synchronisations). They can be automatically computed using the following rules:

- Each element $f \in \mathcal{F}_d$ corresponds to a PN-process. Places supported by f are *states*, and transitions that are connected only to states of f become *local actions*. Please note that local actions should have one input state and one output state only.
- Each place that is not supported by any element $f \in \mathcal{F}_d$ corresponds to a *channel place*.
- Each transition that has more than one input state and more than one output state is a *synchronisation*.

The result of such an operation for partition $\mathcal{F}_{d_1}$ in our example is shown in Figure 21.5. It clearly outlines the structure that is convenient for the generated prototype.

When the model is coloured, it is also necessary to compute the set of local variables for each PN-process. Each PN-process instance (deduced from the initial number of tokens in the net) will have its own copy of these variables. These variables constitute the *PN-process context*.

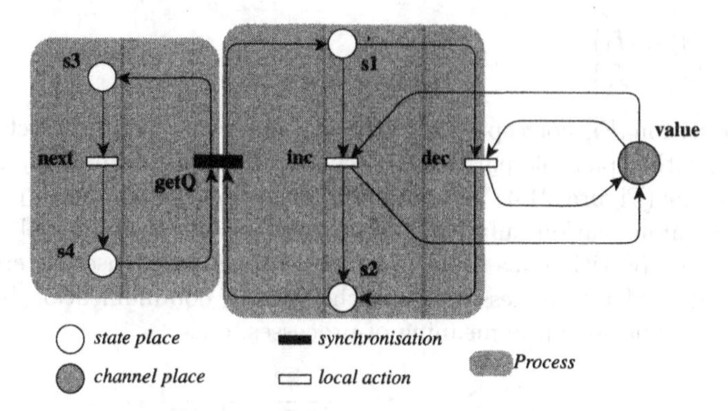

Fig. 21.5. Prototype objects computed for partition $\mathcal{F}_{d_1}$

Computation of the PN-process context corresponds to an interpretation of the variables found in arc valuations. This is a complex operation that must deal with a large number of configurations, especially when there is a lack of writing rules for arc valuations.

Figure 21.6 presents two examples illustrating some of the interpretation problems that have to be solved. Model A illustrates the overloading of variables. Thus x, z and u are equivalent (as are y, t, and v). a is another potential context variable. It is obvious that the context of the only PN-process of model A should be either variable a (typed by domain D) or a set of two variables (x and y, respectively typed by classes *c1* and *c2*, for example). This choice is difficult to automate when PN-processes are complex and deal with state places having discrete colour domains.

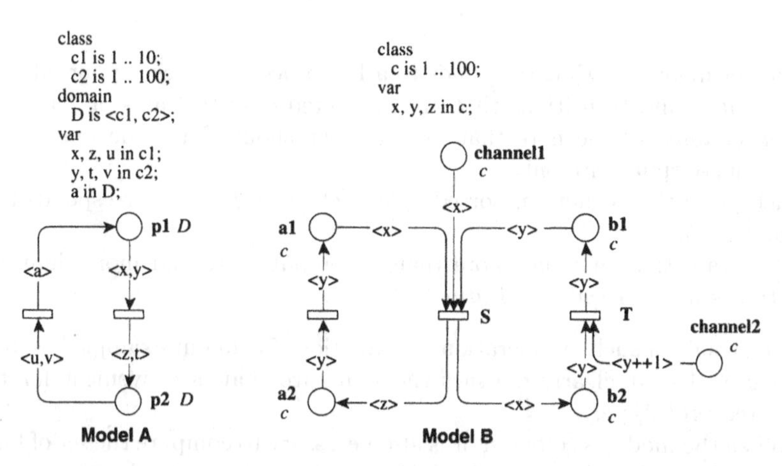

Fig. 21.6. Examples of nets for which the automatic computation of context variables is difficult

Model B illustrates the problem that occurs when there is data exchange by means of a synchronisation (here transition S) or a complex selection of tokens from a channel place (here transition T). In the case of model B, the conclusion could be:

- In PN-process *a1 / a2*, the context should be variable x.
- In PN-process *b1 / b2*, the context should be variable y.
- There is a transmission of information from PN-process *a1 / a2* to PN-process *b1 / b2* via synchronisation S.
- Action T requires a token from *channel2* that respects the condition *predecessor* ("context variable y") = "token from *channel2*".

Such a conclusion is also difficult to compute automatically. A solution could be the use of coloured invariants but they are difficult to compute and to interpret[10]. However, some interesting aspects of such an extension are discussed in [BP93, Pey93, BP94]. The solution proposed in [KEK94] to ease the computation of the PN-process context is the definition of writing conventions.

21.2.5 Speeding Up the Algorithm

The algorithm presented deals with a combination of N elements ($N = |\mathcal{F}_p|$) and is thus quite slow (complexity in $n!$). We can speed up the algorithm by discarding a large number of solutions:

1. Let $\mathcal{F}_i$ be a subset of $\mathcal{F}_p$ that verifies condition 21.2, and f an element of $\mathcal{F}_p / f \notin \mathcal{F}_i$.
   $\forall \mathcal{F}_j = \mathcal{F}_i \cup f$, $\mathcal{F}_j$ verifies condition 21.2.
2. Let $\mathcal{F}_i$ be a subset of $\mathcal{F}_p$ that does not verify condition 21.3, and f an element of $\mathcal{F}_p / f \notin \mathcal{F}_i$.
   $\forall \mathcal{F}_k = \mathcal{F}_i \cup f$, $\mathcal{F}_k$ does not verify condition 21.3.

According to property 2, it is useless to investigate supersets of a set $\mathcal{F}_i$ that does not verify condition 21.3 because they will be discarded anyway. According to property 1, it is also useless to evaluate supersets of a set $\mathcal{F}_i$ that already verifies condition 21.2 because they will either fail or introduce useless synchronisations.

These optimisations have been implemented in the CPN/Tagada[11] tool that is a part of the CPN-AMI version 1.3 environment [MT94]. They provide very good speed-ups except when the analysed model has structural symmetries. These symmetries generate many solutions. Thus, much smaller subsets of "obviously bad solutions" are cut off.

[10] The only algorithm to produce a generative family of coloured P-semiflows is the one described in [CHP93]. It runs only on unary coloured nets without guards. As far as we know, there is no systematic way to compute such invariants from any type of coloured Petri nets.

[11] Tagada stands for **T**ranslation, **A**nalysis and **G**eneration of **ADA** code.

21.2.6 Net Transformation When the Algorithm Fails

For space reason, we present here only a few important transformations that lead to a partitionable net. More detailed information and an extension of the algorithm may be found in [Kor92].

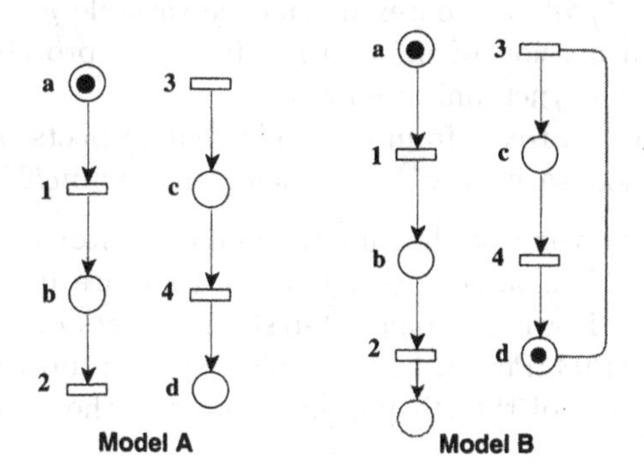

Fig. 21.7. First transformation that enables condition 21.2 in a model. Added items (places, arcs, and tokens) are displayed in grey.

The failure of condition 21.2 means that there are pending places and transitions. There are two possible origins:

- At least one transition without a predecessor or without a successor. A very simple procedure is to add a place after a transition without a successor and to connect a transition without an input place to the end of a "sequence". The first transformation can be done automatically; the second requires semantic information that should be provided by the system designer. Figure 21.7 presents an example of such a transformation. Model *A* cannot be processed using our algorithm while model *B* can.
- At least one dynamic PN-process creation. In this case, forcing the PN-process to be statically instantiated provides good results. An implicit place is added (and marked). The corresponding P-semiflow appears and the condition no longer fails. An example of such a transformation is given in Figure 21.8. Model *A* cannot be processed using our algorithm. Model *B* can be processed but it will be difficult to detect the end of execution for PN-process c/e using the technique previously proposed (when a PN-process instance reaches a place without a successor). Model *C* is another possible transformation that allows such a detection of the end of execution, but it cannot be processed automatically because semantic knowledge is required.

The failure of condition 21.3 means that there are "shared sequences". This configuration corresponds to fork/join sequences as shown in Figure 21.9. In this example, two PN-processes share state places *a* and *f*, and thus

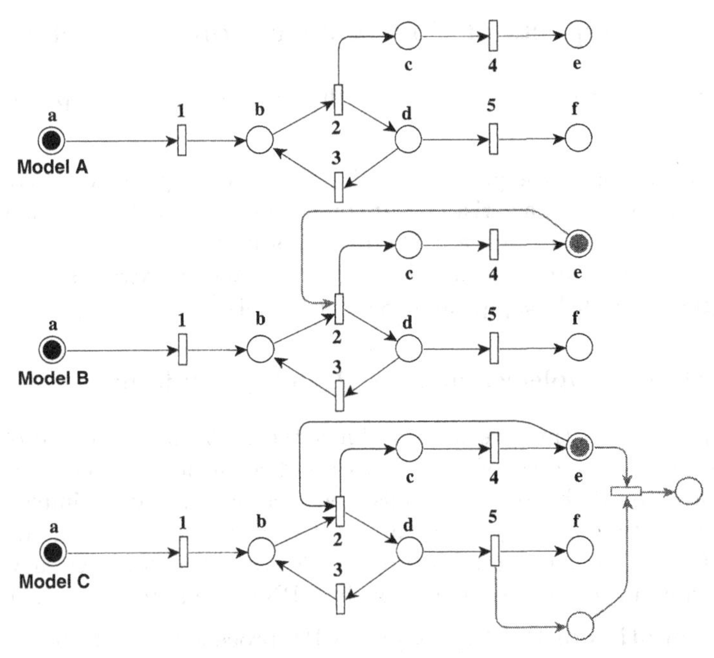

Fig. 21.8. Second transformation that enables condition 21.2 in a model. Added items (places, arcs, and tokens) are displayed in grey.

the algorithm we propose fails. Duplication of these places (including their initial marking) produces model B that is partitionable with our technique. Such a transformation can be performed automatically but produces many synchronisations (transitions *1*, *2*, and *5* in our example).

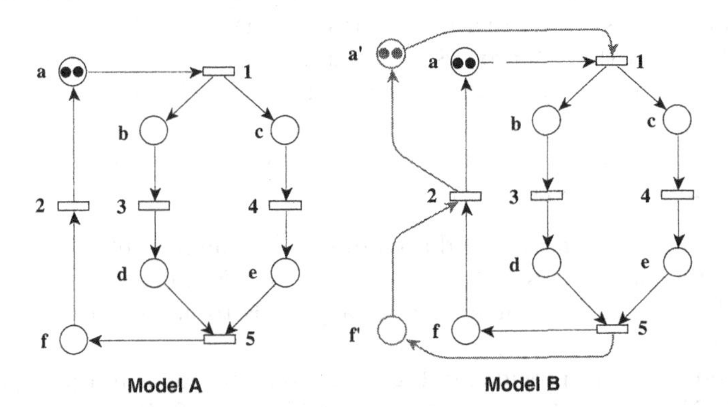

Fig. 21.9. First transformation that enables condition 21.3 in a model. Added items (places, arcs, and tokens) are displayed in grey.

21.3 Some Aspects of Code Generation from Petri Nets

Code generation from Petri nets deals with two separate (but complementary) aspects:

- Implementation of the prototype objects in order to get efficient code
- Computation of characteristics that could lead to a "good" allocation strategy for a given hardware architecture. It is useful to map a software architecture to a hardware one (Section 21.3.3) and to extract some helpful configurations such as pipelines (Section 21.3.4).

21.3.1 On the Implementation of Prototype Objects

In the previous section, we have outlined the following prototype objects: PN-processes (composed of state places and local actions), channels, and synchronisations. This section details how such objects can be implemented in an imperative language that handles parallelism management (such as Ada or C/C++ under Unix). Most of the strategies proposed here have been implemented in the code generator of the CPN/Tagada tool ([KEK94]).

Implementation of PN-Processes. A PN-process is a model of a sequential state machine that aims to become a thread or a process managed by the operating system. Each sequential state machine is a model of a program task implemented as a case included in a loop ([Man90]). The loop ends when a state place without a successor is reached. Each case alternative corresponds to a state of the machine:

- A local action (for non-shared transitions)
- Connection to the synchronisation manager (for shared transitions)
- A state, if it has more than one successor. This is an optimisation that eliminates useless states relating two transitions. Then the next state after a transition in the PN-process automaton is the next transition. In that case, the corresponding code should be a (random?) choice of one of the successors. Concurrent evaluation of all pre-conditions is also possible but may introduce unfair executions when one transition pre-condition is much more complex than the others.

PN-processes are instantiated according to the number of tokens initially found in the state places. Dynamic creation of PN-processes is difficult to consider because such configuration usually leads to non-partitionable Petri net models.

The code segment provided in Figure 21.10 corresponds to the implementation of PN-process *s1*/*s2* in the model of Figure 21.2. Please note:

- According to the optimisation previously outlined, states *s1* and *s2* are not translated because they do not correspond to a choice point (they only relate two sequential transitions).

- Context (variables *op* and *i*) is implemented through local variables.
- Variables *finish* and *c_state* are used to control the current state and the end of the PN-process.

```
PN-process s1/s2
   // declaration of local variables
   type PN-process_state is (tr_inc, tr_getQ);
   variable op : what;
   variable i : user_id;
   finish : boolean := false
   c_state : PN-process_state;

begin
   // behaviour of the state machine
   initialisation phase that gets from the prototype manager
   the values of op, i and c_state
   loop
      case c_state is
         when tr_inc =>
            treatment associated with tr_inc
         when tr_getQ =>
            treatment associated with tr_getQ (connection to the
            synchronisation server etc.)
      end case
   until finish
   termination phase that advises the prototype manager of the
   PN-process end
end
```

Fig. 21.10. Typical algorithm of a PN-process

Evaluation of a local action predicate (if any) has to be split: 1) it evaluates the part that involves context variables first, 2) if necessary, it contacts the channel places manager which evaluates the part that involves channel places. A process should not itself evaluate the part involving channel places in order to minimise conflicts when they occur in more than one transition guard. If the pre-condition fails, the process instance should be suspended until the channel place manager wakes it up. If the transition occurs behind a choice point, a time-out should be managed in order to avoid deadlocks.

The code segment provided in Figure 21.11 shows how to implement a local action when it takes place behind a state that is a link between two transitions. Note that variable *c_state* is the one declared in the previous algorithm.

The code segment provided in Figure 21.12 shows how to implement a local action when it takes place behind a state that is a choice point between a set of transitions. A time-out is now managed in order to avoid deadlocks (the *a priori* choice of a transition to evaluate may not be good).

```
begin
   if condition on context variable = true then
       send query to the channel place manager  (if any)
       wait for answer
       execute code associated with the transition
       update context variables
       send "channel postcondition" (if any) to the
       channel place manager
       c_state := next state
   else
      signal deadlock for this process
   end if
end
```

Fig. 21.11. Typical algorithm of a local action not located behind a choice point (state with more than one successor)

```
begin
   if local condition = true then
       send query to the channel place manager (if any)
       wait for answer until time-out
       if answer OK then
          execute code associated with the transition
          update context variables
          send "channel postcondition" (if any) to the
          channel place manager
          c_state := next state
       else
          c_state := previous state
       end if;
   else
      c_state := previous state
   end if
end
```

Fig. 21.12. Typical algorithm of a local action located behind a choice point (state with more than one successor)

The code associated with a synchronisation is very similar to that for a local transition except that no associated procedure or post-condition construction is executed by the process. It will contact the synchronisation manager and wait for a positive answer (with a time-out if the synchronisation is behind a choice point). The synchronisation manager takes care of pre-conditions and post-conditions.

Implementation of the Synchronisation Manager. A centralised implementation of the synchronisation server is not of interest because it disables a possible distribution of synchronisations over a set of processors. Moreover, synchronisations have nothing in common; each one corresponds to a distinct communication mechanism. Considering this the synchronisation manager is quite easy to distribute.

Each synchronisation is managed by one process[12] that manages its current context. It accepts messages from clients. These clients are instances of the PN-processes that share the corresponding transition.

When a client sends a connection query, the process evaluates it. If at least one instance of each PN-process involved is present and respects the transition condition, it fires the transition (i.e. executes the corresponding transition and generates the post-condition) and wakes up the corresponding clients.

If a synchronisation has input or output channels, the corresponding query (token demand or production) is delegated to the channel place manager.

Implementation of the Channel Place Manager. The implementation of the channel place manager raises some problems that are very close to those of databases:

- Generation of tokens (data) is easy to handle. A complex query involving token production in many places may be split into elementary queries that deal with one place.
- In contrast, consumption of tokens is more complex, especially when there are conditions to satisfy. In the model of Figure 21.13, the evaluation of the pre-condition associated with transition T depends on the set of required values respecting a complex condition.

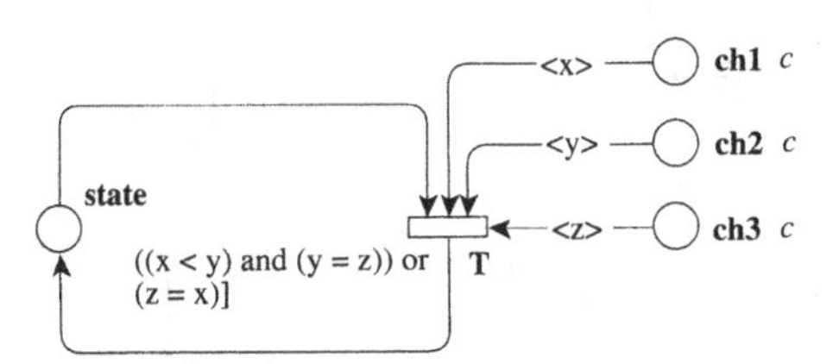

Fig. 21.13. Example of model containing a complex precondition (on transition T)

For these reasons, it is very difficult to evaluate separately the marking of each place.

This conclusion leads to a centralised proposal. The channel place manager is a database-like query evaluator. While the problem is related to the evaluation itself, it seems possible to replicate the query evaluator on each site. Places and their content should then be able to migrate from one site to another. When a query is sent to one of the channel place managers, it calls the required places and performs the evaluation when they are on the site. This solution raises the following problems:

[12] Potential replication of such a process to avoid potential bottlenecks is discussed in Section 21.3.3.

- Ordering of queries should be total. This can be solved using either the Lamport clock ([Lam78]) or the vectorial clock ([Mat89]) synchronisation algorithm.
- Migration should be managed carefully in order to avoid situations such as:
  - The loss of a migration demand that arrives during the migration of the corresponding channel;
  - The never-ending chase of a migration demand that always arrives just after the departure of the corresponding channel.

  Such situations can be solved using a commitment after each channel migration.

The main drawback of this implementation is that the number of messages exchanged cannot be bounded. A small mistake in the mapping of processes to a set of processors may lead to very poor performance due to unnecessary migrations.

A second distributed implementation is based on the following observation: *Sets of channels that belong to a query should never be separated.* Each manager is responsible for a set of channels that never move. Non-local queries are then remotely evaluated. This solution does not generate a large number of messages but its efficiency depends on the net structure.

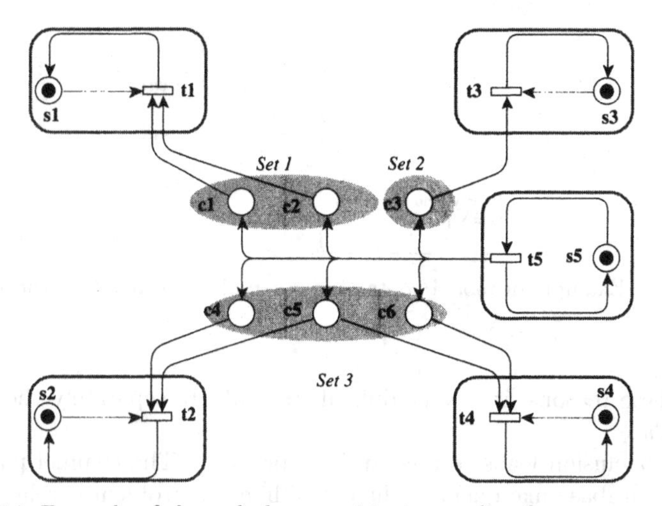

Fig. 21.14. Example of channel place partitioning to distribute management

Figure 21.14 illustrates such a partitioning of channel places. The model contains five outlined processes that communicate through six channels that may be grouped into three sets:

- $set_1 = \{c1, c2\}$: they are both involved in the pre-condition of transition *t1*;

- $set_2 = \{c3\}$: it is involved in the pre-condition of *t3*;
- $set_3 = \{c4, c5, c6\}$: *c4* and *c5* are involved in the pre-condition of *t2* and must be grouped. However, *c5* is also involved in the pre-condition of *t4* with channel *c6*. In order to evaluate both types of pre-condition queries, these three channels have to be grouped.

Recall that post-conditions (in this case, that for transition *t5*) do not have any influence on these sets. The channel manager of this model may be distributed over up to three different hosts if this policy is selected.

21.3.2 Prototype and Execution Environment

Another key issue is connecting the generated prototype to its execution environment. This is important when the prototype is not used to speed up a simulation process as in [PG92].

A solution is to consider a "prototypable" Petri net $\mathcal{N} = \mathcal{N}_p \cup \mathcal{N}_e$ where $\mathcal{N}_p \cap \mathcal{N}_e = I$ is a set of objects (places and/or transitions) that constitute the *interface* between $\mathcal{N}_p$ and $\mathcal{N}_e$. We assume that:

- $\mathcal{N}_p$ corresponds to the system to be prototyped. The description is very precise in order to get all the details of the implementation.
- $\mathcal{N}_e$ corresponds to the execution environment. The description is not as precise as that for $\mathcal{N}_p$. It is the *abstraction* of the environment execution of the system. This means that it mainly describes the interaction with the system, not the way in which the environment is implemented.

$\mathcal{N}$ is important for validation purposes. $\mathcal{N}_p$ is used to generate code and $\mathcal{N}_e$ to deduce interfaces between the generated code and the execution environment (Figure 21.15). Thus, the partitioning algorithm is applied to $\mathcal{N}_p$ only after extraction of $\mathcal{N}_e$, which is used to build an appropriate interface. The prototype is finally linked to the environment interface. Abstractions are reusable specification components that can be stored in a library. A similar strategy is used in CORBA with IDL [MZ95].

Such a technique may also be used for separate modelling and code generation. The system is modelled through n submodels that are connected by mean of specific interfaces. A team may work on a given module using the abstraction of other modules (for the team, they are pieces of the environment execution).

A critical point is to decide what type of interaction should be provided between the prototype and its execution environment. In [KE91], interaction is considered by means of channel places only while [Bré93] prefers a communication by means of transitions. Such choices arise from the underlying software or hardware architecture: distributed Ada in [KE91] (channel places are treated like RPCs), a network of transputers in [Bré93] (shared transitions correspond to process communication in Occam). Such a choice determines the way submodels should be connected, and may depend on a preferred implementation strategy (according to the target language constraints)

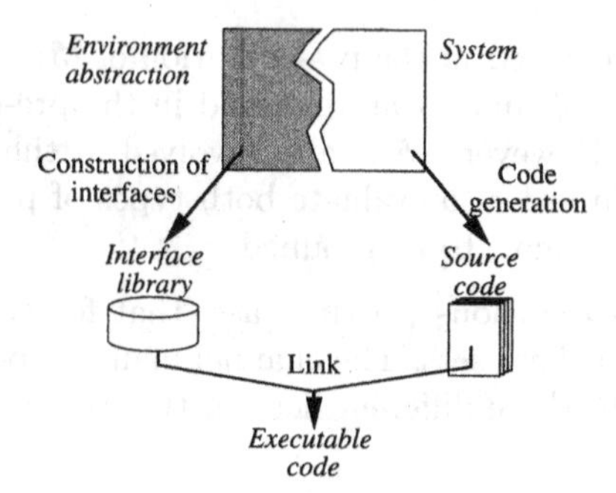

Fig. 21.15. Code generation procedure that integrates relations of produced programs with their execution environment

21.3.3 Mapping Processes onto a Given Architecture

Both target hardware and application architecture must be exploited to model distributed systems and their environment. Focusing only on application architecture could cause a loss of the time gained by automatic code generation for a distributed environment.

Thus, the prototype should be as flexible as possible to allow experimentation over distinct hardware architectures (as a network of interconnected machines or a multi-processor machine). The mapping can be done either in the code or separately in a configuration file. The former is relevant when the target language is dedicated to the execution architecture (e.g. Occam in [Bré93]). The latter is easier to modify (no recompilation or code generation needed after a modification). We will focus on the latter.

To map software onto a given hardware, it is necessary to be able to formalise both. We will first investigate some work that focuses on the hardware description and then show how a software architecture can be deduced from the partitioning obtained in the hybrid approach (Section 21.2). Finally, an example of mapping is discussed.

The Hardware Model. Dealing with both fine- and large-grain descriptions is difficult without a common hardware formalism. A multi-processor or a network of interconnected workstations should differ only by the characteristics of the medium. An ethernet brain connecting two workstations and a bus connecting two processors are not equal in term of throughput and reliability.

Distributed systems such as CONIC [MKS89] and parallel programming languages such as DURRA [BDW+92], ARGUS [Lis88], and EMERALD [JLHB88] provide a simple network description framework. Usually only configuration and reconfiguration management are taken into account.

However, in [ST92] a very interesting network description language called SySl (System Structure language) is proposed. SySl enables the description of systems with hardware, software, and documentation components. Encapsulation of system classes sharing some features is provided, as well as dependencies between components. Genericity is used to describe each object and its properties. For instance, a workstation has characteristics such as processor, memory, and communication interface. Each attribute value is specified when a system class is instantiated.

GATOS [FS93] focuses on the software application structure and works directly on the executable prototype. Dynamic load balancing and fault tolerance are automatically provided. Distribution of parallel applications among heterogeneous hosts and migration of processes are both possible. A configuration file describing the application architecture and the user constraints must be created. The information about the underlying hardware, needed to provide an automatic distributed execution, is not given by the system designer.

In [Fer90], software systems and hardware architectures are both modelled with the same formalism. A program resource mapping net (PRM net) is a modular concept used for the modelling of parallel programs with a Petri net (called the P-net). All resources and their dependencies are also described by means of a Petri net (named the R-net). The mapping between P-nets and R-nets is graphically performed by the designer.

Parallel Proto [Bur91] introduces a graphical hardware description, which can be coupled with the application model to simulate placement effects. The location of objects may be performed either manually or randomly. There is a predefined library that includes some well-known architectures such as Intel Hypercube and Encore Multimax. However, the tool uses this information only for simulation purposes and no code is produced.

The hardware formalism proposed in [EKK94] defines two types of hardware objects: H-Machines and H-Links. They are used to create a graph, where H-Machines (nodes) are linked with H-Links (arcs). Hardware objects are defined by means of attributes and their description may be graphical or textual.

H-Links represent a network cable, a transputer channel, or any other communication link. Characteristics are described by a set of attributes such as:

- Identifier
- Type: channel (local to a motherboard), local (local network), or distant (external network)
- Effective throughput
- Reliability (maximum bound for message delivery).

H-Machines are virtual workstations. Here are some attributes needed to complete their description:

- Identifier
- A list of H-link identifiers

- Technical information such as the current state (alive or dead), the host machine address, and the average load
- The machine characteristics (CPU type, reliability, speed, memory, disk), needed to manage disparities between executable formats and incompatibilities in execution speeds.

The host machine address and its related links must be declared to allow automatic location. The other parameters are necessary only for heuristic computation and statistics purposes.

The Software Model. The software model is composed of elements that can be distributed such as processes and communication mechanisms. Communication mechanisms are as relevant as processes because they generally hide data or managers. This is the case for both the synchronisations and the channels which we get from our partitioning procedure.

It is relevant to consider a hierarchical description of the system architecture in order to have a recursive integration of distribution criteria. To do this, [EKK94] introduces the notion of **D**istributed **V**irtual **S**oftware **C**omponents (DVSCs) that are either software-executable units or resources. DVSCs are connected together and may contain sub-DVSCs.

A DVSC contains the following information:

- Identifier
- Type: decomposable (it is a set of sub-DVSCs that can be assigned to distinct processors) or indivisible (all DVSC components should remain grouped on the same host)
- Behaviour: active (for processes such as a PN-process) or passive (for data or resources such as a channel place)
- Attachments: these may be user-defined constraints that have to be respected or the names of hosts or characteristics that are required (for example, an executable file format).
- Ability to replicate: this is a boolean value that is set to true when the object can be replicated to distinct hosts without changing the behaviour of the system. Replicability of such DVSCs implies that each copy is independent of the others (in our software model, each instance of a replicable process could be assigned to distinct processors).
- Internal description: the contained DVSCs if there are any.

DVSCs could be directly derived from the prototype objects that are computed from the Petri net model:

- Each process becomes an active DVSC that can be replicated (each instance may be located on distinct H-Machines).
- In the case of a distributed management of synchronisations, each is considered to be an active DVSC (*a priori* not replicable). Otherwise, all synchronisations are grouped into one indivisible DVSC that includes sub-DVSCs.

- In the case of a centralised implementation, the channel manager is one indivisible active DVSC. If the first distributed policy is selected, there is one passive DVSC (*a priori* not replicable) per channel. If the second distributed strategy is implemented, each set of channels is a passive DVSC (*a priori* not replicable).

DVSCs are connected with links. Each link has a weight that expresses the relative cost of an access or its frequency[13]. Such information can be deduced from the communications between the two DVSCs. For example, communication through a channel should be weighted less than a synchronisation. It is possible that two distinct policies lead to different values. This is a way to customise the mapping procedure.

Mapping the Prototype Onto a Target Architecture: An Example. Based on both the hardware model and the software model, it is possible to compute a "good" allocation of prototype components (DVSCs) to processors of a distributed architecture (an H-machine).

The partitioning $\mathcal{F}_{d_2}$ of our example (Figure 21.2) is used to illustrate the procedure that is proposed in this section. We assume that the synchronisation manager is distributed as described in Section 21.3.1. We also assume that the channel place manager is implemented using the second distributed strategy (partition of channels depending on the pre-conditions in which they are involved).

In this example, it is not necessary to take into account the implementation policies discussed in Section 21.3.1 because there is only one synchronisation and one channel place. Thus, all proposed implementations lead to the same prototype structure (Figure 21.16). Let us suppose that the valuation of edges, for instance around *getQ*, are computed after the code generation strategy of a given tool. These values could correspond to a usage rate (here, one per connection to the communication mechanism).

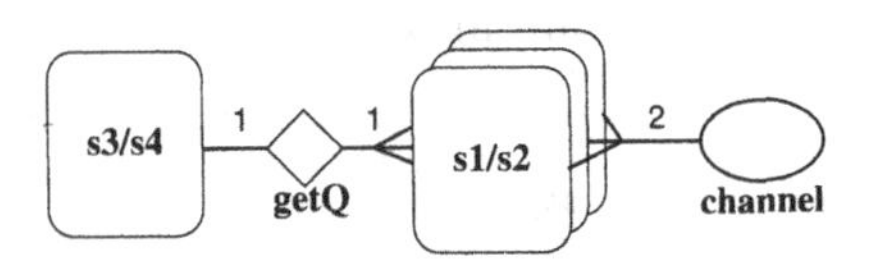

Fig. 21.16. Initial software description of the prototype

Let us further suppose that there is no constraint from the system designer forcing the allocation of any DVSC to a given host (such as an executable file format). Some observations can be done according to the Petri net model.

[13] Evaluation of the message rate can be done after some executions of the prototype.

- Shared transition *getQ* manages a rendezvous between the instance of *s3/s4* and the three instances of *s1/s2*. Such an entity is indivisible because no replication rule can be found.
- Channel place *value* and the instances of process *s3/s4* do share a common "key":
  – variable *i* having colour *user_id* for *s3/s4*;
  – the second field of tokens (having colour *user_id*) for channel *value*.
  It is also useful to notice that both transitions *inc* and *dec* (they both have *value* as a pre-condition) require in their guard, equality between context variable *i* and the second field of *value*'s tokens.

Based on these observations, we finally get from the initial software architecture of Figure 21.16 six DVSCs that are presented in Figure 21.17:

- One PN-process *s3/s4* instance having the following characteristics:

```
type                      = indivisible,
behaviour                 = active,
attachment                = empty,
replication               = no,
internal description      = empty.
```

- One synchronisation server for *getQ* having the following characteristics:

```
type                      = indivisible,
behaviour                 = active,
attachment                = empty,
replication               = no,
internal description      = empty.
```

- Three PN-process *s1/s2* instances having the following characteristics:

```
type                      = indivisible,
behaviour                 = active,
attachment                = channel containing token i,
replication               = no,
internal description      = empty.
```

- One set of channels containing *value* having the following characteristics:

```
type                      = indivisible,
behaviour                 = passive,
attachment                = PN-process s1/s2 number "second field of the token",
replication               = upon colour user_id,
internal description      = empty.
```

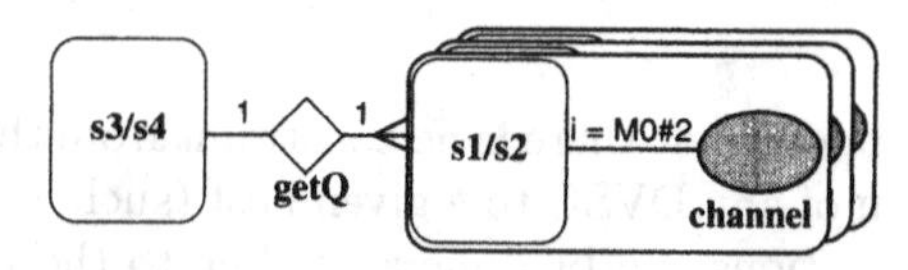

Fig. 21.17. Software architecture of the prototype after analysis

Let us now assume that the hardware architecture is composed of a set of workstations connected by a Ethernet bus[14]. Each machine has the same characteristics.

The maximum number of machines required (five) corresponds to the number of computed DVSCs. Table 21.3 shows a set of possible locations computed according to the previously deduced software architecture. Table 21.2 defines the names of elementary DVSCs to be placed. By convention, a_i is an active DVSC and p_i a passive DVSC. Subscript information (if any) corresponds:

- for active entities, to the initial value for the context (it is then a PN-process instance)
- for passive entities, to the initial information (it is then the marking of a channel place).

Please note that, according to the remarks we have made about the potential replication of channel place *value*, it corresponds to three elementary DVSCs (p_1, p_2, and p_3). Each is a replica of *value* having the appropriate initial marking.

In Table 21.3, the allocation of passive entities p_2 and p_3 on machine 4 (line 5) means that there is one copy of channel *value* that contains both tokens $\langle 40, 2 \rangle$ and $\langle 100, 3 \rangle$.

Table 21.2. Correspondence table for DVSC

Location entity name	Corresponding DVSC
a_1	$s3/s4._{op=plus}$
a_2	$getQ$
a_3	$s1/s2._{i=1,op=?}$
a_4	$s1/s2._{i=2,op=?}$
a_5	$s1/s2._{i=3,op=?}$
p_1	$value._{M_0=\{\langle 200,1\rangle\}}$
p_2	$value._{M_0=\{\langle 40,2\rangle\}}$
p_3	$value._{M_0=\{\langle 100,3\rangle\}}$

21.3.4 Place Invariants and Pipeline Detection

Pipelines are also an interesting configuration to detect because they may provide a significant speed-up when spread over a set of hosts. In some cases, some of the invariants that were discarded may outline a pipeline in a Petri net model. The model of Figure 21.18 is obviously composed of five PN-processes. It has six p-semiflows:

[14] Please note that such a configuration generates no routing problem.

Table 21.3. Possible mapping from one up to five hosts

number of machines	machine 1	machine 2	machine 3	machine 4	machine 5
1 host	a_1, a_2, a_3, a_4, a_5, p_1, p_2, p_3				
2 hosts	a_1, a_2	a_3, a_4, a_5, p_1, p_2, p_3			
3 hosts	a_1	a_2	a_3, a_4, a_5, p_1, p_2, p_3		
4 hosts	a_1	a_2	a_3, p_1	a_4, a_5, p_2, p_3	
5 hosts	a_1	a_2	a_3, p_1	a_4, p_2	a_5, p_3

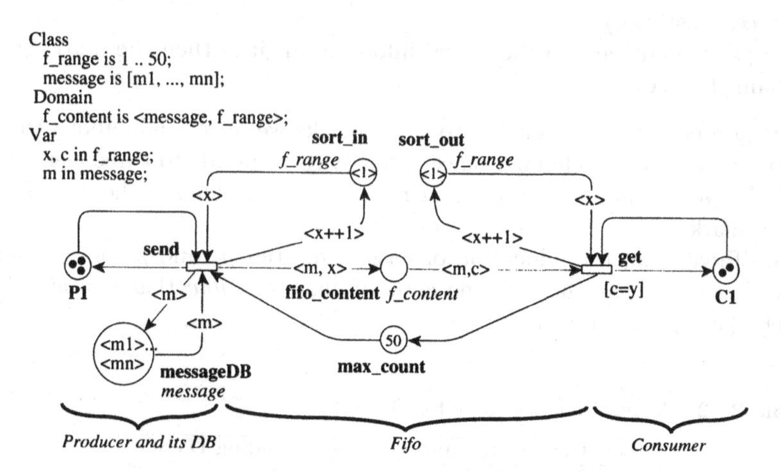

Fig. 21.18. Petri net model that can be implemented in a pipeline

1. $sf_1 = p1$,
2. $sf_2 = p2$,
3. $sf_3 = p3$,
4. $sf_4 = p4$,
5. $sf_5 = p5$,
6. $sf_6 = c1 + c2 + c3 + c4 + c5$.

There are two partitions: $\mathcal{F}_{d_1} = \{sf_1, sf_2, sf_3, sf_4, sf_5\}$ and $\mathcal{F}_{d_2} = \{sf_6\}$. Let us consider $\mathcal{F}_{d_1}$. It discards invariant sf_6 that corresponds to a data-flow. Let us observe that it is the only PN-process of another partition ($\mathcal{F}_{d_2}$).

The problem is that such an configuration is not systematic. If we suppress place *c1* in the model of Figure 21.18, sf_6 disappears and no extra information is provided. The pipeline however remains. The conclusion is that unselected p-semiflows, as well as unselected partitions, may contain useful information for PN-process allocation.

21.4 Code Generation from a High-Level Net

In the previous sections, after presenting a general view of code generation from Petri nets, we focused on a way to produce efficient code. It is obtained after a partitioning of the Petri net model into prototype objects that prefigure the generated code structure.

We also outlined how the Petri net properties were able to help us to extract some relevant information to produce distributed code. In particular, place invariants should be used for the partitioning (see Section 21.2) or to detect pipelines (see Section 21.3.4).

However, although they have may desirable features in the context of validation, we do not think that Petri nets are a good entry point for efficient code generation. Let us illustrate this with an example: a system composed of two PN-processes (P and C) communicating by means of a Unix fifo (pipe) that can contain a maximum of 50 messages. Figure 21.19 shows such a system. The submodel that corresponds to the fifo (places *sort_in*, *sort_out*, *fifo_content*, and *max_count*) complicates the model and introduces many potential PN-processes. It is then difficult to avoid complexity in the generated code. A complex set of entities will be developed from parts that model the fifo although it is a very simple object to model. Searching for specific Petri net patterns (such as the one for a fifo in our example) is not a solution because there are many ways to model such behaviour.

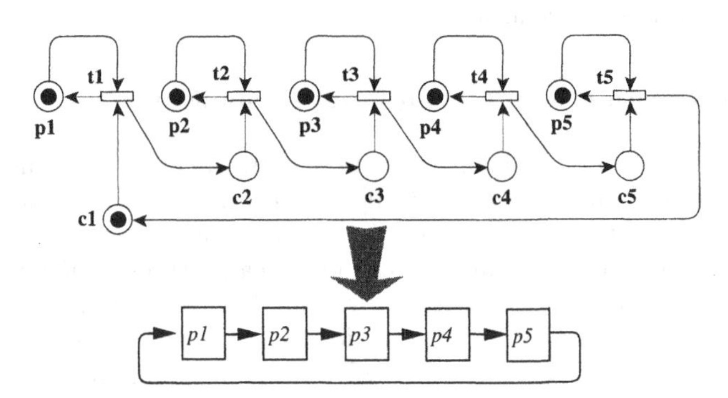

Fig. 21.19. Producer/Consumer model through a Unix fifo

However, while Petri nets enable the computation of relevant information, they also hide some important high-level information. From the code generation point of view, there are two problems:

- Petri nets may be useful for very detailed modelling as well as for high-level modelling. The meaning and precision of a model is decided by the designer. No tool can compute such information.
- There is a lack of structure and thus no guidelines for verification or identification of some specific configurations.

Hence, interpretation of Petri net specifications is quite difficult because relevant semantical information depends on the description level selected by a designer. To compensate for the lack of Petri net structure, two research strategies are usually considered:

- Integration of new high-level features, such as hierarchy: this approach enables modelling of more complex systems but raises theoretical problems. Some properties can no longer be computed. Not all properties extractable from non-hierarchical nets can be computed either. Various proposals for this approach can be found in [BFR92, SB94, Bak96, Bas95, LK95, Val98, Lak96].
- Association between Petri nets and another high-level representation that provides structuring capabilities. The idea is to produce a flat Petri net from a semi-formal model. The Petri net specification could be enormous but is hidden from the end user. Properties of this model are translated back into the terms of the high-level formalism. Proposals for this approach are discussed in [DG90, Pal91, KEK95, DK96].

Elements of both approaches were presented in Section 10.3 but they mainly dealt with modelling. Elements of net execution by simulation were presented in Chapter 20. This section describes how an association between Petri nets and another high-level formalism can be relevant and can preserve both advantages (computation of formal properties and optimised implementation of high-level mechanisms).

21.4.1 Association with a High-Level Formalism

Some early work proposed an association between Petri nets and an pre-existing object-oriented model ([DG90]). However, such an approach can not provide a solution that covers all the high-level formalism capabilities because they are too rich. Some mechanisms cannot be easily transformed into simple Petri nets and thus the computation cannot be fully automated. In their paper, the authors indicate that a modelling strategy should be respected. This idea is of particular interest for solving the modelling problem and can be expressed in:

- The definition of restrictions in the high-level formalism as in [Pal91]
- The definition of a high-level formalism that is dedicated to the modelling philosophy, as in [BE94, KEK95, DK96].

21.4.2 An Example of Work Based on a Pre-Existing High-Level Formalism

A very interesting proposal involving the HOOD specification is investigated in [Pal91]. It proposes HOOD/PNO, a more "natural" method for building large models. The main idea is to use both formalisms: HOOD to decompose the system into objects and then Petri nets to model the system behaviour and deduce properties. This method is best suited for the design of flexible manufacturing systems involving *physical objects* (for example, a robot or a conveyer) driven by a distributed application.

The method has five steps:

- Describing the system environment
- Deciding which physical objects of the PN-process to prototype
- Attaching each physical object to its class
- Describing each class
- Deciding on software and object classes.

When the system is decomposed into modules, the behaviour of such objects is modelled using P/T nets. Each net is contained in a module. Modules are connected. Three types of communications are proposed:

- Strongly or loosely synchronous
- Asynchronous
- Timed-out queries.

Code generation of Ada programs is performed using the information provided by both HOOD and Petri nets. HOOD gives information that is suitable for detecting genericity units, visibility clauses, and public parts of a unit. Petri nets give information about the behaviour of the unit.

Such an approach has three major advantages:

- A complex partitioning algorithm is no longer needed because it is outlined by the designer in the HOOD specification.
- Code generation benefits from HOOD information.
- The complete model may be generated and validated using a "flat" Petri net made of object behaviour connected together according to the communication schemes that are available.

So, both HOOD and Petri nets benefit from this association. HOOD acquires some validation capabilities and Petri nets are structured. A non-Petri-net-expert designer may use such a method without the difficulty of managing big models. The size of each Petri net specification (into a HOOD object) is quite small and thus easy to understand and maintain.

21.4.3 An Example of a High-Level Formalism Dedicated to Code Generation: H-COSTAM

Let us now investigate one dedicated formalism: H-COSTAM (**H**ierarchical **CO**mmunicating **STA**te **M**achine model) [KEK95]. This formalism does not aim at the description of any type of application. It focuses on the modelling of distributed applications for code generation purposes.

It is basically nothing more than an "encapsulation" of coloured Petri nets which supports the following features:

- A typed communication model is proposed in order to meet the needs of distributed systems. Both asynchronous and synchronous communication are provided.
  Asynchronous communication is performed by means of *passive media* that may behave as follows:
  - FIFO: message order is preserved;
  - LIFO: message order is inverted;
  - random: message order is not preserved[15].

  Synchronous communication is provided by means of *active media.* So far, only one mechanism has been investigated: multi-rendezvous[16].
- Processes are the elementary units composed of:
  - a sequential automaton;
  - a set of private variables (each process instance has its own copy);
  - a typed interface to other units that corresponds to communication links.
- Dynamic process creation in order to enable an adequate modelling of distributed systems
- Hierarchy to enable large-scale specification through modularity
- Strong typing to enable verifications as soon as possible and then to derive colour domains safely
- Genericity to reduce the size of a specification and enhance the reusability of modules.

Links to Petri nets are preserved but not only for validation purposes. The computation of "good" properties could lead to optimisation of the system, especially when considering process allocation over a distributed hardware architecture. For example, properties such as the bound of places may be useful for dimensioning some resources in the prototype. Techniques discussed in Section 21.3.3 can be also applied to the Petri net model produced.

It is also possible to ease the calculus of such properties by having distinct and adapted transformations to Petri nets. The procedure is illustrated in Figure 21.20. For example, evaluation entity replication and resource dimensioning do not depend on the same Petri net model. However, both models come from the same high-level description.

Figure 21.21 corresponds to the H-COSTAM version of the system modelled in Figure 21.19. The model is now split into three modules (called *pages*). In this specification, the main page belongs to a *macro-level* description and describes only the structure of the system. It is composed of two processes (*Prod* and *Cons*) that share a fifo link (same behaviour as a Unix pipe). Process *Prod* also has access to a database of messages. Please note that it is possible to define types and constants.

[15] Random links effectively have the same behaviour as channel places.

[16] See footnote 7 on page 438.

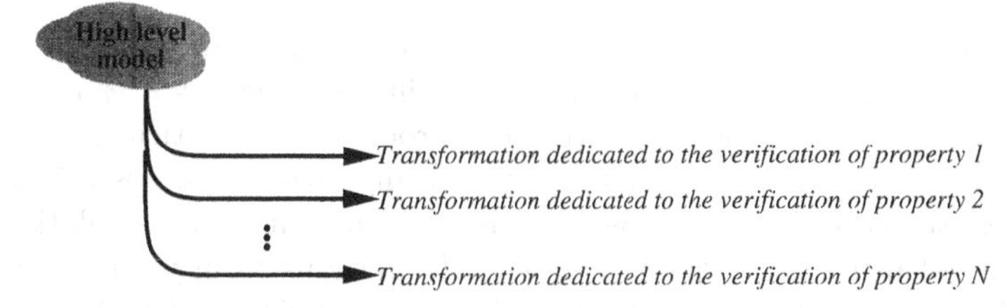

Fig. 21.20. Distinct and property-oriented transformations to Petri nets

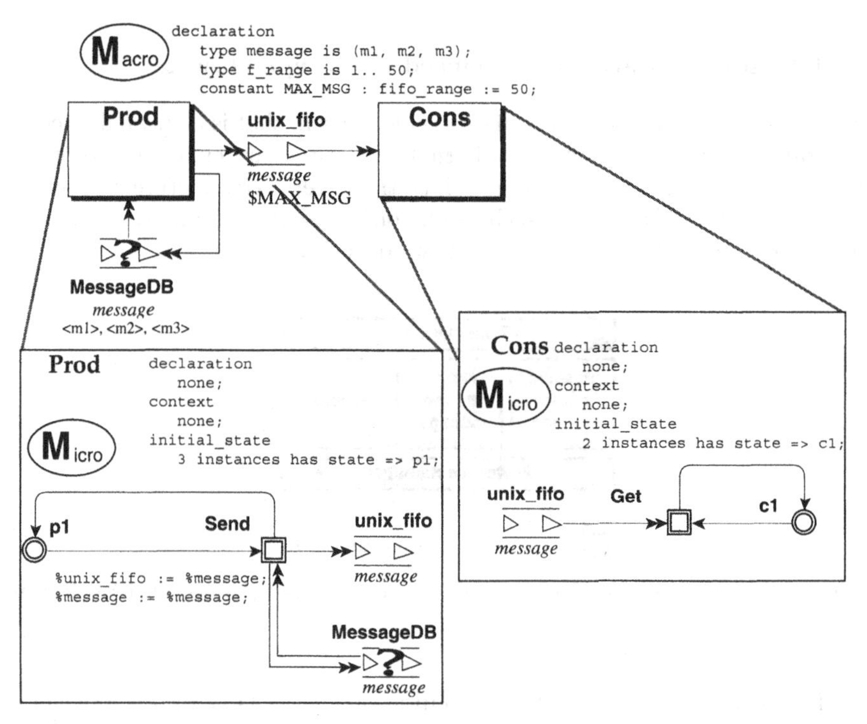

Fig. 21.21. The H-COSTAM version of the producer/consumer model presented in Figure 21.19

Each elementary component (here, boxes *Prod* and *Cons*) is separately defined in other pages. An elementary process should have a *micro-level* description that defines its behaviour by means of an automaton. Both macro and micro pages may introduce new declarative items (types or constants). A page also inherits declarations from higher levels. Pages that are hierarchically enclosed in other communicate through interfaces that are communication media (for example, *unix_fifo*). In the micro-level page describing *Prod*, no local declaration is performed but types *messages* and *fifo_range*, as well as constant *MAX_MSG*, are visible.

Transformation of this specification into Petri nets produces the model shown in Figure 21.19. Please note that this example involves neither genericity nor dynamic process creation (all processes are statically instantiated).

Such models are quite easy to transform because the possible operations (composition, communication, variable assignment or manipulation) can be expressed using Petri net terms. Transformations from more complex high-level formalisms (such as Shlaer & Mellor in [LK95] or HOOD in [Pal91]) are more difficult if we want to preserve some analysis capabilities. This means that only subsets of such high-level formalisms can be exploited.

21.4.4 Implementation of Enhanced Prototype Objects

The main advantage of using a high-level formalism that is dedicated to code generation, such as H-COSTAM, is that its entities are closer to prototype objects than are Petri net entities. Thus, the generic architecture associated with a high-level formalism, such as the one presented in Figure 21.1, only has to be slightly adapted to fit the new concepts.

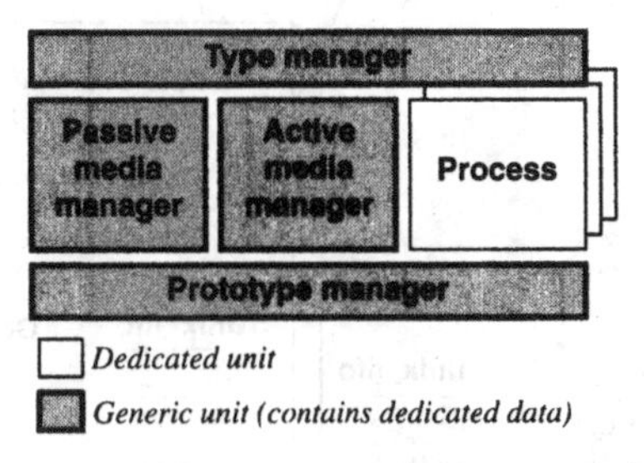

Fig. 21.22. An example of adapted generic architecture (for H-COSTAM in this case)

Figure 21.22 presents a possible adaptation for H-COSTAM of the generic architecture proposed in Figure 21.1. The passive media manager is an extended version of the channel manager able to handle distinct communication behaviour (for FIFO, LIFO, and random links).

Genericity can be supported using the corresponding feature on target programming languages (such as Ada or C++). Otherwise, a rewriting mechanism similar to the one for macro functions can be implemented.

Dynamic process creation should be taken care of by the prototype manager itself. Specific messages are dedicated to this purpose. A process that has to create another process sends a message to the prototype manager which then creates it. Thus, a safe separation between user-defined functions and the prototype runtime is ensured. Such an implementation can be either centralised or distributed.

21.5 Conclusion

In this chapter, after presenting some key issues of code generation from Petri nets, we focused on a particular code generation technique which has been implemented and tested. The approach presented is well suited to the implementation of message-passing-based distributed software systems.

Code generation is a technique that cannot be considered as a final goal. This procedure is part of a *prototyping* process involving several other operations such as: modelling of a system, evaluation of the system model (by simulation or verification), evaluation of the final application (whether or not it has been automatically generated).

It is difficult to consider the Petri net level as an entry point for code generation. It is quite satisfactory in some cases (flexible manufacturing systems for example), however it may not be so accurate for other applications (distributed systems). Here are the two major problems of Petri nets for code generation:

- the lack of high-level structures (such as FIFOs, LIFOs, and multicast mechanisms) and
- the lack of structuring.

The first point is very technical (it is difficult to obtain optimised code and many simple software objects correspond to complex submodels). The second point prevents a modelling/prototyping approach of systems based on design patterns ([Coa92, Oza96]), and makes modelling of large systems more complex.

However, solutions to these problems are being investigated:

- Extension of the Petri net formalism ([SB94, Lak96])
- Association of Petri nets with another structuring formalism (object-oriented representations are good candidates) as in [DG90, Pal91, Hul97]
- Encapsulation of Petri nets using a structured language that tracks useful information for code generation ([KEK95, Hul97]).

The first approach is more likely a long-term solution in which Petri net extensions are traded with the capability of computing formal properties. In contrast, the other two solutions propose a shorter-term solution to this problem.

However, a Petri-net-based approach is valuable in the design of (even complex) systems because it enables the use of formal validation techniques. Such a property gradually becomes a very important point for checking the safety (structural, behavioural, etc.) of applications. The validation of large complex applications should be a major issue of the new century.

22. Conclusion

Validation can be seen as one of the central tasks of systems engineering. It provides the means to check whether the described, planned, or built system fulfils the expectations of the user, customer, or client. These expectations cover all aspects of a system, be it static or dynamic in nature. In this part of the book some critical parts of systems engineering have been described and some major areas of Petri nets have been presented, namely prototyping, net execution, and code generation.

Chapter 19 discussed prototyping as an approach that is a quite natural way to perform systems engineering, especially in the early phases of the development process. In combination with Petri nets, prototyping is very powerful since the operational semantics can be used to execute the nets and therefore provides a good insight into the behaviour of the model developed so far. In several areas the prototype can also directly be used as the final system. However, this depends on many different factors. In this part of the book the use of models dominates, and these are transformed into prototypes.

The importance of supporting approaches in software engineering by tools is well known. Petri nets in particular are well known for their considerable requirements with respect to hardware and software support. In recent years enormous progress has been made, since the graphical interfaces can be built much more easily today than some years ago. The necessary support is now widely available and the great potential of current hardware and software is used more and more by Petri net tool developers.

In Chapter 20 a distributed implementation of Petri nets in a simulation framework was presented. Some technical algorithms, which give a good insight into the underlying ideas, were discussed. Simulation takes place within a certain context. The underlying system components, such as the operating system viewed as a technical aspect or the overall architecture viewed as a conceptual aspect, and their relevance for the simulation were mentioned.

Chapter 21 about code generation presented some key issues that are related to Petri nets. Fundamental questions related to automatically generating code from Petri nets must be solved. The chapter presented an approach that has been implemented and used in several experiments. Prototype objects were deduced by complex algorithms and problems related to colours were mentioned. Furthermore, the importance of a proper integration into an overall prototyping approach was stressed.

Many questions related to the validation of complex systems have been left open. However, an approach based on Petri nets, and its potential within software engineering have been presented. The operational semantics of Petri nets allows for a specific way of prototyping. Models that directly cover central complex issues in an application domain are built and then transformed into a prototype. This prototype can be either the interpreted net itself or generated code. The possibility for users, customers, and developers to choose in a flexible manner the right and appropriate way to develop complex applications is one of the important advantages of Petri nets. Relations to Java, CORBA, Enterprise Java Beans (EJB), etc. – as indicated in Chapter 20 – are of practical importance. The commercial sector requires appropriate approaches and modelling techniques, which can be found in this book. A detailed discussion of some specific application areas will come in the next part (Part V) where some specific approaches for the use of Petri nets in general will be demonstrated.

Part V

Application Domains

23. Introduction

23.1 Putting Petri Nets to Work

Petri nets have existed for over thirty years. Especially in the last decade, Petri nets have been put into practice extensively. Thanks to several useful extensions and the availability of computer tools, Petri nets have become a mature tool for modelling and analysing industrial systems. This part describes how and when Petri nets can be used to model and analyse a variety of systems in application domains ranging from logistics to office automation.

Since the introduction of the classical Petri net by Carl Adam Petri in the '60s, Petri nets have been used to model and analyse all kinds of processes with applications ranging from protocols, hardware, and embedded systems to flexible manufacturing systems, user interaction, and business processes. In the last two decades the classical Petri net has been extended with colour, time, and hierarchy. These extensions facilitate the modelling of complex processes where data and time are important factors. There are several reasons for using Petri nets:

- *Formal semantics*
  A process/system specified in terms of a Petri net has a clear and precise definition, because the semantics of the classical Petri net and several enhancements (colour, time, hierarchy) has been defined formally.
- *Graphical nature*
  Petri nets are a graphical language. As a result, Petri nets are intuitive and easy to learn. The graphical nature also supports the communication with end-users.
- *Expressiveness*
  Petri nets support all the primitives needed to model processes. All the constructs that are needed are present.
- *Properties*
  In the last three decades many people have investigated the basic properties of Petri nets. The firm mathematical foundation allows for reasoning about these properties. As a result, there is a great deal of common knowledge, in the form of books and articles, about this modelling technique.
- *Analysis*
  Petri nets are marked by the availability of many analysis techniques.

Clearly, this is an important factor in favour of the use of Petri nets. These techniques can be used to prove properties (safety properties, invariance properties, deadlock, etc.) and to calculate performance measures (response times, waiting times, occupation rates, etc.). In this way it is possible to evaluate alternative designs.

- *Vendor independence*
  Petri nets provide a tool-independent framework for modelling and analysing processes. (This is in contrast to the techniques promoted by vendors of CASE, WFM, ERP, and simulation tools.) Petri nets are not based on a software package from a specific vendor and do not cease to exist if a new version is released or when one vendor takes over another vendor.

23.2 Domains of Application

In this part, we focus on three application domains: (flexible) manufacturing, telecommunication, and workflow management.

23.2.1 Manufacturing

Factory automation is probably one of the oldest application domains of Petri net theory. Since the early '70s, Petri nets have been used to model and analyse manufacturing systems. Most applications in the field of manufacturing deal with discrete production systems. The number of applications in the field of continuous production is limited. The discrete nature of Petri nets causes some problems with respect to the modelling of continuous flows of materials (e.g. paper mills and oil refineries). Note that many production systems are hybrid, i.e. continuous production flows are made discrete by producing in batches. In this part, we focus on manufacturing systems from a strictly discrete perspective.

A *manufacturing system* is composed of a *physical subsystem* and a *control subsystem*. The physical subsystem is composed of physical components such as conveyors, robots, buffers, and work stations. The control subsystem controls the physical subsystem in order to organise and optimise the production process. When modelling a manufacturing system, both subsystems and their relations need to be specified. Note that the control subsystem has several levels: planning, scheduling, coordination, and local control.

Flexible manufacturing systems (FMS) in particular appear to be an interesting area of application. These systems are characterised by flexible, concurrently operating, and mainly automated elements, such as a production controller, a machine, an automated guided vehicle, and a conveyor. This results in high productivity, short throughput times, and a high degree of diversity in output (i.e. the resulting products).

There are several reasons for using Petri nets in the domain of (flexible) manufacturing. Petri nets allow for the modelling of resource sharing, conflicts, mutual exclusion, concurrency, and non-determinism. Moreover, the well-defined semantics allows for both qualitative and quantitative analysis. In particular, Petri net theory can help to detect potential deadlocks and construct control policies for deadlock prevention.

23.2.2 Workflow Management

In former times, information systems were designed to support the execution of individual tasks. Today's information systems need to support the business processes at hand. It no longer suffices to focus on just the tasks. The information system also needs to control, monitor, and support the logistical aspects of a business process. In other words, the information system also has to manage the flow of work through the organisation. Many organisations with complex business processes have identified the need for concepts, techniques, and tools to support the management of workflows. Based on this need the term *workflow management* was born.

Until recently there were no generic tools to support workflow management. As a result, parts of the business process were hard-coded in the applications. For example, an application to support task X triggers another application to support task Y. This means that one application knows about the existence of another application. This is undesirable, because every time the underlying business process is changed, applications need to be modified. Moreover, similar constructs need to be implemented in several applications and it is not possible to monitor and control the entire workflow. Therefore, several software vendors recognised the need for *workflow management systems*. A workflow management system (WFMS) is a generic software tool which allows for the definition, execution, registration, and control of workflows. At the moment many vendors offer workflow management systems. This shows that the software industry recognises the potential of workflow management tools.

The main purpose of a workflow management system is the support of the definition, execution, registration, and control of *processes*. Because processes are a dominant factor in workflow management, it is important to use an established framework for modelling and analysing workflow processes. In this part we show the application to the workflow domain of a framework based on Petri nets. Petri nets are a good candidate for the foundation of a unified workflow theory.

23.2.3 Telecommunications

Telecommunications has become a dominant factor in today's information society. People and industry rely on telecommunications systems in order to exchange information. Over the last decade the number of services has increased. The ability to forward calls and GSM cellular phones are examples of the new services offered. The statement "The Network is the Computer!" illustrates the emerging role of telecommunications services. A telecommunications system is no longer only a network that transports data and implements protocols. It also provides advanced services. Therefore, the focus is shifting from protocol engineering to service engineering.

A *telecommunications system* consists of two subsystems: a *transport subsystem* and a *processing subsystem*. The transport subsystem is the network, i.e. the communication resources. The processing subsystem is the set of computing resources and programs that control and manage the transport network on the one hand, and that implement the communication software on the other hand. The complexity of the two subsystems and the interaction between them may lead to design errors and performance problems. Therefore, a rigorous approach to modelling and analysis needs to be implemented. In this part, an approach based on Petri nets is presented.

23.2.4 Other Application Domains

There are several other application domains where Petri nets have turned out to be a useful design/analysis tool:

- Distributed software systems
- Logistics (materials handling, production logistics, physical distribution)
- Multi-processor systems
- Software engineering
- Asynchronous circuits
- (Distributed) protocols
- Hardware/software architectures
- Embedded systems
- User interfaces.

It is important to be aware of the trend that today's enterprises are focusing on business processes. Therefore, enterprises are in need of a good formalism for modelling and analysing these processes. There are several reasons for the increased interest in business processes. First of all, management philosophies such as Business Process Re-engineering (BPR) and Continuous Process Improvement (CPI) stimulated organisations to become more aware of business processes. Secondly, today's organisations need to deliver a broad range of products and services. As a result, the number of processes inside organisations has increased. Consider for example mortgages. A decade ago there were just a few types of mortgages, whereas now numerous types are available. Not only has the number of products and services increased, but also their lifetime of has decreased in the last three decades. As a result, today's businesses processes are also subject to frequent changes. Moreover, the complexity of these processes has increased considerably. All these changes in the environment of the information system in an average organisation, have made business processes an important issue in the development of information systems. Therefore, there is a clear need for techniques to model and analyse processes. Clearly, Petri nets are a good candidate.

24. Flexible Manufacturing Systems*

24.1 A Brief Overview of the Domain

A manufacturing system involves manufacturing activity which, as defined in [VN92], is "the transformation process by which raw material, labour, energy and equipment are brought together to produce high-quality goods". A manufacturing system is composed of two main subsystems:

- *The physical subsystem*, composed of the physical resources (hardware components) such as conveyors, robots, buffers, work stations, etc.
- *The control subsystem*, also called Decision Making Subsystem (DMS) in [SV89], which determines how to use the physical subsystem in order to organise and optimise the production process.

Usually, manufacturing transformation processes are classified into continuous (chemical and oil industries, for instance) and discrete (consumer goods and computer industries, for instance). According to the type of transformations to be carried out during the manufacturing process, discrete manufacturing systems are classified into assembly and non-assembly processing. The assembly processes combine several components to obtain a different product, while the non-assembly processes concern the transformation (machining, moulding, painting, etc.) of raw materials.

In order to address some problems related to mass manufacturing systems (very efficient for large production of a small number of products, but inflexible when faced with a changing market), and in parallel with the developments in computer and automation technologies, a new type of production system appeared: the Flexible Manufacturing System (FMS). Using the definition in [PHB93], an FMS is "a computer-controlled configuration of semi-independent work stations and a material handling system (MHS) designed to efficiently manufacture more than one part type from low to medium volumes". The adjective "flexible" indicates the ability of the system to respond effectively to changes in the system. These changes can be internal, breakdowns or quality problems for instance, or external, changes in the design and demand for instance. In [BDR+84] eight different types of flexibility are summarised:

---

* Author: J. Ezpeleta

- machine flexibility (which refers to the time required to change the machines necessary to produce a new type of part),
- process flexibility (related to the mixture of jobs that the system can produce simultaneously),
- product flexibility (the ability to produce new types of products),
- routing flexibility (the ability to route parts via several routes),
- volume flexibility (the ability to operate at different production volumes),
- expansion flexibility (the ability to expand the system in a modular way),
- operation flexibility (the ability to interchange the ordering of several operations for each part type), and
- production flexibility (the set of part types that the system can produce).

Figure 24.1 depicts a typical plant of an FMS ([VN92]). The global coordinating system communicates, via a local area network, with the controllers of each cell. Each of these cell controllers is in charge of the control of the programmable controllers (PC) that are in charge of the control of each of the physical hardware components in the cell. As will be detailed later, the complexity of these systems makes the hierarchical organisation of the control system necessary.

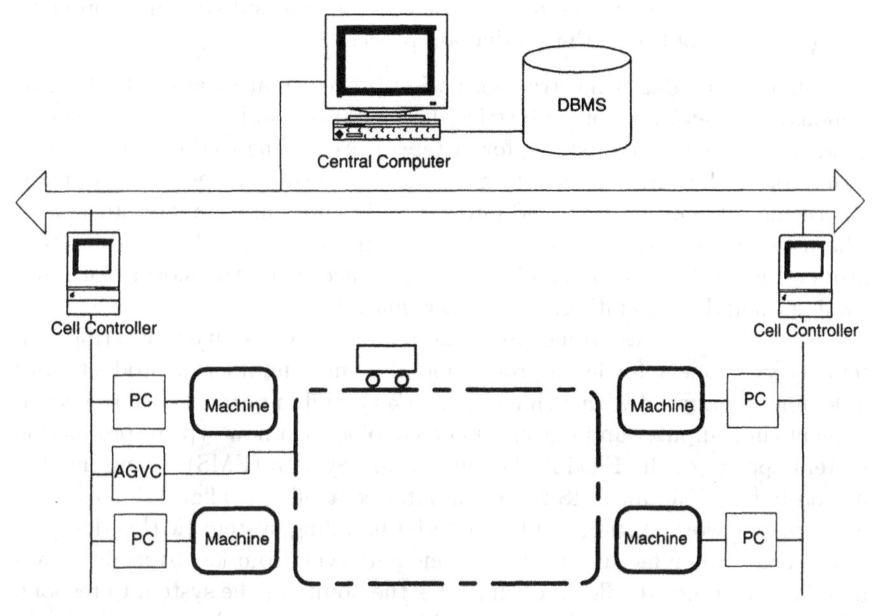

Fig. 24.1. An abstract view of a Flexible Manufacturing System

FMS hardware components are typically a set of work stations; an automated material handling system (conveyors, industrial robots, automated guided vehicles, etc.) allowing a flexible routing of parts through the different work stations; a load/unload station for the entry/exit of parts; some storage means for the work-in-process part storage; some (local and central) tool magazines; and a computer control system that is usually organised in a hierarchical way.

To introduce these systems in a more detailed way, let us present, in an intuitive and informal way, a small FMS.

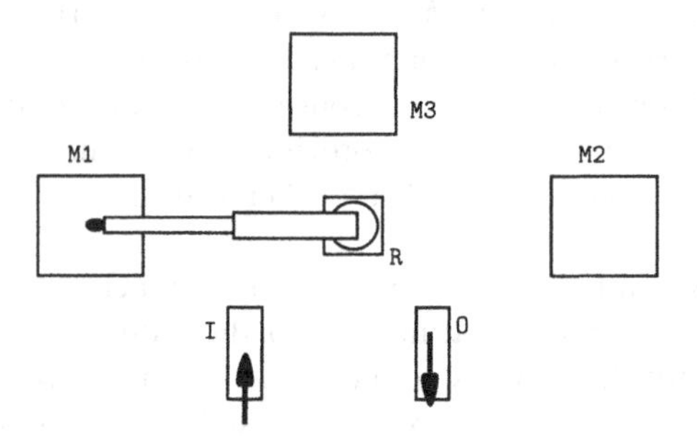

Fig. 24.2. A small manufacturing cell

Consider the manufacturing cell whose physical layout is depicted in Figure 24.2. The cell is composed of three machines, $M1$,$M2$, and $M3$, and a robot R, whose role is to load and unload the machines. The robot can also pick up parts from conveyor I, where parts arrive in the system, and unload parts into conveyor O, where the parts processed in the cell are unloaded. Let us assume that the flexible machines can carry out different operations on the incoming parts. Let us also assume that $M1$ can process three parts at a time, while machines $M2$ and $M3$ can process only two parts at a time. In what follows, we call the elements composing the cell (machines, stores, robots, buffers, etc.) "resources".

Finally, we consider that in this cell two different types of parts must be processed. Parts of type one must be processed first in either machine $M1$ or $M3$ and then in machine $M2$; parts of type two must be processed first in $M2$ and then in $M1$ (at the moment, we are not considering what kind of processing operation must be carried out in each machine and for each type of part).

This system exhibits some important characteristics that are common to almost all FMSs ([ZD93]):

- *It is event-driven.* The system behaviour consists of a discrete state space where a change in the state occurs when certain events are triggered (a new part enters or leaves the cell, a machine loads a part, etc.).
- *It is asynchronous.* Some events in the system occur in an asynchronous way: the end of the processing of a part in machine $M1$ is asynchronous (in time) with respect to the loading of a new part in machine $M2$.

- *It has sequential relations.* Some events must occur in a sequential way. For a part to be unloaded from machine $M1$, this machine must have been previously loaded and the processing of the part must be finished.
- *It has concurrency.* The processing of a part in $M1$ and a second part in $M2$ can be done concurrently, and these two actions do not interact with each other.
- *It has conflicts.* A part of type one that has been held by the robot can be loaded either into $M1$ or into $M3$ (assuming that both machines have free slots for new parts). So a decision must be taken.
- *It has non-determinism.* As a consequence of conflicts, some non-determinism can appear. In the previous situation, we cannot *a priori* predict which action will be taken: either the part is loaded into $M1$ or into $M3$.
- *It has deadlocks.* In the case in which all three machines are fully busy and the robot holds a raw part that must be loaded into one of the machines, the system is in a (total) deadlock situation: no action can be executed since no machine can be unloaded (the robot is busy) and the robot cannot release the part (since it has to go to a machine).
- *Mutual exclusion.* Let us consider the processes corresponding to the processing of a part of type one and a part of type two. These processes cannot be simultaneously in the state "the part is being held by the robot". So this state implies a mutual exclusion for these two processes.

We can conclude that the design of manufacturing systems is a very complex task: many different elements have to be combined, and many different aspects must be taken into account. This complexity has raised two important needs:

1. the design of the production control system in a hierarchical way,
2. the use of formal methods in order to validate the system.

As summarised in [SV89], the DMS is usually split into the following levels:

- *Planning.* This considers both the whole plant and the estimated demand. It considers the production on a long time horizon, establishing the way in which the products needed will be produced during this time interval.
- *Scheduling.* Going down in the DMS hierarchy, this level establishes when each operation on each product must be carried out.
- *Global coordination.* This level must have an updated state of the workshop and must also make real-time decisions taking into consideration the state of each resource and the state of the parts being processed.
- *Subsystem coordination.* The global coordination system can be decomposed into modules specialised for the coordination and supervising of subsystems: a transport system, a robot, a buffer, etc.
- *Local control.* This is the lowest level of the hierarchy, and it is in charge of the interaction with sensors and other low-level components.

The second important need was the use of formal methods. As stated in [JMSW95], the use of a formal framework has some important benefits:

1. In the process of formalising the system requirements omissions, ambiguities, and contradictions can be discovered.
2. A formal method can allow automatic system development.
3. Mathematical methods can be applied to verify system correctness.
4. A formally verified subsystem can be incorporated into larger systems with greater confidence.
5. Different designs for the same system can be compared.

However, the use of different formalisms (e.g. Markov chains, queuing networks or simulation for performance evaluation, mathematical programming for planning, Petri nets for modelling and analysis) for the different problems generates a "Babel Tower" where communication among people working at different stages in the design process is very difficult ([ST97]).

As proposed in [ST97], a good solution is to use a *family of formalisms* which, sharing the basic principles, allows the transformation (in an automatic way if possible) from one formalism into another. The family of Petri net formalisms is a good choice for the manufacturing system environment. This family has the following advantages ([SV89, ZD93]):

1. easy representation of concurrency, resource sharing, conflicts, mutual exclusion, and non-determinism
2. application of top-down and bottom-up design methodologies, and the possibility of having different levels of abstraction of the system
3. ability to generate control code directly from the Petri net model
4. a well-defined semantics that allows qualitative and quantitative analysis for the system validation
5. a graphical interface that allows an intuitive view of the system.

The use of Petri nets in manufacturing systems has been extensively dealt with in literature (see [SV89, ST97] for a large set of references) and many relevant text books have appeared in the last few years: [VN92, DHP+93, ZD93, DAJ95, PX96]. Petri nets have been used in all aspects of the design and operation of FMSs: modelling and verification, performance analysis, scheduling, control, and monitoring.

The present chapter studies some problems related to the design and control of discrete non-assembly FMSs using Petri nets as a family of formal models. Here, we focus on a class of problems that arise at the global coordination level.

The chapter is divided into three main sections. The first one shows how some Petri net elements (tokens, places, transitions, and arcs) can be mapped to FMS concepts. The second one deals with the problem of system modelling. This is not a simple task. Computer-aided design tools are used to make models as well-structured as possible. The section presents a modelling methodology for a wide class of systems. This modelling methodology relies on a clear differentiation between the model of the system layout and the models of the types of parts to be produced. From these inputs, and in an automatic way, a coloured Petri net can be obtained. The section shows the input data models and also explains the process from its first step to the statement of the final model.

As stated previously, one of the advantages of formal models is that system properties can be studied in the model. The second part of the chapter shows how the structure of the Petri net model can be used to deal with one of the main problems in automated manufacturing systems, the deadlock problem. First, we present the place/transition models corresponding to the coloured models obtained by the modelling methodology. Secondly, it is shown how deadlock problems can be characterised in terms of Petri net structural elements called siphons (also called structural deadlocks in the literature). The structural deadlock characterisation is used to get a control policy for deadlock prevention, and this control policy is also implemented by means of Petri net elements (i.e. the addition of some new arcs and places to the model).

Throughout the chapter the same "toy example" will be used. More interesting and complex models can be found in the literature. The aim of this chapter is to show the use of Petri nets (both ordinary and coloured) for modelling and analysing flexible manufacturing systems. In this sense, this chapter is not a survey of all the different approaches that have been proposed for the use of Petri nets in this domain; it presents just one of them. In [ST97] a complete set of references related to this subject can be found. The chapter considers only qualitative aspects of the domain. For quantitative aspects, the reader is referred to [VN92, DHP+93, DAJ95].

24.2 Using Petri Nets in FMS

To get an insight into the use of Petri nets in the domain considered, we will present a series of models corresponding to some basic components of FMSs, such as machines and buffers or stores.

- Figure 24.3 depicts abstract models for three different transportation systems. In the three cases the interactions with the rest of the system are represented by means of the transitions tI and tO, which model the loading and unloading of parts in the module.
  Figure 24.3a is a model of a buffer (also a store) with capacity for k parts. Notice that if we take $k = 1$ this PN can also model a robot for instance. Figure 24.3b is a model of a FIFO queue with capacity for three parts: there are three positions that are accessed sequentially in an ordered way. Finally, Figure 24.3c models a LIFO module. This module represents the set of states that can be reached, but not the firing sequences. Notice that nothing prevents the sequence $(t_{12}t_{21})^*$ which, of course, must not be

allowed. In all the examples we introduce, nothing is said about control; we are concentrating only on the modelling of the structure of the component. In all these cases we are assuming that the time necessary for the execution of the operations related to each transition is negligible.

- Figure 24.4a shows a model of a reliable machine (breakdown is not considered). When the part is loaded into the machine (transition tLM is fired) the processing starts and, once the machine has finished, a token is put in place pAP and then transition tUM can be fired. Notice that in this model two different types of transitions appear. Thick black transitions represent "immediate" actions (here, immediate transitions model system actions whose time execution is negligible); square white transitions model system actions whose execution time can be modelled by means of a probability distribution function. Usually, this function is taken to be an exponential, and the λ parameter which appears near the transition is the firing rate ($1/\lambda$ is then the mean time needed for the processing of the part).
  Figure 24.4b shows a model of the same machine, but here the possibility of a breakdown has been considered. In this model, in order to load or unload a part it is necessary that the machine be in the OK state (there are arcs joining tLM and the OK place and also tUM and the OK place). The machine breaks down with a rate λ_f and is repaired with a rate λ_r.
- Finally, Figure 24.5 shows a model of an unreliable assembly machine. Notice that in order to start the assembly, it is necessary to have loaded a part into $pT1$ and another one into $pT2$. Here λ is the time needed to carry out the assembly. The model for a disassembly machine is almost the same: it suffices to reverse the arcs related to transitions $tL1$, $tL2$, tA, and tU.

As we have seen, when using ordinary Petri nets for the modelling of flexible manufacturing systems, the main Petri net elements (places, transitions, arcs, and tokens) can have different meanings:

- A place can be used for the modelling of different elements. 1) *States in which a part that is being processed can stay.* Let us consider, for instance, place pBP in Figure 24.4a. It represents a part of a given type that has been loaded into the machine and is being processed there, whereas place pAP is used to model a part in the same machine for which processing has already finished so it is ready to be unloaded. 2) *A partial state of a resource.* Place kM in Figure 24.4a models the free state of the machine; hence, this place does not contain "physical" items, but is used instead for a "logical" interpretation.
- A transition usually models a sequence of system actions that changes the state of some system elements. For instance, transition tA in Figure 24.5 models the sequence of system actions by which the parts modelled by the tokens in places $pT1$ and $pT2$ are assembled in order to produce a new product (modelled by means of the new token that is put into place

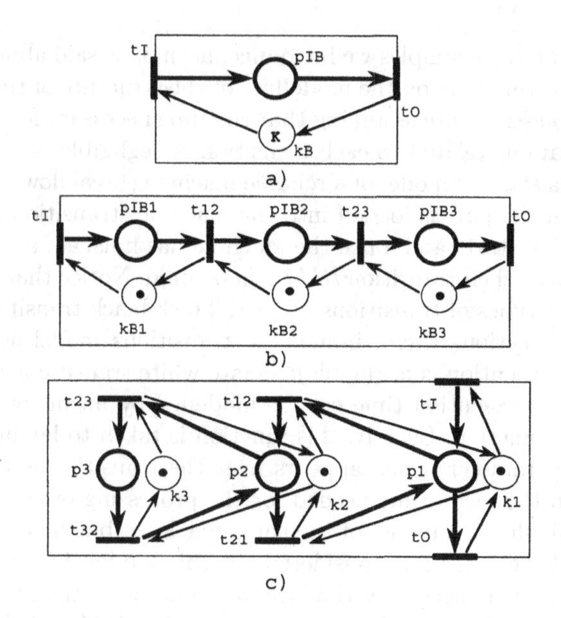

Fig. 24.3. a) A generic model for a storage device; b) a model for a FIFO device; c) a model for a LIFO device

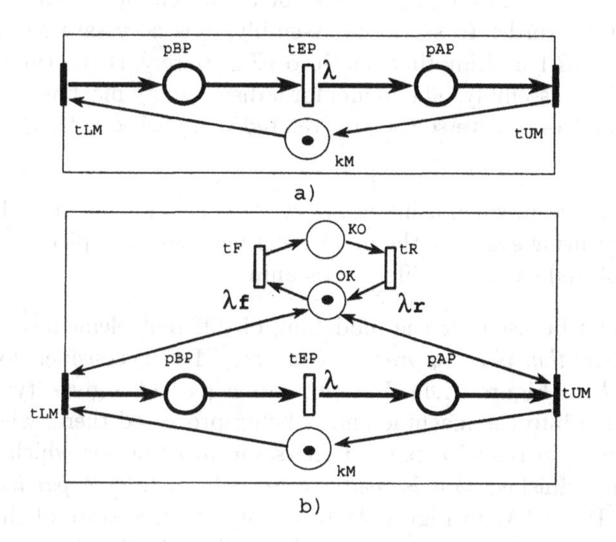

Fig. 24.4. a) A model for a reliable machine; b) a model for an unreliable machine

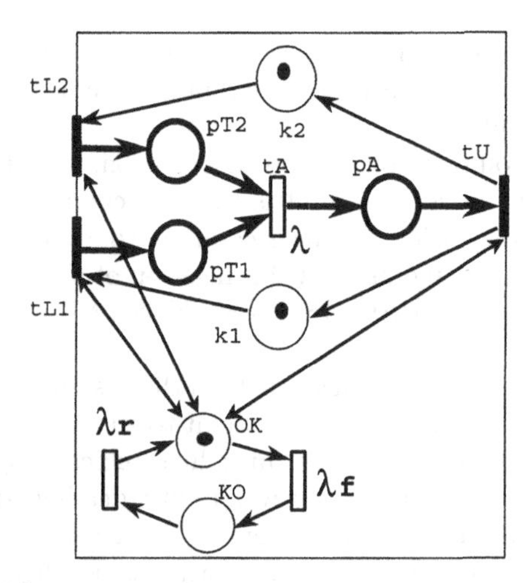

Fig. 24.5. A model for an unreliable assembly machine

pA). Another action that is usually modelled by means of the firing of a transition is the movement of a part between two different locations in the system (for instance, transition $t12$ in Figure 24.3). Also, a transition can model the change in the state of a system resource, as is the case for transition tF in Figure 24.4b: it models a breakdown of the machine modelled in the figure.

- Usually, an arc models either a pre-condition or the flow of parts among resources. The arc joining transition kM and transition tLM in Figure 24.4a is an example of the first kind of arc. It models the need for a free position in the machine in order to load a new part. Arcs from $pT1$ and $pT2$ to transition tA in Figure 24.5 fall into the second class. They model two elements that are withdrawn from the two buffers. The arc joining pA and transition tU in the same figure also belongs to the second class. It models the flow of an assembled element to the output of the assembly machine.
- Tokens can also have different meanings. In Figure 24.4a, the token in place kM models the availability of the machine (the machine is non-busy), whereas a token in place pIB in Figure 24.3a represents a product that is stored in the buffer. In the case of coloured Petri nets, a token can carry a great deal of information, as will be shown later on.

As stated previously, one of the main problems when dealing with real applications is the complexity of the model. From the design point of view, different approaches have been adopted:

- Hierarchical/compositional approach: The idea behind these approaches is the modelling of the systems in an structured way. Using the first approach (also called top-down) the modelling is carried out in several steps. At each step more detailed elements are considered. In general, the process

consists of the replacement of some net elements (place, transition, path, sub-net) by some sub-net in which the replaced elements have been refined ([Val79, SM83, ZD91]).
When using the compositional approach (bottom-up) the global model is obtained by means of simpler models that are combined using some composition mechanism: fusion of places common to a set of submodules (and modelling the same elements), synchronisation of a set of transitions, and fusion of common paths ([ACA78, NV85, KB86, Val90a, BDC92, Feh93]); see Section 9.2.
Although the two approaches below help in the design process, both present one important drawback: it is very difficult to ensure that in the modelling process (either compositionally or in a hierarchical way) desired system properties (such as boundedness, reversibility, deadlock-freeness, liveness) are preserved from one step to the next. This means that, for instance, we can have two live modules whose composition is non-live. The same is true of a hierarchical approach. It can happen that at a given abstraction level the system behaviour is live, but when a new refinement is given the new "view" of the system is not live.
To cope with this problem two different solutions have been adopted:
1. The kit of refinement/composition mechanisms is restricted. This means that composition of modules or refinement must be done only when some special conditions hold.
2. The work is restricted to some special subclasses such as free-choice nets ([ES90]), marked graphs ([Mur89]), modules synchronised by means of (restricted) message passing ([RTS95]) or (restricted) resource sharing ([ECM95, JD95]).

However, in both cases the modelling power is decreased.

- High-level Petri net approach: High-level Petri nets, and coloured Petri nets ([Jen94], Section 4.3) as a particular case, are a very useful tool for modelling complex systems in which different components have analogous behaviour. One of the main advantages of this class of nets is the compactness and the clarity of the models generated ([CMS87, MMS87, VMS88, GBK88a, GBK88b, EM92, Jaf92]). However, usually, they have the drawback that it is difficult to analyse properties.
- Object-Oriented (OO) and Artificial Intelligence (AI) approaches (Section 10.3): Much work on the use of Petri nets in manufacturing systems has tried to extend the capacity of Petri nets for the modelling of systems with the capacity of AI techniques for reasoning about properties. Here too different approaches have been adopted. In some papers, e.g. [BE86, CCG85], elements of AI are used to implement and control the Petri net.
Other work, such as [VA87, SACV87, Rib88, VM94], uses AI elements to implement the Petri net (tokens or places as frames and transitions as rules, for instance), and uses the semantics of the underlying Petri net for simulation and control of the system.

The use of one of these approaches does not exclude the use of another. For instance, we can adopt both a hierarchical approach ([HJS90]) and a compositional approach ([Che91]) in order to obtain a coloured Petri net model.

However, once we have obtained a Petri net model for the system we want to study, what Petri net properties are interesting for our model? Let us enumerate some important behavioural properties. It is important to notice that some of the following properties are related: one property can be deduced from others.

- *Reachability.* From the model point of view, this property determines whether a given (vector) marking is reachable from the initial marking. From the real system point of view, this property indicates whether a system state is reachable from the initial configuration. It can be used to answer questions such as the following: Is it possible to reach a state where machine M is processing two parts while robot R is busy and machine M' is free? Is it possible to reach a state in which buffer B is full? The answers to a set of well-defined questions can be used to establish a correct system design. Notice that if, for instance, the answer to the second question is NO, then the designer can decide whether to use a buffer with less capacity, which could make the system less expensive. A second related property is *coverability.* From the Petri net point of view, this determines whether a reachable marking is greater than or equal to another given marking. From this kind of property, less complete information can be obtained; but this information can be used in a similar way to that provided by reachability properties.
- *Boundedness.* This property determines whether the number of tokens in a given place is always smaller than or equal to a given constant k (see Section 15.2.1). Usually in FMS domains, using the possible meanings of a place as stated previously, all places must be bounded. Thus, if in the analysis of the Petri net model we realise that a place is not bounded, the model is, perhaps, incorrect. However, if the model is correct and a place is detected to be unbounded, some overflow problems may arise. A related property is *safeness* (1-boundedness).
- *Reversibility* (see Section 15.2.4). When verified, this property determines that the initial state can be reached from each reachable state. In the application domain considered, this property means that each possible erroneous situation has been considered by means of some error recovery strategy. These erroneous situations include the case of system deadlocks and the case of resource failures.
- *Deadlock-freeness/liveness* (see Section 15.2.2). These properties will be discussed in more detail in Section 24.3.3.

As has already been intuitively shown by the models of components in an FMS, Petri nets have also been used for performance evaluation of FMSs. To do this, the notion of time has been added to the Petri net models. Introducing time constraints is necessary if we want to consider performance evaluation or the scheduling of real-time control problems.

Usually, time has been introduced in one of two different ways: either associated with places or with transitions. The second way is more natural since transitions usually model system activities (which need some time to be executed). In this approach, time is considered as follows: a transition can fire some time after it is enabled with respect to the number of tokens in its input places. This time can be either deterministic, as in *timed Petri nets*, or random, as in *stochastic Petri nets* (see [VN92, DHP+93, DAJ95] for a clear introduction to these concepts).

In FMS domains the different quantitative measures that can be obtained from the Petri net model have specific and clear meanings: probability of a resource being non-busy, mean number of parts in a machine or buffer, mean waiting time of parts in an input buffer, production rates of parts, mean time of parts in in-process states, etc.

In this chapter we will concentrate on qualitative analysis using structural methods.

24.3 A Design Approach

In this section we will present a particular approach to the design and control of FMSs using Petri nets. As stated in Section 24.1 many different approaches have been adopted. The reader is referred to the literature cited in this chapter for a comprehensive study of the different approaches. The presentation of the method is carried out in an intuitive way following a simple example. A formal presentation can be found in [EC97].

This section is organised as follows. First, we introduce the place/transition model corresponding to the system in Figure 24.2; then we show how an equivalent model can be obtained in an automatic way; finally, we show that structural analysis of the Petri net can be used to establish a control policy for deadlock prevention in order to ensure good behaviour.

24.3.1 An Intuitive Introduction to a Class of Nets

Let us consider the model in Figure 24.4a once again. If we are not interested in performance evaluation we can model each action by means of an immediate transition. This change gives a simpler model. For instance, let us consider a general model of a reliable machine of the figure. If we apply a reduction rule (see Section 15.1), the path pBP, tEP, pAP in the figure can be replaced by a unique place, giving an equivalent model[1]. This approach

[1] When talking about "equivalence" we must specify the type of equivalence. Here, as will be stated later, we are interested in liveness properties. In this case the

will be used below. Since only one kind of transition will be considered, all transitions will be drawn as white rectangles.

Let us consider the manufacturing cell, shown in Figure 24.2, that was described in Section 24.1. Each part belongs to a different part type. The type of the part establishes the correct sequences of operations. In a first step, these sequences are established in terms of transformations to be carried out on parts. For the cell, these sequences of operations are transformed into sequences of pairs *(resource,operation)* which establish, for each operation, the resource where the operation has to be carried out. Each part type can be modelled by an acyclic graph. Figure 24.6 represents the operation graphs corresponding to two different process plans. Parts of type $W1$ have to be processed first in either machine $M1$ or $M3$, and then in machine $M2$. Parts of type $W2$ have to be processed in machine $M2$ and then machine $M1$. Since parts must be loaded (unloaded) into (from) the system, each process plan needs more information than provided by the operation graph. Thus a process plan must be completed with two sets. The first one represents the system actions that load parts of the corresponding type into the system. The second set represents the system actions that unload parts of the corresponding type from the system. So, in the example, we define $W1 = \langle G1, I1, O1 \rangle$ where $I1 = \{fromI\}$ and $O1 = \{toO\}$; and $W2 = \langle G2, I2, O2 \rangle$ where $I2 = \{fromI\}$ and $O2 = \{toO\}$.

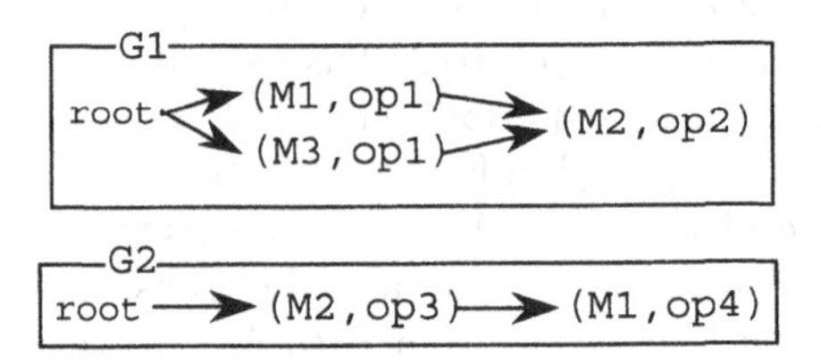

Fig. 24.6. The models of process plans for two types of parts to be processed in the system in Figure 24.2

Each process plan model has an initial node *root* (as shown in Figure 24.6) that models the raw state of parts. The other nodes correspond to the label of the transformation resources which the part can visit during its processing.

From a process point of view, let us show how the processing of parts of type 2 is carried out. The sequence of steps that one part of this type must execute is as follows. The part is held by the robot, loaded into machine $M2$, held once again by the robot, loaded into machine $M1$, held a third time by the robot, and finally unloaded from the system. These different states are modelled in Figure 24.7b by means of the thick places. A place $p0(2)$ (called the *idle state place*) has been added to introduce a notion of repetitive process,

transformation maintains the liveness of both models: the original one and the transformed one.

modelling the repetitive nature of the processing of different parts of the same type. The initial marking of this place establishes the maximum number of parts of type 2 that are allowed to be concurrently processed in the system. Notice that if the initial marking is large enough (as in the example), this idle place becomes implicit (see Section 15.1.2), and has no effect on the model behaviour.

The transitions in this figure model the system actions that perform the state changes of this type of part. The net belongs to the class S^2P in [ECM95] and, essentially, is the same as a *job sub-net* in [HC94b]. It is usually required that all the cycles of the S^2P (and analogous classes) contain the idle state place. This implies that no cyclic behaviour is allowed during the processing of a given part: once the processing of a part has started, the part cannot change its state infinitely often without terminating its processing.

Notice that we have one of these nets for each type of part to be processed. How can these nets be obtained? The process is as follows. Let us classify the set of system resources into two classes: those resources that make some transformation of parts, called *processors* (e.g. lathes, milling machines, saws, grinders) and those which do not transform the parts, called *handlers* (e.g. robots, stores, buffers, conveyors). Notice that since in the operation graphs only part transformations are established, these nodes are always labelled with processors. Let us concentrate once again on parts of type 2. A part of this type, once loaded into the system, must be driven to $M2$ from one of its corresponding loading actions (established by $I2$). Thus all the possibilities for driving the part from the input to $M2$ using only handlers (the first transformation on this part must be performed in $M2$) must be computed. According to the plant layout depicted in Figure 24.2, the only possibility is that the part is held by R and loaded into $M2$. This means that we need an intermediate state (the part is held by R), and also the transitions modelling the flow of the part from $fromI$ to R and from R to $M2$. In this way we obtain the path $fromI(2,s)R(2,s)toM2(2,s)M2(2,M2)$ in Figure 24.7b. Now we must consider the arc $(M2,M1)$ in the operation graph. The part must be driven from $M2$ to $M1$ using only handlers. The only way to do that is to use R once again. Thus the path $fromM2(2,M2)R(2,M2)toM1(2,M2)M1(2,M1)$ is added to the model. From $M2$, the part must be unloaded. So, we must find all the possibilities for the part to be driven to the output of the system using only handlers. The only possibility is that the part is held a third time by the robot. Notice that this process must be repeated for each processing sequence taken from the operation graph of each part. It is also important to point out that in this process both the system layout and the process plans are involved. It can also happen that some operation sequences established by the operation graph are not executable because of the layout architecture (no path joining two machines M_a and M_b exists, when the arc (M_a,M_b) belongs to some operation graph). This justifies the following definition ([EC97]): a process plan is *executable* for a given architecture if for each arc $(\langle p_1, op_1\rangle, \langle p_2, op_2\rangle)$ in the operations graph there exists at least one path from p_1 to p_2 using only handlers. In what follows, we will call the places that are generated during this process *state places*, in order to distinguish them from the places that model the resource capacity constraints, called *resource places*, which will be introduced below.

At this level, the system resources that are used in the processing of parts have not yet been considered. This means that the constraints which the resources impose on the concurrent processing of parts have not been considered. So it is necessary to model these constraints. In order to deal with them, a place is added for each system resource: one place for each machine, with initial marking equal to the number of parts that the machine can process concurrently, and one place for the robot, with initial marking equal to one (we have assumed that the robot can hold one part at a time). The loading of a part into a machine requires at least one of the machine positions to be free (an arc from the resource place to the transition modelling the system actions that load a part into the machine is added). In contrast, the unloading of a part from a machine increases by one the number of non-busy positions in the machine. Thus an arc is added from a transition modelling the unloading of a machine to the resource place. The net in Figure 24.7b depicts the whole model corresponding to the processing of parts of type 2. In the same way, the net in Figure 24.7a models the processing of parts of type 1. These two nets belong to a special class of nets, called S^2PR in [ECM95].

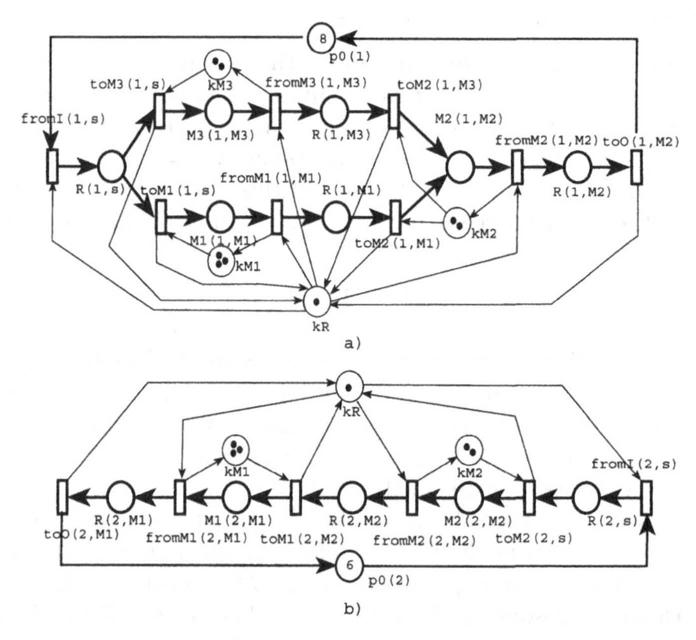

Fig. 24.7. The models corresponding to the processing of the two types of parts under consideration

Finally, the interactions among different types of parts must be considered. The complete system model corresponds to a fusion of the places that the models of the two types of parts have in common, i.e. the places modelling the system resources (in the example considered, places $kM1, kM2, kM3$, and kR). This is quite natural: the interaction of the processing of different parts is done by means of the system resources since all the parts in the system must compete for the same resources. Figure 24.13 depicts the final model once the composition of the sub-models corresponding to the types of parts has been carried out. This net belongs to a class of nets called S^3PR in [ECM95]. This class is analogous to the notion of *production sequence* in [BK90] or *production Petri net* in [HC94b].

24.3.2 Automation of the Modelling Process

In this section we will show that it is possible to adopt a more abstract view of the system, and that this view allows us to obtain easily the Petri net model presented previously. First, we show how the plant layout can be modelled by means of a place/transition Petri net. Secondly, we consider the models of the process plans as introduced above. Finally, we show that the two models can be integrated to obtain the complete model. This final model, which can be obtained in an automatic way from the inputs (the model of the plant layout and the models of the process plans), will be a CPN.

As stated previously, from an abstract point of view, the state of a resource can be modelled by means of two places:

1. The *resource capacity place*, modelling the remaining capacity of the resource for new parts. In the case of multiple copies of identical resources, the marking of this place models the number of copies of the resource that are not engaged in a processing operation.
2. The *resource state place.* Each token in this place models a part that is using either the resource or a copy of the resource in the case of multiple copies of identical resources.

For instance, consider machine $M1$ in Figure 24.8. This machine is modelled by places $M1$ and $kM1$. When considering a state reachable from a given initial state, the tokens in place $M1$ model the parts that are being processed in the machine. The tokens in $kM1$ model the parts that can still be loaded into machine $M1$. Notice that the sum of the number of tokens in $M1$ and the number of tokens in $kM1$ must always be equal to three, the capacity of machine $M1$ that we have assumed.

Let us now show how the possibility of part flow among resources is modelled. Let us consider the resource places R and $M1$. Since the physical layout allows the flow of a part from R to $M1$, transition *toM*1 is added between these two resources. Also, since this flow is from R to $M1$, an arc from R to *toM*1 and an arc from *toM*1 to $M1$ are also added. Since the capacity constraints must also be considered, two more arcs are added: one from $kM1$ to $toM1$ and one from $toM1$ to kR. This must be done for every pair of resources that are directly connected.

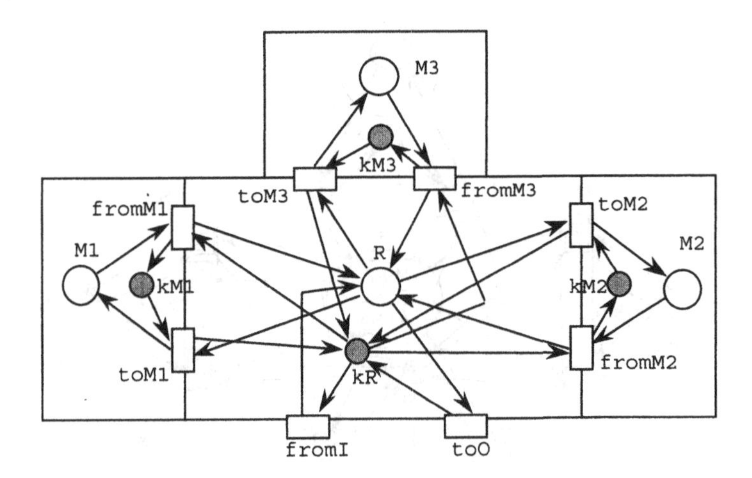

Fig. 24.8. PN layout model of the cell from Figure 24.2

Given the previous considerations, the PN model of the cell considered is depicted in Figure 24.8. Robot R is modelled by places R and kR; machine $M1$ by places $M1$ and $kM1$; machine $M2$ by places $M2$ and $kM2$; and machine $M3$ by places $M3$ and $kM3$ (places whose name starts with 'k' are capacity places). In this PN each directed path between places R, $M1$, $M2$, and $M3$ avoiding capacity places $kM1, kM2, kM3$, and kR models a possible path which a part can follow inside the cell. Since machine $M1$ can process three parts at a time, the initial marking must be $\mathbf{m_0}[kM1] = 3$, while for the other machines $\mathbf{m_0}[kM2] = \mathbf{m_0}[kM3] = 2$ and for the robot $\mathbf{m_0}[kR] = 1$.

In the next step we need to integrate the model of the cell layout with the models of the process plans in Figure 24.6. The modelling of the state of a part in the system is carried out as follows. Each part in the Petri net is modelled by a token. The token has two components, so it will be modelled by a *coloured token*. The first component identifies the part type, i.e. its process plan. The second component identifies the last node of the process plan model which the part has visited during its processing. Let us consider, for instance, a raw part of type $W1$, as considered in Figure 24.6. The part is modelled by a token $\langle W1, root\rangle$ when it is in the system and no transformation has been carried out on it. When the part has already visited machine $M1$, and not yet machine $M2$, the part is modelled by a token $\langle W1, (M1, op1)\rangle$ (since, $(M1, op1)$ is currently the last node of the operation graph "visited" by the part). When the part has been processed in machine $M2$, it is modelled by the token $\langle W1, (M2, op2)\rangle$. Since $(M2, op2)$ is one of the "leaves" of its operations graph, we understand that the processing of the part in the system is finished, and so the part must leave the system.

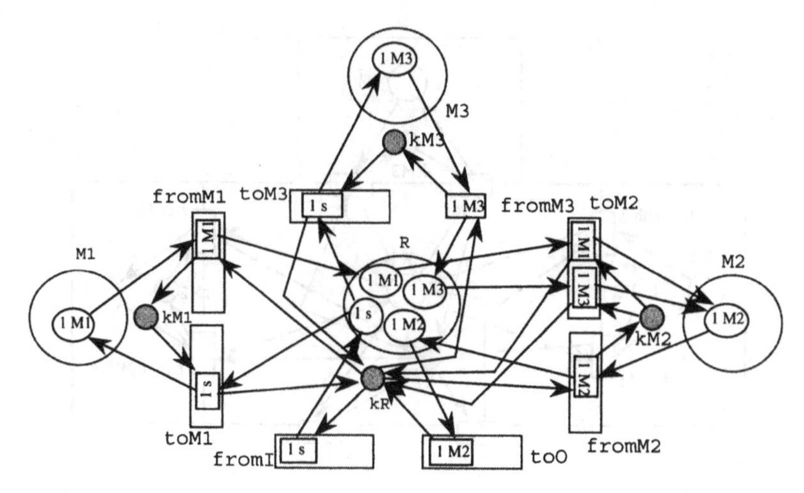

Fig. 24.9. Partial PN model considering only parts of type $W1$

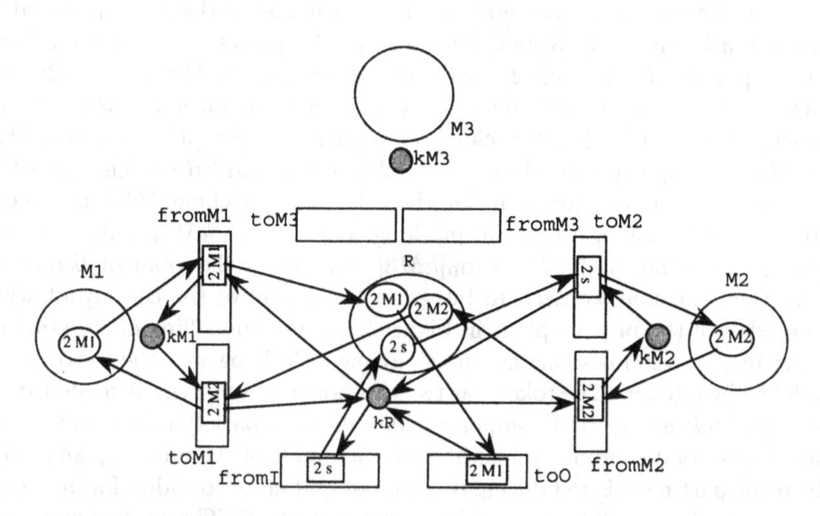

Fig. 24.10. Partial PN model considering only parts of type $W2$

The PN models in Figures 24.9 and 24.10 indicate what parts of type $W1$ and $W2$ respectively supply to the system PN model. In order to make the figures more readable, the operation component does not appear. Thus $(M1, op1)$ is represented as $M1$, while the process plan $W1$ is represented as 1. For the same reason, the *root* node is represented by the letter s. Notice that if an idle state place is added to the net in Figure 24.9 we have exactly the same net as in Figure 24.7a. This is also true for the nets in Figures 24.10 and 24.7b.

In the final model, the different "small" transitions will be modelled by the colour domains of transitions in the global PN model, while "small" places will be modelled by colour domains of places. The arcs joining a place p (transition t) and a transition t (place p) will be modelled by a function defined over the colour domain of transition t and whose images belong to the colour domain of place p. For instance, the colour domain of place $M1$ in the (coloured) global model will be $cd(M1) = \{\langle W2,(M1,op4)\rangle,\langle W1,(M1,op1)\rangle\}$, the colour domain of transition $toM1$ will be $cd(toM1) = \{\langle W1,root\rangle,\langle W2,(M2,op3)\rangle\}$, and the colour domain of capacity places will be the "neutral colour". In this case, $cd(kR) = cd(kM1) = \{\bullet\}$.

The function labelling the arcs connecting the previous places and transitions will be the following. $\mathbf{Post}[M1,toM1] = S_{M1}$ is defined from $cd(toM1)$ to $cd(M1)$ as:

- $S_{M1}(\langle W1,root\rangle) = \langle W1,(M1,op1)\rangle$
- $S_{M1}(\langle W2,(M2,op3)\rangle) = \langle W2,(M1,op4)\rangle$
- $\mathbf{Post}[kM1,toM1] = \mathbf{Pre}[kR,toM1] = \langle\bullet\rangle$
- $\mathbf{Pre}[R,toM1] = Id$

where $\langle\bullet\rangle$ represents the constant function that always returns the neutral colour ("neutral function") and Id is a symbolic representation of the Identity function in its "liberal" meaning; i.e. $Id(x) = x$, even if the origin and final sets are not the same. Figure 24.11 shows the arcs and functions related to transition $toM1$ that the CPN model would have. Figure 24.12 shows the final CPN model for the example considered. The other functions are as follows:

- $S_{M2}(\langle W1,(M1,op1)\rangle) = \langle W1,(M2,op2)\rangle$
- $S_{M2}(\langle W1,(M3,op1)\rangle) = \langle W2,(M1,op4)\rangle$
- $S_{M2}(\langle W2,root\rangle) = \langle W2,(M2,op3)\rangle$
- $S_{M3}(\langle W1,root\rangle) = \langle W1,(M3,op1)\rangle$

We have shown in an intuitive way how the CPN model can be obtained from the input data considered. In [ECS93] algorithms that obtain this coloured model in an automatic way, together with their complexity, are presented.

24.3.3 Using Structural Analysis for System Control

Structural elements like p-semiflows and t-semiflows (see Section 5.2.2) have been widely used to get information from the model. In the example considered (this is also valid for all nets belonging to the S^3PR class), much information about model correctness is given. Let us now consider, once again, the Petri net in Figure 24.13.

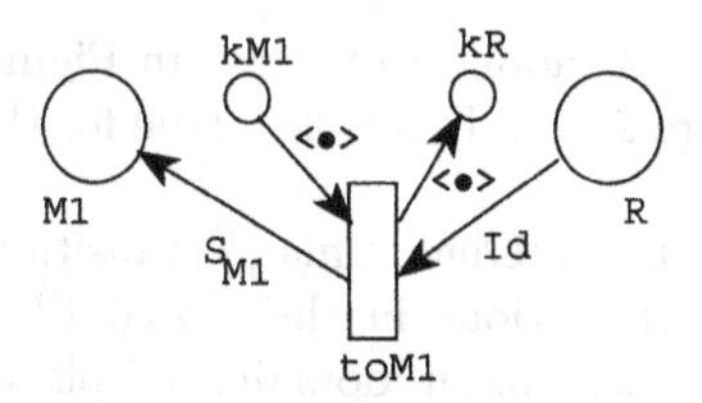

Fig. 24.11. A (partial) view of the arcs and functions surrounding transition *toM*1

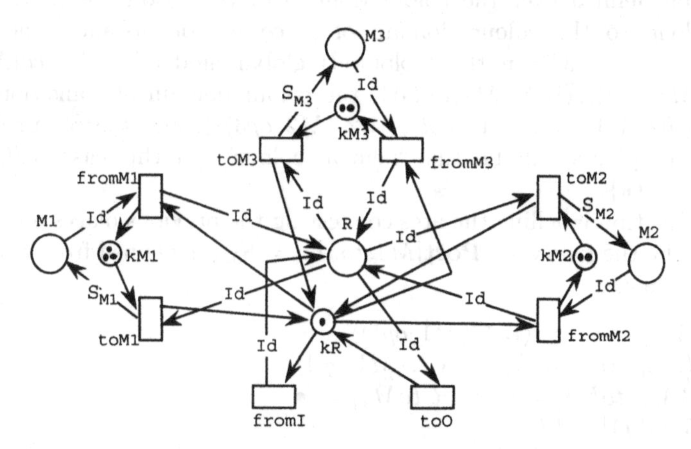

Fig. 24.12. The coloured Petri net obtained applying the proposed methodology. All arcs related to capacity places must be labelled $\langle\bullet\rangle$.

- It is easy to prove that we have two kinds of (minimal) p-semiflows. For each resource, the sum of the number of tokens in the resource and its holders is always equal to the initial marking of the resource. A state place is a *holder* of a resource r if the resource is used in this state. For instance, $M1(1, M1)$ is a holder of the resource $M1$ since the marking of $M1$ decreases when a token enters $M1(1, M1)$ (i.e. place $M1(1, M1)$ "uses" $M1$). Notice also that when a token leaves $M1(1, M1)$, the marking of $M1$ is increased. The set of holders of a resource r is denoted as $H(r)$[2]. $\{r\} \cup H(r)$ induces the following (minimal) p-semiflow: at each reachable marking $\mathbf{m}$, $\mathbf{m}[r] + \sum_{p \in H(r)} = \mathbf{m_0}[r]$. In our example we have:
  - $H(M1) = \{M1(1, M1), M1(2, M1)\}$, which induces the p-semiflow $\mathbf{m}[kM1] + \mathbf{m}[M1(1, M1)] + \mathbf{m}[M1(2, M1)] = \mathbf{m_0}[kM1] = 3$. What is the interpretation of this p-semiflow? Notice that tokens in places $M1(1, M1)$ and $M1(2, M1)$ model parts that are being processed in machine $M1$. The p-semiflow states that the number of parts in $M1$ plus the number

[2] For a set of resources S, we extend the definition of set of holders as follows: $H(S) = \bigcup_{r \in S} H(r)$

of free positions in $M1$ is always 3. This is a necessary condition for our model to be correct.

- $H(M2) = \{M2(2,M2), M2(1,M2)\}$, which induces the p-semiflow $\mathbf{m}[kM2] + \mathbf{m}[M2(1,M2)] + \mathbf{m}[M2(2,M2)] = \mathbf{m_0}[kM2] = 2$.
- $H(M3) = \{M3(1,M3)$, which induces the p-semiflow $\mathbf{m}[kM3] + \mathbf{m}[M3(1,M3)] + \mathbf{m}[kM3] = \mathbf{m_0}[kM3] = 2$.
- $H(R) = \{R(1,s), R(1,M1), R(1,M3), R(1,M2), R(2,s), R(2,M2), R(2,M1)\}$, which induces the p-semiflow $\mathbf{m}[kR] + \mathbf{m}[R(1,s)] + \mathbf{m}[R(1,M1)] + \mathbf{m}[R(1,M3)] + \mathbf{m}[R(1,M2)] + \mathbf{m}[R(2,s)] + \mathbf{m}[R(2,M2)] + \mathbf{m}[R(2,M1)] = \mathbf{m_0}[kR] = 1$.

There is a second type of p-semiflow: for each S^2P, at each reachable marking, the sum of the number of tokens in its places is equal to the initial marking of the idle place. For the example we have:

- $\mathbf{m}[p0(1)] + \mathbf{m}[R(1,s)] + \mathbf{m}[M3(1,M3)] + \mathbf{m}[R(1,M3)] + \mathbf{m}[M1(1,M1)] + \mathbf{m}[R(1,M1)] + \mathbf{m}[M2(1,M2)] + \mathbf{m}[R(1,M2) = \mathbf{m_0}[p0(1)]$
- $\mathbf{m}[p0(2)] + \mathbf{m}[R(2,s)] + \mathbf{m}[M2(2,M2)] + \mathbf{m}[R(2,M2)] + \mathbf{m}[M1(2,M1)] + \mathbf{m}[R(2,M1)] = \mathbf{m_0}[p0(2)]$

The general interpretation of the p-semiflows is easy. p-semiflows of the first type state the correctness of the model with respect to the resources. This means: *1)* A resource can be neither created nor destroyed. *2)* At each reachable state, the sum of the available free positions/copies of each resource and the parts that use it is always equal to the total capacity of the resource. p-semiflows of the second type establish correctness with respect to the types of parts. The initial marking of the idle places establishes, for each type of part, the maximal number allowed to be concurrently processed in the system. The p-semiflow for a type of part states that the total number of parts that are concurrently processed plus the number of parts of this type that can still be accepted is constant.

- It is also very easy to prove that each cycle of each S^2P forms a t-semiflow. The interpretation of these t-semiflows is intuitive: each t-semiflow establishes a possible processing sequence for a part. This means that when the firing of a t-semiflow corresponding to an S^2P is completed, the processing of a part of this type has been finished.

  When the processing of all the parts inside the system finishes (every t-semiflow, once started, is completed), the initial state of the system is reached. Considering parts of type 1, for instance, we have two t-semiflows related to it. The first one, σ_1, is as follows: $\sigma_1[from(1,s)] = 1$, $\sigma_1[toM1(1,s)] = 1$, $\sigma_1[fromM1(1,M1)] = 1$, $\sigma_1[toM2(1,M1)] = 1$, $\sigma_1[fromM2(1,M2)] = 1$, and $\sigma_1[toO(1,M2)] = 1$ with $\sigma_1[t] = 0$ for any other transition.

  Analogously, the second one is the following: $\sigma_2[from(1,s)] = 1$, $\sigma_2[toM3(1,s)] = 1$, $\sigma_2[fromM3(1,M3)] = 1$, $\sigma_2[toM2(1,M3)] = 1$, $\sigma_2[fromM2(1,M2)] = 1$, and $\sigma_2[toO(1,M2)] = 1$ with $\sigma_2[t] = 0$ for any other transition. Given C the net incidence matrix, it is verified that

$C \cdot \sigma_i = \mathbf{0}$, $i = 1, 2$. Notice that the firing of any of the two previous t-semiflows models the completion of the processing of a part of type one.

Now, we will concentrate on another kind of structural element, the siphons (see Section 15.3), and show that, for this class of nets, the siphons are related to system liveness. In Petri net theory there are two main concepts related to the existence of system activities. The first is the concept of deadlock-freeness, while the second is the concept of liveness. Let us now consider these concepts in our application domain.

- *Deadlock-freeness.* Recall (Section 15.2.2) that a Petri net system (i.e. a Petri net with an initial marking) is said to be deadlock-free if at each reachable marking there exists at least one transition that is enabled. In our application domain this means that it is always possible to make *some* production activity (for instance, executing a new step in the production sequence of a part, or introducing a new part into the system).
- *Liveness.* Deadlock-freeness is not enough for this domain: it is possible to have a part of the system that can always run correctly, but also another part of the system that is in a deadlock. For instance, it is possible to have one type of part being correctly processed, as well as other parts whose processing has been started but cannot be finished. Thus deadlock-freeness is not strong enough for highly automated systems; liveness is a stronger property. A Petri net system is said to be *live* if from each reachable marking it is always possible to fire *any* transition (see Section 15.2.2). In the application domain considered, this means that it is always possible to execute the system actions modelled by any transition. As a consequence the processing of each part, once started, can always be finished: the transitions "driving" a token (modelling a part) to the system output can be fired. Thus the processing of the part can be finished. This also means that if there are always new raw materials, their processing can be carried out.

In some cases, e.g. free-choice nets ([ES90]), the previous properties are equivalent. But this is not the case for the class of nets we are considering.

When building automated systems, deadlock problems are very important issues. If we want a system to be highly automated, we must deal with deadlock ([BK90, ECM95, VN92]). As stated above, a deadlock indicates that the processing of a part has been started, but cannot be finished. Therefore, the part can stay in the system for a long period of time (until some recovery strategy is applied). During this time, the part is using system resources and the system performance decreases. In systems where deadlocks can appear, two different approaches have been adopted: the deadlock prevention/avoidance approach and the deadlock detection and recovery approach. In the first approach a deadlock prevention/avoidance control policy is applied in such a way that the system evolutions are controlled in order to ensure that no deadlock is reached. In the second approach, when a deadlock is detected a recovery strategy is applied to change the system state into a non-deadlocked state.

For the general class of nets we are considering, different control policies can be found in [BK90, VN92, ECM95, XHC96]. Let us show how the structure of the net allows us to establish a deadlock characterisation which can be applied to get a control policy for deadlock prevention. Let us consider the net in Figure 24.13. From the initial marking shown in the figure the firing of sequence $\sigma = (\mathit{fromI}(1,s)\mathit{toM}1(1,s))^3\mathit{fromI}(2,s)\mathit{toM}2(2,s)\mathit{fromM}2(2,M2)$ yields a marking $\mathbf{m}$ ($\mathbf{m_0}[\sigma\rangle\mathbf{m}$) such that $\mathbf{m}[kM2] = \mathbf{m}[kM3] = 2$, $\mathbf{m}[M1(1,M1)] = 3$, $\mathbf{m}[R(2,M2)] = 1$, $\mathbf{m}[p_1^0] = 3$, $\mathbf{m}[p_2^0] = 5$, and $\mathbf{m}[p] = 0$ for any other place p. Notice that this state is a deadlock: the parts modelled by the tokens in place $M1(1,M1)$ cannot change the state. This means that these three parts will remain in machine $M1$ (forever if nothing is done!). The same goes for the part modelled by the token in place $R(2,M2)$. A question arises: Is there any information in the Petri net structure allowing the characterisation of deadlock situations? The answer is "yes". For the considered marking $\mathbf{m}$ the set of heavily shaded places in Figure 24.13, $S = \{R(1,s), R(1,M3), R(1,M1), R(1,M2), R(2,M1), R(2,s), kM1, kR\}$, is a siphon and it is unmarked. Remember that one of the most important behavioural properties of a siphon is that once it becomes unmarked, it remains unmarked. Hence no transition in $S^\bullet$ can now fire. So, neither $\mathit{fromM}1(1,M1)$ nor $\mathit{toM}1(2,M2)$ can fire, and the tokens considered will remain in their places. Therefore, the processing of the parts considered cannot be finished.

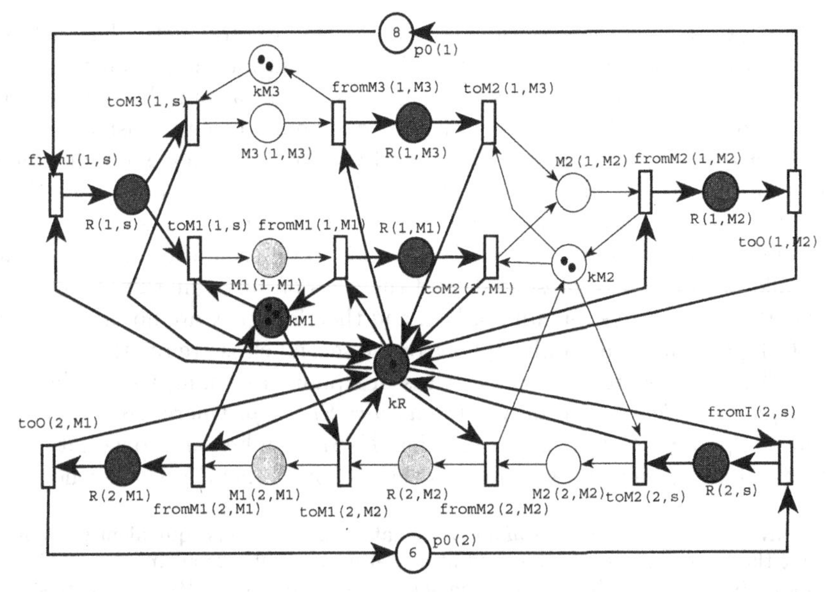

Fig. 24.13. A S^3PR

The following theorem establishes the liveness characterisation.

Theorem 24.3.1 ([ECM95]). *Let $\langle \mathcal{N}, \mathbf{m_0} \rangle$ be a marked S^3PR, let $\mathbf{m} \in \mathrm{RS}(\mathcal{N}, \mathbf{m_0})$, and let $t \in T$ be a dead*[3] *transition for* $\mathbf{m}$. *Then, there exists a reachable marking $\mathbf{m}' \in \mathrm{RS}(\mathcal{N}, \mathbf{m})$ and a (minimal) siphon S such that* $\mathbf{m}'(S) = 0$.

Therefore we can deduce the following corollary.

Corollary 24.3.2 ([ECM95]). *Let $\langle \mathcal{N}, \mathbf{m_0} \rangle$ be a marked S^3PR. Then, $\langle \mathcal{N}, \mathbf{m_0} \rangle$ is live if and only if for every reachable marking $\mathbf{m} \in \mathrm{RS}(\mathcal{N}, \mathbf{m_0})$ and every (minimal) siphon S,* $\mathbf{m}(S) \neq 0$.

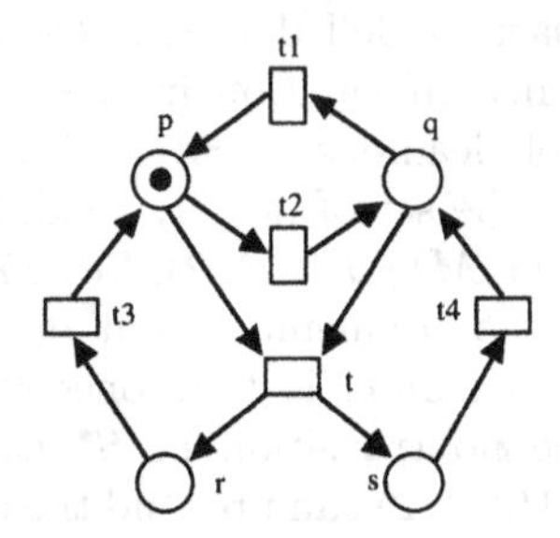

Fig. 24.14. The property *a dead transition implies an empty siphon* is not true for general nets: t is dead, but no siphon is empty.

This liveness characterisation is not true for general nets. The net in Figure 24.14 is a clear example: transition t is dead for the marking shown. However, the only siphon in the net, $\{p, q, r, s\}$, is always marked. Below we will see how this deadlock characterisation can be used to establish a control policy for deadlock prevention. The aim of the control policy is to add some constraints to the system in such a way that no deadlock state is reached.

Let us distinguish two classes of minimal siphons: those which are the support of a p-semiflow and those which are not. Considering the set of minimal p-semiflows (previously presented) and the class of initial markings, siphons of the first class always remain marked, and therefore they are not involved in deadlock problems. Thus only siphons of the second class are related to deadlocks. We will refer to this second class of siphons as "dangerous siphons". A *dangerous siphon* S can be written as $S = S_R \uplus S_P$, where $S_R = S \cap P_K$, $S_P = S \setminus S_R = S \cap P_S$[4]. The set of holders $H(S_R)$ can be partitioned into two subsets: those holders that belong to the siphon S (heavily shaded holders

[3] To say that a transition is *dead* for a reachable marking $\mathbf{m}$ is equivalent to saying that the transition cannot be fired at any state reachable from $\mathbf{m}$.

[4] For a given S^3PR, P_K denotes the set of resource places, P_S the set of state places and P_0 the set of idle states.

in Figure 24.13) and those that do not (light shaded places in Figure 24.13). Notice that, for a token to enter one of these holders, a token needs to have been previously "stolen" from the siphon. For instance, the firing of transition $toM1$ decreases the marking of siphon S by one token. This means that a token in place $M1(2, M1)$ implies one token fewer in $kM1$, and so one token fewer in S.

The control policy for deadlock prevention established in [ECM95] uses this property. For each dangerous siphon, a structurally implicit place (see Section 15.1.2) is added to ensure that at any reachable state, the number of tokens in the system that can reach siphon holders which "steal" tokens from the siphon considered is smaller than the initial marking of the siphon. In this way it is ensured that the marking of the siphon is always ≥ 1, i.e. the siphon cannot be emptied.

For the siphon considered, the control policy will add a place S_1 (see Figure 24.15) such that ${}^\bullet S_1 = \{fromM1(2, M1), fromM1(1, M1), toM3(1, s)\}$ and $S_1{}^\bullet = \{fromI(1, s), fromI(2, s)\}$. Since $\mathbf{m_0}[S_1] = 4$, it is enough to set $\mathbf{m_0}[S_1] = 3$ (for short, we also call $\mathbf{m_0}$ the initial marking of the extended net) to ensure that S cannot be emptied. Of course, any value $\mathbf{m_0}[S_1] \in \{1, 2, 3\}$ will be valid. However, we take the largest in order to have as much parallelism as possible using this control strategy. Notice that the addition of this new place generates a new p-semiflow: for each reachable marking $\mathbf{m}$ of the controlled net, we have $\mathbf{m}[S_1] + \mathbf{m}[R(1, s)] + \mathbf{m}[M1(1, M1)] + \mathbf{m}[R(2, s)] + \mathbf{m}[M2(2, M2)] + \mathbf{m}[R(2, M2)] + \mathbf{m}[M1(2, M1)] = 3$. From this invariant relation it is deduced that $\mathbf{m}[M2(2, M2)] + \mathbf{m}[R(2, M2)] + \mathbf{m}[M1(2, M1)] + \mathbf{m}[M1(1, M1)] \leq 3$, and then, since no more than three tokens can be stolen from the siphon, it cannot become unmarked.

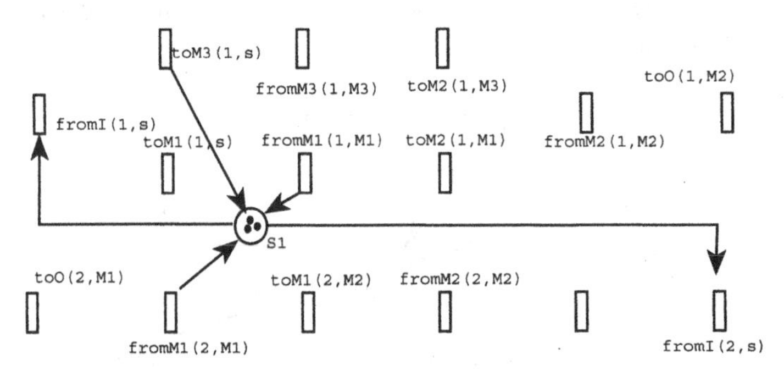

Fig. 24.15. The part of the control policy for deadlock prevention generated by the siphon in Figure 24.13

The same goes for the remaining dangerous siphons of the example. These are the following:

$$
\begin{aligned}
S_2 &= \{R(1,s), R(1,M2), R(2,M2), R(2,M1), kM2, kR\} \\
S_3 &= \{R(1,s), R(1,M2), R(2,M1), kM1, kM2, kR\} \\
S_4 &= \{R(1,M3), R(1,M1), R(1,M2), R(2,s), R(2,M1), kM1, kM3, kR\} \\
S_5 &= \{R(1,M2), R(2,M1), kM1, kM2, kM3, kR\}
\end{aligned}
$$

The added elements are the following:

$$
\begin{aligned}
S_2{}^{\bullet} &= \{fromI(1,s), fromI(2,s)\} \\
{}^{\bullet}S_2 &= \{fromM2(2,M2), fromM2(1,M2)\} \\
S_3{}^{\bullet} &= \{fromI(1,s), fromI(2,s)\} \\
{}^{\bullet}S_3 &= \{fromM1(2,M1), fromM2(1,M2)\} \\
S_4{}^{\bullet} &= \{fromI(1,s), fromI(2,s)\} \\
{}^{\bullet}S_4 &= \{fromM1(2,M1), fromM1(1,M1), fromM3(1,M3)\} \\
S_5{}^{\bullet} &= \{fromI(1,s), fromI(2,s)\} \\
{}^{\bullet}S_5 &= \{fromM1(2,M1), fromM2(1,M2)\}
\end{aligned}
$$

Let us now return to the coloured Petri net model of the example in order to integrate the deadlock prevention control policy. To do this, a new place called CP (Control Place) is added to the coloured model. The colour domain of this place is a set bijective with the set $\{S_1, \ldots, S_k\}$ of control places to the underlying S^3PR model added by the control policy. Let $cd(CP) = \{V_1, \ldots, V_k\}$ be such a set. The arcs which the control policy has added to the underlying S^3PR are represented by the arcs and functions which must be added to the final coloured model.

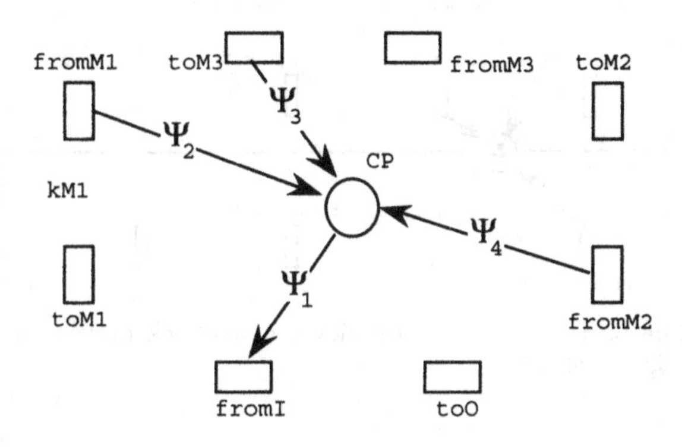

Fig. 24.16. Elements added by the control policy

The elements added in Figure 24.16 are as follows:

$$
\begin{aligned}
cd(CP) &= \{V_1, V_2, V_3, V_4, V_5\} \\
\widehat{\mathbf{m_0}}[CP] &= 3V_1 + 2V_2 + 5V_3 + 7V_4 + 7V_5 \\
\Psi_1(\langle W1, root\rangle) &= V_1 + V_2 + V_3 + V_4 + V_5 \\
\Psi_1(\langle W2, root\rangle) &= V_1 + V_2 + V_3 + V_4 + V_5 \\
\Psi_2(\langle W2, (M1, op4)\rangle) &= V_1 + V_3 + V_4 + V_5 \\
\Psi_2(\langle W1, (M1, op1)\rangle) &= V_1 + V_4 \\
\Psi_3(\langle W1, root\rangle) &= V_1 \\
\Psi_4(\langle W1, (M2, op2)\rangle) &= V_2 + V_3 + V_5 \\
\Psi_4(\langle W2, (M2, op3)\rangle) &= V_2
\end{aligned}
$$

A question arises. In the previous sections no constraint has been imposed on the system layout. However, in the definition of an S^3PR a termination property has been imposed (see the introduction of S^2P in Section 24.3). So, in order to ensure that the underlying system belongs to the S^3PR class we need to constrain the system layout to acyclic handling. This means that in the layout model no cycle is possible using only handlers. This ensures that there is no cycle without transformation, and then in the underlying model each production sequence eventually reaches the idle state. From the application domain point of view, Flexible Manufacturing Cells and Flexible Manufacturing Lines (as considered in [PHB93]) correspond to this class of acyclic handling systems. However, systems where the layout contains a carousel do not fit into this class: a carousel allows parts that can complete cycles with no transformation, which would violate the termination property imposed on the S^2P.

It must be pointed out that the control policy applied is maximally permissive ("maximally permissive" means that only markings related to deadlocks are prevented by the control policy). For the general case of the systems under consideration, to find a maximally permissive control policy remains an open problem. However, for some restricted cases solutions have been found ([XHC96]).

24.4 Conclusion

The chapter has shown how Petri nets can be applied to Flexible Manufacturing Systems. This is a domain whose complexity and inherent concurrency requires the use of formal methods to deal with very important problems, such as the design and control problems, in order to synthesise the software that ensures a correct system behaviour.

With respect to the first problem, we have introduced a design methodology which, given the input data describing both the structure of the system architecture and the logic of the processing of different types of parts, obtains in an automatic way a coloured Petri net model of the entire system. The use of a high-level Petri net model has the advantage of compactness. Also, the model obtained by this methodology shows in a clear way the structure of the system: the skeleton of the net has the same form as the configuration of the hardware components. Since the different processing sequences are modelled by colour domains of places and transitions and the functions labelling the arcs, the introduction/withdrawal of types of parts does not change the "look" of the model.

With respect to the control problem, we have studied the ordinary place/transition nets corresponding to the coloured models synthesised by the modelling methodology presented. This has allowed the study of deadlock problems for these systems from a structural perspective. One of the advantages of Petri net models is that they allow the study of certain properties using structural techniques, avoiding the computation of the reachability graph and, indeed, avoiding the state-space explosion problem. Unfortunately, structural techniques characterising liveness have not been developed for general Petri net models, but only for special subclasses (e.g. state machines, marked graphs, free-choice nets, choice-free systems). However, the special syntactic structure of the class corresponding to the systems we are considering allowed us to establish a deadlock characterisation. As has been shown, this characterisation was used to establish a control policy for deadlock prevention which constrains system evolutions to ensure the liveness of the controlled system.

From the material in the chapter, some general conclusions can be drawn:

- Petri nets are a family of formalisms well suited for application to FMS environments.
- The results obtained by the general Petri net theory are not sufficient for dealing with all the problems that arise in application domains. Therefore, it is necessary to develop specific new results for these domains. However, Petri nets provide a powerful framework which allows such developments.
- These specific results must not be *ad hoc* for each problem, but must concentrate on certain modelling/programming paradigms. For the class of nets considered in the present chapter, we apply the case of sequential processes using monitors (in a restricted way). Natural extensions to this case, applicable to general manufacturing systems, operating systems, databases etc., use monitors in a general way as well as communication through buffers ([Sou91, RTS95, TCE99]).
- The use of coloured Petri nets, and high-level Petri nets in general, has some important advantages for the modelling. However, we are faced with another problem: symbolic processing of these nets is not complete, and so must be developed. The steps given for some subclasses of coloured Petri nets make this approach look promising.

25. Workflow Systems*

25.1 An Overview of the Domain

In this chapter an approach is outlined to support the *definition of business processes with Petri nets*. Today, business processes are defined using specific business applications called *Workflow Management Systems (WFMS)*. They support *workflow management* which can be broadly defined as office logistics. A definition of WFMSs is given by the Workflow Management Coalition (WFMC) [WMC94]: "A Workflow Management System is the computerised execution of a process definition."

This definition contains two key phrases. The first is *process definition* and the second is *computerised execution*. They define workflow management, as shown in Figure 25.1, and are elaborated upon below.

Business Processes: Recently many works have appeared on the subject of business process optimisation, including business process re-engineering and total quality management. Although these writings lack a uniform approach in their definitions of business processes and improvements, most observe a difference between *ad hoc* and *structured* business processes.
The former refers to business processes that largely involve informal communication. These processes are typically supported by groupware applications that enable all participants to access and update a shared information base.
The latter type of business processes is supported by WFMSs because they can be structured. A structured business process can be designed before it is put into practice. During the design, the ordering of tasks and operations is defined. In this chapter the concepts of structured business processes are discussed.

Computerised execution: A Workflow Management System (WFMS) is a generic software package that supports the definition, management, and execution of workflow management. Although it needs a common information infrastructure, as do Groupware tools, a WFMS has an explicitly

* Authors: W. van der Aalst, M. van de Graaf

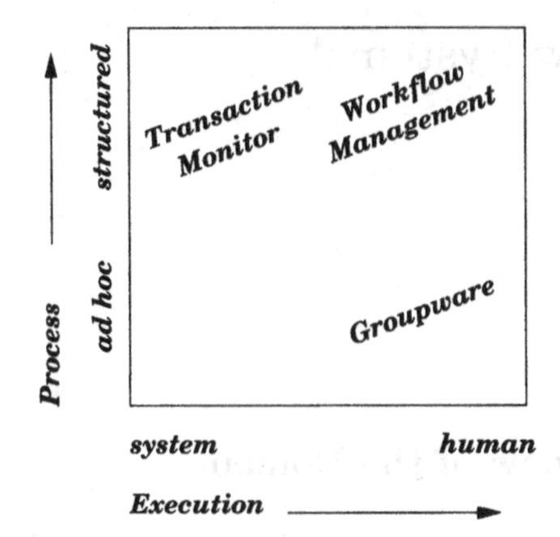

Fig. 25.1. Workflow Management defined according to process and execution characteristics

defined process definition for describing the ordering of activities within a business process. The pre- and post-conditions of activities can be defined and coordination of work can take place via this WFMS. *Humans* execute their work in close cooperation with the WFMS, which appears to them as an electronic mail basket containing their work. The WFMS helps humans to collect applications and data for their work. When the execution is dominated by *systems* there is little or no human labour. Examples are money transfers and inventory management processes. Although not discussed here, notice that there is a continuous drive towards executing business processes with systems rather than humans.

WFMSs aim to define business processes in order to support their computerised execution. To implement this functionality, the WFMC defined a WFMS architecture that is now widely adopted by all vendors of WFMS (and related) products. The WFMS architecture is shown in Figure 25.2. A brief overview of the WFMS architecture is given below.

- **Workflow Engine:** The workflow engine is the heart of a WFMS. It maintains all the data about available and running processes. It interacts with invoked applications and contacts humans by means of dedicated workbaskets.
- **Process Definition Tools:** The tools that are available support the graphical design of a business process. Other tools support the definition of the organisation in terms of resources. These tools are integrated to link resources to specific activities in the process definition. Ideally the process definitions can be analysed for the fulfilment of structural as well as dynamic properties. However, currently most vendors offer tools for simulation with a restricted set of parameters.

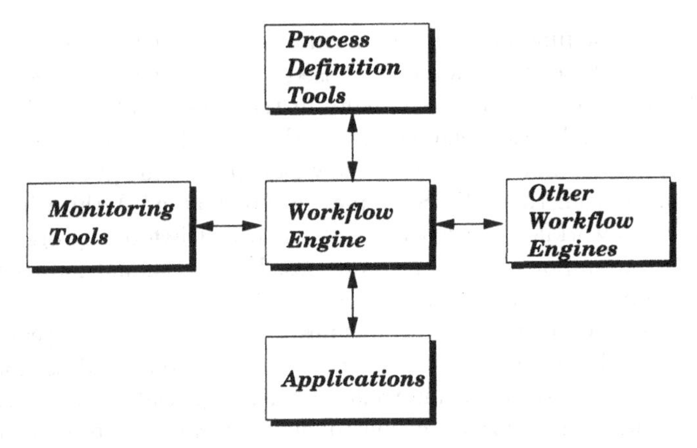

Fig. 25.2. Workflow Management Systems architecture

- **Monitoring Tools:** A monitoring tool allows one to query the detailed results from the workflow management database. All data from the business processes are stored in it for analysis. A ideal monitoring tool can give selective views of performance in the past and handle predictions. Standalone monitoring tools for WFMSs are only useful when the WFMSs are standardised. Currently, most WFMS manufacturers have their own monitoring tool integrated in the product.
- **Applications:** Business processes are made up of many different functions, as stated earlier. These functions, performed with the help of applications, must communicate their state to the workflow engine. A specific set of functions is defined for applications that are *invoked*. This means that they operate under the control of the workflow engine.
- **Other Workflow Engines:** Like DBMSs, the future of business processes will be heterogeneous WFMS-support. Therefore it is important to define a common interface to translate all information at run-time from one WFMS to another.

The WFMS architecture is implemented on different formalisms and platforms by more than 250 vendors. Examples of WFMSs that appear in practice are Staffware (Staffware), FlowMark (IBM), Plexus FloWare (Recognition International Inc.), ActionWorkflow (Action Technologies Inc.), LEU (Verbacom Services) and COSA (Software-Ley) [Act96, Pal97, Sof96, Law97, Bak96]. The process definition tools of the latter two are based on P/T nets. There are also vendors that offer process definition tools only and provide an interface (WFMC standardised) to other WFMSs. Examples of such tools are: Structware (IvyTeam), Income (Promatis), Protos (Pallas Athena) and Meta-Software (Design IDEF).

In this chapter an approach is outlined to support the *definition of business processes with Petri nets*. There are several reasons for this approach. First, since currently more than 250 vendors offer workflow management software ([Aal96]), the functionalities are hard to compare, let alone interchange. In fact, most of these WFMSs are not based on an explicitly defined conceptual model and therefore are incomplete in their specifications. The second need for conceptualisation emerges from the need for businesses to share information across multiple WFMSs. How can they be integrated if the conceptual models from the WFMS engines do not match? A third important reason is that a workflow needs to be proved correct before any organisation will allow it to support their core processes. If these should fail to operate correctly, organisations would cease to exist.

Petri nets have been applied to business processes for many years. Ladd and Tsichritzis [LT80] proposed an office form flow model and used a graph theoretic approach to define interrelated tasks. In that same year, Cook [Coo80] presented the information control net to streamline office procedures. Information control nets were first described by Ellis [Ell79]. Although both used Petri net dialects, Cook describes activities and repositories, such as the transitions and places of the classic Petri net. Later articles such as Voss [Vos87] and Nierstrasz [Nie85] used Petri nets explicitly in the field of office automation. Recent works that have appeared in this field are [WR96, DMEM94, She96, Aal97a, Aal97b, AB97a, AH97, AHV97, Aal98b, Aal98a, Law97].

The elements and properties of business processes are defined using elaborate Petri net constructs, such as those defined in the previous chapters of this book. Colour is added to model complex real-world objects and time is added because business processes have time constraints imposed on them. Although hierarchy is useful for defining business processes, it is not exploited in this chapter; it is however useful when the methodological aspects of constructing business processes with Petri nets are discussed.

This chapter is organised as follows. Section 25.2 elaborates on the main concepts of business processes and computerised execution. The main concepts of business process definitions are given informally. Section 25.3 focuses on formal concepts of WFMSs and shows results for an example business process. This business process serves as a running example throughout this chapter. Section 25.4 shows some Petri net analysis techniques that are applied to workflow management, and Section 25.5 gives a brief summary of a real-world application. A summary of this chapter is given in Section 25.6 which also lists some topics of future research.

25.2 Motivation

Research in the area of Workflow Management Systems (WFMSs) originates from both an application and a technology domain. This is in line with the definition of WFMSs, which states that business processes (application domain) are supported by computerised execution (technology domain). Typical WFMS functionalities are derived from combined developments in these two domains. An example of the former is the need to control business processes to make them more cost effective; an example of the latter is the widespread availability of client/server platforms, large business and common networks, and shared databases.

The application domain is manifested through an increasing need to produce more products with a shorter life-cycle. The only way to do this in a cost-effective manner is to re-use tasks in multiple business processes and to manage the design process. The product variety can be stretched by coupling tasks in different ways, i.e. designing processes in a modular sense. It is expected that a WFMS contributes to a more flexible business process design.

Essential technological functionalities have arisen from the evolution in information systems. Traditionally, information systems consisted of a large number of applications. In the '60s, each application defined its own data structures, user interface and business logic. Of these three, the first to be discarded were the data structures, replaced by Database Management Systems (DBMSs) in the '70s. The DBMSs served as a central repository for applications. This relieved the application builders of extensive data management. Secondly, the user interface was replaced by a User Interface Management System (UIMS). This happened in the '80s. Every application that is built on a certain platform can now use standard UIMS functionality. Figure 25.3 shows this historical perspective.

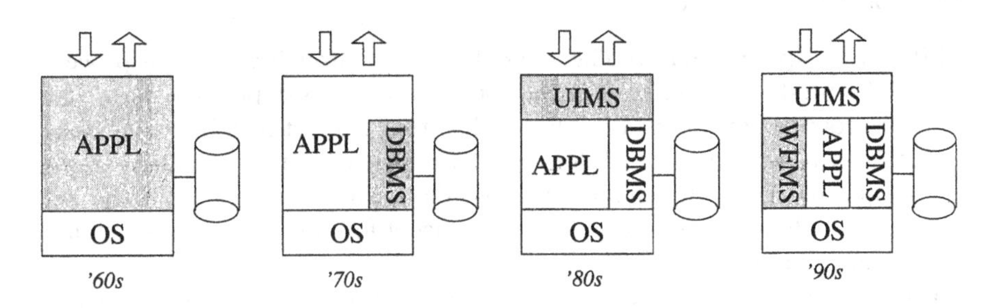

Fig. 25.3. Evolution of information technology

At present the third item, the business logic, is being replaces by a WFMS. This leads to the development of applications that perform only narrowly defined tasks, leaving the control of the business process to dedicated business logic that is defined in a WFMS. Because this is a new development, many vendors, techniques, and concepts are appearing. A similar case in the recent history of information systems is that of the first introduction of DBMSs. Many different techniques for defining a DBMS emerged and application builders had to write just as many interfaces. It was Codd's relational model which ended this situation. From then on each DBMS was specified in a generic manner, displaying just one interface to applications. This introduced the possibility of interchanging data in a uniform way between databases and applications.

The WFMS provides a generic set of functionalities for implementing business logic. It defines which application can start its execution, supplies it with the correct data from a database and handles all extra inputs that fall within the business logic domain, such as the availability of a human or other real-world entity.

At the current stage of WFMS development, two topics are of major interest: 1) constructing a single formal language upon which WFMSs are built, to verify the business process specification, and 2) providing analysis techniques to support business process performance analysis. Petri nets are proposed here to fulfil both requirements. The single formal language also enables the use of a wide range of analysis techniques. These techniques provide the added value of WFMS formalisation with Petri nets.

25.2.1 Formal Language

A common way to define business processes in today's WFMSs is via an event-driven specification. The business processes are specified as events that signal other events. For example, when one activity is completed another is immediately signalled. This approach is useful when the interrelation of activities is supported. It is, however, not the best way to model WFMSs. Some of the reasons are:

Life-cycle specification: The work that flows through a business process is the product under transformation. Think for example of a form that lists your tax or insurance policy. When this product is in between tasks, it is in a certain *state*. A logistical perspective is to view these states as queues. Event-driven specifications do not define states explicitly and therefore lack an essential real-world concept for modelling and specifying business processes.

State-based action: All event-driven specifications assume that after completion of a task the succeeding task can be scheduled immediately. However, it happens frequently in business process scheduling that the subsequent task is not known *a priori*. It has to be chosen from a set of tasks. The choice can depend on many parameters, such as time, messages received from the environment, or related data. When this parameter is not known the product has to wait at a certain state that is monitored by all preceding tasks.

Fail-safe: Workflow Management Systems are embedded in an environment where errors occur. In order to roll-back to a situation before the error occurred, one must know the last proper state of the process. In an event-driven approach, one must maintain a complete history of events in order to get to the event that failed.

Interchangeability: A WFMS has to be able to communicate with other WFMSs while supporting a business process. These different WFMSs have their own span of control. Therefore, they are specified in terms of input/output systems. It is this input/output that is not specified explicitly in an event-driven approach.

The theory of Petri nets is a combination of an event-driven and state-driven specification. When used to specify WFMSs, it provides the functionality that is lacking in current purely event-driven specifications. Moreover, it has good graphical properties, making it a useful candidate for a process definition tool. In the next chapter it is shown that business processes can be mapped to the elements of Petri nets. The result is a unifying conceptual model of business processes, which can be analysed according to different properties.

25.2.2 Analysis Techniques

When business processes are specified in terms of Petri nets, many Petri-net analysis techniques become available. These analysis techniques, which are listed in this book and for example in [Mur89], are applied to the domain of workflow management in Section 25.4. Analysis of workflow management is divided into two areas, known as:

Structural analysis: When a business process is specified many of its properties can be checked based on its structure. The advantage of this feature is the availability of a collection of fast and powerful construction mechanisms to build correct business processes.

Performance analysis: The designer has to test the business process on expected behaviour before the business process is put into practice. With these tests the behaviour of the business process can be analysed according to business properties, such as throughput, occupation rates, and manufacturing costs.

25.3 Design Methodology

The semantics of a Petri net defines the *meaning* of the elements in a specific system. The general concepts of places, transitions, and tokens are translated, or *applied* to specific areas. Without knowing their specific meaning, one would not see the difference between a Petri net for a nuclear plant and that for a detailed specification of operations in a bakery. Furthermore, when the meaning is defined clearly, the analysis techniques of the next section can be placed into a workflow management perspective. Below, the basic concepts for the construction of Workflow Management Systems (WFMSs) are introduced.

An important initial assumption of the design methodology is that a Petri net is constructed for products produced by the business process in *isolation*. This means that they do not interact[1]. When the business process is designed and the behaviour is analysed for multiple products flowing through the business process, they have different colours to separate them.

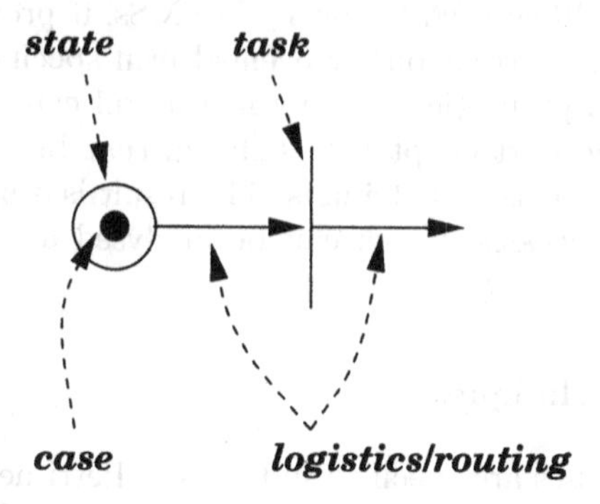

Fig. 25.4. Mapping of business process elements onto a Petri net

The elements that are defined formally in the next section are introduced in Figure 25.4 and described informally below.

Case: The case is the product that is produced by the business process and the main objective of the WFMS is the proper management of this case during its processing. Examples of cases are insurance claims, tax bills, and student registration forms. A case is identified uniquely by its identifier. In the above examples these are the insurance number, social registration number, and student number. The WFMS will manage the life-cycle of each case. It will maintain its attributes, the conditions placed on it, and (an entry to) its data. A *case colour* defines the logistical properties of the case. For example, if the insurance claim is larger than amount x, it should be handled by a senior manager and otherwise by a junior. Another example is that a student who started in year x loses governmental support after year $x + 4$. These attributes can change the state of the case.
Case data is not maintained by the WFMS. A case can exist in a number of documents, files, databases, etc. Only by the unique colour of a case does the WFMS maintain a link to this data. For operational practice this is a necessity.

State: The state of a case in the business process is defined by the states of the Petri net. Examples of relevant states for WFMSs are for the insur-

[1] This aspect implies some special properties for the structural analysis of business processes.

ance cases: `claim_received`, `large_claim`, and `claim_denied`. The student has states such as `receives_support` or `no_support`. These states provide natural implementation of the business process and allow for automatic actions which can be defined on them. They are also the essential feature for synchronising the flow of work, as will be shown later in this chapter.

Task: A task is an elementary action, also referred to as a logical unit of work. Although a task is an elementary action within a business process, it can maintain many subtasks on a lower level. For example, for a supervisor the execution of a task by an employee is a logical unit of work, while the employee views the same task as being composed of a number of tasks ([AH97]).
Examples of tasks are: typing a letter, reviewing a document, filling an insurance claim, or writing an exam. Tasks can be divided into manual, automatic, and semi-automatic (or semi-manual) types. The manual task is executed without the interference of computers, e.g. inspecting damage or lecturing[2]. A task that is executed completely automatically has no human interference. Based on the available data and instructions, an application can create, update, or delete case data. When both human input and applications are needed to execute the task, it is called semi-automatic. Examples of semi-automatic tasks are: the management of documents with a word processor, or the calculation of the annual budgets with spreadsheets.

Logistics: The life-cycle of a case is the path which it takes through the tasks in order to complete. The life-cycle of a case is maintained by the WFMS. Therefore the WFMS sends the case to appropriate tasks and signals the appropriate resources (to be discussed later). This function is referred to as business logistics or routing. The elementary logistical function is called the *sequence*, meaning that one task precedes another task. More elaborate logistical functions enable work to be executed in *parallel* or allow *selective routing*. The former implies that tasks can be executed at the same time and the latter that one task to be executed is chosen from a set of tasks. Repeating a (set of) task(s) is called *iteration*. Logistical functions play an important role in the Petri net constructs that will be defined in the remainder of this section.

The interpretation of Petri nets for Workflow Management Systems (WFMSs) will be discussed in detail. First a task is identified with a transition in Section 25.3.1. Additional triggers to start a task are viewed as extra input conditions that have to be fulfilled. They are exploited in this section. Section 25.3.2 shows the logistical functions that provide the basis for coordination of a case. Finally, Section 25.3.3 shows the equivalent of a case, which

[2] However, experiments with computer-based training have proved to be effective and will be more common in the near future.

is a (set of) tokens. The case concept is defined by a Petri net for WFMSs. Therefore it is discussed after all the elements have been introduced.

25.3.1 Tasks and Transitions

The first Petri net element that is discussed is the transition. A transition represents a transformation of a case between consuming it from an input place and producing it into an output place. The transition therefore is an abstraction of a task executed in a business process.

Normally a token (which represents a case) is modified by applying some function to it. These functions can be seen as applications and therefore pure transitions mimic automatic tasks. As was showed previously, other tasks require human input and therefore these transitions require an external event to occur. In a Petri net these events can be simulated as an extra input condition to a task. For structural analysis they are discarded. They play an important role when the Petri net is interpreted by business employees, simulated by staff employees, and finally implemented. The events show the type of external event that needs consideration.

In Figure 25.5 four types of triggers are shown. These events are extra conditions that have to be satisfied to enable a task. The meaning of each condition is explained below. Note that a detailed description of trigger management and the interaction with tasks is more complex and beyond the scope of this introduction. A more detailed analysis of triggers and their application to WFMSs is given in [Gra96, GK95, MS95].

Message

Time

Automatic

User

Fig. 25.5. Four external triggers placed on tasks

Message: When a task can be executed after another system signalled it, the message icon is placed on the transition. An example of a message trigger is that a task can start when a specific document has arrived in another department. Thus, even if tokens are available at the input place they are not consumed. This introduces the possibility of a delay in the execution (i.e. tasks are not eager) and the input place becomes an input queue. In logistics this queue is termed a *complementary queue* ([Pla96]) since the task needs an additional event from another system in order to complete.

Time: Some tasks start after a certain time has passed. Time-based signals can be given for a single case or for a number of cases. The former can be seen as a time stamp t on a case when it is produced in a place. A task is enabled after $t + \Delta t$. Because this time-based task depends on the time the case is placed in the place, it is called a *relative* time signal. Another relative time signal is implemented by a time stamp t as an attribute of a case. Enabling a task is now controlled by this attribute. An example of this is an arrival date that orders cases.
When the task is enabled at fixed time intervals for a number of cases it is called an *absolute* time signal. An example of this is that order picking starts at 8:00 pm each day. When the task is disabled due to this absolute time trigger a queue of cases appears at its input place. These queues have their equivalent in logistics, where they are termed *platform queues*.

Automatic: In general, when all pre-conditions of an automatic task are satisfied the task fires instantly (i.e. it is eager). Firing an automatic task means that an application program is invoked which transforms the case. The task is non-eager when the availability of this application is limited, e.g. when there is a bound on the number of application licences. To show graphically that there is a resource dependency, the icon is placed on the task.

User: Many tasks require the input of a user to execute. Thus the WFMS has to receive a signal from a user stating that this task can be started. There are many pragmatic solutions which support this type of interaction between users and WFMSs. Basically the concept of *in-baskets*, defined for example in [WMC94, AAM$^+$95], is introduced for communication with (groups of) persons. As with applications and their WFMS interaction, this topic is not further detailed.

Extra task conditions in Petri nets are used to model these triggers, shown in Figure 25.5. In the following section these tasks are structured so as to form a business process. The structuring of tasks follows the logistical primitives that were outlined at the beginning of this section.

25.3.2 Logistics and Transitions

A case flows through a Petri net by means of logistical properties, which are defined below as pre- and post-conditions of tasks (transitions). There are four basic logistical properties, also identified by the WFMC [WMC94]. They are divided into parallel, mutual exclusive, splitting, and joining properties.

First the logistical property that enables parallel execution is depicted. A case is split into a number of case tokens that enable tasks for that case at the same time. Thus multiple tasks are enabled for the same case at the same time and can be executed in unrelated order. The AND-SPLIT provides this parallel split for a single case. All case tokens must at some point be joined again with an AND-JOIN (see Section 25.4). Both AND constructs are shown in Figure 25.6. The upper two nets show the P/T net representation and the lower two the counterpart for WFMS design.

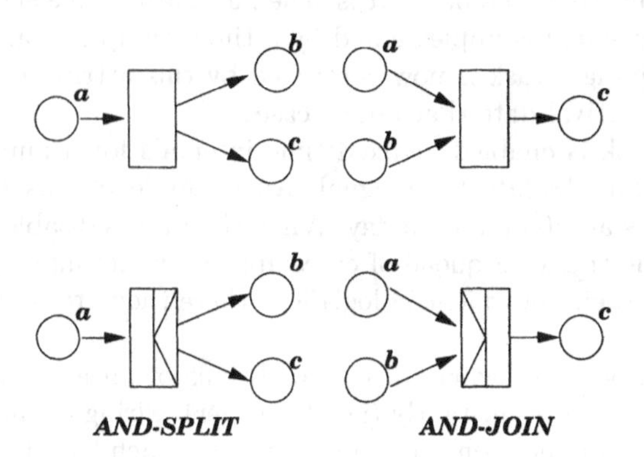

Fig. 25.6. AND-SPLIT and AND-JOIN in Petri nets

This new representation is introduced to make clear to a designer that this is a logistical task, i.e. the case is not transformed. The AND-JOIN is enabled if two tokens from the same case (based on its unique colour when multiple tokens flow through the Petri net) are available in the input; it produces one case token in its output.

A brief note on parallel work is made here. Traditionally cases were limited to sequential routing due to the physical constraints of the case. These have been removed by the widespread availability of shared databases and networks that store cases and enable widespread access to them. An example of parallel case assignment for the insurance claim is that it can be reviewed by an office worker while another employee is checking the actual damage. One need not wait for the other to start, but they must both be finished before the next task for this case can be executed, which introduces interesting new ways of conducting business ([HC94a]).

The mutual exclusion logistical property also has a split, known as the OR-SPLIT, and a join called the OR-JOIN. In Figure 25.7 both are represented in the P/T net representation (upper two figures) and their WFMS counterpart (bottom figures). Again the new representation is introduced to show a designer that a case is only routed and not transformed by this transition.

An OR-SPLIT takes a case from its input and places it in one of its output places based on a logical function. This function is based on the case attributes. With this construct a decision can be implemented. For example, every insurance claim for which the amount x falls within a certain domain *DOM* is handled separately from other claims, as Figure 25.7 shows.

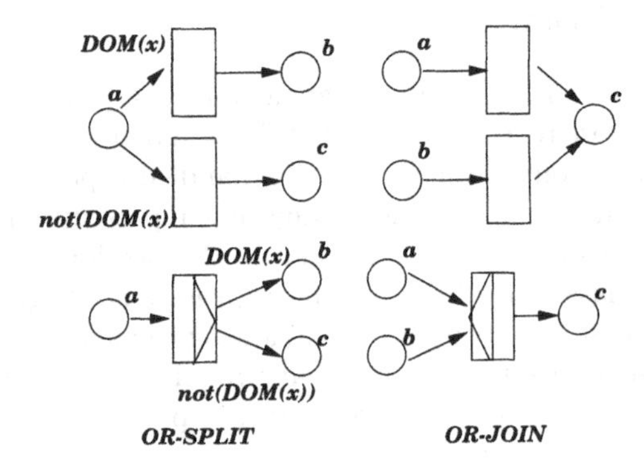

Fig. 25.7. OR-SPLIT and OR-JOIN in Petri nets

At some point the separate paths of the OR-SPLIT have to be merged again and this happens with an OR-JOIN. This transition is empty, i.e. it does not contain any logical function; every case that appears in one of its input places is sent to an output place. No synchronisation on a case attribute is needed because a case is not multiplied by the OR-SPLIT.

The OR-SPLIT is used when the outcome of the decision is known *a priori*. This is called an *explicit OR-SPLIT*. However, sometimes a decision is known only *a posteriori*, which requires an *implicit OR-SPLIT*. Such an OR-SPLIT can be defined on a state in the Petri net.

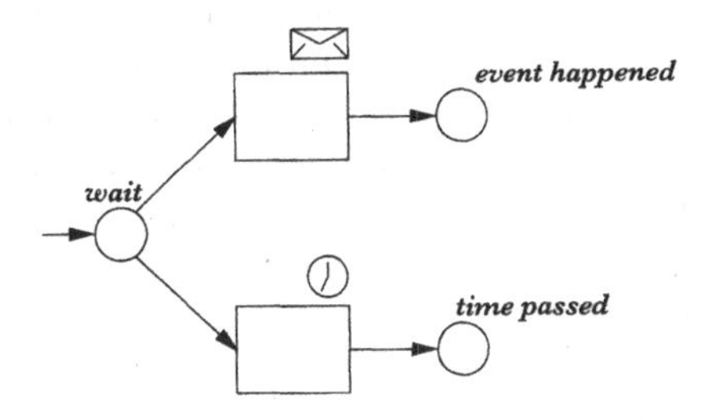

Fig. 25.8. An implicit OR-SPLIT

An example of an implicit OR-SPLIT is shown in Figure 25.8. This example shows two tasks with extra triggers, a message and a clock. The idea is that a case is flowing through the upper path (with the message trigger). However, since time is an important parameter in business processes, a timed action is also placed on it. If the case is not handled by the upper path before time δt (which can be absolute or relative to the case) the case flows through the lower path and is executed by other means. Because the decision is known *a posteriori* an implicit OR-SPLIT is used instead of an explicit OR-SPLIT.

25.3.3 Case and Tokens

Earlier it was mentioned that a case is analogous to a product. In the Petri net the tokens therefore reflect the case that flows and also reflect the features of the case (by the colours). Since a case is handled separately from other cases, the set of tokens in the Petri net at any time represents the state of that single case. Properties of a case are defined in this section. A case c always is an instance of a class type C. The business process is designed to produce cases of type C. A case type contains the following elements: $C = \{A, M, P\}$.

First, A represents a set of colours that provide the logistical values for each case c. The logistical values, defined in the previous section, correctly route a case through the process.

Secondly, M is a set of colours that list case properties used for efficiency of operations. It has information for both *simulation* and *real-time* scheduling. Think for example of information about the added costs of the case and its due dates. Typically this information is maintained in separate databases to obtain aggregate information over a selection of cases. For example, one might be interested in the average cost structure of all cases that have the logistical properties: `amount>1,000, date>01-01-98`.

P is the set of actual product data of the case and in practice it falls outside the WFMS scope. The data undergoes transformation by tasks, i.e. creation, updating, and deletion. Product data is not a primary concern during the formal modelling of WFMSs. However, it is important to *order* the tasks of the business process. A designer has to know where data is created and where it is needed (the pre-condition) before a task can start. Therefore the specification of product data is of importance for the method of designing the formal model. Note that there can be a logistical colour that is also product data $A \cap P \neq \phi$. When the P/T net of a WFMS is put into practice, this overlap causes synchronisation questions to be answered since P resides in another domain than A.

Each case $c \epsilon C$ consists of these three elements. When multiple tokens of a case are distributed over a set of states of the Petri net, together they represent the total state for the case c at a certain point in time. The examples that are used in this chapter will outline the usage of the elements A for routing c through the Petri net, M for simulation and management information, and P for product data.

25.3.4 Case Study: Justice Department

To illustrate the use of Petri nets to model business processes a real world example is presented here. It is an example obtained from the Dutch Justice Department. First the informal description is given and after that the formal model is derived from it. Note that the identified tasks are modelled at a high level of abstraction and have to be detailed in further studies.

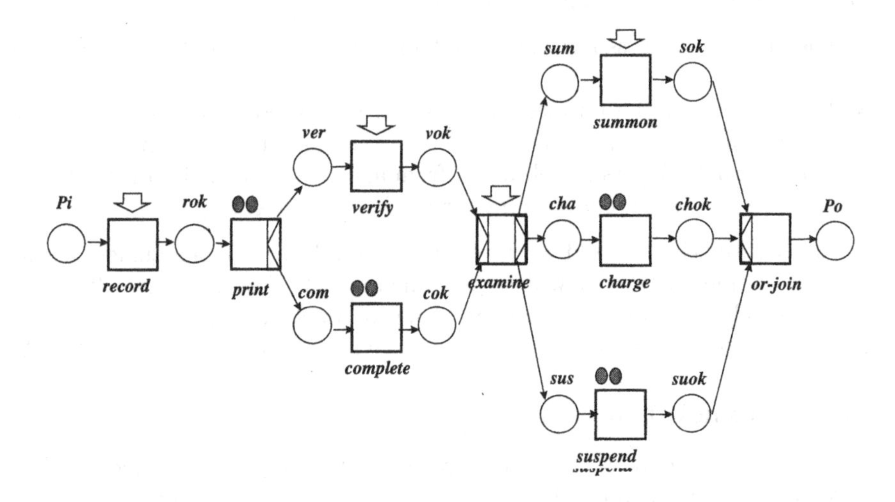

Fig. 25.9. Case study example for a Petri net

> When a criminal offence happens and the police have a suspect an official record is made. This is printed and sent to the secretary of the Justice Department. Extra information about the history of the suspect and some data from the local government are supplied. Meanwhile, the information on the official record is verified by a secretary. When these activities are complete, a prosecutor determines whether the suspect should be summoned or charged, or the case should be suspended.

Based on the analysis of information manipulation in the informal description, a corresponding Petri net is depicted in Figure 25.9[3]. It is shown that a case starts when the task record is enabled by the first case token in P_i. The types of external triggers are shown in the graphical representation. In this case the record can be made only with the input of a user (from the police department). Printing is an automatic task handled by a dedicated application and once completed it enables both verification and completion by the Justice Department. The AND-SPLIT symbol is placed in this task. The examination of the case by the prosecutor can start only after both tasks have been completed (AND-JOIN). This is a manual task and it enables one of three tasks: summon, charge, or suspend. The OR-SPLIT is also placed in

[3] This initial Petri net is derived from the informal description. After analysing its properties we may change the model. The sequence of information manipulation is, however, a process constraint that has to be obeyed in any improved model.

this task. If the enabled task is completed the case is closed and placed in the P_o state. Points of importance are:

- The business process has a start and an end state, called P_i and P_o respectively.
- The logistics of the business process can be placed on a task, as for the `print` task, or as a primary logistic task, such as the `or-join`.
- The external triggers are obtained from another domain (information systems design) and are a given factor for formal modelling.

This case will be used as a running example in the remainder of this chapter. Some parts of it will return when specific issues of modelling and analysis techniques are explained. Several proposals for changing this business process in order or improve the performance are suggested in Section 25.4.2.

25.3.5 Business Process Definition

A Petri net that defines a business process is defined. The fundamental concepts are explained in detail in [Aal95]. These concepts are introduced here and are elaborated upon in the next section.

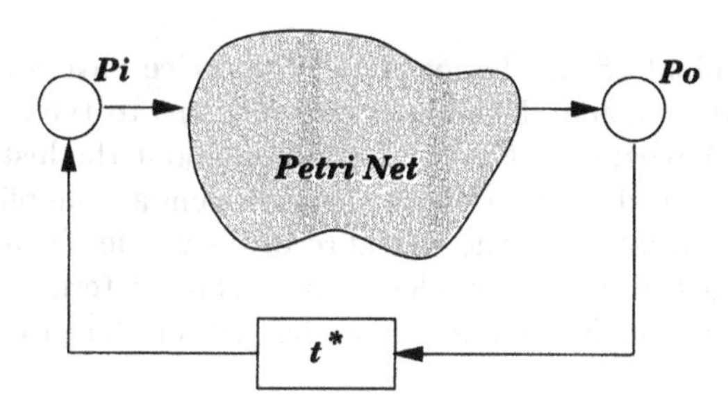

Fig. 25.10. A strongly connected Petri net

Start and end: A formal model of a business process has two special places, p_i and p_o. A case starts at the input place which has the property $\bullet p_i = \emptyset$. This means that this place has no incoming arcs, i.e. it is a *source* place. A case always ends in the output place which has the property $p_o \bullet = \emptyset$, meaning that this place has no outgoing arcs thus representing a *sink* place. In the example of the Justice Department these places are called *start* and *end*.

Strongly connected: If the transition t^* is added to a Petri net connecting p_o with p_i, the Petri net is strongly connected. This property states that every transition (task) t is in a direct path leading from p_i to p_o via t^*. This requirement is added to avoid dangling tasks, i.e. tasks which do not contribute to the processing of a case. Figure 25.10 shows a Petri net with the t^* added.

25.4 Workflow Analysis

Building a WFMS using the Petri net concepts described above enables a wide variety of Petri net analysis techniques to be used. When WFMSs are designed they are analysed according to process, resource, and case dimensions, see Figure 25.11.

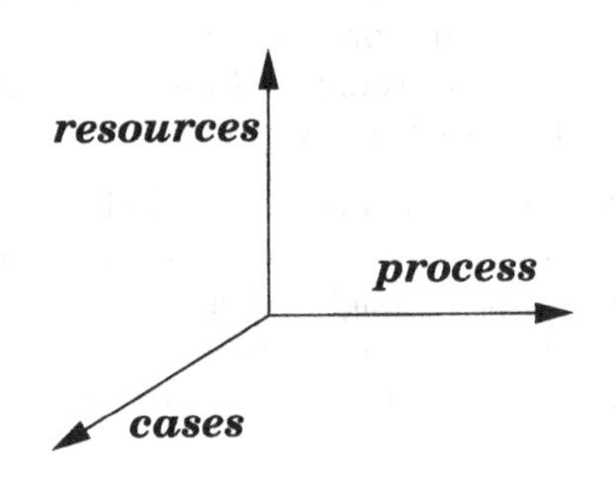

Fig. 25.11. Dimensions of Workflow Management analyses

The first aim of analysis is to guide a WFMS designer in building correct specifications of business processes. Therefore in Section 25.4.1 the structural analysis techniques of Petri nets are described since they cover the *process dimension*. These structural analysis techniques are described for a special marking of cases in the net: only one case token is assumed at p_i, i.e. $M = i$. This means that we can use the analysis techniques developed for P/T nets, since $C_A = \phi$.

Resources are not analysed when the structural properties of the model are proved. Adding resources makes this task computationally harder to solve since resources violate some rules that will be defined for models constructed for single-case Petri nets. The abstraction of resources can safely be done because we assume intelligent resources (i.e. humans). Therefore, it is always possible to recover from a deadlock ([Lom80]). Furthermore, for some structural analysis techniques the availability of resources would not even make sense, as will be shown.

The second aim of analysis techniques is to help a designer find an optimal solution for modelling tasks *and* resources. Guidelines are proposed for the choice of different correct models and their resource distributions. This is called dynamic or performance analysis, and it is discussed in Section 25.4.2. Since the solution space is limited to constraints from the outside world (the availability of resources, the product structure of the case, etc.) only a local optimum can be found.

The third dimension is the *case dimension*. The structural properties are proved for single-case business processes, whereas dynamic analysis introduces multiple cases into a business process to analyse its behavioural aspects.

25.4.1 Structural Analysis

An interesting feature of Petri nets in the design of WFMSs is that design errors can be detected before a specification is put into execution in the workflow engine. As explained earlier, when a model is made, the structural analysis techniques apply for each *individual* case (meaning $M = i$).

Structural analysis resembles the analysis of a business process because it is assumed that tokens which represent a single case do not interfere with tokens from any other case. The example of the Justice Department case will be used to clarify these topics when needed.

Properties: Safeness. A business process is designed to produce a certain case type. Therefore, each case instance is manipulated in isolation. This implies that the effect of every single token that is assigned to p_i can be analysed. This token in p_i results in a number of tokens that are distributed over the Petri net according to the logistical properties that were defined in the previous section.

Since a single case is analysed, it makes no sense to have multiple tokens in the same condition at the same time. A condition is either true or false. Thus, a Petri net designed for a single case must be safe, i.e. it can hold a maximum of one case token in each place at the same time. In the Justice Department example it makes no sense to enable a task twice at the same time for the same record. This would mean that for one and the same record two similar tasks have to be executed at the same time. This undesirable situation is prevented by proving the safeness property for the model.

Boundedness. Since a business process has to be safe for each case it is 1-bounded. Safeness is a special case of boundedness, meaning that a maximum of one case token per place is allowed.

Reachability. A business process is designed to deliver output based on some input. A general check after designing a business process therefore is that a case that started in place p_i results in a token in place p_o, the end-state. Place p_o may be unreachable if errors occur in the design, such as deadlocks by the omission of certain transitions. The possibility of certain tasks that are enabled at the same time is also shown with this type of analysis.

A designer also wants to know which sequences of tasks can appear. In this way the causal effects that exist in a business process can be analysed. Thus, given a marking M' of the business process (for a single case), is a certain marking M'' reachable? This analysis helps a designer to find causal effects and also helps users in a later stage to query their own business process. The question that can be answered is: if a certain task is executed, is it still possible to execute another specific task?

Deadlock. A deadlock occurs when a business process is not terminated and no transitions (tasks) are enabled. In practice this can occur when a critical resource is not available, but again, we abstract from resources in this section.

For a single case, a deadlock can occur when the logistical properties are not applied correctly. Suppose a designer places an OR-SPLIT in the model and (by mistake) synchronises both branches with an AND-JOIN for completion. This example is given in Figure 25.12.

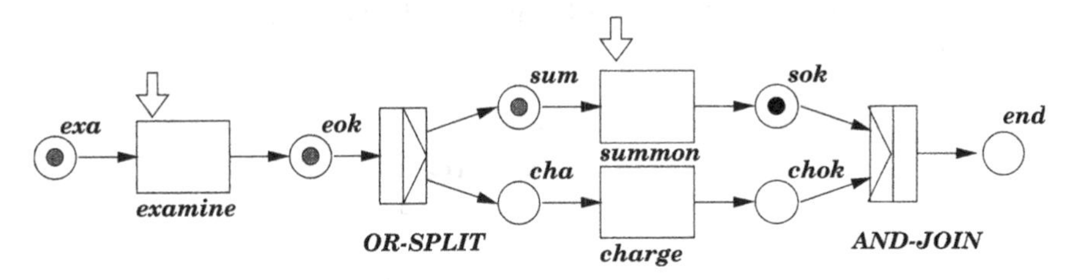

Fig. 25.12. Deadlock caused by incorrect place synchronisation

In this example the path of the tokens is shown by depicting them in grey. After the examination is completed (`eok`), either a summon or a charge task can start according to the OR-SPLIT. In the example the first option is chosen and completed, and a token is placed in (`sok`). Now the AND-JOIN can not fire, since a token has to arrive in `chok` as well, and a deadlock is the result. Therefore p_o is not reached and the business process can not terminate. In the following a subclass of Petri nets is proposed that is based on correct synchronisation of tasks.

Liveness. A designer of a WFMS wants a guarantee that a case token always enables a task in the business process model. The task can fire immediately (it is eager) because resources are not considered here. Thus, once a token is generated in p_i, there has to be a firing sequence σ that leads to this task. The t^* property ensures that from each marking of the Petri net all other markings are reachable. Even when an OR-SPLIT enables `task-a` it is always possible to enable `task-a` again or enable `task-b`.

In the Justice Department case, once an official record is made, all tasks in the process can be enabled. Some tasks are enabled simultaneously, while others are disjunctive. Via the task t^* there is always a path leading an initial marking to all possible markings of this business process.

Soundness. The last property discussed here is called soundness. A business process has to terminate by placing a token in p_o and at the same time leave all other places empty. This means that each task is disabled from that time on. This requirement is placed on business processes because it is illegal for a business process to be ended and at the same time have some tasks enabled. This would mean that one case token, started in p_i, results in multiple end tokens in p_o. To implement soundness in business processes the underlying formal model must therefore prevent this situation from happening.

In the Justice Department case the violation of this property would introduce a WFMS that allows the possibility of assigning a criminal offence that has already formally ended in for example a summons, to tasks such as the printing of the record. Violating this property therefore implies a violation of proper logistics and presumably the data manipulation: the printed record is needed before it can be handled.

The foregoing properties of business processes have to be analysed in the formal model. The following techniques are introduced to guarantee the properties.

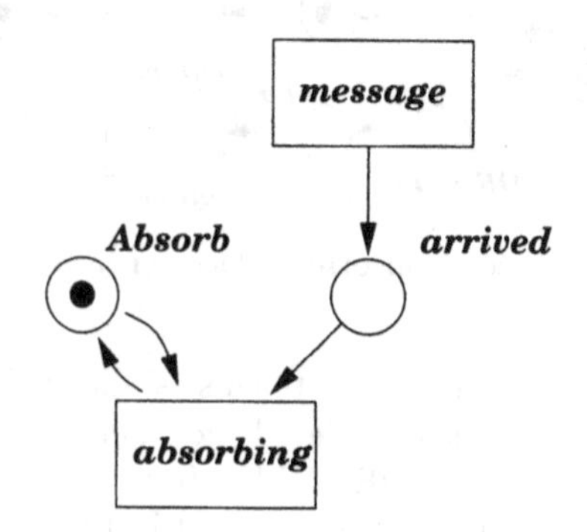

Fig. 25.13. A token absorber is a trap

Techniques: Siphons and Traps. Formally defined, a *siphon* is a set of places S which is a subset of all places P in the model and $\bullet S \subseteq S\bullet$. Thus, when the case leaves this siphon it can never enter again. Similarly, a set of places S which is a subset of all places P in the model is a *trap* when $S\bullet \subseteq \bullet S$.

The techniques of siphons and traps can be used to analyse the Petri net that specifies a WFMS. Both are found for initial markings of the Petri net $M = i$. In Figure 25.13 an example of the use of a trap is shown. In this case, a designer wants to get rid of external messages. This situation sometimes occurs in practice. Places `Absorb` and `arrived` have a set of input tasks: $\bullet S = \{message, absorbing\}$ and a set of output tasks $S\bullet = \{absorbing\}$. This makes $S\bullet \subset \bullet S$. Note that place `Absorb` is also safe.

An example of the use of siphons is shown in Figure 25.14. When a case starts flowing through the business process, all other places are empty. Therefore, when a siphon is found in this situation it means that those places remain empty. Their usage can then be questioned. In the example two places $S1$ and $S2$ are highlighted that are supposed to make up a siphon. A designer of this business process is now supported by a system-executed check on the business process, and can focus on $S1$ and $S2$ to remodel the business process.

Reachability Graph. The reachability graph is a basic analysis technique for the problem of reachability. In the foregoing the properties of a good Petri net for business processes were described. A business process that is constructed with Petri nets can be analysed with a reachability graph.

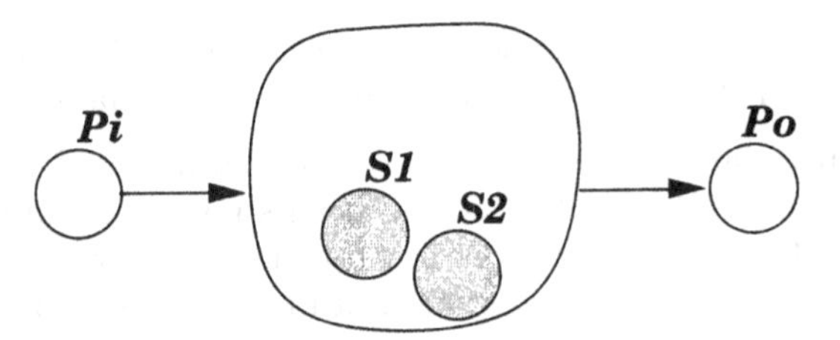

Fig. 25.14. Siphons that show parts of a business process which will remain empty

In this topic an initial marking is needed similar to that for a single case which enters the business process in p_i. Then a reachability graph is constructed to analyse the properties of the Petri net. In Figure 25.15 a simple graph is shown for the example from Figure 25.12 with one case token assigned to place `exa` that enables the task *examine*. Each place that can hold a case token after this state is depicted in the graph.

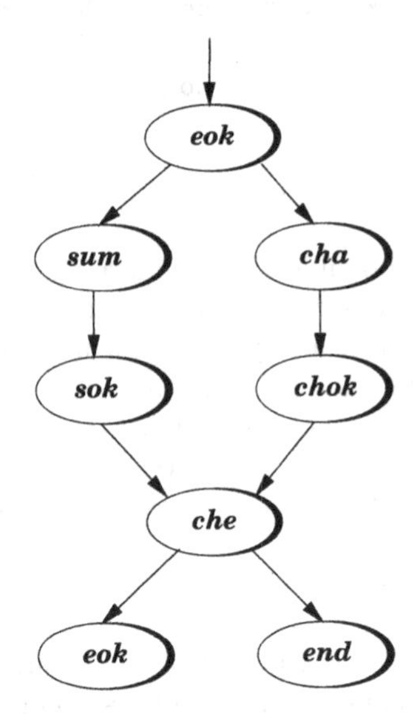

Fig. 25.15. Reachability graph of a single case

This reachability graph shows that there exist two paths from `eok` to `end`, either via `sum`, `sok`, i.e. via the task *summon*, or via places `cha`, `chok` executing the task *charge*.

For this simple Petri net the reachability graph shows some properties discussed previously. However if the number of tasks grows, i.e. a complex business process is modelled, the reachability graph can help a designer to find possible flaws in the model. Note that a typical reachability graph can be very large for these complex business processes. A reachability graph analysis of a partial Petri net is then a practical solution to this problem.

Invariants. Invariants, either place or transition, are used to analyse stable properties of the Petri net. Place invariants give information about the weighted sum of tokens in a Petri net, and transition invariants show which firing sequences lead to its initial state. Both are discussed for business processes.

A case is always started with one token placed in p_i. Then the case can be split into many tokens that signal a task to start. However, there are execution paths in a business process that have their tasks sequentially linked. An important design implication for place invariants is that the token sum on this path always equals 1, i.e. in a sequential flow of tasks there is always one and only one task enabled.

The Justice Department example of Figure 25.9 has the following place invariants when one token in p_i is assumed:

1. P_i `+ rok + ver + vok + sum + sok + cha + chok + sus + suok +` P_o `= 1`
2. P_i `+ rok + com + cok + sum + sok + cha + chok + sus + suok +` P_o `= 1`
3. `ver + vok - com - cok = 0`
4. `com + cok - ver - vok = 0`

On one path it is not possible to have more than one case token, i.e. the possibility of two case tokens in the same place is excluded. As mentioned, this safeness or 1-bounded property serves an important practical goal: for one case a task can not be enabled more than once at a time.

The transition invariants help a designer to list all possible execution paths from the start of the business process to the end. Possible dangling tasks can be detected and the designer gets an overview of the allocation of tasks.

The transition invariants of the Justice Department example show that only three tasks are case dependent; the rest have to be executed for all cases. This is important information when the business process is large and statements about the number and size of its components must be given. In the Justice Department case of Figure 25.9 the transition invariants are (assuming t^*):

1. *record + print + verify + complete + examine + summon + or-join*
2. *record + print + verify + complete + examine + charge + or-join*
3. *record + print + verify + complete + examine + suspend + or-join*

These analysis techniques help a designer to construct proper business processes. In the following some subclasses are introduced that preserve these properties and analysis techniques but also have good computational properties. Limiting the model for a business process to such a subclass is therefore advisable; however, some circumstantial design requirements need elaborate models that fall outside the scope of these subclasses.

Subclasses: Well-structured. The first subclass of Petri nets ensures a correct specification regarding the *logistical properties*. As mentioned, the logistical properties that are used (AND and OR) always come in pairs, i.e. after each SPLIT there is a JOIN of the same type. A well-structured Petri net has its logistical properties correctly paired. Further, the rules that apply are the same as those for the use of braces. For example, an AND-SPLIT can be followed by a Petri net as long as it is closed by an AND-JOIN. In this Petri net new logistical properties can be nested.

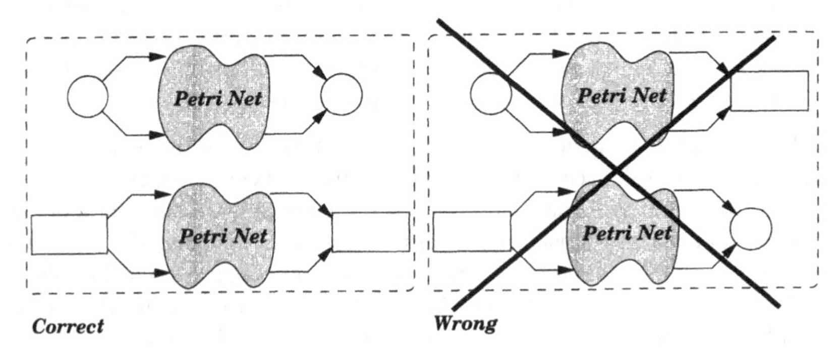

Fig. 25.16. Examples of well-structured nets

A design error occurs when this synchronisation property is violated. When an OR-SPLIT is closed by an AND-JOIN (upper right corner), the resulting specification leads to a deadlock (see also Figure 25.12). When an AND-SPLIT is synchronised by an OR-JOIN (lower right corner) the resulting specification violates the safeness property. One case enters the process and multiple case tokens will leave it. Every task that is linked to this specification receives multiple case tokens of exactly the same real-world customer order. This is an unwanted situation.

The well-structured property therefore helps to design good processes in terms of synchronisation of all branches that occur.

Free-Choice Petri Nets. A second subclass that is recommended for business process design is the (extended) free-choice Petri net, which has either $p_a\bullet = p_b\bullet$ or $p_a \bullet \cap p_b\bullet = \emptyset$. Free-choice Petri nets have been studied for example in [DE95], because they seem to be a good compromise between expressive power and the need for analysis.

Once a case token in a free-choice business process reaches p_o it means that all other places are empty. The main result is that a decision on the logistics of a case is not biased by previously (not) executed tasks.

Both subclasses ensure the expressive power to define business processes. The advantage of both subclasses is that the important property of **soundness** can be proven for a business process in polynomial time. This is an important aspect if the designer wants to check the specification in an interactive manner.

25.4.2 Dynamic Analysis

Dynamic analysis is used when a new business process is analysed for a *number* of cases that have to be handled. It must deliver a statement of the implementation of business performance goals. In analogy with industrial design, the behaviour of the business process should *conform to the specifications*. To achieve this a designer wants to test different options and choose an optimal procedure for a given situation. The behaviour of a business process is tested with Petri net techniques on the following properties.

Simulation. The behaviour of a business process is investigated by analysing, for example, accumulated waiting time, service time, throughputs, and costs. Recent studies have shown that on average more than 95% of the time a case spends in a business process is 'wait-time' leaving the reminder for the activities. Simulation is a practical way to obtain information about the behaviour of a business process. Although it lacks a mathematical proof, the results of simulation can be of great value when used correctly.

Simulation is used to analyse the behaviour of a specification for different combinations of market demand, structure, and resource management. The question to be answered is: what is the throughput[4] of a certain combination, and is that desirable or should improvements be made? The throughput is the total time it takes to get an individual case from p_i to p_o. The following parameters are required during simulation:

- The expected market demand, used to generate a specific number of cases that arrive with a certain distribution. This is a statistical function based on historical or other data.
- The structure of the specification in terms of a business process definition. Different structures change the throughput, e.g. a sequential execution of tasks is probably more time-consuming than parallel execution.
- The required resources and their expected operation time per activity transition.

Why are resources important? Since resources provide input to tasks they are therefore another pre-condition from a conceptual point of view ([AHH95]). If resources, such as humans, forms, computers, or telephones are analysed during simulation, they are added to a model as outlined in Figure

[4] Throughput is often mentioned as an important performance indicator for the dynamic behaviour of business processes.

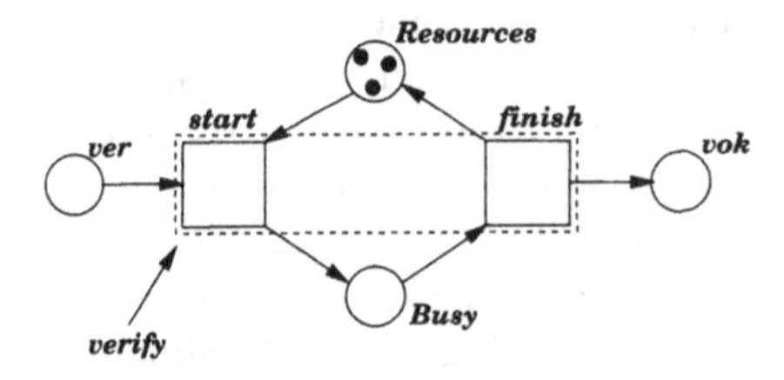

Fig. 25.17. Resources modelled in a Petri net

25.17. Here, task verify has a *start* that is enabled when there is a case in its logical in-basket (`ver`) and a proper resource is available (`Resources`). During the execution of this task the resource is kept in `Busy`. It is now no longer available for other incoming cases. Once completed, i.e. when *finish* fires, the resource is put back and is available for a new case. The case that was handled is placed in the out-basket (`vok`).

The use of resources decreases the ability to analyse the business process when one resource place can enable multiple tasks. It violates the free-choice property and solutions will therefore be harder to find.

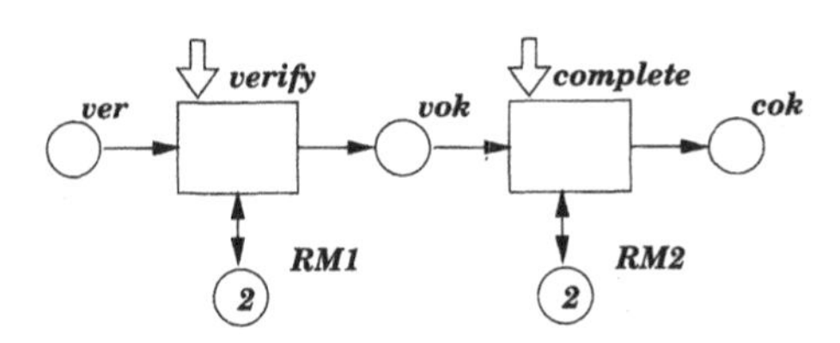

Fig. 25.18. Scenario-1

Next some examples illustrate the use of simulation in business process design. Suppose a designer creates *scenario-1*, shown in Figure 25.18. This is the Justice Department example in its initial state with two tasks *verify* and *complete* that are specified to be sequential. This is the **structure** that is specified. The **market demand** is stated here as 24 cases that arrive in p_i per hour (according to a Poisson process). Both tasks *verify* and *complete* have two **resources** available, shown in their Resource Managers (*RM1, RM2*). The operation time is 4 minutes for each task and is independent of the resource.

When this situation is simulated, a large number of cases are generated according to the market demand. The simulation shows that the average throughput of scenario-1 is 22.2 minutes, meaning that on average a case will spend 22.2 minutes in the model from its start at p_i to completion in p_o. The actual operation time of this model is 8 minutes when each case can be handled without delays. Thus scenario-1 has a large amount of *overhead* time, i.e. time other than execution time spent on the case.

A designer therefore wants to test other possibilities in the search for an optimal solution with these constraints. In scenario-2 another model is tested that executes these tasks in parallel, while both the market demand and the resources are kept constant. Thus, a change in structure is made based on the AND-SPLIT concept, shown in Figure 25.19. Simulating this model shows that the throughput is decreased to 15.0 minutes, an improvement of 7.2 minutes over scenario-1. Parallel execution of tasks is one of the main contributions of workflow management in decreasing the throughput of cases, since tasks have no physical limitation on their ordering (in contrast to the processes based on physical document flows). Therefore this result is interesting, however if other parameters are changed further performance improvement may be possible.

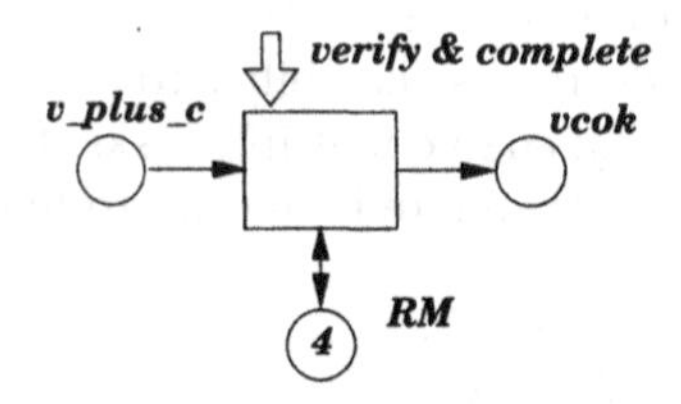

Fig. 25.19. Scenario-2

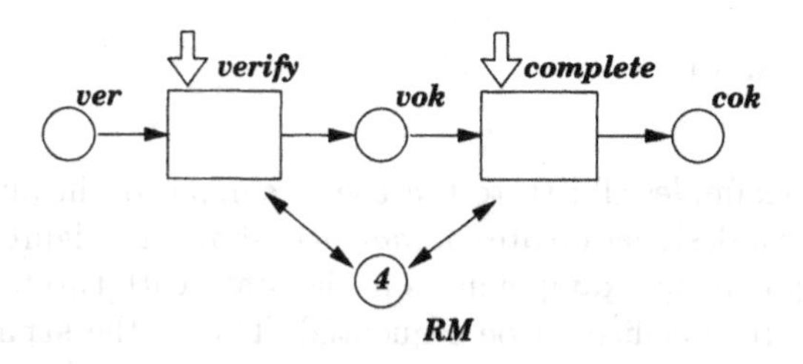

Fig. 25.20. Scenario-3

In scenario-3, shown in Figure 25.20, the market demand and sequential structure are kept constant and the resources are assigned differently. In this case, the resources can perform either activity and the total operation time is the same. Simulation shows that the throughput is decreased to 14.0 minutes. In this scenario the resources are *flexible* as opposed to the *specialised* resources of the previous scenarios.

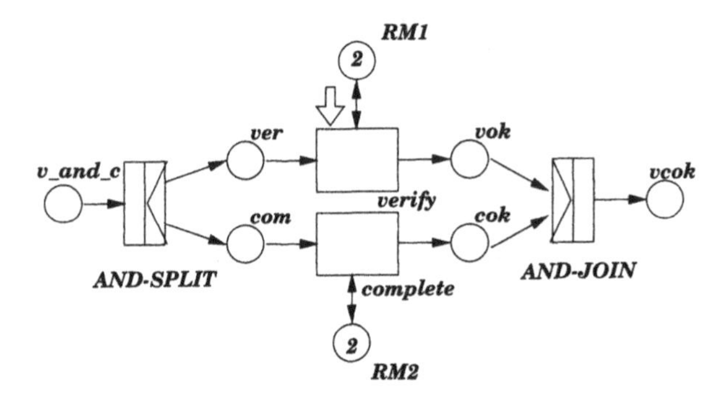

Fig. 25.21. Scenario-4

This scenario introduces the possibility of scenario-4, shown in Figure 25.21. In this scenario the two tasks are executed in combination. This eliminates set-up time between verification and completion and decreases the total operation time to 7 minutes. The elimination of set-up time is also a major time saver, however, when tasks are integrated the ability to switch between tasks is lost and this loss of flexibility can introduce a scheduling problem. When the resources are assigned to tasks as displayed in scenario-4, the throughput is decreased to 9.5 minutes on a total operation time of 7 minutes.

It has been shown that the initial business process model can change either in structure or resource management in order to achieve better dynamic behaviour. The changes in structure are facilitated by the use of different logistical properties. These transformations can be applied automatically without changing the properties of the initial business process, as stated in [Aal95]. The resource management policy influences performance. As showed, general resources allow a lower throughput than specialised resources. The choice between generalised resources and specialised resources follows from organisational debate and normally is a given input for the designer. However, suggestions for changing this resource distribution can be given based on expected performance indicators that are found using simulation.

Table 25.1. Results from the simulation studies

Scenario	Description	throughput	service time	waiting time
1	sequential	22.2	8.0	14.2
2	parallel	15	4(8)	11(7)
3	compose	9.5	7	2.5
4	flexible resources	14	8	6
5	triage	27.8	6.6(12)	21.2(15.8)

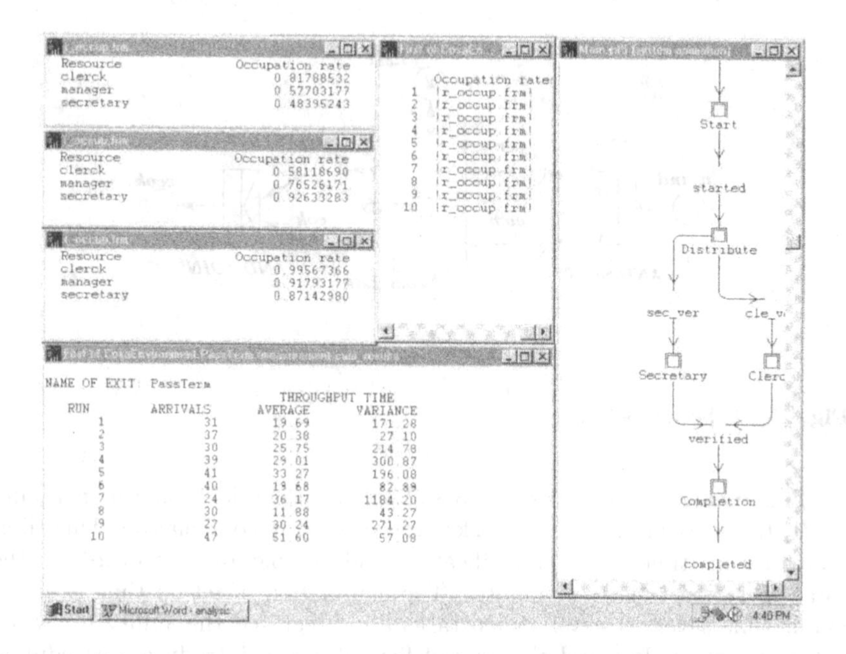

Fig. 25.22. Simulation screen for scenario-5

The final example of simulation uses a *token colour* to make a distinction between *difficult* and *standard* cases. The business process is shown in Figure 25.22 and splits the task *verify* into two different subtasks. The figure also shows some windows obtained from a simulation run using the business process definition tool from COSA with ExSpect as simulation environment. The business process is explained first, followed by some details shown in the simulation screens.

Cases are generated in the first task, called *Start*. This task is added to our example to set the simulation parameters for the throughput, resources etc. In the following task, *Distribute*, 25% of all cases that pass are marked as *difficult* and the remainder as *standard*. The difficult cases are sent to *Secretary* and will take on average 8 minutes to process. The standard cases flow to *Clerck* and will take on average 2.6 minutes to process. The OR-SPLIT is thus based on the case colour defined in M, stating its level of difficulty.

The branches are merged using an OR-JOIN and then *Completion*, which will last about 4 minutes, is enabled. For task *Clerck* one resource called *clerck* is available. This is not shown in the model; it is placed in the data specifying each task. Similarly, *Secretary* has one resource called *secretary* and *Completion* has two resources called *manager*.

The simulation windows show the following information. The window on the right shows a part of the business process model that is explained in the foregoing. The circles are the states and the squares represent the tasks. During simulation one can see cases moving through the model from state to state. The three upper left windows show resource occupation rates for the subruns of this simulation. Using the simulator in this way, one can zoom into the performance of subprocesses (using the hierarchy) and change the resource distribution before performing a re-run. Underneath the resource occupation windows, a window displays for each subrun the number of cases that reached the end of the business process (`completed`) and the average time that these cases spent *in* the process (throughput).

The result of this simulation is that the use of triage is not as straightforward as one may expect. Because the resources that operate on the subtasks of *Verify* are specialised, the total throughput of the process increases. Changing the level of difficulty and the resource sharing can change the performance. Better solutions are found interactively by rerunning the simulation with different parameters.

Scheduling Principles. In Figure 25.4 the relation between a case, a task, and logistics is shown. When a designer of a workflow specifies a business process, priority rules have to be used to assign cases to tasks. These priority rules specify which case is handled by a resource at a certain time. The priority of a case can be adjusted by employees, and cases are ordered according to this priority. The order depends on the scheduling principle that is used and this can differ for each resource.

Which scheduling principle is preferred is not known *a priori*. Simulation can be used to test different scheduling principles when all other parameters are kept constant. Note that the scheduling principle may differ for each resource manager. This topic is not further addressed here. Below some scheduling principles are listed, based on a complete overview given in [Hau89].

First-In-First-Out (FIFO): The most simple and effective scheduling principle is to assign a resource to the first case available. The cases will thus be ordered by arrival at the task. This is an attractive rule since no additional information is needed to make a decision and it is therefore easy to implement.

Last-In-First-Out (LIFO): If the scheduling principle chooses the last case arrived, one speaks of LIFO queuing. In some situations this can lead to a higher service-level, i.e. more cases reach p_o before their due date (at the cost of some other cases which are delayed longer).

First-At-Shop-First-Served (FASFS): When resources must be assigned to tasks based on their ordering at their entrance in p_i, the FASFS base priority rule is chosen. Here a resource is assigned to the oldest case in the business process. FASFS can be considered as an overall FIFO scheduling principle for the process.

Shortest-Processing-Time (SPT): If the amount of labour required for each case is known *a priori*, the case with the smallest processing time gets the resource first. In general, the average throughput decreases when this base priority rule is used. It is based on the assumption that the throughput of a case should be proportional to the expected processing time. If the case with the longest processing time gets the resources first, the rule is called Longest-Processing-Time (LPT).

Shortest-Rest-Processing-Time (SRPT): If the amount of labour required for a case is decreased by the amount of processing time already consumed, the result is an ordering of cases by shortest remaining processing time. This scheduling principle minimises the work in progress. Note that operations have to occur on the management data of a case, which makes this solution computationally more demanding.

Earliest-Due-Date (EDD): If the due date on which the case has to be ready (i.e. has to be in place p_o) is entered in the management information part, the policy can be to choose the case with the earliest due date.

These scheduling principles have an impact on the total throughput of a business process. Combinations are found in many ways, e.g. with resource management functions or with transitions that balance the workload by assigning the case to groups of people with the lowest level of work in progress. However, typical Workflow Management Systems are constrained in their scheduling of resources by other considerations:

- Function separation: When two tasks for one case are not allowed to be handled by the same person, this constraint is applied. It limits the number of available resources. In the financial industry in particular, this is an important constraint.
- Case workers: Opposed to function separation is this constraint which applies when sequential tasks must be handled by the same person. This reduces set-up time and leads to a reduction of time and errors.
- Special assignments: Similar to the case worker constraint, this constraint assigns people to particular tasks for particular cases. For example, follow-ups of the cases handled by person p are also handled by p.

Animation. When a model is animated, some real-world aspects are visualised during simulation. The use of animation is especially helpful when the model must be validated by end users. Thus, animation helps to clarify the underlying formal definitions.

Within the field of workflow management, the following aspects are generally seen during animation:

Forms flow through a Petri net instead of the token 'bullets'. When different colours are assigned to forms, a different priority is assigned to them.

People are substituted in and managed by a resource manager. This resources manager assigns resources when cases appear. Resource managers can manage different classes of people, for example clerks, typists, and office managers.

Places and arcs are removed from a model for animation. One sees forms 'flowing' from one activity to the next without the enabling places in between. If places are visualised it is done to show a queue.

Colours and signs denote performance indicators such as wait times and occupation rates. The colours and signs change depending on values of the model. For example, the colour of a resource manager is red when understaffed and blue when overstaffed. When the number of tokens in a place has reached a maximum it can change shape.

Plant layouts are scanned and placed in the background of the screen. Each location now gets either its queues (places) or tasks (transitions) assigned. If the animation options above are added, the screen come to life.

Although the display of the elements is conceptually irrelevant, it is important for validating the model. Note that system engineers have a different perspective on the workflow management world than participants from the user organisation. Animation can bridge the gap between these groups to obtain a vision shared by both designers and users.

25.5 Lessons Learned: The Sagitta-2000 Case

Both the technological and organisational advantages of workflow management concepts have been recognised by the Dutch Customs Department. In 1994 the Customs Department started the development of a nationwide information system for handling customs declarations, known as *Sagitta-2000*. A primary goal of this project is to define the core customs declaration processes and implement them in workflow software. The estimated number of customs declarations per year is 10 million, with some peak days when more than seventy thousand declarations must be handled. The number of direct customs officers is well over five thousand. The complexity of this type of process (an average customs declaration can take up to fifty activities to complete), combined with a frequent change in government regulations for customs declarations, makes *Sagitta-2000* a very ambitious project.

The approach taken was to separate information on the logistics of declarations from information on task execution. A Workflow Management System (WFMS) incorporates logistical data and was therefore a possible solution from the start of the project.

To allow implementation on any future platform, the business processes were modelled with a diagramming technique based on Petri nets. The designer team consists of system engineers, management consultants, staff members, and end users, and the current number is close to one hundred. The number of business process models constructed is very large, and each model is very complex. Although the concepts of tasks, cases, and resources are used there is much specific terminology for each model. Figure 25.23 shows

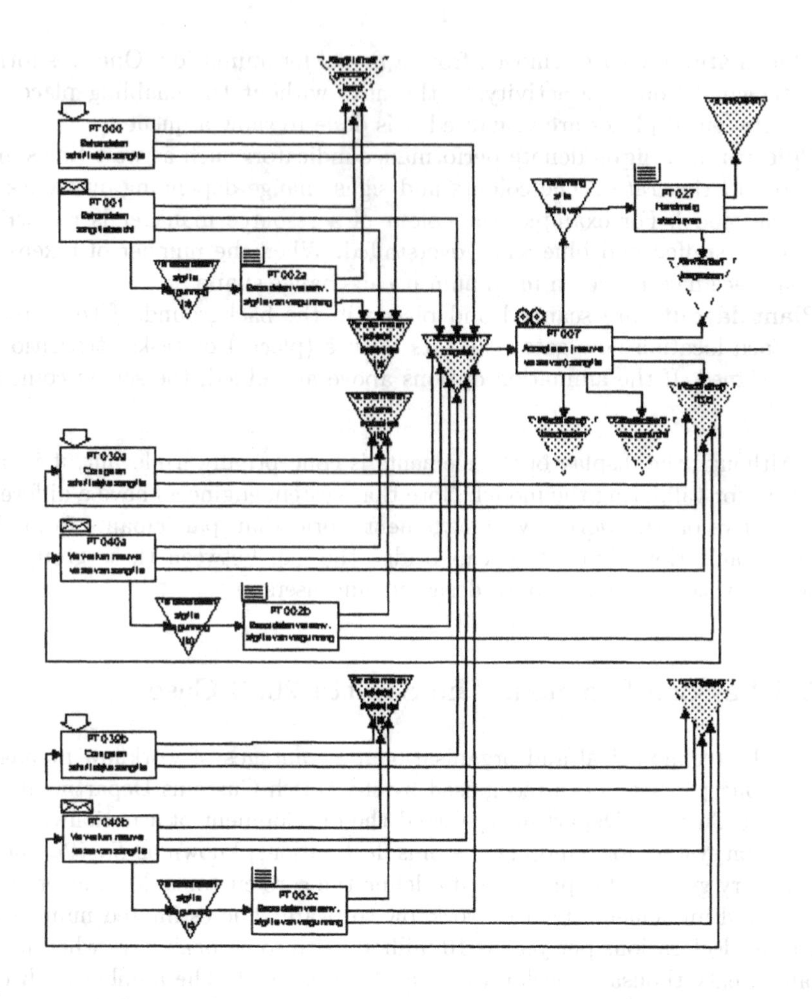

Fig. 25.23. An example of the Sagitta-2000 workflow model

an example of the diagramming technique, obtained from the *Sagitta-2000* project.

This part of the process shows the ordering of tasks for checking individual customer declarations. The descriptions are given in Dutch. The meaning of the elements is given below.

1. Tasks are modelled as transitions. To close the gap between conceptual modelling and implementation, the activities are defined with a *trigger type*. This trigger is used to couple the supporting application in a later stage. The number of the task corresponds to the mainframe application which supports the task.

2. Tasks are assigned to work-stores. These work-stores correspond directly to places, since they enable activities when filled with a token. They are depicted by a triangle for other reasons.
3. Routing of tasks is based on the building blocks presented in the WFMC Glossary [WMC94]. Thus the modelling technique shows ANDs, ORs, SPLITs, and JOINs.

During the modelling process the selection of WFMSs to support the implementation was made. The complexity of the process led tor a client/server architecture. After careful evaluation it became clear that both local Workflow Management Systems, based on client/server technology, and the mainframe workflow engine had to be Petri-net-based. Although standard client/server-software based on Petri nets is available, a Petri-net-based mainframe workflow engine is being developed since no available product has the required functionality. At present the mainframe workflow engine is being built according to the Petri net concept described above.

This practical experience supports the perspective of Petri-net-based standardisation of workflow management concepts. These are preliminary results from the *Sagitta-2000* project which is not yet complete. When the project is finished the results concerning cost savings, development time, performance indicators, etc. will be published.

25.6 Conclusion

The concepts that define workflow management can be specified by Petri nets. This provides an essential function for verifying the correctness of Workflow Management Systems. The easy translation of workflow concepts, such as tasks, cases, logistics, and resources, to the Petri net model enables both specification and the use of a wide range of analysis techniques. The analysis techniques are valuable if the translation makes sense to the people designing and using the Workflow Management System. Three reasons to use Petri nets for workflow management are summarised here:

1. **Formal semantics combined with a graphical nature.** The formal semantics guides designers to build business processes that are correct. This is a necessity because the specifications are used to execute the business process. The user organisation therefore must also participate in the design of the business process. Many users appreciate the Petri net approach and in the *Sagitta-2000* case it helped both engineers and large user groups to communicate their goals.
2. **State-based instead of event-based.** The explicit representation of states plays an important role in the design of WFMSs. Besides the clear meaning in relation to real-world entities, such as mail-baskets or batches, it also provides an essential control property. When it is not known *a*

priori where a case should be routed to, its state defines all possible alternatives, as shown in Figure 25.8. It is not possible to represent this control property with event-based techniques.

3. **Abundance of analysis techniques.** Analysis techniques are available to check properties of the specification based on its structure. Therefore errors are reported to the designer before the specification is put into practice. When dynamic analysis techniques are used, the specification is tested on different (expected) market demands, structures, and resource management. The abundance of analysis techniques enables the design of a correct business process that also has expected dynamic properties.

These three reasons summarise the need for Petri-net-based Workflow Management Systems. This research will be extended in many ways, such as: adaptation of workflow process models to new (enhanced) models without violating correctness; enhanced modelling techniques for defining tasks in a generic way; and using simulation to find new and better ways to execute business processes. The Petri-net-based approach has been proved to be an enabling technology for workflow management.

26. Telecommunications Systems*

26.1 Overview of the Domain

A telecommunications system enables communication between remote entities in order to exchange information. It is composed of two subsystems, namely the transport network and the processing system. The transport network is the set of communication resources that enable information transfer between the communicating entities. The processing system is the set of computing resources and programs that control and manage the transport network on the one hand, and that implement the communication software on the other hand ([SKL94]). This software is designed according to syntactic and semantic rules called protocols. The development of these protocols constitutes protocol engineering, which can be defined as the specialisation of software engineering for communication software. Protocol engineering encompasses the set of techniques, methods, and tools that enable the production and exploitation of communication software with an industrial control ([Raf95]). This process of production is organised according to the classical stages of the development cycle, namely specification, design, implementation, deployment, and maintenance.

In order to ensure software quality, validation and formal verification are required. Validation consists of verifying the global coherence of the specification and design, while verification consists of formal proof of expected structural and behavioural properties. Validation and verification must be applied to each step of the development process, since the later an error is detected, the harder its correction ([Boe84]).

In the case of protocol engineering, these activities are strongly dependent on the complexity of the telecommunications system. Examples of this complexity are the system size, the hardware and software heterogeneity, the real-time aspect since the system has to operate and to be available permanently, and the guarantee of a correct operation. Moreover, the evolution of the telecommunications system is also a factor of complexity because it extends the system functionality. From the provision of voice-transport-oriented service, i.e. the basic call service, telecommunications services have evolved to include today services such as freephone or call forwarding. Nowadays,

* Author: M.-P. Gervais

a telecommunications system is a service network. As the number of these telecommunications services grows, and increasingly complex interactions appear among them, the architecture of the telecommunications system has to be adapted in order to provide adequate support. Therefore a new architecture known as Intelligent Network (IN) has been created.

IN is an architectural concept allowing the rapid introduction of new telecommunications services in any kind of network. Its goal is to provide a flexible and open architecture that facilitates and accelerates service implementation and provisioning in a cost-effective manner, and in a multi-vendor environment. A first set of recommendations concerning the IN, named IN Capability Set 1 (IN-CS1) [ITU93], has been produced by the International Telecommunication Union (ITU), and a second one is ongoing (IN-CS2).

Before detailing the current IN architecture that is used in the remainder of this chapter, it should be mentioned that studies are ongoing to define the long-term IN architecture. The Telecommunications Information Networking Architecture (TINA) is one of the most promising conceptual frameworks ([DNI95]). TINA studies are performed by the TINA Consortium (TINA-C) which is a worldwide consortium of network operators, and telecommunications and computer equipment suppliers. Its aim is to define and validate an open architecture for telecommunications service that provides a set of concepts and principles to be applied in the design, processing, and operation of telecommunications software. Telecommunications services are treated as software-based applications that operate on a distributed computing platform, which hides the details of underlying technologies and distribution concerns. TINA makes use of recent advances in distributed computing, especially the Open Distributed Processing (ODP) standardisation works. ODP is a standard developed by the International Standardization Organization (ISO) and the ITU. Its aim is to provide a conceptual framework for rigorously specifying a distributed system architecture in heterogeneous environments. ODP makes use of object-based and client-server approaches and defines a reference model ([ITU96]). So TINA is a specialisation of this RM-ODP to the telecommunications domain.

26.1.1 The IN Architecture

The IN architecture defined by ITU is in keeping with the Plain Old Telephony System (POTS) context dedicated to voice transport. Its principle is separation between service control and call control. It is composed of functional entities that are mapped onto physical entities. The description here is limited to the entities that are relevant for the case study presented in Section 26.3. According to the separation principle, some functional entities are related to call control, namely the Call Control Function (CCF) and the Service Switching Function (SSF), whereas others are related to service control, namely the Service Control Function (SCF) and the Service Data Function (SDF).

- The Call Control Function (CCF) is responsible for the call processing. It corresponds to the classical functionality of a switch.
- The Service Switching Function (SSF) is the interface between the CCF and the SCF. It enables the CCF to be controlled by the SCF.
- The Service Control Function (SCF) contains the service logic, i.e. the programs performing the service functionality. It is responsible for their execution.
- The Service Data Function (SDF) stores and manages data services, i.e. data related to the services.

Functional entities can be mapped onto physical entities in different ways. Typically, the Service Switching Point (SSP) implements the CCF and the SSF. This is an IN switch. The Service Control Point (SCP), which implements the SCF, is a real-time computer with a very high availability. The Service Data Point (SDP), which implements the SDF, is a database system. However, the SCF and SDF can be implemented together in the SCP (see Figure 26.1). Relations between SSP and SCP are based on the Common Channel Signalling #7 (CCS7) [ITU88].

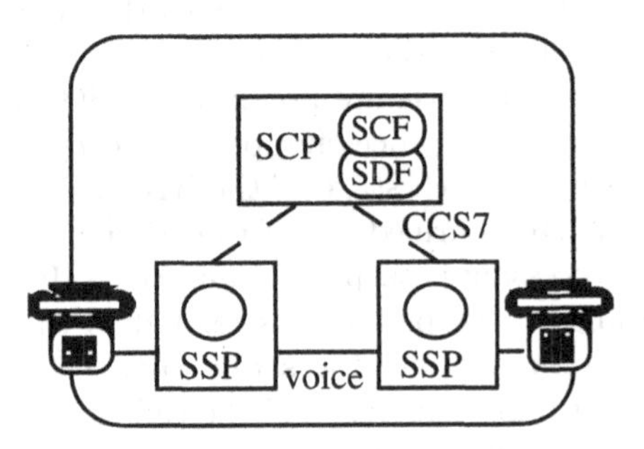

Fig. 26.1. IN simplified architecture

26.1.2 The IN Service Processing

The IN services are built upon the basic call service. This means that they offer advanced functionality based on and enhancing the basic call, and that they follow the basic call processing of the telephony service. Examples of IN services are the freephone, credit card calling, and the virtual private network. An IN service is processed as follows. The user picks up the phone and dials a number. The SSP performs the basic call processing. During this processing there are some points, the Detection Points (DPs), at which the SSP can detect that the call contains an IN service demand. Then it suspends the basic call processing and transfers the control to the SCP, which performs the service logic. Then the SCP sends back instructions concerning further processing of the call in the SSP.

The SSP is then a switch that is able to detect a service demand through the DP processing. DPs are located at stages in the basic call processing where events can be detected. Several types of DPs are defined. Here, the description is limited to the types of DPs that enable the SSP to identify an IN service demand[1] and to request the SCP for invoking the corresponding service logic. Several DPs of this type are defined depending on the stage at which they occur in the basic call processing. They are identified by a number and a name. For example DP3, named Analysed_Information, corresponds to the point located at the stage where the SSP has analysed the dialled number. The number dialled by a user can be an ordinary phone number or a specific number corresponding to an IN service demand. For instance, a freephone number is a logical number identifiable by a special prefix (e.g. 800 in the USA or 0800 in France and Germany). Then DP3 processing enables the SSP to recognise such special numbers.

DPs are characterised with attributes that determine their types. Moreover, a DP has a state, and some criteria are associated with a DP. The state of a DP is armed or disarmed. A DP must be armed in order to be processed. If it is not, the SSP continues the call processing. The DP is armed through service provisioning. The criteria are conditions that must be met in order to suspend the basic call processing and to notify the SCP that the DP was encountered. Examples of criteria are prefixes (e.g. 800) and specific calling or called numbers. So the DP processing performed by the SSP for a given call consists of checking, at each stage of the basic call processing at which a DP is located, if the DP is armed and if the criteria are met. In the latter case, the basic call processing is suspended and the SSP transfers control to the SCP by providing it with parameters related to the requested service. Examples of these parameters are the service key, i.e. an identifier of the requested service, the calling number, and the called number.

The SCP is responsible for executing the program corresponding to a service, namely the Service Logic Program (SLP). The SCP can be dedicated to one service, i.e. it performs only the SLP related to this service, or it can implement several SLPs corresponding to different services. The SCP is composed of several modules, including a library of SLPs, but in order to simplify the description, it will be considered that an SSP request is processed as follows: Based on the service key contained in the SSP request, the SCP calls the appropriate SLP to execute the service logic. During this execution, a dialogue may be needed with the user to get information such as an authentication code. In this case, the SCP controls the SSP by giving it orders related to the actions needed for establishing such a dialogue. At the end of execution, the SLP returns a result to the SCF, which it uses to transmit an order to the SSP concerning the basic call processing. Three types of orders are identified:

[1] These DPs are named Trigger Detection Point-Requests (TDP-R) in the IN terminology.

- Continue with same data: the service execution has not modified the initial data. The SSP can resume the basic call processing with the same data, i.e. the calling number and the called number.
- Continue with new data: the service execution has provided new data by modifying the initial ones. So the SSP has to resume the basic call processin, taking into account these new data.
- Clear the call: following the service execution, the call must be cleared by the SSP. This response is provided for example in case of errors occurring during the service execution.

26.1.3 Conclusion

IN constitutes an important evolution of telecommunications system functionality since it is a service-oriented architecture. A telecommunications system is no longer simply a network that transports information and implements protocols. It now provides advanced services, i.e. applications that improve the basic call. So it is no longer concerned only with protocol engineering, but also with service engineering.

The next section examines the way in which formal methods are used in such an architecture.

26.2 Motivation

Formal methods have been used ever since the activity of designing protocols appeared. In fact, a formal description is required to express the system functions unambiguously, to analyse and to validate the description in order to detect design errors, and to develop software tools resulting from the formalism. Generally speaking, two types of formal techniques can be identified ([Jua95]):

- The models represent the main mechanisms but do not totally represent a specification. On the other hand, an exhaustive analysis of the represented mechanisms (verification) is feasible. Petri nets are an example of such models.
- The languages represent all of the mechanisms, but do not enable an exhaustive analysis of the specification, instead just simulation or test sequences. Examples of languages are Estelle [BD87] and LOTOS [BB88] developed at ISO by the Formal Description Techniques working group or SDL [Bra96] developed at ITU-T. It should be noticed that in the telecommunications context, the term Formal Description Techniques (FDTs) is commonly used to refer to these three languages.

These two types are not opposed, but each of them addresses a specific scope of concerns. Modelling enables the verification of the basic mechanisms of communication, especially synchronisation. For that purpose, Petri nets have been widely used in the protocol engineering area since they have a very good expressive power for specifying parallelism, synchronisation, and causality ([Dia82, JAD84]). Moreover, they enable verification properties related to control and synchronisation. On the other hand, using FDTs enables the representation and testing of implementation-oriented mechanisms for which modelling would be not suitable. Thus FDTs are also widely used in the protocol engineering field.

SDL is a Specification and Description Language standardised by ITU. The basis for the description of behaviour is communicating Extended State Machines that are represented by processes. Communication is represented by signals and can take place among processes, or between processes and the environment of the system model. Data are represented by algebraic abstract types. Estelle is close to SDL since it also uses Extended State Machines communicating through FIFO channels, but the language Pascal is used to represent data. LOTOS is based on the CCS (Calculus of Communicating Systems) formalism and the abstract types of data ACT ONE.

As mentioned in the previous section, due to the evolution of telecommunications systems, service engineering is an emerging domain of interest. The aspects related to telecommunications service provision pose new problems, especially in terms of compatibility and interactions among services. When a new service is introduced into an operational telecommunications system, one must be able to evaluate its impact on the system and the existing services. A question to be addressed is: "Are the features of this new service compatible with the existing ones, or do they create conflicts?" So the specification and design of the services require undertaking validation and formal verification to manage their quality assurance.

For specifying telecommunications services, SDL has proved to be a popular language. However, object-oriented (OO) approaches are being used more and more. According to the ODP reference model, a telecommunications service is designed as a set of interacting objects. Actually, the most recent version of SDL integrates the OO paradigm. However, both SDL and the OO approach are hampered by the lack of analysis tools for evaluating and validating the dynamic aspects of the system at the specification stage. They do not provide validation and verification features to evaluate the correctness and compliance of the model. Motivation to use Petri nets is the need for the designer to have a validation and verification of his specification. Some versions of Petri nets adapted to object-oriented features are proposed, such as interacting modular nets and hierarchical nets. Combination of the OO paradigm and Petri net concepts produces a modelling tool that has profitable aspects of both. So the use the Object Formalism Class (OF-Class), presented in Section 10.3 of Part II, and which integrates the Petri net proving power with an OO methodology is proposed.

The next section is devoted to a description of a case study illustrating the design methodology of an intelligent network telecommunications system into which a new service must be introduced.

26.3 Design Methodology

We address the problem of the introduction of a new telecommunications service into an operational telecommunications system. The example is an Intelligent Network (IN) telecommunications system that provides the Call Forwarding Unconditional (CFU) service and into which a new service is added, namely the Terminating Call Screening (TCS) service. With such an example, it is possible to illustrate one of the benefits of formalising application creation,i.e. the ability to detect errors that can occur during such a process. It is considered that a telecommunications service is a distributed application composed of a collection of interacting objects. The description of the environment supporting the application is also object-based. This allows one to have a modular representation of system components.

In order to be able to validate and to verify a telecommunications system described with an OO approach, the design methodology used is the Object Formalism Class (OF-Class) already mentioned. Our example is composed of three steps. Firstly a basic telecommunications system that provides the most elementary service, namely the basic call, is modelled. Then this model is modified to introduce the Intelligent Network (IN) capability, which is the ability to detect an IN service demand, and a telecommunications system that provides the IN Call Forwarding Unconditional (CFU) service is specified. Finally this model is made richer by the introduction of the Terminating Call Screening (TCS) service in order to get an IN Telecommunications System.

Once a syntactically correct OF-Class model of this system is obtained, it is automatically transformed into OF-CPNs. This process is recalled at the end of this section.

26.3.1 The OF-Class Model of a Basic Telecommunications System

A telecommunications system provides the users with the ability to communicate. The users manipulate phones that are attached to the network, which is composed of lines and switches (Figure 26.2). The telecommunications system operation is based on the exchange of messages or signals. The specification of such a system allows for the validation of these exchanges and for verification that the system behaviour is in conforms to the requirements.

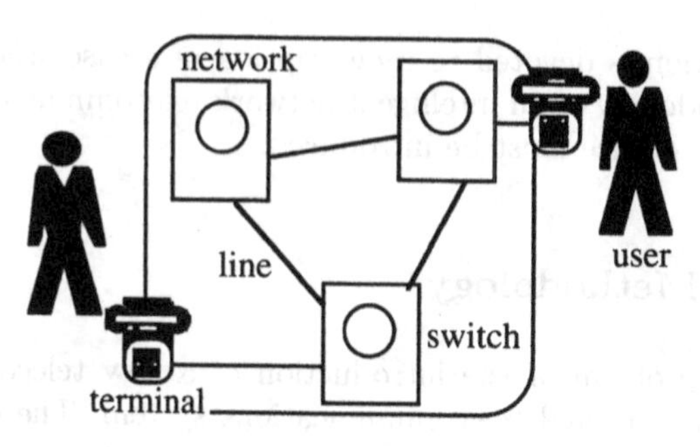

Fig. 26.2. Global view of the system

- The **user** initiates or receives calls, i.e. he[2] is the caller or the callee. The user behaviour is described independently of the telecommunications system since the user does not belong to the system. He can pick up the phone and dial digits if he is the caller, or pick up the phone when it rings if he is a callee. Another action is to hang up the phone either by choice or because he receives a busy tone.
- The **phone**, also named the **terminal**, is the interface between the user and the network. It represents the access point to the network for the user: from the terminal, the user can send or receive calls. The terminal must provide the user with the ability to perform the actions mentioned above.
- The **network** is a graph composed of switches and lines. The switch enables a connection between two users. It must be available in order to capture a call demand coming from a terminal. It must also control the establishment and release of the connections, and manage the lines.
- The **line** is represented by the messages and signals exchanged between the terminal and the switch. One line per terminal is modelled.

Communication between two users encompasses three steps: establishment, transfer, and termination. The communication progress described here corresponds to a simplified scenario (Figure 26.3). It is assumed that the network is composed of a single switch and two lines. Hence, no routing function is taken into account by the switch. Other assumptions are that a party corresponding to the dialled number always exists, and that the user dials correctly. So the switch does not manage dialling errors. Moreover, charging aspects are omitted. Finally, a unified process for call termination is adopted as explained below.

- The establishment phase deals with the connection of two users, the caller and the callee. The caller picks up the phone, which sends an off-hook signal to the switch. The switch answers by sending a dial tone signal. Once the user hears the dial tone, he dials the number, which is transmitted

[2] "He" should be read as "he" or "she" throughout this chapter and indeed throughout the book.

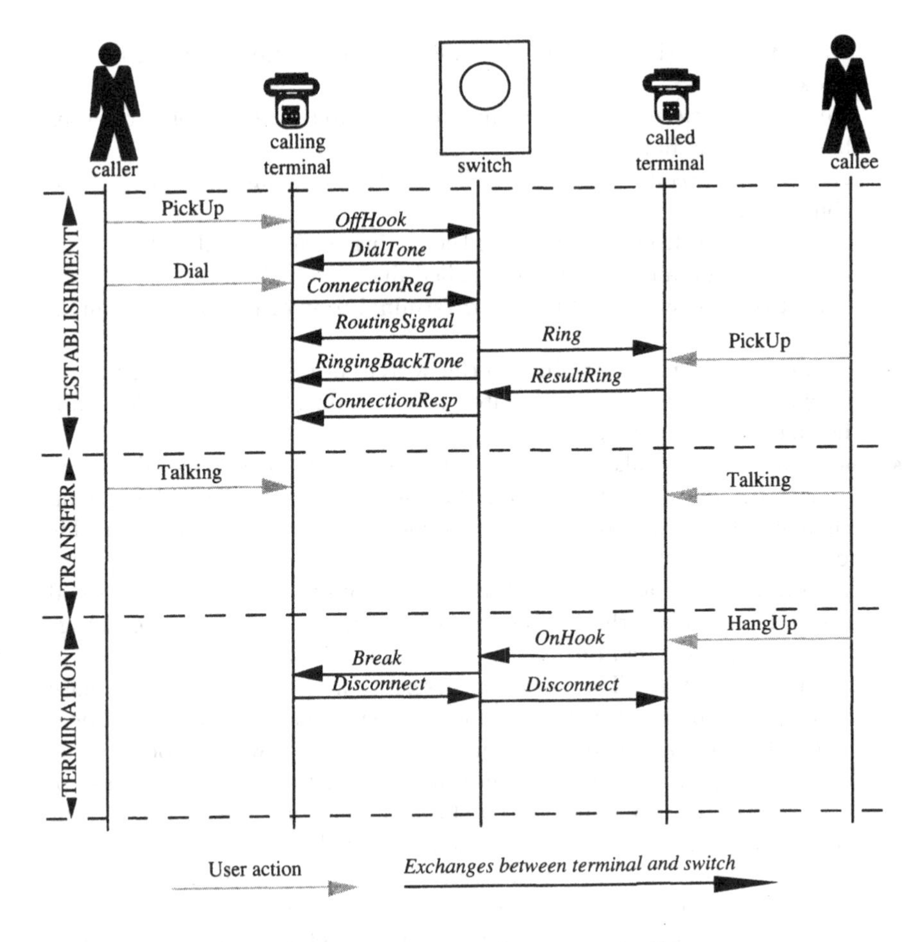

Fig. 26.3. An example of the three steps of a communication

by the terminal to the switch. Based on this number, the switch selects the corresponding line while sending back a routing signal to the calling terminal. If the called-line state is free, the switch sends a ring to the called terminal and a ringing back tone to the calling terminal. Once the called user picks up the phone, the switch establishes the connection, i.e. changes the line state. No connection is established if the line is not free, or if the called user does not answer the ringing phone after a period of time. In both cases, the switch sends a busy tone to the calling terminal. The actions of a user during this phase are *pick-up* and *dial*. The exchanges between the terminal and the switch are:

- *off-hook*: from the calling terminal to the switch, following a user request for a call;

- *dial-tone*: from the switch to the calling terminal after receiving the off-hook signal;
- *connection request*: from the calling terminal to the switch, once the user has dialled the number;
- *connection response*: from the switch to the terminal, in response to a connection request;
- *routing signal*: from the switch to the calling terminal while establishing the connection and selecting the called line;
- *ring*: from the switch to the called terminal to inform it of an incoming call;
- *result-ring*: from the terminal to the switch, in response to the ring;
- *ringing back tone*: from the switch to the calling terminal while establishing the connection.

• The transfer phase takes place once the connection is established. It corresponds to conversation between the users, represented by the talking exchange. This phase ends as soon as one of the two users hangs up his phone.

• The termination phase deals with the release of the connection. In normal cases, it begins when the transfer phase terminates because a user hangs up. But it can also occur during the establishment phase since the caller can hang up at any time. The following description of the termination phase is a simplified one, since it does not deal with the asymmetrical aspect of the call termination. It is assumed here that the switch processes the termination in the same way whether the caller or the callee first hangs up. So once one of the two users hangs up his terminal, an on-hook signal is sent by the terminal to the switch. This one looks for the other party of the call. This party exists if there is a connection, i.e. if the user hangs up during the transfer phase. Then the switch sends to this party a break command. This party must hang up, and then the switch releases the connection. If there is no connection, i.e. the caller hangs up during the establishment phase, then only the calling-line state has to be changed by the switch. So the action of a user during this phase is *hangup*. The exchanges between the terminal and the switch during this phase are:
  - *on-hook*: from the terminal to the switch, following a user request for a call termination;
  - *break*: from the switch to a party involved in a call to alert it that the other party has hung up;
  - *disconnect*: from the switch to the terminal that has requested a call termination and from the other terminal to the switch.

The OF-Class model for the telecommunications system described above contains two OF-Classes, namely the OF-Class `Terminal` and the OF-Class `TSwitch`. The user is not modelled since he does not belong to the system. His interaction with the system will be treated as an external entity that constitutes the environment of the system. The OF-Classes interact through their interfaces in order to perform the system function. The set of possible interactions between a terminal and a switch reflects the communication progress and is expressed through operational, signal, and stream interfaces, which constitute the macro-level description of the OF-Class. Moreover, each of these OF-Classes owns resources. For example, a switch manages a set of lines, and a set of connections that result from the ongoing association of two lines. Finally, some triggers can be identified for an OF-Class to enable automated processing such as sending or receiving signals. The micro-level description of an OF-Class contains the definition of its instances, resources, operations, and triggers.

The OF-Class `Terminal`. The OF-Class `Terminal` models a terminal, i.e. the means by which the user accesses the system. It interacts with the user and with the switch through an operational interface, a signal interface, and a stream interface. Thus the exchanges described in Figure 26.3 are expressed through these interfaces. Some of these exchanges are expressed as operations while others are signals or streams.

- The operational interface of the OF-Class `Terminal` is the set of the *imported* service from the switch and the *exported* services to the user and to the switch. The service *imported* by this OF-Class is the `OutgoingCall` service, by which a calling terminal can request the switch for a connection with another terminal. It includes the operation `ConnectionReq`.
  The services *exported* by the OF-Class `Terminal` are defined as follows:
  - The `UseTerminal` service is used by the user to place a call or to answer a call. It is composed of four operations that reflect the actions performed by the user: the `PickUp` operation, the `Dial` operation, the `Talk` operation, and the `HangUp` operation. The usage pattern of the service describes the allowed chaining of operations and is illustrated in Figure 26.4. The user picks up the terminal to initiate a call or to receive a call. If he is the callee, he talks immediately, else he must first dial, then talk. He hangs up to terminate the call. He can also hang up at any time once he has picked up. The invocation mode of the `UseTerminal` service is synchronous, i.e. blocking for all the operations.
  - The `EndConnection` service is used by the switch to alert a party that the conversation is broken because the other party has already hung up. The service is composed of the `Break` operation. The invocation mode of this service is synchronous.
  - The `IncomingCall` service is used by the switch to alert the called party. It is composed of the `Ring` operation. Its invocation mode is asynchronous because the switch does not need to block until the return from `Ring` operation. It can perform other tasks meanwhile.
- The signal interface of the OF-Class `Terminal` is the set of signals exchanged between the terminal and the switch. Two signals are synchronously transmitted by the terminal, namely the `OffHook` signal and

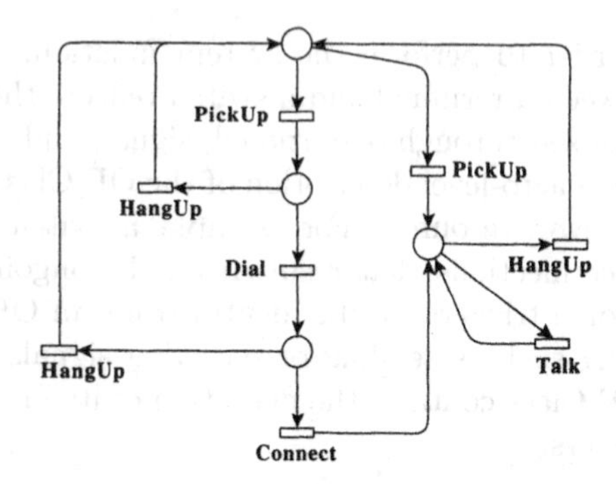

Fig. 26.4. The usage pattern of the `UseTerminal` service

the `OnHook` signal. The signals expected by the terminal are `DialTone`, `RoutingSignal`, `RingingBackTone`, and `Disconnect`.

- The stream interface of the OF-Class `Terminal` is the voice flow exchanged between the users during a conversation.
- The instances of the OF-Class `Terminal` used correspond to the simplified scenario described above. Two instances are declared, identified by a number that is assumed to be the terminal number and identifier of the switch to which the terminal is attached.
- The resources of the OF-Class `Terminal` are the state of the terminal `tstate` and the identity of the switch to which the terminal is attached. The evolution of the terminal state during a communication is illustrated in Figure 26.5. The default state is `disconnected`. When the user picks up the phone to initiate a call, the state shifts to the `calling` value. Then the user dials a number, so the state shifts to the `dialling` value. If the connection can be established with the other party by the switch, then the state of the calling terminal shifts to the `connected` value. When the user picks up the phone because it is ringing, the state shifts from the `ringing` value to the `connected` value. It is possible to shift from any of the previous states to the state `disconnected`.
- The trigger `CallRequest` of the OF-Class `Terminal` is used by the terminal as soon as its state shifts to the `calling` value, to wait for the signals `RoutingSignal` and `RingingBackTone` sent by the switch.

The OF-Class `TSwitch`. The OF-Class `TSwitch` models the switch, which must be permanently available to detect the user requests coming through the terminals that are attached to it. Its macro-level description is related to the macro-level description of the OF-Class `Terminal`. In fact, the operational interface of the OF-Class `TSwitch` contains the services that it *imports* from the `Terminal`, namely the `IncomingCall` and the `EndConnection` services. These services correspond to those exported by the OF-Class `Terminal`. This interface also contains the `OutgoingCall` service that the OF-Class `TSwitch` *exports* to the `Terminal` (which *imports* it). The signal interface is composed

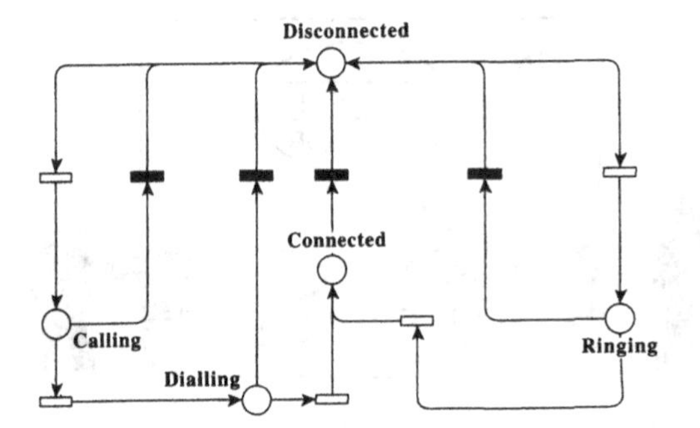

Fig. 26.5. The states of a terminal

of the signals provided by the switch, namely `DialTone`, `RoutingSignal`, and `RingingBackTone`, and of those expected by the switch, i.e. `OffHook` and `OnHook` signals. At the micro-level, the resources described are the set of lines managed by the switch, and the ongoing connections that result from the association of two lines. In order to simplify the description, a line is characterised by an identifier and a state. The line identifier corresponds to the terminal number. In our example, the line state is `free`, `calling`, or `called`. The switch is then able to identify the calling line from the called line in a given connection and in the set of lines. Two triggers are defined. The trigger `OpenLine` expresses the ability of the switch to detect a call demand, i.e. an `OffHook` signal coming from a terminal. The trigger `CloseLine` enables the switch to permanently detect a call termination demand, i.e. an `OnHook` signal coming from a terminal. The switch has to release the connection related to this terminal if such a connection exists (i.e. the user has hung up after the transfer phase, and not during the establishment phase).

In the next section, the SSF capability, i.e. the ability to detect an IN service demand contained in a call, is added to the switch. The OF-Class `TSwitch` is then modified to take into account this capability.

26.3.2 The OF-Class Model of a CFU Telecommunications System

A CFU Telecommunications System is a system that provides the Call Forwarding Unconditional service according to the intelligent network principle. In this case, the telecommunications system is composed of physical entities that implement the needed functional entities, namely the CCF/SSF, SCF, and SDF entities. As a simplification, the SCF and the SDF are grouped together (see Figure 26.6).

The CFU Service. The CFU service enables a user to forward his incoming calls to another terminal. Only the operational part of this service is taken into account; it encompasses several aspects such as operator procedures (e.g. adding or removing a subscriber) and user procedures. The description is

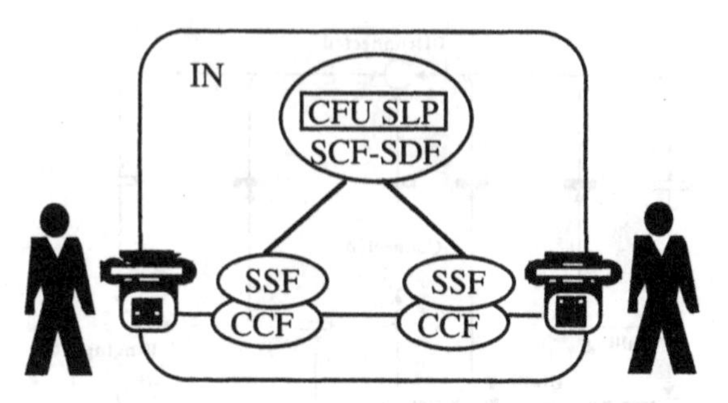

Fig. 26.6. The CFU telecommunications system

focused on the user procedures, namely activation, deactivation, and use. A CFU subscriber who wants to forward his calls must first activate the service. The activation procedure is performed by first dialling a special code, and then dialling a terminal number to which the calls will be forwarded; this is named the forwarded-to number. If a user dials the CFU subscriber number, then his call will be forwarded to this forwarded-to number. Thus no specific procedure is required of the user calling the CFU subscriber. Only the system is involved, detecting that such a basic call demand is in fact a CFU processing demand. When the subscriber wants to deactivate the service, he must dial another special code. The model described below details only the CFU service part corresponding to the processing of a CFU number demand.

The CFU Number Demand Processing. A call to a CFU subscriber is processed as illustrated in Figure 26.7. When a user dials a terminal number, the CCF/SSF has to detect if it corresponds to a CFU subscriber number or not. This is done by the Detection Point (DP) processing included in the basic call processing. If the CCF/SSF recognises a CFU subscriber number, it suspends the basic call processing and addresses a request to the SCF by providing it with the corresponding service key. Based on this information, the SCF invokes the appropriate service logic, i.e. the CFU processing, by providing the dialled number. The CFU processing determines if the service is activated for this subscriber, and if so, retrieves the Forwarded-To number. The result of this execution is then provided to the SCF. Depending on this result, the SCF has to transmit an order to the CCF/SSF. Three types of orders are identified:

- Continue with the same data: the CCF/SSF can resume the basic call processing with the same data, i.e. the calling-line identity (`callingLineID`) and the called-line identity (`calledPartyNumber`).
- Continue with new data: the service execution provides new data by modifying the initial ones (e.g. the forwarded-to number replaces the called number). So the CCF/SSF has to resume the basic call processin, taking into account these new data.

- Clear the call: following the service execution, the call must be cleared by the CCF/SSF. This response is provided, for example, in case of errors occurring during the service execution. It is also used to terminate a service demand for which no connection request is required, for example the activation or deactivation of a service. In such cases, once the service is performed, it is assumed that the call can be cleared.

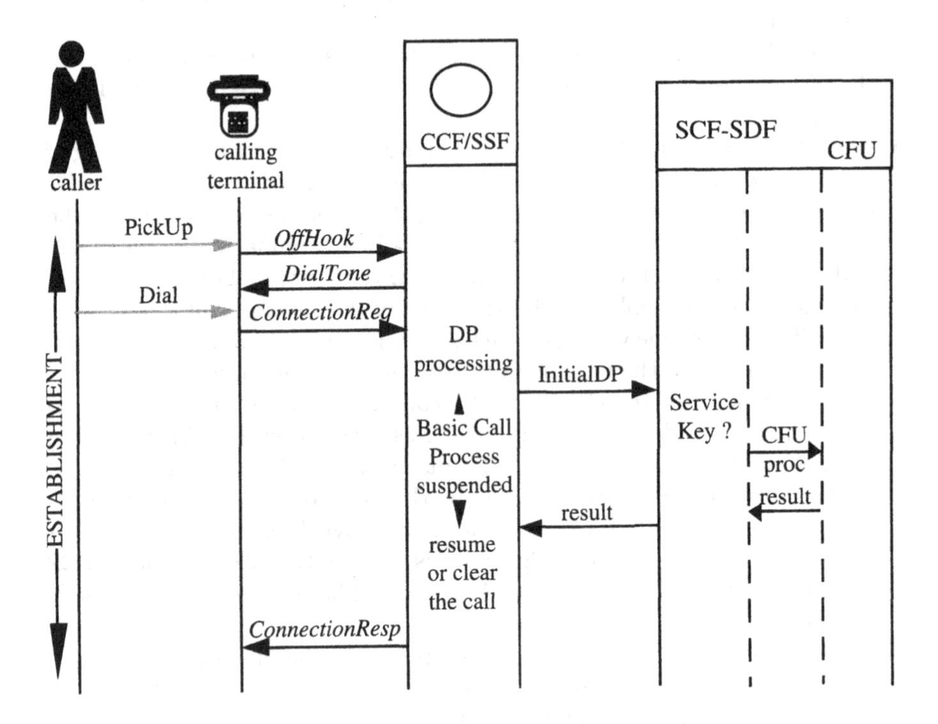

Fig. 26.7. The CFU number demand processing

The OF-Class Model. The OF-Class model of such a system contains the same OF-Class `Terminal` as the previous model, a new OF-class `SSF` corresponding to the modified OF-Class `TSwitch` and two other new OF-Classes, namely the `SCF` and `OFCallForwarding` OF-Classes.

- The OF-Class `SSF` corresponds to an OF-Class `TSwitch` that has the ability to detect an IN demand, i.e. to process detection points (DPs). A simplified DP processing is considered: only the DP enabling the processing associated with the acceptance of a call on the terminating side is taken into account, i.e. DP12. Actually, before delivering the call to the called terminal, the CCF/SSF determines if it is a basic call or if it is an IN service demand. For this, it is assumed that an armed DP12 and the corresponding service key, namely the CFU number demand, are associated with the

CFU subscriber number. This is achieved by using a table that associates the armed DPs of a service with the user profile identified by a terminal number. This table `tPointTable` is defined as an SSF resource. So any call with call parameters corresponding to a CFU subscriber number is identified as an IN service demand and is suspended while a request is addressed to the SCF for service processing. The CCF/SSF will continue the basic call processing according to the SCF order related to the end of the service execution.
So the OF-Class `SSF` imports from the OF-Class `SCF` the service `StartUp` that is composed of the operation `InitialDP`. The service invocation mode is synchronous.

- The OF-Class `SCF` models the entity that is responsible for the processing of the SSF requests. When the SSF invokes operation `InitialDP`, it provides to the SCF one parameter `initialDPArg` that contains the identifier of the service `serviceKey`, the identity of the calling terminal `callingLineID`, and the identity of the called terminal `calledPartyNumber`. Once again, the description is simplified. The `serviceKey` is a discriminant that enables the SCF to determine which service logic program is relevant to the request. The SCF transfers the request to this service logic program.
The macro-level of the OF-Class `SCF` contains only an operational interface. The *imported* service of this OF-Class is the service `CallForwarding` composed of three operations: `CF_Activation`, `CF_Deactivation`, and `CF_Processing`. This service is used by the SCF to invoke the appropriate service logic program. Its invocation mode is synchronous. The OF-Class `SCF` exports the service `StartUp`, which is used by the `SSF`. The `InitialDP` operation is the sole operation of the service. The service invocation mode is synchronous.
The micro-level description is very simple: only the operation bodies are needed. The SCF manages no resource because it works only on supplied data and does not need to have persistent states.
It should be noted that in the IN recommendations, the SCF is the entity performing the services. Thus it is composed of several modules, including the service logic programs. In the present model, this characteristic is expressed by the definition of two OF-Classes. One is the OF-Class `SCF` which represents the module that receives an SSF request, analyses it, and invokes the appropriate service logic program. The other is the OF-Class `OFCallForwarding` which represents the service logic itself. Since a service logic program is a part of the SCF, the OF-Class `SCF` includes the OF-Class `OFCallForwarding`.
- The OF-Class `OFCallForwarding` represents the service logic for the CFU service. As mentioned above, it is included in the OF-Class `SCF`. Its macro-level description contains the *exported* service `CallForwarding` with its three operations `CF_Activation`, `CF_Deactivation`, and `CF_Processing`, corresponding to the user procedures. The usage pattern expresses the

dependency relation between the CF_Activation and CF_Deactivation operations, since the CF_Activation operation must always precede the CF_Deactivation operation. On the other hand, the CF_Processing operation can be invoked independently of the two others.
It should be noticed that it is assumed that grouping the SCF and the SDF can be grouped together (see Figure 26.7). Since the OF-Class SCF models the SDF too, one resource is defined at the micro-level. This is the list of subscribers who have activated the CFU service, CF_activeList. Only the CF_Processing operation is specified in detail. It performs CFU number demand processing.

26.3.3 The OF-Class Model of an IN Telecommunications System

The system now modelled is the CFU Telecommunications System in which a new service is introduced, namely the Terminating Call Screening service (TCS), in order to obtain an IN Telecommunications System. It is considered that the same SCF performs the two services (Figure 26.8). Actually, whether we have a single SCF performing the two services, or two SCFs, each dedicated to the execution of one service, has no impact on the activities of validation and verification of the system realised by the formalism. The proving process enables us to validate the correctness of the specification in terms of expected functionality, and to verify its structural and behavioural properties. Thus it is independent of the assumptions made in the specification itself. Its role is precisely to evaluate whether or not, given these assumptions, the specification is correct. Introducing the TCS service then requires the modification of the SSF and SCF OF-Classes and the addition of a new OF-Class OFTerminatingCallScreening.

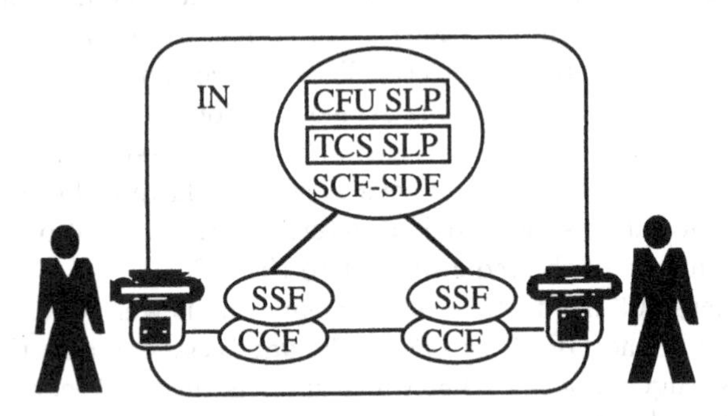

Fig. 26.8. The IN telecommunications system

The Terminating Call Screening Service. The TCS service enables a user to prevent some calls from being delivered to his terminal. This means that the subscriber has to provide the calling numbers which he does not want to accept, and when an incoming call is presented to the switch from one of these numbers, the switch stops this call. As was done for the CFU service, the focus is on the user procedures of the TCS service, especially the processing of a TCS number demand. The subscriber has to activate his service by a specific procedure `TCS_Activation`. Then his incoming calls will be filtered according to the given list of denied callers. When the subscriber wants to deactivate the service, he performs the `TCS_Deactivation` procedure. The subscriber can also manage his list of denied numbers, i.e. add or remove a number by using the `TCS_AddNumber` or `TCS_RemoveNumber` procedures.

The TCS Number Demand Processing. The TCS number demand processing is very close to CFU processing, since it is based on the call presentation to the called terminal. If the called number corresponds to a TCS subscriber, then the CCF/SSF detects it through DP12 processing, suspends the basic call processing, and addresses an `InitialDP` request to the SCF by providing the `initialDPArg`, i.e. the `serviceKey`, the `callingLineID`, and the `calledPartyNumber`.

Based on the service key, the SCF invokes the TCS service logic program. It checks whether the service is activated for this subscriber, and if so, it checks whether or not the calling terminal is authorised to make the call. As for the response, the SCF will deliver to the CCF/SSF the order to resume or to clear the call.[3]

The OF-Class Model. To introduce a new service, the CFU Telecommunications System requires some modifications.

- The OF-Class `SSF` has to be modified to be able to detect a TCS service demand and to initiate a corresponding request to the SCF. Obviously DP12 has to be armed. As in the previous model, the SSF resource `tPointTable` is used to associate an armed DP of a service with the user profile identified by the terminal number. So this table has to be modified by adding the association corresponding to the TCS service. This table represents the list of services for which DP12 is armed and for which the CCF/SSF has to request from the SCF the execution of the corresponding service logic programs during a given call. If several services have to be invoked in the same call, the table is also used to determine the invocation priority and then the sequence of the requests sent to the SCF.
- The OF-Class `SCF` has to take into account the new service. So it will include not only the OF-Class `OFCallForwarding`, but also the OF-Class `OFTerminatingCallScreening`, which models the TCS service logic. Into the operational interface of the OF-Class `SCF` is added the *imported* service `TerminatingCallScreening`, composed of the five operations corresponding to user procedures, `TCS_Activation`, `TCS_Deactivation`,

[3] Once again, the procedure is simplified by the omission of the announcement to the caller that he is not authorised to place his call.

TCS_AddNumber, TCS_RemoveNumber, and TCS_Processing. The usage pattern expresses the dependency relation between the TCS_Activation and TCS_Deactivation operations, since the TCS_Activation operation must always precede the TCS_Deactivation operation. The same precedence relation exists between the TCS_AddNumber and TCS_RemoveNumber operations. On the other hand, the TCS_Processing operation can be invoked independently of the others. This service is synchronously invoked. At the micro-level, the service keys for the TCS service and enabling the SCF to determine which service logic program is relevant to the SSF request are added.
- The OF-Class OFTerminatingCallScreening represents the service logic program of the TCS service. As mentioned above, it is included in the OF-Class SCF. Its macro-level description contains the *exported* service TerminatingCallScreening with its five operations. Resources are the list of subscribers who have activated their TCS service, TCS_activeList, and for each subscriber, the list of his denied numbers, deniedList. Only the operation TCS_Processing is specified in detail.

26.3.4 From OF-Class to OF-CPN: The Principles of the Transformation

Once a syntactically correct OF-Class model is obtained, it is automatically transformed into OF-CPNs, as described in section 10.3. This process is recalled here for the sake of clarity. The principles of this transformation, which are based on the method developed by Heiner [Hei92] are:

- Each OF-Class (the source element) is transformed into one OF-CPN (the target element). This means that composition links are not taken into account. Thus a composite OF-Class is considered as a set of flat OF-Classes, without any hierarchical relationship.
- For a given OF-Class, its declaration section determines the type system of its target OF-CPN.
- The elementary unit of an OF-Class, i.e. the micro-level description is transformed into the target Petri net. Actions are mapped onto transitions or onto a sub-net, whereas resources, parameters, and variables are mapped onto places. Domains of these places are determined by the modelled data types. Arcs describe the effect of actions on resources, variables, and parameters. Arc valuations determine the semantics of transformations.
- The interfacing units (macro-level description) contain offered and required services. They are transformed into state machines controlling the interactions between the elementary units.

Based on these principles, each OF-CPN is generated from the OF-Class model of the IN Telecommunications System. They are not represented here because of their size and complexity. Table 26.1 below illustrates the complexity of the net models by giving the number of places, transitions, and arcs for each net obtained by transformation of the corresponding OF-Class. The specification size, i.e. number of lines, of the OF-Class is also given as an indication. The next section illustrates the transformation on one component, namely `Terminal`.

Table 26.1. Complexity of net models

OF-Class	Number of Lines	Number of Places	Number of Transitions	Number of Arcs
Terminal	152	78	57	196
SSF	203	79	65	227
SCF	99	24	28	91
OFCallForwarding	58	48	28	90
OFTerminating-CallScreening	89	29	17	52

26.3.5 From OF-Class to OF-CPN: Illustration of the Transformation

An OF-CPN is built from a set of Petri nets produced from the OF-Class specification according to the principles given in Section 26.3.4. Each usage pattern gives a net which enforces valid sequences between the operations. Figure 26.4 illustrates the usage pattern of the terminal, which determines the correct behaviour of a telephone user. The places in this net express the sequencing constraints. The operations are transformed into nets according to the principles above. The places which model the resources are connected to the actions which modify them. A couple of places are added to model the incoming parameters and the outgoing results of the operation.

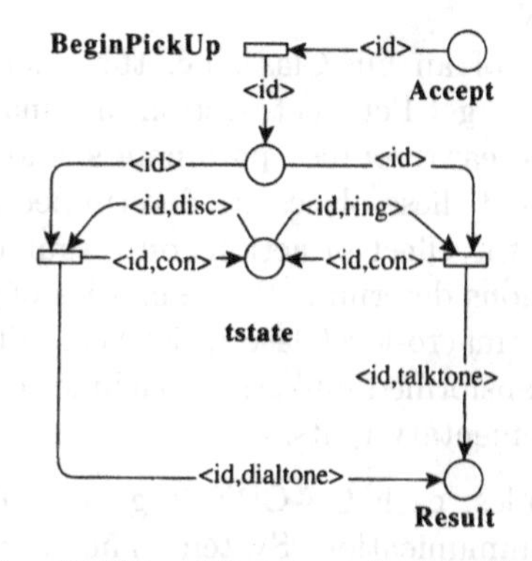

Fig. 26.9. The operation PickUp of `Terminal` in OF-Class and OF-CPN

These nets are used to refine the usage pattern and produce from it the OF-CPN model. Actually, each transition of the usage pattern models an operation, i.e. a sequence of elementary and elaborated actions. It is therefore refined by the net modelling this operation (see Figure 26.9 for the example of `PickUp`).

The procedure sketched above shows a progressive way to produce an OF-CPN from an OF-Class. A simple case for the OF-Class `Terminal` with only one usage pattern is illustrated in Figure 26.10. In the case of many different usage patterns, they are first combined according to rules which are skipped here for the sake of space.

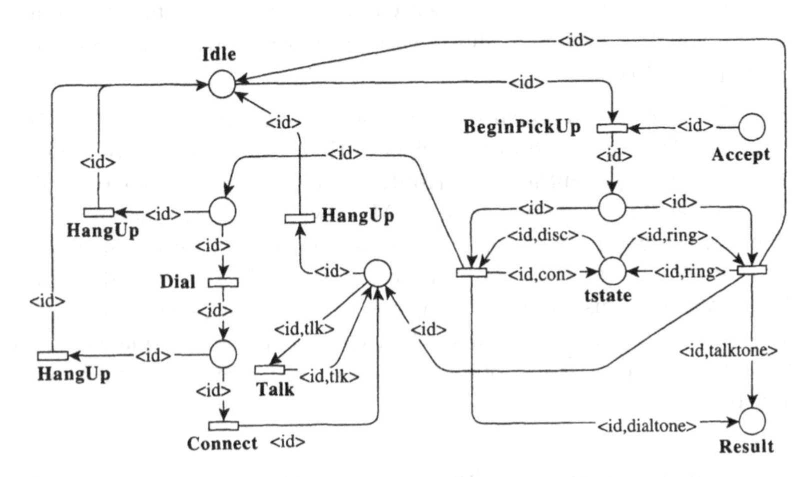

Fig. 26.10. Part of the OF-CPN synthesised from OF-Class `Terminal`

26.4 Analysis

26.4.1 Overview of Analysis with Petri Nets in the Area of Telecommunications Systems

Many analysis techniques and methods have been developed for the formalism of Petri nets (see Part III). Almost all these techniques and methods can be applied to telecommunications systems. First of all, the direct executability of nets allows a validation approach based on simulation and animation. Telecommunications systems modelled directly with nets can be simulated. If nets are used only for verification and validation needs (as for the modelling approach used in this chapter), the nets provide a basis for animation of the models because of their tight correspondence.

Structural analysis such as invariant computation allows us to validate the usage of resources and the sequence of actions in net-based telecommunications system models. For instance, a given telephone line in a telecommunications system has a given number of possible states which can be matched by a P-semiflow. The actions a subscriber can perform using a telephone can also be matched by repetitive sequences of transitions. There are two basic approaches in invariant-based analysis and modelling:

- Models are built and their invariants computed. These invariants are then used to corroborate those expected in the models.
- Invariants are modelled and afterwards the models are refined in such a way that these invariants are preserved. This approach puts constraints on the refinement process but allows us to produce models with correct and precise properties.

Net models for telecommunications systems can also support analysis of their performance. Deterministic and stochastic time can be associated with the transitions. For exponentially distributed stochastic time, the reachability graph is shown to be equivalent to a Markov chain if the net is bounded (see [Dia82, Mur89]). One interesting feature of nets in telecommunications is that they allow a state-/transition-based approach to their models. Reachability graphs allow us to build and check the global state space of a model. Properties can be expressed and verified by model-checking means. For the expression of the properties, one can use, for instance, temporal logic (see Section 14.1).

26.4.2 Analysis of the IN Model: Detection of Feature Interaction

The purpose of this analysis is twofold:

- It enables the designer to establish that the specification of the system is correct in terms of safety and reliability.
- It assists him/her to check and validate some specific scenarios suspected to produce feature interactions.

The technique used here is the analysis of the reachability graph. Once each OF-CPN has been obtained by the transformation of its source OF-Class, it is extended by including the usage patterns of its offered services. In this case, all OF-CPNs are interfaced by places. Therefore, the place-fusion method is chosen for the analysis. This method is the most suitable because state information on the SSF is required. The initial configuration of the system is determined as follows. Three users A, B, and C are considered. A has subscribed to the CFU and TCS services and has activated them. A forwards his calls to B's terminal and denies calls from C. B is idle and C initiates a call to A.

First the reachability graph for each OF-CPN is computed according to the initial configuration and the exhaustive abstraction of the environment.

Exhaustive abstraction of the environment means marking with all their possible values the interface places that model incoming results and signals. It contains the expected results of interactions with other OF-CPNs as well as a representation of the part of the system that is not explicitly modelled. This abstraction enables an OF-CPN to guarantee a correct operation provided that the environment operates correctly. It gives an implicit model of the environment based on the rely/guarantee principle ([Col93]). An aggregated equivalent for each OF-CPN with its environment abstraction is computed. Below we give the aggregated equivalent of the OF-CPN deduced from OF-Class `Terminal` (Figure 26.11). Its initial configuration is C initiating a call to A. This is represented by a request for each of the operations of the OF-Class `Terminal`. The sequencing between these requests is achieved by the usage pattern of the OF-Class. For the sake of space and readability, the value of the resources at each state is not shown in detail.

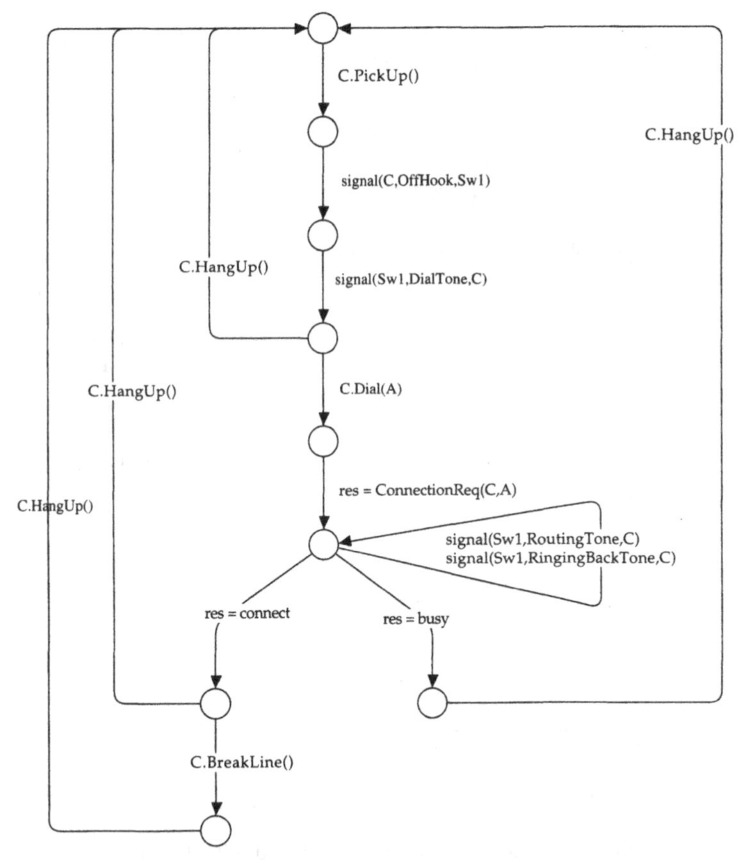

Fig. 26.11. Aggregated equivalent of the OF-Class `Terminal`

The aggregated equivalent of an OF-CPN is a restriction of the reachability graph to actions that modify the interface places. Thus it models the part of the behaviour that is observable rather than the internal part that modifies only resource variables and parameters.

Such equivalents are considered for each of the other OF-Classes, namely **SSF**, **SCF**, **OFCallForwarding**, and **OFTerminatingCallScreening**. Each equivalent (except the **SSF** equivalent) is transformed back into smaller Petri nets that will be combined with the OF-CPN modelling the CCF/SSF by place-fusion. The reachability graph for the net modelling the **SSF** interfaced with other equivalents is illustrated in Figure 26.12. It shows that a communication can be established between A and C, thereby violating the TCS functionality. In fact, the precedence relationship between the invocation of the CFU and TCS services in the same call, established by the way the table is managed, is not correct.

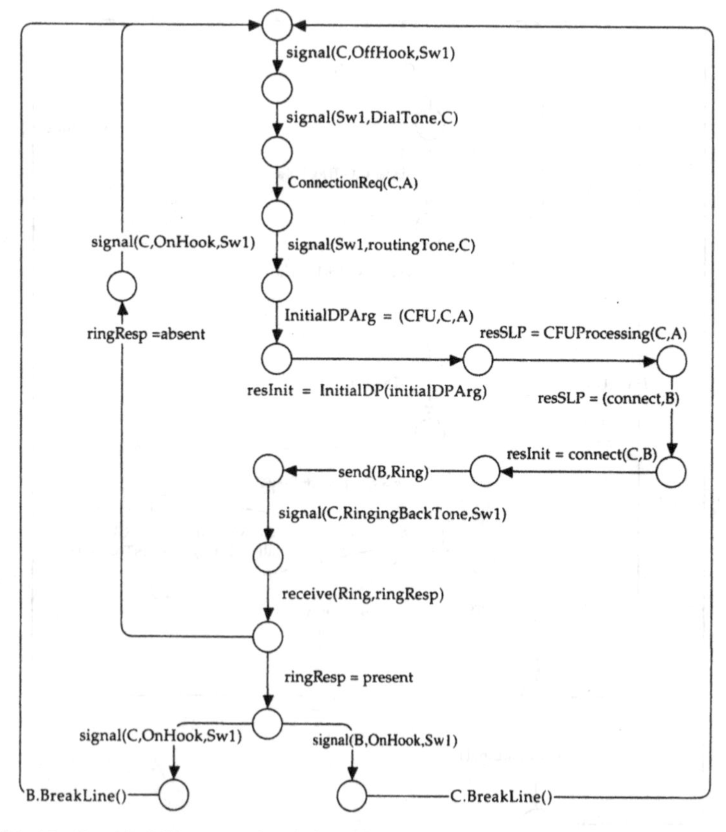

Fig. 26.12. Reachability graph of the SSF interfaced with the aggregated equivalents of Terminal, TCS, and CFU

In a correct operation, when a call is addressed to a subscriber that has subscribed to the CFU and TCS services, if the services are activated and if the caller belongs to the `deniedList` of the subscriber, then the call must not be forwarded. So the first service to be invoked must be the TCS service, followed by the CFU service. The invocation priority is determined by the table `tPointTable` and so depends on the management policy of this resource. This policy is the combination of an updating policy and a consulting policy, and both must be consistent with each other in order to guarantee that no interactions between services will occur. Thus introducing a service requires us to modify the resource `tPointTable` and to build a new one, according to the management policy. Once the table is built, our formalism enables us to determine the correctness of the construction, by verifying the precedence relationship between the invocations.

A very simple resource management policy has been chosen, namely FIFO management. Hence the oldest service created in the system, namely the CFU service, is invoked first. The verification of the table construction has shown that the management policy is not correct.

The IN-CS1 principles have been applied, namely single-ended and single point-of-control services. The service invocations of SSF are sequential and blocking, i.e. the basic call process is suspended. This is expressed in the specification by the synchronous invocation mode of the service `StartUp` (cf. the DP processing in the OF-Class `SSF`). The OF-Class formalism enables non-blocking and parallel invocations of services imported from different OF-Classes. So a scenario in which several points of control are involved can be described with the OF-Class formalism, i.e. an OF-Class `SSF` can invoke several OF-Classes `SCF` in a parallel and non-blocking way. In this case, several services can be invoked at the same time. If a precedence relationship must be respected between services, it is no longer expressed as an invocation priority, but as a priority of result collecting and processing. As for priority between invocations, the specification must state a priority policy for result collecting and processing. The proving tool is then able to validate that the policy does not exhibit interaction between services.

The OF-Class formalism is linked with another one called H-COSTAM, which is mostly dedicated to code generation with optimisation features and was presented in Chapter 21. The translation from one formalism to the other is described in [DK96]. The implementation of a tool for this translation is ongoing. So it will become possible to undertake code generation from an OF-Class specification. The pair of formalisms allows one to trace properties proved on the specification to the final implementation. The OF-Class formalism provides a very detailed specification level from which final code could be easily derived in an automated way. But it is useful to trace and check properties via the H-COSTAM model.

26.5 Conclusion

This chapter has illustrated the benefits of using formal methods, namely Petri nets, in the context of telecommunications systems. The emerging domain of service engineering needs formal support. The new challenge for the providers is the ability to specify and to design new services which can be integrated them into an operational system without altering the existing service operation.

The proving toolset described in this chapter enables a service designer to validate and to verify a service specification. The proposed approach, based on a coupling of OO and Petri net paradigms, can be incorporated in a standard software production environment without too much effort. The shift from the initial OO model to the OF-Class model can be made easier by means of design patterns. The transformation from the OF-Class to the OF-CPN model is fully automated and supported by a tool. A prototype of the approach described here has been developed and integrated in the AMI environment, which is a framework dedicated to the formalisation of software development throughout its life-cycle.

The case study developed in the chapter uses this proving toolset for the validation and verification of an intelligent network system that provides call forwarding and terminating call screening services. Applying the tool gives an OF-CPN model which is analysed. It reveals an error due to an incorrect resource management policy. This example clearly establishes the suitability of the proving toolset. More generally, it illustrates that Petri nets have great potential as formal methods for the telecommunications service engineering area.

27. Conclusion

27.1 Common Modelling Problems

The three application domains have common characteristics with respect to the modelling of their processes. Most of the processes considered in this part are *case-based*, i.e. every piece of work is executed for a specific *case*. Examples of cases are a mortgage, a telephone call, a production order, an insurance claim, a tax declaration, an order, or a request for information. Cases are often generated by an external customer. However, it is also possible that a case is generated by another department within the same organisation (internal customer). The goal of the processes in the three application domains is to handle cases as efficiently and effectively as possible.

Cases are handled by executing *tasks* (operations, process steps) in a specific order. The *process definition* (procedure) specifies which tasks need to be executed and in what order. Since tasks are executed in a specific order, it is useful to identify *conditions* which correspond to causal dependencies between tasks. A condition holds or does not hold (true or false). Each task has pre- and post-conditions: the pre-conditions should hold before the task is executed, and the post-conditions should hold after execution of the task. Many cases can be handled by following the same process definition. As a result, the same task has to be executed for many cases. A task which needs to be executed for a specific case is called a *work item*. An example of a work item is: Execute task 'send refund form to customer' for case 'complaint sent by customer Baker'.

Most work items are executed by a *resource*. A resource is either a machine (e.g. a conveyor belt, a robot, or a router) or a person (participant, worker, employee). Resources are allowed to deal with specific work items. To facilitate the allocation of work items to resources, resources are grouped into classes. A *resource class* is a group of resources with similar characteristics. There may be many resources in the same class, and a resource may be a member of multiple resource classes. A work item which is being executed by a specific resource is called an *activity*. If we take a photograph of a process, we see cases, work items, and activities. Work items link cases and processes. Activities link cases, processes, and resources.

Figure 27.1 shows that a case-based process has three dimensions: (1) the case dimension, (2) the process dimension, and (3) the resource dimension.

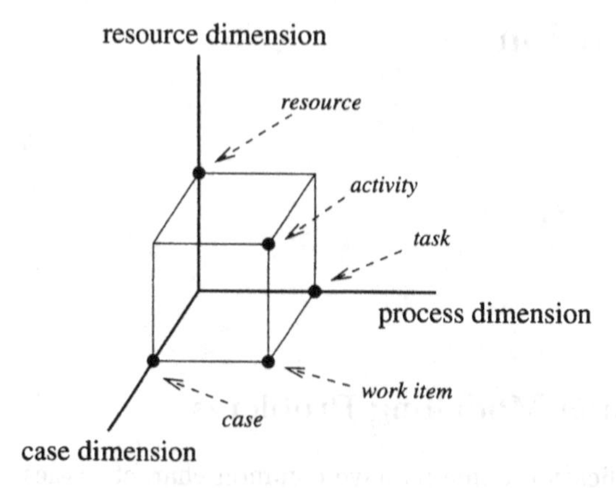

Fig. 27.1. A three-dimensional view of a process

The case dimension signifies the fact that all cases are handled individually. In principle, cases do not directly influence each other. Clearly they influence each other indirectly via the sharing of resources and data. In the process dimension, the procedure, i.e. the tasks and the routing through these tasks, is specified. In the resource dimension, the resources are grouped into classes. We can visualise the state of a process as a number of dots in the three-dimensional view shown in Figure 27.1. Each dot represents either a work item (case + task) or an activity (case + task + resource).

The fact that the three-dimensional view of a process presented in Figure 27.1 makes sense in the three application domains presented in this part illustrates that from a modelling point of view the three domains are not that different! These application domains share *design patterns* which transcend each of the domains. Future research should aim at capturing these design patterns so that they can be made available to people applying Petri nets.

27.2 Shared Analysis Results

To support the analysis of processes in manufacturing, telecommunications, and business, many approaches have been developed. There has been a massive invasion of quantitative techniques to aid decision making at all levels. Techniques provided by operations research (OR) include linear programming, Markovian analysis, dynamic programming, and critical path analysis. These OR techniques can be used for scheduling, risk analysis, project planning, plant location, etc. However, for each level we need different techniques and models. Moreover, there is often no relation between the efforts performed at the operational, tactical, and strategical levels.

Another problem related to the use of mathematical techniques such as linear programming, Markovian analysis, dynamic programming, etc. is the fact that it is often necessary to remodel the situation if another type of question needs to be answered. For example, to use another method of analysis we may have to remodel the situation.

To solve these problems one should use a 'solver-independent' medium for the modelling of the problem situation. This medium is used to make a concise blueprint of the process/system. This blueprint may be used at different levels of decision making and can be used as a starting point for various means of analysis.

The solver-independent medium proposed in this book is Petri nets. A specification in terms of a (high-level) Petri net provides a concise and solver-independent description of the process at hand, i.e. a blueprint. Unlike more traditional approaches, the emphasis is on a natural representation rather than the method(s) of analysis. The result of the modelling process, i.e. the specification, can be used as a starting point for all kinds of analysis, i.e. the specification provides an interface between the modelling activities and the analysis activities. Since the specification does not depend upon the method of analysis, it is possible to do several kinds of analysis without having to remodel the entire system. Sometimes, additional analysis activities are the price paid for this flexibility, i.e. the specification has to be transformed into a model which can be analysed using a suitable method of analysis. Therefore, it is essential to support these transformations with tools that allow for an easy transition between the specification, the analytical model, and the analysis results.

The use of Petri nets as a modelling language allows for the *unification* of analysis techniques over a number of application domains. Consider, for example, techniques for deadlock detection: they can be applied in any of the application domains presented in this part.

References

[Aal95] W.M.P. van der Aalst. A class of Petri nets for modelling and analysing business processes. Computing Science Report 95/26, Eindhoven University of Technology, Department of Mathematics and Computing Science, Eindhoven, The Netherlands, August 1995.

[Aal96] W.M.P. van der Aalst. Three good reasons for using a Petri-net-based workflow management system. In S. Navathe and T. Wakayama, editors, *Proceedings of the International Working Conference on Information and Process Integration in Enterprises (IPIC'96)*, pages 179–201. MIT, Cambridge, Massachusetts, 1996.

[Aal97a] W.M.P. van der Aalst. Designing workflows based on product structures. In K. Li, S. Olariu, Y. Pan, and I. Stojmenovic, editors, *Proceedings of the 9th IASTED International Conference on Parallel and Distributed Computing Systems*, pages 337–342. IASTED/Acta Press, Anaheim, CA, USA, 1997.

[Aal97b] W.M.P. van der Aalst. Verification of workflow nets. In Azéma and Balbo [AB97b], pages 407–426.

[Aal98a] W.M.P. van der Aalst. The application of Petri nets to workflow management. *The Journal of Circuits, Systems and Computers*, pages 21–66, 1998.

[Aal98b] W.M.P. van der Aalst. Three good reasons for using a Petri-net-based workflow management system. In T. Wakayama, S. Kannapan, C. Khoong, S. Navathe, and J. Yates, editors, *Information and Process Integration in Enterprises: Rethinking Documents*, The Kluwer International Series in Engineering and Computer Science, pages 161–182 (Chapter 10). Kluwer Academic Publishers, Norwell, 1998.

[AAM+95] G. Alonso, D. Agrawal, C. Mohan, R. Günthör, A. El Abbadi, and M. Kamath. Exotica/FMQM: a persistent message based architecture for distributed workflow management. In *Proceedings of the IFIP Working Conference on Information Systems in Decentralised Organisations*, 1995.

[AB97a] W.M.P. van der Aalst and T. Basten. Life-cycle inheritance: A Petri-net-based approach. In Azéma and Balbo [AB97b], pages 62–81.

[AB97b] P. Azéma and G. Balbo, editors. *Application and Theory of Petri Nets 1997, Proceedings of the 18th International Conference (ICATPN'97), Toulouse, France*, volume 1248 of *Lecture Notes in Computer Science*. Springer-Verlag, 1997.

[ABC+95] M. Ajmone-Marsan, G. Balbo, G. Conte, S. Donatelli, and G. Franceschinis. *Modelling with Generalized Stochastic Petri Nets*. John Wiley and Sons, 1995.

[ABH+97] R. Alur, R.K. Brayton, T.A. Henzinger, S. Qadeer, and S.K. Rajamani. Partial-order reduction in symbolic state space exploration. In Grumberg [Gru97], pages 340–351.

[ACA78] T. Agerwala and Y. Choed-Amphai. A synthesis rule for concurrent systems. In *Proceedings of the 15th Design Automation Conference*, pages 305–311, Las Vegas, NV, USA, June 1978.

[Act96] Action Technologies. *ActionWorkflow Enterprise Series 3.0 User Guide.* Action Technologies, Inc., Alameda, 1996.

[AG96] K. Arnold and J.A. Gosling. *The Java Programming Language.* Addison-Wesley, Reading, MA, USA, 1996.

[AH93] S. Asur and S. Hufnagel. Taxonomy of rapid-prototyping methods and tools. In RSP93 [RSP93], pages 42–56.

[AH97] W.M.P. van der Aalst and K.M. van Hee. *Workflow Management: Models, Methods and Systems (in Dutch).* Academic Service, Schoonhoven, 1997.

[AHH95] W.M.P. van der Aalst, K.M. Hee, and G.J. Houben. Formal description of workflow using a Petri based approach. *Petri Nets*, 1995.

[AHI98] K. Ajami, S. Haddad, and J.M. Ilié. Exploiting symmetry in linear temporal model checking: One step beyond. In *Tools and Algorithms for the Construction and Analysis of Systems, Proceedings of the 4th International Conference, TACAS'98, part of Theory and Practice of Software (ETAPS'98), Lisbon, Portugal*, volume 1384 of *Lecture Notes in Computer Science*, pages 52–67. Springer-Verlag, 1998.

[AHV97] W.M.P. van der Aalst, D. Hauschildt, and H.M.W. Verbeek. A Petri-net-based tool to analyze workflows. In Farwer et al. [FMS97], pages 78–90.

[AK86] K.R. Apt and D.C Kozen. Limits for automatic verification of finite-state concurrent systems. *Information Processing Letters*, 22(6):307–309, 1986.

[Ake78] S.B. Akers. Binary decision diagrams. *IEEE Transactions on Computers*, C-27(6):509–516, 1978.

[And91] G. Andrews. *Concurrent Programming: Principles and Practice.* Benjamin/Cummings, 1991.

[APP94] F. Andersen, K.D. Petersen, and J.S. Pettersson. Program verification using HOL-UNITY. In J.J. Joyce and C.-J.H. Segar, editors, *Higher Order Logic Theorem Proving and Its Applications. Proceedings of the 6th International Workshop, HUG'93, Vancouver, B.C., Canada*, volume 780 of *Lecture Notes in Computer Science*, pages 1–15. Springer-Verlag, 1994.

[AS85] B. Alpern and F.B. Schneider. Defining liveness. *Information Processing Letters*, 21(4):181–185, 1985.

[AS87] B. Alpern and F.B. Schneider. Recognizing safety and liveness. *Distributed Computing*, 2(3):117–126, 1987.

[ATP91] *Application and Theory of Petri Nets, Proceedings of the 12th International Conference (ICATPN'91), Gjern, Denmark*, 1991.

[BA99] T. Basten and W.M.P. van der Aalst. Inheritance of behavior. Computing Science Report 99/17, Eindhoven University of Technology, Department of Mathematics and Computing Science, Eindhoven, The Netherlands, November 1999.

[Bak96] Bakkenist Management Consultants. *ExSpect 5.0 User Manual*, 1996.

[Bak97] Bakkenist Management Consultants. *ExSpect 6 User Manual*, 1997. Information available at `http://www.exspect.com/`.

[Bas95] R. Bastide. Approaches in unifying Petri nets and the object-oriented approach. In *1st Workshop on Object-Oriented Programming and Models of Concurrency, Turin, Italy*, 1995.

[Bas98] T. Basten. *In Terms of Nets: System Design with Petri Nets and Process Algebra.* PhD thesis, Eindhoven University of Technology, Department of Mathematics and Computing Science, Eindhoven, The Netherlands, December 1998.

[Bau97] F. Bause. Analysis of Petri nets with a dynamic priority method. In Azéma and Balbo [AB97b], pages 215–234.

[BB88] T. Bolognesi and E. Brinksma. Introduction to the ISO specification language LOTOS. *Computer Networks and ISDN Systems*, 14(1):25–59, 1988.

[BBC+99] B. Barras, S. Boutin, C. Cornes, J. Courant, Y. Coscoy, D. Delahaye, D. de Rauglaudre, J.-Ch. Filliatre, E. Giménez, H. Herbelin, G. Huet, H. Laulhère, C. Muñoz, C. Murthy, C. Parent-Vigouroux, P. Loiseleur, Ch. Paulin-Mohring, A. Sabi, and B. Werner. *The Coq Proof Assistant - Reference Manual, Version 6.3.1, Coq Project*, December 1999. `http://pauillac.inria.fr/coq`.

[BC93] H. Bachatène and J.M. Couvreur. A reference model for modular colored Petri nets. In *IEEE International Conference on Systems, Man and Cybernetics, Proceedings Vol. II, Le Touquet, France*, pages 724–729, 1993.

[BC94] J.L. Briz and J.M. Colom. Implementation of weighted place/transition nets based on linear enabling functions. In Valette [Val94a], pages 99–118.

[BCAP95] G. Bruno, A. Castella, R. Agarwal, and M.P. Pescarmona. CAB: an environment for developing concurrent applications. In De Michelis and Diaz [DMD95], pages 141–160.

[BCC+86] B. Berthomieu, N. Choquet, C. Colin, B. Loyer, J.M. Martin, and A. Mauboussin. Abstract data nets: Combining Petri nets and abstract data types for high level specifications of distributed systems. In *Proceedings of the Seventh European Workshop on Application and Theory of Petri Nets, Oxford, UK*, pages 25–48, 1986.

[BCDC96] E. Battiston, A. Chizzoni, and F. De Cindio. Modeling a cooperative development environment with CLOWN. In *Proceedings of the 2nd Workshop on Object-Oriented Programming and Models of Concurrency, Osaka, Japan*, 1996. Workshop at ICATPN'96, see [BR96].

[BCDE90] E. Best, L. Cherkasova, J. Desel, and J. Esparza. Characterization of home states in free choice systems. Berichte 7/90, Hildesheimer Informatik, Hildesheim, Germany, July 1990.

[BCE94] H. Bachatène, J.M. Couvreur, and P. Estraillier. Specification of compositional active objects using modular colored nets. In C. Girault, editor, *International IFIP Conference: Applications in Parallel and Distributed Computing*, pages 205–214. North-Holland, 1994.

[BCL+94] J.R. Burch, E.M. Clarke, D.E. Long, K.L. McMillan, and D.L. Dill. Symbolic model checking for sequential circuit verification. *IEEE Transactions on Computer-Aided Design of Integrated Circuits and Systems*, 13(4):401–424, 1994.

[BCM+90] J.R. Burch, E.M. Clarke, K.L. McMillan, D.L. Dill, and L.J. Hwang. Symbolic model checking: 10^{20} states and beyond. In *Proceedings of the 5th IEEE Symposium on Logic in Computer Science, LICS'90*, 1990. Also *Information and Computation*, 98(2):142–170, 1992.

[BD87] S. Budkowski and P. Dembinski. An introduction to Estelle. *Computer Networks and ISDN Systems*, 4:3–23, 1987.

[BD97] M. Bidoit and M. Dauchet, editors. *TAPSOFT'97: Theory and Practice of Software Development, Proceedings of the 7th International Joint Conference CAAP/FASE, Lille, France, April 1997*, volume 1214 of *Lecture Notes in Computer Science*. Springer-Verlag, 1997.

[BDC92] L. Bernardinello and F. De Cindio. A survey of basic net models and modular subclasses. In Rozenberg [Roz92], pages 304–351.

[BDCM88] E. Battiston, F. De Cindio, and G. Mauri. OBJSA nets: A class of high level nets having objects as domains. In Rozenberg [Roz88], pages 20–43.

[BDKP91] E. Best, R. Devillers, A. Kiehn, and L. Pomello. Concurrent bisimulations in Petri nets. *Acta Informatica*, 28(3):231–264, 1991.

[BDR+84] J. Browne, D. Dubois, K. Rathmill, S.P. Sethi, and K.E. Stecke. Classification of flexible manufacturing systems. *The FMS Magazine*, 2(2), 1984.

[BDW+92] M.R. Barbacci, D.L. Doubleday, M.J. Weinstock, M.J. Gardner, and R.W. Lichotta. Building fault tolerant distributed applications with Durra. In *International IEEE Workshop on Configurable Distributed Systems*, 1992.

[BE86] G. Bruno and A. Elia. Operational specification of process control systems: Execution of Prot nets using OPS5. In *Proceedings of the 10th World IFIP Congress*, Dublin, 1986.

[BE94] H. Bachatène and P. Estraillier. Composing objects: state of the art and definition of a general support for specification composition. In *IFIP WG 10.3 Concurrent Systems, International Conference on Applications in Parallel and Distributed Computing, Caracas, Venezuela*, 1994.

[Ber86] G. Berthelot. Checking properties of nets using transformations. In G. Rozenberg, editor, *Advances in Petri Nets 1985*, volume 222 of *Lecture Notes in Computer Science*, pages 19–40. Springer-Verlag, 1986.

[Ber87] G. Berthelot. Transformations and decompositions of nets. In Brauer et al. [BRR87b], pages 359–376.

[Bes87] E. Best. Structure theory of Petri nets: the free choice hiatus. In Brauer et al. [BRR87b], pages 168–205.

[Béz87] J. Bézivin. Some experiments in object-oriented simulation. In *Proc. OOPSLA'87*, pages 394–405. ACM Press, 1987.

[BFR92] D. Buchs, J. Flumet, and P. Racloz. Producing prototypes from CO-OPN specifications. In RSP92 [RSP92], pages 77–93.

[BG91] D. Buchs and N. Guelfi. A concurrent object-oriented Petri nets approach for system specification. In ATPN91 [ATP91], pages 432–454.

[BG92] C. Brown and D. Gurr. Refinement and simulation of nets – a categorical characterisation. In Jensen [Jen92a], pages 76–92.

[BG94] C. Brown and D. Gurr. Temporal logic and categories of Petri nets. Technical report, University of Sussex, School of Cognitive and Computer Sciences, 1994.

[BG96] B. Best and B. Grahlmann. PEP – more than a Petri net tool. In T. Margaria and B. Steffen, editors, *Tools and Algorithms for the Construction and Analysis of Systems, Proceedings of the 2nd International Workshop, TACAS'96, Passau, Germany*, volume 1055 of *Lecture Notes in Computer Science*, pages 397–401. Springer-Verlag, 1996.

[BGV91] W. Brauer, R. Gold, and W. Vogler. A survey of behaviour and equivalence preserving refinements of Petri nets. In Rozenberg [Roz91a], pages 1–46.

[BH73] P. Brinch Hansen. *Operating System Principles.* Prentice Hall, Englewood Cliffs, NJ, 1973.

[BH99] T. Basten and J. Hooman. Process algebra in PVS. In W.R. Cleaveland, editor, *Tools and Algorithms for the Construction and Analysis of Systems, Proceedings of the 5th International Conference, TACAS'99, Amsterdam, The Netherlands*, volume 1579 of *Lecture Notes in Computer Science*, pages 270–284. Springer-Verlag, 1999.

[BJM97] A. Bouhoula, J.-P. Jouannaud, and J. Meseguer. Specification and proof in membership equational logic. In Bidoit and Dauchet [BD97], pages 67–92.

[BK90] Z. Banaszak and B. Krogh. Deadlock avoidance in flexible manufacturing systems with concurrently competing process flows. *IEEE Transactions on Robotics and Automation*, 6(6):724–734, 1990.

[BKK97] P. Borovansky, C. Kirchner, and H. Kirchner. Rewriting as a unified specification tool for logic and control: The ELAN language. In *Proceedings of International Workshop on Theory and Practice of Algebraic Specifications, ASF+SDF'97, Amsterdam.* Workshops in Computing, Springer-Verlag, 1997.

[BKKZ92] R. Budde, K. Kautz, K. Kuhlenkamp, and H. Züllighoven. *Prototyping - An Approach to Evolutionary System Development.* Springer-Verlag, 1992.

[BKMZ84] R. Budde, K. Kuhlenkamp, L. Mathiassen, and H. Züllighoven, editors. *Approaches to Prototyping.* Springer-Verlag, Berlin, 1984.

[BM86] G. Bruno and G. Marchetto. Process translatable Petri nets for the rapid prototyping of process control systems. *IEEE Transactions on Software Engineering*, SE-12(2):346–357, 1986.

[BMMV93] J.A. Bañares, P.R. Muro-Medrano, and J.L. Villarroel. Taking advantages of temporal redundancy in high level Petri nets implementations. In Marsan [Mar93], pages 32–48.

[Boe81] B.W. Boehm. *Software Engineering Economics.* Prentice Hall, Englewood Cliffs, NJ, 1981.

[Boe84] B.W. Boehm. Software life cycle factors. In R.V. Vick and C.V. Ramamoorthy, editors, *Handbook of Software Engineering*, pages 494–518. Van Nostrand Reinhold, 1984.

[Boe86] B.W. Boehm. A spiral model of software development and enhancement. *ACM SIGSOFT Software Engineering Notes*, 11(4):22–42, 1986.

[Bou88] L. Bougé. On the existence of symmetric algorithms to find leaders in networks of communicating sequential processes. *Acta Informatica*, 25:179–201, 1988.

[BP93] F. Bréant and J.F. Peyre. A new prototyping method of massively parallel applications using colored Petri nets. *Transputer Research and Applications*, 6:83–98, 1993.

[BP94] F. Bréant and J.F. Peyre. An improved massively parallel implementation of colored Petri nets specifications. In *IFIP WG 10.3 Working Conference on Programming Environments for Massively Parallel Distributed Systems, Ascona, Switzerland*, 1994.

[BR96] J. Billington and W. Reisig, editors. *Application and Theory of Petri Nets 1996, Proceedings of the 17th International Conference (ICATPN'96), Osaka, Japan*, volume 1091 of *Lecture Notes in Computer Science.* Springer-Verlag, 1996.

[Bra83] G.W. Brams. *Réseaux de Petri: Théorie et Pratique (2 vols.).* Masson, Paris, 1983.

[Bra90] J.C. Bradfield. Proving temporal properties of Petri nets. In Rozenberg [Roz91b], pages 29–47.

[Bra92] J.C. Bradfield. *Verifying Temporal Properties of Systems.* Birkhäuser, 1992.

[Bra96] R. Braek. SDL basics. *Computer Networks and ISDN Systems*, 28(12): 1585–1602, 1996.

[Bré90] F. Bréant. Tapioca: OCCAM rapid prototyping from Petri nets. In *Proceedings of the Vth Jerusalem Conference on Information Technology, Jerusalem, Israel*, 1990.

[Bré93] F. Bréant. *Décomposition de réseaux de Petri colorés. Modélisation d'architectures parallèles. Application au protypage sur des réseaux.* Thèse de doctorat, Université Pierre & Marie Curie (Paris VI), France, 1993.

[Bri87] E. Brinksma. A theory for the derivation of tests. In S. Aggarwal, editor, *Proceedings of the 8th International Conference on Protocol Specification, Testing and Verification.* North-Holland, 1987.

[Bro89] C. Brown. Relating Petri nets to formulae of linear logic. Technical report, Department of Computer Science, University of Edinburgh, 1989.

[BRR87a] W. Brauer, W. Reisig, and G. Rozenberg, editors. *Petri Nets: Applications and Relationships to other Models of Concurrency. Advances in Petri Nets 1986. Proceedings of an Advanced Course, Bad Honnef, Germany, Part II*, volume 255 of *Lecture Notes in Computer Science*. Springer-Verlag, 1987.

[BRR87b] W. Brauer, W. Reisig, and G. Rozenberg, editors. *Petri Nets: Central Models and their Properties. Advances in Petri Nets 1986. Proceedings of an Advanced Course, Bad Honnef, Germany, Part I*, volume 254 of *Lecture Notes in Computer Science*. Springer-Verlag, 1987.

[Bry86] R.E. Bryant. Graph-based algorithms for boolean function manipulation. *IEEE Transactions on Computers*, C-35(6):677–691, 1986.

[Bry92] R.E. Bryant. Symbolic boolean manipulation with ordered binary decision diagrams. *ACM Computing Surveys*, 24(3):293–318, 1992.

[BS69] K.A. Bartlett and R.A. Scantlebury. A note on reliable full-duplex transmission over half-duplex links. *Communications of the ACM*, 12:260–261, 1969.

[BSS91] K. Birman, A. Schiper, and P. Stephenson. Lightweight causal and atomic group multicast. *ACM Transactions on Computer Systems*, 9(3):272–314, 1991.

[BSV94] F. Balarin and A.L. Sangiovanni-Vincentelli. On the automatic computation of network invariants. In D.L. Dill, editor, *Proceedings of the 6th International Conference on Computer Aided Verification (CAV'94), Stanford, CA, USA*, volume 818 of *Lecture Notes in Computer Science*, pages 234–246. Springer-Verlag, 1994.

[BT80] J. Bergstra and J. Tucker. Characterization of computable data types by means of a finite equational specification method. In J.W. de Bakker and J. van Leeuwen, editors, *Automata, Languages and Programming, 7th Colloquium ICALP'80*, volume 85 of *Lecture Notes in Computer Science*, pages 76–90. Springer-Verlag, 1980.

[BT87] E. Best and P. S. Thiagarajan. Some classes of live and safe Petri nets. In K. Voss, H.J. Genrich, and G. Rozenberg, editors, *Concurrency and Nets – Advances in Petri Nets*, pages 71–94. Springer-Verlag, 1987.

[BT95] R.L. Bagrodia and Y.K. Tsay. Deducing fairness properties in UNITY logic – a new completeness result. *ACM Transactions on Programming Languages and Systems*, 17(1):17–27, 1995.

[Bur90] C. Burns. PROTO – a software requirements specification, analysis and validation tool. In RSP90 [RSP90], pages 196–203.

[Bur91] C. Burns. Parallel PROTO – a prototyping tool for analyzing and validating sequential and parallel processing software requirements. In RSP91 [RSP91], pages 151–160.

[Bur93] C. Burns. REE – a requirements engineering environment for analyzing & validating software and system requirements. In RSP93 [RSP93], pages 188–193.

[BV84] E. Best and K. Voss. Free choice systems have home states. *Acta Informatica*, 21:89–100, 1984.

[BV95a] T. Basten and M. Voorhoeve. An algebraic semantics for hierarchical P/T nets. Computing Science Report 95/35, Eindhoven University of Technology, Department of Mathematics and Computing Science, Eindhoven, The Netherlands, December 1995. An extended abstract appeared as [BV95b].

[BV95b] T. Basten and M. Voorhoeve. An algebraic semantics for hierarchical P/T nets (extended abstract). In De Michelis and Diaz [DMD95], pages 45–65.

[BW90] J.C.M. Baeten and W.P. Weijland. *Process Algebra*, volume 18 of *Cambridge Tracts in Theoretical Computer Science*. Cambridge University Press, Cambridge, UK, 1990.

[Cas91] V. Cassigneul. S.A.O. presentation. Technical Report 463.097/91, Aérospatiale, Toulouse, France, 1991.

[CC99] P. Cousot and R. Cousot. Refining model checking by abstract interpretation. *Automated Software Engineering Journal*, 6(1):69–95, 1999.

[CCG85] E. Castelain, D. Corbeel, and J. Gentina. Comparative simulations of control processes described by Petri nets. In *Proceedings of the IEEE COMPINT'85 Conference*, 1985.

[CCS90] J. M. Colom, J. Campos, and M. Silva. On liveness analysis through linear algebraic techniques. In *Proceedings of the AGM of Esprit BRA 3148 (DEMON)*, 1990.

[CCS91] J. Campos, G. Chiola, and M. Silva. Properties and performance bounds for closed free choice synchronized monoclass queueing networks. *IEEE Transactions on Automatic Control*, 36(12):1145–1155, 1991.

[CDE+99] M. Clavel, F. Duran, S. Eker, P. Lincoln, N. Marti-Oliet, J. Meseguer, and J. Quesada. *Maude: Specification and Programming in Rewriting Logic.* SRI International, 1999. Available via Internet at `http://maude.csl.sri.com`.

[CDF91a] G. Chiola, S. Donatelli, and G. Franceschinis. On parametric P/T nets and their modelling power. In ATPN91 [ATP91].

[CDF91b] G. Chiola, S. Donatelli, and G. Franceschinis. Priorities, inhibitor arcs, and concurrency in P/T nets. In ATPN91 [ATP91].

[CDFH91] G. Chiola, C. Dutheillet, G. Franceschinis, and S. Haddad. On well-formed coloured nets and their symbolic reachability graph. In Jensen and Rozenberg [JR91], pages 373–396.

[CDFH93] G. Chiola, C. Dutheillet, G. Franceschinis, and S. Haddad. Stochastic well-formed colored nets and symmetric modeling applications. *IEEE Transactions on Computers*, C-42(11):1343–1360, 1993.

[CELM96] M. Clavel, S. Eker, P. Lincoln, and J. Meseguer. Principles of Maude. *Electronic Notes in Theoretical Computer Science*, 4, 1996.

[CES86] E.M. Clarke, E.A. Emerson, and J. Sistla. Automatic verification of finite-state concurrent systems using temporal logic specification. *ACM Transactions on Programming Languages and Systems*, 8(2):244–263, 1986.

[CG87] E.M. Clarke and O. Grumberg. Avoiding the state explosion problem in temporal logic model checking algorithms. In *Proceedings of the 6th ACM Symposium on Principles of Distributed Computing, Vancouver, B.C., Canada*, pages 244–303, 1987.

[CGJ95] E.M. Clarke, O. Grumberg, and S. Jha. Verifying parameterized networks using abstraction and regular languages. In *Proceedings of the 6th International Conference on Concurrency Theory, Philadelphia, PA, USA*, volume 962 of *Lecture Notes in Computer Science*, pages 365–407. Springer-Verlag, 1995.

[Cha94] K.M. Chandy. Properties of concurrent programs. *Formal Aspects of Computing*, 6(6):607–619, 1994.

[Che91] G. Chehaibar. Use of reentrant nets in modular analysis of colored nets. In Jensen and Rozenberg [JR91], pages 596–617.

[CHEP71] F. Commoner, A. W. Holt, S. Even, and A. Pnueli. Marked directed graphs. *Journal of Computer and System Sciences*, 5:72–79, 1971.

[Chi91] G. Chiola. GreatSPN 1.5: software architecture. In *5th Conference on Modelling Techniques and Tools for Computer Performance Evaluation, Turin, Italy*, 1991.

[CHP93] J.M. Couvreur, S. Haddad, and J.F. Peyre. Generative families of positive invariants in coloured nets sub-classes. In Rozenberg [Roz93], pages 51–70.

[CK97a] S. Christensen and L.M. Kristensen. State space analysis of hierarchical coloured Petri nets. In Farwer et al. [FMS97].

[CK97b] P. Collette and E. Knapp. A foundation for modular reasoning about safety and progress properties of state-based concurrent programs. *Theoretical Computer Science*, 183:253–279, 1997.

[Cla98] Manuel Clavel. *Reflection in General Logics and in Rewriting Logic, with Applications to the Maude Language.* PhD thesis, University of Navarre, 1998.

[CM88] K.M. Chandy and J. Misra. *Parallel Program Design – A Foundation.* Addison-Wesley, 1988.

[CMS87] J.M. Colom, J. Martínez, and M. Silva. Packages for validating discrete production systems modeled with Petri nets. In P. Borne and S.G. Tzafestas, editors, *Applied Modelling and Simulation of Technological Systems*, pages 529–536. Elsevier Science Publishers B.V. (North-Holland), 1987.

[Coa92] P. Coad. Object oriented patterns. *Communications of the ACM*, 35(9):153–159, 1992.

[Col89] J.M. Colom. *Análisis estructural de Redes de Petri, programación lineal y geometría convexa.* PhD thesis, Universidad de Zaragoza, Departamento de Ingeniería Eléctrica e Informática, June 1989.

[Col93] P. Colette. Application of the composition principle to unity-like specifications. In Gaudel and Jouannaud [GJ93], pages 230–242.

[Coo80] C.L. Cook. Streamlining office procedures, an analysis using the information control nets. In *Proceedings of the AFIPS Working Conference*, 1980.

[Cou91] J.M. Couvreur. The general computation of flows for coloured nets. In Rozenberg [Roz91b], pages 204–223.

[CP92] S. Christensen and L. Petrucci. Towards a modular analysis of coloured Petri nets. In Jensen [Jen92a], pages 113–133.

[CP96] J.M. Couvreur and D. Poitrenaud. Model checking based on occurrence net graph. In R. Gotzhein and J. Bredereke, editors, *Proceedings of the Formal Description Techniques IX, Theory, Application and Tools, FORTE/PSTV'96, Kaiserslautern, Germany*, pages 380–395. Chapman & Hall, 1996.

[CS89] J.M. Colom and M. Silva. Improving the linearly based characterization of P/T nets. In De Michelis [DM89], pages 52–73.

[CS90a] J.M. Colom and M. Silva. Convex geometry and semiflows in P/T nets. A comparative study of algorithms for computation of minimal P-semiflows. In Rozenberg [Roz91a], pages 79–112.

[CS90b] J.M. Colom and M. Silva. Improving the linearly based characterization of P/T nets. In Rozenberg [Roz91a], pages 113–145.

[CSV86] J.M. Colom, M. Silva, and J.L. Villarroel. On software implementation of Petri nets and colored Petri nets using high-level concurrent languages. In *Proceedings of the 7th Workshop on Application and Theory of Petri Nets, Oxford, UK*, pages 207–241, 1986.

[CVWY90] C. Courcourbetis, M. Vardi, P. Wolper, and M. Yannakakis. Memory efficient algorithms for the verification of temporal properties. In *Proceedings of the 2nd International Workshop on Computer Aided Verification (CAV'90)*, volume 30 of *DIMACS*. North-Holland, 1990.

[CVWY92] C. Courcoubetis, M. Vardi, P. Wolper, and M. Yannakakis. Memory efficient algorithms for the verification of temporal properties. *Formal Methods in System Design*, 1:275–288, 1992.

[DAJ95] A.A. Desrochers and R.Y. Al-Jaar. *Application of Petri Nets in Manufacturing Systems: Modeling, Control and Performance Analysis.* IEEE Computer Society Press, 1995.

[DC90] A. Dollas and V. Chi. Rapid system prototyping in academic laboratories of the 1990's. In RSP90 [RSP90], pages 38–47.

[DDBD94] E. Dubois, Ph. Du Bois, and F. Dubru. Animating formal requirements specifications of cooperative information systems. In *Proceedings of the 2nd International Conference on Cooperative Information Systems, Toronto, Canada*, pages 216–225, May 1994.

[DE93] J. Desel and J. Esparza. Reachability in cyclic extended free choice nets. *Theoretical Computer Science*, 114:93–118, 1993.

[DE95] J. Desel and J. Esparza. *Free Choice Petri Nets*, volume 40 of *Cambridge Tracts in Theoretical Computer Science.* Cambridge University Press, Cambridge, UK, 1995.

[DE96] A. Diagne and P. Estraillier. Formal specification and design of distributed system. In *Proceedings of the 1st International IFIP Workshop on Formal Methods for Object-based Open Distributed Systems, Paris, France*, 1996.

[Des92] J. Desel. A proof of the rank theorem for extended free choice nets. In Jensen [Jen92a], pages 134–153.

[DG90] R. Di Giovanni. Petri nets and software engineering: HOOD nets. In Rozenberg [Roz91b], pages 123–138.

[DGK+92] J. Desel, D. Gomm, E. Kindler, R. Walter, and B. Paech. Bausteine eines kompositionalen Beweiskalküls für netzmodellierte Systeme. SFB-Bericht 342/16/92 A, Technische Universität München, Institut für Informatik, 1992.

[DHP+93] F. DiCesare, G. Harhalakis, J. M. Proth, M. Silva, and F. B. Vernadat. *Practice of Petri Nets in Manufacturing.* Chapman & Hall, 1993.

[Dia82] M. Diaz. Modeling and analysis of communication and cooperation protocols using Petri net based models. *Computer Networks*, 6:419–441, 1982.

[Dij65] E.W. Dijkstra. Solution of a problem in concurrent programming control. *Communications of the ACM*, 8(9):569, 1965.

[Dij68] E.W. Dijkstra. Co-operating sequential processes. In F. Genuys, editor, *Programming Languages*, pages 43–112. Academic Press N.Y., 1968.

[DJ90] N. Dershowitz and J.-P. Jouannaud. Rewrite systems. In J. van Leeuwen, editor, *Handbook of Theoretical Computer Science, B: Formal Methods and Semantics*, pages 243–320. North-Holland, 1990.

[DJS92] A. Desrochers, H. Jungnitz, and M. Silva. An approximation method for the performance analysis of manufacturing systems based on GSPNs. In *Proceedings of the Third International Conference on Computer Integrated Manufacturing and Automation Technology (CIMAT'92)*, pages 46–55. IEEE Computer Society Press, 1992.

[DK96] A. Diagne and F. Kordon. A multi formalisms prototyping approach from formal description to implementation of distributed systems. In *7th IEEE International Workshop on Rapid System Prototyping, Thessaloniki, Greece*, pages 102–107. IEEE Comp. Soc. Press, 1996.

[DM89] G. De Michelis, editor. *Application and Theory of Petri Nets 1989, Proceedings of the 10th International Conference (ICATPN'89), Bonn, Germany*, 1989.

[DM96] J. Desel and A. Merceron. Vicinity respecting homomorphisms for abstracting system requirements. Technical Report 337, Institut AIFB, Universität Karlsruhe, 1996.

[DMD95] G. De Michelis and M. Diaz, editors. *Application and Theory of Petri Nets 1995, Proceedings of the 16th International Conference (ICATPN'95), Turin, Italy*, volume 935 of *Lecture Notes in Computer Science.* Springer-Verlag, 1995.

[DMEM94] G. De Michelis, C. Ellis, and G. Memmi. Computer supported cooperative work, Petri nets and related formalisms. In Valette [Val94a].

[DMM89] P. Degano, J. Meseguer, and U. Montanari. Axiomatizing net computations and processes. In *Proceedings of the 4th IEEE Symposium on Logic in Computer Science, LICS'89*, pages 175–185. IEEE Computer Society Press, 1989.

[DNI95] F. Dupuy, G. Nilsson, and Y. Inoué. The TINA consortium: Toward networking telecommunications information services. *IEEE Communications Magazine*, 33(11):78–83, 1995.

[DNV90] R. De Nicola and F. Vaandrager. Action versus State Based Logics for Transition Systems. In I. Guessarian, editor, *Semantics of Systems of Concurrent Processes, LITP Spring School, Proceedings*, volume 469 of *Lecture Notes in Computer Science*, pages 407–419. Springer-Verlag, 1990.

[DoD83] DoD. *Reference Manual for the Ada programming language (ANSI/MIL-STD 1815A)*. AJPO, 1983.

[DS80] E.W. Dijkstra and C.S. Scholten. Termination detection for diffusing computations. *Information Processing Letters*, 11(1):1–4, 1980.

[DS98] J. Desel and M. Silva, editors. *Application and Theory of Petri Nets 1998, Proceedings of the 18th International Conference (ICATPN'98), Lisbon, Portugal*, volume 1420 of *Lecture Notes in Computer Science*. Springer-Verlag, 1998.

[DT98] C. Daws and S. Tripakis. Model checking of real-time reachability properties using abstractions. In *Tools and Algorithms for the Construction and Analysis of Systems, Proceedings of the 4th International Conference, TACAS'98, part of Theory and Practice of Software (ETAPS'98), Lisbon, Portugal*, volume 1384 of *Lecture Notes in Computer Science*, pages 313–329. Springer-Verlag, 1998.

[Dur99] F. Durán. *A Reflective Module Algebra with Applications to the Maude Language*. PhD thesis, University of Malaga, 1999.

[Dut92] C. Dutheillet. *Symétries dans les Réseaux Colorés. Définition, Analyse et Application à l'Evaluation de Performances*. Thèse de doctorat, Université Pierre & Marie Curie (Paris VI), France, 1992.

[EC97] J. Ezpeleta and J.M. Colom. Automatic synthesis of colored Petri nets for the control of FMS. *IEEE Transactions on Robotics and Automation*, 13(3):327–337, 1997.

[ECM95] J. Ezpeleta, J.M. Colom, and J. Martínez. A Petri net based deadlock prevention policy for flexible manufacturing systems. *IEEE Transactions on Robotics and Automation*, 11(2):173–184, 1995.

[ECS93] J. Ezpeleta, J.M. Couvreur, and M. Silva. A new technique for finding a generating family of siphons, traps and st-components. Application to colored Petri nets. In Rozenberg [Roz93], pages 126–147.

[EG92] P. Estraillier and C. Girault. Applying Petri net theory to the modelling, analysis and prototyping of distributed systems. In R. Zurawski, editor, *2nd IEEE International Workshop on Emerging Technologies and Factory Automation - State of the Art and Future Directions, Cairns, Australia*, pages 239–260. Elsevier, 1992.

[EH86] E.A. Emerson and J.Y. Halpern. "Sometimes" and "not never" revisited: On branching versus linear time. *Journal of the ACM*, 33(1):151–178, 1986.

[EKK94] W. El Kaim and F. Kordon. An integrated framework for rapid system prototyping and automatic code distribution. In *5th IEEE International Workshop on Rapid System Prototyping, Grenoble*, pages 52–61. IEEE Comp. Soc. Press, 1994.

[EL87] E.A. Emerson and C.L. Lei. Modalities for model checking: branching time strikes back. *Science of Computer Programming*, 8:275–306, 1987.

[Ell79] C.A. Ellis. Information control nets: A mathematical model for office information flow. In *Proceedings of the ACM Conference on Simulation, Modeling and Measurement of Computer Systems*, August 1979.

[EM85] H. Ehrig and B. Mahr. *Fundamentals of Algebraic Specification 1: Equations and Initial Semantics*, volume 6 of *EATCS Monographs on Theoretical Computer Science*. Springer-Verlag, 1985.

[EM92] J. Ezpeleta and J. Martínez. Petri nets as a specification language for manufacturing systems. In J.C. Gentina and S.G. Tzafestas, editors, *Robotics and Flexible Manufacturing Systems*, pages 427–436. Elsevier Science Publishers B.V. (North-Holland), 1992.

[Enc95] E. Encrenaz. *Une Méthode de Vérification de Propriétés de Programmes VHDL Basée sur des Modèles Formels de Réseaux de Petri.* Thèse de doctorat, Université Pierre & Marie Curie (Paris VI), France, 1995.

[Eng91] J. Engelfriet. Branching processes of Petri nets. *Acta Informatica*, 28:575–591, 1991.

[ERW96] J. Esparza, S. Romer, and Vogler W. An improvement of McMillan's unfolding algorithm. In *Tools and Algorithms for the Construction and Analysis of Systems, Proceedings of the 2nd International Workshop, TACAS'96, Passau, Germany*, volume 1055 of *Lecture Notes in Computer Science*, pages 87–106. Springer-Verlag, 1996.

[ES90] J. Esparza and M. Silva. On the analysis and synthesis of free choice systems. In Rozenberg [Roz91a], pages 243–286.

[ES91] J. Esparza and M. Silva. Circuits, handles, bridges and nets. In Rozenberg [Roz91a], pages 210–242.

[Esp92a] J. Esparza. Model checking using net unfoldings. Technical Report 14/92, Hildesheimer Informatik-Fachbericht, 1992.

[Esp92b] J. Esparza. A solution to the covering problem for 1-bounded conflict-free Petri nets using linear programming. *Information Processing Letters*, 41:313–319, 1992.

[Esp93] J. Esparza. Model checking using net unfoldings. In Gaudel and Jouannaud [GJ93], pages 613–628.

[Esp94] J. Esparza. Model checking using net unfoldings. *Science of Computer Programming*, 23:151–195, 1994.

[EW90] U. Engberg and G. Winskel. Petri nets as models for linear logic. Technical report, Computer Science Department, Aarhus University, Denmark, 1990.

[Far96] B. Farwer. Relating object systems to formulae of infinitary linear logic. Talk given at the Third Seminar on Algebra, Logic, and Geometry in Informatics (ALGI 3), Tokyo, 1996.

[Far98] B. Farwer. Towards linear logic Petri nets – From P/T-nets to object systems. Technical Report FBI-HH-B-211/98, Fachbereich Informatik, Universität Hamburg, 1998.

[Far99a] B. Farwer. *Linear Logic Based Calculi for Object Petri Nets.* Dissertation, Universität Hamburg, Fachbereich Informatik, 1999. Also published by Logos Verlag, Berlin, 2000.

[Far99b] B. Farwer. A linear logic view of object Petri nets. *Fundamenta Informaticae*, 37(3):225–246, 1999.

[Far99c] B. Farwer. Towards a linear logic based calculus for structural modifications of Petri nets. In H.-D. Burkhard, L. Czaja, H.-S. Nguyen, and P. Starke, editors, *Concurrency Specification and Programming (CSP'99), Proceedings*, pages 47–58. University of Warsaw, 1999.

[Far00a] B. Farwer. A multi-region linear logic based calculus for dynamic Petri net structures. *Fundamenta Informaticae*, 43(1–4):61–79, 2000.

[Far00b] B. Farwer. Relating formalisms for non-object-oriented object Petri nets. In H.-D. Burkhard, L. Czaja, and P. Starke, editors, *Concurrency Specification and Programming (CSP'2000), Proceedings*, pages 53–64. Informatik-Bericht Nr. 140, Vol. 1, Humboldt-Universität, Berlin, 2000.

[Feh93] R. Fehling. A concept of hierarchical Petri nets with building blocks. In Rozenberg [Roz93], pages 148–168.

[Fer90] A. Ferscha. Modeling mappings of parallel computations onto parallel architecture with the prm-net model. In C. Girault and M. Cosnard, editors, *Decentralized Systems. Proceedings of the IFIP-WG 10.3 Working Conference on Decentralized Systems, Lyon, France*, pages 349–362. North-Holland, 1990.

[Fin79] S.G. Finn. Resynch procedures and a fail-safe network protocol. *IEEE Transactions on Communications*, COM-27(6):840–845, 1979.

[Fin93] A. Finkel. The minimal coverability graph for Petri nets. In Rozenberg [Roz93], pages 210–243.

[Flo84] C. Floyd. A systematic look at prototyping. In Budde et al. [BKMZ84].

[FMS97] B. Farwer, D. Moldt, and M.O. Stehr, editors. *Proceedings of the Workshop on Petri Nets in Systems Engineering (PNSE'97)*, Hamburg, September 1997. Universität Hamburg (FBI-HH-B-205/97).

[Fra86] N. Francez. *Fairness.* Texts and Monographs in Computer Science. Springer-Verlag, New York, 1986.

[FS93] B. Folliot and P. Sens. Load balancing and fault tolerance management in distributed systems. In *5th European Workshop on Dependable Computing, Dependability, Decentralization, and Distribution (EWDC-5), Lisbon, Portugal*, 1993.

[GBK88a] J.C. Gentina, J.P. Bourey, and M. Kapusta. Coloured adaptive structured Petri nets (FMS design). *Systems*, 1(1):39–47, 1988.

[GBK88b] J.C. Gentina, J.P. Bourey, and M. Kapusta. Coloured adaptive structured Petri nets. Part II. *Systems*, 1(2):103–109, 1988.

[GD94] J.A. Goguen and R. Diaconescu. An Oxford survey of order sorted algebra. *Mathematical Structures in Computer Science*, 4(3):363–392, 1994.

[Gen88] H.J. Genrich. Equivalence transformations of Pr/Tr nets. In *Proceedings of the 9th European Workshop on Application and Theory of Petri Nets, Venezia, Italy*, 1988.

[GHJV95] E. Gamma, R. Helm, R. Johnson, and J. Vlissides. *Design Patterns.* Addison-Wesley, Reading, MA, 1995.

[GHT91] K. Grönbœk, A. Hviid, and R. Trigg. APPLBUILDER – an object oriented application generator supporting rapid prototyping. In *International Conference on Software Engineering & its Applications, Toulouse, France*, 1991.

[Gir87] J.-Y. Girard. Linear logic. *Theoretical Computer Science*, 50:1–102, 1987.

[GJ93] M.C. Gaudel and J.P. Jouannaud, editors. *TAPSOFT'93: Theory and Practise of Software Development, Proceedings of the 4th International Joint Conference CAAP/FASE, Orsay, France*, volume 688 of *Lecture Notes in Computer Science.* Springer-Verlag, 1993.

[GK95] M.C.A. van de Graaf and M. Kersten. Formalising business rules for long transaction management. In *Proceedings of the ERCIM Workshop on Database Issues and Cooperative Systems*, August 1995.

[GL79] H.J. Genrich and K. Lautenbach. The analysis of distributed systems by means of predicate/transition-nets. In G. Kahn, editor, *Semantics of Concurrent Computation, Evian, France*, volume 70 of *Lecture Notes in Computer Science*, pages 123–146. Springer-Verlag, 1979.

[GL81] H.J. Genrich and K. Lautenbach. System modelling with high-level Petri nets. *Theoretical Computer Science*, 13:109–136, 1981.

[God90a] P. Godefroid. Using partial orders to improve automatic verification methods. In *Proceedings of Computer Aided Verification (CAV'90)*, volume 30 of *DIMACS*, pages 321–340. North-Holland, 1990.

[God90b] P. Godefroid. Using partial orders to improve automatic verification methods (extended abstract). In E.M. Clarke and R.P. Kurshan, editors, *Proceedings of the 2nd International Workshop on Computer Aided Verification (CAV'90), New Brunswick, NJ, USA*, volume 531 of *Lecture Notes in Computer Science*, pages 176–185. Springer-Verlag, 1990.

[God94] P. Godefroid. *Partial-Order Methods for the Verification of Concurrent Systems, an Approach to the State-Explosion Problem.* Thèse de doctorat, Université de Liège, 1994. Also published as volume 1032 of *Lecture Notes in Computer Science*, Springer-Verlag, 1996.

[GP94] J.F. Groote and A. Ponse. Proof theory for μCRL: A language for processes with data. In D.J. Andrews, J.F. Groote, and C.A. Middelburg, editors, *Semantics of Specification Languages, Proceedings of the International Workshop SoSL, Utrecht, The Netherlands*, pages 232–251. Workshops in Computing, Springer-Verlag, 1994.

[GPC95] F. Girault and B. Pradin-Chézalviel. Petri Nets Theory Enhanced by Linear Logic. Manuscript, 1995.

[GPVW93] R. Gerth, D. Peled, M. Vardi, and P. Wolper. Simple on-the-fly automatic verification of linear temporal logic. In *Proceedings of the 15th International Conference on Protocol Specification, Testing and Verification, Warsaw, Poland*, 1993.

[GR89] Goldberg and Robson. *Smalltalk-80: The Language and its Implementation.* Addison-Wesley, 1989.

[Gra96] M.C.A van de Graaf. Active databases supporting workflow. Master's thesis, University of Amsterdam, April 1996.

[Gru97] O. Grumberg, editor. *Computer Aided Verification, Proceedings of the 9th International Conference (CAV'97), Haifa, Israel*, volume 1254 of *Lecture Notes in Computer Science.* Springer-Verlag, 1997.

[GS92] S. German and A.P. Sistla. Reasoning about systems with many processes. *Journal of the ACM*, 39:675–735, 1992.

[GVC95] F. García-Vallés and J.M. Colom. A boolean approach to the state machine decomposition of Petri nets with obdds. In *IEEE International Conference on Systems, Man and Cybernetics, Vancouver, B.C., Canada*, pages 3451–3456, 1995.

[GW91a] P. Godefroid and P. Wolper. A partial approach to model checking. In *Proceedings of the 6th IEEE Symposium on Logic in Computer Science, LICS'91, Amsterdam, Holland*, pages 406–415, 1991.

[GW91b] P. Godefroid and P. Wolper. Using partial orders for the efficient verification of deadlock freedom and safety properties. In Larsen and Skou [LS91], pages 332–342. Also published in *Formal Methods in System Design*, 2(2): 149–164, 1993.

[GW94] P. Godefroid and P. Wolper. A partial approach to model checking. *Information and Computation*, 110(2):305–326, 1994.

[GW96] R.J. van Glabbeek and W.P. Weijland. Branching time and abstraction in bisimulation semantics. *Journal of the ACM*, 43(3):555–600, 1996.

[GWM$^+$92] J.A. Goguen, T. Winkler, J. Meseguer, K. Futatsugi, and J.-P. Jouannaud. Introducing OBJ. In J.A. Goguen, D. Coleman, and R. Gallimore, editors, *Applications of Algebraic Specification Using OBJ.* Cambridge University Press, 1992.

[Hac72] M. Hack. Analysis of production schemata by Petri nets. Master's thesis, MIT, Cambridge, MA, USA, 1972. Corrections in Project MAC, *Computation Structures Note* 17, 1974.

[Hac74] M. Hack. Extended state-machine allocatable nets (esma), an extension of free choice Petri net results. Technical Report Memo 78-1, MIT, Project MAC, Computation Structures Group, 1974.

[Hac75] M. Hack. *Decidability Questions for Petri Nets.* PhD thesis, MIT, Cambridge, MA, December 1975. Also published as Technical Report 161, Lab. for Computer Science, June 1976.

[Had89] S. Haddad. A reduction theory for colored Petri nets. In Rozenberg [Roz89], pages 209–235.

[Hal91] H. Hallmann. A process model for prototyping. In *International Conference on Software Engineering & Its Applications, Toulouse, France*, 1991.

[Hau87] D. Hauschildt. A Petri net implementation. Technical report, Fachbereich Informatik, Universität Hamburg, Germany, 1987.

[Hau89] R. Haupt. A survey of priority rule-based scheduling. *OR-Spektrum*, 11:3–16, 1989.

[HBVB97] G. Hilderink, J. Broenink, W. Vervoort, and A. Bakkers. Communicating Java threads. In *Proc. Parallel Programming and Java, Twente, The Netherlands*, April 1997.

[HC88] S. Haddad and J.M. Couvreur. Towards a general and powerful computation of flows for parameterized coloured nets. In *Proceedings of the 9th European Workshop on Application and Theory of Petri Nets*, 1988.

[HC94a] M. Hammer and J. Champy. *Business Process Reengineering.* Cambridge University Press, Cambridge, UK, 1994.

[HC94b] F.S. Hsieh and S.C. Chang. Deadlock avoidance controller synthesis for flexible manufacturing systems. *IEEE Transactions on Robotics and Automation*, 10(2):196–209, 1994.

[HC96] B. Heyd and P. Crégut. A modular coding of UNITY in Coq. In J. von Wright, J. Grundy, and J. Harrison, editors, *Theorem Proving in Higher Order Logics. 9th International Conference, TPHOLs'96, Turku, Finland*, volume 1125 of *Lecture Notes in Computer Science*, pages 251–266. Springer-Verlag, 1996.

[Hei92] M. Heiner. Petri net based software validation, prospects and limitations. Technical Report TR92-022, International Computer Science Institute, Berkeley, CA, USA, March 1992.

[HG86] S. Haddad and C. Girault. Algebraic structure of flows of a regular coloured net. In *Proceedings of the 7th European Workshop on Application and Theory of Petri Nets, Oxford, UK*, 1986.

[HGP92] G.J. Holzmann, P. Godefroid, and D. Pirotin. Coverage preserving reduction strategies for the reachability analysis. In R.J. Linn and M.U. Uyar, editors, *Proceedings of the 7th International Conference on Protocol Specification, Testing and Verification, Florida, USA*, pages 349–364. IFIP, North-Holland, 1992.

[HHK96] R.H. Hardin, Z. Har'El, and R.P. Kurshan. Cospan. In R. Alur and T. A. Henzinger, editors, *Proceedings of the 8th International Conference on Computer Aided Verification (CAV'96), New Brunswick, NJ, USA*, volume 1102 of *Lecture Notes in Computer Science*, pages 423–427. Springer-Verlag, 1996.

[HI88] S. Hekmatpour and D. Ince. *Software Prototyping, Formal Methods and VDM.* Addison-Wesley, Wokingham, 1988.

[HITZ95] S. Haddad, J.M. Ilié, M. Taghelit, and B. Zouari. Symbolic reachability graph and partial symmetries. In De Michelis and Diaz [DMD95], pages 238–257.

[HJJJ84] P. Huber, A.M. Jensen, L.O. Jepsen, and K. Jensen. Towards reachability trees for high-level Petri nets. In Rozenberg [Roz85], pages 215–233.

[HJS90] P. Huber, K. Jensen, and R.M. Shapiro. Hierarchies in coloured Petri nets. In Rozenberg [Roz91a], pages 313–341.

[HM95] C. Hanen and A. Munier. Cyclic scheduling on parallel processors: an overview. In P. Chrétienne, E.G. Coffman, J.K. Lenstra, and Z. Liu, editors, *Scheduling Theory and Its Applications*, pages 193–226. John Wiley and Sons, 1995.

[HO80] G. Huet and D. Oppen. Equations and rewrite rules: A survey. In R.V. Book, editor, *Formal Language Theory: Perspectives and Open Problems*, pages 349–405. Academic Press, 1980.

[Hoa69] C.A.R. Hoare. An axiomatic basis for computer programming. *Communications of the ACM*, 12(10):576–583, 1969.

[Hoa74] C.A.R. Hoare. Monitors: an operating system structuring concept. *Communications of the ACM*, 17(10):549–557, 1974.

[Hoa85] C.A.R. Hoare. *Communicating Sequential Processes*. Prentice Hall, Englewood Cliffs, NJ, 1985.

[Hol85] G.J. Holzmann. Tracing protocols. *AT&T Technical Journal*, 64(12):2413–2434, 1985.

[Hol87] G.J. Holzmann. Automated protocol validation in ARGOS: Assertion proving and scatter searching. *IEEE Transactions on Software Engineering*, SE-13(6):683–696, 1987.

[Hol88] G.J. Holzmann. An improved protocol reachability analysis technique. *Software Practice and Experience*, 18(2):137–161, 1988.

[Hol96] G.J. Holzmann. Early fault detection tools. In *Tools and Algorithms for the Construction and Analysis of Systems*, volume 1055 of *Lecture Notes in Computer Science*, pages 1–13. Springer-Verlag, 1996.

[Hol97] G.J. Holzmann. The spin model checker. *IEEE Transactions on Software Engineering*, SE-23(5):279–295, 1997.

[HSV91] K.M. van Hee, L.J. Somers, and M. Voorhoeve. A formal framework for dynamic modelling of information systems. In H.G. Sol and K.M. van Hee, editors, *Dynamic Modelling of Information Systems*, pages 227–236. Elsevier Science Publishers, Amsterdam, 1991.

[Hul97] G.-J. Hulaas. *An Incremental Prototyping Methodology for Distributed Systems Based on Formal Specifications*. PhD thesis, Ecole Polytechnique Fédérale de Lausanne, Switzerland, 1997.

[HV87] D. Hauschildt and R. Valk. Safe states in banker-like resource allocation problems. *Information and Computation*, 75:232–263, 1987.

[IA97] J.M. Ilié and K. Ajami. Model checking through symbolic reachability graph. In Bidoit and Dauchet [BD97], pages 213–224.

[IR93] J.M. Ilié and O. Rojas. On well-formed nets and optimizations in enabling tests. In Marsan [Mar93], pages 300–318.

[ITU88] ITU-T. Recommendation Q700. Introduction to CCITT signalling system 7, 1988.

[ITU93] ITU-T. Recommendations Serie Q12xx. Intelligent Network, 1993.

[ITU96] ITU-T. Recommendations X901 & ISO IS 10746-1. Basic reference model of open distributed processing, Part 1: Overview and guide to use, 1996.

[JAD84] G. Juanole, B. Algayres, and J. Dufau. On communication protocol modelling and design. In Rozenberg [Roz85], pages 267–287.

[Jaf92] M.A. Jafari. An architecture for a shop-floor controller using colored Petri nets. *The International Journal of Flexible Manufacturing Systems*, 4(4):159–181, 1992.

[Jan94] P. Jancar. Decidability questions for bisimilarity of Petri nets and some related problems. In *Proceedings of the STACS'94, Caen, France*, volume 775 of *Lecture Notes in Computer Science*, pages 581–592. Springer-Verlag, 1994.

[JCHH91] K. Jensen, S. Christensen, P. Huber, and M. Holla. *Design/CPN. A Reference Manual.* Meta Software Corporation, 125 Cambridge Park Drive, Cambridge, Massachusetts 02140, USA, 1991. Information available at `http://www.daimi.au.dk/designCPN`.

[JD95] M.D. Jeng and F. DiCesare. Synthesis using resource control nets for modeling shared-resource systems. *IEEE Transactions on Robotics and Automation*, 11(3):317–327, 1995.

[Jen81] K. Jensen. Coloured Petri nets and the invariant-method. *Theoretical Computer Science*, 14:317–336, 1981.

[Jen92a] K. Jensen, editor. *Application and Theory of Petri Nets 1992, Proceedings of the 13th International Conference (ICATPN'92), Sheffield, UK*, volume 616 of *Lecture Notes in Computer Science.* Springer-Verlag, 1992.

[Jen92b] K. Jensen. *Coloured Petri Nets. Basic Concepts, Analysis Methods and Practical Use. Volume 1: Basic Concepts.* EATCS Monographs on Theoretical Computer Science. Springer-Verlag, 1992. 2nd edition 1996, 2nd corr. printing 1997.

[Jen94] K. Jensen. *Coloured Petri Nets. Basic Concepts, Analysis Methods and Practical Use. Volume 2: Analysis Methods.* EATCS Monographs on Theoretical Computer Science. Springer-Verlag, 1994.

[Jen97] K. Jensen. *Coloured Petri Nets. Basic Concepts, Analysis Methods and Practical Use. Volume 3: Practical Use.* EATCS Monographs on Theoretical Computer Science. Springer-Verlag, 1997.

[JJ89] C. Jard and T. Jeron. On-line model checking for finite linear temporal logic specifications. In J. Sifakis, editor, *Automatic Verification Methods for Finite State Systems, Proceedings of the International Workshop, Grenoble, France*, volume 407 of *Lecture Notes in Computer Science*, pages 189–196. Springer-Verlag, 1989.

[JJ91] C. Jard and T. Jeron. Bounded-memory algorithms for verification on-the-fly. In Larsen and Skou [LS91], pages 192–202.

[JLHB88] E. Jul, H. Levy, N. Hutchinson, and A. Black. Fine grained mobility in the Emerald system. *ACM Transactions on Computer Systems, 6(1)*, 1988.

[JMSW95] S.B. Joshi, E.G. Mettala, J.S. Smith, and R.A. Wysk. Formal models for control of flexible manufacturing cells: Physical and system mode. *IEEE Transactions on Robotics and Automation*, 11(4):558–570, 1995.

[Joh88] C. Johnen. Algorithmic verification of home spaces in P/T systems. In *IMACS 1988, Proceedings of the 12th World Congress on Scientific Computation*, pages 491–493, 1988.

[JR91] K. Jensen and G. Rozenberg, editors. *High-Level Petri Nets - Theory and Application.* Springer-Verlag, 1991.

[Jua95] G. Juanole. Réseaux de Petri et communications. In G. Juanole, A. Serhrouchni, and D. Seret, editors, *Réseaux de communication et conception de protocoles*, pages 266–287. Hermès, 1995.

[JV80] M. Jantzen and R. Valk. Formal properties of place/transition nets. In W. Brauer, editor, *Net Theory and Applications*, volume 84 of *Lecture Notes in Computer Science*, pages 165–212. Springer-Verlag, 1980.

[JV87] E. Jessen and R. Valk. *Rechensysteme - Grundlagen der Modellbildung.* Springer-Verlag, Berlin, 1987.

[Kan94] M.I. Kanovich. Petri nets, horn programs, linear logic, and vector games. In M. Hagiya and J.C. Mitchell, editors, *Theoretical Aspects of Computer Software, International Symposium TACS'94, Sendai, Japan*, volume 789 of *Lecture Notes in Computer Science*, pages 642–666. Springer-Verlag, 1994.

[KB86] B.H. Krogh and C.L. Beck. Synthesis of place/transition nets for simulation and control of flexible manufacturing systems. In *Proceedings of the IFIP Symposium on Large Scale Systems, Zurich, Switzerland*, 1986.

[KE91] F. Kordon and P. Estraillier. Complex systems rapid prototyping and environment abstraction. In RSP91 [RSP91], pages 34–46.

[KEC90] F. Kordon, P. Estraillier, and R. Card. Rapid Ada prototyping: principles and example of a complex application. In *9th IEEE International Conference on Computers and Communications, Phoenix, AZ, USA*, pages 453–460. IEEE Comp. Soc. Press, 1990.

[KEK94] F. Kordon and W. El Kaim. CPN/Tagada User Guide. Technical report, Laboratoire MASI, Université Pierre & Marie Curie (Paris VI), France, 1994.

[KEK95] F. Kordon and W. El Kaim. H-COSTAM: a hierarchical communicating state-machine model for generic prototyping. In *6th IEEE International Workshop on Rapid System Prototyping, Chapel Hill, North Carolina, USA*, pages 131–137. IEEE Comp. Soc. Press, 1995.

[Kel76] R.M. Keller. Formal verification of parallel programs. *Communications of the ACM*, 19(7):371–384, 1976.

[Kin95a] E. Kindler. Invariants, composition, and substitution. *Acta Informatica*, 32(4):299–312, 1995.

[Kin95b] E. Kindler. *Modularer Entwurf verteilter Systeme mit Petrinetzen.* Bertz Verlag, Berlin, 1995.

[KJ87] F. Krückeberg and M. Jaxy. Mathematical methods for calculating invariants in Petri nets. In Rozenberg [Roz87], pages 104–131.

[KKTT96] A. Kondratyev, M. Kishinevsky, A. Taubin, and S. Ten. A structural approach for the analysis of Petri nets by reduced unfoldings. In Billington and Reisig [BR96], pages 346–365.

[Klo92] J.W. Klop. Term rewriting systems. In S. Abramsky, D. Gabbay, and T. Maibaum, editors, *Handbook of Logic in Computer Science*, volume 2, pages 1–117. Oxford University Press, 1992.

[KM66] R.M. Karp and R.E. Miller. Properties of a model for parallel computations: Determinacy, termination, queueing. *SIAM Journal on Applied Mathematics*, 14(6):1390–1411, 1966.

[KM69] R.M. Karp and R.E. Miller. Parallel program schemata. *Journal of Computer Sciences*, 3:147–195, 1969.

[KM89] R.P. Kurshan and K. McMillan. A structural induction theorem for processes. In *Proceedings of the 8th Annual ACM Symposium on Principles of Distributed Computing, Alberta, Canada*, pages 239–247, 1989.

[KMM+97] Y. Kesten, O. Maler, M. Marcus, A. Pnueli, and E. Shahar. Symbolic model checking with rich assertional languages. In Grumberg [Gru97], pages 424–435.

[Kna94] E. Knapp. Soundness and completeness of UNITY logic. In P.S. Thiagarajan, editor, *Foundations of Software Technology and Theoretical Computer Science*, volume 880 of *Lecture Notes in Computer Science*, pages 378–389. Springer-Verlag, 1994.

[Kor92] F. Kordon. *Prototypage de systèmes parallèlles à partir de réseaux de Petri colorés, application au langage Ada dans un environnement centralisé ou réparti.* Thèse de doctorat, Université Pierre & Marie Curie (Paris VI), France, 1992.

[Kor94] F. Kordon. Proposal for a generic prototyping approach. In *IEEE Symposium on Emerging Technologies and Factory Automation, Tokyo, Japan*, pages 396–403. IEEE Comp. Soc. Press, 1994.

[KP91] F. Kordon and J.F. Peyre. Process decomposition for rapid prototyping of parallel systems. In *6th International Symposium on Computer and Information Science, Kener, Antalya, Turkey*, 1991.

[KP92] S. Katz and D. Peled. Verification of distributed programs using representative interleaving sequences. *Distributed Computing*, 6:107–120, 1992.

[KR96] E. Kindler and W. Reisig. Algebraic system nets for modelling distributed algorithms. *Petri Net Newsletter*, 51:16–31, December 1996.

[KV98] E. Kindler and H. Völzer. Flexibility in algebraic nets. In Desel and Silva [DS98], pages 345–384.

[Lak95] C.A. Lakos. From colored Petri nets to object Petri nets. In De Michelis and Diaz [DMD95], pages 278–297.

[Lak96] C. Lakos. The consistent use of names and polymorphism in the definition of objects Petri nets. In Billington and Reisig [BR96], pages 380–399.

[Lam74] L. Lamport. A new solution of Dijkstra's concurrent programming problem. *Communications of the ACM*, 17(8):453–455, 1974.

[Lam77] L. Lamport. Proving the correctness of multiprocess programs. *IEEE Transactions on Software Engineering*, SE-3(2):126–143, 1977.

[Lam78] L. Lamport. Time, clocks, and the ordering of events in a distributed system. *Communications of the ACM*, 21(7):558–564, 1978.

[Lam80] L. Lamport. "Sometimes" is sometimes "not never". In *Proceedings of the 7th Annual ACM Symposium on Principles of Programming Languages*, pages 174–185, January 1980.

[Lau87] K. Lautenbach. Linear algebraic techniques for place/transition nets. In Brauer et al. [BRR87b], pages 142–167.

[Law97] P. Lawrence, editor. *Workflow Handbook 1997, Workflow Management Coalition.* John Wiley and Sons, New York, 1997.

[LC94] C. Lakos and S. Christensen. A general systematic approach to arc extensions for coloured Petri nets. In Valette [Val94a], pages 338–357.

[Lea96] D. Lea. *Concurrent Programming in Java.* Addison-Wesley, Reading, MA, USA, 1996.

[Lil96] J. Lilius. $O(PN)^2$: An object-oriented Petri net programming notation. In F. G. Agha, De Cindio and A. Yonezawa, editors, *Proceedings of the 2nd Workshop on Object-Oriented Programming and Models of Concurrency, Osaka, Japan*, 1996. Workshop at ICATPN'96, see [BR96].

[Lio96] J.-L. Lions. Rapport de la Commission d'enquête ARIANE 501 Failure, July 1996. ESA/CNES, Paris, communiqué de presse conjoint ESA-CNES.

[Lis88] B. Liskov. Distributed programming in Argus. *Communications of the ACM*, 31(3):300–312, 1988.

[LK95] C. Lakos and C. Keen. An open software engineering environment based on object Petri nets. Technical report, Dept. of Computer Science, University of Tasmania, Hobart, Tasmania, Australia, 1995.

[LLRKS93] E.L. Lawler, J.K. Lenstra, A.H.G. Rinnooy Kan, and D.B. Shmoys. Sequencing and scheduling: algorithms and complexity. In S.C. Graves, A.H.G. Rinnooy Kan, and P.H. Zipkin, editors, *Handbook of Operations Research and Management Science*, pages 445–522. North-Holland, Amsterdam, 1993.

[Lom80] D.B. Lomet. Subsystems of processes with deadlock avoidance. *IEEE Transactions on Software Engineering*, SE-6(5):562–632, 1980.

[LP85] O. Lichtenstein and A. Pnueli. Checking that finite state concurrent programs satisfy their linear specification. In *Proceedings of the 12th ACM Symposium of Principles of Programming Languages, New Orleans, Louisiana, USA*, pages 97–107, January 1985.

[LP90] R. Lintulampi and P. Pulli. Graphics based prototyping of real-time systems. In RSP90 [RSP90], pages 128–137.

[LPS81] D. Lehmann, A. Pnueli, and J. Stavi. Impartiality, justice and fairness. In S. Even and O. Kariv, editors, *Automata, Languages and Programming, Eigth Colloquium ICALP'81, Acre (Akko), Israel*, volume 115 of *Lecture Notes in Computer Science*, pages 264–277. Springer-Verlag, 1981.

[LR78] L.H. Landweber and E.L. Robertson. Properties of conflict-free and persistent Petri nets. *Journal of the ACM*, 25(3):352–364, 1978.

[LS91] K.G. Larsen and A. Skou, editors. *Proceedings of the 3rd Workshop on Computer Aided Verification (CAV'91), Aalborg, Denmark*, volume 575 of *Lecture Notes in Computer Science*. Springer-Verlag, 1991.

[LSB93] Luqi, M. Shing, and J. Brockett. Real-time scheduling for software prototyping. In RSP93 [RSP93], pages 150–163.

[LSS92] P. Lincoln, A. Scedrov, and N. Shankar. Decision problems for propositional linear logic. *Annals of Pure and Applied Logic*, 56:239–311, 1992.

[LSY94] J. Li, I. Suzuki, and M. Yamashita. A new structural induction theorem for rings of temporal Petri nets. *IEEE Transactions on Software Engineering*, SE-20(2):115–126, 1994.

[LT80] I. Ladd and D. Tschritzis. An office form flow model. In *Proceedings of the AFIPS Working Conference*, 1980.

[Luq89] Luqi. Software evolution through rapid prototyping. *IEEE Computer*, 22, 1989.

[Luq92] Luqi. Computer aided system prototyping. In RSP92 [RSP92], pages 50–57.

[Mac91] R. Mackenthun. Using temporal logic to develop correct coloured nets. Fachbereichsmitteilung FBI-HH-M-227/93, Universität Hamburg, Fachbereich Informatik, 1991.

[Mac98] R. Mackenthun. A method to develop mutex-algorithms (work in progress), 1998.

[Man90] D. Mange. *Systèmes Microprogrammés: Une Introduction au Magiciel.* Presses Polytechniques et Universitaires Romandes, 1990.

[Mar93] M.A. Marsan, editor. *Application and Theory of Petri Nets 1993, Proceedings of the 14th International Conference (ICATPN'93), Chicago, Illinois, USA*, volume 691 of *Lecture Notes in Computer Science*. Springer-Verlag, 1993.

[Mat89] F. Mattern. Virtual time and global states of distributed systems. In M. Cosnard et al., editor, *Parallel and Distributed Algorithms*, pages 215–226. North-Holland, 1989.

[Maz87] A. Mazurkiewicz. Trace theory. In Brauer et al. [BRR87a], pages 279–324.

[McM92] K.L. McMillan. Using unfoldings to avoid the state explosion problem in the verification of asynchronous circuits. In G. v. Bochmann and D.K. Probst, editors, *Proceedings of the 4th International Conference on Computer Aided Verification (CAV'92), Montreal, Canada*, volume 663 of *Lecture Notes in Computer Science*, pages 164–177. Springer-Verlag, 1992.

[McM93] K.L. McMillan. *Symbolic Model Checking.* Kluwer, 1993.

[McM95] K.L. McMillan. Trace theoretic verification of asynchronous circuits using unfoldings. In P. Wolper, editor, *Proceedings of the 7th International Conference on Computer Aided Verification (CAV'95), Liège, Belgium*, volume 939 of *Lecture Notes in Computer Science*, pages 180–195. Springer-Verlag, 1995.

[Meh84] K. Mehlhorn. *Data Structures and Algorithms 2: Graph Algorithms and NP-Completeness*, volume 2 of *EATCS Monographs on Theoretical Computer Science*. Springer-Verlag, Berlin, 1984.

[Mem78] G. Memmi. *Fuites et Semi-flots dans les Réseaux de Petri.* Thèse de doctorat, Université Pierre & Marie Curie (Paris VI), France, December 1978.

[Mes92] J. Meseguer. Conditional rewriting logic as a unified model of concurrency. *Theoretical Computer Science*, 96:73–155, 1992.

[Mes93] J. Meseguer. A logical theory of concurrent objects and its realization in the Maude language. In G. Agha, P. Wegner, and A. Yonezawa, editors, *Research Directions in Concurrent Object-Oriented Programming*. MIT Press, 1993.

[Mes98a] J. Meseguer. Personal communication, 6th May 1998.

[Mes98b] J. Meseguer. Membership algebra as a semantic framework for equational specification. In F. Parisi-Presicce, editor, *Recent Trends in Algebraic Development Techniques, 12th International Workshop, WADT '97, Tarquinia, Italy, June 3-7, 1997, Selected Papers*, volume 1376 of *Lecture Notes in Computer Science*, pages 18–61. Springer-Verlag, 1998.

[MG85] J. Meseguer and J.A. Goguen. Initiality, induction and computability. In M. Nivat and J.C. Reynolds, editors, *Algebraic Methods in Semantics*, pages 459–540. Cambridge Univ. Press, 1985.

[MGK89] S.C. Murphy, P. Gunningberg, and J.P.J. Kelly. Implementing protocols with multiple specifications : Experiences with Estelle, LOTOS and SDL. In E. Brinksma, G. Scollo, and C.A. Vissers, editors, *Proceedings of the 9th International Conference on Protocol Specification, Testing and Verification, Enschede, The Netherlands*. North-Holland, 1989.

[Mil89] R. Milner. *Communication and Concurrency*. Prentice Hall, Englewood Cliffs, NJ, 1989.

[Mis95] J. Misra. A logic for concurrent programming: Safety and progress. *Journal of Computer and Software Engineering*, 3(2):251–257, 1995.

[MKM86] T. Murata, N. Komoda, and K. Matsumoto. A Petri nets based factory automation controller for flexible and maintainable control specification. *IEEE Transactions on Industrial Electronics*, 33, 1986.

[MKS89] J. Magee, J. Kramer, and M. Sloman. Constructing distributed systems in conic. *IEEE Transactions on Software Engineering*, SE-15(6), 1989.

[MM88] J. Meseguer and U. Montanari. Petri nets are monoids: A new algebraic foundation for net theory. Technical Report SRI-CSL-88-3, SRI International, Computer Science Laboratory, Stanford, CA, USA, 1988.

[MM90] J. Meseguer and U. Montanari. Petri nets are monoids. *Information and Computation*, 88(2):105–155, 1990.

[MMS87] J. Martínez, P. Muro, and M. Silva. Modeling, validation and software implementation of production systems using high level Petri nets. In *Proceedings of the IEEE International Conference on Robotics and Automation*, pages 1180–1184, Raleigh, NC, USA), 1987.

[MN95] O. Müller and T. Nipkow. Combining model checking and deduction for I/O-automata. In E. Brinksma et al., editors, *Tools and Algorithms for the Construction and Analysis of Systems. Selected Papers of the 1st International Workshop, TACAS'95, Aarhus, Denmark*, volume 1019 of *Lecture Notes in Computer Science*, pages 1–16. Springer-Verlag, 1995.

[Mol96] D. Moldt. *Höhere Petrinetze als Grundlage für Systemspezifikationen*. Dissertation, Universität Hamburg, Fachbereich Informatik, August 1996.

[MOM91] N. Martí-Oliet and J. Meseguer. From Petri nets to linear logic through categories: A survey. *International Journal of Foundations of Computer Science*, 2(4):297–399, 1991.

[MP89] Z. Manna and A. Pnueli. The anchored version of the temporal framework. In J. W. de Bakker et al., editors, *Linear Time, Branching Time and Partial Order in Logics and Models of Concurrency*, volume 354 of *Lecture Notes in Computer Science*, pages 201–284. Springer-Verlag, 1989.

[MP92] Z. Manna and A. Pnueli. *The Temporal Logic of Reactive and Concurrent Systems – Specification*. Springer-Verlag, New York, 1992.

[MR80] G. Memmi and G. Roucairol. Linear algebra in net theory. In W. Brauer, editor, *Net Theory and Applications*, volume 84 of *Lecture Notes in Computer Science*, pages 213–223. Springer-Verlag, Berlin, 1980.

[MR93] A. Mostefaoui and M. Raynal. Causal multicasts in overlapping groups: Towards a low cost approach. Technical Report TR93-09, ESPRIT Basic Research Project BROADCAST, August 1993.

[MS94] S.J Mellor and S. Shlaer. A deeper look at ... at execution and translation. *Journal of Object-Oriented Programming*, 7(3):24–26, 1994.

[MS95] P. McBrien and A.H. Seltveit. Coupling process models and business rules. In *Proceedings of the IFIP Working Conference on Information Systems in Decentralised Organisations*, 1995.

[MT94] MARS-Team. *The CPN-AMI environment version 1.3.* Lip6 Lab, Institut Blaise Pascal, Université Pierre & Marie Curie (Paris VI), France, June 1994.

[Mur83] K.G. Murty. *Linear Programming.* John Wiley and Sons, New York, 1983.

[Mur89] T. Murata. Petri nets: Properties, analysis and applications. *Proceedings of the IEEE*, 77(4):541–580, 1989.

[MV87] G. Memmi and J. Vautherin. Analysing nets by the invariant method. In Brauer et al. [BRR87a], pages 300–336.

[MV91] R. Mackenthun and R. Valk. Verifying coloured nets in UNITY-style. Fachbereichsmitteilung FBI-HH-M-222/93, Universität Hamburg, Fachbereich Informatik, 1991.

[MZ95] T. Mowbray and R. Zahavu. *The Essential CORBA: Systems Integration Using Distributed Objects.* John Wiley and Sons, 1995.

[NHS83] R. Nelson, L. Haibt, and P. Sheridan. Casting Petri nets into programs. *IEEE Transactions on Software Engineering*, SE-9(5):590–602, 1983.

[Nie85] O.M. Nierstrasz. Message flow analysis. In D. Tsichritzis, editor, *Office Automation: Concepts and Tools*, chapter 12, pages 283–314. Springer-Verlag, 1985.

[NM94] M. Notomi and T. Murata. Hierarchical reachability graph of bounded Petri nets for concurrent-software analysis. *IEEE Transactions on Software Engineering*, SE-20(5):325–336, 1994.

[NPW80] M. Nielsen, G. Plotkin, and G. Winskel. Petri nets, events structures and domains, Part I. *Theoretical Computer Science*, 13(1):85–108, 1980. Previously published in *Semantics of Concurrent Computation.* Volume 70 of *Lecture Notes in Computer Science*, pages 266–284, Springer-Verlag, 1979.

[NRKT89] G.L. Nemhauser, A.H. Rinnooy Kan, and M.J. Todd. *Optimization*, volume 1 of *Handbook in Operations Research and Management Science.* North-Holland, Amsterdam, 1989.

[NV85] Y. Narahari and N. Viswanadham. A Petri net approach to the modelling and analysis of flexible manufacturing systems. *Annals of Operations Research*, 3:449–472, 1985.

[NW88] G.L. Nemhauser and L.A. Wolsey. *Integer and Combinatorial Optimization.* John Wiley and Sons, 1988.

[Obe81] H. Oberquelle. More on the readability of net representations. *Petri Nets and Related System Models*, GI-Newsletter No. 9:5–8, October 1981.

[Old91] E.R. Olderog. *Nets, Terms and Formulas.* Oxford University Press, 1991.

[Oza96] C. Ozanne. *Conception d'applications client-serveur: modèles d'architecture fonctionnelle et opérationnelle.* Thèse de doctorat, Université Pierre & Marie Curie (Paris VI), France, 1996.

[Pac92a] J. Pachl. A simple proof of a completeness result for leads-to in the UNITY logic. *Information Processing Letters*, 41, 1992.

[Pac92b] J. Pachl. A simple proof of a completeness result for leads-to in the UNITY logic (corrigendum). *Information Processing Letters*, 44, 1992.

[Pae95] P. Paeppinghaus. On the logic of UNITY. *Theoretical Computer Science*, 139(1):27–67, 1995.

[Pal91] M. Paludetto. *Sur la commande de procédés industriels: une méthodologie basée objets et réseaux de Petri.* Thèse de doctorat, Université Paul Sabatier, Toulouse, France, 1991.

[Pal97] Pallas Athena. *Protos User Manual.* Pallas Athena BV, Plasmolen, The Netherlands, 1997.

[Pau99] L. C. Paulson. Mechanizing UNITY in Isabelle. Technical Report 467, Computer Laboratory, University of Cambridge, UK, May 1999.

[Pel93] D. Peled. All from one, one for all: on model checking using representatives. In C. Courcoubetis, editor, *Proceedings of the 5th International Conference on Computer Aided Verification (CAV'93), Elounda, Greece*, volume 697 of *Lecture Notes in Computer Science*, pages 409–423. Springer-Verlag, 1993.

[Pel94] D. Peled. Combining partial order reductions with on-the-fly model-checking. In D.L. Dill, editor, *Proceedings of the 6th International Conference on Computer Aided Verification (CAV'94), Stanford, CA, USA*, volume 818 of *Lecture Notes in Computer Science*, pages 377–390. Springer-Verlag, 1994.

[Pet62] C.A. Petri. *Kommunikation mit Automaten.* Dissertation, Technische Universität Darmstadt, 1962. Institut für Instrumentelle Mathematik, Schriften des IIM Nr. 2 (in German), 1962. English translation in: Griffiss Air Force Base, New York, Technical Report RADC-TR-65–377, Vol. 1, 1966.

[Pet81] J.L. Peterson. *Petri Net Theory and the Modeling of Systems.* Prentice Hall, New York, 1981.

[Pet96] C.A. Petri. Nets, time and space. *Theoretical Computer Science*, 153:3–48, 1996.

[Pey93] J.-F. Peyre. *Résolution paramétrée de systèmes linéaires; applications au calcul d'invariants positifs dans les réseaux colorés, à la validation formelle et à la génération de code.* Thèse de doctorat, Université Pierre & Marie Curie (Paris VI), France, 1993.

[PG92] M. Pezzè and C. Ghezzi. Cabernet: a customizable environment for the specification and analysis of real-time systems. Technical report, Dipartimento di Elettronica e dell'Informazione, Politecnico di Milano, Italy, 1992.

[PHB93] H.T. Papadopoulos, C. Heavy, and J. Browne. *Queuing Theory in Manufacturing Systems Analysis and Design.* Chapman & Hall, London, 1993.

[Pla96] E.A.H. Platier. *A Logistical Perspective Upon Business Processes.* PhD thesis, Eindhoven University of Technology, 1996.

[Pnu81] A. Pnueli. The temporal semantics of concurrent programs. *Theoretical Computer Science*, 13:45–60, 1981.

[Poi96] D. Poitrenaud. *Graphes de Processus Arborescents pour la Vérification de Propriétés de Systèmes Concurrents.* Thèse de doctorat, Université Pierre & Marie Curie (Paris VI), France, December 1996.

[PP94] D. Peled and A. Pnueli. Proving partial-order properties. *Theoretical Computer Science*, 126:143–182, 1994.

[PRCB94] E. Pastor, O. Roig, J. Cortadella, and R.M. Badia. Petri net analysis using boolean manipulation. In Valette [Val94a], pages 416–435.

[PRS92] L. Pomello, G. Rozenberg, and C. Simone. A survey of equivalence notions for net based systems. In Rozenberg [Roz92], pages 410–472.

[Pru98] D. Prun. *Méthodologie de conception de composants logiciels coopératifs: une approche pour l'observation, la mise au point et la maintenance évolutive d'applications réparties.* Thèse de doctorat, Université Pierre & Marie Curie (Paris VI), France, 1998.

[PSL80] M. Pease, R. Shostak, and L. Lamport. Reaching agreement in the presence of faults. *Journal of the ACM*, 27(2):228–234, 1980.

[PTVdB90] T.A. Petersen, D.A. Thomae, and D.E. Van den Bout. The anyboard: a rapid-prototyping system for use in teaching digital circuit design. In RSP90 [RSP90], pages 25–32.

[PWW96] D. Peled, T. Wilke, and P. Wolper. An algorithmic approach for checking closure properties of ω-regular languages. In U. Montanari and V. Sassone, editors, *Proceedings of the 7th International Conference on Concurrency Theory, Pisa, Italy*, volume 1119 of *Lecture Notes in Computer Science*, pages 596–626. Springer-Verlag, 1996.

[PX96] J.M. Proth and X. Xie. *Petri Nets. A Tool for Design and Management of Manufacturing Systems.* John Wiley and Sons, 1996.

[Raf95] O. Rafiq. Historique et problématique des réseaux informatiques. In G. Juanole, A. Serhrouchni, and D. Seret, editors, *Réseaux de communication et conception de protocoles*, pages 11–33. Hermès, 1995.

[Rao95] J.R. Rao. *Extensions of the UNITY Methodology: Compositionality, Fairness and Probability in Parallelism*, volume 908 of *Lecture Notes in Computer Science.* Springer-Verlag, 1995.

[Ray86] M. Raynal. *Algorithms for Mutual Exclusion.* North Oxford Academic Publishers Ltd, 1986.

[RBP+91] J. Rumbaugh, M. Blaha, W. Premeralani, F. Eddy, and W. Lorensen. *Object-Oriented Modeling and Design.* Prentice Hall, Englewood Cliffs, NJ, 1991.

[Rei83] W. Reisig. *System Design Using Petri Nets.* Springer-Verlag, Berlin, 1983.

[Rei85a] W. Reisig. *Petri Nets – An Introduction*, volume 4 of *Springer EATCS Monographs in Theoretical Computer Science.* Springer-Verlag, Berlin, 1985.

[Rei85b] W. Reisig. Petri nets with individual tokens. *Theoretical Computer Science*, 41:185–213, 1985.

[Rei91] W. Reisig. Petri nets and algebraic specifications. *Theoretical Computer Science*, 80:1–34, 1991.

[Rei92] W. Reisig. Combining Petri nets and other formal methods. In Jensen [Jen92a], pages 24–44.

[Rei98] W. Reisig. *Elements of Distributed Algorithms: Modeling and Analysis with Petri Nets.* Springer-Verlag, Berlin, 1998.

[Rib88] S. Ribaric. Knowledge representation scheme based on Petri net theory. *International Journal of Pattern Recognition and Artificial Intelligence*, 2(4):691–700, 1988.

[Röm96] S. Römer. An efficient algorithm for the computation of unfoldings of finite and safe Petri nets (on efficiently implementing McMillan's unfolding algorithm). Technical report, Technische Universität München, Institut für Informatik, 1996.

[Roz85] G. Rozenberg, editor. *Advances in Petri Nets 1984*, volume 188 of *Lecture Notes in Computer Science.* Springer-Verlag, 1985.

[Roz87] G. Rozenberg, editor. *Advances in Petri Nets 1987*, volume 266 of *Lecture Notes in Computer Science.* Springer-Verlag, 1987.

[Roz88] G. Rozenberg, editor. *Advances in Petri Nets 1988*, volume 340 of *Lecture Notes in Computer Science.* Springer-Verlag, 1988.

[Roz89] G. Rozenberg, editor. *Advances in Petri Nets 1989*, volume 424 of *Lecture Notes in Computer Science.* Springer-Verlag, 1989.

[Roz91a] G. Rozenberg, editor. *Advances in Petri Nets 1990*, volume 483 of *Lecture Notes in Computer Science.* Springer-Verlag, 1991.

[Roz91b] G. Rozenberg, editor. *Advances in Petri Nets 1991*, volume 524 of *Lecture Notes in Computer Science.* Springer-Verlag, 1991.

[Roz92] G. Rozenberg, editor. *Advances in Petri Nets 1992*, volume 609 of *Lecture Notes in Computer Science.* Springer-Verlag, 1992.

[Roz93] G. Rozenberg, editor. *Advances in Petri Nets 1993*, volume 674 of *Lecture Notes in Computer Science*. Springer-Verlag, 1993.

[RS93] J.K. Rho and F. Somenzi. Automatic generation of network invariants for the verification of iterative sequential systems. In C. Courcoubetis, editor, *Proceedings of the 5th International Conference on Computer Aided Verification (CAV'93), Elounda, Greece*, volume 697 of *Lecture Notes in Computer Science*, pages 123–137. Springer-Verlag, 1993.

[RSP90] *1st IEEE International Workshop on Rapid System Prototyping, Research Triangle Park Institute, North Carolina, USA*. IEEE Comp. Soc. Press, 1990.

[RSP91] *2nd IEEE International Workshop on Rapid System Prototyping, Research Triangle Park Institute, North Carolina, USA*. IEEE Comp. Soc. Press, 1991.

[RSP92] *3rd IEEE International Workshop on Rapid System Prototyping, Research Triangle Park Institute, North Carolina, USA*. IEEE Comp. Soc. Press, 1992.

[RSP93] *4th IEEE International Workshop on Rapid System Prototyping, Research Triangle Park Institute, North Carolina, USA*. IEEE Comp. Soc. Press, 1993.

[RTS95] L. Recalde, E. Teruel, and M. Silva. On well-formedness analysis: The case of deterministic systems of sequential processes. In J. Desel, editor, *Proceedings of the International Workshop on Structures in Concurrency Theory, Berlin, Germany*, pages 279–293. Workshops in Computing, Springer-Verlag, 1995.

[RTS96] L. Recalde, E. Teruel, and M. Silva. Modeling and analysis of cooperating processes with Petri nets. Research report, Dep. Informática e Ingeniería de Sistemas, Universidad de Zaragoza, Spain, 1996.

[RV87] W. Reisig and J. Vautherin. An algebraic approach to high level Petri nets. In *Proceedings of the 8th European Workshop on Application and Theory of Petri Nets*, pages 51–72. Universidad de Zaragoza, Spain, 1987.

[SACV87] A. Sahraoui, H. Atabakhche, M. Courvoisier, and R. Valette. Joining Petri nets and knowledge based systems for monitoring purposes. In *Invited Sessions: Petri Nets and Flexible Manufacturing Systems, IEEE International Conference on Robotics and Automation, Raleigh, USA*, pages 1160–1165, 1987.

[San91] B. A. Sanders. Eliminating the substitution axiom from UNITY logic. *Formal Aspects of Computing*, 3:189–205, 1991.

[Sas95] V. Sassone. On the category of Petri net computations. In P.D. Mosses, M. Nielsen, and M.I. Schwartzbach, editors, *TAPSOFT'95: Theory and Practice of Software Development, Proceedings of the 6th International Joint Conference CAAP/FASE, Aarhus, Denmark*, volume 915 of *Lecture Notes in Computer Science*, pages 334–348. Springer-Verlag, 1995.

[SB93] C. Sibertin-Blanc. A client-server protocol for the composition of Petri nets. In Marsan [Mar93], pages 377–396.

[SB94] C. Sibertin-Blanc. Cooperative nets. In Valette [Val94a], pages 471–490.

[SC88] M. Silva and J.M. Colom. On the computation of structural synchronic invariants in P/T nets. In Rozenberg [Roz88], pages 386–417.

[Sce90] A. Scedrov. A brief guide to linear logic. *Bulletin of the EATCS*, 41:154–165, 1990.

[Sce93] A. Scedrov. A brief guide to linear logic. In G. Rozenberg and A. Salomaa, editors, *Current Trends in Theoretical Computer Science*, pages 377–394. World Scientific, 1993. Revised version of [Sce90].

[Sch95] K. Schmidt. Symmetry calculation. In *Proceedings of the Workshop Concurrency, Specification, and Programming*, pages 147–161, Warsaw, Poland, 1995.

[SG89] Z. Shtadler and O. Grumberg. Network grammars, communication behaviors and automatic verification. In J. Sifakis, editor, *International Workshop on*

Automatic Verification Methods for Finite State Systems, Grenoble, France, volume 407 of *Lecture Notes in Computer Science*, pages 151–165. Springer-Verlag, 1989.

[Sha93] A.U. Shankar. An introduction to assertional reasoning for concurrent systems. *ACM Computing Surveys*, 25(3):226–262, 1993.

[She96] A. Sheth. Workflow process automation in information systems: State-of-the-art and future directions. In *Proceedings of the NSF Workshop*, May 1996.

[Sil81] M. Silva. Sur le concept de macroplace et son utilisation pour l'analyse des réseaux de Petri. *R.A.I.R.O. Automatique/Systems Analysis and Control*, 15(4):335–345, 1981.

[Sil85] M. Silva. *Las redes de Petri en la Automática y la Informática.* Editorial AC, Madrid, 1985.

[Sil89] M. Silva. Logical controllers. In *Proceedings of the IFAC Symposium on Low Cost Automation*, pages 157–166, 1989.

[SKL94] J.B. Stefani, R. Kung, and Y. Lepetit. L'architecture à long terme SERENITE. *L'Echo des recherches*, No. 157:45–48, 1994.

[SM83] I. Suzuki and T. Murata. A method for stepwise refinements and abstraction of Petri nets. *Journal of Computer and System Sciences*, 27:51–76, 1983.

[SM99] M.-O. Stehr and J. Meseguer. Pure type systems in rewriting logic. In *Proceedings of LFM'99: Workshop on Logical Frameworks and Meta-languages, Paris, France*, 1999.

[Sof96] Software-Ley. *COSA User Manual.* Software-Ley GmbH, Pulheim, Germany, 1996.

[Som96] I. Sommerville. *Software Engineering.* International Computer Science Series. Addison-Wesley, Wokingham, 5th edition, 1996.

[Sou89] Y. Souissi. Compositions of nets via a communication medium. In De Michelis [DM89].

[Sou91] Y. Souissi. Deterministic systems of sequential processes: a class of structured Petri nets. In ATPN91 [ATP91], pages 62–81.

[Spr98] C. Sprenger. A verified model checker for the modal μ-calculus in coq. In Desel and Silva [DS98], pages 167–183.

[SRI00] SRI International. *PVS home page*, 2000. `http://www.csl.sri.com/sri-csl-pvs.html`.

[ST92] I. Sommerville and R. Thomson. Configuration specification using a system structure language. In *International IEEE Workshop in Configurable Distributed Systems*, 1992.

[ST97] M. Silva and E. Teruel. Petri nets for the design and operation of manufacturing systems. *European Journal of Control*, 3(3):182–199, 1997.

[Sta91] P. Starke. Reachability analysis of Petri nets using symmetries. *Systems Analysis Modelling Simulation*, 8:293–303, 1991.

[Sta93] M.G. Staskauskas. Formal derivation of concurrent programs: an example from industry. *IEEE Transactions on Software Engineering*, SE-19:503–528, 1993.

[Ste98a] M.-O. Stehr. Assertional reasoning and temporal logic for system verification. Lecture at the MATCH Advanced Summer School on Systems Engineering, September 14-22, Jaca, Spain, 1998.

[Ste98b] M.-O. Stehr. Embedding UNITY into the calculus of inductive constructions. Technical Report FBI-HH-B-214/98, Universität Hamburg, Fachbereich Informatik, 1998.

[Sti96] C. Stirling. Modal and temporal logics for processes. In Faron Moller and Graham Birtwistle, editors, *Logics for Concurrency – Structure versus Automata*, volume 1043 of *Lecture Notes in Computer Science*, pages 149–237. Springer-Verlag, 1996.

[Suz88] I. Suzuki. Proving properties of a ring of finite-state machines. *Information Processing Letters*, 28(4):213–214, 1988.

[SV82] M. Silva and S. Velilla. Programmable logic controllers and Petri nets. In *International Symposium IFAC-IFIP on Software Computer Control, Madrid, Spain*, 1982.

[SV89] M. Silva and R. Valette. Petri nets and flexible manufacturing. In Rozenberg [Roz89], pages 374–417.

[Tau88] D. Taubner. On the implementation of Petri nets. In Rozenberg [Roz88], pages 418–439.

[TCE99] F. Tricas, J.M. Colom, and J. Ezpeleta. A solution to the problem of deadlocks in concurrent systems using Petri nets and integer linear programming. In *11th European Simulation Symposium and Exhibition Simulation in Industry*, Erlangen-Nuremberg, 1999.

[TCS93] E. Teruel, J.M. Colom, and M. Silva. Linear analysis of deadlock-freeness of Petri net models. In *Proceedings of the European Control Conference, ECC'93, Groningen, The Netherlands*, pages 513–518, 1993.

[TCS97] E. Teruel, J. M. Colom, and M. Silva. Choice-free Petri nets: A model for deterministic concurrent systems with bulk services and arrivals. *IEEE Transactions on Systems, Man and Cybernetics. Part A: Systems and Humans*, 27(1):73–83, 1997.

[Tel91] G. Tel. *Topics in Distributed Algorithms.* Cambridge University Press, 1991.

[Tel94] G. Tel. *Introduction to Distributed Algorithms.* Cambridge University Press, 1994. 2nd edition, 2001.

[Thi87] P.S. Thiagarajan. Elementary net system. In Brauer et al. [BRR87a], pages 26–59.

[Thu85] E. Thuriot. Le synchroniseur coloré: une approche pour la mise en oeuvre des systèmes multitâches. Master's thesis, Université Paul Sabatier, Toulouse, France, 1985.

[Tiu94] M. Tiusanen. Symbolic, symmetry and stubborn set searches. In Valette [Val94a], pages 511–530.

[Tro92] A.S. Troelstra. *Lectures on Linear Logic.* CSLI Lecture Notes No. 29, Stanford, CA, USA, 1992.

[Tro93] A.S. Troelstra. Tutorial on linear logic. In P. Schröder-Heister and K. Došen, editors, *Substructural Logics. Workshop Tübingen 1990.* Oxford Science Publications, 1993.

[TS93] E. Teruel and M. Silva. Liveness and home states in Equal Conflict systems. In Marsan [Mar93], pages 415–432.

[TS94] E. Teruel and M. Silva. Well-formedness of Equal Conflict systems. In Valette [Val94a], pages 491–510.

[TS96] E. Teruel and M. Silva. Structure theory of Equal Conflict Systems. *Theoretical Computer Science*, 153(1-2):271–300, 1996.

[TV84] P.S. Thiagarajan and K. Voss. A fresh look at free choice nets. *Information and Control*, 61(2):85–113, 1984.

[VA87] R. Valette and H. Atabakhche. Petri nets for sequence constraint propagation in knowledge based approaches. In K. Voss, H.J. Genrich, and G. Rozenberg, editors, *Concurrency and Nets – Advances in Petri Nets*, pages 555–570. Springer-Verlag, 1987.

[Val79] R. Valette. Analysis of Petri nets by stepwise refinements. *Journal of Computer and System Sciences*, 18(1):35–46, 1979.

[Val87a] R. Valette. Nets in production systems. In Brauer et al. [BRR87a], pages 191–217.

[Val87b] R. Valk. Modelling of task flow in systems of functional units. Technical Report FBI-HH-B-124/87, Fachbereich Informatik, Universität Hamburg, 1987.

[Val88] A. Valmari. Error detection by reduced reachability graph generation. In *Proceedings of the 9th European Workshop on Application and Theory of Petri Nets*, pages 95–112, Venice, Italy, 1988.

[Val89] A. Valmari. Stubborn sets for reduced space generation. In De Michelis [DM89]. Volume 2, pages 1–22.

[Val90a] K.P. Valavanis. On the hierarchical modeling analysis and simulation of flexible manufacturing systems with extended Petri nets. *IEEE Transactions on Systems, Man and Cybernetics*, 20(1):94–110, 1990.

[Val90b] A. Valmari. A stubborn attack on state explosion. In E.M. Clarke and R.P. Kurshan, editors, *Proceedings of the 2th International Conference on Computer Aided Verification (CAV'90), New Brunswick, NJ, USA*, volume 531 of *Lecture Notes in Computer Science*, pages 156–165. Springer-Verlag, 1990.

[Val90c] A. Valmari. Stubborn sets for reduced state space generation. In Rozenberg [Roz91a], pages 491–515. Revised and extended version of [Val89].

[Val91a] R. Valk. Modelling concurrency by task/flow EN systems. In *3rd Workshop on Concurrency and Compositionality, Goslar, Germany*, volume 191 of *GMD-Studien*, pages 207–215, St. Augustin, Bonn, 1991. Gesellschaft für Mathematik und Datenverarbeitung.

[Val91b] A. Valmari. Stubborn sets of coloured Petri nets. In ATPN91 [ATP91].

[Val92] A. Valmari. A stubborn attack on state explosion. *Formal Methods in System Design*, 1(4):297–322, 1992.

[Val93a] R. Valk. Bridging the gap between place- and Floyd-invariants with applications to preemptive scheduling. In Marsan [Mar93], pages 433–452.

[Val93b] R. Valk. Extending S-invariants for coloured and selfmodifying nets. Technical Report FBI-HH-B-93-165, Universität Hamburg, Fachbereich Informatik, 1993.

[Val93c] A. Valmari. On-the-fly verification with stubborn sets. In C. Courcoubetis, editor, *Proceedings of the 5th International Conference on Computer Aided Verification (CAV'93), Elounda, Greece*, volume 697 of *Lecture Notes in Computer Science*, pages 397–408. Springer-Verlag, 1993.

[Val94a] R. Valette, editor. *Application and Theory of Petri Nets 1994, Proceedings of the 15th International Conference (ICATPN'94), Zaragoza, Spain*, volume 815 of *Lecture Notes in Computer Science*. Springer-Verlag, 1994.

[Val94b] A. Valmari. Compositional analysis with place-bordered subnets. In Valette [Val94a], pages 531–547.

[Val95] R. Valk. Petri nets as dynamical objects. In *Proceedings of the 1st International Workshop on Object-Oriented Programming and Models of Concurrency, Turin, Italy*, June 1995.

[Val96a] R. Valk. How to define markings in object systems. *Petri Net Newsletter*, 50, 1996.

[Val96b] R. Valk. On processes of object Petri nets. Technical Report FBI-HH-B-185/96, Fachbereich Informatik, Universität Hamburg, 1996.

[Val98] R. Valk. Petri nets as token objects – an introduction to elementary object nets. In Desel and Silva [DS98], pages 1–25.

[Val00] R. Valk. Relating different semantics for object Petri nets. Technical Report FBI-HH-B-266/00, Fachbereich Informatik, Universität Hamburg, 2000.

[VAM96] F. Vernadat, P. Azéma, and F. Michel. Covering step graphs. In Billington and Reisig [BR96], pages 516–535.

[Vau85] J. Vautherin. *Un Modèle Algébrique, Basé sur les Réseaux de Petri, pour l'Etude des Systèmes Parallèles.* Thèse de Docteur Ingenieur, Univ. de Paris-Sud, Centre d'Orsay, June 1985.

[Vau87] J. Vautherin. Parallel systems specifications with coloured Petri nets and algebraic specifications. In Rozenberg [Roz87], pages 293–308.

[VB90] R. Valette and B. Bako. Software implementation of Petri nets and compilation of rule-based systems. In Rozenberg [Roz91b], pages 296–316.

[VCBA83] R. Valette, M. Courvoisier, J.M. Bigou, and J. Alburkerque. A Petri net based programmable logic controller. In *1st International Conference on Computer Application in Production and Engineering, Amsterdam*, 1983.

[Ver94a] I. Vernier. *Graphe symbolique paramétré de réseaux de Petri et logique temporelle.* Thèse de doctorat, Université Pierre & Marie Curie (Paris VI), France, December 1994.

[Ver94b] I. Vernier. Parameterized evaluation of CTL-X formulae. In H.J. Ohlbach, editor, *Proceedings of the 1st International Conference on Temporal Logic, ICTL-94 Workshop, Bonn, Germany*, pages 22–31. Max-Planck-Institut für Informatik, MPI-I-94-230, June 1994.

[Ver95] I. Vernier. Symbolic verification of parallel programs. Technical Report 95/03, IBP - MASI, February 1995.

[Ver96] I. Vernier. Symbolic executions of symmetrical parallel programs. In *Proceedings of the 4th Euromicro Workshop on Parallel and Distributed Processing*, pages 327–334, Braga, Portugal, January 1996.

[VHHP95] K. Varpaaniemi, J. Halme, K. Hiekkanen, and T. Pyssysalo. Prod reference manual. Technical Report B 13, Helsinki University of Technology, Digital Systems Laboratory, Espoo, Finland, 1995.

[Vin81] J. Vinkowski. Protocol of accessing overlapping sets of resources. *Information Processing Letters*, 12(5):239–243, 1981.

[Vir94] P. Viry. Rewriting: An effective model of concurrency. In C. Halatsis, D. Maritsas, G. Philokyprou, and S. Theodoridis, editors, *PARLE'94 - Parallel Architectures and Languages Europe, 6th International PARLE Conference, Athens, Greece, July 1994, Proceedings*, volume 817 of *Lecture Notes in Computer Science*, pages 648–660. Springer-Verlag, 1994.

[VJ85] R. Valk and M. Jantzen. The residue of vector sets with applications to decidability problems in Petri nets. *Acta Informatica*, 21:643–674, 1985.

[VM94] J.L. Villarroel and P.R. Muro. Using Petri net models at the coordination level for manufacturing systems control. *Robotics and Computer-Integrated Manufacturing*, 1(11):41–50, 1994.

[VM97] F. Vernadat and F. Michel. Covering step graphs preserving failure semantics. In Azéma and Balbo [AB97b], pages 253–270.

[VMS88] J.L. Villarroel, J. Martínez, and M. Silva. GRAMAN: a graphic system for manufacturing system modelling and simulation. In *Proceedings of the IMACS Symposium on System Modelling and Simulation, Cetraro, Italy*, pages 311–316, 1988.

[VN92] N. Viswanadham and Y. Narahari. *Performance Modeling of Automated Manufacturing Systems.* Prentice Hall, Englewood Cliffs, NJ, 1992.

[Vog89] W. Vogler. Live and bounded free choice nets have home states. *Petri Net Newsletter*, 32:18–21, 1989.

[Von90] R. Vonk. *Prototyping.* Prentice Hall, Englewood Cliffs, NJ, 1990.

[Vos87] K. Voss. Nets in office automation. In Brauer et al. [BRR87a], pages 234–257.

[VW86] M.Y. Vardi and P. Wolper. An automata-theoretic approach to automatic program verification. In *Proceedings of the First Annual Symposium on Logic in Computer Science, LICS'86, Cambridge, MA, USA*, pages 322–331. IEEE Computer Society Press, 1986.

[Wad93] P. Wadler. A taste of linear logic. In A.M. Borzyszkowski and S. Sokolowski, editors, *Mathematical Foundations of Computer Science 1993, Proceedings of the 18th International Symposium, MFCS '93, Gdansk, Poland*, volume 711 of *Lecture Notes in Computer Science*, pages 185–210. Springer-Verlag, 1993.

[Wal95] R. Walter. *Petrinetzmodelle verteilter Algorithmen. Beweistechnik und Intuition.* Volume 2 of Edition Versal, Bertz Verlag, Dissertation, Humboldt-Universität zu Berlin, 1995.

[Wec92] W. Wechler. *Universal Algebra for Computer Scientists*, volume 25 of *EATCS Monographs on Theoretical Computer Science*. Springer-Verlag, 1992.

[Win87] G. Winskel. Event structures. In Brauer et al. [BRR87a], pages 325–392.

[Win90] G. Winskel. A compositional proof system on a category of labelled transition systems. *Information and Computation*, 87:2–57, 1990.

[WL89] P. Wolper and V. Lovinfosse. Verifying properties of large sets of processes with network invariants. In *Proceedings of the International Workshop on Automatic Verification Methods for Finite State Systems Grenoble, France*, volume 407 of *Lecture Notes in Computer Science*, pages 68–80. Springer-Verlag, 1989.

[WL93] P. Wolper and D. Leroy. Reliable hashing compaction without collision detection. In C. Courcoubetis, editor, *Proceedings of the 5th International Conference on Computer Aided Verification (CAV'93), Elounda, Greece*, volume 697 of *Lecture Notes in Computer Science*, pages 59–70. Springer-Verlag, 1993.

[WMC94] Workflow reference model. Technical report, Workflow Management Coalition, Brussels, 1994.

[WR96] M. Wolf and U. Reimer. Workshop on adaptive workflow. In *Proceedings of the 1st International Conference on Practical Aspects of Knowledge Management (PAKM'96)*, 1996.

[WW96] B. Willems and P. Wolper. Partial-order methods for model checking: From linear time to branching time. In *Proceedings of the 11th Annual Symposium on Logic in Computer Science, LICS'96, New Brunswick, NJ, USA*, pages 294–303, New Brunswick, NJ, USA, 1996.

[WWV+97] M. Weber, R. Walter, H. Völzer, T. Vesper, W. Reisig, S. Peuker, E. Kindler, J. Freiheit, and J. Desel. DAWN: Petrinetzmodelle zur Verifikation verteilter Algorithmen. Informatik-Bericht 88, Humboldt-Universität zu Berlin, 1997.

[XHC96] K.Y. Xing, B.S. Hu, and H.X. Chen. Deadlock avoidance policy for Petri-net modeling of flexible manufacturig systems with shared resources. *IEEE Transactions on Automatic Control*, 41(2):289–295, 1996.

[Zak96] N. Zakhama. *Le prototypage hétérogène formel de systèmes distribués, une perspective pour le développement continu.* Thèse de doctorat, Université Pierre & Marie Curie (Paris VI), France, 1996.

[ZD91] M. Zhou and F. DiCesare. Parallel and sequential mutual exclusions for Petri net modelling of manufacturing systems with shared resources. *IEEE Transactions on Robotics and Automation*, 7(4), 1991.

[ZD93] M. Zhou and F. DiCesare. *Petri Net Synthesis for Discrete Event Control of Manufacturing Systems.* Kluwer Academic Publishers, 1993.

[Zha99] W. Zhang. *State-Space Search – Algorithms, Complexity, Extensions, and Applications.* Springer-Verlag, New York, 1999.

Index